STUDENT'S SOLUTIONS MANUAL

CARRIE GREEN

INTRODUCTORY AND INTERMEDIATE ALGEBRA THROUGH APPLICATIONS

SECOND EDITION

Geoffrey Akst

Borough of Manhattan Community College/City University of New York

Sadie Bragg

Borough of Manhattan Community College/City University of New York

Boston San Francisco New York
London Toronto Sydney Tokyo Singapore Madrid
Mexico City Munich Paris Cape Town Hong Kong Montreal

Reproduced by Pearson Addison-Wesley from electronic files supplied by the author.

Publishing as Pearson Addison-Wesley, 75 Arlington Street, Boston, MA 02116.

ISBN-13: 978-0-321-55670-7
ISBN-10: 0-321-55670-4

1 2 3 4 5 6 BB 11 10 09 08

Table of Contents

Chapter R

PREALGEBRA REVIEW

Pretest

1. $7 \cdot 7 \cdot 7 = 7^3$

2. $2^4 \cdot 5^2 = (2 \cdot 2 \cdot 2 \cdot 2) \cdot (5 \cdot 5) = 400$

3. $16 - 2^3 = 16 - 8 = 8$

4. $$\begin{aligned} 20 - 2(3 - 1^2) &= 20 - 2(3 - 1) \\ &= 20 - 2(2) \\ &= 20 - 4 \\ &= 16 \end{aligned}$$

5. 12
Factors: 1, 2, 3, 4, 6, 12

6. $20 = 2 \cdot 2 \cdot 5 = 2^2 \cdot 5$

7. $$\begin{aligned} \frac{16}{24} &= \frac{2 \cdot 2 \cdot 2 \cdot 2}{2 \cdot 2 \cdot 2 \cdot 3} \\ &= \frac{\cancel{2} \cdot \cancel{2} \cdot \cancel{2} \cdot 2}{\cancel{2} \cdot \cancel{2} \cdot \cancel{2} \cdot 3} \\ &= \frac{2}{3} \end{aligned}$$

8. $$\begin{aligned} \frac{3}{10} + \frac{1}{10} &= \frac{3+1}{10} \\ &= \frac{4}{10} \\ &= \frac{2}{5} \end{aligned}$$

9. $$\begin{aligned} 10\frac{1}{3} - 2\frac{5}{6} &= 9\frac{4}{3} - 2\frac{5}{6} \\ &= 9\frac{8}{6} - 2\frac{5}{6} \\ &= 7\frac{3}{6} \\ &= 7\frac{1}{2} \end{aligned}$$

10. $\frac{2}{9} \cdot \frac{2}{3} = \frac{2 \cdot 2}{9 \cdot 3} = \frac{4}{27}$

11. $$\begin{aligned} 4 \div 1\frac{1}{2} &= \frac{4}{1} \div \frac{3}{2} \\ &= \frac{4}{1} \cdot \frac{2}{3} \\ &= \frac{8}{3}, \text{ or } 2\frac{2}{3} \end{aligned}$$

12. 9.013
Nine and thirteen thousandths

13. $3.072 \approx 3.1$

14. $$\begin{array}{r} 7 \\ 4.01 \\ +\ 9.3003 \\ \hline 20.3103 \end{array}$$

15. $$\begin{array}{r} 8.00 \\ -\ 2.34 \\ \hline 5.66 \end{array}$$

16. $$\begin{array}{r} 923 \\ \times\ \ 41 \\ \hline 923 \\ 3692 \\ \hline 37843 \end{array}$$

$9.23 \times 4.1 = 37.843$

17. $0.235 \times 100 = 23.5$

18. $0.25\overline{)0.045} \Rightarrow 25\overline{)4.5}$, quotient 0.18

$$\begin{array}{r} 25 \\ \hline 200 \\ 200 \\ \hline 0 \end{array}$$

19. $\frac{31}{1000} = 0.0031$

20. $7\% = 0.07$

21.
$$\begin{array}{r} 3.000 \\ -\ 1.316 \\ \hline 1.684 \end{array}$$

$1.684 \approx 2$

\$2 million

22. $\frac{1}{2}\cdot\frac{2}{3}=\frac{1\cdot\cancel{2}}{\cancel{2}\cdot 3}=\frac{1}{3}$

$\frac{1}{3}$ mi

23. $0.05\overline{)0.1} \Rightarrow 5\overset{2}{\overline{)10}}$

Two times

24. $0.1 = 10\%$

25. $95\% = 0.95$

R.1 Exponents and Order of Operations

Practice

1. $2\cdot2\cdot2\cdot2\cdot2=2^5$

2. $7^2\cdot2^4=(7\cdot7)\cdot(2\cdot2\cdot2\cdot2)$
$=49\cdot16$
$=784$

3. $1{,}000{,}000{,}000=10^9$

4. $8\div2+4\cdot3=4+12=16$

5. $(4+1)^2-4\cdot6=5^2-4\cdot6$
$=25-4\cdot6$
$=25-24$
$=1$

6. $\begin{pmatrix}84+85+88+92+80+96\\ +150+175+100+95+75+80\end{pmatrix}\div12$
$=(1200)\div12$
$=100$

No, the monthly average bill this year was \$100.

R.2 Factors, Primes, and Least Common Multiples

Practice

1. $10=2\cdot5$
$25=5\cdot5=5^2$
$LCM=2\cdot5^2=50$

2. $20=2\cdot2\cdot5=2^2\cdot5$
$36=2\cdot2\cdot3\cdot3=2^2\cdot3^2$
$60=2\cdot2\cdot3\cdot5=2^2\cdot3\cdot5$
$LCM=2^2\cdot3^2\cdot5=180$

3. $6=2\cdot3$
$8=2\cdot2\cdot2=2^3$
$LCM=2^3\cdot3=24$
24 hr

R.3 Fractions

Practice

1. $3\frac{2}{9}=\frac{(9\times3)+2}{9}$
$=\frac{27+2}{9}$
$=\frac{29}{9}$

2. $\frac{8}{3}=3\overset{2}{\overline{)8}}$
$\begin{array}{r}\underline{6}\\2\end{array}$

$\frac{8}{3}=2\frac{2}{3}$

3. $\frac{24}{30}=\frac{2\cdot2\cdot2\cdot3}{2\cdot3\cdot5}$
$=\frac{\cancel{2}\cdot2\cdot2\cdot\cancel{3}}{\cancel{2}\cdot\cancel{3}\cdot5}$
$=\frac{4}{5}$

4. $\frac{7}{15}+\frac{3}{15}=\frac{7+3}{15}$
$=\frac{10}{15}$
$=\frac{2}{3}$

5. $\frac{19}{20}-\frac{11}{20}=\frac{19-11}{20}$
$=\frac{8}{20}$
$=\frac{2}{5}$

6. $\frac{11}{12}+\frac{3}{4}=\frac{11}{12}+\frac{9}{12}$
$=\frac{20}{12}$
$=\frac{5}{3}$, or $1\frac{2}{3}$

7. $\frac{4}{5}-\frac{1}{2}=\frac{8}{10}-\frac{5}{10}=\frac{3}{10}$

8. $4\frac{5}{8}=4\frac{5}{8}$
$+3\frac{1}{2}=+3\frac{4}{8}$
$7\frac{9}{8}$
$=7+\frac{9}{8}=7+1\frac{1}{8}=8\frac{1}{8}$

9. $15\frac{1}{12}$
$-9\frac{11}{12}$

$15\frac{1}{12}=14+1+\frac{1}{12}=14+\frac{12}{12}+\frac{1}{12}=14\frac{13}{12}$

$14\frac{13}{12}$
$-9\frac{11}{12}$
$5\frac{2}{12}=5\frac{1}{6}$

10. $\frac{1}{2}\cdot\frac{3}{4}=\frac{1\cdot 3}{2\cdot 4}=\frac{3}{8}$

11. $\frac{7}{10}\cdot\frac{5}{11}=\frac{7\cdot \cancel{5}^{1}}{\cancel{10}_{2}\cdot 11}=\frac{7}{22}$

12. $3\frac{3}{4}\cdot 2\frac{1}{10}=\frac{15}{4}\cdot\frac{21}{10}$
$=\frac{\cancel{15}^{3}\cdot 21}{4\cdot \cancel{10}_{2}}$
$=\frac{63}{8}$, or $7\frac{7}{8}$

13. $\frac{3}{4}\div\frac{1}{8}=\frac{3}{4}\cdot\frac{8}{1}$
$=\frac{3\cdot \cancel{8}^{2}}{\cancel{4}_{1}\cdot 1}$
$=6$

14. $6\div 3\frac{3}{4}=\frac{6}{1}\div\frac{15}{4}$
$=\frac{6}{1}\cdot\frac{4}{15}$
$=\frac{\cancel{6}^{2}\cdot 4}{1\cdot \cancel{15}_{5}}$
$=\frac{8}{5}$, or $1\frac{3}{5}$

15. $1-\frac{3}{5}=\frac{5}{5}-\frac{3}{5}=\frac{2}{5}$

$1000\cdot\frac{2}{5}=\frac{\cancel{1000}^{200}}{1}\times\frac{2}{\cancel{5}}$
$=400$

400 students

R.4 Decimals

Practice

1. a. $0.5=\frac{5}{10}=\frac{1}{2}$

b. $2.073=2\frac{73}{1000}$

2. 4.003
Four and three thousandths

3. $748.0772 = 784.0\underline{7}72 \approx 748.08$

4.
$$\begin{array}{r} 5.92 \\ 35.872 \\ +\ 0.3 \\ \hline 42.092 \end{array}$$

5.
$$\begin{array}{r} 3.800 \\ -\ 2.621 \\ \hline 1.179 \end{array}$$

6.
$$\begin{array}{r} 281 \\ \times\ \ 35 \\ \hline 1405 \\ 843\ \\ \hline 9835 \end{array}$$

$2.81 \times 3.5 = 9.835$

7. $32.7 \times 10{,}000 = 327{,}000$

8.
$$0.15\overline{)2.706} \Rightarrow \begin{array}{r} 18.04 \\ 15\overline{)270.6} \\ \underline{270}\ \ \ \\ 6\ \ \ \\ \underline{0}\ \ \ \\ 60 \\ \underline{60} \\ 0 \end{array}$$

9. $0.86 \div 1000 = 0.00086$

10.
$$72.7\overline{)112.0} \Rightarrow \begin{array}{r} 1.54 \\ 727\overline{)1120.00} \\ \underline{727}\ \ \ \ \ \\ 3930\ \ \\ \underline{3635}\ \ \\ 2950 \\ \underline{2908} \\ 42 \end{array}$$

$1.54 \approx 1.5$

R.5 Percent Conversions

Practice

1. $7\% = \dfrac{7}{100}$

2. $5\% = 0.05$

3. $0.025 = 2.5\%$

4. $\dfrac{1}{4} = \dfrac{25}{100} = 0.25 = 25\%$

5. $40\% = 0.40$, or 0.4

Exercises

1. $6 \cdot 6 \cdot 6 \cdot 6 \cdot 6 = 6^5$

3. $2 \cdot 2 \cdot 10 \cdot 10 \cdot 10 = 2^2 \cdot 10^3$

5. $5^2 \cdot 10^3 = 25 \cdot 1000 = 25{,}000$

7. $6 + 3^2 = 6 + 9 = 15$

9.
$$\begin{aligned} 2 + 18 \div 3(9-7) &= 2 + 18 \div 3(2) \\ &= 2 + 18 \div 6 \\ &= 2 + 3 \\ &= 5 \end{aligned}$$

11.
$$\begin{aligned} \frac{4^2 + 8}{9 - 3} &= \frac{16 + 8}{6} \\ &= \frac{24}{6} \\ &= 4 \end{aligned}$$

13. 150
Factors: 1, 2, 3, 5, 6, 10, 15, 25, 30, 50, 75, 150

15. 23
Prime

17. 51
Composite

19. $42 = 2 \cdot 3 \cdot 7$

21. $48 = 2 \cdot 2 \cdot 2 \cdot 2 \cdot 3 = 2^4 \cdot 3$

23. $6 = 2 \cdot 3$
$8 = 2 \cdot 4$
$LCM = 2 \cdot 3 \cdot 4 = 24$

25. $24 = 2 \cdot 2 \cdot 2 \cdot 3 = 2^3 \cdot 3$
$36 = 2 \cdot 2 \cdot 3 \cdot 3 = 2^2 \cdot 3^2$
$72 = 2 \cdot 2 \cdot 2 \cdot 3 \cdot 3 = 2^3 \cdot 3^2$
$LCM = 2^3 \cdot 3^2 = 72$

27. $3\frac{4}{5} = \frac{(5 \times 3) + 4}{5} = \frac{15+4}{5} = \frac{19}{5}$

29. $\frac{23}{4} = 4\overline{)23}$, quotient 5; 20; remainder 3

$\frac{23}{4} = 5\frac{3}{4}$

31. $\frac{14}{28} = \frac{7 \cdot 2}{7 \cdot 2 \cdot 2} = \frac{\cancel{7} \cdot \cancel{2}}{\cancel{7} \cdot \cancel{2} \cdot 2} = \frac{1}{2}$

33. $5\frac{2}{4} = 5\frac{2}{2 \cdot 2} = 5\frac{\cancel{2}}{\cancel{2} \cdot 2} = 5\frac{1}{2}$

35. $\frac{1}{9} + \frac{4}{9} = \frac{1+4}{9} = \frac{5}{9}$

37. $\frac{3}{8} - \frac{1}{8} = \frac{3-1}{8} = \frac{2}{8} = \frac{1}{4}$

39. $\frac{2}{5} + \frac{4}{7} = \frac{14}{35} + \frac{20}{35} = \frac{14+20}{35} = \frac{34}{35}$

41. $\frac{3}{10} - \frac{1}{20} = \frac{6}{20} - \frac{1}{20} = \frac{6-1}{20} = \frac{5}{20} = \frac{1}{4}$

43.
$$\begin{array}{r} 1\frac{1}{8} \\ +5\frac{3}{8} \\ \hline 6\frac{4}{8} = 6\frac{1}{2} \end{array}$$

45.
$$\begin{array}{r} 8\frac{7}{10} \\ +1\frac{9}{10} \\ \hline 9\frac{16}{10} = 10\frac{6}{10} = 10\frac{3}{5} \end{array}$$

47.
$$\begin{array}{r} 9\frac{11}{12} \\ -6\frac{7}{12} \\ \hline 3\frac{4}{12} = 3\frac{1}{3} \end{array}$$

49.
$$\begin{array}{r} 6\frac{1}{10} \\ -4\frac{3}{10} \\ \hline \end{array}$$

$$6\frac{1}{10}=5+1+\frac{1}{10}=5+\frac{10}{10}+\frac{1}{10}=5\frac{11}{10}$$

$$\begin{array}{r} 5\frac{11}{10} \\ -4\frac{3}{10} \\ \hline 1\frac{8}{10}=1\frac{4}{5} \end{array}$$

51.
$$\begin{array}{rcl} 12 & = & 11\frac{2}{2} \\ -5\frac{1}{2} & = & -5\frac{1}{2} \\ \hline & & 6\frac{1}{2} \end{array}$$

53.
$$\begin{array}{rcrcr} 7\frac{1}{2} & = & 7\frac{4}{8} & = & 6\frac{12}{8} \\ -4\frac{5}{8} & = & -4\frac{5}{8} & = & -4\frac{5}{8} \\ \hline & & & & 2\frac{7}{8} \end{array}$$

55. $\frac{2}{3}\cdot\frac{1}{5}=\frac{2\cdot 1}{3\cdot 5}=\frac{2}{15}$

57.
$$\begin{aligned} 1\frac{2}{5}\cdot 10 &= \frac{7}{5}\cdot\frac{10}{1} \\ &= \frac{7\cdot\cancel{10}^{2}}{\cancel{5}_{1}\cdot 1} \\ &= 14 \end{aligned}$$

59.
$$\begin{aligned} 3\frac{1}{4}\cdot 4\frac{2}{3} &= \frac{13}{4}\cdot\frac{14}{3} \\ &= \frac{13\cdot\cancel{14}^{7}}{\cancel{4}_{2}\cdot 3} \\ &= \frac{91}{6} \\ &= 15\frac{1}{6} \end{aligned}$$

61.
$$\begin{aligned} 2\frac{5}{6}\div\frac{1}{2} &= \frac{17}{6}\cdot\frac{2}{1} \\ &= \frac{17\cdot\cancel{2}^{1}}{\cancel{6}_{3}\cdot 1} \\ &= \frac{17}{3} \\ &= 5\frac{2}{3} \end{aligned}$$

63.
$$\begin{aligned} \frac{2}{3}\div 6 &= \frac{2}{3}\cdot\frac{1}{6} \\ &= \frac{\cancel{2}\cdot 1}{3\cdot\cancel{6}_{3}} \\ &= \frac{1}{9} \end{aligned}$$

65.
$$\begin{aligned} 8\div 2\frac{1}{3} &= \frac{8}{1}\div\frac{7}{3} \\ &= \frac{8}{1}\cdot\frac{3}{7} \\ &= \frac{8\cdot 3}{1\cdot 7} \\ &= \frac{24}{7} \\ &= 3\frac{3}{7} \end{aligned}$$

67.
$$\begin{aligned} \left(\frac{3}{4}\right)^2-\frac{3}{8}\div 6 &= \frac{9}{16}-\frac{\cancel{3}^{1}}{8}\cdot\frac{1}{\cancel{6}_{2}} \\ &= \frac{9}{16}-\frac{1}{16} \\ &= \frac{8}{16} \\ &= \frac{1}{2} \end{aligned}$$

69. $0.875=\frac{875}{1000}=\frac{7}{8}$

71. $18.3\underline{5}9$
Hundredths

73. 0.72
Seventy-two hundredths

75. 3.009
Three and nine thousandths

77. $7.\underline{3}1 \approx 7.3$

79. $4.3\underline{8}68 \approx 4.39$

81.
$$\begin{array}{r} 8.2 \\ 3.91 \\ +\ 6 \\ \hline 18.11 \end{array}$$

83.
$$\begin{array}{r} 3.800 \\ -1.927 \\ \hline 1.873 \end{array}$$

85.
$$\begin{array}{r} 728 \\ \times\quad 4 \\ \hline 2912 \end{array}$$

$7.28 \times 0.4 = 2.912$

87. $2.71 \cdot 1000 = 2710$

89.
$$\begin{array}{r} 0.0015 \\ 4\overline{)0.0060} \\ \underline{4} \\ 20 \\ \underline{20} \\ 0 \end{array}$$

91. $2.4\overline{)12} \Rightarrow 24\overline{)120}$
$$\begin{array}{r} 5 \\ 24\overline{)120} \\ \underline{120} \\ 0 \end{array}$$

93. $7.1 + 0.5^2 = 7.1 + 0.25 = 7.35$

95. $75\% = \frac{75}{100} = \frac{3}{4}$

97. $106\% = 1\frac{6}{100} = 1\frac{3}{50}$

99. $6\% = 0.06$

101. $150\% = 1.5$

103. $0.31 = 31\%$

105. $0.0145 = 1.45\%$

107. $\frac{1}{10} = 0.1 = 10\%$

109. $\frac{4}{5} = 0.8 = 80\%$

111. $(67 + 72 + 78 + 70 + 65 + 77 + 82) \div 7$
$= (511) \div 7$
$= 73$

73°F

113. $\frac{1}{2} + \frac{1}{4} = \frac{2}{4} + \frac{1}{4} = \frac{3}{4}$

$\frac{3}{4}$ mi

115. $27 \cdot \frac{2}{3} = \frac{\cancel{27}^{9}}{1} \cdot \frac{2}{\cancel{3}_{1}} = 18$

$18

117.

$(9.87 \div 3) - (7.96 \div 4) = 3.29 - 1.99 = 1.30$

$1.30

119. $15\% = \frac{15}{100} = \frac{3}{20}$

Chapter R Posttest

1. $8^2 \cdot 2^3 = 8 \cdot 8 \cdot 2 \cdot 2 \cdot 2 = 512$

2. $11 \cdot 2 + 5 \cdot 3 = 22 + 15 = 37$

3. 20
Factors: 1, 2, 4, 5, 10, 20

4. $3\frac{1}{4} = \frac{(4 \times 3) + 1}{4}$
$= \frac{12 + 1}{4}$
$= \frac{13}{4}$

5. $\frac{10}{36} = \frac{2 \cdot 5}{2 \cdot 2 \cdot 3 \cdot 3}$

$= \frac{\not{2} \cdot 5}{\not{2} \cdot 2 \cdot 3 \cdot 3}$

$= \frac{5}{18}$

6. $\frac{5}{8} + \frac{7}{8} = \frac{5+7}{8}$

$= \frac{12}{8}$

$= \frac{3}{2}$

$= 1\frac{1}{2}$

7. $7\frac{7}{8} + 4\frac{1}{6} = \frac{63}{8} + \frac{25}{6}$

$= \frac{189}{24} + \frac{100}{24}$

$= \frac{289}{24}$

$= 12\frac{1}{24}$

8. $\frac{4}{9} - \frac{3}{10} = \frac{40}{90} - \frac{27}{90} = \frac{13}{90}$

9.
$$\begin{array}{rcrcr} 12\frac{1}{4} & = & 12\frac{10}{40} & = & 11\frac{50}{40} \\ -8\frac{3}{10} & = & -8\frac{12}{40} & = & -8\frac{12}{40} \\ \hline & & & & 3\frac{38}{40} = 3\frac{19}{20} \end{array}$$

10. $\frac{3}{4} \cdot \frac{4}{5} = \frac{3 \cdot \not{4}}{\not{4} \cdot 5} = \frac{3}{5}$

11. $\frac{2}{3} \div \frac{1}{3} = \frac{2}{3} \cdot \frac{3}{1}$

$= \frac{2 \cdot \not{3}}{\not{3} \cdot 1}$

$= 2$

12. $7 \div 3\frac{1}{5} = \frac{7}{1} \div \frac{16}{5}$

$= \frac{7}{1} \cdot \frac{5}{16}$

$= \frac{7 \cdot 5}{1 \cdot 16}$

$= \frac{35}{16}$, or $2\frac{3}{16}$

13. 2.396
Two and three hundredths ninety-six thousandths

14.
$$\begin{array}{r} 5.2 \\ 3 \\ +\ 8.002 \\ \hline 16.202 \end{array}$$

15.
$$\begin{array}{r} 10.00 \\ -\ 3.01 \\ \hline 6.99 \end{array}$$

16.
$$\begin{array}{r} 502 \\ \times\ 89 \\ \hline 4518 \\ 4016\ \\ \hline 44678 \end{array}$$

$5.02 \times 8.9 = 44.678$

17. $2.07 \times 1000 = 2070$

18. $\frac{0.05}{100} = 0.0005$

19. $\frac{1}{8} = 0.125 = 12.5\%$

20. $0.7 = 70\%$

$= \frac{70}{100}$

$= \frac{7}{10}$

21. $3 \div \frac{1}{2} = \frac{3}{1} \div \frac{1}{2}$

$= \frac{3}{1} \cdot \frac{2}{1}$

$= 6$

6 times

22. $1.\underline{7}7 \approx 1.8$

23.

$$\begin{array}{r} 305 \\ \times\ 205 \\ \hline 1525 \\ 6100 \\ \hline 62525 \end{array}$$

$30.5 \times 20.5 = 625.25 \approx 625.3$

625.3 sq m

24. $0.6\overline{)\ \ 9}$ with quotient 0.125 $\Rightarrow 6\overline{)90}$ with quotient 15

$$\begin{array}{r} \underline{90} \\ 0 \end{array}$$

15 times

25. $100\% - 16\% - 21\% - 43\% = 20\% = 0.2$

Chapter 1

REAL NUMBERS AND ALGEBRAIC EXPRESSIONS

Pretest

1. $+\$2000$

2. yes

3. Number line from -3 to 3 with a point at $-\frac{1}{2}$.

4. 5

5. $\left|-\frac{2}{3}\right|=\frac{2}{3}$

6. -31 is to the left of -1, so $-31<-1$.

7. Commutative Property of Addition.

8. $9+(-4)+2+(-9)$
$=9+2+(-4)+(-9)$
$=11+(-13)$
$=-2$

9. $3(-7)-5$
$=-21-5$
$=-26$

10. $\frac{4}{1}$, or 4

11. $(-72)\div(-8)=9$

12. Ten less than the product of three and n. Answers may vary.

13. $-6\cdot6\cdot6\cdot6=-6^4$

14. $2x-4y+8$
$=2(-2)-4(3)+8$
$=-4-12+8$
$=-8$

15. $3n-7+n$
$=3n+n-7$
$=(3+1)n-7$
$=4n-7$

16. $-5(2-x)+9x$
$=-10+5x+9x$
$=-10+(5+9)x$
$=-10+14x$
$=14x-10$

17. $5895-(-156)$
$=5895+156$
$=6051$ m

18. $\frac{L}{3}$, or $\frac{1}{3}L$

19. $P=2l+2w$
$=2(12)+2(3)$
$=24+6$
$=30$ cm

20. -4 is to the left of $-3, -1$, and $+2$.
He had the lowest score in the third round.

1.1 Real Numbers and the Real-Number Line

Practice

1. The number is question is below 0 (negative). -5°F

2. Number line from -4 to 4 with points at -1.7 and $\frac{5}{4}$.

3. **a.** 41

   b. $\frac{8}{9}$

   c. -1.7

   d. $\frac{2}{5}$

4. **a.** $\left|\frac{1}{2}\right| = \frac{1}{2}$

b. $|0| = 0$

c. $|-9| = 9$

d. $-|-3| = -(3) = -3$

5. **a.** True, because -2 is to the left of -1.

b. False, because 0 is to the right of -5.

c. True, because $\frac{10}{4}$ is equal to $\frac{5}{2}$.

d. True, because 0.3 is to the right of 0.

e. False, because -2.4 is to the left of 1.6.

6.

−2.4 −1.6 $-\frac{1}{2}$ 3

−3 −2 −1 0 1 2 3

$3, -\frac{1}{2}, -1.6$, and -2.4

7. Since $-92 < -52 < 0$, the Caspian Sea has the lowest elevation.

8. **a.**

B.C. A.D.

A B D C F E G

2000 1000 0 1000 2000

Babylonian Algebra (1700) Pythagoreans (500) Euclid (300) Diophantus (250) Al-Khowarizmi (825) Bhaskara (1100) Modern symbols (1500)

b. A, B, D, C, F, E, and G.

Exercises

1. The symbol $\varnothing$ represents the empty set.

3. The whole numbers consist of 0 and the natural numbers.

5. The set of integers is $\{..., -4, -3, -2, -1, 0, +1, +2, +3, +4, ...\}$.

7. Real numbers that cannot be written as the quotient of two integers are called irrational numbers.

9. The rational and the irrational numbers together make up the real numbers.

11. -5 km

13. -22.5°C

15. $-\$160$

17. −4 −3 −2 −1 0 1 2 3 4

19. −4 −3 −2 −1 0 1 2 3 4

21. −4 −3 −2 −1 0 1 2 3 4

23. −4 −3 −2 −1 0 1 2 3 4

25. Integers, Rational Numbers, Real Numbers

27. Rational Numbers, Real Numbers

29. Whole Numbers, Integers, Rational Numbers, Real Numbers

31. 3

33. 0

35. 3.5

37. $|-4| = 4$

39. $|0| = 0$

41. $|-4.6| = 4.6$

43. $-\left|\frac{1}{2}\right| = -\left(\frac{1}{2}\right) = -\frac{1}{2}$

45. 4 and -4

47. Impossible; absolute value is always positive or zero.

49. True, because -7 is to the left of -5.

51. True, because -1 is to the left of 2.5.

53. True, because 0 is to the right of $-1\frac{1}{4}$.

55. 0 is to the right of -1, so $0 > -1$.

57. -1.5 is to the right of -2, so $-1.5 > -2$.

59. 2.5 is equal to $2\frac{1}{2}$, so $2.5 = 2\frac{1}{2}$.

61. $|-4| = 4$ and $|4| = 4$, so $|-4| = |4|$.

63. $|-7.1| = 7.1$ and 6.2 is to the left of 7.1, so $6.2 < |-7.1|$

65.

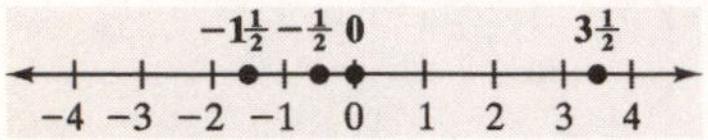

$3\frac{1}{2}, 0, -\frac{1}{2}, -1\frac{1}{2}$

67.

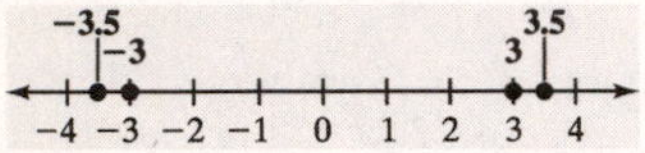

$3.5, 3, -3, -3.5$

69. The value is a loss (negative). $-\$53$

71. Rational Numbers, Real Numbers

73. $-|-1.5| = -(1.5) = -1.5$

75. False, because -5 is to the right of $-5\frac{1}{3}$.

77. $-|3| = -3$ and $|-3| = 3$. Since -3 is to the left of 3, then $-|3| < |-3|$.

79. $-\$200 > -\$2000$, so he is better off financially today.

81. $-64.8^\circ C < -64.3^\circ C < -54.5^\circ C$, so the coldest temperature is $-64.8^\circ C$.

83. **a.** Sirius is brighter as seen from Earth.

b. -13

c.

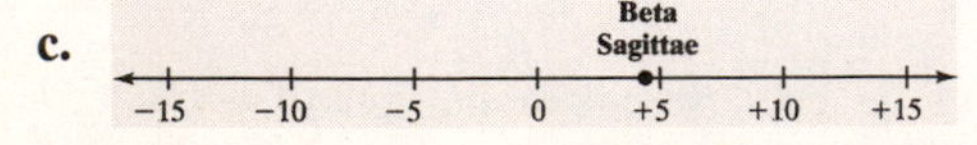

85. **a.** Socrates

b. Pythagoras

c. Euclid

d. Woods

1.2 Addition of Real Numbers

Practice

1. $2 + (-3) = -1$

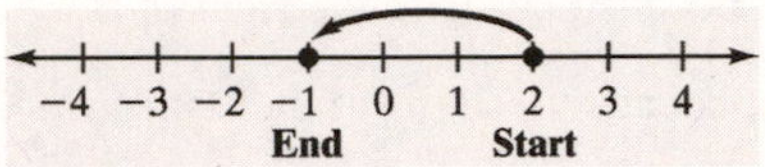

2. $-5 + 5 = 0$

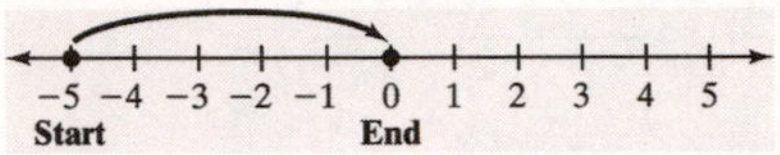

3. $0 + 0.5 = 0.5$

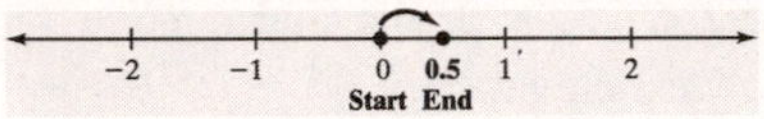

4. $-13 + (-18) = -(13 + 18) = -31$

5. $10.1 + (-6.6) = 10.1 - 6.6 = 3.5$

6. $-\frac{1}{3} + \frac{1}{3} = 0$

7. $-\frac{7}{18} + \frac{1}{9}$

$= -\frac{7}{18} + \frac{2}{18}$

$= -\frac{5}{18}$

8. $\$37.50 + \$2 + (-\$1) + (-\$2)$

$= \$39.50 + (-\$3)$

$= \$36.50$

9. $6.002 + (-9.37) + (-0.22)$

$= 6.002 + (-9.59)$

$= -3.588$

Exercises

1. To add real numbers a and b on a number line, start at a and move to the left if b is negative.

3. Additive inverses are also called opposites.

5. $4+(-3)=1$

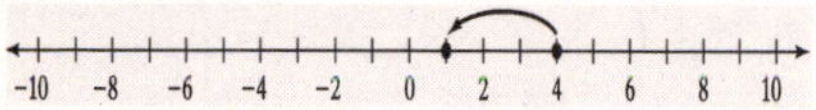

7. $8+(-8)=0$

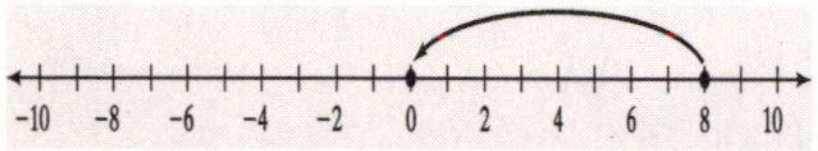

9. $-5\frac{1}{2}+10=4\frac{1}{2}$

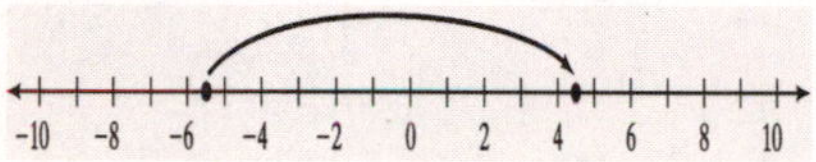

11. 18, because $-18+18=0$.

13. 0, because $0+0=0$.

15. $24+(-1)$
$=24-1$
$=23$

17. $-10+5$
$=-(10-5)$
$=-5$

19. $10+(-6)$
$=10-6$
$=4$

21. $-50+(-30)$
$=-(50+30)$
$=-80$

23. $-10+2$
$=-(10-2)$
$=-8$

25. $-18+18=0$

27. $5.2+(-0.9)$
$=5.2-0.9$
$=4.3$

29. $-0.2+0.8$
$=0.8-0.2$
$=0.6$

31. $-9.6+3.9$
$=-(9.6-3.9)$
$=-5.7$

33. $(-9.8)+(-6.5)$
$=-(9.8+6.5)$
$=-16.3$

35. $-\frac{1}{2}+\left(-\frac{1}{2}\right)$
$=-\left(\frac{1}{2}+\frac{1}{2}\right)$
$=-\frac{2}{2}$
$=-1$

37. $\frac{4}{15}+\left(-\frac{2}{3}\right)$
$=\frac{4}{15}+\left(-\frac{10}{15}\right)$
$=-\frac{6}{15}$
$=-\frac{2}{5}$

39. $-1\frac{3}{5}+2$
$=-\frac{8}{5}+\frac{10}{5}$
$=\frac{2}{5}$

41. $2\frac{1}{3}+\left(-1\frac{1}{2}\right)$
$=\frac{7}{3}+\left(-\frac{3}{2}\right)$
$=\frac{14}{6}+\left(-\frac{9}{6}\right)$
$=\frac{5}{6}$

43. $-24+(25)+(-89)$
$=25+(-24)+(-89)$
$=25+(-113)$
$=-88$

45. $-0.4+(-2.6)+(-4)$
$=-3+(-4)$
$=-7$

47. $107+(-97)+(-45)+23$
$=107+23+(-97)+(-45)$
$=130+(-142)$
$=-12$

49. $-2.001+(0.59)+(-8.1)+10.756$
$=0.59+10.756+(-2.001)+(-8.1)$
$=11.346+(-10.101)$
$=1.245$

51. $2+(-9)=-7$

-10 -8 -6 -4 -2 0 2 4 6 8 10

53. $53+(-38)=15$

55. $-8.5+4.8=-3.7$

57. $-4.1+2.3+1.5$
$=-1.8+1.5$
$=-0.3$

59. $-2+7$
$=7-2$
$=5$

$5°$ above $0°$ $(+5)$

61. $\$132{,}000+(-\$148{,}000)$
$=-(\$148{,}000-\$132{,}000)$
$=-\$16{,}000$
Lost $16,000 (–$16,000)

63. $3+(-3)+(-7)+11$
$=0+(-7)+11$
$=11-7$
$=4$
The San Francisco 49ers won by 4 points.

65. $\$371.25+(-\$71.33)+(-\$51.66)+\$35$
$=\$371.25+\$35+(-\$71.33)+(-\$51.66)$
$=\$406.25+(-\$122.99)$
$=\$283.26$
Yes, he will have $283.26 in the account.

67. $32{,}000+(-700)$
$=32{,}000-700$
$=31{,}300$
$31,300 ft

1.3 Subtraction of Real Numbers

Practice

1. $4-(-1)$
$=4+1$
$=5$

2. $-12-(-15)$
$=-12+15$
$=3$

3. $8-12$
$=8+(-12)$
$=-4$

4. $-8.1-7.6$
$=-8.1+(-7.6)$
$=-15.7$

5. $5-(-8)-(-15)$
$=5+8-(-15)$
$=13+15$
$=28$

6. $4+(-6)-(-11)+8$
$=-2-(-11)+8$
$=-2+11+8$
$=9+8$
$=17$

7. $770-(-100)$
$=770+100$
$=870$
Paper is invented 870 years earlier than wood block printing.

Exercises

1. $25-8=17$

3. $-24-7$
$=-24+(-7)$
$=-31$

5. $(-19)-25$
$=(-19)+(-25)$
$=-44$

7. $52-(-19)$
$=52+19$
$=71$

9. $60-95$
$=60+(-95)$
$=-35$

11. $-34-(-2)$
$=-34+2$
$=-32$

13. $16-(-16)$
$=16+16$
$=32$

15. $0-45$
$=0+(-45)$
$=-45$

17. $-31-31$
$=-31+(-31)$
$=-62$

19. $22-(-22)$
$=22+22$
$=44$

21. $200-(-800)$
$=200+800$
$=1000$

23. $6-7.42$
$=6+(-7.42)$
$=-1.42$

25. $-7.3-(0.5)$
$=-7.3+(-0.5)$
$=-7.8$

27. $(-5.6)-(-5.6)$
$=-5.6+5.6$
$=0$

29. $8.6-(-1.7)$
$=8.6+1.7$
$=10.3$

31. $-\frac{1}{3}-\frac{5}{6}$
$=-\frac{2}{6}-\frac{5}{6}$
$=-\frac{7}{6}$

33. $-12-\frac{1}{4}$
$=-12+\left(-\frac{1}{4}\right)$
$=-12\frac{1}{4}$

35. $4\frac{3}{5}-\left(-1\frac{1}{2}\right)$

$=4\frac{3}{5}+1\frac{1}{2}$

$=\frac{23}{5}+\frac{3}{2}$

$=\frac{46}{10}+\frac{15}{10}$

$=\frac{61}{10}$

$=6\frac{1}{10}$

37. $3+(-6)-(-15)$

$=-3-(-15)$

$=-3+15$

$=12$

39. $8-10+(-5)$

$=8+(-10)+(-5)$

$=-2+(-5)$

$=-7$

41. $-9+(-4)-9+4$

$=-13-9+4$

$=-13+(-9)+4$

$=-22+4$

$=-18$

43. $9-12-18$

$=9+(-12)-18$

$=-3-18$

$-3+(-18)$

$=-21$

45. $-10.722+(-3.913)-8.36-3.492$

$=-14.635-8.36-3.492$

$=-14.635+(-8.36)-3.492$

$=-22.995-3.492$

$=-22.995+(-3.492)$

$=-26.487$

47. $17-(-31)$

$=17+31$

$=48$

49. $(-23)-(-15)$

$=(-23)+(-15)$

$=-38$

51. $\frac{5}{6}-\left(-\frac{3}{4}\right)$

$=\frac{5}{6}+\frac{3}{4}$

$=\frac{10}{12}+\frac{9}{12}$

$=\frac{19}{12}$

$=1\frac{7}{12}$

53. $18+(-13)-(-9)$

$=5-(-9)$

$=5+9$

$=14$

55. $2002-(-776)$

$=2002+776$

$=2778$

Approximately 28 centuries

57. $4000-(-5000)$

$=4000+5000$

$=9000$ ft

59. $-\$5291-(-\$281,330)$

$=-\$5291+\$281,330$

$=\$276,039$

61. $10,152-(-184)$

$=10,152+184$

$=10,336$ ft

63. **a.** Radon

$-61.8-(-71)$
$=-61.8+71=9.2^{\circ}$

Neon

$-246-(-248.7)$
$=-246+248.7=2.7^{\circ}$

Bromine

$58.8-(-7.2)$
$=58.8+7.2=66^{\circ}$

b. Bromine is liquid in the widest range of temperatures.

c. Bromine is liquid at 0°C.

1.4 Multiplication and Division of Real Numbers

Practice

1. $-1(-100)=100$

2. **a.** $\left(-\frac{2}{3}\right)(-12)$

$=-\frac{2}{\not{3}_1}\cdot\frac{\not{-12}^{4}}{1}$

$=8$

b. $\left(-\frac{1}{3}\right)\left(\frac{5}{9}\right)=-\frac{5}{27}$

c. $(-0.4)(-0.3)=0.12$

d. $2.5(-1.9)=-4.75$

e. $0\cdot(-2.8)=0$

f. $1\cdot\frac{2}{3}=\frac{2}{3}$

3. $-8(4)(-2)$
$=-32(-2)$
$=64$

4. $4(-25)-(-2)(36)$
$=-100-(-72)$
$=-100+72$
$=-28$

5. $-5(-9+15)$
$=-5(6)$
$=-30$

6. $10-[5(3+1)]$
$=10-[5(4)]$
$=10-(20)$
$=-10$

7. $10-(5.3)(3)$
$=10-15.9$
$=-5.9$

About -5.9; the rock is moving downward at a velocity of 5.9 ft/sec.

8. **a.** $40\div(-5)=-8$

b. $\frac{-42}{-6}=7$

c. $\frac{-5}{10}=-\frac{1}{2}$

d. $\frac{-6.3}{9}=-0.7$

e. $\frac{-24}{-0.4}=60$

9. **a.** $\frac{0}{-2}=0$

b. $-2\div0$ is not possible because $-2\div0$ is undefined.

c. $0\div7=0$

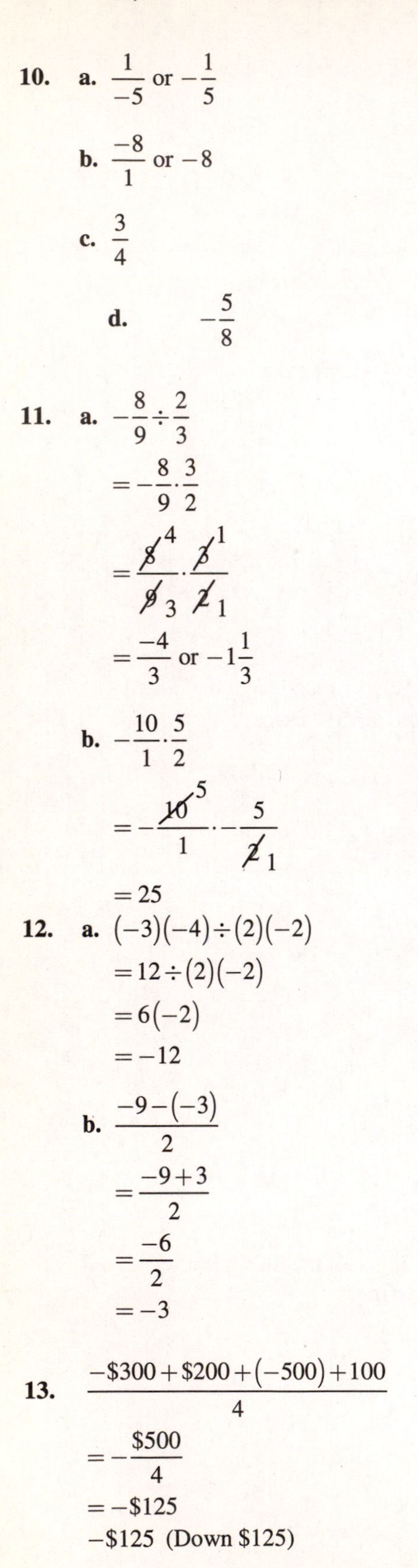

10. **a.** $\frac{1}{-5}$ or $-\frac{1}{5}$

b. $\frac{-8}{1}$ or -8

c. $\frac{3}{4}$

d. $-\frac{5}{8}$

11. **a.** $-\frac{8}{9}\div\frac{2}{3}$

$= -\frac{8}{9}\cdot\frac{3}{2}$

$= \frac{\cancel{8}^4}{\cancel{9}_3}\cdot\frac{\cancel{3}^1}{\cancel{2}_1}$

$= \frac{-4}{3}$ or $-1\frac{1}{3}$

b. $-\frac{10}{1}\cdot\frac{5}{2}$

$= -\frac{\cancel{10}^5}{1}\cdot-\frac{5}{\cancel{2}_1}$

$= 25$

12. **a.** $(-3)(-4)\div(2)(-2)$

$= 12\div(2)(-2)$

$= 6(-2)$

$= -12$

b. $\frac{-9-(-3)}{2}$

$= \frac{-9+3}{2}$

$= \frac{-6}{2}$

$= -3$

13. $\frac{-\$300+\$200+(-500)+100}{4}$

$= -\frac{\$500}{4}$

$= -\$125$

$-\$125$ (Down \$125)

Exercises

1. The product of two real numbers with different signs is <u>negative</u>.

3. The <u>multiplicative identity</u> property tells us that the product of any number and 1 is the original number.

5. To divide real numbers, first divide the <u>absolute values</u>.

7. Any real number *a* divided by zero is <u>undefined</u>.

9. The product of any real number and its multiplicative inverse is <u>one</u>.

11. Multiplicative inverse property

13. No

15. No

17. No

19. $6(-2) = -12$

21. $-7(-3) = 21$

23. $-12\left(\frac{1}{4}\right) = \frac{-\cancel{12}^3}{1}\cdot\frac{1}{\cancel{4}_1} = -3$

25. $-\frac{1}{3}\cdot\frac{4}{9} = -\frac{4}{27}$

27. $\left(1\frac{1}{3}\right)\left(-\frac{4}{9}\right) = \left(\frac{4}{3}\right)\left(-\frac{4}{9}\right) = -\frac{16}{27}$

29. $-1.5(-0.6) = 0.9$

31. $1.2(-50) = -60$

33. $3(-2)(-20)$

$= -6(-20)$

$= 120$

35. $-15(-3)(0)$
$=45(0)$
$=0$

37. $-6(1)(-2)(-3)(-4)$
$=-6(-2)(-3)(-4)$
$=12(-3)(-4)$
$=-36(-4)$
$=144$

39. $-4(5)(-6)(1)$
$=5(1)(-4)(-6)$
$=5(24)$
$=120$

41. $\left(-\frac{1}{3}\right)\left(-\frac{1}{3}\right)\left(-\frac{1}{3}\right)=\frac{1}{9}\left(-\frac{1}{3}\right)=-\frac{1}{27}$

43. $-6.24(0.08)(-1.97)$
$=-0.4992(-1.97)$
$=0.983424$
≈ 0.98

45. $-7+3(-2)-10$
$=-7+(-6)-10$
$=-13-10$
$=-23$

47. $-3-5(-6)$
$=-3-(-30)$
$=-3+30$
$=27$

49. $\frac{3}{5}(-15)-6$
$=\frac{3}{\not{5}_1}\cdot\frac{-\not{15}^3}{1}-6$
$=-9-6$
$=-15$

51. $-5\cdot(-3+4)$
$=-5\cdot(1)$
$=-5$

53. $-6-[3(5-9)]$
$=-6-[3(-4)]$
$=-6-(-12)$
$=-6+12$
$=6$

55. **a.** $-4(-2)-3$
$=8-3$
$=5$

b. $-4(-1)-3$
$=4-3$
$=1$

c. $-4(0)-3$
$=0-3$
$=-3$

d. $-4(1)-3$
$=-4-3$
$=-7$

e. $-4(2)-3$
$=-8-3$
$=-11$

57. **a.** $-\frac{2}{1}$ or -2

b. $\frac{1}{5}$

c. $-\frac{4}{3}$ or $-1\frac{1}{3}$

d. $\frac{5}{16}$

e. $\frac{1}{-1}$ or -1

59. $-8\div(-1)=8$

61. $-63\div 7=-9$

63. $\frac{0}{-9}=0$

65. $-2500\div(100)=-25$

67. $-200 \div (-8) = 25$

69. $-64 \div (-16) = 4$

71. $\frac{-25}{-5} = 5$

73. $\frac{-2}{16} = -\frac{1}{8}$ or -0.125

75. $\frac{10}{-20} = -\frac{1}{2}$ or -0.5

77. $\frac{4}{5} \div \left(-\frac{2}{3}\right)$
$= \frac{\cancel{4}^2}{5} \cdot \left(-\frac{3}{\cancel{2}_1}\right)$
$= -\frac{6}{5}$

79. $8 \div \left(-\frac{1}{4}\right)$
$= \frac{8}{1} \cdot \left(-\frac{4}{1}\right)$
$= -32$

81. $2\frac{1}{2} \div (-20)$
$= \frac{\cancel{5}^1}{2} \cdot \left(-\frac{1}{\cancel{20}_4}\right)$
$= -\frac{1}{8}$

83. $(-3.5) \div 7 = -0.5$

85. $10 \div (-0.5) = -20$

87. $\frac{-7.2}{0.9} = -\frac{72}{9} = -8$

89. $\frac{-3}{-0.3} = \frac{30}{3} = 10$

91. $(-15.5484) \div (-6.13) \approx 2.54$

93. $-0.8385 \div (0.715) \approx -1.17$

95. $-16 \div (-2)(-2)$
$= 8(-2)$
$= -16$

97. $(3-7) \div (-4)$
$= (-4) \div (-4)$
$= 1$

99. $\frac{2+(-6)}{-2} = \frac{-4}{-2} = 2$

101. $(4-6) \div (1-5)$
$= -2 \div (-4)$
$= \frac{1}{2}$

103. $-56 \div 7 - 4 \cdot (-3)$
$= -8 - (-12)$
$= -8 + 12$
$= 4$

105. $-4\left(\frac{1}{2}\right)(-2) \div \left(-\frac{1}{8}\right)$
$= -2 - 2 \cdot (-8)$
$= -2 - (-16)$
$= -2 + 16$
$= 14$

107. $-2.8(-1.3) = 3.64$

109. $3(-5)(1)(-4)(-2)$
$= 3(1)(-5)(-4)(-2)$
$= 3(-40)$
$= -120$

111. a. $-5(2)+4$
$=-10+4$
$=-6$

b. $-5(1)+4$
$=-5+4$
$=-1$

c. $-5(0)+4$
$=0+4$
$=4$

d. $-5(-1)+4$
$=5+4$
$=9$

e. $-5(-2)+4$
$=10+4$
$=14$

113. $-5-6(-2)+(-3)$
$=-5-(-12)+(-3)$
$=-5+12+(-3)$
$=7+(-3)$
$=7-3$
$=4$

115. $-4\frac{1}{2}\div 3$
$=-\frac{9}{2}\div\frac{3}{1}$
$=-\frac{\cancel{9}^{3}}{2}\cdot\frac{1}{\cancel{3}_{1}}$
$=-\frac{3}{2}$

117. $(-0.72)\div(-6)=0.12$

119. $-65\div(-13)=5$

121. $-12\div(5-7)$
$=-12\div(-2)$
$=6$

123. $2\cdot(-5)=-10$
$-\$10$ (lost $\$10$)

125. $3(4)+2(1)+4(-3)+1(0)$
$=12+2+(-12)+0$
$=2+0$
$=2$
The team scored 2 more points than its opponents $(+2)$.

127. $-3\cdot 5=-15$
-15 in. (dropped 15 in.)

129. a. $3(30)+\frac{1}{2}(-450)$
$=90-225$
$=-135$ calories

b. $\frac{1}{2}(-288)+2(125)$
$=-144+250$
$=106$ calories

c. $3(80)+2(210)+2(-612)+1(-288)$
$=240+420-1224-288$
$=660-1224-288$
$=-564-288$
$=-852$ calories

131. $-130\div 5=-26$

133. $-47{,}355\div 10=-4735.5$
A decrease of about 4736 people per year (-4735.5)

135. $-24\div 6=-4$
An average loss of 4 yards

137. $-\$72{,}000\div 12=-\$6000$
Expenses averaged $\$6000$ per month $(-\$6000)$.

139. $\frac{2+0+(-7)+(-11)+1}{5}$
$=\frac{-15}{5}$
$=3$
Yes.

1.5 Properties of Real Numbers

Practice

1. a. $-3+5=5+(-3)$

   b. $b+3a=3a+b$

2. a. $(-8)(2)=(2)(-8)$

   b. $-4n=n(-4)$

3. a. $[8+(-1)]+2=8+[(-1)+2]$

   b. $(x+3y)+z=x+(3y+z)$

4. a. $[(-3)(5)](-2)=(-3)[(5)(-2)]$

   b. $(3)(-6n)=[3(-6)]n$

5. $$\begin{aligned}&-8+(-4)+3+(-8)\\&=-8+(-4)+(-8)+3\\&=-12+(-8)+3\\&=-20+3\\&=-17\end{aligned}$$

6. a. $$\begin{aligned}&(-6)(-1)(4)(-5)\\&=(-6)(-1)(-5)(4)\\&=6(-5)(4)\\&=-30(4)\\&=-120\end{aligned}$$

   b. $$\begin{aligned}&(3)(-4)(2)(-1)(6)\\&=[(3)(2)(6)][(-4)(-1)]\\&=[(6)(6)](4)\\&=(36)(4)\\&=144\end{aligned}$$

7. a. $-5+0=-5$

   b. $0+6y=6y$

   c. $(1)(-2)=-2$

   d. $(-5x)(1)=-5x$

8. a. 2

   b. $\frac{2}{3}$

   c. $-y$

9. a. 5

   b. $\frac{1}{2}$

   c. $-\frac{1}{5}$

   d. $-\frac{7}{2}$

10. a. $(-2)(9+4.3)=(-2)(9)+(-2)(4.3)$

    b. $0.2(a+b)=0.2a+0.2b$

    c. $(2-p)\cdot q=2q-pq$

11. $4\cdot\left(\frac{1}{4}x\right)=\left(4\cdot\frac{1}{4}\right)x$ **a.** The associative property of multiplication

    $=1x$ **b.** The multiplicative inverse property

    $=x$ **c.** The multiplicative identity property

12. Because 100 = 1, the return is $1n = n$ dollars by the multiplicative identity property.

Exercises

1. The <u>commutative property</u> of multiplication states that we get the same product when we multiply two numbers in any order.

3. A product with an odd number of negative factors is <u>negative</u>.

5. The <u>multiplicative identity property</u> states that the product of a number and 1 is the number itself.

7. If $a\cdot\frac{1}{a}=1$ and $\frac{1}{a}\cdot a=1$, then a and $\frac{1}{a}$ are multiplicative inverses, or <u>reciprocals</u>, of each other.

9. $3.7+2=2+3.7$

11. $[(-1)+(-6)]+7=(-1)+[(-6)+7]$

13. $-3+0=-3$

15. $3(1+9)=3\cdot1+3\cdot9$

17. $(2+7)\cdot5=5(2+7)$

19. $2a+2b=2(a+b)$

21. Commutative property of multiplication.

23. Commutative property of addition.

25. Distributive property.

27. Multiplicative identity property

29. Additive inverse property.

31. Associative property of addition.

33. Associative property of multiplication.

35. $8+\underline{(-3)+(-5)}=8+(-8)=0$ or

$\underline{8+(-3)}+(-5)=5+(-5)=0$

37. Rearrange the numbers by sign.

$2+(-3.8)+9.13+(-1)$

$=\underline{2+9.13}+\underline{(-3.8)+(-1)}$

$=11.13+(-4.8)$

$=6.33$

39. $\underline{(-7)(-2)}(3)=(14)(3)=42$

41. $\underline{(-2)(-2)}(-2)=(4)(-2)=-8$

43. $\underline{(-5)(-7)}(-2)(10)=\underline{(35)(-2)}(10)$

$=(-70)(10)=-700$

45. The additive inverse of 2 is -2.

47. The additive inverse of -7 is 7.

49. The multiplicative inverse of 7 is $\frac{1}{7}$.

51. The multiplicative inverse of -1 is

$\frac{1}{-1}=-1.$

53. $(-4)(2+5)=(-4)(2)+(-4)(5)$

55. $(x+10)\cdot3=3\cdot x+3\cdot10=3x+30$

57. $-(a+6b)=(-1)(a)+(-1)(6b)$

$=-a+(-6b)$ or $-a-6b$

59. $n(n-2)=n\cdot n-n\cdot2=n^2-2n$

61. **a.** The commutative property of multiplication.
b. The associative property of multiplication.
c. Multiplication of real numbers.

63. **a.** The associative property of multiplication.
b. The multiplicative inverse property.
c. The multiplicative identity property.

65. 0.2

67. Associative property of multiplication.

69. Multiplicative inverse property.

71. Yes, by the commutative property of addition. (Distance traveled on bus) + (distance walked)
= (distance walked) + (distance traveled on bus).

73. Yes, by the distributive property.
$r(p+q)=rp+rq$

75. Using the additive identity property: $p+0=p$, his weight at the end of the week is p lb.

77. Yes, by the commutative property of multiplication: $\frac{1}{2}\cdot9\cdot12=\frac{1}{2}\cdot12\cdot9$. The calculations give the same result.

1.6 Algebraic Expressions, Translations, and Exponents

Practice

1. **a.** 3

   b. 1

2. **a.** $\frac{1}{3}$ of p

   b. the difference between 9 and x

   c. s divided by -8

   d. n plus -6

   e. the product $\frac{3}{8}$ and m

3. **a.** twice x minus the product of 3 and y

   b. 4 plus $3m$

   c. 5 times the difference between a and b

   d. the difference between r and s divided by the sum of r and s

4. **a.** $\frac{1}{6}n$

   b. $n+(-5)$

   c. $m-(-4)$

   d. $\frac{100}{x}$

   e. $-2y$

5. **a.** $m+(-n)$

   b. $5y-11$

   c. $\frac{m+n}{mn}$

   d. $-6(x+y)$

6. $60(\text{m}+1)\text{words}$

7. **a.** $-6^2=-(6)(6)=-36$

   b. $(-6)^2\cdot(-3)^2$
   $=(-6)(-6)(-3)(-3)$
   $=(36)(9)$
   $=324$

8. $2(2)(2)(2)(-5)(-5)=2^4(-5)^2$

9. **a.** $-x\cdot x\cdot x\cdot x\cdot x=-x^5$

   b. $2m\cdot m\cdot m\cdot n\cdot n\cdot n\cdot n=2m^3n^4$

10. $\frac{10\,\text{hrs}}{2\,\text{hrs}}=5$

    The population after 10 hr was $343x$, or 3^5x.

Exercises

1. In algebra, a(n) <u>variable</u> can be used as an unknown quantity.

3. A(n) <u>algebraic expression</u> consists of one or more terms, separated by addition signs.

5. The algebraic expression $\frac{x}{7}$ can be translated as <u>the quotient of x and 7</u>.

7. In the expression x^a, x is called the <u>base</u>.

9. 1

11. 3

13. 2

15. 3 plus t

17. 4 less than x

19. 7 times r

21. the quotient of a and 4

23. the product of $\frac{4}{5}$ and w

25. the sum of negative 3 and z

27. twice n plus 1

29. 4 times the quantity x minus y

31. 1 minus 3 times x

33. the product of a and b divided by the sum of a and b

35. twice x minus 5 times y

37. $x+5$

39. $d-4$

41. $-6a$

43. $y+(-15)$

45. $\frac{1}{8}k$

47. $\frac{m}{n}$

49. $a-2b$

51. $4z+5$

53. $12(x-y)$

55. $\frac{b}{a-b}$

57. $-3^2=-(3)(3)=-9$

59. $(-3)^3\cdot(-4)^2$
$=(-3)(-3)(-3)(-4)(-4)$
$=-27\cdot16$
$=-432$

61. $-2(-2)(-2)(4)(4)=(-2)^3\cdot(4)^2$

63. $6(6)(-3)(-3)(-3)=6^2\cdot(-3)^3$

65. $3(n)(n)(n)=3n^3$

67. $-4a\cdot a\cdot a\cdot b\cdot b=-4a^3b^2$

69. $-y\cdot y\cdot y=-y^3$

71. $10a\cdot a\cdot a\cdot b\cdot b\cdot c=10a^3b^2c$

73. $-x\cdot x\cdot y\cdot y\cdot y=-x^2y^3$

75. 8 times the quantity w minus y

77. the product of s and r divided by the difference between r and s

79. $a\cdot a(-b)(-b)=a^2(-b)^2$

81. $-3^2(-2)^2$
$=-(3)(3)(-2)(-2)$
$=-(9)(4)$
$=-36$

83. $x+2y$

85. $90°+x°+y°$

87. $\frac{30,000}{p}$ dollars

89. $(t+x)$ dollars

91. $\frac{30\text{ years}}{10\text{ years}}=3$ time periods;
$(2^3\cdot5000)$ dollars

93. $s\cdot s=s^2$

95. $10,000\left(\frac{1}{20}\right)(20-n)$ dollars or
$500(20-n)$ dollars

97. The entire area of the floor is ab sq ft. The area of the rug is cd sq ft. The area of the floor not covered by the rug is $(ab-cd)$ sq ft.

1.7 Simplifying Algebraic Expressions

Practice

1. **a.** Terms: m and $-3m$; Like

   b. Terms: $5x$ and 7; Unlike

   c. Terms: $2x^2y$ and $-3xy^2$; Unlike

   d. Terms: m, $2m$, and $-4m$; Like

2. **a.** $5x+x=(5+1)x$
   $=6x$

   b. $-5y-y=(-5-1)y$
   $=-6y$

   c. $a-3a+b=(1-3)a+b$
   $=-2a+b$

   d. $-9t+3t+6t=(-9+3+6)t$
   $=0\cdot t$
   $=0$

3. **a.** $y^2-3y^2=(1-3)y^2$
   $=-2y^2$

   b. $7a^2b+3ab^2$
   The terms are not like terms.
   This cannot be simplified.

   c. $4xy^2-xy^2=(4-1)xy^2$
   $=3xy^2$

4. $3(y-4)+2=3y-12+2$
   $=3y-10$

5. $-(2a-3b)=-1\cdot 2a+(-1)(-3b)$
   $=-2a+3b$

6. $5y-6-(y-5)=5y-6-y+5$
   $=4y-1$

7. $(y+3)-3(y+7)=y+3-3y-21$
   $=-2y-18$

8. $10-[4y+3(2y-1)]=10-[4y+6y-3]$
   $=10-[10y-3]$
   $=10-10y+3$
   $=-10y+13$

9. $5c+12(c-40)=5c+12c-480$
   $=17c-480$

Exercises

1. The <u>coefficient</u> of the term $-x$ is -1.

3. The <u>distributive property</u> states that for any real numbers, a, b, and c, $a\cdot(b+c)=a\cdot b+a\cdot c$.

5. When removing parentheses preceded by a(n) <u>negative</u> sign, change all the terms in parentheses to the opposite sign.

7. 7

9. -5

11. 1

13. -1

15. -0.1

17. $\frac{2}{3}$

19. 2; -5

21. Terms: $2a$ and $-a$; like

23. Terms: $5p$ and 3; unlike

25. Terms: $4x^2$ and $-6x^2$; like

27. Terms: x^2 and $7x^3$; unlike

29. $3x+7x=(3+7)x$
    $=10x$

31. $-10n-n=(-10-1)n$
    $=-11n$

33. $20a-10a+4a=(20-10+4)a$
$=14a$

35. $3y-y+2=(3-1)y+2$
$=2y+2$

37. $8b^3+b^3-9b^3=(8+1-9)b^3$
$=0\cdot b^3$
$=0$

39. $-b^2+ab^2$ The terms are not like terms. This cannot be simplified.

41. $3r^2t^2+r^2t^2=(3+1)r^2t^2$
$=4r^2t^2$

43. $3x^2y-5xy^2$ The terms are not like terms. This cannot be simplified.

45. $2(x+3)-4=2x+6-4$
$=2x+2$

47. $(7x+1)+(2x-1)=7x+1+2x-1$
$=9x$

49. $-(3y-10)=-1(3y-10)$
$=-3y+10$

51. $5x-3-(x+6)=5x-3-x-6$
$=4x-9$

53. $-4(n-9)+3(n+1)$
$=-4n+36+3n+3$
$=-n+39$

55. $x-4-2(x-1)+3(2x+1)$
$=x-4-2x+2+6x+3$
$=(1-2+6)x+(-4+2+3)$
$=5x+1$

57. $7+3[x-2(x-1)]=7+3[x-2x+2]$
$=7+3[-x+2]$
$=7-3x+6$
$=-3x+13$

59. $10-3[4a+8-3a]=10-3[a+8]$
$=10-3a-24$
$=-3a-14$

61. $4pq^2-6q^2$ The terms are not like terms. This cannot be simplified.

63. Terms: $3a$ and $3a^2$; unlike

65. 1

67. $y-3-4(y-2)+2(3y+1)$
$=y-3-4y+8+6y+2$
$=(1-4+6)y+(-3+8+2)$
$=3y+7$

69. $x^\circ+x^\circ+40^\circ=2x^\circ+40^\circ$

71. $d+2(d+4)=d+2d+8$
$=3d+8$
$(3d+8)$ dollars

73. $n+(n+1)+(n+2)=n+n+1+n+2$
$=(1+1+1)n+(1+2)$
$=3n+3$

75. $0.05x+0.04(1000-x)=0.05x+40-0.04x$
$=0.01x+40$
$(0.01x+40)$ dollars

1.8 Translating and Evaluating Algebraic Expressions

Practice

1. a. $25+m$
$=25+(-10)$
$=15$

b. $-3xy$
$=-3(-2)(5)$
$=6(5)$
$=30$

2. **a.** $5a-2c$

$=5(2)-2(-4)$

$=10+8$

$=18$

b. $2(d-b)$

$2(5-(-3))$

$2(5+3)$

$=2(8)$

$=16$

c. $3cd^2$

$=3(-4)(5)^2$

$=3(-4)(25)$

$=-12(25)$

$=-300$

d. $2a^3+4b^2$

$2(2)^3+4(-3)^2$

$=2(8)+4(9)$

$=16+36$

$=52$

3. **a.** $\frac{x-2z}{y}$

$=\frac{-5-2(1)}{-3}$

$=\frac{-5-2}{-3}$

$=\frac{-7}{-3}$

$=\frac{7}{3}$

b. $\frac{x-z}{x+y}$

$=\frac{-5-1}{-5+(-3)}$

$=\frac{-6}{-8}$

$=\frac{3}{4}$

c. $(-y)^4$

$=\left[-(-3)^4\right]$

$=(3)^4$

$=(3)(3)(3)(3)$

$=81$

d. $-y^4$

$=-(-3)^4$

$=-(-3)(-3)(-3)(-3)$

$=-81$

4. $\frac{(-0.5)^2+(0.3)^2+(0.2)^2}{3}$

$=\frac{0.25+0.09+0.04}{3}$

$=\frac{0.38}{3}$

≈ 0.1

5. $F=\frac{9}{5}C+32$

6. $d=rt$

$=50(1.6)$

$=80$ mi

The distance d is 80 mi.

7. **a.** $K=C+273$

b. $K=C+273$

$=-6+273$

$=267$

Exercises

1. $b-5=3-5$

$=-2$

3. $-2ac=-2(4)(-2)$

$=(-8)(-2)$

$=16$

5. $-2a^2=-2(4)^2$

$=-2(16)$

$=-32$

7. $2a-15=2(4)-15$
$=8-15$
$=-7$

9. $a+2c=4+2(-2)$
$=4+(-4)$
$=0$

11. $2(a-c)=2(4-(-2))$
$=2(4+2)$
$=2(6)$
$=12$

13. $-a+b^2=-4+(3)^2$
$=-4+9$
$=5$

15. $3a^2-c^3=3(4)^2-(-2)^3$
$=3(16)-(-8)$
$=48+8$
$=56$

17. $\frac{a+b}{b-a}=\frac{4+3}{3-4}$
$=\frac{7}{-1}$
$=-7$

19. $\frac{3}{5}(a+b+c)^2=\frac{3}{5}(4+3+(-2))^2$
$=\frac{3}{5}(7+(-2))^2$
$=\frac{3}{5}(5)^2$
$=\frac{3}{5}\cdot\frac{25}{1}$
$=15$

21. $2w^2-3x+y-4z$
$=2(-0.5)^2-3(2)+(-3)-4(1.5)$
$=2(0.25)-3(2)+(-3)-4(1.5)$
$=0.5+(-6)+(-3)+(-6)$
$=-5.5+(-3)+(-6)$
$=-8.5+(-6)$
$=-14.5$

23. $w-7z-\frac{1}{4}(x-6y)$
$=-0.5-7(1.5)-\frac{1}{4}(2-6(-3))$
$=-0.5-7(1.5)-\frac{1}{4}(2+18)$
$=-0.5-7(1.5)-\frac{1}{4}(20)$
$=-0.5+(-10.5)+(-5)$
$=-11+(-5)$
$=-16$

25. $\frac{-10xy}{(w-z)^2}$
$=\frac{-10(2)(-3)}{(-0.5-1.5)^2}$
$=\frac{-10(2)(-3)}{(-2)^2}$
$=\frac{-10(2)(-3)}{4}$
$=\frac{60}{4}$
$=15$

27. $2x+5$
$2(0)+5=0+5=5$
$2(1)+5=2+5=7$
$2(2)+5=4+5=9$
$2(-1)+5=-2+5=3$
$2(-2)+5=-4+5=1$

29. $y-0.5$
$0-0.5=-0.5$
$1-0.5=0.5$
$2-0.5=1.5$
$3-0.5=2.5$
$4-0.5=3.5$

31. $-\frac{1}{2}x$

$-\frac{1}{2}(0)=0$

$-\frac{1}{2}(2)=-1$

$-\frac{1}{2}(4)=-2$

$-\frac{1}{2}(-2)=1$

$-\frac{1}{2}(-4)=2$

33. $\frac{n}{2}$

$\frac{2}{2}=1$

$\frac{4}{2}=2$

$\frac{6}{2}=3$

$\frac{-2}{2}=-1$

$\frac{-4}{2}=-2$

35. $-g^2$

$-(0)^2=0$

$-(1)^2=-1$

$-(2)^2=-4$

$-(-1)^2=-1$

$-(-2)^2=-4$

37. a^2+2a-2

$(0)^2+2(0)-2=0+0-2=-2$

$(1)^2+2(1)-2=1+2-2=1$

$(2)^2+2(2)-2=4+4-2=6$

$(-1)^2+2(-1)-2=1-2-2=-3$

$(-2)^2+2(-2)-2=4-4-2=-2$

39.
$$\begin{aligned}C&=\frac{5}{9}(F-32)\\&=\frac{5}{9}(-4-32)\\&=\frac{5}{9}(-36)\\&=-20^\circ\end{aligned}$$

41.
$$\begin{aligned}P&=2l+2w\\&=2\left(2\frac{1}{2}\right)+2\left(1\frac{1}{4}\right)\\&=2\left(\frac{5}{2}\right)+2\left(\frac{5}{4}\right)\\&=5+\frac{5}{2}\\&=\frac{10}{2}+\frac{5}{2}\\&=\frac{15}{2}\\&=7\frac{1}{2}\text{ ft}\end{aligned}$$

43.
$$\begin{aligned}C&=\pi d\\&\approx(3.14)(100)\\&=314\text{ m}\end{aligned}$$

45.
$$\begin{aligned}6e^2&=6(1.5\text{ cm})^2\\&=6(2.25\text{cm}^2)\\&=13.5\text{ cm}^2\end{aligned}$$

47.
$$\begin{aligned}&y^2-2z+\frac{1}{3}(2x-w)\\&=(-2)^2-2(0.5)+\frac{1}{3}(2(-1.5)-3)\\&=(-2)^2-2(0.5)+\frac{1}{3}(-3-3)\\&=(-2)^2-2(0.5)+\frac{1}{3}(-6)\\&=4-2(0.5)+\frac{1}{3}(-6)\\&=4-1-2\\&=1\end{aligned}$$

49. $-2x+4$

$-2(0)+4=0+4=4$

$-2(1)+4=-2+4=2$

$-2(2)+4=-4+4=0$

$-2(-1)+4=2+4=6$

$-2(-2)+4=4+4=8$

51. $$\begin{aligned} b^2-3a^2 &= (-4)^2-3(3)^2 \\ &= 16-3(9) \\ &= 16-27 \\ &= -11 \end{aligned}$$

53. $$\begin{aligned} A &= \frac{1}{2}bh \\ &= \frac{1}{2}(7\text{ in.})(3\text{ in.}) \\ &= \frac{21}{2}\text{ in.}^2 \\ &= 10\frac{1}{2}\text{ in.}^2 \text{ or } 10.5\text{ in}^2 \end{aligned}$$

55. $A=\dfrac{a+b+c}{3}$

57. $P=2(l+w)$

59. $E=mc^2$

61. $l=0.4w+25$

63. $$\begin{aligned} S &= \frac{1}{2}gt^2 \\ &= \frac{1}{2}(32)(2)^2 \\ &= \frac{1}{2}(32)(4) \\ &= 16(4) \\ &= 64 \end{aligned}$$

The object falls 64 ft.

65. **a.** $m=\dfrac{100(s-c)}{c}$

$$\begin{aligned} &\frac{100(s-c)}{c} \\ &= \frac{100(\$8.75-\$6.25)}{\$6.25} \\ &= \frac{100(\$2.50)}{\$6.25} \\ &= \frac{\$250}{\$6.25} \\ &= 40 \end{aligned}$$

b. 40%

Chapter 1 Review Exercises

1. The value is above sea level (positive).
+3 mi

2. The value is a withdrawal (negative).
−$160

3.

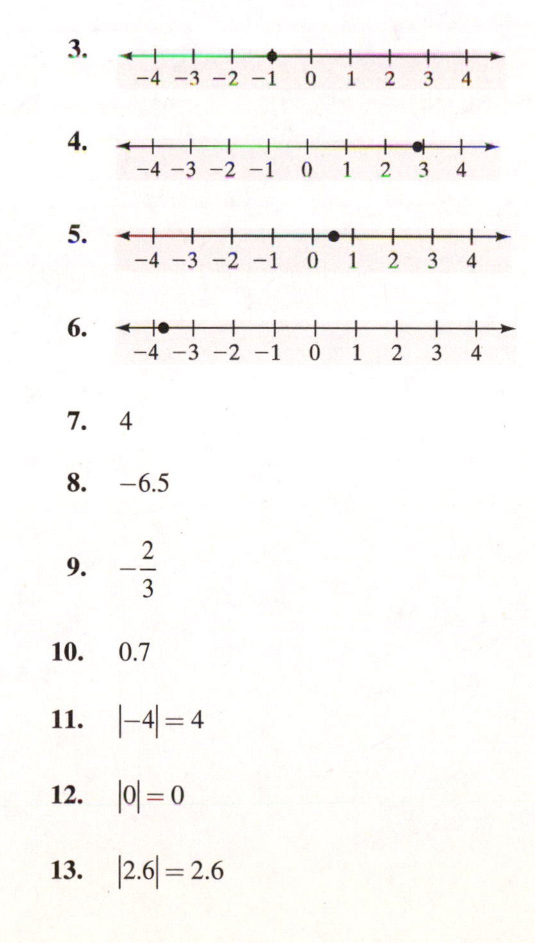

4.

5.

6.

7. 4

8. −6.5

9. $-\dfrac{2}{3}$

10. 0.7

11. $|-4|=4$

12. $|0|=0$

13. $|2.6|=2.6$

14. $\left|-\frac{5}{9}\right| = \frac{5}{9}$

15. True, because -7 is to the left of -5.

16. False, because -1 is to the left of 3.

17. $-4+(-1)=-5$

−5 −4 −3 −2 −1 0 1 2 3 4 5

18. $3+(-7)=-4$

−5 −4 −3 −2 −1 0 1 2 3 4 5

19. Another term for opposites is additive inverses.

20. Additive inverse property

21. $9+(-9)=0$

22. $4+(-2)=2$

23. $-3+5=2$

24. $(-3)+(-2)=-5$

25. $$\begin{aligned}-3+7+(-89)&=4+(-89)\\&=-85\end{aligned}$$

26. $$\begin{aligned}-2+5.3+12&=3.3+12\\&=15.3\end{aligned}$$

27. $12-3=9$

28. $$\begin{aligned}36-47&=36+(-47)\\&=-11\end{aligned}$$

29. $$\begin{aligned}-52-3&=-52+(-3)\\&=-55\end{aligned}$$

30. $$\begin{aligned}2-5&=2+(-5)\\&=-3\end{aligned}$$

31. $$\begin{aligned}-19-8&=-19+(-8)\\&=-27\end{aligned}$$

32. $$\begin{aligned}24-(-3)&=24+3\\&=27\end{aligned}$$

33. $$\begin{aligned}8-(-8)&=8+8\\&=16\end{aligned}$$

34. $$\begin{aligned}0-5&=0+(-5)\\&=-5\end{aligned}$$

35. $$\begin{aligned}6-7.42&=6+(-7.42)\\&=-1.42\end{aligned}$$

36. $$\begin{aligned}-9-\left(-\frac{3}{8}\right)&=-9+\frac{3}{8}\\&=-\frac{72}{8}+\frac{3}{8}\\&=-\frac{69}{8}\\&=-8\frac{5}{8}\end{aligned}$$

37. $$\begin{aligned}2+(-4)-(-7)&=2+(-4)+7\\&=-2-(-7)\\&=-2+7\\&=5\end{aligned}$$

38. $$\begin{aligned}-3-(-1)+12&=-3+1+12\\&=-2+12\\&=10\end{aligned}$$

39. Multiplicative inverse property

40. Multiplication property of zero

41. $2(-5)=-10$

42. $-3\cdot 7=-21$

43. $-60\cdot 90=-5400$

44. $-8(-300)=2400$

45. $(-2.7)(-10)=27$

46. $$\begin{aligned}\left(\frac{3}{4}\right)\left(-\frac{1}{3}\right)&=\frac{\cancel{3}^{1}}{4}\left(-\frac{1}{\cancel{3}_{1}}\right)\\&=-\frac{1}{4}\end{aligned}$$

47. $5(-4)(-300) = -20(-300)$
$= 6000$

48. $(-1)(-12)(3) = 12(3)$
$= 36$

49. $-8 + 3(-2) - 9 = -8 - 6 - 9$
$= -14 - 9$
$= -23$

50. $3 - 2(-3) - (-5) = 3 + 6 + 5$
$= 9 + 5$
$= 14$

51. $-9 - 5(-7) = -9 + 35$
$= 26$

52. $20 - 3(-6) = 20 + 18$
$= 38$

53. $-4(-2 + 5) = -4(3)$
$= -12$

54. $(-12 + 6)(-1) = (-6)(-1)$
$= 6$

55. $-\frac{3}{2}$

56. $\frac{1}{8}$

57. $-30 \div (-10) = 3$

58. $6 \div (-1) = -6$

59. $-\frac{11}{5} = -2\frac{1}{5}$

60. $\frac{4}{5} \div \left(-\frac{2}{3}\right) = \frac{4}{5} \cdot \left(-\frac{3}{2}\right)$
$= \frac{\cancel{4}^{2}}{5} \cdot \left(-\frac{3}{\cancel{2}_{1}}\right)$
$= -\frac{6}{5}$
$= -1\frac{1}{5}$

61. $-16 \div 2(-4)$
$= -8(-4)$
$= 32$

62. $(9 - 23) \div (-13 + 6)$
$= -14 \div (-7)$
$= 2$

63. $\frac{3 + (-1)}{-2}$
$= \frac{2}{-2}$
$= -1$

64. $\frac{5(7 - 3)}{-8 - 2}$
$= \frac{5(4)}{-8 - 2}$
$= \frac{20}{-8 - 2}$
$= \frac{20}{-10}$
$= -2$

65. $(-3) + 8 - 2 \cdot (-4)$
$= -3 + 8 + 8$
$= 5 + 8$
$= 13$

66. $10 \div (-2) + (-3) \cdot 5$
$= -5 + (-15)$
$= -20$

67. $3 + 9 = 9 + 3$

68. $(-3)(1 + 9) = (-3)(1) + (-3)(9)$

69. $(3x+y)+z=3x+(y+z)$
Associative property of addition.

70. $-(x+1)+(x+1)=0$
Additive inverse property.

71. $10+(-2)+(-1)=8+(-1)=7$

72. $(-4)(-5)(-2)(4)=(20)(-2)(4)$
$=(-40)(4)=-160$

73. The additive inverse of 4 is –4.

74. The multiplicative inverse of $\frac{2}{3}$ is $\frac{3}{2}$.

75. $3(a-4b)=3(a)-3(4b)=3a-12b$

76. $-(x-5)=(-1)(x-5)$
$=(-1)(x)-(-1)(5)$
$=-x+5$

77. 3

78. 2

79. 1

80. 4

81. the sum of negative 6 and w

81. the product of negative $\frac{1}{3}$ and x

83. 6 more than negative 3 times n

84. 5 times the quantity p minus q

85. $x-10$

86. $\frac{1}{2}s$

87. $\frac{p}{q}$

88. $R-2V$

89. $6(4n-2)$

90. $\frac{-4a}{5b+c}$

91. $-3(-3)(-3)(-3)=(-3)^4$

92. $-5(-5)(-5)(3)(3)=(-5)^3 3^2$

93. $4(x)(x)(x)=4x^3$

94. $-5\cdot a\cdot a\cdot b\cdot b\cdot b\cdot c=-5a^2b^3c$

95. $\frac{3}{4}$

96. $4x+10x-2y=(4+10)x-2y$
$=14x-2y$

97. $3x^2-x^2-4x^2=(3-1-4)x^2$
$=-2x^2$

98. $2r^2t^2-r^2t^2=(2-1)r^2t^2$
$=r^2t^2$

99. $2(a-5)+1=2a-10+1$
$=2a-9$

100. $-(3x+2)=-1(3x+2)$
$=-3x-2$

101. $-3x-5-(x+10)=-3x-5-x-10$
$=(-3-1)x+(-5-10)$
$=-4x-15$

102. $(2a-4)+2(a-5)-3(a+1)$
$=2a-4+2a-10-3a-3$
$=(2+2-3)a+(-4-10-3)$
$=a-17$

103. $30+c=30+(-1)$
$=29$

104. $-\frac{4}{9}b=-\frac{4}{9}(5)$
$=-\frac{20}{9}$

105. $-5a^2 = -5(2)^2$
$= -5(4)$
$= -20$

106. $10(b-c) = 10(5-(-1))$
$= 10(5+1)$
$= 10(6)$
$= 60$

107. $\frac{1-a}{c} = \frac{1-2}{-1}$
$= \frac{-1}{-1}$
$= 1$

108. $4a^2 - 4ab + b^2$
$= 4(2)^2 - 4(2)(5) + (5)^2$
$= 4(4) - 4(2)(5) + 25$
$= 16 - 40 + 25$
$= -24 + 25$
$= 1$

109. The value represents a gain (positive).
$+\$700$

110. The value represents a loss (negative).
$-\$7000$

111. $-\$0.50(4) = -\$2.00$

112. $-4-(-7) = -4+7 = -3$
Exothermic $(+3°\text{C})$

113. $I - 0.5h$ degrees

114. $F = \frac{9}{5}C + 32$
$= \frac{9}{5}(-10) + 32$
$= -18 + 32$
$= 14°$

115. More money was flowing from the Netherlands.

116. $\text{P} = 2l + 2w$
$= 2(2w) + 2w$
$= 4w + 2w$
$= 6w$

117. Tue: $\$57.04 - \$57.19 = -\$0.15$
Wed: $\$55.91 - \$57.04 = -\$1.13$
Thur: $\$56.06 - \$55.91 = +\$0.15$
Fri: $\$57.65 - \$56.06 = +\$1.59$

118. $-61.8 - (-71)$
$= -61.8 + 71$
$= 71 - 61.8$
$= 9.2°$
The boiling point is $9.2°$ higher.

119. $3^3 \cdot 10$

120. $S = 0.5N + 26$
$= 0.5(90) + 26$
$= 45 + 26$
$= 71$ ft

121. $3 > -4$, so a temperature of $3°$ is warmer.

122. $[0.05x + 0.07(600 - x)]$ dollars

123. $-496 + 90 = -406$
Approximately 406 B.C.

124. $\frac{-\$60{,}000}{-\$20{,}000} = 3$

125. $I = \frac{P}{E}$
$= \frac{2300}{115}$
$= 20$ amperes

126. $\$410 + \$900 + (-\$720) + 2(-\$300)$
$= \$1310 + (-\$720) + (-\$600)$
$= \$590 + (-\$600)$
$= -\$10$
The account is overdrawn by $10 (-10).

127. $x+(32-20)y=x+12y$
$(x+12y)$ dollars

128. $f+s+s+s+s=f+4s$
$(f+4s)$ students

Chapter 1 Posttest

1. The value represents a loss (negative).
$-10,000$

2. Yes, $\frac{2}{5}$ is rational

3.
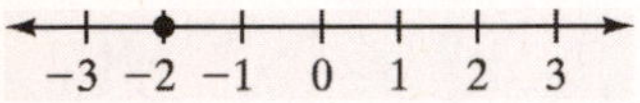

4. -7

5. $|-3.5|=3.5$

6. True, because 1 is to the right of -4.

7. $10+(-3)=10-3=7$

8. $2+(-3)+(-1)+5=-1+(-1)+5$
$=-2+5$
$=3$

9. $4+(-1)(-6)=4+6=10$

10. $\frac{1}{12}$

11. $-15\div 0.3=-50$

12. $x+2y$

13. $-5(-5)(-5)=(-5)^3$

14. $3a+b-c=3(-1)+0-2$
$=-3+0-2$
$=-5$

15. $4y+3-7y+10y+1$
$=(4-7+10)y+(3+1)$
$=7y+4$

16. $8t+1-2(3t-1)=8t+1-6t+2$
$=(8-6)t+(1+2)$
$=2t+3$

17. $d+0.05d=10.5d$
$1.05d$ dollars

18. $\$50,000-(-\$20,000)=\$50,000+\$20,000$
$=\$70,000$
An improvement of $70,000

Chapter 2

SOLVING LINEAR EQUATIONS AND INEQUALITIES

Pretest

1.
$$7-2x=3x-11$$
$$7-2(4)=3(4)-11$$
$$7-8=12-11$$
$$-1\neq 1$$
No, 4 is not a solution

2.
$$n+2=-6$$
$$n+2-2=-6-2$$
$$n=-8$$

3.
$$\frac{y}{-5}=1$$
$$-5\cdot\frac{y}{-5}=-5\cdot 1$$
$$y=-5$$

4.
$$-n=8$$
$$-1\cdot -n=-1\cdot 8$$
$$n=-8$$

5.
$$\frac{2}{3}x-3=-9$$
$$\frac{2}{3}x-3+3=-9+3$$
$$\frac{2}{3}x=-6$$
$$\frac{3}{2}\cdot\frac{2}{3}x=\frac{3}{2}\cdot -6$$
$$x=-9$$

6.
$$4x-8=-10$$
$$4x-8+8=-10+8$$
$$4x=-2$$
$$\frac{1}{4}\cdot 4x=\frac{1}{4}\cdot -2$$
$$x=-\frac{1}{2}$$

7.
$$6-y=-5$$
$$6-6-y=-5-6$$
$$-y=-11$$
$$-1\cdot -y=-11\cdot -1$$
$$y=11$$

8.
$$9x+13=7x+19$$
$$9x+13-13=7x+19-13$$
$$9x=7x+6$$
$$9x-7x=7x-7x+6$$
$$2x=6$$
$$\frac{1}{2}\cdot 2x=\frac{1}{2}\cdot 6$$
$$x=3$$

9.
$$-2(3n-1)=-7n$$
$$-6n+2=-7n$$
$$-6n+7n+2=-7n+7n$$
$$n+2=0$$
$$n+2-2=0-2$$
$$n=-2$$

10.
$$14x-(8x-13)=12x+3$$
$$14x-8x+13=12x+3$$
$$6x+13=12x+3$$
$$6x-12x+13=12x-12x+3$$
$$-6x+13=3$$
$$-6x+13-13=3-13$$
$$-6x=-10$$
$$\frac{-6x}{-6}=\frac{-10}{-6}$$
$$x=\frac{5}{3}$$

11.
$$v-5u=w$$
$$v-5u+5u=w+5u$$
$$v=w+5u$$

12.
$$36x=9$$
$$x=.25$$
$$=25\%$$

13.
$$0.6x = 12$$
$$\frac{0.6x}{0.6} = \frac{12}{0.6}$$
$$x = 20$$

14.

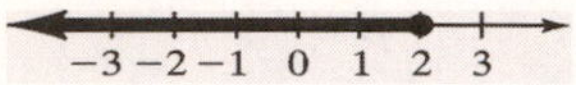

15.
$$x + 3 > 3$$
$$x + 3 - 3 > 3 - 3$$
$$x > 0$$

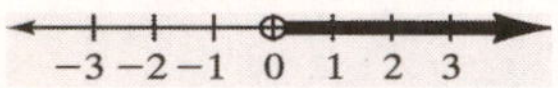

16.
$$30x = 360$$
$$\frac{1}{30} \cdot 30x = \frac{1}{30} \cdot 360$$
$$x = 12$$
12 min

17.
$$100 + 70c = 1500$$
$$100 - 100 + 70c = 1500 - 100$$
$$70c = 1400$$
$$\frac{70c}{70} = \frac{1400}{70}$$
$$c = 20$$
20 centerpieces

18.
$$E = \frac{1}{2}mv^2$$
$$\frac{2}{v^2} \cdot E = \frac{2}{v^2} \cdot \frac{1}{2}mv^2$$
$$\frac{2E}{v^2} = m$$
$$m = \frac{2E}{v^2}$$

19.
$$0.08(2x) + 0.05x = 420$$
$$0.16x + 0.05x = 420$$
$$0.21x = 420$$
$$\frac{0.21x}{0.21} = \frac{420}{0.21}$$
$$x = 2000$$
\$4000 was invested at 8%, and \$2000 was invested at 5%.

20.
$$10 + 3x > 55$$
$$10 - 10 + 3x > 55 - 10$$
$$3x > 45$$
$$\frac{3x}{3} > \frac{45}{3}$$
$$x > 15$$
Option A is a better deal if the member used the gym than 15 hours per month $(x > 15)$.

2.1 Solving Linear Equations: The Addition Property

Practice

1.
$$5x - 4 = 2x + 5$$
$$5(4) - 4 = 2(4) + 5$$
$$20 - 4 = 8 + 5$$
$$16 \neq 13$$
No, 4 is not a solution.

2.
$$5(x + 3) = 3x - 1$$
$$5(-8 + 3) = 3(-8) - 1$$
$$5(-5) = -24 - 1$$
$$-25 = -25$$
Yes, -8 is a solution.

3.
$$y - 12 = -7$$
$$y - 12 + 12 = -7 + 12$$
$$y = 5$$

4.
$$-2 = n + 15$$
$$-2 - 15 = n + 15 - 15$$
$$-17 = n$$
$$n = -17$$

5.
$$5 = 4.9 - (-x)$$
$$5 = 4.9 + x$$
$$5 - 4.9 = 4.9 - 4.9 + x$$
$$0.1 = x$$
$$x = 0.1$$

6.
$$9.68 + x = 24.56$$
$$9.68 - 9.68 + x = 24.56 - 9.68$$
$$x = 14.88$$
14.88g

Exercises

1. A(n) equation is a mathematical statement that two expressions are equal.

3. Equations that have the same solutions are equivalent equations.

5. The addition property of equality tells us that for any real numbers a, b, and c, if $a = b$, then $a + c = b + c$.

7. a. $3x + 13 = -11$
$3(-8) + 13 = -11$
$-24 + 13 = -11$
$-11 = -11$
True

b. $28 - x = 7 - 4x$
$28 - 7 = 7 - 4(7)$
$28 - 7 = 7 - 28$
$21 \neq -21$
False

c. $2(x - 3) = 12$
$2(9 - 3) = 12$
$2(6) = 12$
$12 = 12$
True

d. $12x - 2 = 6x + 2$
$12\left(\frac{2}{3}\right) - 2 = 6\left(\frac{2}{3}\right) + 2$
$8 - 2 = 4 + 2$
$6 = 6$
True

9. Subtract 4 (or add -4).

11. Add 1.

13. Subtract 3.5 (or add -3.5).

15. Add $2\frac{1}{5}$.

17. $y + 9 = -14$
$y + 9 - 9 = -14 - 9$
$y = -23$
Check:
$-23 + 9 = -14$
$-14 = -14$

19. $t - 4 = -4$
$t - 4 + 4 = -4 + 4$
$t = 0$
Check:
$0 - 4 = -4$
$-4 = -4$

21. $9 + a = -3$
$9 - 9 + a = -3 - 9$
$a = -12$
Check:
$9 + (-12) = -3$
$-3 = -3$

23. $z - 4 = -10$
$z - 4 + 4 = -10 + 4$
$z = -6$
Check:
$-6 - 4 = -10$
$-10 = -10$

25. $12 = x + 12$
$12 - 12 = x + 12 - 12$
$0 = x$
$x = 0$
Check:
$12 = 0 + 12$
$12 = 12$

27. $-6 = t - 12$
$-6 + 12 = t - 12 + 12$
$6 = t$
$t = 6$
Check:
$-6 = 6 - 12$
$-6 = -6$

29.
$$\begin{aligned} -4 &= -10 + r \\ -4 + 10 &= -10 + 10 + r \\ 6 &= r \\ r &= 6 \end{aligned}$$
Check:
$$\begin{aligned} -4 &= -10 + 6 \\ -4 &= -4 \end{aligned}$$

31.
$$\begin{aligned} -15 + n &= -2 \\ -15 + 15 + n &= -2 + 15 \\ n &= 13 \end{aligned}$$
Check:
$$\begin{aligned} -15 + 13 &= -2 \\ -2 &= -2 \end{aligned}$$

33.
$$\begin{aligned} x + \frac{2}{3} &= -\frac{1}{3} \\ x + \frac{2}{3} - \frac{2}{3} &= -\frac{1}{3} - \frac{2}{3} \\ x &= -1 \end{aligned}$$
Check:
$$\begin{aligned} -1 + \frac{2}{3} &= -\frac{1}{3} \\ -\frac{3}{3} + \frac{2}{3} &= -\frac{1}{3} \\ -\frac{1}{3} &= -\frac{1}{3} \end{aligned}$$

35.
$$\begin{aligned} 8 + y &= 4\frac{1}{2} \\ 8 - 8 + y &= 4\frac{1}{2} - 8 \\ y &= \frac{9}{2} - \frac{16}{2} \\ y &= -\frac{7}{2} \\ y &= -3\frac{1}{2} \end{aligned}$$
Check:
$$\begin{aligned} 8 + \left(-3\frac{1}{2}\right) &= 4\frac{1}{2} \\ \frac{16}{2} - \frac{7}{2} &= 4\frac{1}{2} \\ \frac{9}{2} &= 4\frac{1}{2} \\ 4\frac{1}{2} &= 4\frac{1}{2} \end{aligned}$$

37.
$$\begin{aligned} m + 2.4 &= 5.3 \\ m + 2.4 - 2.4 &= 5.3 - 2.4 \\ m &= 2.9 \end{aligned}$$
Check:
$$\begin{aligned} 2.9 + 2.4 &= 5.3 \\ 5.3 &= 5.3 \end{aligned}$$

39.
$$\begin{aligned} -2.3 + t &= -5.9 \\ -2.3 + 2.3 + t &= -5.9 + 2.3 \\ t &= -3.6 \end{aligned}$$
Check:
$$\begin{aligned} -2.3 + (-3.6) &= -5.9 \\ -5.9 &= -5.9 \end{aligned}$$

41.
$$\begin{aligned} a - (-35) &= 30 \\ a + 35 &= 30 \\ a + 35 - 35 &= 30 - 35 \\ a &= -5 \end{aligned}$$
Check:
$$\begin{aligned} -5 - (-35) &= 30 \\ -5 + 35 &= 30 \\ 30 &= 30 \end{aligned}$$

43.
$$\begin{aligned} m - \left(-\frac{1}{4}\right) &= -\frac{1}{4} \\ m + \frac{1}{4} &= -\frac{1}{4} \\ m + \frac{1}{4} - \frac{1}{4} &= -\frac{1}{4} - \frac{1}{4} \\ m &= -\frac{2}{4} \\ m &= -\frac{1}{2} \end{aligned}$$
Check:
$$\begin{aligned} -\frac{1}{2} - \left(-\frac{1}{4}\right) &= -\frac{1}{4} \\ -\frac{2}{4} + \frac{1}{4} &= -\frac{1}{4} \\ -\frac{1}{4} &= -\frac{1}{4} \end{aligned}$$

45.
$$\begin{aligned} y + 2.932 &= 48.11 \\ y + 29.32 - 29.32 &= 4.811 - 2.932 \\ y &= 1.879 \\ y &\approx 1.88 \end{aligned}$$

47.
$$x+2=12$$
$$x+2-2=12-2$$
$$x=10$$
Check:
$$10+2=12$$
$$12=12$$

49.
$$n-4=21$$
$$n-4+4=21+4$$
$$n=25$$
Check:
$$25-4=21$$
$$21=21$$

51.
$$x+(-3)=-1$$
$$x-3=-1$$
$$x-3+3=-1+3$$
$$x=2$$
Check:
$$2+(-3)=-1$$
$$-1=-1$$

53.
$$n+7=11$$
$$n+7-7=11-7$$
$$n=4$$
Check:
$$4+7=11$$
$$11=11$$

55. d

57. a

59. Subtract $\frac{2}{3}$ (or add $-\frac{2}{3}$).

61. **a.**
$$6x+15=-11$$
$$6(-4)+15=11$$
$$-24+15=-11$$
$$-9\neq-11$$
False

b.
$$4x-21=2x-3$$
$$4(9)=2(9)-3$$
$$36-21=18-3$$
$$15=15$$
True

c.
$$-2(x-5)=4$$
$$-2(7-5)=4$$
$$-2(2)=4$$
$$-4\neq4$$
False

d.
$$5-8x=-4x+2$$
$$5-8\left(\frac{3}{4}\right)=-4\left(\frac{3}{4}\right)+2$$
$$5-6=-3+2$$
$$-1=-1$$
True

63.
$$x-2.5=-3.8$$
$$x-2.5+2.5=-3.8+2.5$$
$$x=-1.3$$
Check:
$$x-2.5=-3.8$$
$$-1.3-2.5=-3.8$$
$$-3.8=-3.8$$

65.
$$-19=m+4$$
$$-19-4=m+4-4$$
$$-23=m$$
$$m=-23$$
Check:
$$-19=-23+4$$
$$-19=-19$$

67.
$$7+n=2\frac{2}{3}$$
$$7-7+n=2\frac{2}{3}-7$$
$$n=\frac{8}{3}-\frac{21}{3}$$
$$n=-\frac{13}{3}$$
$$n=-4\frac{1}{3}$$
Check:
$$7+\left(-4\frac{1}{3}\right)=2\frac{2}{3}$$
$$\frac{21}{3}-\frac{13}{3}=2\frac{2}{3}$$
$$\frac{8}{3}=2\frac{2}{3}$$
$$2\frac{2}{3}=2\frac{2}{3}$$

69.
$$x+10=44$$
$$x+10-10=44-10$$
$$x=34 \text{ mph}$$
Check:
$$34+10=44$$
$$44=44$$

71.
$$x-190=370$$
$$x-190+190=370+190$$
$$x=560 \text{ cal}$$
Check:
$$560-190=370$$
$$370=370$$

73.
$$h-170=215$$
$$h-170+170=215+170$$
$$h=385 \text{ m}$$
Check:
$$385-170=215$$
$$215=215$$

75.
$$x+118.5=180$$
$$x+118.5-118.5=180-118.5$$
$$x=61.5°$$
Check:
$$61.5+118.5=180$$
$$180=180$$

2.2 Solving Linear Equations: The Multiplication Property

Practice

1.
$$\frac{y}{3}=21$$
$$3\cdot\frac{y}{3}=3\cdot 21$$
$$y=63$$
Check:
$$\frac{63}{3}=21$$
$$21=21$$

2.
$$7y=63$$
$$\frac{7y}{7}=\frac{63}{7}$$
$$y=9$$
Check:
$$7(9)=63$$
$$63=63$$

3.
$$-x=10$$
$$\frac{-x}{-1}=\frac{10}{-1}$$
$$x=-10$$
Check:
$$-(-10)=10$$
$$10=10$$

4.
$$-11.7=-0.9z$$
$$\frac{-11.7}{-0.9}=\frac{-0.9z}{-0.9}$$
$$13=z$$
$$z=13$$
Check:
$$-11.7=-0.9(13)$$
$$-11.7=-11.7$$

5.
$$\frac{6}{7}y=-12$$
$$\frac{7}{6}\cdot\frac{6}{7}y=\frac{7}{6}\cdot -12$$
$$y=-14$$
Check:
$$\frac{6}{7}(-14)=-12$$
$$-12=-12$$

6.
$$\frac{1}{4}x=189.50$$
$$4\cdot\frac{1}{4}x=4\cdot 189.50$$
$$x=758$$
The total bill was $758.
Check:
$$\frac{1}{4}(758)=189.50$$
$$189.50=189.50$$

7. $60t = 139$

$$\frac{60t}{60} = \frac{139}{60}$$

$$t \approx 2.3$$

It will take about 2.3 hr.

Check:

$$60(2.3) \approx 139$$

$$138 \approx 139$$

Exercises

1. Multiply by 3.

3. Divide by -5.

5. Divide by -2.2.

7. Multiply by $\frac{4}{3}$.

9. Multiply by $-\frac{2}{5}$.

11. $6x = -30$

$$\frac{6x}{6} = \frac{-30}{6}$$

$$x = -5$$

Check:

$$6(-5) = -30$$

$$-30 = 30$$

13. $\frac{n}{2} = 9$

$$2 \cdot \frac{n}{2} = 2 \cdot 9$$

$$n = 18$$

Check:

$$-8(-1) = 8$$

$$8 = 8$$

15. $\frac{a}{4} = 1.2$

$$4 \cdot \frac{a}{4} = 4 \cdot 1.2$$

$$a = 4.8$$

Check:

$$\frac{4.8}{4} = 1.2$$

$$1.2 = 1.2$$

17. $-5x = 2.5$

$$\frac{-5x}{-5} = \frac{2.5}{-5}$$

$$x = -0.5$$

Check:

$$-5(-0.5) = 2.5$$

$$2.5 = 2.5$$

19. $42 = -6c$

$$\frac{42}{-6} = \frac{-6c}{-6}$$

$$-7 = c$$

$$c = -7$$

Check:

$$42 = -6(-7)$$

$$42 = 42$$

21. $11 = -\frac{r}{2}$

$$-2 \cdot 11 = -2 \cdot \left(-\frac{r}{2}\right)$$

$$-22 = r$$

$$r = -22$$

Check:

$$11 = \frac{-22}{-2}$$

$$11 = 11$$

23. $\frac{5}{6}x = 10$

$$\frac{6}{5} \cdot \frac{5}{6}x = \frac{6}{5} \cdot 10$$

$$x = 12$$

Check:

$$\frac{5}{6}(12) = 10$$

$$10 = 10$$

25. $-\frac{2}{5}y = 1$

$$-\frac{5}{2} \cdot \left(-\frac{2}{5}y\right) = -\frac{5}{2} \cdot 1$$

$$y = -\frac{5}{2}$$

Check:

$$-\frac{2}{5}\left(-\frac{5}{2}\right) = 1$$

$$1 = 1$$

27. $\frac{3}{4}n = 6$

$\frac{4}{3}\cdot\frac{3}{4}n = \frac{4}{3}\cdot 6$

$n = 8$

Check:

$\frac{3(8)}{4} = 6$

$\frac{24}{4} = 6$

$6 = 6$

29. $\frac{4}{3}c = -4$

$\frac{3}{4}\cdot\frac{4}{3}c = \frac{3}{4}\cdot(-4)$

$c = -3$

Check:

$\frac{4(-3)}{3} = -4$

$\frac{-12}{3} = -4$

$-4 = -4$

31. $\frac{x}{2.4} = -1.2$

$2.4\cdot\frac{x}{2.4} = 2.4\cdot(-1.2)$

$x = -2.88$

Check:

$\frac{-2.88}{2.4} = -1.2$

$-1.2 = -1.2$

33. $-2.5a = 5$

$\frac{-2.5a}{-2.5} = \frac{5}{-2.5}$

$a = -2$

Check:

$-2.5(-2) = 5$

$5 = 5$

35. $\frac{2}{3}y = \frac{4}{9}$

$\frac{3}{2}\cdot\frac{2}{3}y = \frac{3}{2}\cdot\frac{4}{9}$

$y = \frac{2}{3}$

Check:

$\frac{2}{3}\cdot\frac{2}{3} = \frac{4}{9}$

$\frac{4}{9} = \frac{4}{9}$

37. $\frac{x}{-1.515} = 1.515$

$(-1.515)\left(\frac{x}{-1.515}\right) = (-1.515)\cdot(1.515)$

$x \approx -2.30$

39. $-3.14x = 21.4148$

$\frac{-3.14x}{-3.14x} = \frac{21.4148}{-3.14}$

$x \approx -6.82$

41. $-4x = 56$

$\frac{-4x}{-4} = \frac{56}{-4}$

$x = -14$

Check:

$-4(-14) = 56$

$56 = 56$

43. $\frac{n}{0.2} = 1.1$

$0.2\cdot\frac{n}{0.2} = 0.2\cdot 1.1$

$n = 0.22$

Check:

$\frac{0.22}{0.2} = 1.1$

$1.1 = 1.1$

45. $\frac{x}{-3} = 20$

$(-3)\cdot\left(\frac{x}{-3}\right) = (-3)\cdot(20)$

$x = -60$

Check:

$\frac{-60}{-3} = 20$

$20 = 20$

47. $\frac{1}{6}x = 2\frac{4}{5}$

$6\cdot\frac{1}{6}x = 6\cdot\frac{14}{5}$

$x = \frac{84}{5}$

Check:

$\frac{1}{6}\left(\frac{84}{5}\right) = 2\frac{4}{5}$

$\frac{14}{5} = \frac{14}{5}$

49. c

51. a

53. $\frac{2n}{7} = 4$

$\frac{7}{2}\cdot\frac{2n}{7} = \frac{7}{2}\cdot 4$

$n = 14$

Check:

$\frac{2(14)}{7} = 4$

$\frac{28}{7} = 4$

$4 = 4$

55. $-\frac{y}{3.8} = -0.3$

$(-3.8)\cdot\left(-\frac{y}{3.8}\right) = (-3.8)\cdot(-0.3)$

$y = 1.14$

57. $\frac{x}{5} = 2$

$5\cdot\frac{x}{5} = 5\cdot 2$

$x = 10$

Check:

$\frac{10}{5} = 2$

$2 = 2$

59. Divide by -5.2.

61. $0.02x = 10.5$

$\frac{0.02x}{0.02} = \frac{10.5}{0.02}$

$x = 525$

525 yr

Check:

$0.02(525) = 10.5$

$10.5 = 10.5$

63. $70r = 3348$

$\frac{70r}{70} = \frac{3348}{70}$

$r \approx 48$

48 mph

Check:

$70(48) \approx 3348$

$3360 \approx 3348$

65. $0.05c = 20$

$\frac{0.05c}{0.05} = \frac{20}{0.05}$

$c = 400$

400 copies

Check:

$0.05(400) = 20$

$20 = 20$

67. $\frac{2}{3}x = 800{,}000$

$\frac{3}{2}\cdot\frac{2}{3}x = \frac{3}{2}\cdot 800{,}000$

$x = 1{,}200{,}000$

\$1,200,000

Check:

$\frac{2}{3}(1{,}200{,}000) = 800{,}000$

$800{,}000 = 800{,}000$

69. $\frac{1}{5}d = 1000$

$5 \cdot \frac{1}{5}d = 5 \cdot 1000$

$d = 5000$

5000 m

Check:

$\frac{1}{5}(5000) = 1000$

$1000 = 1000$

71. $7.50t = 187.50$

$\frac{7.50}{7.50}t = \frac{187.50}{7.50}$

$t = 25$

25 hr

Check:

$7.50(25) = 187.50$

$187.50 = 187.50$

73. $12x = 10{,}020$

$\frac{12x}{12} = \frac{10{,}020}{12}$

$x = 835$

\$835 per month

Check:

$12(835) = 10{,}020$

$10{,}020 = 10{,}020$

2.3 Solving Linear Equations by Combining Properties

Practice

1. $2y + 1 = 9$

$2y + 1 - 1 = 9 - 1$

$2y = 8$

$\frac{2y}{2} = \frac{8}{2}$

$y = 4$

Check:

$2(4) + 1 = 9$

$8 + 1 = 9$

$9 = 9$

2. $\frac{c}{5} - 1 = 8$

$\frac{c}{5} - 1 + 1 = 8 + 1$

$\frac{c}{5} = 9$

$5 \cdot \frac{c}{5} = 5 \cdot 9$

$c = 45$

Check:

$\frac{45}{5} - 1 = 8$

$9 - 1 = 8$

$8 = 8$

3. $-6b - 5 = 13$

$-6b - 5 + 5 = 13 + 5$

$-6b = 18$

$\frac{-6b}{-6} = \frac{18}{-6}$

$b = -3$

4. $8n + 10n = 24$

$18n = 24$

$\frac{18n}{18} = \frac{24}{18}$

$n = \frac{4}{3}$

5. $5 - t - t = -1$

$5 - 2t = -1$

$5 - 5 - 2t = -1 - 5$

$-2t = -6$

$\frac{-2t}{-2} = \frac{-6}{-2}$

$t = 3$

Check:

$5 - 3 - 3 = -1$

$-1 = -1$

6.
$$\begin{aligned}
3f-12&=-f-15\\
3f+f-12&=-f+f-15\\
4f-12&=-15\\
4f-12+12&=-15+12\\
4f&=-3\\
\frac{4f}{4}&=\frac{-3}{4}\\
f&=-\frac{3}{4}
\end{aligned}$$

7.
$$\begin{aligned}
-5(z+6)&=z\\
-5z-30&=z\\
-5z+5z-30&=z+5z\\
-30&=6z\\
\frac{-30}{6}&=\frac{6z}{z}\\
-5&=z
\end{aligned}$$

Check:
$$\begin{aligned}
-5(-5+6)&=-5\\
-5(1)&=-5\\
-5&=-5
\end{aligned}$$

8.
$$\begin{aligned}
2(t-3)-3(t-2)&=t+8\\
2t-6-3t+6&=t+8\\
-t&=t+8\\
-t-t&=t-t+8\\
-2t&=8\\
\frac{-2t}{-2}&=\frac{8}{-2}\\
t&=-4
\end{aligned}$$

9.
$$\begin{aligned}
4[5y-(y-1)]&=7(y-2)\\
4[5y-y+1]&=7(y-2)\\
4[4y+1]&=7(y-2)\\
16y+4&=7y-14\\
16y-7y+4&=7y-7y-14\\
9y+4&=-14\\
9y+4-4&=-14-4\\
9y&=-18\\
\frac{9y}{9}&=\frac{-18}{9}\\
y&=-2
\end{aligned}$$

10.
$$\begin{aligned}
12{,}000-1100x&=6500\\
12{,}000-12{,}000-1100x&=6500-12{,}000\\
1100x&=-5500\\
\frac{-1100x}{-1100}&=\frac{-5500}{-1100}\\
x&=5
\end{aligned}$$

The car will have a value of \$6500 in 5 years.

11.
$$\begin{aligned}
60x&=50(x+0.5)\\
60x&=50x+25\\
60x-50x&=50x-50x+25\\
10x&=25\\
\frac{10x}{10}&=\frac{25}{10}\\
x&=2.5
\end{aligned}$$

The express train will catch up with the local train in 2.5 hr, or $2\frac{1}{2}$ hr.

12.
$$\begin{aligned}
\frac{d}{25}+\frac{d}{30}&=5.5\\
\frac{6d}{150}+\frac{5d}{150}&=5.5\\
\frac{11d}{150}&=5.5\\
\frac{150}{11}\cdot\frac{11d}{150}&=\frac{150}{11}\cdot 5.5\\
d&=75
\end{aligned}$$

75 mi

13.
$$\begin{aligned}
2(x-10)+x&=40\\
2x-20+x&=40\\
3x-20+20&=40+20\\
3x&=60\\
3x&=60\\
\frac{3x}{3}&=\frac{60}{3}\\
x&=20
\end{aligned}$$

20 mph

Exercises

1.
$$\begin{aligned} 3x-1&=8\\ 3x-1+1&=8+1\\ 3x&=9\\ \frac{3x}{3}&=\frac{9}{3}\\ x&=3 \end{aligned}$$
Check:
$$\begin{aligned} 3(3)-1&=8\\ 9-1&=8\\ 8&=8 \end{aligned}$$

3.
$$\begin{aligned} 9t+17&=-1\\ 9t+17-17&=-1-17\\ 9t&=-18\\ \frac{9t}{9}&=\frac{-18}{9}\\ t&=-2 \end{aligned}$$
Check:
$$\begin{aligned} 9(-2)+17&=-1\\ -18+17&=-1\\ -1&=-1 \end{aligned}$$

5.
$$\begin{aligned} 20-5m&=45\\ 20-20-5m&=45-20\\ -5m&=25\\ \frac{-5m}{-5}&=\frac{25}{-5}\\ m&=-5 \end{aligned}$$
Check:
$$\begin{aligned} 20-5(-5)&=45\\ 20+25&=45\\ 45&=45 \end{aligned}$$

7.
$$\begin{aligned} \frac{n}{2}-1&=5\\ \frac{n}{2}-1+1&=5+1\\ \frac{n}{2}&=6\\ 2\cdot\frac{n}{2}&=2\cdot6\\ n&=12 \end{aligned}$$
Check:
$$\begin{aligned} \frac{12}{2}-1&=5\\ 6-1&=5\\ 5&=5 \end{aligned}$$

9.
$$\begin{aligned} \frac{x}{5}+15&=0\\ \frac{x}{5}+15-15&=0-15\\ \frac{x}{5}&=-15\\ 5\cdot\frac{x}{5}&=5\cdot(-15)\\ x&=-75 \end{aligned}$$
Check:
$$\begin{aligned} \frac{-75}{5}+15&=0\\ -15+15&=0\\ 0&=0 \end{aligned}$$

11.
$$\begin{aligned} 3-t&=1\\ 3-3-t&=1-3\\ -t&=-2\\ \frac{-t}{-1}&=\frac{-2}{-1}\\ t&=2 \end{aligned}$$
Check:
$$\begin{aligned} 3-2&=1\\ 1&=1 \end{aligned}$$

13.
$$\begin{aligned} -8-b&=11\\ -8+8-b&=11+8\\ -b&=19\\ \frac{-b}{-1}&=\frac{19}{-1}\\ b&=-19 \end{aligned}$$
Check:
$$\begin{aligned} -8-(-19)&=11\\ -8+19&=11\\ 11&=11 \end{aligned}$$

15. $\frac{2}{3}x-9=17$

$\frac{2}{3}x-9+9=17+9$

$\frac{2}{3}x=26$

$\frac{3}{2}\cdot\frac{2}{3}x=\frac{3}{2}\cdot 26$

$x=39$

Check:

$\frac{2}{3}(39)-9=17$

$26-9=17$

$17=17$

17. $\frac{4}{5}r+20=-20$

$\frac{4}{5}r+20-20=-20-20$

$\frac{4}{5}r=-40$

$\frac{5}{4}\cdot\frac{4}{5}r=\frac{5}{4}\cdot(-40)$

$r=-50$

Check:

$\frac{4}{5}(-50)+20=-20$

$-40+20=-20$

$-20=-20$

19. $3y+y=-8$

$4y=-8$

$\frac{4y}{4}=\frac{-8}{4}$

$y=-2$

Check:

$3(-2)+(-2)=-8$

$-6-2=-8$

$-8=-8$

21. $7z-2z=-30$

$5z=-30$

$\frac{5z}{5}=\frac{-30}{5}$

$z=-6$

Check:

$7(-6)-2(-6)=-30$

$-42+12=-30$

$-30=-30$

23. $28-a+4a=7$

$28+3a=7$

$28-28+3a=7-28$

$3a=-21$

$\frac{3a}{3}=\frac{-21}{3}$

$a=-7$

Check:

$28-(-7)+4(-7)=7$

$28+7+-28=7$

$7=7$

25. $1=1-6t-4t$

$1=1-10t$

$1-1=1-1-10t$

$0=-10t$

$\frac{0}{-10}=\frac{-10t}{-10}$

$0=t$

$t=0$

Check:

$1=1-6(0)-4(0)$

$1=1$

27. $3y+2=-y-2$

$3y+y+2=-y+y-2$

$4y+2=-2$

$4y+2-2=-2-2$

$4y=-4$

$\frac{4y}{4}=\frac{-4}{4}$

$y=-1$

Check:

$3(-1)+2=-(-1)-2$

$-3+2=1-2$

$-1=-1$

29.
$$5r-4=2r+6$$
$$5r-2r-4=2r-2r+6$$
$$3r-4=6$$
$$3r-4+4=6+4$$
$$3r=10$$
$$\frac{3r}{3}=\frac{10}{3}$$
$$r=\frac{10}{3}$$

Check:
$$5\left(\frac{10}{3}\right)-4=2\left(\frac{10}{3}\right)+6$$
$$\frac{50}{3}-4=\frac{20}{3}+6$$
$$\frac{50}{3}-\frac{12}{3}=\frac{20}{3}+\frac{18}{3}$$
$$\frac{38}{3}=\frac{38}{3}$$

31.
$$4(x+7)=7+x$$
$$4x+28=7+x$$
$$4x-x+28=7+x-x$$
$$3x+28=7$$
$$3x+28-28=7-28$$
$$3x=-21$$
$$\frac{3x}{3}=\frac{-21}{3}$$
$$x=-7$$

Check:
$$4((-7)+7)=7+(-7)$$
$$4(0)=0$$
$$0=0$$

33.
$$5(y-1)=2y+1$$
$$5y-5=2y+1$$
$$5y-2y-5=2y-2y+1$$
$$3y-5=1$$
$$3y-5+5=1+5$$
$$3y=6$$
$$\frac{3y}{3}=\frac{6}{3}$$
$$y=2$$

Check:
$$5(2-1)=2(2)+1$$
$$5(1)=4+1$$
$$5=5$$

35.
$$3a-2(a-9)=4+2a$$
$$3a-2a+18=4+2a$$
$$a+18=4+2a$$
$$a-2a+18=4+2a-2a$$
$$-a+18=4$$
$$-a+18-18=4-18$$
$$-a=-14$$
$$\frac{-a}{-1}=\frac{-14}{-1}$$
$$a=14$$

Check:
$$3(14)-2(14-9)=4+2(14)$$
$$3(14)-2(5)=4+2(14)$$
$$42-10=4+28$$
$$32=32$$

37.
$$5(2-t)-(1-3t)=6$$
$$10-5t-1+3t=6$$
$$9-2t=6$$
$$9-9-2t=6-9$$
$$-2t=-3$$
$$\frac{-2t}{-2}=\frac{-3}{-2}$$
$$t=\frac{3}{2}$$

Check:
$$5\left(2-\left(\frac{3}{2}\right)\right)-\left(1-3\left(\frac{3}{2}\right)\right)=6$$
$$5\left(\frac{4}{2}-\frac{3}{2}\right)-\left(\frac{2}{2}-\frac{9}{2}\right)=6$$
$$5\left(\frac{1}{2}\right)-\left(-\frac{7}{2}\right)=6$$
$$\frac{5}{2}+\frac{7}{2}=6$$
$$\frac{12}{2}=6$$
$$6=6$$

39.
$$\begin{aligned}
2y-3(y+1) &= -(5y+3)+y \\
2y-3y-3 &= -5y-3+y \\
-y-3 &= -4y-3 \\
-y+4y-3 &= -4y+4y-3 \\
3y-3 &= -3 \\
3y-3+3 &= -3+3 \\
3y &= 0 \\
\frac{3y}{3} &= \frac{0}{3} \\
y &= 0
\end{aligned}$$

Check:
$$\begin{aligned}
2(0)-3((0)+1) &= -(5(0)+3)+0 \\
0-3 &= -3+0 \\
-3 &= -3
\end{aligned}$$

41.
$$\begin{aligned}
2[3z-5(2z-3)] &= 3z-4 \\
2(3z-10z+15) &= 3z-4 \\
2(-7z+15) &= 3z-4 \\
-14z+30 &= 3z-4 \\
-14z-3z+30 &= 3z-3z-4 \\
-17z+30 &= -4 \\
-17z+30-30 &= -4-30 \\
-17z &= -34 \\
\frac{-17z}{-17} &= \frac{-34}{-17} \\
z &= 2
\end{aligned}$$

Check:
$$\begin{aligned}
2[3(2)-5(2(2)-3)] &= 3(2)-4 \\
2[6-5(4-3)] &= 6-4 \\
2[6-5(1)] &= 2 \\
2(6-5) &= 2 \\
2(1) &= 2 \\
2 &= 2
\end{aligned}$$

43.
$$\begin{aligned}
-8m-[2(11-2m)+4] &= 9m \\
-8m-(22-4m+4) &= 9m \\
-8m-(26-4m) &= 9m \\
-8m-26+4m &= 9m \\
-4m-26 &= 9m \\
-4m-9m-26 &= 9m-9m \\
-13m-26 &= 0 \\
-13m-26+26 &= 0+26 \\
-13m &= 26 \\
\frac{-13m}{-13} &= \frac{26}{-13} \\
m &= -2
\end{aligned}$$

Check:
$$\begin{aligned}
-8(-2)-[2(11-2(-2))+4] &= 9(-2) \\
16-[2(11+4)+4] &= -18 \\
16-[2(15)+4] &= -18 \\
16-(30+4) &= -18 \\
16-34 &= -18 \\
-18 &= -18
\end{aligned}$$

45.
$$\begin{aligned}
\frac{y}{0.87}+2.51 &= 4.03 \\
\frac{y}{0.87}+2.51-2.51 &= 4.03-2.51 \\
\frac{y}{0.87} &= 1.52 \\
0.87\cdot\frac{y}{0.87} &= 0.87\cdot 1.52 \\
y &\approx 1.32
\end{aligned}$$

47.
$$\begin{aligned}
7.37n+4.06 &= -1.98+6.55 \\
7.37n+1.98+4.06 &= -1.98+1398n+6.55 \\
9.35n+4.06 &= 6.55 \\
9.35n+4.06-4.06 &= 6.55-4.06 \\
9.35n &= 2.49 \\
\frac{9.35n}{9.35} &= \frac{2.49}{9.35} \\
n &\approx 0.27
\end{aligned}$$

49. a

51. d

53. $5-5x+x=21$
$$5-4x=21$$
$$5-5-4x=21-5$$
$$-4x=16$$
$$\frac{-4x}{-4}=\frac{16}{-4}$$
$$x=-4$$
Check:
$$5-5(-4)+(-4)=21$$
$$5+20-4=21$$
$$21=21$$

55. $4z+3(5-z)=-(z-3)+8$
$$4z+15-3z=-z+3+8$$
$$z+15=-z+11$$
$$z+z+15=-z+z+11$$
$$2z+15=11$$
$$2z+15-15=11-15$$
$$2z=-4$$
$$\frac{2z}{2}=\frac{-4}{2}$$
$$z=-2$$
Check:
$$4(-2)+3[5-(-2)]=-[(-2)-3]+8$$
$$-8+3(5+2)=-(-5)+8$$
$$-8+3(7)=5+8$$
$$-8+21=13$$
$$13=13$$

57. $-2.31y+0.14=-9.23$
$$-2.31y+0.14-0.14=-9.23-0.14$$
$$-2.31y=-9.37$$
$$\frac{-2.31y}{-2.31}=\frac{-9.37}{-2.31}$$
$$y\approx 4.06$$

59. $50+120x=1010$
$$50-50+120x=1010-50$$
$$120x=960$$
$$\frac{120x}{120}=\frac{960}{120}$$
$$x=8$$
The student is carrying 8 credits.

61. $x+2x=3690$
$$3x=3690$$
$$\frac{3x}{3}=\frac{3690}{3}$$
$$x=1230$$
One candidate received 1230 votes; the other received 2460 votes.

63. $3+2(t-1)=9$
$$3+2t-2=9$$
$$2t+1=9$$
$$2t+1-1=9-1$$
$$2t=8$$
$$\frac{2t}{2}=\frac{8}{2}$$
$$t=4$$
The car was parked in the garage for 4 hr.

65. $0.02x+0.01(5000-x)=85$
$$0.02x+50-0.01x=85$$
$$0.01x+50=85$$
$$0.01x+50-50=85-50$$
$$0.01x=35$$
$$\frac{0.01x}{0.01}=\frac{35}{0.01}$$
$$x=3500$$
3500 large postcards and 1500 small postcards can be printed.

67. $24\left(t+\frac{1}{3}\right)=36t$
$$24t+8=36t$$
$$24t-24t+8=36t-24t$$
$$8=12t$$
$$\frac{8}{12}=\frac{12t}{12}$$
$$\frac{2}{3}=t$$

It took $\frac{2}{3}$ hr, or 40 min, to catch the bus.

69.
$$\begin{aligned}27r+27(r+2)&=432\\27r+27r+54&=432\\54r+54&=432\\54r+54-54&=432-54\\54r&=378\\\frac{54r}{54}&=\frac{378}{54}\\r&=7\end{aligned}$$
One snail is crawling at a rate of 7 cm/min, the other is crawling at a rate of 9 cm/min.

71.
$$\begin{aligned}2r+2(r+4)&=212\\2r+2r+8&=212\\4r+8&=212\\4r+8-8&=212-8\\4r&=204\\\frac{4r}{4}&=\frac{204}{4}\\r&=51\end{aligned}$$
The speed of the slower truck is 51 mph.

2.4 Solving Literal Equations and Formulas

Practice

1.
$$\begin{aligned}q&=1-p\\q-1&=1-1-p\\q-1&=-p\\(-1)(q-1)&=(-1)(-p)\\1-q&=p\\p&=1-q\end{aligned}$$

2.
$$\begin{aligned}3r-s&=t\\3r-s+s&=t+s\\3r&=t+s\\\frac{3r}{3}&=\frac{t+s}{3}\\r&=\frac{t+s}{3}\end{aligned}$$

3.
$$\begin{aligned}\frac{4x}{5a}&=c\\5a\cdot\frac{4x}{5a}&=5a\cdot c\\4x&=5ac\\\frac{4x}{4}&=\frac{5ac}{4}\\x&=\frac{5ac}{4}\end{aligned}$$

4. **a.**
$$\begin{aligned}y&=mx+b\\y-b&=mx+b-b\\y-b&=mx\\\frac{y-b}{m}&=\frac{mx}{m}\\\frac{y-b}{m}&=x\\x&=\frac{y-b}{m}\end{aligned}$$

b.
$$\begin{aligned}x&=\frac{10-7}{-3}\\x&=\frac{3}{-3}\\x&=-1\end{aligned}$$

$$\begin{aligned}A&=P(1+rt)\\A&=P+Prt\\A-P&=P-P+Prt\\A-P&=Prt\\\frac{A-P}{Pt}&=\frac{Prt}{Pt}\\\frac{A-P}{Pt}&=r\end{aligned}$$

5. **a.** $r=\dfrac{A-P}{Pt}$

b.
$$\begin{aligned}r&=\frac{2100-2000}{(2000)(2)}\\r&=\frac{100}{4000}\\r&=0.025 \text{ or } 2.5\%\end{aligned}$$

6. a. $$A = \frac{1}{2}bh$$
$$\frac{2}{b} \cdot A = \frac{2}{b} \cdot \frac{1}{2}bh$$
$$\frac{2A}{b} = h$$
$$h = \frac{2A}{b}$$

b. $$h = \frac{2A}{b}$$
$$h = \frac{2(63)}{9}$$
$$h = 14 \text{ in.}$$

7. a. $$A = \frac{1}{2}h(b+B)$$

b. $$2 \cdot A = 2 \cdot \frac{1}{2}h(b+B)$$
$$2A = h(b+B)$$
$$2A = hb + hB$$
$$2A - hB = hb + hB - hB$$
$$2A - hB = hb$$
$$\frac{2A - hB}{h} = \frac{hb}{h}$$
$$b = \frac{2A - hB}{h}$$

c. $$b = \frac{2(32) - 4(11)}{4}$$
$$b = \frac{64 - 44}{4}$$
$$b = \frac{20}{4}$$
$$b = 5 \text{ cm}$$

Exercises

1. A literal equation is an equation involving two or more variables.

3. The solution to a literal equation is an algebraic expression.

5. $$y + 10 = x$$
$$y + 10 - 10 = x - 10$$
$$y = x - 10$$

7. $$d - c = 4$$
$$d - c + c = 4 + c$$
$$d = 4 + c$$

9. $$-3y = da$$
$$\frac{-3y}{a} = \frac{da}{a}$$
$$\frac{-3y}{a} = d$$
$$d = \frac{-3y}{a}$$

11. $$\frac{1}{2}n = 2p$$
$$2 \cdot \frac{1}{2}n = 2 \cdot 2p$$
$$n = 4p$$

13. $$a = \frac{1}{2}xyz$$
$$2 \cdot a = 2 \cdot \frac{1}{2}xyz$$
$$2a = xyz$$
$$\frac{2a}{xy} = \frac{xyz}{xy}$$
$$\frac{2a}{xy} = z$$
$$z = \frac{2a}{xy}$$

15. $$3x + y = 7$$
$$3x + y - y = 7 - y$$
$$3x = 7 - y$$
$$\frac{3x}{3} = \frac{7 - y}{3}$$
$$x = \frac{7 - y}{3}$$

17. $$3x + 4y = 12$$
$$3x - 3x + 4y = 12 - 3x$$
$$4y = 12 - 3x$$
$$\frac{4y}{4} = \frac{12 - 3x}{4}$$
$$y = \frac{12 - 3x}{4}$$

19.
$$y-4t=0$$
$$y-4t+4t=0+4t$$
$$y=4t$$

21.
$$-5b+p=r$$
$$-5b+p-p=r-p$$
$$-5b=r-p$$
$$\frac{-5b}{-5}=\frac{r-p}{-5}$$
$$b=\frac{p-r}{5}$$

23.
$$h=2(m-2l)$$
$$h=2m-4l$$
$$h-2m=2m-2m-4l$$
$$h-2m=-4l$$
$$\frac{h-2m}{-4}=\frac{-4l}{-4}$$
$$\frac{2m-h}{4}=l$$
$$l=\frac{2m-h}{4}$$

25.
$$I=prt$$
$$\frac{I}{pt}=\frac{prt}{pt}$$
$$\frac{I}{pt}=r$$
$$r=\frac{I}{pt}$$

27.
$$d=rt$$
$$\frac{d}{t}=\frac{rt}{t}$$
$$\frac{d}{t}=r$$
$$r=\frac{d}{t}$$

29.
$$P=a+b+c$$
$$P-a=a-a+b+c$$
$$P-a=b+c$$
$$P-a-c=b+c-c$$
$$P-a-c=b$$
$$b=P-a-c$$

31.
$$C=\pi d$$
$$\frac{C}{\pi}=\frac{\pi d}{\pi}$$
$$\frac{C}{\pi}=d$$
$$d=\frac{C}{\pi}$$

33.
$$P=I^2R$$
$$\frac{P}{I^2}=\frac{I^2R}{I^2}$$
$$\frac{P}{I^2}=R$$
$$R=\frac{P}{I^2}$$

35.
$$A=\frac{a+b+c}{3}$$
$$3\cdot A=3\cdot\frac{a+b+c}{3}$$
$$3A=a+b+c$$
$$3A-b=a+b-b+c$$
$$3A-b=a+c$$
$$3A-b-c=a+c-c$$
$$3A-b-c=a$$
$$a=3A-b-c$$

37.
$$S=a+(n-1)d$$
$$S-(n-1)d=a+(n-1)d-(n-1)d$$
$$a=S-dn+d$$

39.
$$3x+y=6$$
$$3x-3x+y=6-3x$$
$$y=6-3x$$

Check:
$$y=6-3(-12)$$
$$y=6+36$$
$$y=42$$

41.
$$3x-7=y$$
$$3x-7+7=y+7$$
$$3x=y+7$$
$$\frac{3x}{3}=\frac{y+7}{3}$$
$$x=\frac{y+7}{3}$$

Check:
$$x=\frac{5+7}{3}$$
$$x=\frac{12}{3}$$
$$x=4$$

43.
$$-\frac{1}{3}y=x$$
$$-3\left(-\frac{1}{3}y\right)=-3\cdot x$$
$$y=-3x$$

Check:
$$y=-3\left(\frac{1}{2}\right)$$
$$y=-\frac{3}{2}$$

45.
$$ax+by=c$$
$$ax+by-by=c-by$$
$$ax=c-by$$
$$\frac{ax}{a}=\frac{c-by}{a}$$
$$x=\frac{c-by}{a}$$

Check:
$$x=\frac{(2)-(3)(-4)}{1}$$
$$x=\frac{2+12}{1}$$
$$x=\frac{14}{1}$$
$$x=14$$

47.
$$V=\pi r^2h$$
$$\frac{V}{\pi r^2}=\frac{\pi r^2h}{\pi r^2}$$
$$\frac{V}{\pi r^2}=h$$
$$h=\frac{V}{\pi r^2}$$

49.
$$m=\frac{2}{5}abc$$
$$\frac{5}{2}\cdot m=\frac{5}{2}\cdot\frac{2}{5}abc$$
$$\frac{5}{2}m=abc$$
$$\frac{5m}{2ac}=\frac{abc}{ac}$$
$$\frac{5m}{2ac}=b$$

51.
$$4w+9z=3$$
$$4w-4w+9z=3-4w$$
$$9z=3-4w$$
$$\frac{9z}{9}=\frac{3-4w}{9}$$
$$z=\frac{3-4w}{9}$$

53.
$$7x+2y=-20$$
$$7x+2y-2y=-20-2y$$
$$7x=-20-2y$$
$$\frac{7x}{7}=\frac{-20-2y}{7}$$
$$x=\frac{-20-2y}{7}$$

Check:
$$x=\frac{-20-2(4)}{7}$$
$$x=\frac{-20-8}{7}$$
$$x=\frac{-28}{7}$$
$$x=-4$$

55.
$$K=\frac{V}{T}$$
$$K\cdot T=\frac{V}{T}\cdot T$$
$$KT=V$$
$$V=KT$$

57. **a.** $C = \frac{w}{150} \cdot A$

b.
$$C = \frac{w}{150} \cdot A$$
$$\frac{150}{w} \cdot C = \frac{150}{w} \cdot \frac{w}{150} \cdot A$$
$$\frac{150C}{w} = A$$
$$A = \frac{150C}{w}$$

59. **a.** $m = \frac{t}{5}$

b. $5 \cdot m = 5 \cdot \frac{t}{5}$
$$5m = t$$
$$t = 5m$$

c. $t = (5)(2.5)$
$$t = 12.5$$
The thunder will be heard in 12.5 sec.

61. **a.** $C = 2\pi r$

b. $\frac{C}{2\pi} = \frac{2\pi r}{2\pi}$
$$\frac{C}{2\pi} = r$$
$$r = \frac{C}{2\pi}$$

c. $r \approx \frac{(5)}{2(3.14)}$
$$r \approx \frac{5}{6.28}$$
$$r \approx 0.08 \text{ ft.}$$

2.5 Solving Equations Involving Percent

Practice

1.
$$8 = 0.4 \cdot x$$
$$\frac{8}{0.4} = \frac{0.4x}{0.4}$$
$$20 = x$$

2.
$$0.12 \cdot x = 3$$
$$\frac{0.12x}{0.12} = \frac{3}{0.12}$$
$$x = 25$$
$25 million

3. $0.23 \cdot 45 = 10.35$
10.35 m

4.
$$x = (0.50 - 0.24)2{,}860{,}000$$
$$x = (0.26)2{,}860{,}000$$
$$x = 743{,}600$$

5.
$$14 = x \cdot 16$$
$$\frac{14}{16} = \frac{16x}{16}$$
$$x = 87.5$$
$87\frac{1}{2}\%$

6.
$$x - 42 = 13$$
$$\frac{42x}{42} = \frac{13}{42}$$
$$x = .3095$$
31% of the presidents had been vice president.

7.
$$300x = 361 - 300$$
$$300x = 61$$
$$\frac{300x}{300} = \frac{61}{300}$$
$$x = 0.203$$
The number of nursing homes increased by 20.3%.

8. $x \cdot 300 = 300 - 230$
$300x = 70$
$\frac{300x}{300} = \frac{70}{300}$
$x \approx 0.23$

$x \cdot 2250 = 2250 - 1750$
$2250x = 500$
$\frac{2250x}{2250} = \frac{500}{2250}$
$x \approx 0.22$

The stock index dropped about 23% in the 1929 crash and about 22% in the 1987 crash. So the stock index dropped more in 1929.

9. $r = \frac{I}{pt}$
$r = \frac{130}{2000(1)}$
$r = 0.065$
6.5%

10. $0.1(2x) - 0.1x = 700$
$0.2x - 0.1x = 700$
$0.1x = 700$
$\frac{0.1x}{0.1} = \frac{700}{0.1}$
$x = 7000$

She invested \$7000 in a mutual fund and \$14,000 in bonds.

11. $0.2(30) + x = 0.25(30 + x)$
$6 + x = 7.5 + 0.25x$
$6 + x - 0.25x = 7.5 + 0.25 - 0.25$
$6 + 0.75x = 7.5$
$6 - 6 + 0.75x = 7.5 - 6$
$0.75x = 1.5$
$\frac{0.75x}{0.75} = \frac{1.5}{0.75}$
$x = 2$

2g

Exercises

1. In solving the percent problem "30% of what number is 7?", "what number" represents the <u>base</u>.

3. In translating the problem" What percent of 36 is 9?", "of" becomes <u>times</u>.

5. $0.75 \cdot 8 = 6$

7. $1.00 \cdot 23 = 23$

9. $0.41 \cdot 7 \text{ kg} = 2.87 \text{ kg}$

11. $0.08 \cdot \$500 = \$40$

13. $0.125 \cdot 32 = 4$

15. $0.25x = 8$
$\frac{0.25x}{0.25} = \frac{8}{0.25}$
$x = 32 \text{ sq in.}$

17. $\$12 = 0.1x$
$\frac{\$12}{0.1} = \frac{0.1x}{0.1}$
$\$120 = x$

19. $20 = 0.1x$
$\frac{\$20}{0.1} = \frac{0.1x}{0.1}$
$\$200 = x$

21. $3.5 = 2x$
$\frac{3.5}{2} = \frac{2x}{2}$
$1.75 = x$

23. $0.05l = 23$
$\frac{0.05l}{0.05} = \frac{23}{0.05}$
$l = 4600 \text{ m}$

25. $50 = x \cdot 80$
$\frac{50}{80} = \frac{80x}{50}$
$x = 0.625$
62.5%

27. $5 = x \cdot 15$
$\frac{5}{15} = \frac{15x}{15}$
$\frac{1}{3} = x$
$33\frac{1}{3}\%$

29. $10 = x \cdot 8$

$\frac{10}{8} = \frac{8x}{8}$

$125 = x$

125%

31. $x \cdot 5 = \frac{1}{2}$

$\frac{1}{5} \cdot 5x = \frac{1}{5} \cdot \frac{1}{2}$

$x = \frac{1}{10} = 0.1$

10%

33. $2.5 = x \cdot 4$

$\frac{2.5}{4} = \frac{4x}{4}$

$0.625 = x$

62.5%

35. $x = 0.35 \cdot \$400$

$x = \$140$

37. $x \cdot 50 = 20$

$\frac{50x}{50} = \frac{20}{50}$

$x = 0.4$

40%

39. $0.7 \cdot x = 14$

$\frac{0.7x}{0.7} = \frac{14}{0.7}$

$x = 20$

41. $0.001 \cdot 35 = x$

$0.035 = x$

43. $\$8 = x \cdot \$240$

$\frac{\$8}{\$240} = \frac{\$240x}{\$240}$

$0.03\overline{3} = x$

$x = 0.03\overline{3}$

$3\frac{1}{3}\%$

45. $3 = 0.2 \cdot x$

$\frac{3}{0.2} = \frac{0.2x}{0.2}$

$x = 15$

15 oz

47. $\$14 = x \cdot \$8$

$\frac{\$14}{\$8} = \frac{\$8x}{\$8}$

$1.75 = x$

175%

49. $0.8 \cdot x = 96$

$\frac{0.8x}{0.8} = \frac{96}{0.8}$

$x = 120$

51. $0.08 \cdot 120 = 9.6$

53. $21 = 0.2 \cdot x$

$\frac{21}{0.2} = \frac{0.2x}{0.2}$

$105 = x$

55. Faculty: $0.75 \cdot 160 = 120$

Staff: $0.25 \cdot 160 = 40$

$120 - 40 = 80$

There are 80 more faculty than staff.

57. $\frac{9}{12} = 4x$

$\frac{1}{4} \cdot \frac{9}{12} = \frac{1}{4} \cdot 4$

$\frac{9}{48} = x$

$x = \frac{9}{48}$

$x = 0.1875$

18.75%

59. $8 = 0.25 \cdot x$

$\frac{8}{0.25} = \frac{0.25x}{0.25}$

$32 = x$

There are 32 employees

61. $x \cdot \$250{,}000 = \$50{,}000$

$$\frac{\$250{,}000x}{\$250{,}000} = \frac{\$50{,}000}{250{,}000}$$

$$x = 0.20$$

20%

63. $1.5 = 0.08x$

$$\frac{1.5}{0.08} = \frac{0.08x}{0.08}$$

$18.75 = x$

The work force is 18.75 million people.

65. $0.6 \cdot 90 = 54$

54 tables

67. $\dfrac{4000 - 1800}{4000}$

$$= \frac{2200}{4000}$$

$= 0.55$

55% ; No

69. $I = Prt$

$I = \$2000(0.05)(1)$

$I = \$100$

71. $I = Prt$

$I = \$5000(0.05)(1)$

$I = \$250$

73. $x \cdot 0.08 + (34{,}000 - x) \cdot 0.10 = 3000$

$$0.08 + 3400 - 0.10x = 3000$$

$$3400 - 0.02x = 3000$$

$$3400 - 3400 - 0.02x = 3000 - 3400$$

$$-0.02x = -400$$

$$\frac{-0.02x}{-0.02} = \frac{-400}{-0.02}$$

$$x = \$20{,}000$$

$\$34{,}000 - \$20{,}000 = \$14{,}000$

\$20,000 was invested at 8% and 14,000 was invested at 10%.

75. $\$20{,}000(0.08) + x(0.05) = \$2100$

$$\$1600 + 0.05x = \$2100$$

$$\$1600 - \$1600 + 0.05x = \$2100 - \$1600$$

$$0.05x = \$500$$

$$\frac{0.05x}{0.05} = \frac{\$500}{0.05}$$

$$x = \$10{,}000$$

\$10,000 was invested at 5%

77. The ratio of lemon juice to olive oil is 20%, so if x cups of oil are added we have:

$$\frac{\frac{1}{3}}{x+1} = 0.2$$

$$3 \cdot \frac{\frac{1}{3}}{x+1} = 3 \cdot 0.2$$

$$\frac{1}{x+1} = 0.6$$

$$x + 1 = \frac{1}{0.6} = \frac{5}{3}$$

$$x = \frac{2}{3} \text{ cups of additional olive oil}$$

79. $x(0.4) + 30(0.04) = (x + 30)(0.10)$

$$0.4x + 1.2 = 0.1x + 3$$

$$0.4x + 1.2 - 1.2 = 0.1x + 3 - 1.2$$

$$0.4x = 0.1x + 1.8$$

$$0.4x - 0.1x = 0.1x - 0.1x + 1.8$$

$$0.3x = 1.8$$

$$\frac{0.3x}{0.3} = \frac{1.8}{0.3}$$

$$x = 6$$

6 oz

2.6 Solving Linear Inequalities

Practice

1. $\frac{1}{2}x - 2 < -1$

$$\frac{1}{2}(4) - 2 < -1$$

$$2 - 2 < -1$$

$$0 < -1$$

No, 4 is not a solution.

2.

−5 −4 −3 −2 −1 0 1 2 3 4 5

3.

−5 −4 −3 −2 −1 0 1 2 3 4 5

4.

−5 −4 −3 −2 −1 0 1 2 3 4 5

5.

$$\begin{aligned} n+5&>4 \\ n+5-5&>4-5 \\ n&>-1 \end{aligned}$$

−5 −4 −3 −2 −1 0 1 2 3 4 5

6.

$$\begin{aligned} x-4&\le 1\frac{1}{2} \\ x-4+4&\le 1\frac{1}{2}+4 \\ x&\le 5\frac{1}{2} \end{aligned}$$

−2 −1 0 1 2 3 4 5 6 7 8

7.

$$\begin{aligned} 4x+5&\ge 3x-2 \\ 4x+5-5&\ge 3x-2-5 \\ 4x&\ge 3x-7 \\ 4x-3x&\ge 3x-3x-7 \\ x&\ge -7 \end{aligned}$$

−8 −7 −6 −5 −4 −3 −2 −1 0 1 2

8.

$$\begin{aligned} \frac{x}{3}&\le 1 \\ 3\cdot\frac{x}{3}&\le 3\cdot 1 \\ x&\le 3 \end{aligned}$$

−5 −4 −3 −2 −1 0 1 2 3 4 5

9.

$$\begin{aligned} -3x&>15 \\ \frac{-3x}{-3}&<\frac{15}{-3} \\ x&<-5 \end{aligned}$$

−11 −10 −9 −8 −7 −6 −5 −4 −3 −2 −1

10.

$$\begin{aligned} 10&>5x-7x \\ 10&>-2x \\ \frac{10}{-2}&<\frac{-2x}{-2} \\ -5&<x \\ x&>-5 \end{aligned}$$

−8 −7 −6 −5 −4 −3 −2 −1 0 1 2

11.

$$\begin{aligned} -6&\ge 3z+4-z \\ -6&\ge 2z+4 \\ -6-4&\ge 2z+4-4 \\ -10&\ge 2z \\ \frac{-10}{2}&\ge\frac{2z}{2} \\ -5&\ge z \\ z&\le -5 \end{aligned}$$

12.

$$\begin{aligned} 7x-(9x+1)&>-5 \\ 7x-9x-1&>-5 \\ -2x-1&>-5 \\ -2x-1+1&>-5+1 \\ -2x&>-4 \\ \frac{-2x}{-2}&<\frac{-4}{-2} \\ x&<2 \end{aligned}$$

13.

$$\begin{aligned} (x+3)+(x+2)+x&\ge 14 \\ 3x+5&\ge 14 \\ 3x+5-5&\ge 14-5 \\ 3x&\ge 9 \\ \frac{3x}{3}&\ge\frac{9}{3} \\ x&\ge 3 \end{aligned}$$

The perimeter will be greater than or equal to 14 in. for any value of x greater than or equal to 3.

14.

$$\begin{aligned} 15(8.50)+7.50t&\ge 300 \\ 127.5+7.5t&\ge 300 \\ 127.5-127.5+7.5t&\ge 300-127.5 \\ 7.5t&\ge 172.5 \\ \frac{7.5t}{7.5}&\ge\frac{172.5}{7.5} \\ t&\ge 23 \end{aligned}$$

You should work at least 23 hr on the second job.

Exercises

1. A(n) inequality is any mathematical statement containing $<, \le, >, \ge$, or $\ne$.

3. In the graph of the inequality $x>-7$, the circle is open.

5. When the same number is added to or subtracted from each side of an inequality, the direction of the inequality is unchanged.

7. According to the multiplication property of inequalities, if $a \geq b$ and c is negative, the $ac \leq bc$.

9. a. $8-3x>5$

$8-3(1)>5$

$8-3>5$

$5>5$

False

b. $4x-7 \leq 2x+1$

$4(4)-7 \leq 2(4)+1$

$16-7 \leq 8+1$

$9 \leq 9$

True

c. $6(x+6)<-9$

$6(-7+6)<-9$

$6(-1)<-9$

$-6<-9$

False

d. $8x+10 \geq 12x+15$

$8\left(-\frac{3}{4}\right)+10 \geq 12\left(-\frac{3}{4}\right)+15$

$-6+10 \geq -9+15$

$4 \geq 6$

False

11.

−2 −1 0 1 2 3 4 5 6 7 8

13.

−8 −7 −6 −5 −4 −3 −2 −1 0 1 2

15.

−5 −4 −3 −2 −1 0 1 2 3 4 5

17.

−5 −4 −3 −2 −1 0 1 2 3 4 5

19.

−5 −4 −3 −2 −1 0 1 2 3 4 5

21.

−5 −4 −3 −2 −1 0 1 2 3 4 5

23.

−5 −4 −3 −2 −1 0 1 2 3 4 5

25.

−5 −4 −3 −2 −1 0 1 2 3 4 5

27. $v+2<-5$

$v+2-2<-5-2$

$v<-7$

−15 −14 −13 −12 −11 −10 −9 −8 −7 −6 −5

29. $y-5>-5$

$y-5+5>-5+5$

$y>0$

−5 −4 −3 −2 −1 0 1 2 3 4 5

31. $y+2 \leq 5.5$

$y+2-2 \leq 5.5-2$

$y \leq 3.5$

−5 −4 −3 −2 −1 0 1 2 3 4 5

33. $v-17 \leq -15$

$v-17+17 \leq -15+17$

$v \leq 2$

−5 −4 −3 −2 −1 0 1 2 3 4 5

35. $-2 \geq x-4$

$-2+4 \geq x-4+4$

$2 \geq x$

$x \leq 2$

−5 −4 −3 −2 −1 0 1 2 3 4 5

37. $\frac{1}{3}a<-1$

$3 \cdot \frac{1}{3}a<3 \cdot (-1)$

$a<-3$

−10 −9 −8 −7 −6 −5 −4 −3 −2 −1 0

39. $-5y>10$

$-\frac{1}{5}(-5y)<-\frac{1}{5} \cdot (10)$

$y<-2$

−10 −9 −8 −7 −6 −5 −4 −3 −2 −1 0

41.
$$\begin{aligned} 2x &\geq 0 \\ \frac{2x}{2} &\geq \frac{0}{2} \\ x &\geq 0 \end{aligned}$$

−5 −4 −3 −2 −1 0 1 2 3 4 5

43.
$$\begin{aligned} \frac{-3}{4}a &\geq 3 \\ -\frac{4}{3}\cdot\frac{-3a}{4} &\leq -\frac{4}{3}\cdot 3 \\ a &\leq -4 \end{aligned}$$

−12 −11 −10 −9 −8 −7 −6 −5 −4 −3 −2

45.
$$\begin{aligned} 6 &\leq -\frac{2}{3}n \\ -\frac{3}{2}\cdot 6 &\geq \frac{-3}{2}\cdot\left(-\frac{2}{3}\right)n \\ -9 &\geq n \\ n &\leq -9 \end{aligned}$$

−18 −17 −16 −15 −14 −13 −12 −11 −10 −9 −8

47.
$$\begin{aligned} \frac{n}{3}+2 &> 3 \\ \frac{n}{3}+2-2 &> 3-2 \\ \frac{n}{3} &> 1 \\ 3\cdot\frac{n}{3} &> 3\cdot 1 \\ n &> 3 \end{aligned}$$

49.
$$\begin{aligned} 3x-12 &\leq 6 \\ 3x-12+12 &\leq 6+12 \\ 3x &\leq 18 \\ \frac{1}{3}\cdot 3x &\leq \frac{1}{3}\cdot 18 \\ x &\leq 6 \end{aligned}$$

51.
$$\begin{aligned} -21-3y &> 0 \\ -21+21-3y &> 0+21 \\ -3y &> 21 \\ -\frac{1}{3}\cdot(-3y) &< -\frac{1}{3}\cdot 21 \\ y &< -7 \end{aligned}$$

53.
$$\begin{aligned} 5n-11 &\geq 2n+28 \\ 5n-11+11 &\geq 2n+28+11 \\ 5n &\geq 2n+39 \\ 5n-2n &\geq 2n-2n+39 \\ 3n &\geq 39 \\ \frac{3n}{3} &\geq \frac{39}{3} \\ n &\geq 13 \end{aligned}$$

55.
$$\begin{aligned} -4m+8 &\leq -3m+1 \\ -4m+8-8 &\leq -3m+1-8 \\ -4m &\leq -3m-7 \\ -4m+3m &\leq -3m+3m-7 \\ -m &\leq -7 \\ \frac{-m}{-1} &\geq \frac{-7}{-1} \\ m &\geq 7 \end{aligned}$$

57.
$$\begin{aligned} -7x+4x+23 &< 2 \\ -3x+23 &< 2 \\ -3x+23-23 &< 2-23 \\ -3x &< -21 \\ \frac{-3x}{-3} &> \frac{-21}{-3} \\ x &> 7 \end{aligned}$$

59.
$$\begin{aligned} -3(z+5) &> -15 \\ -3z-15 &> -15 \\ -3z-15+15 &> -15+15 \\ -3z &> 0 \\ \frac{-3z}{-3} &< \frac{0}{-3} \\ z &< 0 \end{aligned}$$

61.
$$\begin{aligned} 0.5(2x+1) &\geq 3x \\ \frac{0.5(2x+1)}{0.5} &\geq \frac{3x}{0.5} \\ 2x+1 &\geq 6x \\ 2x-2x+1 &\geq 6x-2x \\ 1 &\geq 4x \\ \frac{1}{4} &\geq \frac{4x}{4} \\ \frac{1}{4} &\geq x \\ x &\leq 0.25 \end{aligned}$$

63.
$$\begin{aligned}
2(x-2)-3x &\geq -1\\
2x-4-3x &\geq -1\\
-x-4 &\geq -1\\
-x-4+4 &\geq -1+4\\
-x &\geq 3\\
\frac{-x}{-1} &\leq \frac{3}{-1}\\
x &\leq -3
\end{aligned}$$

65.
$$\begin{aligned}
7y-(9y+1) &< 5\\
7y-9y-1 &< 5\\
-2y-1 &< 5\\
-2y-1+1 &< 5+1\\
-2y &< 6\\
\frac{-2y}{-2} &> \frac{6}{-2}\\
y &> -3
\end{aligned}$$

67.
$$\begin{aligned}
0.4(5x+1) &\geq 3x\\
\frac{0.4(5x+1)}{0.4} &\geq \frac{3x}{0.4}\\
5x+1 &\geq 7.5x\\
5x+1-1 &\geq 7.5x-1\\
5x &\geq 7.5x-1\\
5x-7.5x &\geq 7.5x-7.5x-1\\
-2.5x &\geq -1\\
\frac{-2.5x}{-2.5} &\leq \frac{-1}{-2.5}\\
x &\leq 0.4
\end{aligned}$$

69.
$$\begin{aligned}
5x+1 &< 3x-2(4x-3)\\
5x+1 &< 3x-8x+6\\
5x+1 &< -5x+6\\
5x+1-1 &< -5x+6-1\\
5x &< -5x+5\\
5x+5x &< -5x+5x+5\\
10x &< 5\\
\frac{10x}{10} &< \frac{5}{10}\\
x &< \frac{1}{2}
\end{aligned}$$

71.
$$\begin{aligned}
3+5n &\leq 6(n-1)+n\\
3+5n &\leq 6n-6+n\\
3+5n &\leq 7n-6\\
3-3+5n &\leq 7n-6-3\\
5n &\leq 7n-9\\
5n-7n &\leq 7n-7n-9\\
-2n &\leq -9\\
\frac{-2n}{-2} &\geq \frac{-9}{-2}\\
n &\geq 4.5
\end{aligned}$$

73.
$$\begin{aligned}
-\frac{4}{3}x-16 &> x+\frac{1}{3}x\\
-\frac{4}{3}x-16 &> \frac{4}{3}x\\
-\frac{4}{3}x+\frac{4}{3}x-16 &> \frac{4}{3}x+\frac{4}{3}x\\
16 &> \frac{8}{3}x\\
\frac{3}{8}\cdot(16) &> \frac{3}{8}\cdot\frac{8}{3}x\\
-6 &> x\\
x &< -6
\end{aligned}$$

75.
$$\begin{aligned}
0.2y &> 1500+2.6y\\
0.2y-2.6y &> 1500+2.6y-2.6y\\
-2.4y &> 1500\\
\frac{-2.4y}{-2.4} &< \frac{1500}{-2.4}\\
y &< -625
\end{aligned}$$

77. d

79. d

81.

−2 −1 0 1 2 3 4 5

83. **a.** $4-2x<-1$

$4-2(2)<-1$

$4-4<-1$

$0<-1$

False

b. $3x-5\geq 21-2x$

$3(6)-5\geq 21-2(6)$

$18-5\geq 21-12$

$13\geq 9$

True

c. $-(2x+4)\leq 12$

$-[2(-8)+4]\leq 12$

$-(-16+4)\leq 12$

$-(12)\leq 12$

$12\leq 12$

True

d. $8x-5>4x-8$

$8\left(-\frac{1}{2}\right)-5>4\left(-\frac{1}{2}\right)-8$

$-4-5>-2-8$

$-9>10$

True

85.

$-\frac{2}{3}m\geq 4$

$-\frac{3}{2}\cdot\left(-\frac{2}{3}m\right)\leq-\frac{3}{2}\cdot 4$

$m\leq-6$

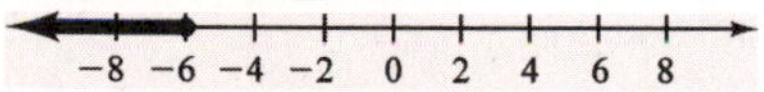

87. $3(a+4)\geq 2(3a-2)$

$3a+12\geq 6a-4$

$3a+12-12\geq 6a-4-12$

$3a\geq 6a-16$

$3a-6a\geq 6a-6a-16$

$-3a\geq-16$

$\frac{-3a}{-3}\leq\frac{-16}{-3}$

$a\leq\frac{16}{3}$

89. $8x+7-3x<32$

$5x+7<32$

$5x+7<32$

$5x+7-7<32-7$

$5x<25$

$\frac{5x}{5}<\frac{25}{5}$

$x<5$

91. $\frac{81+85+91+x}{4}>85$

$\frac{257+x}{4}>85$

$4\cdot\frac{257+x}{4}>4\cdot 85$

$257+x>340$

$257-257+x>340-257$

$x>83$

The student must score above 83.

93.

$\frac{250+250+150++130+180+x}{6}\geq 200$

$\frac{960+x}{6}\geq 200$

$6\cdot\frac{960+x}{6}\geq 6\cdot 200$

$960+x\geq 1200$

$960-960+x\geq 1200-960$

$x\geq 240$

The store must make at least $240.

95. $0.50+0.10x\geq 2$

$0.50-0.50+0.10x\geq 2-0.50$

$0.10x\geq 1.50$

$\frac{0.10x}{0.10}\geq\frac{1.50}{0.10}$

$x\geq 15$

Each call lasts at least 15 min.

97.
$$\begin{aligned}1000+1500h &> 1500+1200h\\1000-1000+1500h &> 1500-1000+1200h\\1500h &> 500+1200h\\1500h-1200h &> 500+1200h-1200h\\300h &> 500\\\frac{300h}{300} &> \frac{500}{300}\\h &> \frac{5}{3}\\h &> 1\frac{2}{3}\end{aligned}$$

He should accept the deal is he sells 2 or more houses each month.

99.
$$\begin{aligned}200-2.5x &< 180\\200-200-2.5x &< 180-200\\-2.5x &< -20\\\frac{-2.5x}{-2.5} &> \frac{-20}{-2.5}\\x &> 8\end{aligned}$$

He will weigh less than 180 lbs after 8 mos.

Chapter 2 Review Exercises

1.
$$\begin{aligned}5x+3 &= 7-4x\\5(2)+3 &= 7-7(2)\\10+3 &= 7-8\\13 &\neq -1\end{aligned}$$

-1 is not a solution.

2.
$$\begin{aligned}4x-15 &= 5(x-3)\\4(0)-15 &= 5(0-3)\\0-15 &= 15(-3)\\-15 &= -15\end{aligned}$$

0 is a solution.

3.
$$\begin{aligned}x-3 &= -12\\x-3+3 &= -12+3\\x &= -9\end{aligned}$$

Check:
$$\begin{aligned}-9-3 &= -12\\-12 &= -12\end{aligned}$$

4.
$$\begin{aligned}t+10 &= 8\\t+10-10 &= 8-10\\t &= -2\end{aligned}$$

Check:
$$\begin{aligned}-2+10 &= 8\\8 &= 8\end{aligned}$$

5.
$$\begin{aligned}-9 &= a+5\\-9-5 &= a+5-5\\-14 &= a\\a &= -14\end{aligned}$$

Check:
$$\begin{aligned}-9 &= -14+5\\-9 &= -9\end{aligned}$$

6.
$$\begin{aligned}4 &= n-7\\4+7 &= n-7+7\\11 &= n\\n &= 11\end{aligned}$$

Check:
$$\begin{aligned}4 &= 11-7\\4 &= 4\end{aligned}$$

7.
$$\begin{aligned}y-(-3.1) &= 11\\y+3.1 &= 11\\y+3.1-3.1 &= 11-3.1\\y &= 7.9\end{aligned}$$

Check:
$$\begin{aligned}7.9-(-3.1) &= 11\\7.9+3.1 &= 11\\11 &= 11\end{aligned}$$

8.
$$\begin{aligned}r+4.8 &= 20\\r+4.8-4.8 &= 20-4.8\\r &= 15.2\end{aligned}$$

Check:
$$\begin{aligned}15.2+4.8 &= 20\\20 &= 20\end{aligned}$$

9.
$$\begin{aligned}\frac{x}{3} &= -2\\3\cdot\frac{x}{3} &= 3\cdot(-2)\\x &= -6\end{aligned}$$

Check:
$$\begin{aligned}\frac{-6}{3} &= -2\\-2 &= -2\end{aligned}$$

10.
$$\frac{z}{2}=-5$$
$$2\cdot\frac{z}{2}=2\cdot(-5)$$
$$z=-10$$
Check:
$$\frac{-10}{2}=-5$$
$$-5=-5$$

11.
$$2x=-20$$
$$\frac{2x}{2}=\frac{-20}{2}$$
$$x=-10$$
Check:
$$2(-10)=-20$$
$$-20=-20$$

12.
$$-5d=15$$
$$\frac{-5d}{-5}=\frac{15}{-5}$$
$$d=-3$$
Check:
$$-5(-3)=15$$
$$15=15$$

13.
$$-y=-4$$
$$\frac{-y}{-1}=\frac{-4}{-1}$$
$$y=4$$
Check:
$$-(4)=-4$$
$$-4=-4$$

14.
$$-x=3$$
$$\frac{-x}{-1}=\frac{3}{-1}$$
$$x=-3$$
Check:
$$-(-3)=3$$
$$3=3$$

15.
$$20.5=0.5n$$
$$\frac{20.5}{0.5}=\frac{0.5n}{0.5}$$
$$41=n$$
$$n=41$$
Check:
$$20.5=0.5(41)$$
$$20.5=20.5$$

16.
$$30=-0.2r$$
$$\frac{30}{-0.2}=\frac{-0.2r}{-0.2}$$
$$r=-150$$
Check:
$$30=-0.2(-150)$$
$$30=30$$

17.
$$\frac{2t}{3}=-6$$
$$\frac{3}{2}\cdot\frac{2}{3}t=\frac{3}{2}\cdot(-6)$$
$$t=-9$$
Check:
$$\frac{2(-9)}{3}=-6$$
$$-6=-6$$

18.
$$\frac{5y}{6}=-10$$
$$\frac{6}{5}\cdot\frac{5y}{6}=\frac{6}{5}\cdot(-10)$$
$$y=-12$$
Check:
$$\frac{5(-12)}{6}=-10$$
$$-10=-10$$

19.
$$2x+1=7$$
$$2x+1-1=7-1$$
$$2x=6$$
$$\frac{2x}{2}=\frac{6}{2}$$
$$x=3$$
Check:
$$2(3)+1=7$$
$$6+1=7$$
$$7=7$$

20.
$$-t-4=5$$
$$-t-4=5$$
$$-t-4+4=5+4$$
$$-t=9$$
$$\frac{-t}{-1}=\frac{9}{-1}$$
$$t=-9$$
Check:
$$-(-9)-4=5$$
$$9-4=5$$

21.
$$\frac{a}{2}-3=-10$$
$$\frac{a}{2}-3+3=-10+3$$
$$\frac{a}{2}=-7$$
$$2\cdot\frac{a}{2}=2\cdot(-7)$$
$$a=-14$$
Check:
$$\frac{-14}{2}-3=-10$$
$$-7-3=-10$$
$$-10=-10$$

22.
$$\frac{r}{3}-6=12$$
$$\frac{r}{3}-6+6=12+6$$
$$\frac{r}{3}=18$$
$$3\cdot\frac{r}{3}=3\cdot 18$$
$$r=54$$
Check:
$$\frac{54}{3}-6=12$$
$$18-6=12$$
$$12=12$$

23.
$$-y+7=-2$$
$$-y+7-7=-2-7$$
$$-y=-9$$
$$\frac{-y}{-1}=\frac{-9}{-1}$$
$$y=9$$
Check:
$$-9+7=-2$$
$$-2=-2$$

24.
$$-2t+3=1$$
$$-2t+3-3=1-3$$
$$-2t=-2$$
$$\frac{-2t}{-2}=\frac{-2}{-2}$$
$$t=1$$
Check:
$$-2(1)+3=1$$
$$-2+3=1$$
$$1=1$$

25.
$$4x-2x-5=7$$
$$2x-5=7$$
$$2x-5+5=7+5$$
$$2x=12$$
$$\frac{2x}{2}=\frac{12}{2}$$
$$x=6$$
Check:
$$4(6)-2(6)-5=7$$
$$24-12-5=7$$
$$12-5=7$$
$$7=7$$

26.
$$3y-y+12=6$$
$$2y+12=6$$
$$2y+12-12=6-12$$
$$2y=-6$$
$$\frac{2y}{2}=\frac{-6}{2}$$
$$y=-3$$
Check:
$$3(-3)-(-3)+12=6$$
$$-9+3+12=6$$
$$-6+12=6$$
$$6=6$$

27.
$$\begin{aligned} z+1&=-2z+10\\ z+1-1&=-2z+10-1\\ z&=-2z+9\\ z+2z&=-2z+2z+9\\ 3z&=9\\ \frac{3z}{3}&=\frac{9}{3}\\ z&=3 \end{aligned}$$

Check:
$$\begin{aligned} 3+1&=-2(3)+10\\ 4&=-6+10\\ 4&=4 \end{aligned}$$

28.
$$\begin{aligned} n-3&=-n+7\\ n-3+3&=-n+7+3\\ n&=-n+10\\ n+n&=-n+n+10\\ 2n&=10\\ \frac{2n}{2}&=\frac{10}{2}\\ n&=5 \end{aligned}$$

Check:
$$\begin{aligned} 5-3&=-5+7\\ 2&=2 \end{aligned}$$

29.
$$\begin{aligned} c&=-2(c+1)\\ c&=-2c-2\\ c+2c&=-2c+2c-2\\ 3c&=-2\\ \frac{3c}{3}&=\frac{-2}{3}\\ c&=-\frac{2}{3} \end{aligned}$$

Check:
$$\begin{aligned} -\frac{2}{3}&=-2\left(-\frac{2}{3}+1\right)\\ -\frac{2}{3}&=-2\left(\frac{1}{3}\right)\\ -\frac{2}{3}&=-\frac{2}{3} \end{aligned}$$

30.
$$\begin{aligned} p&=-(p-5)\\ p&=-p+5\\ p+p&=-p+p+5\\ 2p&=5\\ \frac{2p}{2}&=\frac{5}{2}\\ p&=\frac{5}{2} \end{aligned}$$

Check:
$$\begin{aligned} \frac{5}{2}&=-\left(\frac{5}{2}-5\right)\\ \frac{5}{2}&=-\left(-\frac{5}{2}\right)\\ \frac{5}{2}&=\frac{5}{2} \end{aligned}$$

31.
$$\begin{aligned} 2(x+1)-(x-8)&=-x\\ 2x+2-x+8&=-x\\ x+10&=-x\\ x+10-10&=-x-10\\ x&=-x-10\\ x+x&=-x+x-10\\ 2x&=-10\\ \frac{2x}{2}&=\frac{-10}{2}\\ x&=-5 \end{aligned}$$

Check:
$$\begin{aligned} 2((-5)+1)-((-5)-8)&=-(-5)\\ 2(-4)-(-13)&=-5\\ -8+13&=-5\\ -5&=-5 \end{aligned}$$

32.
$$\begin{aligned}
-(x+2)-(x-4)&=-5x\\
-x-2-x+4&=-5x\\
-2x+2&=-5x\\
-2x+2-2&=-5x-2\\
-2x&=-5x-2\\
-2x+5x&=-5x+5x-2\\
3x&=-2\\
\frac{3x}{3}&=\frac{-2}{3}\\
x&=-\frac{2}{3}
\end{aligned}$$

Check:
$$\begin{aligned}
-\left(\left(-\frac{2}{3}\right)+2\right)-\left(\left(-\frac{2}{3}\right)-4\right)&=-5\left(-\frac{2}{3}\right)\\
-\left(\frac{4}{3}\right)-\left(-\frac{14}{3}\right)&=\frac{10}{3}\\
\frac{10}{3}&=\frac{10}{3}
\end{aligned}$$

33.
$$\begin{aligned}
3[2n-4(n+1)]&=6n-12\\
3(2n-4n-4)&=6n-12\\
3(-2n-4)&=6n-12\\
-6n-12&=6n-12\\
-6n-12+12&=6n-12+12\\
-6n&=6n\\
-6n+6n&=6n+6n\\
0&=12n\\
n&=0
\end{aligned}$$

Check:
$$\begin{aligned}
3[2(0)-4(0+1)]&=6(0)-12\\
3(0-4)&=0-12\\
3(-4)&=-12\\
-12&=-12
\end{aligned}$$

34.
$$\begin{aligned}
-4(2x-6)&=7[x-(3x-1)]\\
-8x+24&=7(x-3x+1)\\
-8x+24&=7(-2x+7)\\
-8x+24&=-14x+7\\
-8x+14x+24&=-14x+14x+7\\
6x+24&=7\\
6x+24-24&=7-24\\
6x&=-17\\
\frac{6x}{6}&=\frac{-17}{6}\\
x&=-\frac{17}{6}
\end{aligned}$$

Check:
$$\begin{aligned}
-4\left[2\left(-\frac{17}{6}\right)-6\right]&=7\left[-\frac{17}{6}-\left(3\left(-\frac{17}{6}\right)-1\right)\right]\\
-4\left(-\frac{34}{6}-6\right)&=7\left[-\frac{17}{6}-\left(-\frac{51}{6}-\frac{6}{6}\right)\right]\\
-4\left(-\frac{70}{6}\right)&=7\left[-\frac{17}{6}-\left(-\frac{57}{6}\right)\right]\\
\frac{280}{6}&=\left(-\frac{17}{6}+\frac{57}{6}\right)\\
\frac{280}{6}&=7\left(\frac{40}{6}\right)\\
\frac{280}{6}&=\frac{280}{6}
\end{aligned}$$

35. $$\begin{aligned}10-[3+(2x-1)]&=3x\\10-(2x+2)&=3x\\10-2x-2&=3x\\8-2x&=3x\\8-2x+2x&=3x+2x\\8&=5x\\\frac{8}{5}&=\frac{5x}{5}\\\frac{8}{5}&=x\\x&=\frac{8}{5}\end{aligned}$$

Check:

$$\begin{aligned}10-\left[3+\left(2\left(\frac{8}{5}\right)-1\right)\right]&=3\left(\frac{8}{5}\right)\\10-\left[3+\left(\frac{16}{5}-\frac{5}{5}\right)\right]&=\frac{24}{5}\\10-\frac{26}{5}&=\frac{24}{5}\\\frac{24}{5}&=\frac{24}{5}\end{aligned}$$

36. $$\begin{aligned}x-[5-(3x-4)]&=-x\\x-(5-3x+4)&=-x\\x-(3x+1)&=-x\\x-3x-1&=-x\\-2x-1&=-x\\-2x+2x-1&=-x+2x\\-1&=x\\x&=-1\end{aligned}$$

Check:

$$\begin{aligned}-1-[5+(3(-1)-4)]&=-(-1)\\-1-[5+(-7)]&=1\\-1-(-2)&=1\\-1+2&=1\\1&=1\end{aligned}$$

37. $$\begin{aligned}a-5b&=2c\\a-5b+5b&=2c+5b\\a&=2c+5b\end{aligned}$$

38. $$\begin{aligned}\frac{2a}{b}&=n\\b\cdot\frac{2a}{b}&=b\cdot n\\2a&=bn\\\frac{2a}{2}&=\frac{bn}{2}\\a&=\frac{bn}{2}\end{aligned}$$

39. **a.** $$\begin{aligned}Ax+By&=C\\Ax+By-By&=C-By\\Ax&=C-By\\\frac{Ax}{x}&=\frac{C-By}{x}\\x&=\frac{C-By}{A}\end{aligned}$$

b. $$\begin{aligned}x&=\frac{0-(-1)(5)}{2}\\x&=\frac{0+5}{2}\\x&=\frac{5}{2}\end{aligned}$$

40. **a.** $$\begin{aligned}A&=\frac{bh}{2}\\2\cdot A&=2\cdot\frac{bh}{2}\\2A&=bh\\\frac{2A}{b}&=\frac{bh}{b}\\\frac{2A}{b}&=h\end{aligned}$$

b. $$\begin{aligned}h&=\frac{2(12)}{4}\\h&=\frac{24}{4}\\h&=6\text{ cm}\end{aligned}$$

41. $$\begin{aligned}0.3\cdot x&=12\\\frac{0.3x}{0.33}&=\frac{12}{0.3}\\x&=40\end{aligned}$$

42. $1.25 \cdot x = 5$

$$\frac{1.25x}{1.25} = \frac{5}{1.25}$$

$$x = 4$$

43. $x \cdot 5 = 8$

$$\frac{5x}{5} = \frac{8}{5}$$

$$x = 1.6$$

160%

44. $x \cdot 5 = 5$

$$\frac{8x}{8} = \frac{5}{8}$$

$$x = 0.625$$

$62\frac{1}{2}\%$

45. $0.85 \cdot \$300 = \$25.50$

46. $0.035 \cdot \$2000 = \$70$

47.

−4 −3 −2 −1 0 1 2 3 4

48.

−6 −5 −4 −3 −2 −1 0 1 2

49.

−2 −1 0 1 2 3 4 5 6

50.

−2 −1 0 1 2 3 4 5 6

51. $y + 1 > 6$

$$y + 1 - 1 > 6 - 1$$

$$y > 5$$

4 5 6 7 8 9 10

52. $-\frac{1}{2}t + 3 \leq 2$

$$-\frac{1}{2}t + 3 - 3 \leq 3 - 3$$

$$\frac{1}{2}t \leq 0$$

−1 0 1 2 3 4 5

53. $8y - 2 \leq 6y + 2$

$$8y - 6y - 2 \leq 6y - 6y + 2$$

$$2y - 2 \leq 2$$

$$2y - 2 + 2 \leq 2 + 2$$

$$2y \leq 4$$

$$\frac{2y}{2} \leq \frac{4}{2}$$

$$y \leq 2$$

−3 −2 −1 0 1 2 3

54. $\frac{1}{2}(8 - 12x) \leq x - 10$

$$4 - 6x \leq x - 10$$

$$4 - 6x - x \leq x - x - 10$$

$$4 - 7x \leq -10$$

$$4 - 4 - 7x \leq -10 - 4$$

$$-7x \leq -14$$

$$\frac{-7x}{-7} = \frac{-14}{-7}$$

$$x \geq 2$$

0 1 2 3 4 5 6

55. $0.5n - 0.3 < 0.2(2n + 1)$

$$0.5n - .03 < 0.4n + .02$$

$$0.5n - 0.4n - .03 < 0.4n - 0.4n + 0.2$$

$$0.1n - 0.3 < 0.2$$

$$0.1n - 0.3 + .03 < 0.2 + 0.3$$

$$0.1n < 0.5$$

$$\frac{0.1}{0.1} = \frac{0.5}{0.1}$$

$$n < 5$$

1 2 3 4 5 6 7

56. $\frac{Btu}{2000} = 8$

$$2000 \cdot \frac{Btu}{2000} = 2000 \cdot 8$$

$$Btu = 16{,}000$$

57. $a + a + a = 180$

$$3a = 180$$

$$\frac{3a}{3} = \frac{180}{3}$$

$$a = 60^\circ$$

58.
$$\$5000 + \$50x = \$12,000$$
$$\$5000 - \$5000 + \$50x = \$12,000 - \$5000$$
$$\$50x = \$7000$$
$$\frac{\$50x}{\$50x} = \frac{\$7000}{\$50x}$$
$$x = 140$$
140 guests

59.
$$180(n-2) = 540$$
$$\frac{180(n-2)}{180} = \frac{540}{180}$$
$$(n-2) = 3$$
$$n - 2 + 2 = 3 + 2$$
$$n = 5$$
5 sides

60.
$$x + (15,360 + x) = 39,210$$
$$2x + 15,360 = 39,210$$
$$2x + 15,360 - 15,360 = 39,210 - 15,360$$
$$2x = 23,850$$
$$\frac{2x}{2} = \frac{23,850}{2}$$
$$x = 11,925$$
One candidate received 11,925; the other received 27,285 votes.

61. **a.** $c = 2(26-10)y$
$c = 2 + 16y$

b.
$$c = 12 + 16y$$
$$c - 12 = 12 - 12 + 16y$$
$$c - 12 = 16y$$
$$\frac{c-12}{16} = \frac{16y}{16}$$
$$\frac{c-12}{16} = y$$
$$y = \frac{c-12}{16}$$

62.
$$45x + 50x = 380$$
$$95x = 380$$
$$\frac{95x}{95} = \frac{380}{95}$$
$$x = 4$$
The trucks will meet 4 hr after departure.

63.
$$40x + 32x = 18$$
$$72x = 18$$
$$\frac{72x}{72} = \frac{18}{72}$$
$$x = \frac{1}{4}$$
10:15 pm

64.
$$\frac{x}{400} - 0.5 = \frac{x}{500}$$
$$\frac{x}{400} - \frac{x}{500} - 0.5 = \frac{x}{500} - \frac{x}{500}$$
$$\frac{5x}{2000} - \frac{4x}{2000} - 0.5 = 0$$
$$\frac{x}{2000} - 0.5 = 0$$
$$\frac{x}{2000} - 0.5 + 0.5 = 0 + 0.5$$
$$\frac{x}{2000} = 0.5$$
$$2000 \cdot \frac{x}{2000} = 2000 \cdot 0.5$$
$$x = 1000$$
1000 mi

65. $\frac{425}{430} \approx 0.99$

Seaver received 99% of the votes cast.

66.
$$\frac{8.8 - 5.7}{8.8}$$
$$= \frac{3.1}{8.8}$$
$$\approx 0.3522$$
35%

67. $0.007(4000) = 28$

28 students

68. $I = prt$

$0.06(\$500) = \$30$

69.
$$\frac{170 - 60}{170}$$
$$= \frac{110}{170}$$
$$\approx 0.65$$
Van Buren's electoral vote count dropped 65%.

70.
$$6(0.60)+x=(6+x)(0.70)$$
$$3.6+x=4.2+0.7x$$
$$3.6+x-0.7x=4.2+0.7x-0.7x$$
$$3.6+0.3x=4.2$$
$$3.6-3.6+0.3x=4.2-3.6$$
$$0.3x=0.6$$
$$\frac{0.3x}{0.3}=\frac{0.6}{0.3}$$
$$x=2$$
2 L

71.
$$\frac{2L}{4(0.10)+x(0.01)=(4+x)(0.05)}$$
$$0.4+0.01x=0.2+0.05x$$
$$0.4+0.01x-0.05x=0.2+0.5x-0.5x$$
$$0.4-0.04x=0.2$$
$$0.4-0.4-0.04x=0.2-0.4$$
$$-0.04x=-0.2$$
$$\frac{-0.04x}{-0.04}=\frac{-0.2}{-0.04}$$
$$x=5$$
5 pt

72. **a.** $p=2.2k$

b. $\frac{p}{2.2}=\frac{2.2k}{2.2}$
$$\frac{p}{2.2}=k$$
$$k=\frac{p}{2.2}$$

73.
$$100{,}000+25b=50b$$
$$100{,}000+25b-25b=50b-25b$$
$$100{,}000=25b$$
$$\frac{100{,}000}{25}=\frac{25b}{25}$$
$$b=4000$$
4000 books

74.
$$\$350+x(3)(\$50)=\$1000$$
$$\$350+\$150x=\$1000$$
$$\$350-\$350+\$150x=\$1000-\$350$$
$$\$150x=\$650$$
$$\frac{\$150x}{\$150}=\frac{\$650}{\$150}$$
$$x=\frac{13}{3}$$
$$x=4\frac{1}{3}$$
Up to 4 songs.

Chapter 2 Posttest

1.
$$3x-4=6(x+2)$$
$$3(-2)-4=6((-2)+2)$$
$$-6-4=6(0)$$
$$-10=0$$
-2 is not a solution.

2.
$$x-1=-10$$
$$x-1+1=-10+1$$
$$x=-9$$
Check:
$$(-9)-1=-10$$
$$-10=-10$$

3.
$$\frac{n}{2}=-3$$
$$2\cdot\frac{n}{2}=2\cdot(-3)$$
$$n=-6$$
Check:
$$\frac{-6}{2}=-3$$
$$-3=-3$$

4.
$$-y=-11$$
$$-1\cdot(-y)=-1\cdot(-11)$$
$$y=11$$
Check:
$$-(11)=-11$$
$$-11=-11$$

5.
$$\frac{3y}{4}=6$$
$$\frac{4}{3}\cdot\frac{3y}{4}=\frac{4}{3}\cdot 6$$
$$y=8$$
Check:
$$\frac{3(8)}{4}=6$$
$$6=6$$

6.
$$2x+5=11$$
$$2x+5-5=11-5$$
$$2x=6$$
$$\frac{2x}{2}=\frac{3}{2}$$
$$x=3$$
Check:
$$2(3)+5=11$$
$$6+5=11$$
$$11=11$$

7.
$$-s+4=2$$
$$-s+4-4=2-4$$
$$-s=-2$$
$$-1\cdot -s=-2\cdot -1$$
$$s=2$$
Check:
$$-2+4=2$$
$$2=2$$

8.
$$10x+1=-x+23$$
$$10x+x+1=-x+x+23$$
$$11x+1=23$$
$$11x+1-1=23-1$$
$$11x=22$$
$$\frac{11x}{11}=\frac{22}{11}$$
$$x=2$$
Check:
$$10(2)+1=-(2)+23$$
$$20+1=-2+23$$
$$21=21$$

9.
$$16a=-4(a-5)$$
$$16a=-4a+20$$
$$16a+4a=-4a+4a+20$$
$$20a=20$$
$$\frac{20a}{20}=\frac{20}{20}$$
$$a=1$$
Check:
$$16(1)=-4(1-5)$$
$$16=-4(-4)$$
$$16=16$$

10.
$$2(x+5)-(x+4)=7x+1$$
$$2x+10-x-4=7x+1$$
$$x+6=7x+1$$
$$x-7x+6=7x-7x+1$$
$$-6x+6=1$$
$$-6x+6-6=1-6$$
$$-6x=-5$$
$$-\frac{-6x}{6}=\frac{-5}{6}$$
$$x=\frac{5}{6}$$
Check:
$$2\left(\frac{5}{6}+5\right)-\left(\frac{5}{6}+4\right)=7\left(\frac{5}{6}\right)+1$$
$$2\left(\frac{35}{6}\right)-\left(\frac{29}{6}\right)=7\left(\frac{5}{6}\right)+1$$
$$\frac{70}{6}-\frac{29}{6}=\frac{35}{6}+\frac{6}{6}$$
$$\frac{41}{6}=\frac{41}{6}$$

11.
$$5n+p=t$$
$$5n-5n+p=t-5n$$
$$p=t-5n$$

12.
$$0.40\cdot x=8$$
$$\frac{0.4x}{0.4}=\frac{8}{0.4}$$
$$x=20$$

13.
$$x\cdot 5=10$$
$$\frac{5x}{5}=\frac{10}{5}$$
$$x=2$$
200%

14.

−3 −2 −1 0 1 2 3

15.

$$-2z \leq 6$$
$$\frac{-2z}{-2} \geq \frac{6}{-2}$$
$$z \geq -3$$

−3 −2 −1 0 1 2 3

16.

$$4 + 1.25x = 16.5$$
$$4 - 4 + 1.25x = 16.5 - 4$$
$$1.25x = 12.5$$
$$\frac{1.25x}{1.25} = \frac{12.5}{1.25}$$
$$x = 10$$

10 mi

17.

$$S = 3L - 21$$
$$S + 21 = 3L - 21 + 21$$
$$S + 21 = 3L$$
$$\frac{S+21}{3} = \frac{3L}{3}$$
$$\frac{S+21}{3} = L$$
$$L = \frac{S+21}{3}$$

18.

$$0.37x = 8{,}400{,}000$$
$$\frac{0.37x}{0.37} = \frac{8{,}400{,}000}{0.37}$$
$$x \approx 23{,}000{,}000$$

23,000,000 operations

19.

$$1.5(x+2) + 1.5x = 33$$
$$1.5x + 3 + 1.5x = 33$$
$$3x + 3 = 33$$
$$3x3 - 3 = 33 - 3$$
$$3x = 30$$
$$\frac{3x}{3} = \frac{30}{3}$$
$$x = 10$$
$$x + 2 = 12$$

10 mph and 12 mph

20.

$$39.99 + 0.79x > 54.99 + 0.59x$$
$$39.99 - 39.99 + 0.79x > 54.99 - 39.99 + 0.59x$$
$$0.79x > 15 + 0.59x$$
$$0.79x - 0.59x > 15 + 0.59x - 0.59x$$
$$0.2x > 15$$
$$\frac{0.2x}{0.2} > \frac{15}{0.2}$$
$$x > 75$$

The monthly cost of Plan A exceeds the monthly cost of Plan B if more than 75 min of calls are made outside the network.

Chapter 3

GRAPHING LINEAR EQUATIONS AND INEQUALITIES; FUNCTIONS

Pretest

1. More than half the cancer patients survived 5 or more years during the periods 1983-1985 and 1995-2001.

2. In 2006, there were approximately 395 thousand ATMs in the United States.

3. **a.**

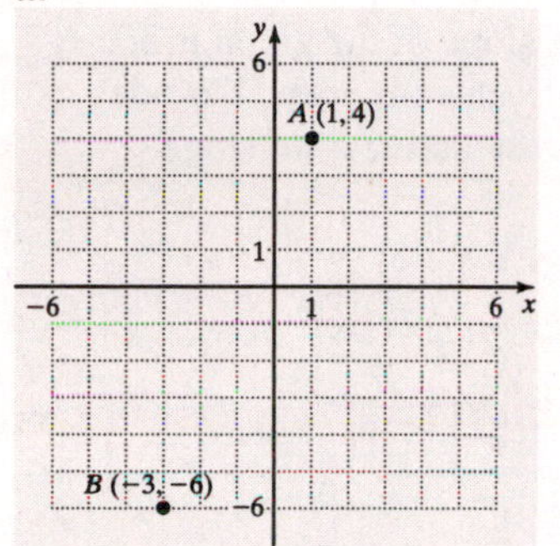

b. IV

4. $m = \dfrac{5-2}{7-1} = \dfrac{3}{6} = \dfrac{1}{2}$

5. $\overleftrightarrow{PQ}$: $m = \dfrac{3-(-1)}{-3-1} = \dfrac{4}{-4} = -1$

$\overleftrightarrow{RS}$: $m = \dfrac{-2-4}{-2-4} = \dfrac{-6}{-6} = 1$

The slope of $\overleftrightarrow{PQ}$ is -1 and the slope of $\overleftrightarrow{RS}$ is 1. $\overleftrightarrow{PQ}$ is perpendicular to $\overleftrightarrow{RS}$, since the product of their slopes is -1.

6. x-intercept: (3, 0); y-intercept: (0, 4)

7.

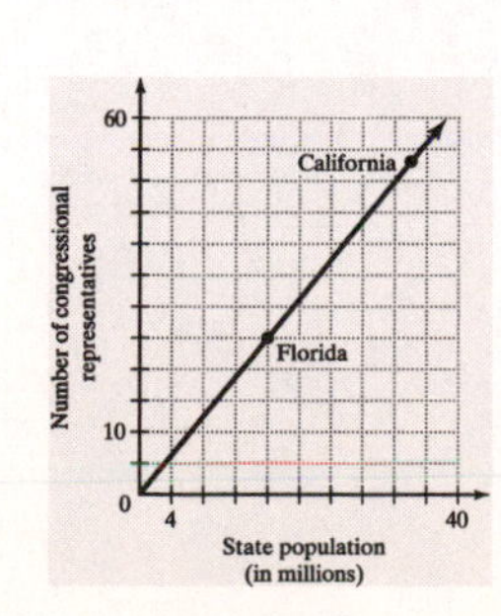

The slope of the line is positive. As the population of a state increases, the number of representatives in Congress from that state increases.

8. Variety A grows faster.

9. $y = 2x - 5$

$y = 2(4) - 5 = 3$

$y = 2(7) - 5 = 9$

$0 = 2x - 5$

$5 = 2x$

$\dfrac{5}{2} = x$

$-1 = 2x - 5$

$4 = 2x$

$2 = x$

10. 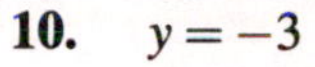

$y = -3$

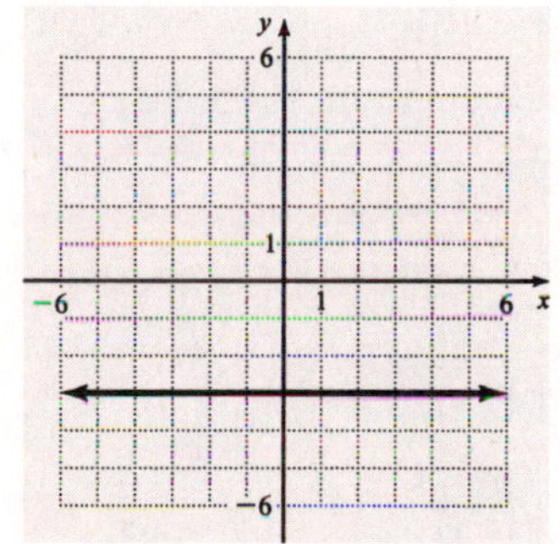

11. 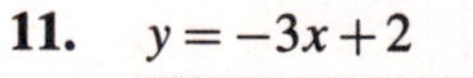

$y = -3x + 2$

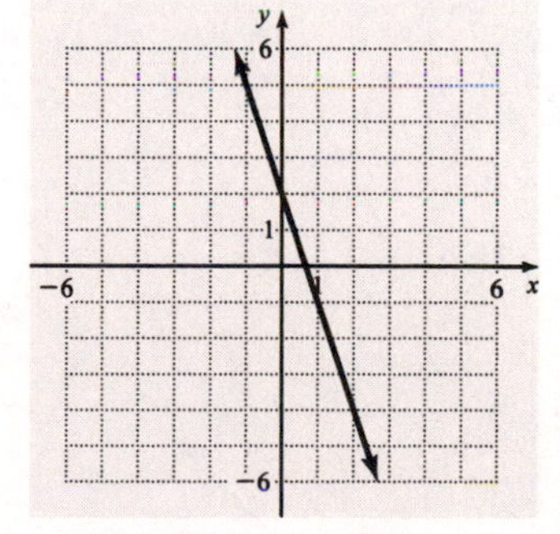

12. $2x - 3y = 6$

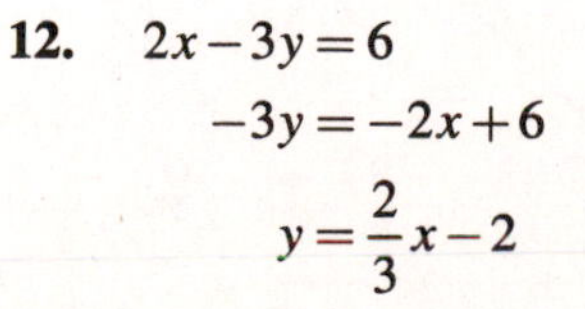

$-3y = -2x + 6$

$y = \dfrac{2}{3}x - 2$

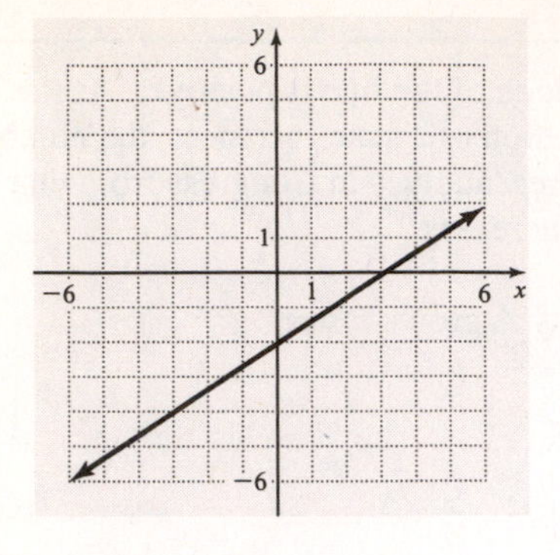

13.
$$5x - y = 8$$
$$-y = -5x + 8$$
$$y = 5x - 8$$

14. Slope-intercept form: $y = 2x + 8$;
point-slope: $y - 8 = 2(x - 0)$

15. $m = \dfrac{1-(-1)}{4-2} = \dfrac{2}{2} = 1$
$$y - 1 = (x - 4)$$
$$y = x - 3$$

Slope-intercept form: $y = x - 3$;
point-slope: $y - 1 = x - 4$

16.

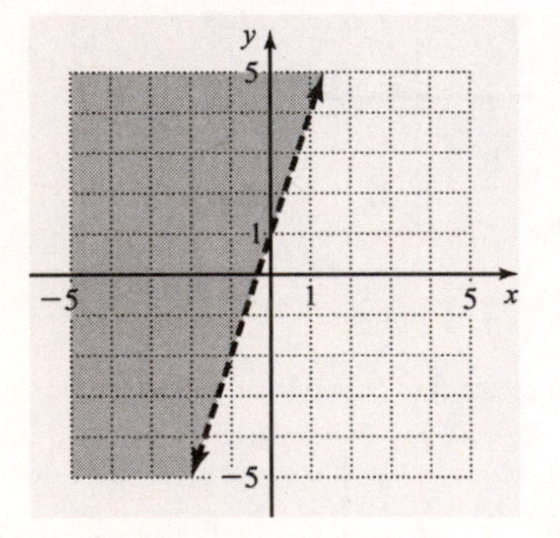

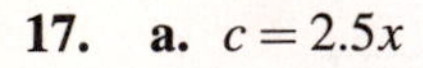

17. **a.** $c = 2.5x$

b.

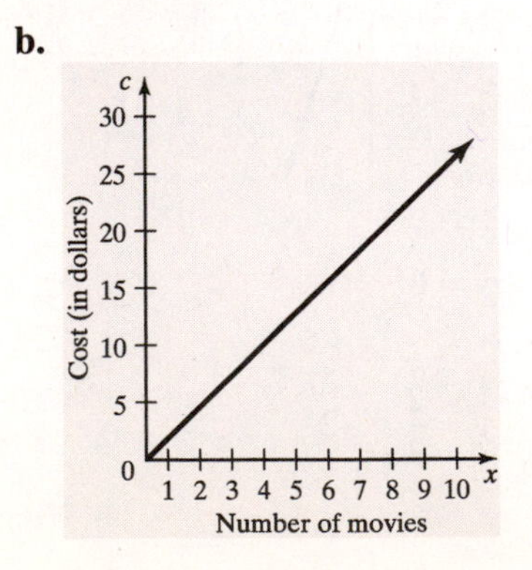

c. The slope of the line is 2.5. It represents the cost of renting a movie.

18. **a.** $d = 50t$

b.

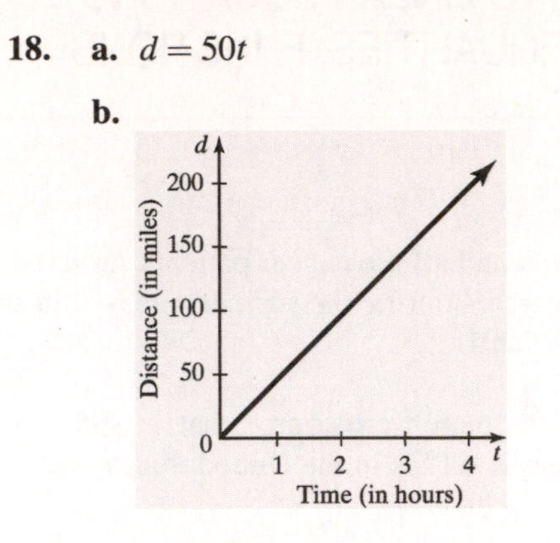

c. The slope of the graph is 50. It represents the speed the sales representative is driving.

19.
$$f(x) = |2x - 8|$$
$$f(-6) = |2(-6) - 8|$$
$$= |-12 - 8|$$
$$= |-20| = 20$$

20. $f(x) = 1 - 1.5x$ for $x \geq -2$

x	$f(x) = 1 - 1.5x$	$(x, f(x))$
−2	$1 - 1.5(-2) = 4$	$(-2, 4)$
0	$1 - 1.5(0) = 1$	$(0, 1)$
2	$1 - 1.5(2) = -2$	$(2, -2)$

Domain: $[-2, \infty)$, Range: $(-\infty, 4]$

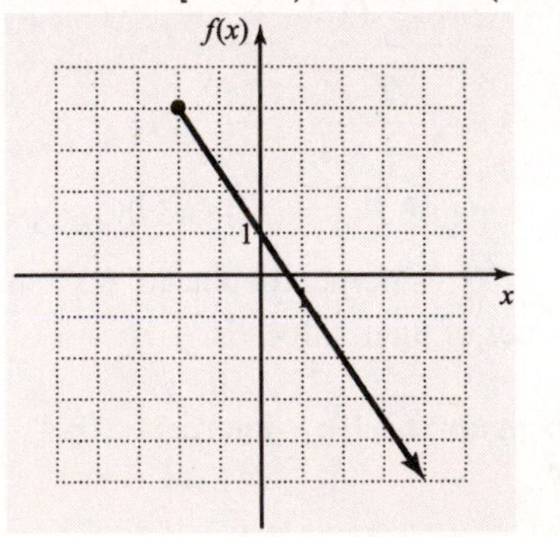

3.1 Introduction to Graphing

Practice

1. **a.** Cattle and claves had the greatest value.

   b. The approximate value was $27 million.

   c. The value of corn is about $21 million and the value of broilers was about $19 million, so the value of corn was about $2 million more than broilers.

2. **a.** July has the highest mean temperature.

   b. The mean temperature is about 27°.

   c. In Chicago, mean temperatures increase from January through July and then decrease from July through December.

3. **a.** Approximately 47,000 children were adopted in 2005.

   b. In 2003 the number of adopted children exceeded the number of children in foster care.

   c. Possible answer: Between 1995 and 2005, the number of children in foster care declined, and the number of children who were adopted increased.

4.

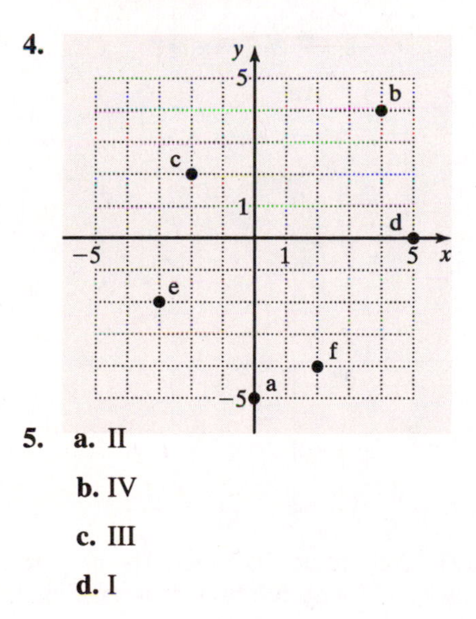

5. **a.** II

   b. IV

   c. III

   d. I

6.

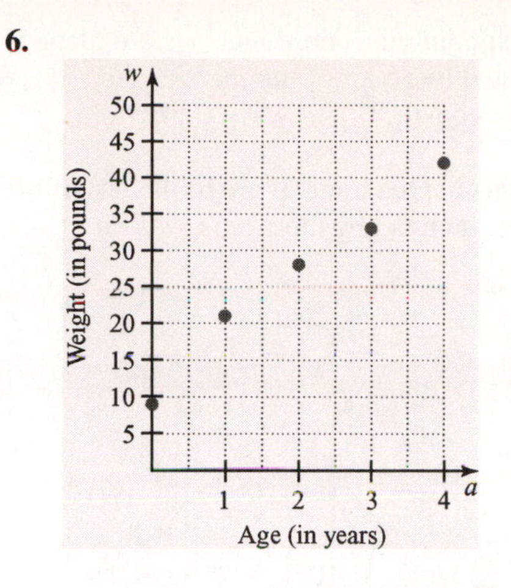

7. **a.**

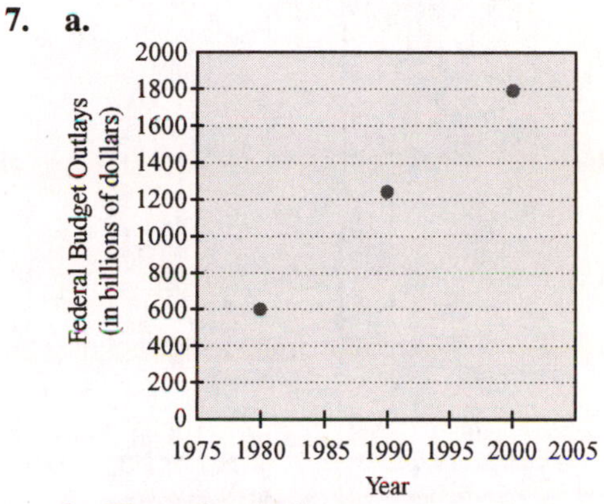

   b. The points indicate an upward trend and appear to lie approximately in a straight line. We can use this trend to estimate the federal budget outlays for years in which the exact data are not available.

8. From *A* to *B* and *B* to *C*, the line segments slant upward to the right, indicating that the runner's heartbeats per minute increase over this period of time. From *C* to *D*, the line segment slants downward to the right, indicating that the runner's heartbeats per minute decrease. Possible story: The runner jogs slowly to warm up (*A* to *B*), then runs more quickly, then jogs slowly (*C* to *D*), and finally rests (*D*).

Exercises

1. On a(n) bar graph, quantities are represented by thin, parallel rectangles.

3. A coordinate plane has two number lines that intersect at a point called the origin.

5. Each point in a coordinate system plane is represented by a pair of numbers called a(n) <u>ordered pair</u>.

7. Points in Quadrant III are to the left of the y-axis and <u>below</u> the x-axis.

9.

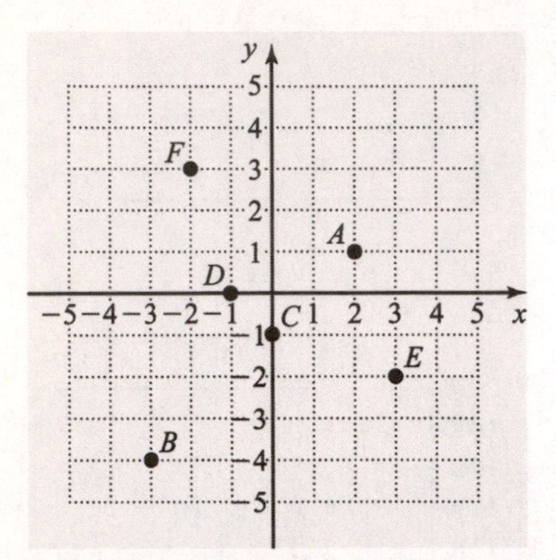

11.

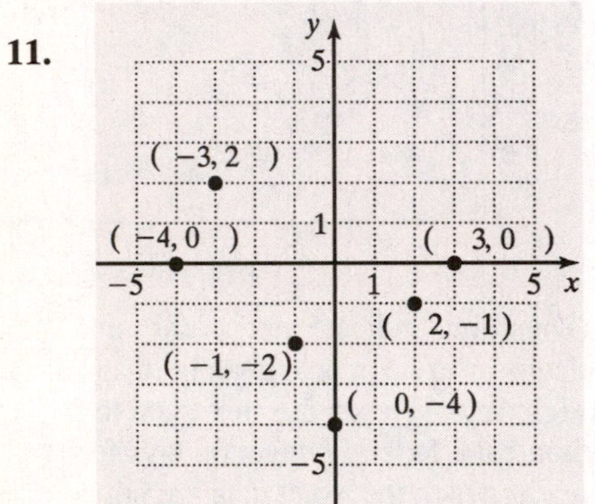

13. III

15. IV

17. I

19. II

21.

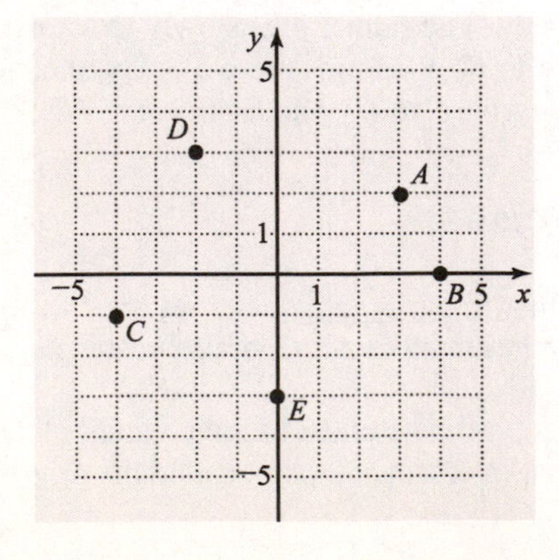

23. II

25. I

27. **a.** Milk, lemon juice, and vinegar are acids.

b. The pH of sea water is about 8.

c. Pure water is neutral.

29. **a.** There were about 90 million subscribers in the year 2000.

b. In 2003 the number of subscribers reached 150 million.

c. Every year the number of cell phone subscribers increased.

31. **a.** $A(20,40)$, $B(52,90)$, $C(76,80)$, $D(90,28)$

b. Students A, B, and C scored higher in English than in mathematics.

33.

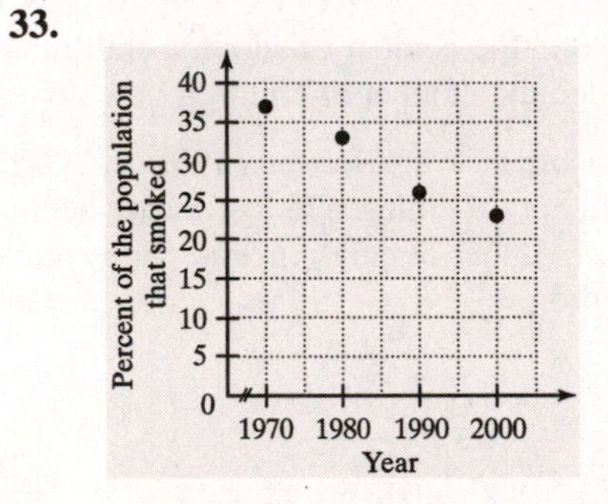

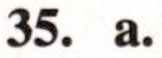

35. **a.**

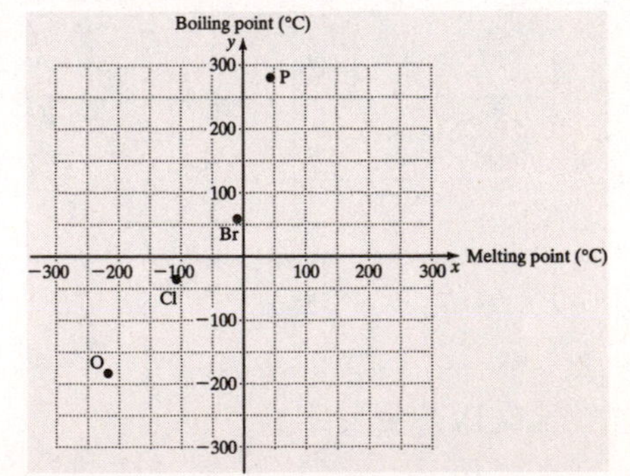

b. The y-coordinate is larger. The pattern shows that for each substance its boiling point is higher than its melting point.

37. The number of senators from a state (2) is the same regardless of the size of the state's population.

39. The graph in (a) could describe this motion. As the child moves away from the wall, the distance from the wall increases (line segment slants upward to the right). When the child stands still, the distance does not change (horizontal line segment). Finally, as the child moves toward the wall, the child's distance from the wall decreases (line segment slants downward to the right).

3.2 Slope

Practice

1. $m=\dfrac{3-2}{4-1}=\dfrac{1}{3}$

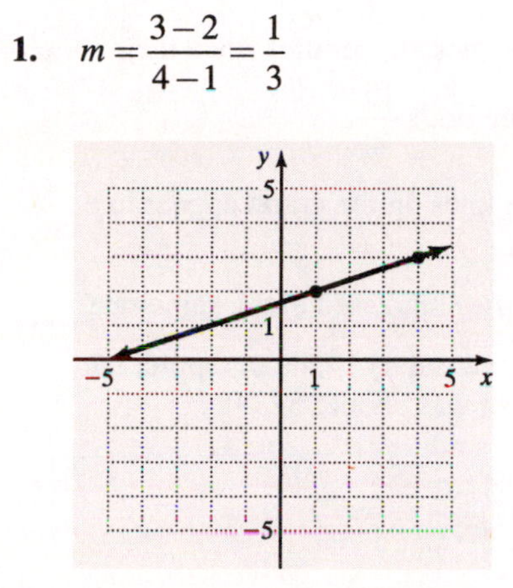

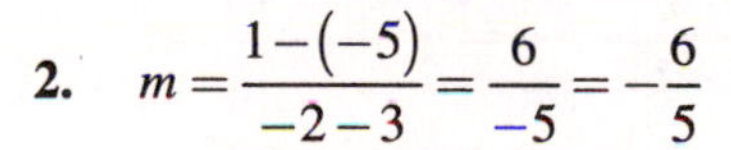

2. $m=\dfrac{1-(-5)}{-2-3}=\dfrac{6}{-5}=-\dfrac{6}{5}$

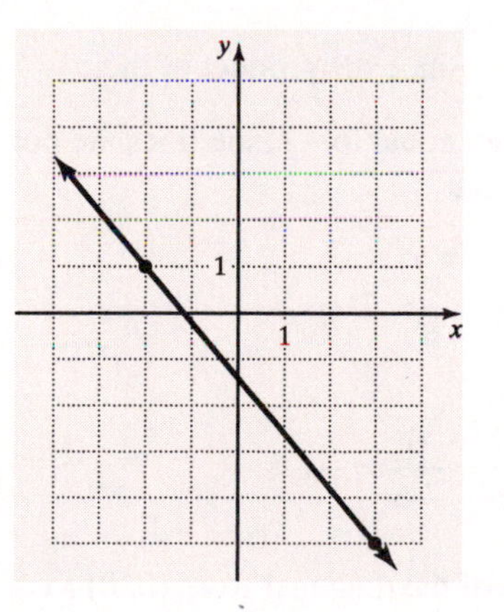

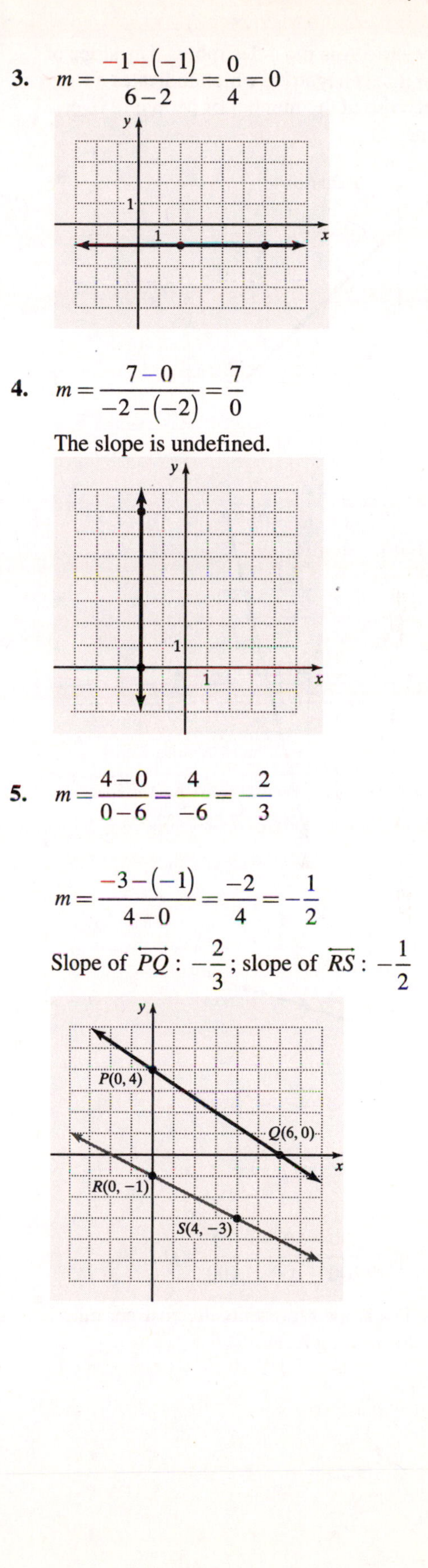

3. $m=\dfrac{-1-(-1)}{6-2}=\dfrac{0}{4}=0$

4. $m=\dfrac{7-0}{-2-(-2)}=\dfrac{7}{0}$

The slope is undefined.

5. $m=\dfrac{4-0}{0-6}=\dfrac{4}{-6}=-\dfrac{2}{3}$

$m=\dfrac{-3-(-1)}{4-0}=\dfrac{-2}{4}=-\dfrac{1}{2}$

Slope of $\overleftrightarrow{PQ}$: $-\dfrac{2}{3}$; slope of $\overleftrightarrow{RS}$: $-\dfrac{1}{2}$

6. Scenario A is most desirable. The slope of the line is negative, which indicates a decrease in the number of people ill over time.

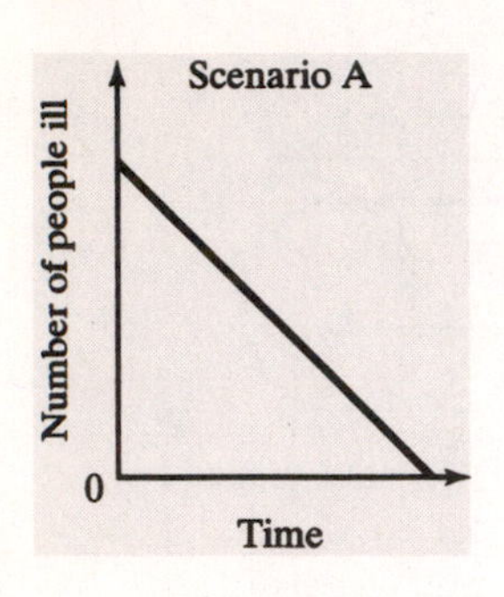

7.

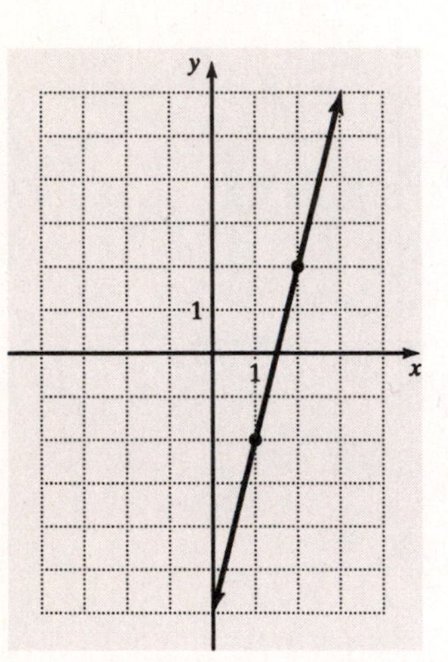

8. a.

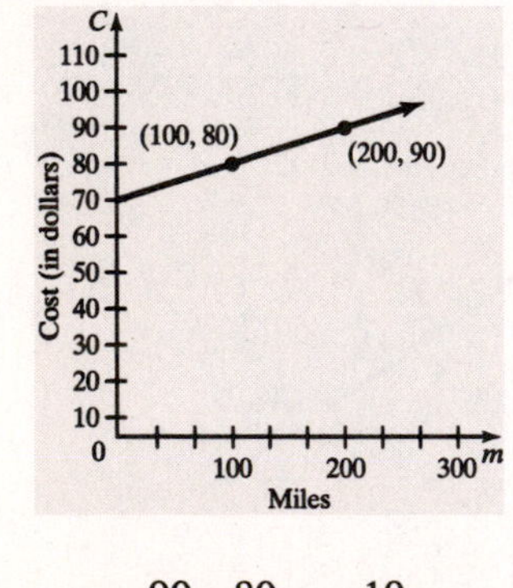

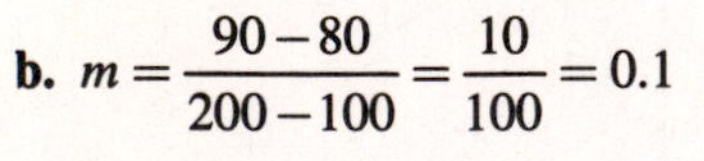

b. $m=\dfrac{90-80}{200-100}=\dfrac{10}{100}=0.1$

c. The slope represents the cost per mile for renting a car.

9. $\overleftrightarrow{EF}: m=\dfrac{-1-4}{4-0}=\dfrac{-5}{4}=-\dfrac{5}{4}$

$\overleftrightarrow{GH}: m=\dfrac{-2-8}{8-0}=\dfrac{-10}{8}=-\dfrac{5}{4}$

The slope of $\overleftrightarrow{EF}$ is $-\dfrac{5}{4}$ and the slope of $\overleftrightarrow{GH}$ is $-\dfrac{5}{4}$. Since their slopes are equal, the lines are parallel.

10. a. Technician: $m=\dfrac{80-65}{6-2}=\dfrac{15}{4}$

Designer: $m=\dfrac{45-30}{6-2}=\dfrac{15}{4}$

Yes, the lines are parallel since their slopes are both $\dfrac{15}{4}$.

b. Yes; the lines on the graph appear to be parallel.

c. The salaries increased at the same rate.

d. The starting salary of the computer lab technician was about $57,000.

11. $\overleftrightarrow{AB}: m=\dfrac{5-3}{2-1}=\dfrac{2}{1}=2$

$\overleftrightarrow{AC}: m=\dfrac{2-3}{-1-1}=\dfrac{-1}{-2}=\dfrac{1}{2}$

The slope of $\overleftrightarrow{AB}$ is 2 and the slope of $\overleftrightarrow{AC}$ is $\dfrac{1}{2}$. Since the product of their slopes is not equal to -1, the lines are not perpendicular.

12. $m=\dfrac{6-0}{6-0}=\dfrac{6}{6}=1$

$m=\dfrac{6-0}{0-6}=\dfrac{6}{-6}=-1$

The slope of the diagonal from (0, 0) to (6, 6) is 1. The slope of the diagonal from (0, 6) to (6, 0) is -1. Since the product of the slopes is -1, the diagonals of the square are perpendicular.

Exercises

1. The slope of a line is also called its <u>rate of change</u>.

3. A line with <u>negative</u> slope is decreasing.

5. A line with zero slope is <u>horizontal</u>.

7. Two nonvertical lines <u>parallel</u> if and only if their slopes are equal.

9. $m = \dfrac{3-0}{2-(-2)} = \dfrac{3}{2+2} = \dfrac{3}{4}$

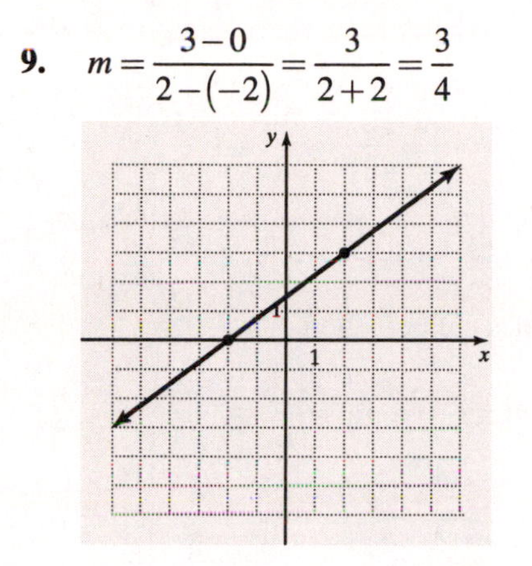

11. $m = \dfrac{-4-1}{6-6} = \dfrac{-5}{0} = \text{undefined}$

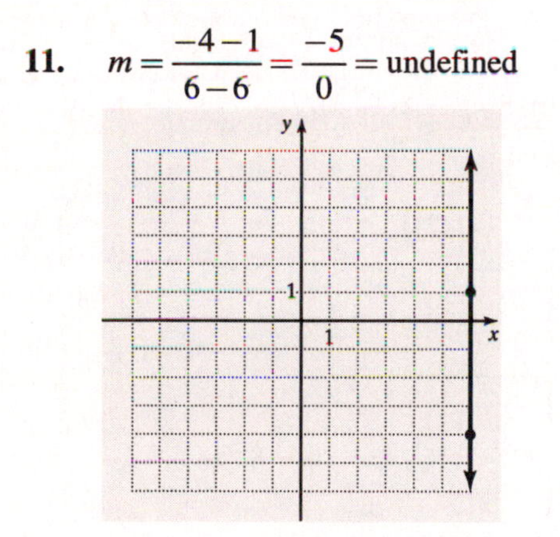

13. $m = \dfrac{-1-1}{3-(-2)} = \dfrac{-2}{5} = -\dfrac{2}{5}$

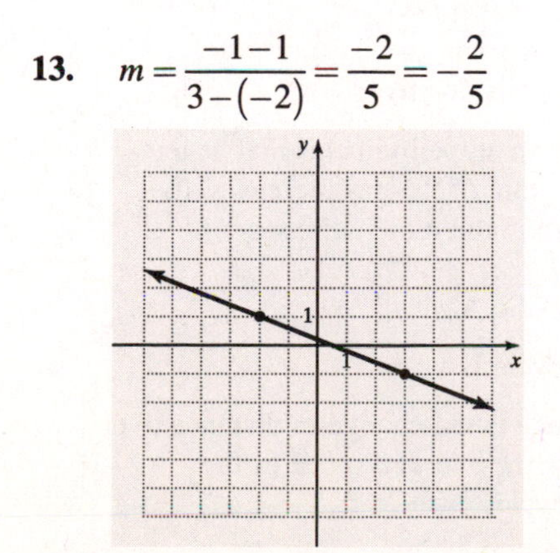

15. $m = \dfrac{-4-(-4)}{3-(-1)} = \dfrac{0}{4} = 0$

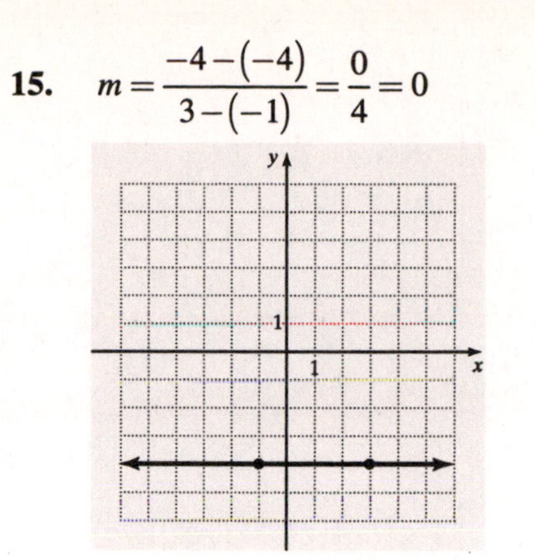

17. $m = \dfrac{3.5-0}{0-0.5} = \dfrac{3.5}{-0.5} = -7$

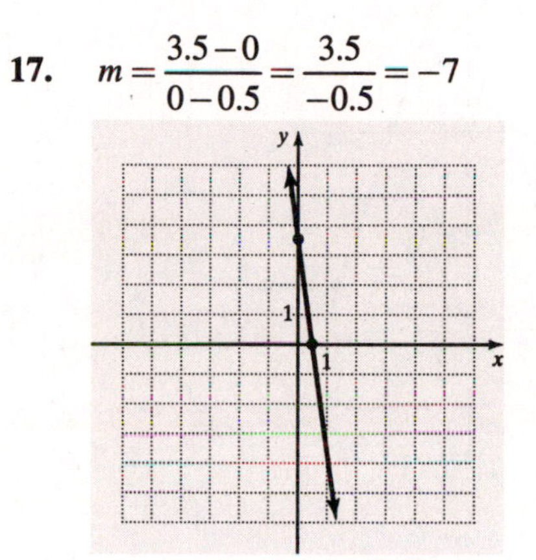

19. $\overleftrightarrow{AB}: m = \dfrac{-3-2}{6-1} = \dfrac{-5}{5} = -1$

$\overleftrightarrow{CD}: m = \dfrac{8-4}{3-1} = \dfrac{4}{2} = 2$

The slope of $\overleftrightarrow{AB}$ is -1. The slope of $\overleftrightarrow{CD}$ is 2.

21.

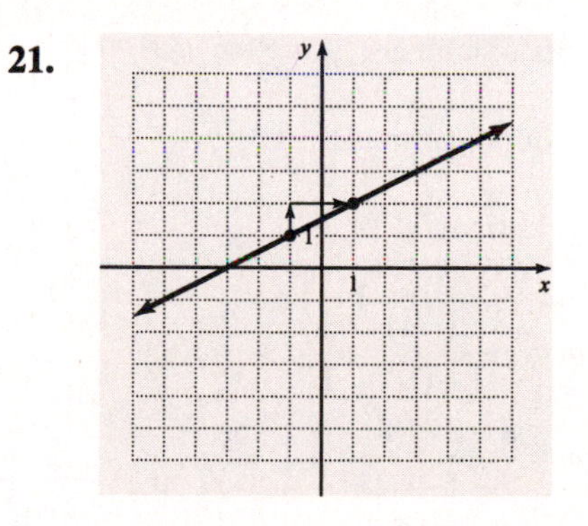

23.

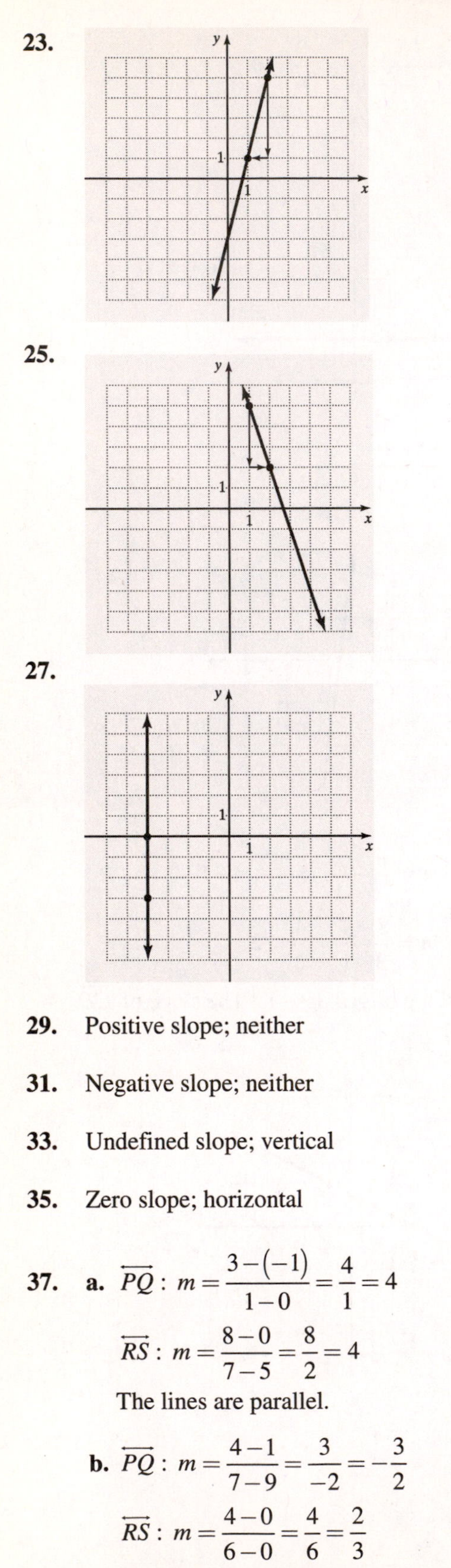

25.

27.

29. Positive slope; neither

31. Negative slope; neither

33. Undefined slope; vertical

35. Zero slope; horizontal

37. **a.** $\overleftrightarrow{PQ}: m = \dfrac{3-(-1)}{1-0} = \dfrac{4}{1} = 4$

$\overleftrightarrow{RS}: m = \dfrac{8-0}{7-5} = \dfrac{8}{2} = 4$

The lines are parallel.

b. $\overleftrightarrow{PQ}: m = \dfrac{4-1}{7-9} = \dfrac{3}{-2} = -\dfrac{3}{2}$

$\overleftrightarrow{RS}: m = \dfrac{4-0}{6-0} = \dfrac{4}{6} = \dfrac{2}{3}$

The lines are perpendicular.

39. Zero slope; horizontal

41.

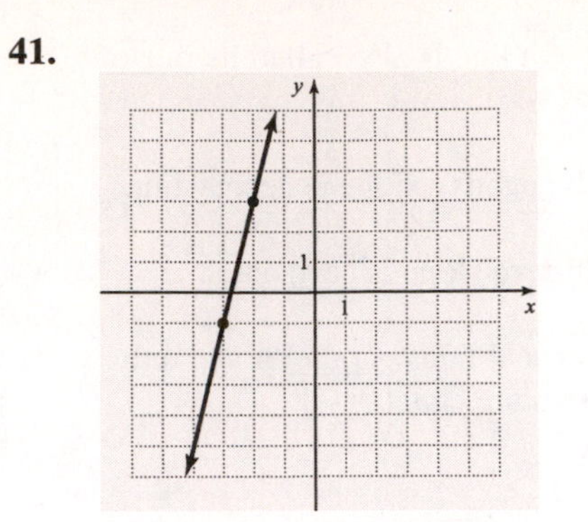

43. **a.** $\overleftrightarrow{AB}: m = \dfrac{-2-2}{5-(-3)} = \dfrac{-4}{8} = -\dfrac{1}{2}$

$\overleftrightarrow{CD}: m = \dfrac{4-(-4)}{5-1} = \dfrac{8}{4} = 2$

The lines are perpendicular.

b. $\overleftrightarrow{AB}: m = \dfrac{-3-5}{5-(-1)} = \dfrac{-8}{6} = -\dfrac{4}{3}$

$\overleftrightarrow{CD}: m = \dfrac{-2-2}{5-2} = \dfrac{-4}{3} = -\dfrac{4}{3}$

The lines are parallel.

45. $m = \dfrac{4-(-2)}{-3-(-3)} = \dfrac{6}{0} =$ undefined

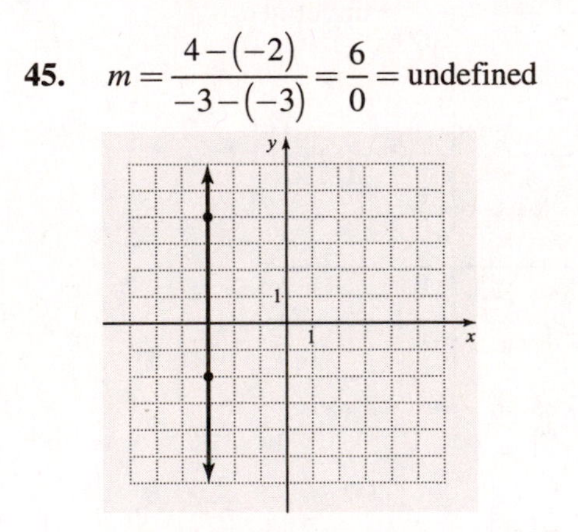

47. **a.** The slope is positive.

b. A positive slope indicates that as the temperature of the gas increases, the pressure in the tube increases.

49. **a.** Motorcycle A

b. Motorcycle B

c. The slope is the change in distance over time, or the average speed of the motorcycles.

51. Landfill A: $m=\frac{8-4}{7-3}=\frac{4}{4}=1$

Landfill B: $m=\frac{6-2}{7-3}=\frac{4}{4}=1$

The slope of each line is 1. Since the slopes of the lines are equal, the landfills are growing at the same rate.

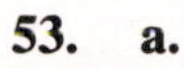

53. **a.**

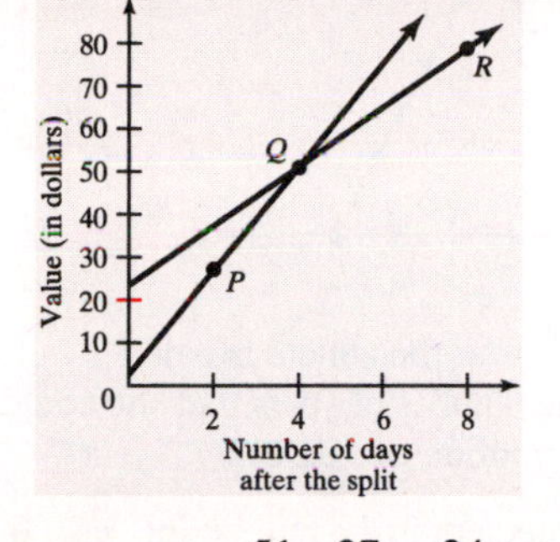

b. $\overleftrightarrow{PQ}$: $m=\frac{51-27}{4-2}=\frac{24}{2}=12$

53. **c.** $\overleftrightarrow{QR}$: $m=\frac{79-51}{8-4}=\frac{28}{4}=7$

The slopes are 12 and 7. Since the slopes are not equal, the rate of increase did change over time.

55. $\overleftrightarrow{AD}$: $m=\frac{2-7}{6-5}=\frac{-5}{1}=-5$

$\overleftrightarrow{BC}$: $m=\frac{3-1}{9-3}=\frac{2}{6}=\frac{1}{3}$

The product of the slopes of the two lines is $-5\cdot\frac{1}{3}$, which is not equal to -1, so $\overleftrightarrow{AD}$ is not the shortest route.

57. **a.** Graph II; As the car travels, its distance increases with time. This implies a positive slope.

b. Graph I; The car is set for a constant speed. The speed of the car does not change over time. This implies a 0 slope.

3.3 Linear Equations and Their Graphs

Practice

1. **a.** $2(3)-3(2)\stackrel{?}{=}5$

$6-6\stackrel{?}{=}5$

$0\neq 5$

It is not a solution.

b. $2(1)-3(-1)\stackrel{?}{=}5$

$2+3\stackrel{?}{=}5$

$5=5$

It is a solution.

2.

x	0	5	-3	$-\frac{1}{2}$	-2
y	1	11	-5	0	-3

3.

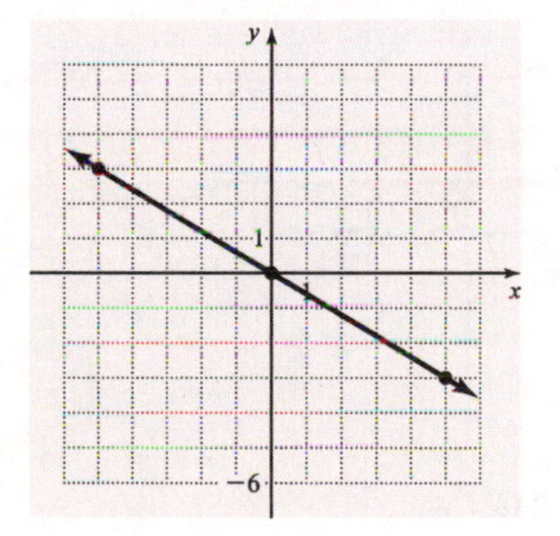

4. **a.**

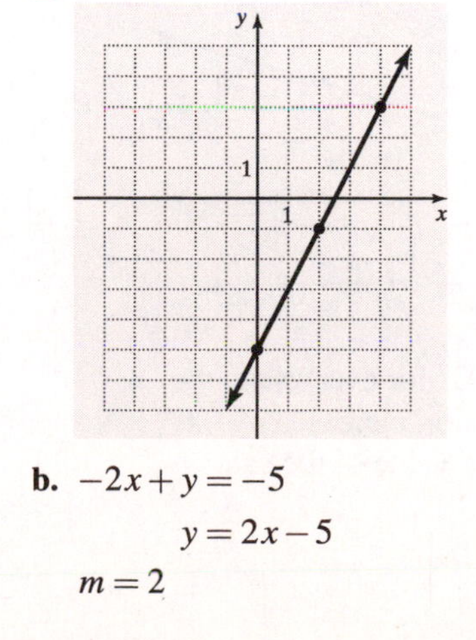

b. $-2x+y=-5$

$y=2x-5$

$m=2$

5. $x-2y=4$

$0-2y=4$

$-2y=4$

$y=-2$

$x-2(0)=4$

$x=4$

x-intercept: $(4,0)$

y-intercept: $(0,-2)$

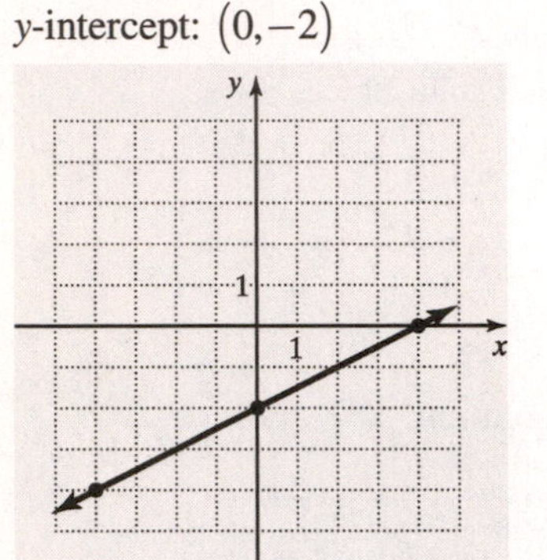

6.

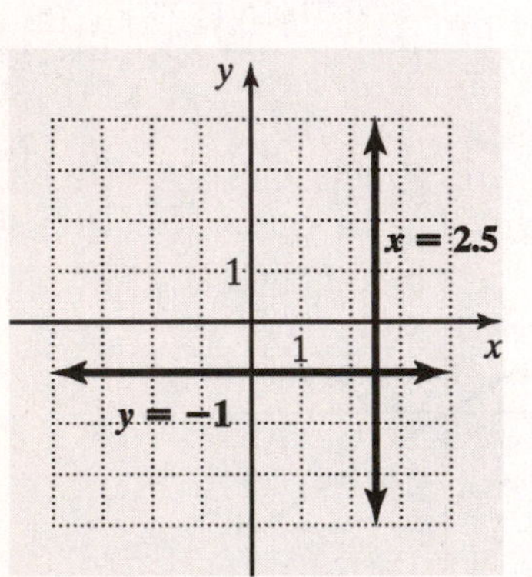

7. **a.** $C=0.03s+40$

b.

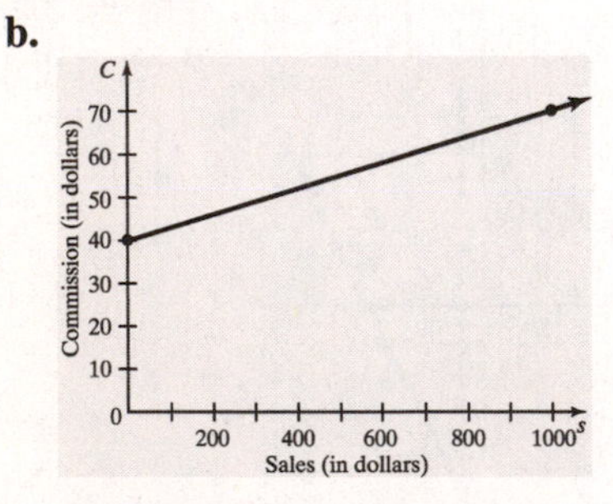

c. $m=0.03$; for every sale, the commission increases by 0.03 times the value of the sale.

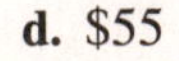

d. $55

8. **a.** $w+2t=10$

b.

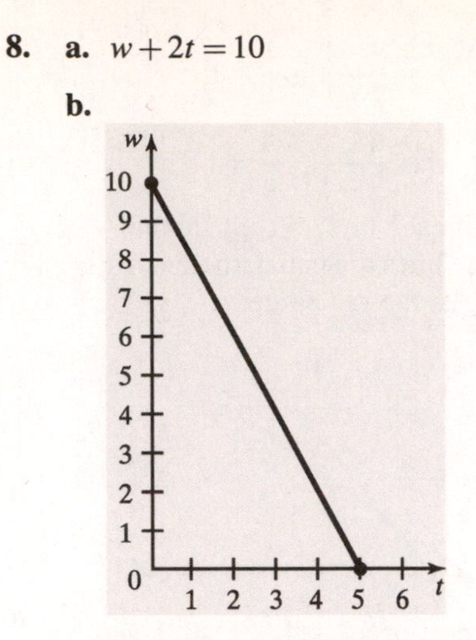

c. For each year the athlete is paid $2 million, the number of years that she could be paid $1 million decreases by 2 years.

d. The t-intercept is the number of years of the contract if she was paid $2 million in each year of the contract. The w-intercept is the number of years of the contract if she was paid $1 million in each year of the contract.

Exercises

1. A solution of an equation in two variables is an ordered pair of numbers that when substituted for the variables makes the equation true.

3. One way of graphing a linear equation in two variables is to plot three points.

5. One way to graph a linear equation using intercepts is to first let $x=0$ and then find the y-intercept.

7. $y=3x-8$

$y=3(4)-8=12-8=4$

$y=3(7)-8=21-8=13$

$0=3x-8$

$8=3x$

$\frac{8}{3}=x$

9.

$$y = 5x$$
$$y = 5(3.5) = 17.5$$
$$y = 5(6) = 30$$
$$\left(\frac{1}{2}\right) = 5x$$
$$\frac{1}{10} = x$$
$$(-8) = 5x$$
$$-\frac{8}{5} = x$$

11.

$$3x + 4y = 12$$
$$3(0) + 4y = 12$$
$$4y = 12$$
$$y = 3$$
$$3(-4) + 4y = 12$$
$$-12 + 4y = 12$$
$$4y = 24$$
$$y = 6$$
$$3x + 4(-3) = 12$$
$$3x - 12 = 12$$
$$3x = 24$$
$$x = 8$$
$$3x + 4(0) = 12$$
$$3x = 12$$
$$x = 4$$

13.

$$y = \frac{1}{3}x - 1$$
$$y = \frac{1}{3}(3) - 1$$
$$y = 1 - 1$$
$$y = 0$$
$$y = \frac{1}{3}(6) - 1$$
$$y = 2 - 1$$
$$y = 1$$
$$y = \frac{1}{3}(-3) - 1$$
$$y = -1 - 1$$
$$y = -2$$
$$-1 = \frac{1}{3}x - 1$$
$$0 = \frac{1}{3}x$$
$$0 = x$$

15.

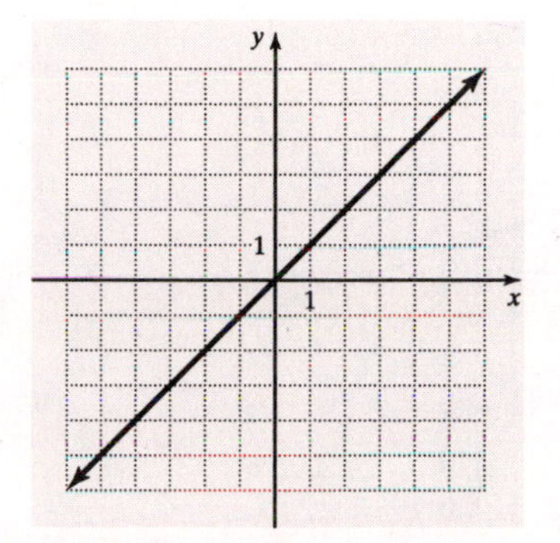

17.

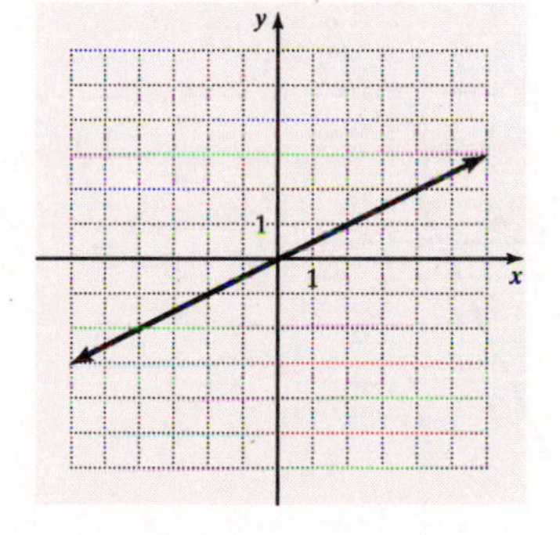

19.

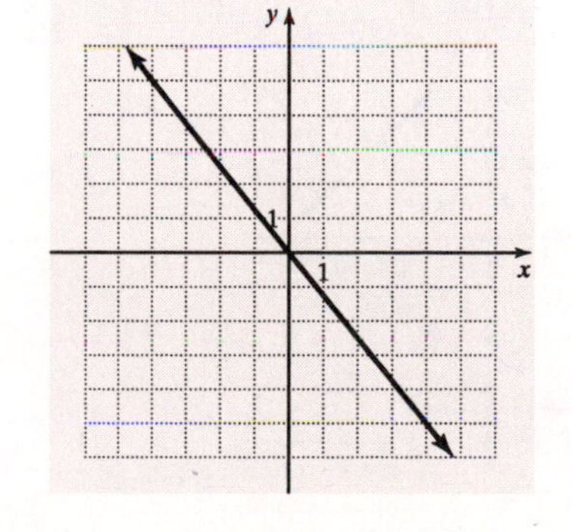

21.

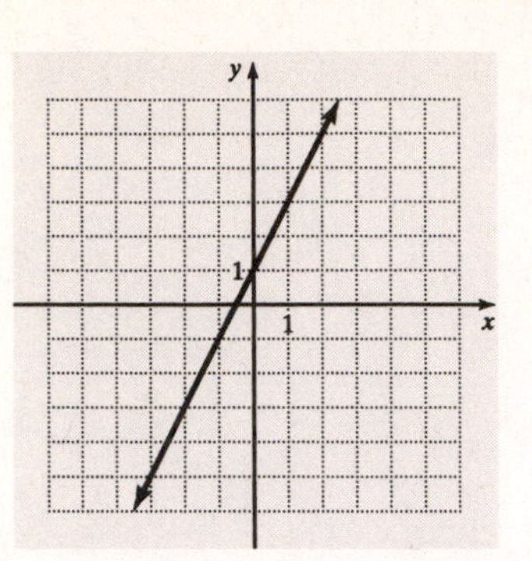

23.

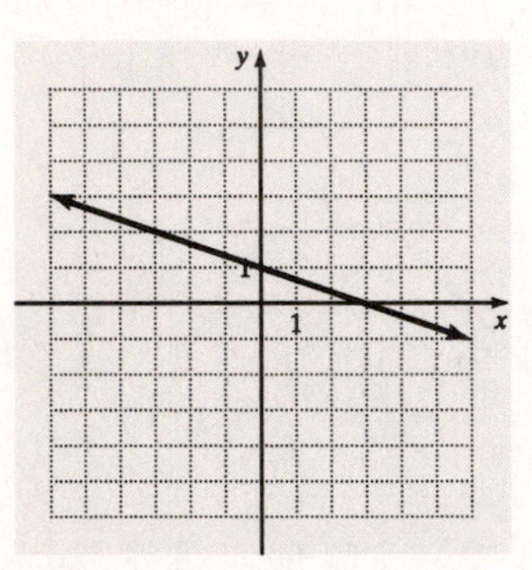

25.

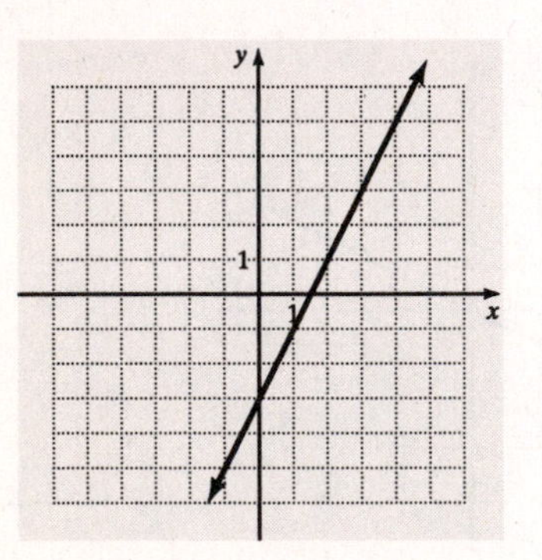

27.

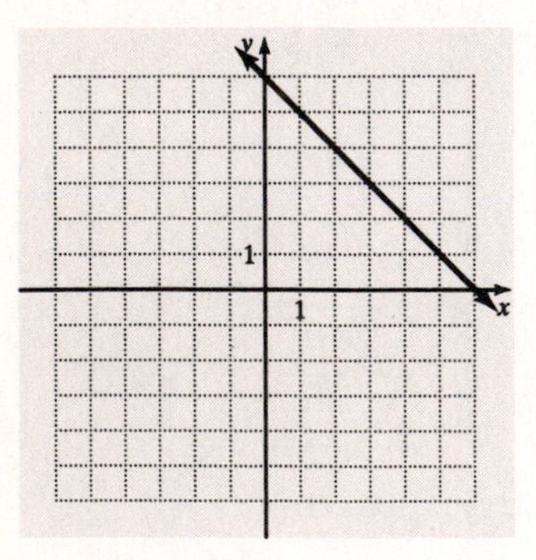

29.

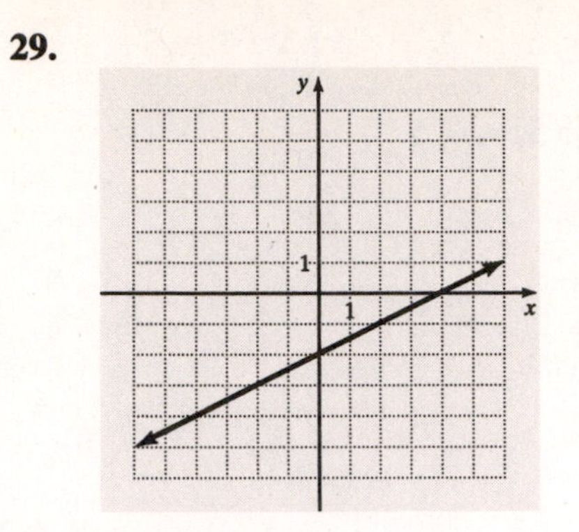

31. $5x+3y=15$

$5(0)+3y=15$ $\qquad$ $5x+3(0)=15$

$3y=15$ $\qquad$ $5x=15$

$y=5$ $\qquad$ $x=3$

x-intercept: $(3,0)$, y-intercept: $(0,5)$

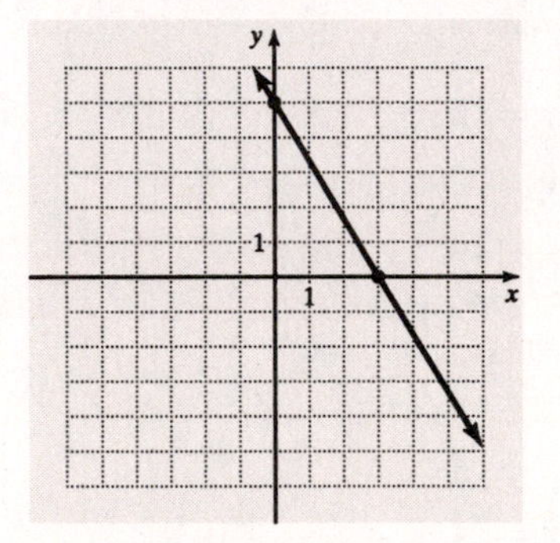

33. $3x-6y=18$

$3(0)-6y=18$

$-6y=18$

$y=-3$

$3x-6(0)=18$

$3x=18$

$x=6$

x-intercept: $(6,0)$, y-intercept: $(0,-3)$

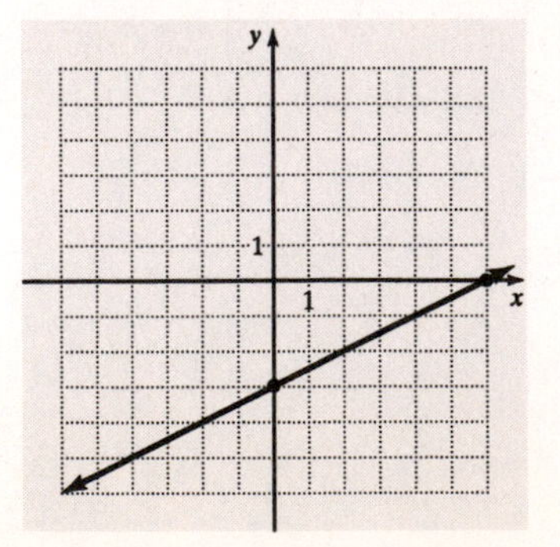

35. $4y-5x=10$

$$4y-5(0)=10$$
$$4y=10$$
$$y=\frac{10}{4}$$
$$y=\frac{5}{2}$$

$$4(0)-5x=10$$
$$-5x=10$$
$$x=-2$$

x-intercept: $(-2,0)$, y-intercept: $\left(0,\frac{5}{2}\right)$

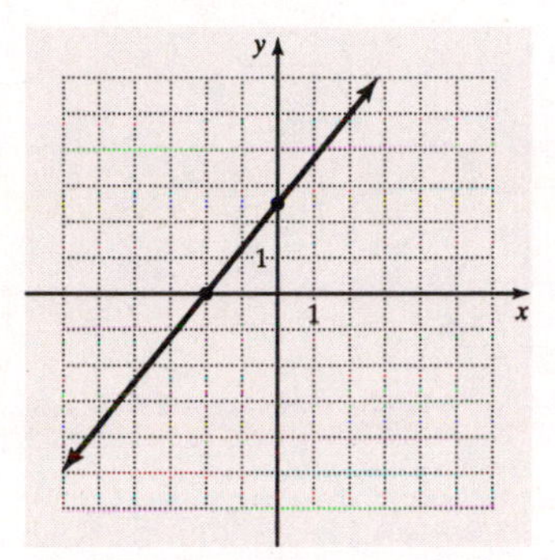

37. $9y+6x=-9$

$$9y-6(0)=-9$$
$$9y=-9$$
$$y=-1$$

$$9(0)+6x=-9$$
$$6x=-9$$
$$x=-\frac{9}{6}$$
$$x=-\frac{3}{2}$$

x-intercept: $\left(-\frac{3}{2},0\right)$, y-intercept: $(0,-1)$

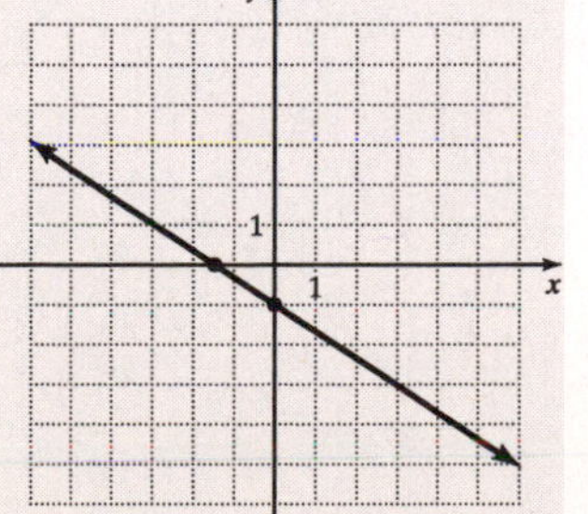

39. $y=\frac{1}{2}x+2$

$$y=\frac{1}{2}(0)+2$$
$$y=2$$

$$0=\frac{1}{2}x+2$$
$$-2=\frac{1}{2}x$$
$$-4=x$$

x-intercept: $(-4,0)$, y-intercept: $(0,2)$

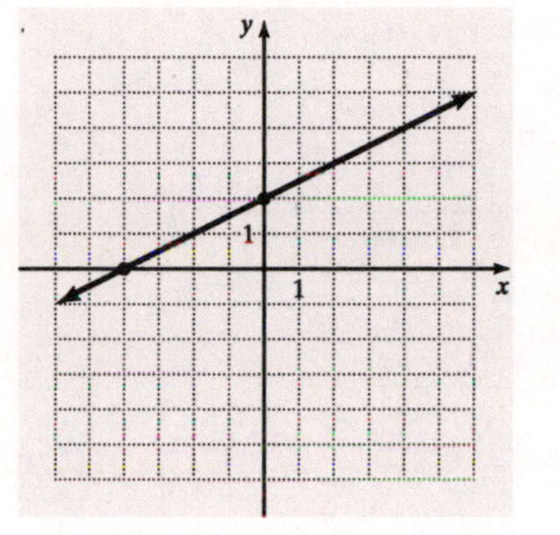

41.

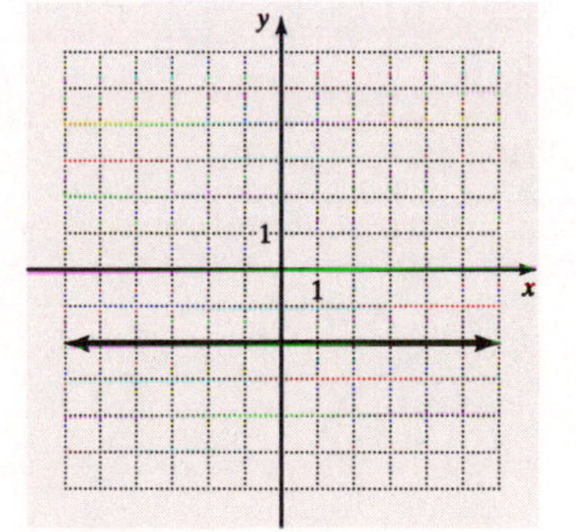

43.

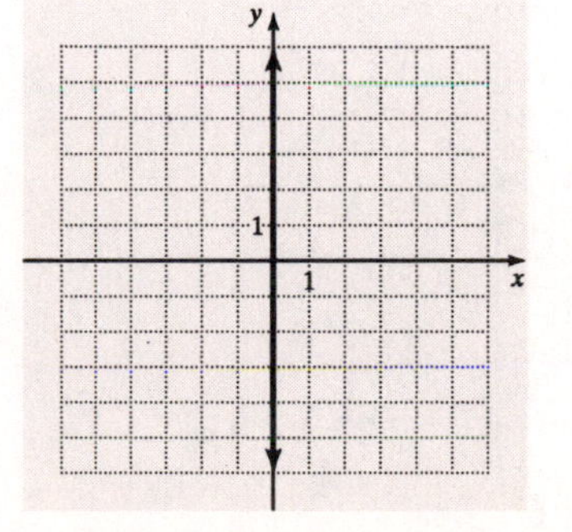

45.

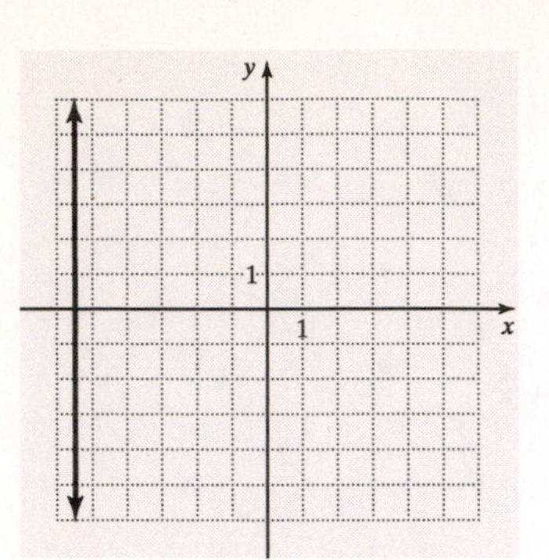

47.

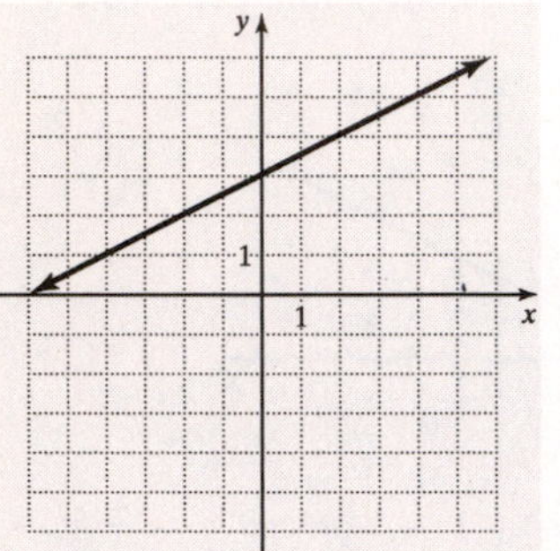

49.

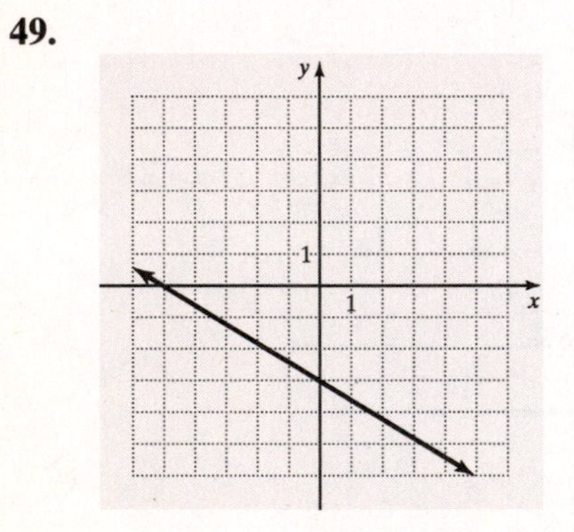

51.

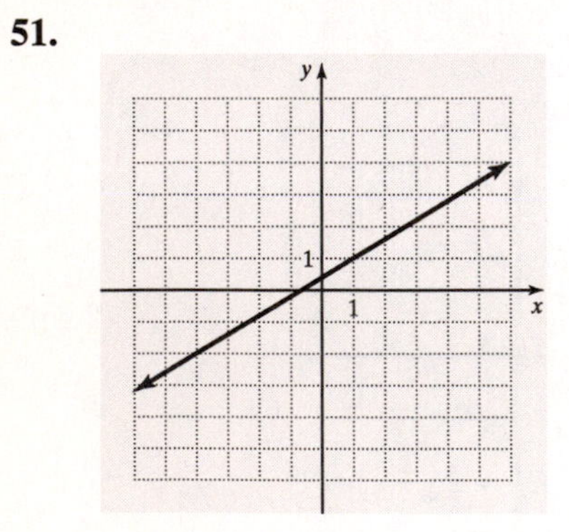

53. $y = -2x + 6$

$y = -2(-3) + 6$
$y = 12$

$y = -2\left(\frac{5}{2}\right) + 6$
$y = -5 + 6$
$y = 1$

$-10 = -2x + 6$
$-16 = -2x$
$8 = x$

$4 = -2x + 6$
$-2 = -2x$
$1 = x$

55. $y = -\frac{1}{2}x + 2$

$y = -\frac{1}{2}(0) + 2$
$y = 2 \qquad (0,2)$

$0 = -\frac{1}{2}x + 2$
$-2 = -\frac{1}{2}x$
$4 = x \qquad (4,0)$

$y = -\frac{1}{2}(-4) + 2$
$y = 2 + 2$
$y = 4 \qquad (-4,4)$

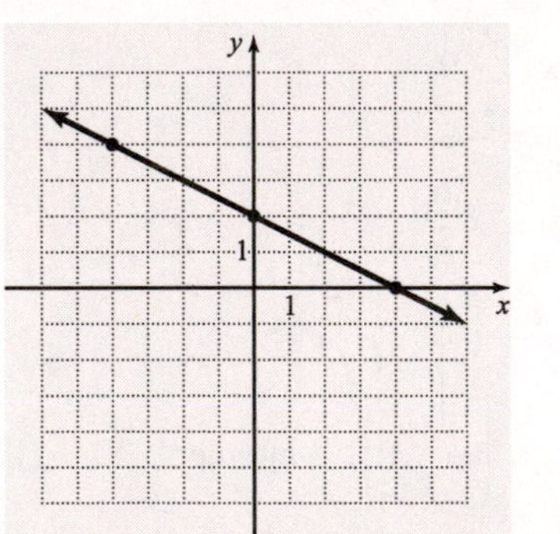

57. $6x-2y=-12$

$$\begin{aligned} 6(0)-2y&=-12 \\ -2y&=-12 \\ y&=6 \qquad (0,6) \end{aligned}$$

$$\begin{aligned} 6x-2(0)&=-12 \\ 6x&=-12 \\ x&=-2 \qquad (-2,0) \end{aligned}$$

$$\begin{aligned} 6(-3)-2y&=-12 \\ -18-2y&=-12 \\ -2y&=6 \\ y&=-3 \qquad (-3,-3) \end{aligned}$$

59. $y=-4x$

$$\begin{aligned} y&=-4(0) \\ y&=0 \qquad (0,0) \end{aligned}$$

$$\begin{aligned} 0&=-4x \\ 0&=x \qquad (0,0) \end{aligned}$$

$$\begin{aligned} y&=-4(1) \\ y&=-4 \qquad (1,-4) \end{aligned}$$

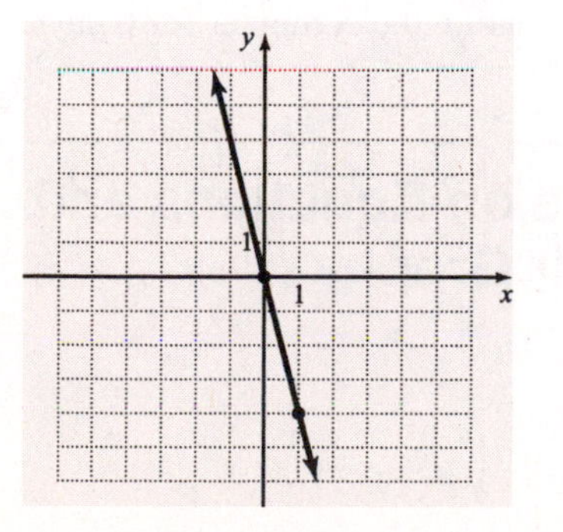

61. $3y+6x=-3$

$$\begin{aligned} 3y+6(0)&=-3 \\ 3y&=-3 \\ y&=-1 \qquad (0,-1) \end{aligned}$$

$$\begin{aligned} 3(0)+6x&=-3 \\ 6x&=-3 \\ x&=\frac{-3}{6} \\ x&=-\frac{1}{2} \qquad \left(-\frac{1}{2},0\right) \end{aligned}$$

$$\begin{aligned} 3y+6(-2)&=-3 \\ 3y-12&=-3 \\ 3y&=9 \\ y&=3 \qquad (-2,3) \end{aligned}$$

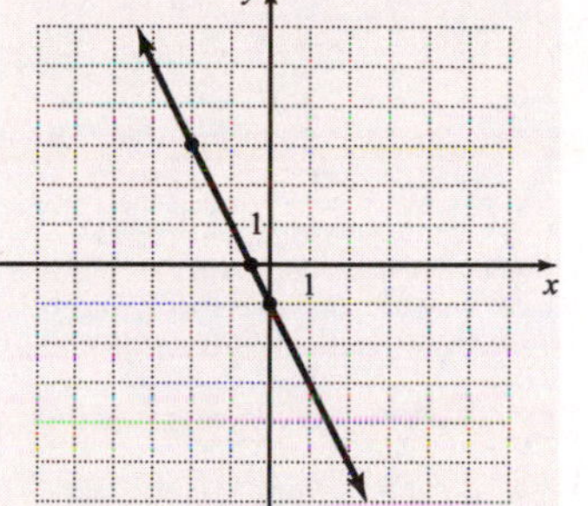

63.

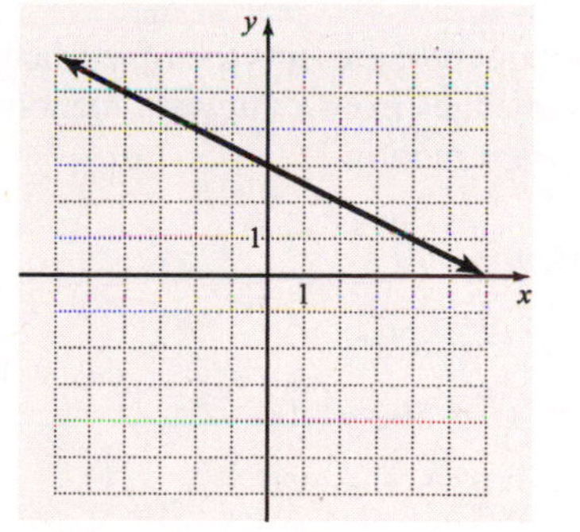

65. **a.** $v = 10 - 32t$

$v = 10 - 32(0) = 10$

$-6 = 10 - 32t$

$-16 = -32t$

$0.5 = t$

$v = 10 - 32(1) = -22$

$v = 10 - 32(1.5) = -38$

$v = 10 - 32(2) = -54$

A positive value of v means that the object is moving upward. A negative value of v means that the object is morning downward.

b.

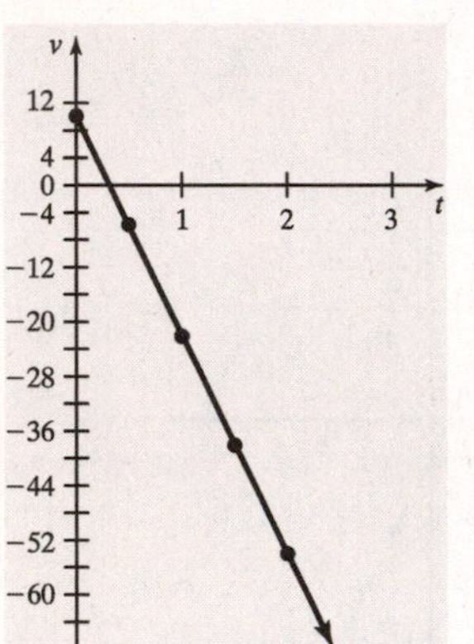

c. The v-intercept is the initial velocity of the object.

d. The t-intercept represents the time when the object changes from an upward motion to a downward motion.

67. **a.** $P = 100m + 500$

b. $P = 100(1) + 500 = 600$

$P = 100(2) + 500 = 700$

$P = 100(3) + 500 = 800$

c.

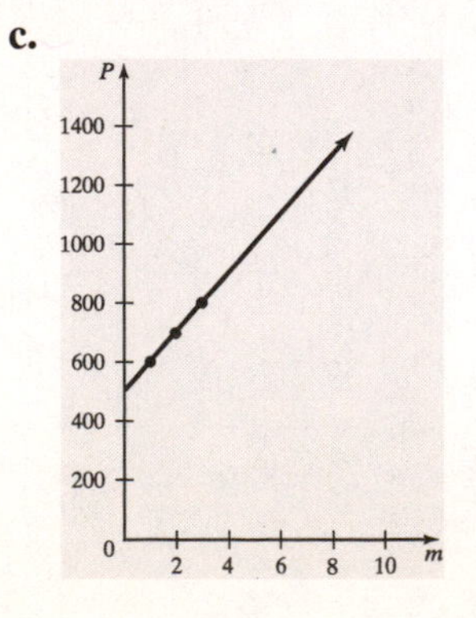

69. **a.** $0.05n + 0.1d = 2$

b. $0.05(0) + 0.1d = 2$

$0.1d = 2$

$d = 20$

$0.05n + 0.1(0) = 2$

$0.05n = 2$

$n = 40$

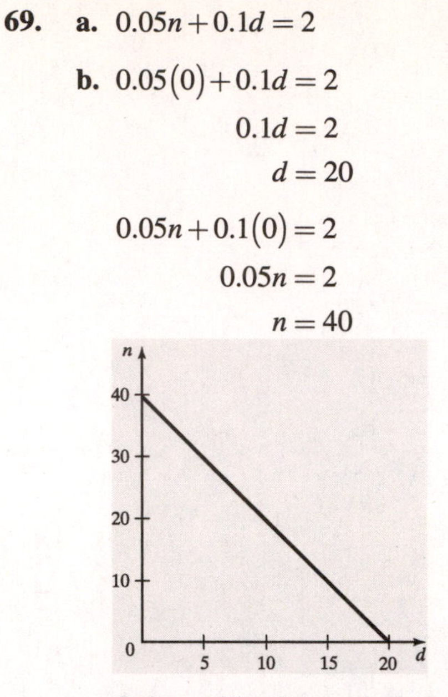

c. Only positive integer values make sense, since you cannot have fractions of nickels or dimes.

71. **a.** $F = 5d + 40$

b. $F = 5(0) + 40 = 40$

$F = 5(8) + 40 = 80$

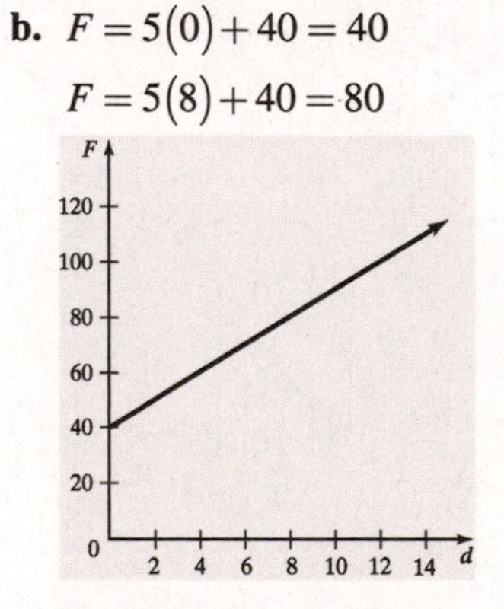

c. The F-intercept represents the fixed cost of renting the computer for 0 days.

3.4 More on Equations and Their Graphs

Practice

1. Slope is -2; y-intercept is (0,3).

2. $3x-2y=4$

$-2y=-3x+4$

$y=\frac{3}{2}x-2$

3. $y-2=4(x+1)$

$y-2=4x+4$

$y=4x+6$

4. Slope is -1; y-intercept is (0,0)

5.

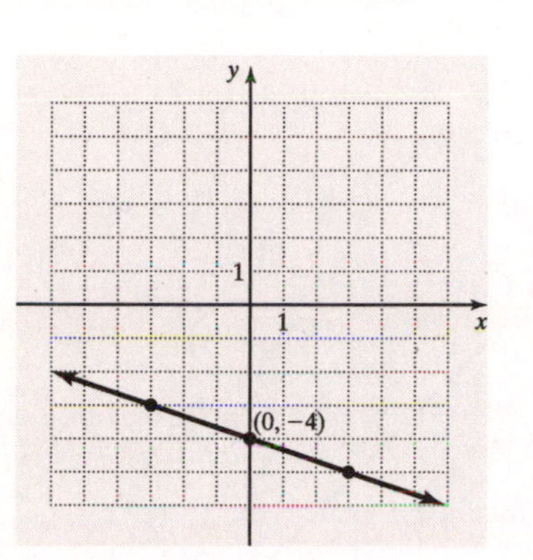

6. $y=1x+2$ or $y=x+2$

7. $y=-2x-1$

8. $y=-\frac{1}{2}x-2$

9. $w=45-3t$

10. $y-0=2(x-7)$

11. $y-7=1(x-7)$ or $y-0=1(x-0)$

12. Point-slope form: $w-27=5(b-4)$;

$w-27=5(b-4)$

$w-27=5b-20$

$w=5b+7$

Slope-intercept form: $w=5b+7$

13.

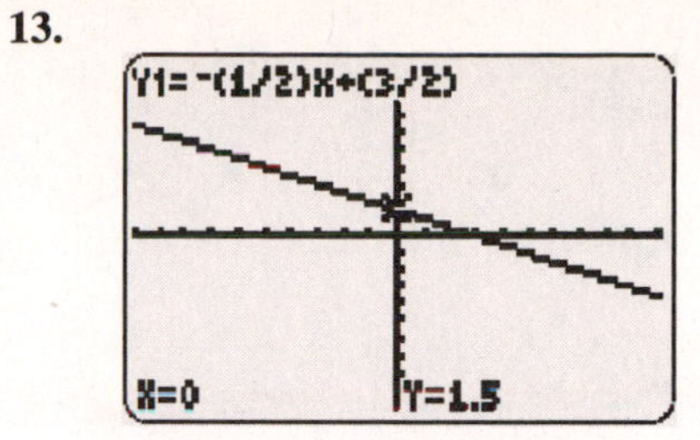

The displayed coordinates of the y-intercept are $x=0$ and $y=1.5$.

Exercises

1. A linear equation is in slope-intercept form if it is written as $y=mx+b$ where m and b are constants.

3. For an equation of a line written in slope-intercept form, $(0,b)$ is the y-intercept of the line.

5.

Equation	Slope, m	y-intercept $(0,b)$	Graph Type	x-intercept
$y=3x-5$	3	$(0,-5)$	/	$\left(\frac{5}{3},0\right)$
$y=-2x$	-2	$(0,0)$	\	$(0,0)$
$y=0.7x+3.5$	0.7	$(0,3.5)$	/	$(-5,0)$
$y=\frac{3}{4}x-\frac{1}{2}$	$\frac{3}{4}$	$\left(0,-\frac{1}{2}\right)$	/	$\left(\frac{2}{3},0\right)$
$6x+3y=12$	-2	$(0,4)$	\	$(2,0)$
$y=-5$	0	$(0,-5)$	–	no x-intercept
$x=-2$	undefined	no y-intercept	|	$(-2,0)$

7. Slope: -1; y-intercept: $(0,2)$

9. Slope: $-\frac{1}{2}$; y-intercept: $(0,0)$

11. $x-y=10$

$-y=-x+10$

$y=x-10$

13. $x + 10y = 10$

$$10y = -x + 10$$
$$y = -\frac{1}{10}x + \frac{10}{10}$$
$$y = -\frac{1}{10}x + 1$$

15. $6x + 4y = 1$

$$4y = -6x + 1$$
$$y = -\frac{6}{4}x + \frac{1}{4}$$
$$y = -\frac{3}{2}x + \frac{1}{4}$$

17. $2x - 5y = 10$

$$-5y = -2x + 10$$
$$y = \frac{-2}{-5}x + \frac{10}{-5}$$
$$y = \frac{2}{5}x - 2$$

19. $y + 1 = 3(x + 5)$

$$y + 1 = 3x + 15$$
$$y = 3x + 15 - 1$$
$$y = 3x + 14$$

21. b

23. a

25. 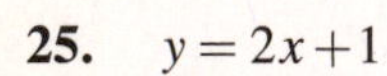 $y = 2x + 1$

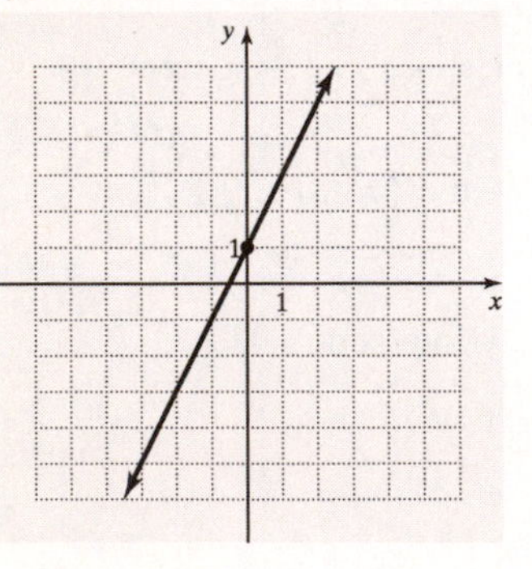

27. $y = -\frac{2}{3}x + 6$

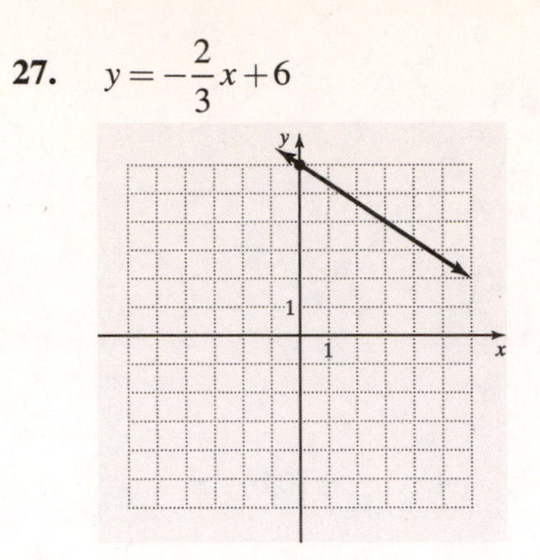

29. $x + y = 1$

$$y = -x + 1$$

31. $y = -\frac{3}{4}x$

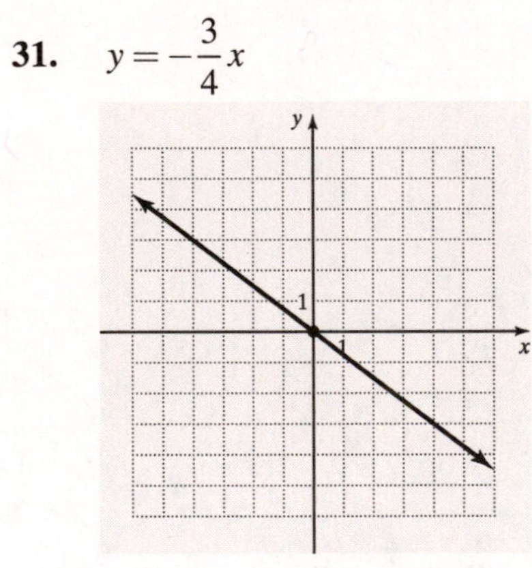

33.
$$\begin{aligned} x+2y&=4\\ 2y&=-x+4\\ y&=-\frac{1}{2}x+\frac{4}{2}\\ y&=-\frac{1}{2}x+2 \end{aligned}$$

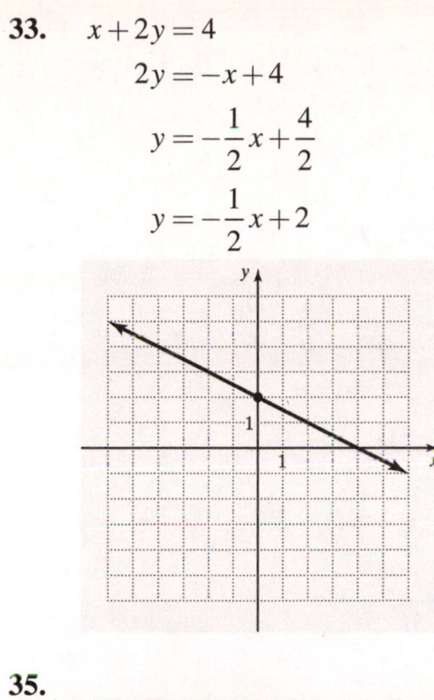

35.

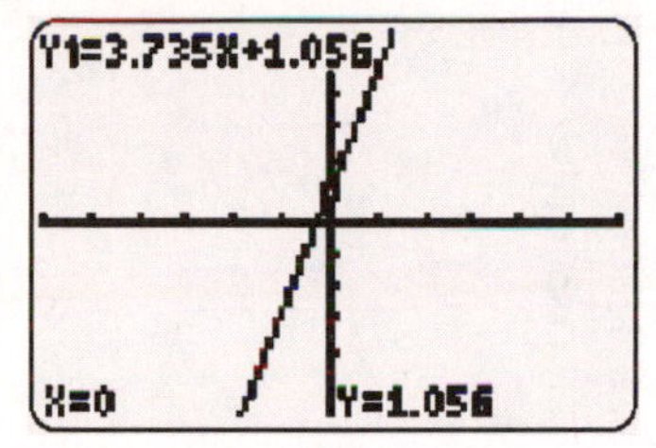

37. $y=3x+7$

39.
$$\begin{aligned} y&=5x+b\\ 0&=5(4)+b\\ 0&=20+b\\ b&=-20 \end{aligned}$$
$$y=5x-20$$

41.
$$\begin{aligned} y-5&=-\frac{1}{2}(x-(-2))\\ y-5&=-\frac{1}{2}(x+2)\\ y-5&=-\frac{1}{2}x-1\\ y&=-\frac{1}{2}x-1+5\\ y&=-\frac{1}{2}x+4 \end{aligned}$$

43. $m=\dfrac{2-1}{1-2}=\dfrac{1}{-1}=-1$
$$\begin{aligned} y-1&=-1(x-2)\\ y-1&=-x+2\\ y&=-x+2+1\\ y&=-x+3 \end{aligned}$$

45. $m=\dfrac{-6-(-5)}{-7-(-1)}=\dfrac{-1}{-6}=\dfrac{1}{6}$
$$\begin{aligned} y-(-5)&=\frac{1}{6}(x-(-1))\\ y+5&=\frac{1}{6}x+\frac{1}{6}\\ y&=\frac{1}{6}x+\frac{1}{6}-5\\ y&=\frac{1}{6}x+\frac{1}{6}-\frac{30}{6}\\ y&=\frac{1}{6}x-\frac{29}{6} \end{aligned}$$

47. $x=-3$

49. $y=0$

51. $m=\dfrac{3-0}{0-(-4)}=\dfrac{3}{4}$
$$\begin{aligned} y-0&=\frac{3}{4}(x-(-4))\\ y&=\frac{3}{4}(x+4)\\ y&=\frac{3}{4}x+3 \end{aligned}$$

53. $y=2$

55. $m=\dfrac{2-(-2)}{-4-0}=\dfrac{4}{-4}=-1$
$$\begin{aligned} y-2&=-1(x-(-4))\\ y-2&=-(x+4)\\ y-2&=-x-4\\ y&=-x-4+2\\ y&=-x-2 \end{aligned}$$

57.

Equation	Slope, m	y-intercept $(0,b)$	Graph Type	x-intercept
$y=-7x+2$	-7	$(0,2)$	\	$\left(\frac{2}{7},0\right)$
$y=4x$	4	$(0,0)$	/	$(0,0)$
$y=2.5x+10$	2.5	$(0,10)$	/	$(-4,0)$
$y=\frac{2}{3}x-\frac{1}{4}$	$\frac{2}{3}$	$\left(0,-\frac{1}{4}\right)$	/	$\left(\frac{3}{8},0\right)$
$5x+4y=20$	$-\frac{5}{4}$	$(0,5)$	\	$(4,0)$
$x=9$	undefined	no y-intercept	|	$(9,0)$
$y=-3.2$	0	$(0,-3.2)$	—	no x-intercept

59.
$$4x-y=5$$
$$-y=-4x+5$$
$$\frac{-y}{-1}=\frac{-4x}{-1}+\frac{5}{-1}$$
$$y=4x-5$$

61. d

63.
$$y=\frac{1}{2}x+b$$
$$1=\frac{1}{2}(4)+b$$
$$1=2+b$$
$$b=-1$$

$$y=\frac{1}{2}x-1$$

65.

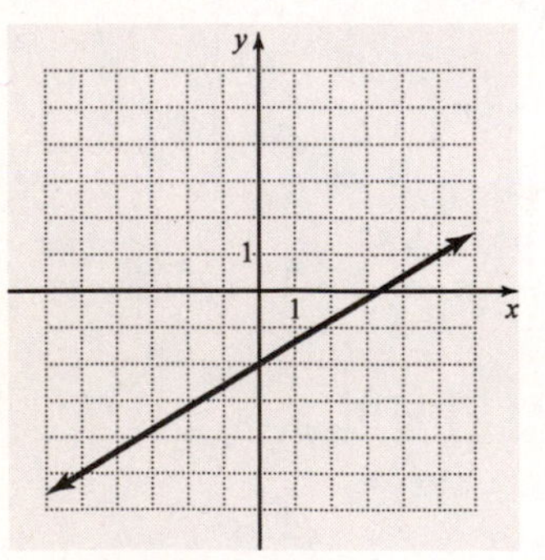

67. $m=\frac{5-0}{0-5}=\frac{5}{-5}=-1$
$$y-5=-1(x-0)$$
$$y-5=-x$$
$$y=-x+5$$

69. **a.** $m=\frac{32-14}{0-(-10)}=\frac{18}{10}=\frac{9}{5}$

b. $F-32=\frac{9}{5}(C-0)$
$$F-32=\frac{9}{5}C$$
$$F=\frac{9}{5}C+32$$

c.
$$212=\frac{9}{5}C+32$$
$$212-32=\frac{9}{5}C$$
$$180=\frac{9}{5}C$$
$$\frac{5}{9}\cdot 180=\frac{5}{9}\cdot\frac{9}{5}C$$
$$100=C$$
Water boils at 100 °C.

71. **a.** $y-3500=4(x-500)$

b. $y-3500=4x-2000$
$$y=4x-2000+3500$$
$$y=4x+1500$$

c. The y-intercept represents the monthly flat fee the utility company charges its residential customers.

73. **a.** $m=\frac{1-2}{5-10}=\frac{-1}{-5}=\frac{1}{5}$

$t-1=\frac{1}{5}(L-5)$ or $L-5=5(t-1)$

b. $L-5=5(t-1)$
$$L-5=5t-5$$
$L=5t$ or $t=\frac{1}{5}L$

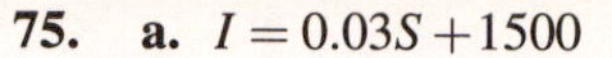

75. **a.** $I = 0.03S + 1500$

b.

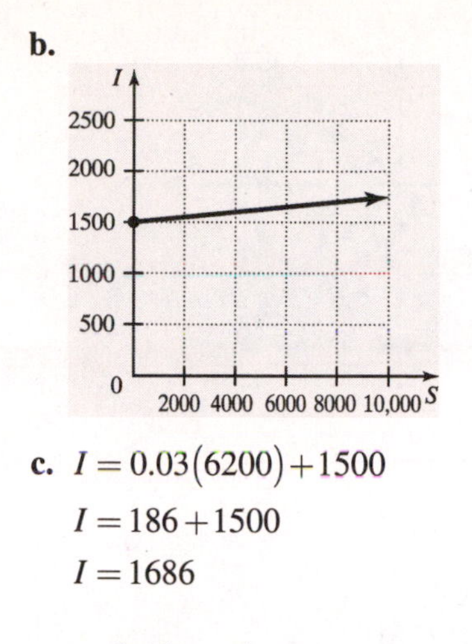

c. $I = 0.03(6200) + 1500$

$I = 186 + 1500$

$I = 1686$

77. $m = \dfrac{2-1}{33-0} = \dfrac{1}{33}$

$P - 1 = \dfrac{1}{33}(d - 0)$

$P = \dfrac{1}{33}d + 1$

79. $m = \dfrac{15-10}{6-0} = \dfrac{5}{6}$

$L - 10 = \dfrac{5}{6}(F - 0)$

$L = \dfrac{5}{6}F + 10$

3.5 Linear Inequalities and Their Graphs

Practice

1. $y < x - 1$

$3 < 1 - 1$

$3 < 0$

No, (1,3) is not a solution to the inequality.

2.

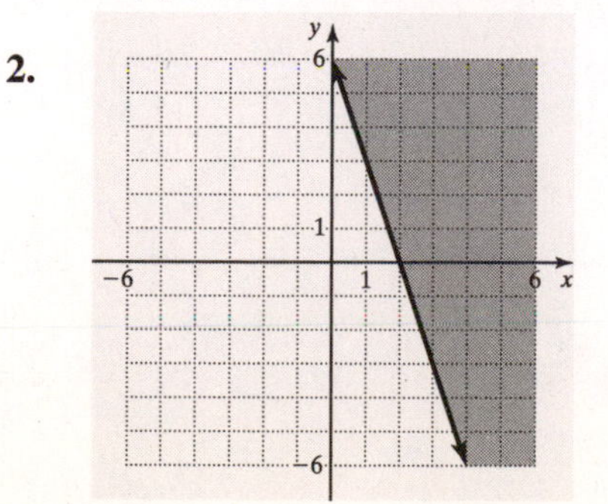

3.

4.

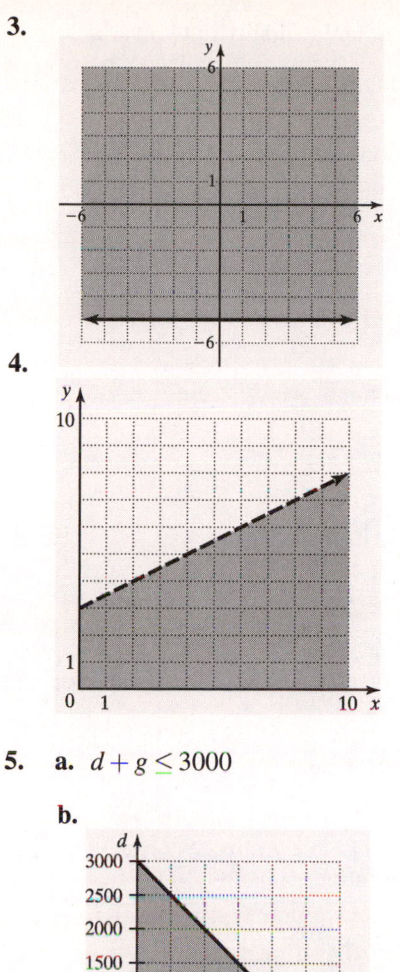

5. **a.** $d + g \le 3000$

b.

c. The d-intercept represents the maximum number of gallons of diesel fuel the refinery can produce if no gasoline is produced. The g-intercept represents the maximum number of gallons of gasoline the refinery can produce if no diesel fuel is produced.

Exercises

1. The graph of a linear inequality in two variables is a half-plane.

3. The graph of an inequality in two variables is the set of all points on the plane whose coordinates satisfy the inequality.

5. A broken boundary line is drawn when graphing a linear inequality that involves the symbol < or the symbol >.

7. $y < 3x$

$0 < 3(0)$

$0 < 0$

No, not a solution

9. $y \geq 2x - 1$

$-2 \geq 2\left(-\frac{1}{2}\right) - 1$

$-2 \geq -1 - 1$

$-2 \geq -2$

Yes, a solution

11. $2x - 3y > 10$

$2(10) - 3(8) > 10$

$20 - 24 > 10$

$-4 > 10$

No, not a solution

13.

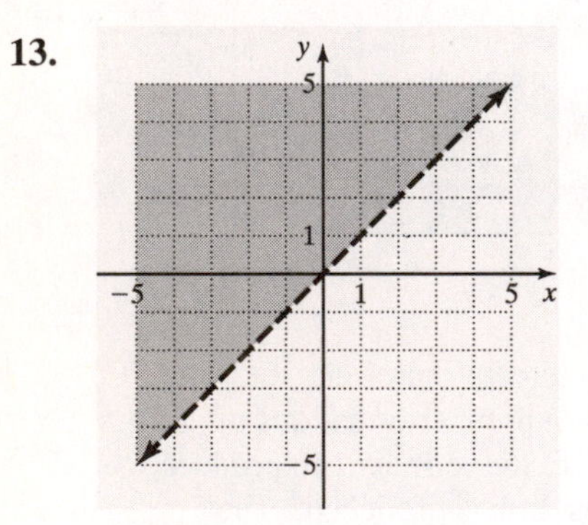

15.

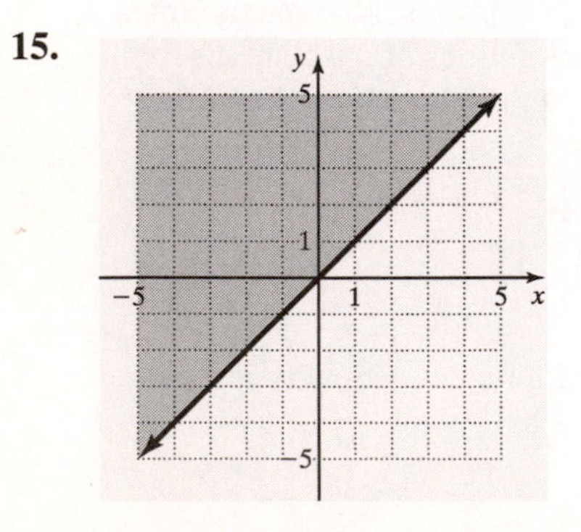

17.

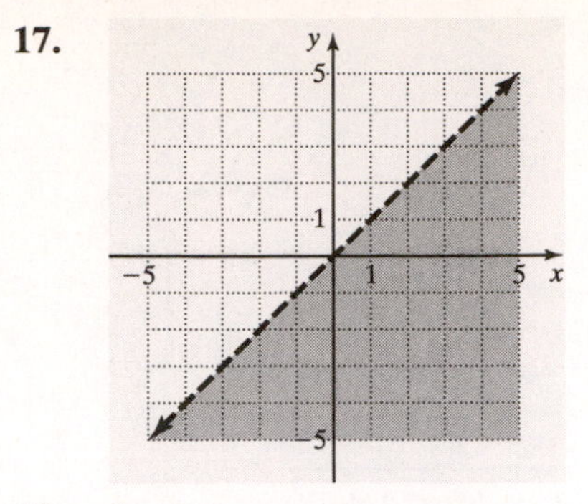

19. d

21. b

23. $x > -5$

Test point: $(0,0)$

$0 > -5$ True

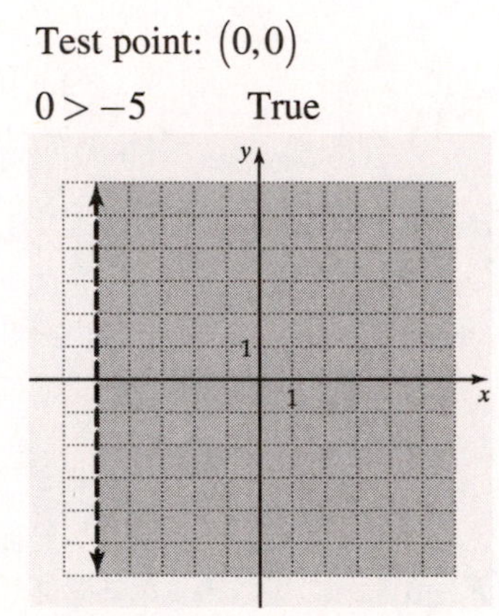

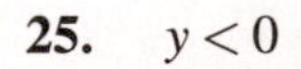

25. $y < 0$

Test point: $(1,1)$

$1 < 0$ False

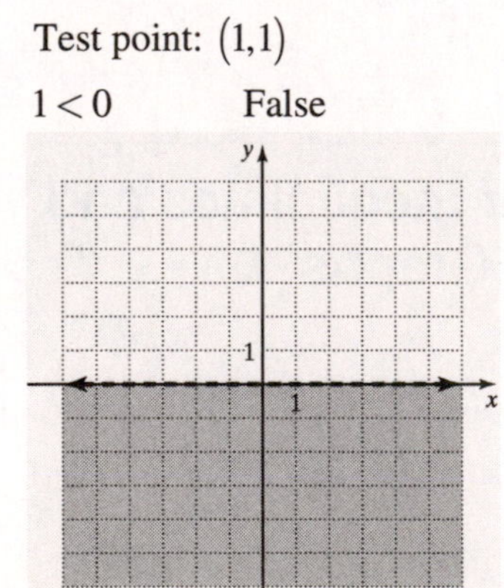

27. $y \le 3x$

Test point: $(1,0)$

$0 \le 3(1)$

$0 \le 3$ True

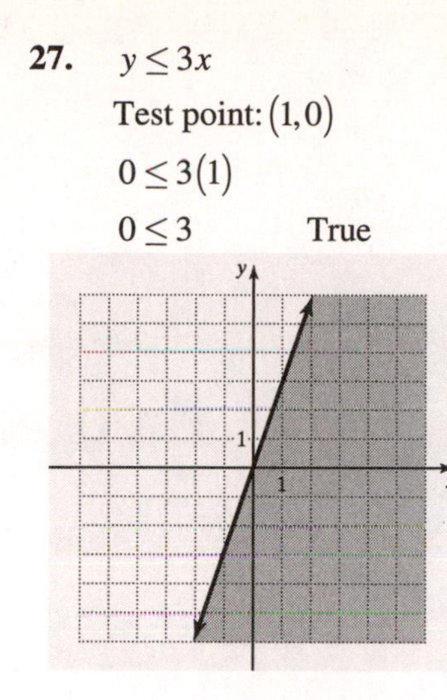

29. $y \ge -2x$

Test point: $(1,0)$

$0 \ge -2(1)$

$0 \ge -2$ True

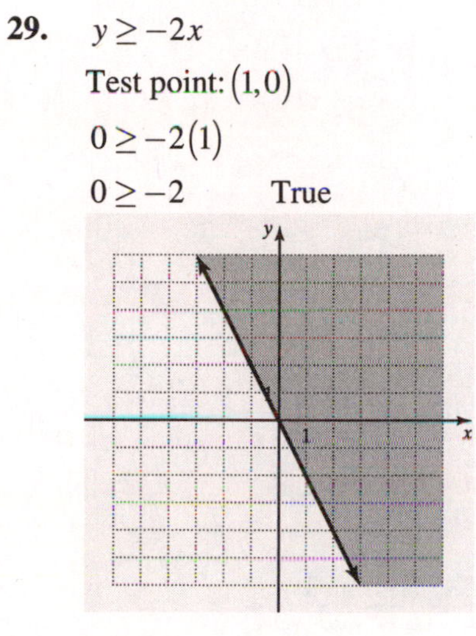

31. 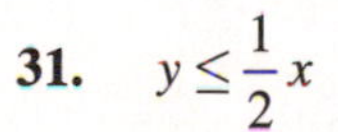
$y \le \frac{1}{2}x$

Test point: $(1,0)$

$0 \le \frac{1}{2}(1)$

$0 \le \frac{1}{2}$ True

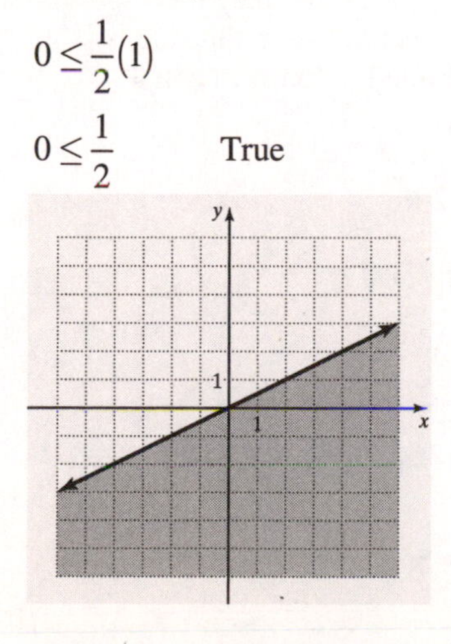

33. $y > 3x + 5$

Test point: $(1,0)$

$0 > 3(1) + 5$

$0 > 3 + 5$

$0 > 8$ False

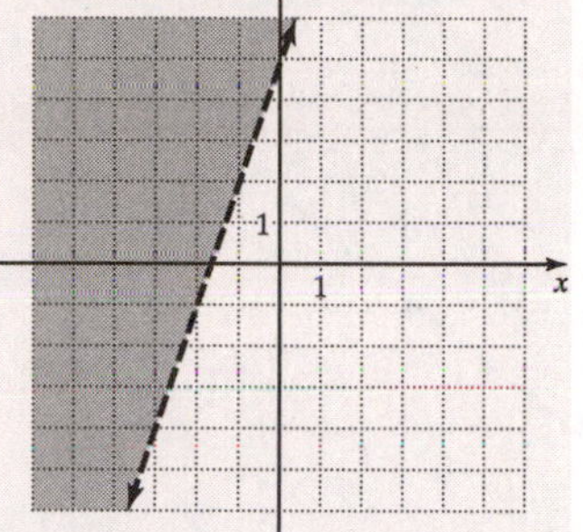

35. $5y - x > 10$

$y > \frac{1}{5}x + 2$

Test point: $(1,0)$

$5(0) - (1) > 10$

$-1 > 10$ False

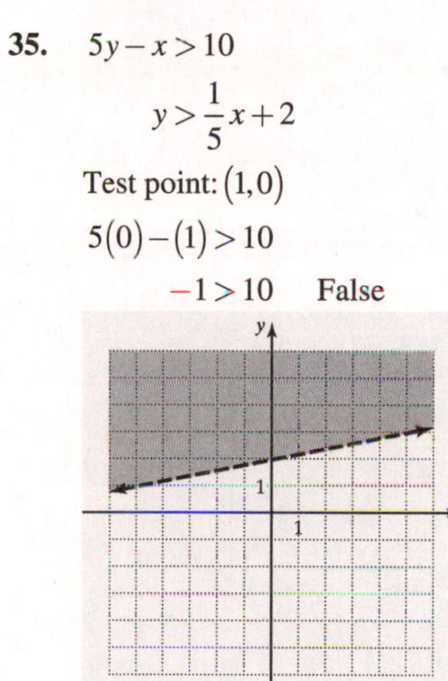

37. $2x - 3y \ge 3$

$y \le \frac{2}{3}x - 1$

Test point: $(0,0)$

$2(0) - 3(0) \ge 3$

$0 \ge 3$ False

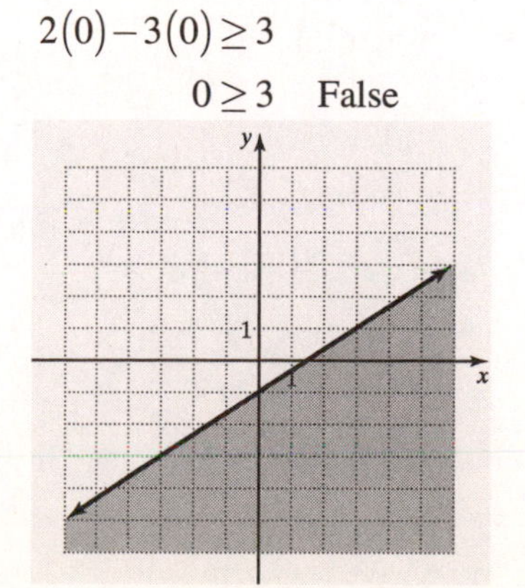

39. $4x-5y\le 10$

$$y\ge \frac{4}{5}x-2$$

Test point: $(1,0)$

$4(1)-5(0)\le 10$

$4\le 10$ True

41. $y>-\frac{1}{2}x+2$

$0>-\frac{1}{2}(4)+2$

$0>-2+2$

$0>0$

No, not a solution

43. b

45. $y<-3$

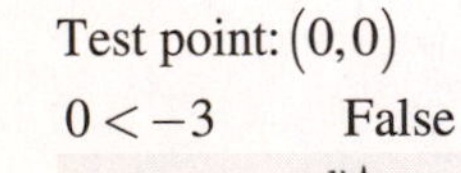

Test point: $(0,0)$

$0<-3$ False

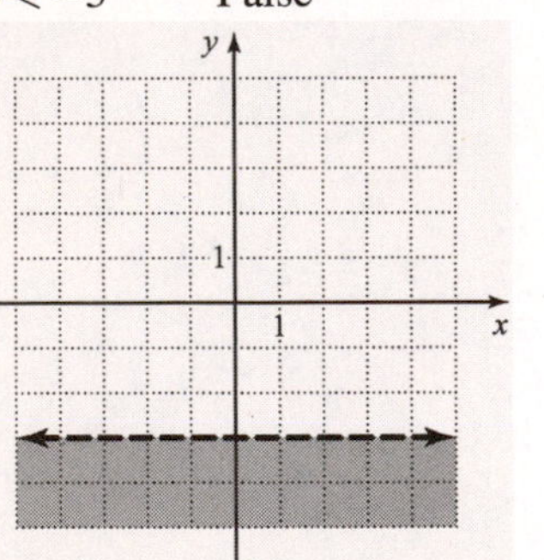

47. $y>-2x-2$

Test point: $(1,0)$

$0>-2(1)-2$

$0>-4$ True

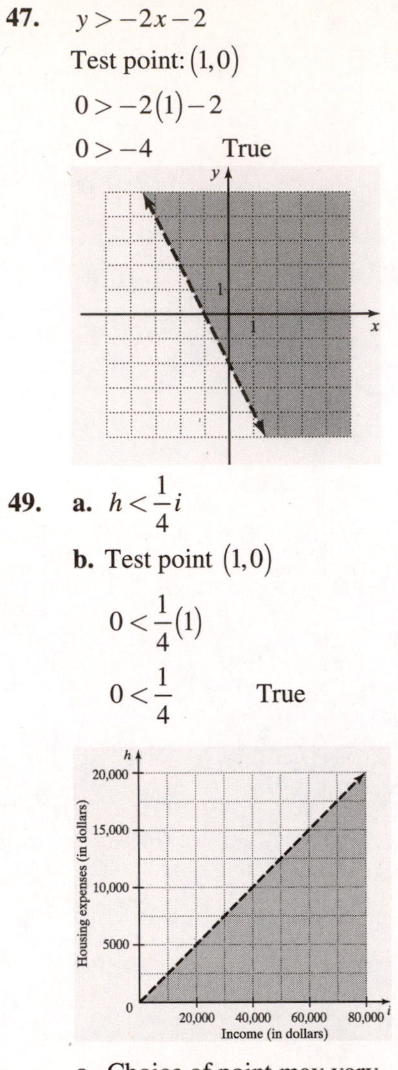

49. **a.** $h<\frac{1}{4}i$

b. Test point $(1,0)$

$0<\frac{1}{4}(1)$

$0<\frac{1}{4}$ True

c. Choice of point may vary.

Possible point: $(20{,}000,\ 2500)$.

The guideline holds since the inequality is true when the values are substituted in to the original inequality.

51. **a.** $x+y\ge 200$

b. Test point $(1,0)$

$(1)+(0)\ge 200$

$1\ge 200$ False

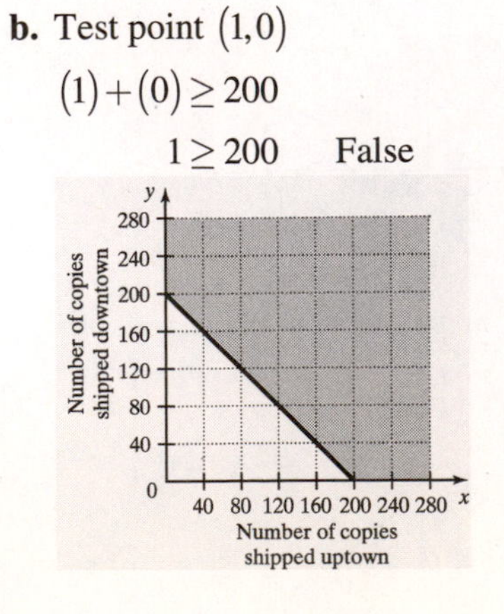

51. **c.** Choice of point may vary.
Possible point: $(200, 60)$.

$$(200)+(60)\geq 200$$
$$260\geq 200 \quad \text{True}$$

53. **a.** $30x+75y\geq 1500$

b. Test point $(1,0)$

$$30(1)+75(0)\geq 1500$$
$$30\geq 1500 \quad \text{False}$$

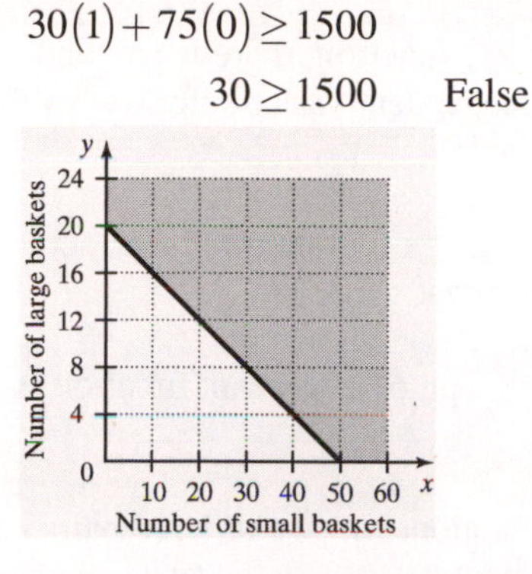

c. Since the point (20, 20) lies in the solution region, selling 20 small and 20 large gift baskets will generate the desired revenue.

55. $60a+100b<300$

$$b<-\frac{3}{5}a+3$$

Test point: $(1,0)$

$$60(1)+100(0)<300$$
$$60<300 \quad \text{True}$$

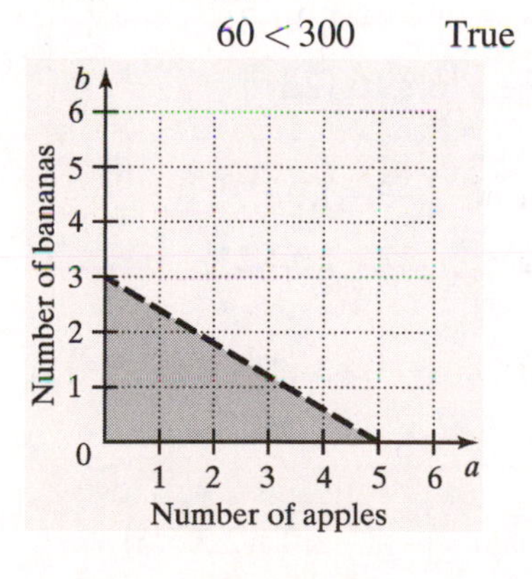

57. **a.** $10w+15m\leq 50{,}000$

$$w<-\frac{3}{2}m+5000$$

57. **b.** Test point: $(1,0)$

$$60(0)+50(1)<300$$
$$50<300 \quad \text{True}$$

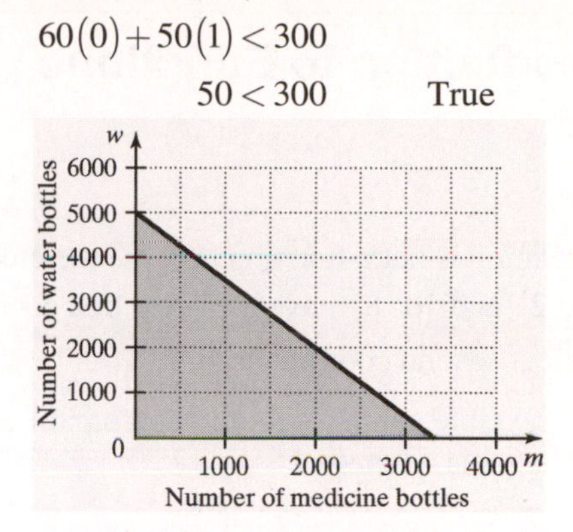

c. Answers may vary. Possible answers:
1500 medicine containers and 1000 bottles of water: (1500, 1000)
300 medicine containers and 500 bottles of water: (300, 500)
1200 medicine containers and 2000 bottles of water: (1200, 2000)

59. **a.** $8x+10y\geq 200$

$$y\geq -\frac{4}{5}x+20$$

b. Test point: $(1,0)$

$$8(1)+10(0)\geq 200$$
$$8\geq 200 \quad \text{False}$$

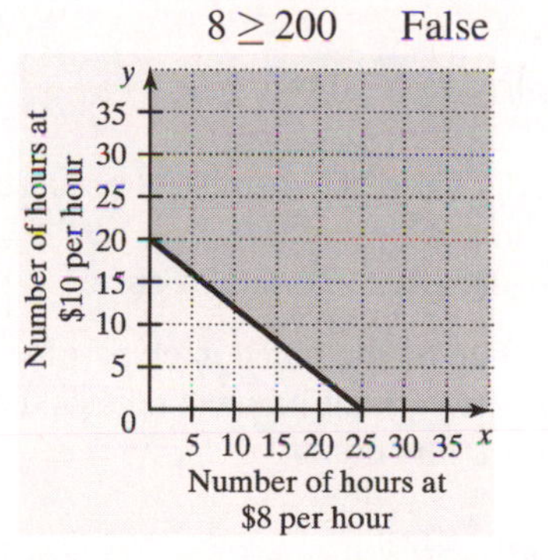

c. Answers may vary. Possible answers:
20 hr at the job paying $8 per hr and 10 hr at the job paying $10 per hr
10 hr at the job paying $8 per hr and 15 hr at the job paying $10 per hr

3.6 Introduction to Functions

Practice

1. **a.** The relation is not a function because the ordered pairs $(0, 2)$ and $(0, 8)$ have the same first coordinate but different second coordinates.

   b. From the table, all the x-coordinates are different so the relation is a function.

2. **a.** Domain: $\{2, 5, 8\}$; range: $\{-6, 0, 3\}$

   b. Domain: $\{1, 2, 5\}$; range: $\{-1, 2, 4\}$

   c. Domain: $\{1, 2, 3, 4, 5\}$;
   range: $\{6.5, 7, 7.5, 8, 9\}$

3. **a.** $g(2)=3(2)-1=6-1=5$

   b. $g(-1)=3(-1)-1=-3-1=-4$

   c. $g(n)=3(n)-1=3n-1$

   d. $g(n+1)=3(n+1)-1$
   $=3n+3-1=3n+2$

4. Domain: $[0,\infty)$; range: $[0,\infty)$

5. **a.** The graph fails the vertical-line test because any vertical line with $x>3$ or $x<6$ intersects the graph more than once.

   b. Any vertical line on the plane intersects the graph at one point, so the vertical-line test tells us that this graph represents a function.

6. **a.** $C(m)=0.1m+40$

   b. $C(200)=0.1(200)+40$
   $=20+40=\$60$

   c.

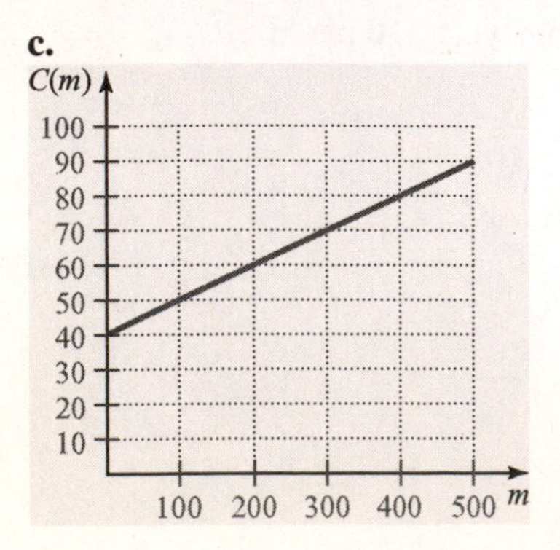

   d. Domain: $[0,\infty)$; range: $[40,\infty)$

   e. The graph in part (c) passes the vertical-line test, showing that the relation is a function.

Exercises

1. A(n) relation is a set of ordered pairs.

3. For any function, there is one and only one value of the dependent variable for each value of the other variable.

5. The range of a function is the set of all values of the dependent variable.

7. The graph of a constant function is a(n) horizontal line.

9. A function. No two ordered pairs have the same first element.

11. Not a function. The ordered pairs $(-2,-3)$ and $(-2,3)$ have the same first element.

13. Not a function. The ordered pairs $(3,4)$ and $(3-4)$ have the same first element.

15. A function. No two ordered pairs have the same first element.

17. Domain: $\{-2,-1,0,1,2\}$
    Range: $\{6,8,10,12,14\}$

19. Domain: $\{-3,-1,0,1,3\}$
    Range: $\{-27,-1,0,1,27\}$

21. Domain: $\{-4,-3,-2,-1,1,2,3,4\}$
    Range: $\{-8,-5,0,7\}$

23. Domain: $\{-4,-2,0,1,2,3.5,5\}$
    Range: $\{1,3,5\}$

25. Domain: $\{-7,-3,-1,2,4\}$
    Range: $\{-4,-2,1,3,7\}$

27. Domain: $\{2000,2001,2002,2003,2004\}$
    Range: $\{20,23,32,34,38\}$

29. **a.** $f(2)=8-5(2)=8-10=-2$

b. $f(-1)=8-5(-1)=8+5=13$

c. $f\left(\frac{3}{5}\right)=8-5\left(\frac{3}{5}\right)=8-3=5$

d. $f(1.8)=8-5(1.8)=8-9=-1$

31. **a.** $g(5)=2.4(5)-7=12-7=5$

b. $g(-2)=2.4(-2)-7=-11.8$

c. $g(a)=2.4(a)-7=2.4a-7$

d. $g(a^2)=2.4(a^2)-7=2.4a^2-7$

33. **a.** $f(0)=\left|\frac{1}{2}(0)+3\right|=|3|=3$

b. $f(-8)=\left|\frac{1}{2}(-8)+3\right|$
$=|-4+3|=|-1|=1$

c. $f(-4t)=\left|\frac{1}{2}(-4t)+3\right|=|-2t+3|$

d. $f(t-6)=\left|\frac{1}{2}(t-6)+3\right|$
$=\left|\frac{1}{2}t-3+3\right|=\left|\frac{1}{2}t\right|$

35. **a.** $h(2)=3(2)^2-6(2)-9$
$=3(4)-12-9$
$=12-12-9=-9$

b. $h(-1)=3(-1)^2-6(-1)-9$
$=3(1)+6-9=0$

c. $h(-n)=3(-n)^2-6(-n)-9$
$=3n^2+6n-9$

d. $h(2n)=3(2n)^2-6(2n)-9$
$=3(4n^2)-12n-9$
$=12n^2-12n-9$

37. **a.** $g(7)=10$

b. $g(-150)=10$

c. $g(t)=10$

d. $g(5-9t)=10$

39. $f(x)=5x-4$

x	$f(x)=5x-4$	(x,y)
0	$5(0)-4=-4$	$(0,-4)$
$\frac{1}{5}$	$5\left(\frac{1}{5}\right)-4=-3$	$\left(\frac{1}{5},-3\right)$
$\frac{4}{5}$	$5\left(\frac{4}{5}\right)-4=0$	$\left(\frac{4}{5},0\right)$

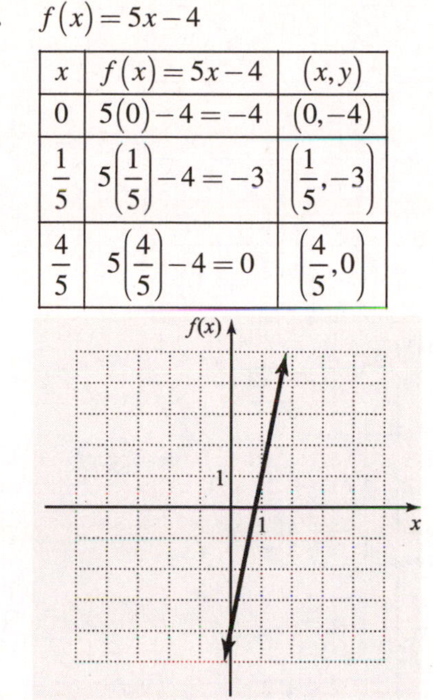

Domain: $(-\infty,\infty)$

Range: $(-\infty,\infty)$

41. $f(x)=-5$

x	$f(x)=-5$	(x,y)
-4	-5	$(-4,-5)$
0	-5	$(0,-5)$
4	-5	$(4,-5)$

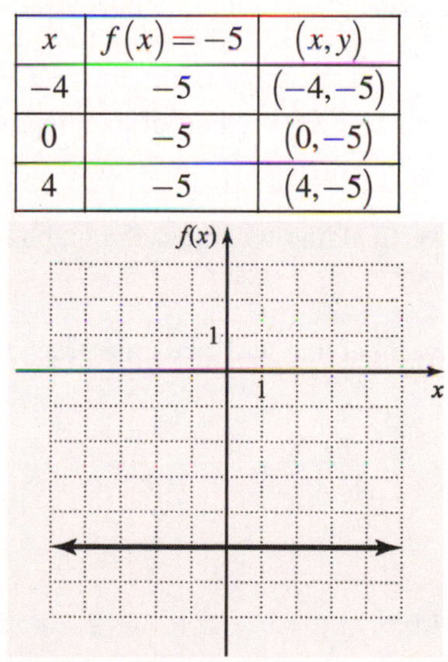

Domain: $(-\infty,\infty)$

Range: $\{-5\}$

43. $f(x) = -\frac{1}{4}x - 1$ for $x \leq 0$

x	$f(x) = -\frac{1}{4}x - 1$	(x, y)
0	$-\frac{1}{4}(0) - 1 = -1$	$(0, -1)$
−2	$-\frac{1}{4}(-2) - 1 = -\frac{1}{2}$	$\left(-2, -\frac{1}{2}\right)$
−4	$-\frac{1}{4}(-4) - 1 = 0$	$(-4, 0)$

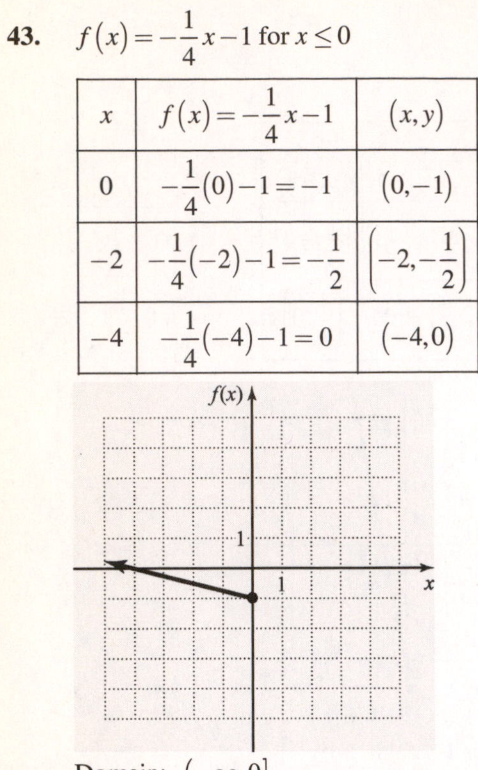

Domain: $(-\infty, 0]$

Range: $[-1, \infty)$

45. A function. No vertical line will cross the graph more than once.

47. No, not a function. The vertical part of the graph fails the vertical line test.

49. A function. No vertical line will cross the graph more than once.

51. A function. No vertical line will cross the graph more than once.

53. A function.

55. A function.

57. Domain: $\{0, 2, 4, 6, 8, 9\}$

Range: $\{-3, -1, 0, 1, 5, 6\}$

59. **a.** $f(0) = 3(0)^2 + (0)$

$= 3(0) + 0$

$= 0 + 0$

$= 0$

59. **b.** $f(-2) = 3(-2)^2 + (-2)$

$= 3(4) + (-2)$

$= 12 - 2$

$= 10$

c. $f(n) = 3(n)^2 + (n)$

$= 3n^2 + n$

d. $f(-4n) = 3(-4n)^2 + (-4n)$

$= 3(16n) + (-4n)$

$= 48n - 4n$

61. **a.** $d(a) = 0.20a$

b. $d(150) = 0.20(150) = 30$

The discount is \$30, so \$30 was saved.

c.

a	$d(a) = 0.20a$	(a, d)
0	$0.20(0) = 0$	$(0, 0)$
200	$0.20(200) = 40$	$(200, 40)$
400	$0.20(400) = 80$	$(400, 80)$

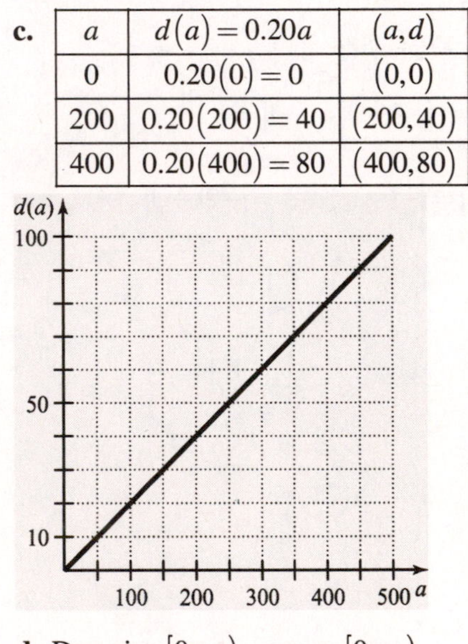

d. Domain: $[0, \infty)$; range: $[0, \infty)$

63. **a.** $V(t) = 22{,}500 - 1875t$

b. $V(6)$ represents the value of the car 6 years after it is purchased.

$V(6) = 22{,}500 - 1875(6)$

$= 22{,}500 - 11{,}250 = 11{,}250$

The value of the car after 6 years is \$11,250.

63. c.

t	$V(t) = 22{,}500 - 1875t$	$(t, V(t))$
0	$22{,}500 - 1875(0) = 22{,}500$	(0, 22,500)
6	$22{,}500 - 1875(6) = 11{,}250$	(6, 11,250)
12	$22{,}500 - 1875(12) = 0$	(12,0)

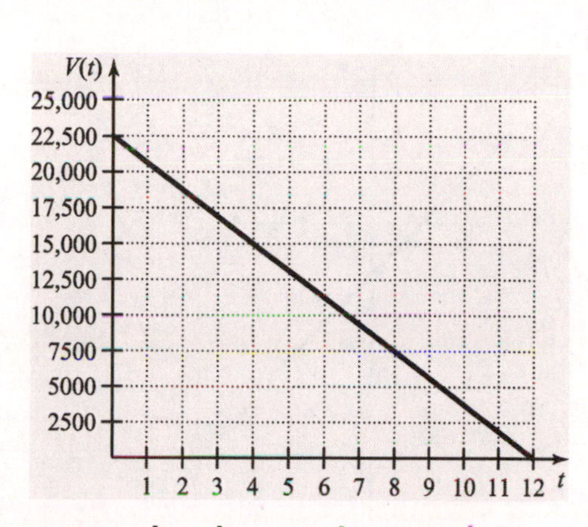

d. Domain: $[0,12]$; range: $[0, 22{,}500]$

65. a. $d(x) = 500 - 50x$

b. $d(2) = 500 - 50(2) = 500 - 100 = 400$

After two weeks the patient's dosage is 400 mg.

c.

x	$d(x) = 500 - 50x$	$(x, d(x))$
0	$500 - 50(0) =$ $500 - 0 = 500$	(0,500)
5	$500 - 50(5) =$ $500 - 250 = 250$	(5,250)
10	$500 - 50(10) =$ $500 - 500 = 0$	(10,0)

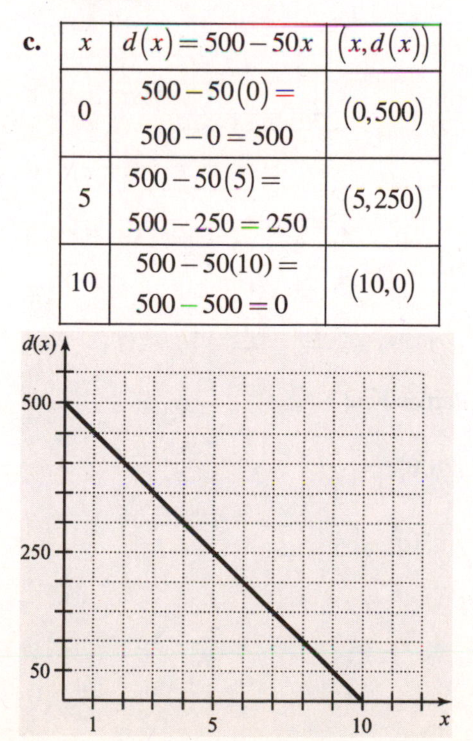

Domain: $[0,10]$

Range: $[0,500]$

Chapter 3 Review Exercises

1.

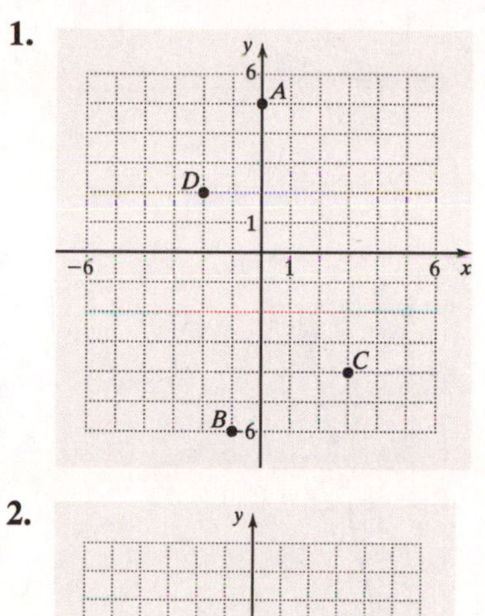

2.

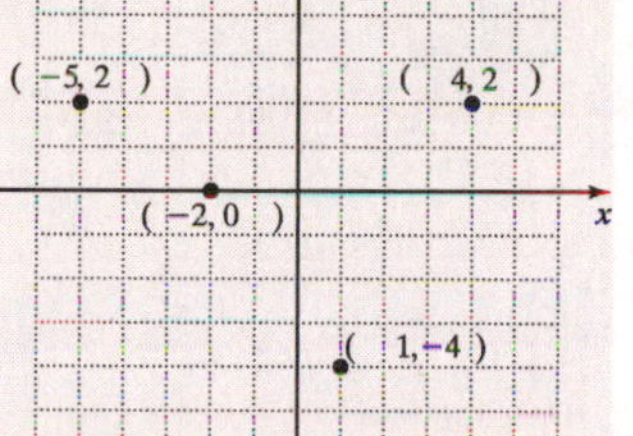

3. IV

4. III

5. $m = \dfrac{5-0}{3-2} = \dfrac{5}{1} = 5$

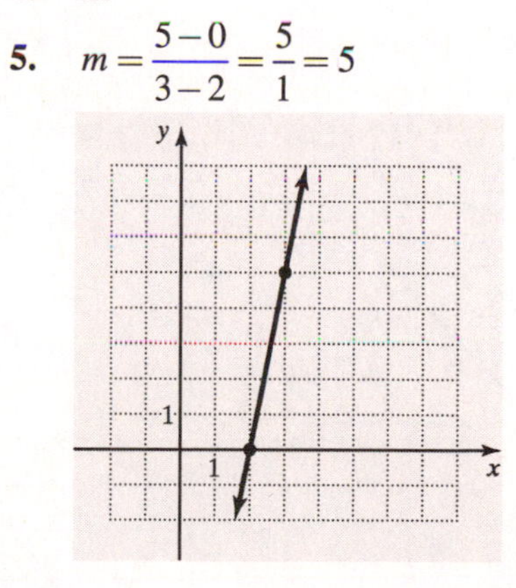

6. $m = \dfrac{7-7}{2-5} = \dfrac{0}{-3} = 0$

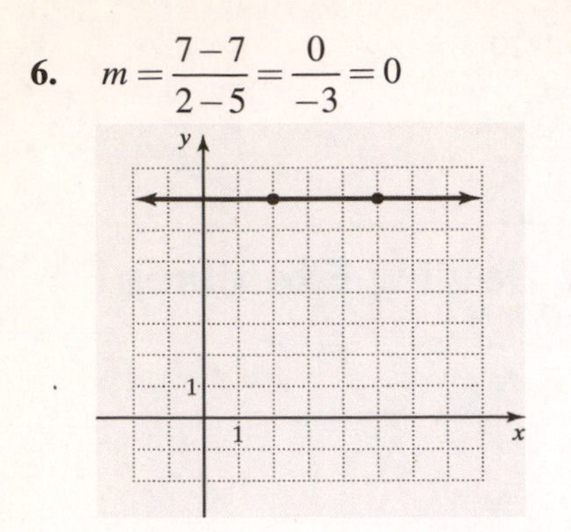

7.

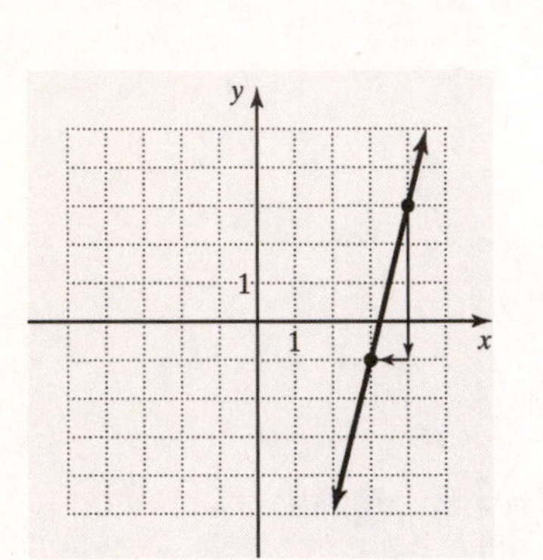

8.

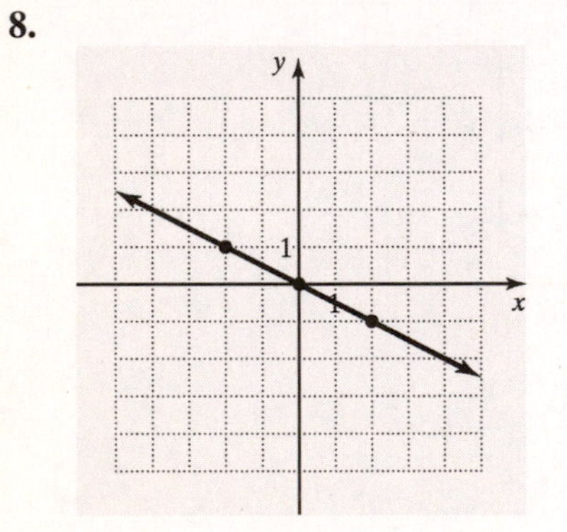

9. Positive slope

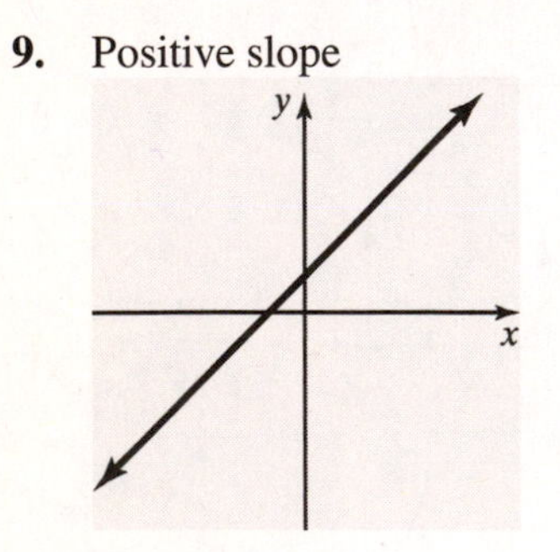

10. Undefined slope

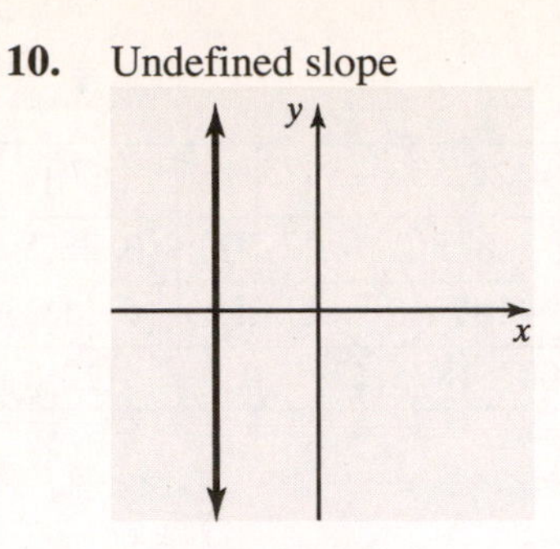

11. Negative slope

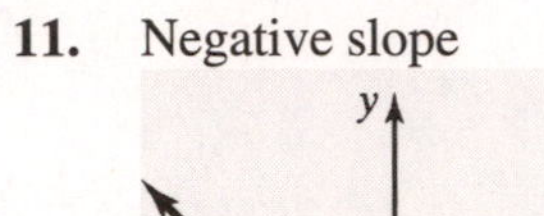
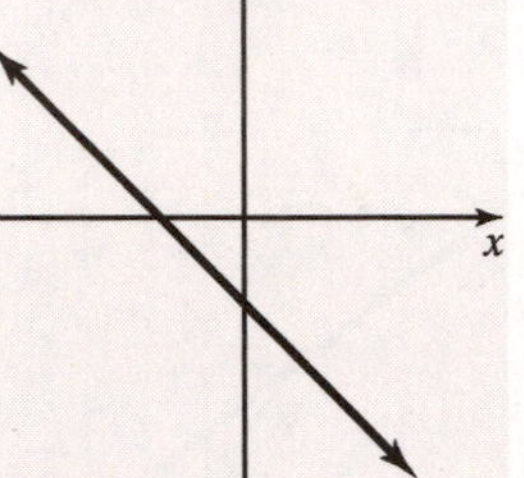

12. Zero slope

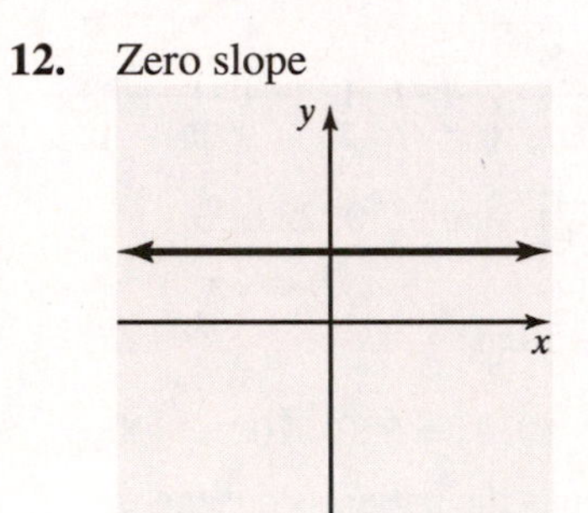

13. $\overleftrightarrow{AB}: \ m = \dfrac{0-0}{3-5} = \dfrac{0}{-2} = 0$

$\overleftrightarrow{CD}: \ m = \dfrac{-2-(-2)}{-3-1} = \dfrac{0}{-4} = 0$

Parallel

14. $\overleftrightarrow{AB}: \ m = \dfrac{9-8}{5-4} = \dfrac{1}{1} = 1$

$\overleftrightarrow{CD}: \ m = \dfrac{-1-(-3)}{0-2} = \dfrac{2}{-2} = -1$

Perpendicular

15. $(30, 0)$

16. $(0, 50)$

17. $y = 2x - 5$

$y = 2(0) - 5 = -5$

$y = 2(1) - 5 = -3$

$0 = 2x - 5$

$5 = 2x$

$\frac{5}{2} = x$

$1 = 2x - 5$

$6 = 2x$

$3 = x$

18. $y = -x + 3$

$y = -(2) + 3 = 1$

$y = -(5) + 3 = -2$

$7 = -x + 3$

$4 = -x$

$-4 = x$

$-5 = -x + 3$

$-8 = -x$

$8 = x$

19. $4x - 3y = -12$

$-3y = -4x - 12$

$y = \frac{4}{3}x + 4$

20. $x + 2y = -6$

$2y = -x - 6$

$y = -\frac{1}{2}x - 3$

21. $y = \frac{1}{2}x$

22. $y = -x + 2$

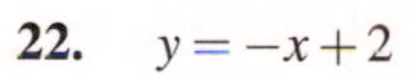

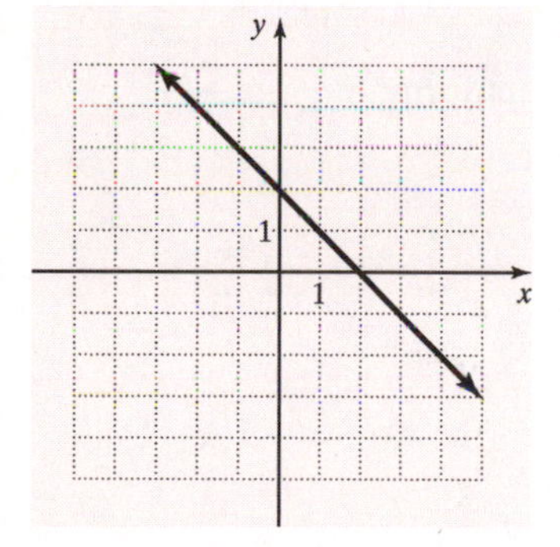

23. $x - y = 10$

$-y = -x + 10$

$y = x - 10$

24. $x + 2y = -1$

$2y = -x - 1$

$y = -\frac{1}{2}x - \frac{1}{2}$

25.

Equation	Slope, m	y-intercept	Type	x-intercept
$y=4x-16$	4	$(0,-16)$	/	$(4,0)$

26.

Equation	Slope, m	y-intercept	Type	x-intercept
$y=-\frac{1}{3}x$	$-\frac{1}{3}$	$(0,0)$	\	$(0,0)$

27. $5x-10y=20$

$$-10y=-5x+20$$
$$y=\frac{1}{2}x-2$$
$$m=\frac{1}{2}$$

The slope of a line perpendicular to this line is -2.

28. $6x-2y=2$

$$-2y=-6x+2$$
$$y=3x-1$$
$$m=3$$

The slope of a line parallel to this line is 3.

29. Point-slope form: $y-5=-(x-3)$

$$y-5=-x+3$$
$$y=-x+8$$

Slope-intercept form: $y=-x+8$

30. $y=0$

31. $m=\frac{5-0}{1-2}=\frac{5}{-1}=-5$

Point-slope form: $y-5=-5(x-1)$

$$y-5=-5x+5$$
$$y=-5x+10$$

Slope-intercept form: $y=-5x+10$

32. $m=\frac{0-1}{-2-3}=\frac{-1}{-5}=\frac{1}{5}$

Point-slope form: $y-1=\frac{1}{5}(x-3)$

$$y-1=\frac{1}{5}x-\frac{3}{5}$$
$$y=\frac{1}{5}x-\frac{3}{5}+1$$
$$y=\frac{1}{5}x-\frac{3}{5}+\frac{5}{5}$$
$$y=\frac{1}{5}x+\frac{2}{5}$$

Slope-intercept form: $y=\frac{1}{5}x+\frac{2}{5}$

33. $m=\frac{3-0}{0-2}=\frac{3}{-2}=-\frac{3}{2}$

$$y=-\frac{3}{2}x+3$$

34. $m=\frac{0-(-3)}{2-0}=\frac{3}{2}$

$$y=\frac{3}{2}x-3$$

35.

$$x+y<1$$
$$(-2)+(7)<1$$
$$5<1$$

No, it is not a solution.

36.

$$2x-3y\geq 14$$
$$2(1)-3(-4)\geq 14$$
$$2+12\geq 14$$
$$14\geq 14$$

Yes, it is a solution.

37.

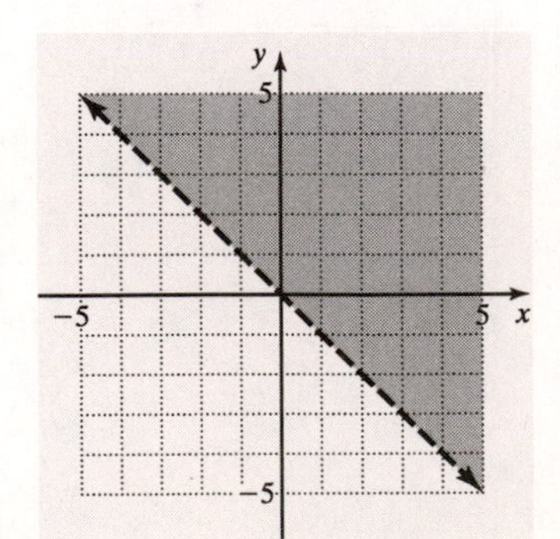

38.

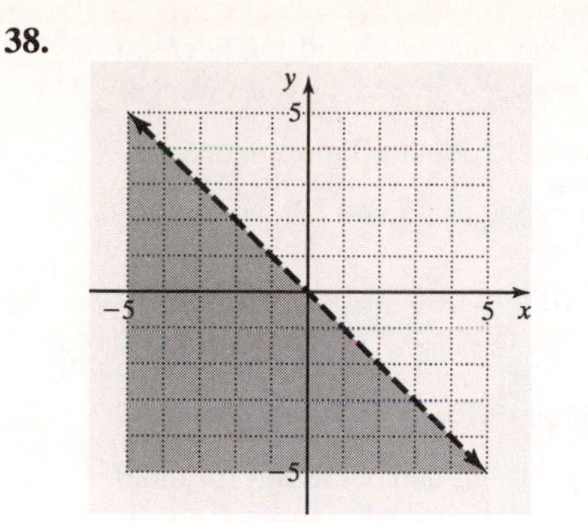

39. $y \le 2x$

Test point: $(1,0)$

$0 \le 2(1)$

$0 \le 2$ True

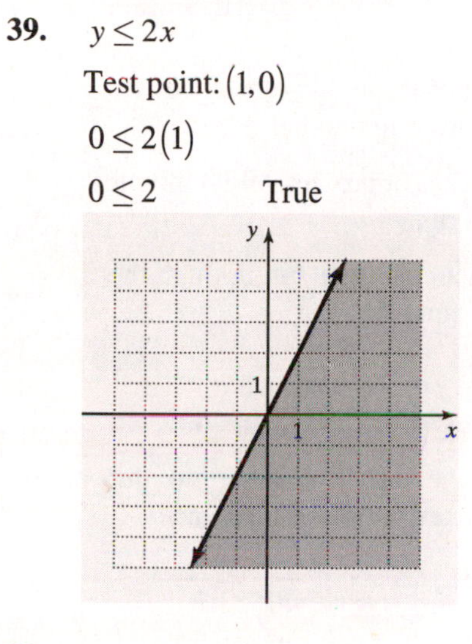

40. $y - x > -1$

Test point: $(0,0)$

$(0) - (0) > -1$

$0 > -1$ True

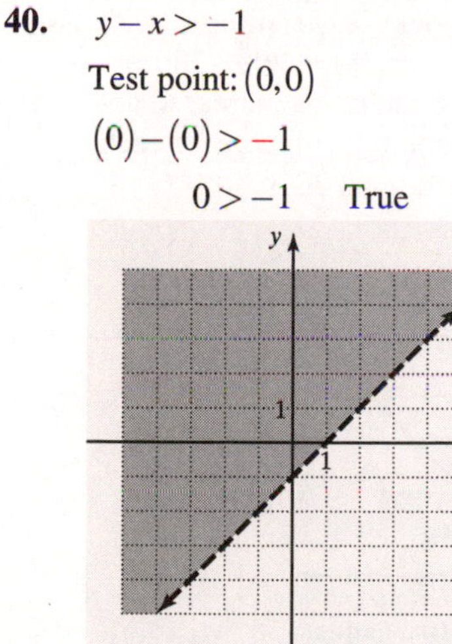

41. A function. No two ordered pairs have the same first element.

42. Not a function (3, –2) and (3, 4) have the same first element and different second elements.

43. Domain: $\{-7,-5,-3,-1\}$.

Range: $\{3\}$.

44. Domain: $\{-6,-4,-3,-2,0,2,3,4,6\}$.

Range: $\{-6,-3,0,1,2,5,8\}$.

45. Domain: $\{-27,-8,-1,0,1,8,27\}$.

Range: $\{-3,-2,-1,0,1,2,3\}$.

46. Domain: $\{20,24,26,30,32,38\}$.

Range: $\{190,228,247,285,304,361\}$.

47. **a.** $f(-9) = \frac{1}{3}(-9) + 6 = -3 + 6 = 3$

b. $f(3.6) = \frac{1}{3}(3.6) + 6 = 1.2 + 6 = 7.2$

c. $f(3a) = \frac{1}{3}(3a) + 6 = a + 6$

d. $f(6a - 12) = \frac{1}{3}(6a - 12) + 6$

$= 2a - 4 + 6 = 2a + 2$

48. **a.** $g(3) = |4(3) - 7| = |12 - 7| = |5| = 5$

b. $g\left(-\frac{3}{4}\right) = \left|4\left(-\frac{3}{4}\right) - 7\right|$

$= |-3 - 7| = |-10| = 10$

c. $g(2n) = |4(2n) - 7| = |8n - 7|$

d. $g\left(\frac{1}{4}n + 1\right) = \left|4\left(\frac{1}{4}n + 1\right) - 7\right|$

$= |n + 4 - 7| = |n - 3|$

49. $f(x) = 4 - \frac{1}{2}x$

x	$f(x) = 4 - \frac{1}{2}x$	(x, y)
0	$4 - \frac{1}{2}(0) = 4$	$(0,4)$
2	$4 - \frac{1}{2}(2) = 3$	$(2,3)$
4	$4 - \frac{1}{2}(4) = 2$	$(4,2)$

49. (continued)

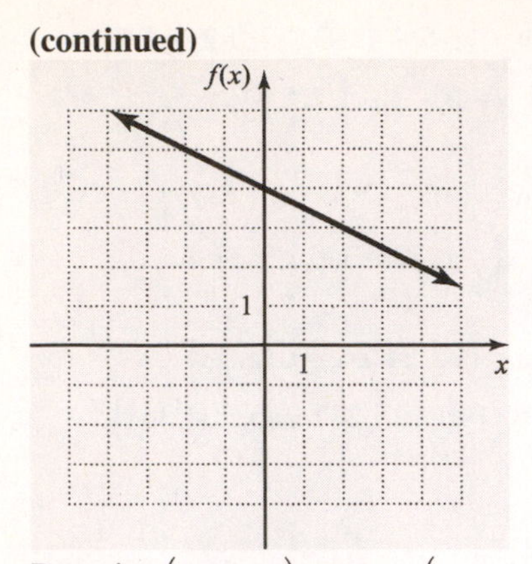

Domain: $(-\infty,\infty)$; range: $(-\infty,\infty)$

50. $g(x)=4x-9$ for $x\geq 1$

x	$g(x)=4x-9$	(x,y)
1	$4(1)-9=-5$	$(1,-5)$
2	$4(2)-9=-1$	$(2,-1)$
3	$4(3)-9=3$	$(3,3)$

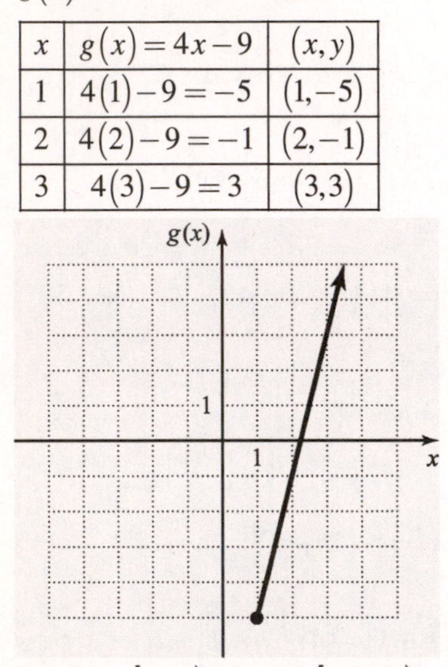

Domain: $[1,\infty)$; range: $[-5,\infty)$

51. No, not a function. The y-axis crosses the graph twice failing the vertical line test.

52. A function. No vertical line will cross the graph more than once.

53. **a.** In 2003 there were approximately 36,000 movie screens.

b. In 2005 there were about 37,750 movie screens and in 2000 there were about 36,250 movie screens. So in 2005 there were about 1,500 more movie screens than in 2000.

c. The number of movie screens in the United States decreased from 2000 to 2001 and then increased each year from 2001 to 2005.

54. **a.** There were about 61 million licensed drivers between 35 and 49 years of age.

b. There were about 20 million drivers between the ages 35 and 39, which is

$\frac{20}{200}=10\%$ of 200 million drivers.

c. It increases for younger age groups, peaks for 40-44, and decreases for older age groups.

55. **a.** The rat runs through the maze in 10 minutes during run number 3.

b. The rat takes approximately 3 minutes on the tenth run.

c. With practice, the rat ran through the maze more quickly.

56. **a.** Insurance expenditures were approximately $700 billion in 2005.

b. In 2000, the approximate ratio of out-of-pocket expenditures to insurance expenditures was about $\frac{180}{405}=\frac{4}{9}$.

c. Both out-of-pocket expenditures (with the exception of the period 2005-2006) and insurance expenditures increased and are expected to continue to increase each year from 2000 through 2010.

57. **a.**

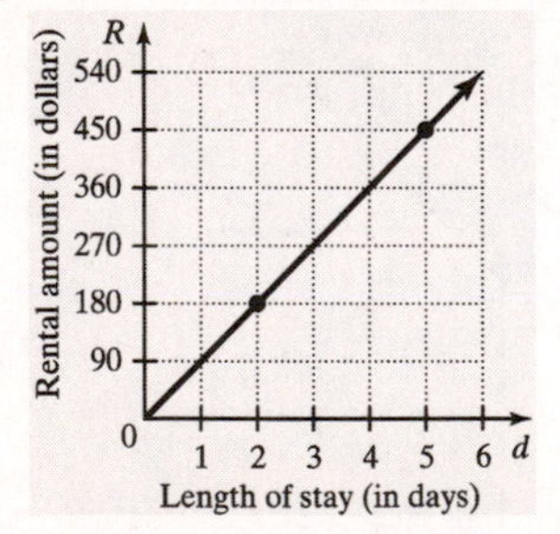

b. The R-intercept is $(0,0)$. The R-intercept means that the cost for renting a room for 0 days is $0.

58. **a.**

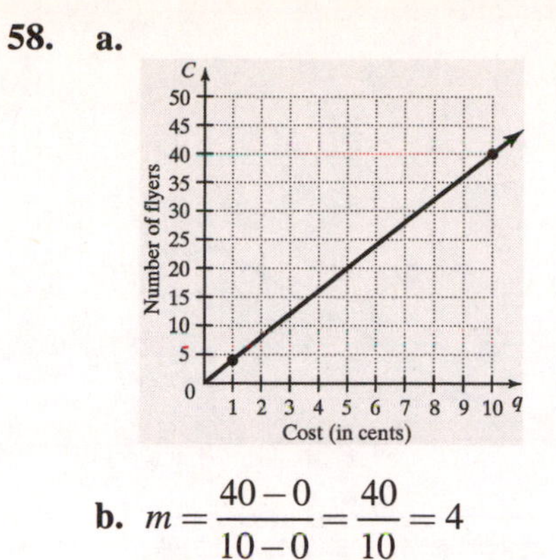

b. $m = \dfrac{40-0}{10-0} = \dfrac{40}{10} = 4$

The slope of the line is 4. The slope represents the rate the print shop charges for each flyer, which is 4 cents.

59. The graph in part (a) could describe this motion. As the child walks toward the wall, the distance between the child and the wall decreases, implying a negative slope. When the child stands still, the distance between the child and the wall remains the same, as indicated by the horizontal line segment. When the child moves toward the wall again, the distance again decreases, implying a negative slope.

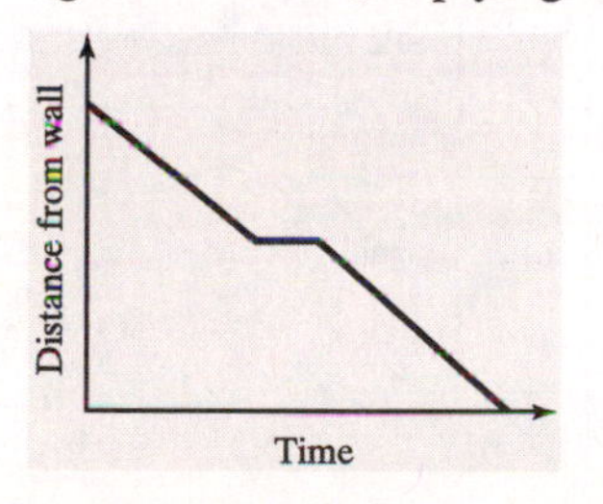

60. In the first part of the flight, the airplane takes off and ascends to a particular altitude (line segment slanting up to the right), then it flies at that same altitude during the second and longest part of the flight (horizontal line segment), and finally in the last part of the flight, it descends and lands (line segment slanting down to the right).

61. **a.** $i = 20{,}000 + 0.09s$

b.

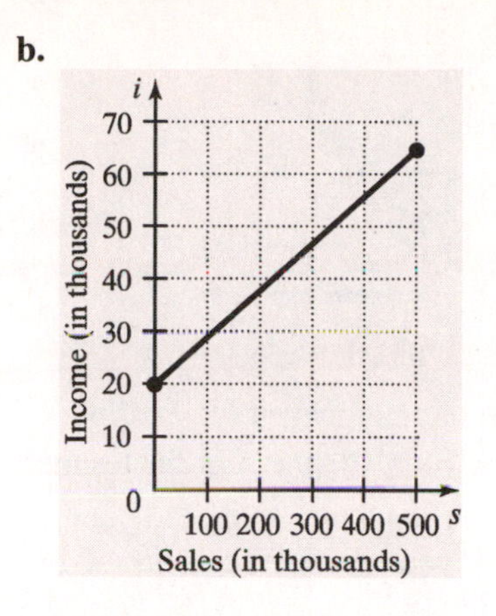

62. **a.**

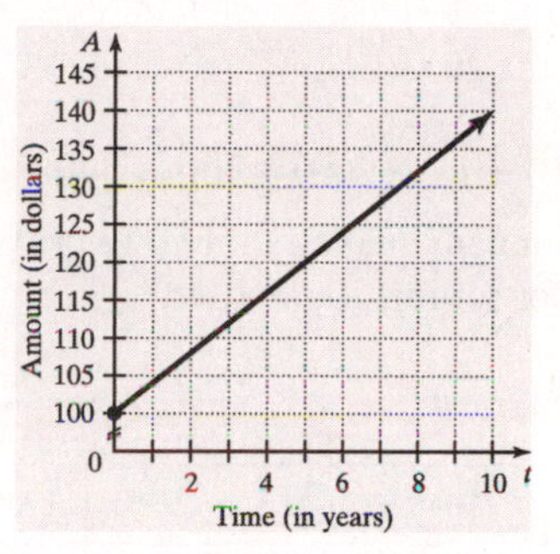

b. The A-intercept of the graph is (0, 100). The A-intercept represents the initial balance in the bank account.

63. **a.** $\dfrac{1}{4}s + \dfrac{1}{2}d < 30$

$s < -2d + 120$

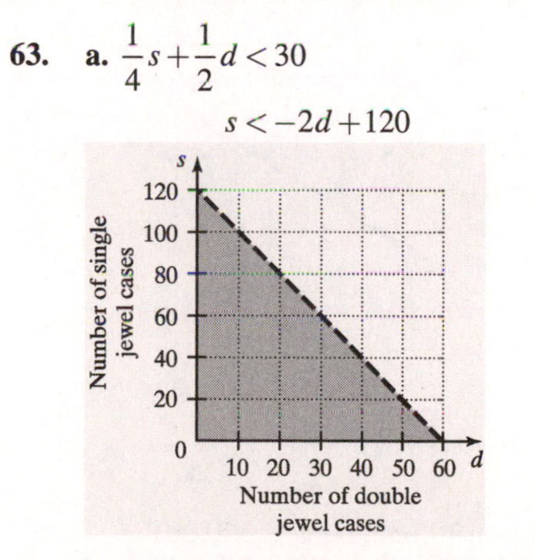

b. Answers may vary. Possible answer: 20 double jewel cases and 30 single jewel cases.

64. **a.**

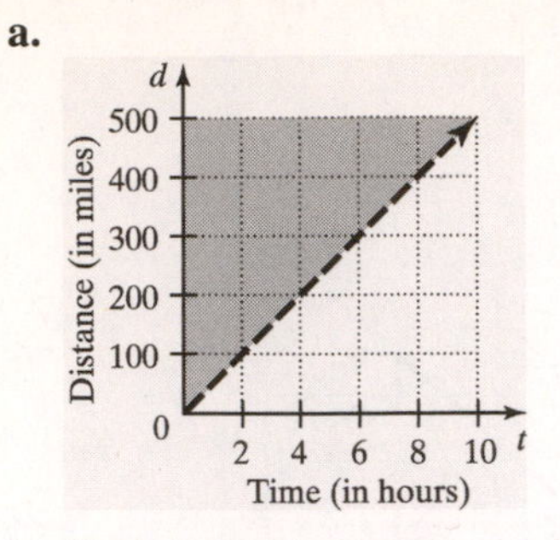

b. Choice of point may vary. Possible answer: $(2, 110)$; the coordinates mean that you caught up and passed your friend if you covered a distance of 110 mi in 2 hr.

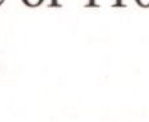

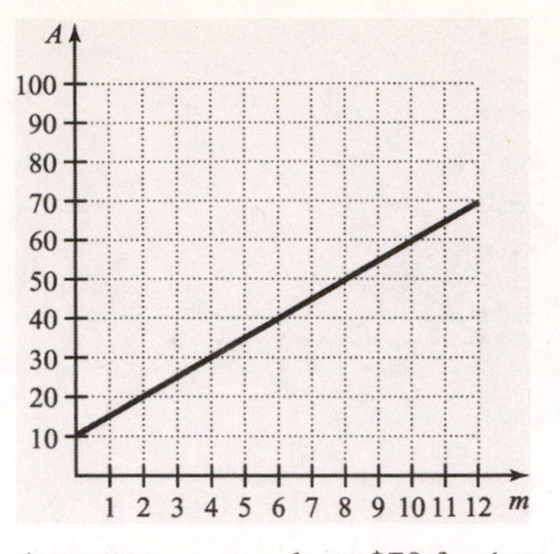

c. A customer pays about \$70 for 1 year of service.

65. **a.** $P(x) = 1.20x + 300$

b. $P(200) = 1.20(200) + 300 = 540$

The company makes a monthly profit of \$540 for selling 200 bottles.

c.

x	$P(x) = 1.20x + 300$	$(x, P(x))$
0	$1.20(0) + 300 = 300$	$(0, 300)$
200	$1.20(200) + 300 = 540$	$(200, 540)$
500	$1.20(500) + 300 = 900$	$(500, 900)$

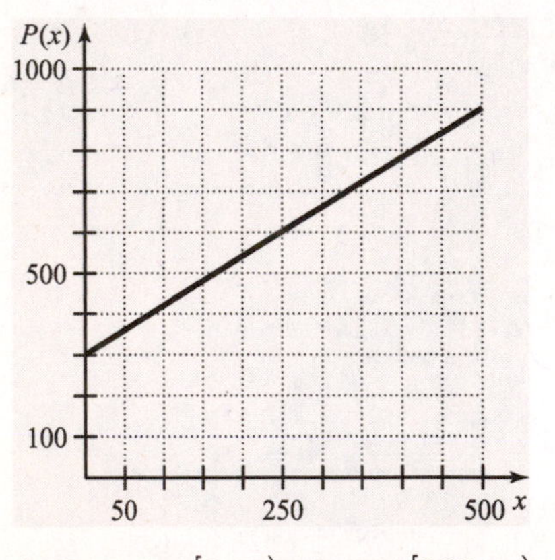

d. Domain: $[0, \infty)$; range: $[300, \infty)$

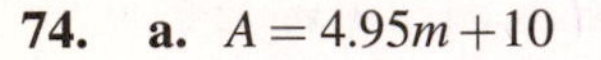

74. **a.** $A = 4.95m + 10$

b.

m	$A = 4.95m + 10$	(x, y)
0	$4.95(0) + 10 = 10$	$(0, 10)$
6	$4.95(6) + 10 = 39.70$	$(6, 39.70)$
10	$4.95(10) + 10 = 59.50$	$(10, 59.50)$

Chapter 3 Posttest

1. In 2004 there were about 13,500,000 students enrolled in public colleges.

2.

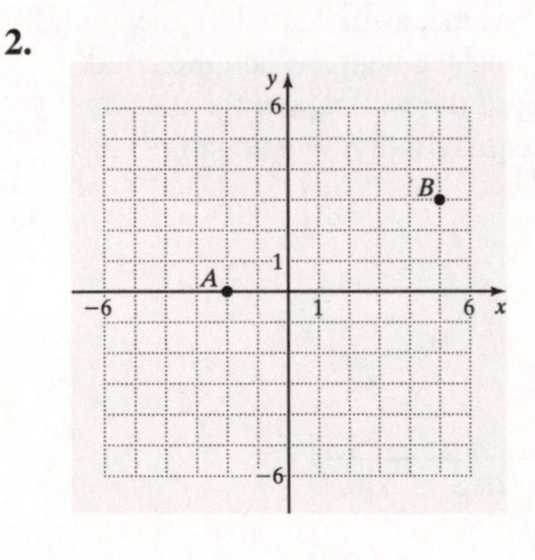

3. II

4. $m = \dfrac{1-(-4)}{8-3} = \dfrac{5}{5} = 1$

5. The graphs are parallel. The slope of $y = 3x + 1$ is 3 and the slope of $y = 3x - 2$ is also 3. Since the slopes of the two lines are equal, their graphs are parallel.

6. $\overleftrightarrow{AB}: \ m = \dfrac{8-1}{2-0} = \dfrac{7}{2}$

$\overleftrightarrow{CD}: \ m = \dfrac{4-6}{7-0} = \dfrac{-2}{7} = -\dfrac{2}{7}$

The slope of $\overleftrightarrow{AB}$ is $\dfrac{7}{2}$. The slope of $\overleftrightarrow{CD}$ is $-\dfrac{2}{7}$. $\overleftrightarrow{AB}$ is perpendicular to $\overleftrightarrow{CD}$, since the product of their slopes is -1.

7. x-intercept: $(-5,0)$; y-intercept: $(0,2)$

8. The slope is positive. As the number of miles driven increases, the rental cost increases.

9. $y = -3x+1$

$y = -3(-3)+1 = 10$

$y = -3(5)+1 = -14$

$0 = -3x+1$

$3x = 1$

$x = \dfrac{1}{3}$

$-2 = -3x+1$

$-3 = -3x$

$1 = x$

10. $y = 2$

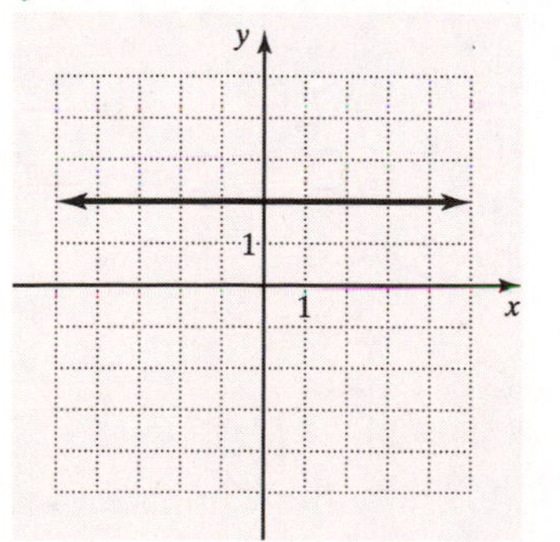

11. $y = -x-3$

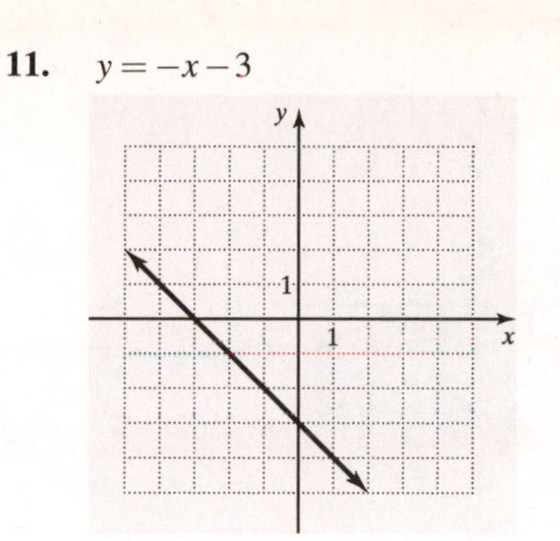

12. $3x-2y = 6$

$-2y = -3x+6$

$y = \dfrac{3}{2}x-3$

13. Slope: 3; y-intercept: $(0,1)$

14. $2x-y = 5$

$-y = -2x+5$

$y = 2x-5$

15. $y = -x-3$

16. $m = \dfrac{5-2}{3-(-4)} = \dfrac{3}{7}$

Point-slope form: $y-5 = \dfrac{3}{7}(x-3)$

$y-5 = \dfrac{3}{7}(x-3)$

$y-5 = \dfrac{3}{7}x - \dfrac{9}{7}$

$y = \dfrac{3}{7}x - \dfrac{9}{7} + \dfrac{35}{7}$

$y = \dfrac{3}{7}x + \dfrac{26}{7}$

Slope-intercept form: $y = \dfrac{3}{7}x + \dfrac{26}{7}$

17.

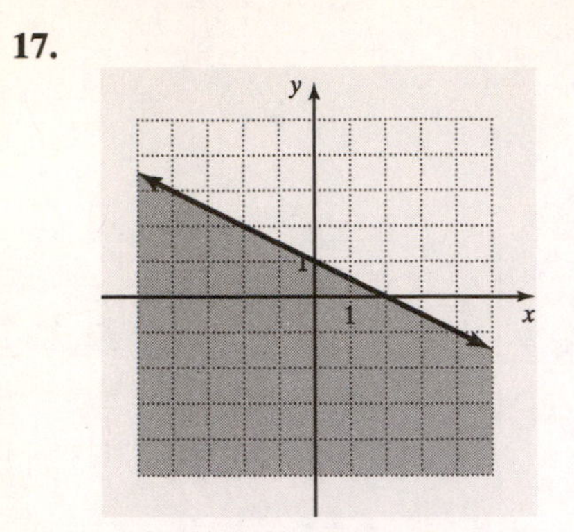

18. 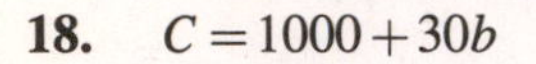 $C = 1000 + 30b$

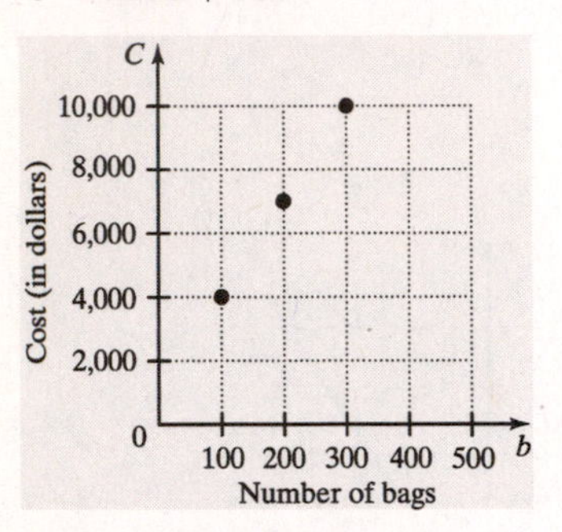

19. $0.04x + 0.08y \geq 500$

Test point $(1,0)$

$0.04(1) + 0.08(0) \geq 500$

$0.04 \geq 500$ False

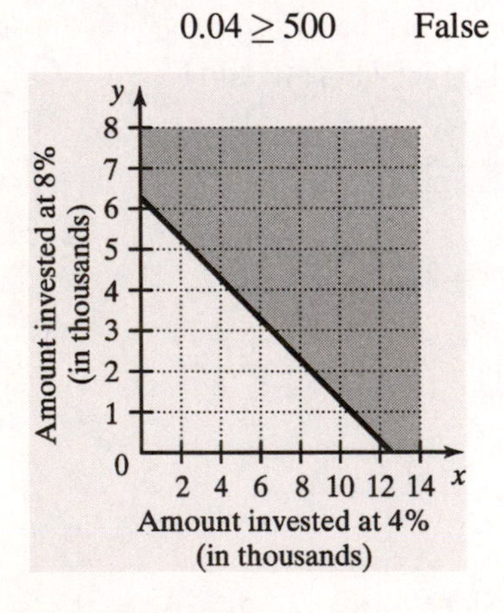

20. $g(x) = -2x + 11$

$g(a+3) = -2(a+3) + 11$

$g(a+3) = -2a - 6 + 11$

$g(a+3) = -2a + 5$

Chapter 4

SYSTEMS OF EQUATIONS AND INEQUALITIES

Pretest

1. **a.**
$$(5)+2(0)=5$$
$$5=5 \quad \text{True}$$
$$5(5)-(0)=-8$$
$$25=-8 \quad \text{False}$$

Not a solution

b.
$$(-1)+2(3)=5$$
$$-1+6=5$$
$$5=5 \quad \text{True}$$
$$5(-1)-(3)=-8$$
$$-5-3=-8$$
$$-8=-8 \quad \text{True}$$

A solution

c.
$$(1)+2(-3)=5$$
$$1-6=5$$
$$-5=5 \quad \text{False}$$
$$5(1)-(-3)=-8$$
$$5+3=-8$$
$$8=-8 \quad \text{False}$$

Not a solution

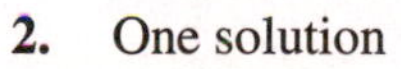

2. One solution

3.

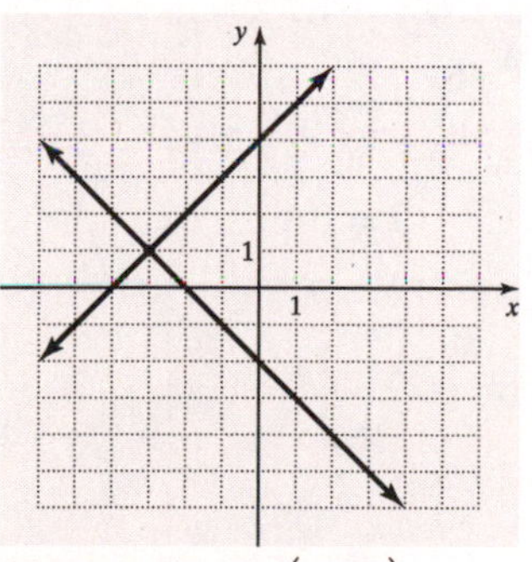

The solution is $(-3,1)$.

4.

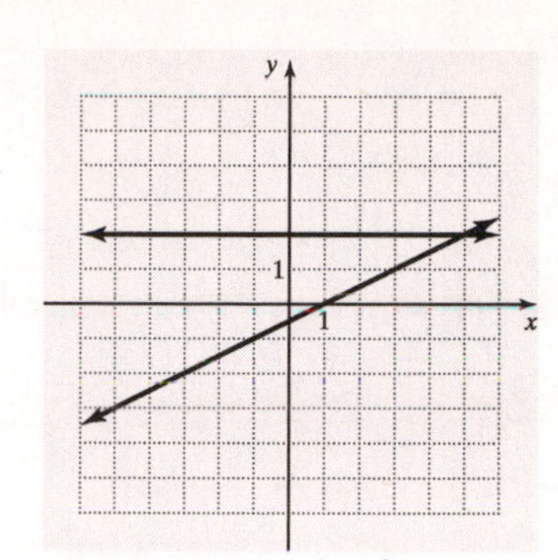

The solution is $(5,2)$.

5.

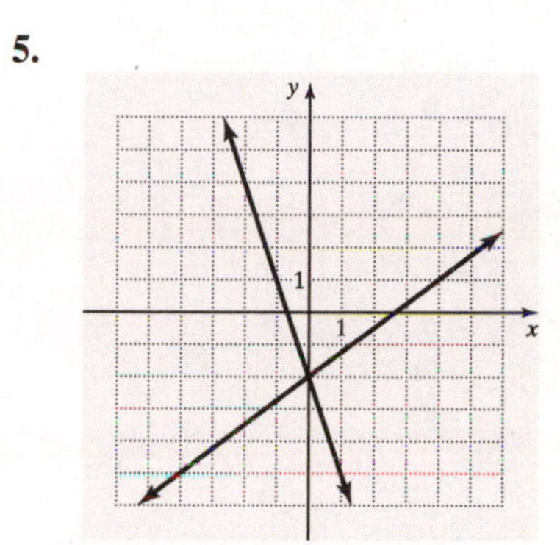

The solution is $(0,-2)$.

6. **(1)**
$$x-2y=7$$
$$x-2(-11-x)=7$$
$$x+22+2x=7$$
$$3x=-15$$
$$x=-5$$

(2)
$$y=-11-(-5)$$
$$y=-6$$

$(-5,-6)$

7. **(1)**
$$x-2y=0$$
$$x=2y$$

(2)
$$7(2y)-4y=10$$
$$14y-4y=10$$
$$10y=10$$
$$y=1$$

(1)
$$x-2(1)=0$$
$$x-2=0$$
$$x=2$$

$(2,1)$

8. **(1)**

$$a+3b=-2$$
$$(2b-7)+3b=-2$$
$$5b-7=-2$$
$$5b=5$$
$$b=1$$

(2)

$$a=2(1)-7$$
$$a=2-7$$
$$a=-5$$
$$a=-5, b=1$$

9.

$$2x+5y=-13$$
$$-2x+6y=-20$$
$$11y=-33$$
$$y=-3$$

$$2x+5(-3)=-13$$
$$2x-15=-13$$
$$2x=2$$
$$x=1$$
$$(1,-3)$$

10.

$$6x-8y=36$$
$$1.5x-2y=96$$

$$6x-8y=36$$
$$-6x+8y=-36$$
$$0=0$$

Infinitely many solutions.

11.

$$3x-7y=-19$$
$$2x+3y=-5$$

$$-6x+14y=38$$
$$6x+9y=-15$$
$$23y=23$$
$$y=1$$

$$3x-7(1)=-19$$
$$3x-7=-19$$
$$3x=-12$$
$$x=-4$$
$$(-4,1)$$

12.

$$-3n+5m=10$$
$$-4m=-2(n+1)$$
$$-4m=-2n-2$$
$$2n-4m=-2$$

$$-6n+10m=20$$
$$6n-12m=-6$$
$$-2m=14$$
$$m=-7$$

$$-3n+5(-7)=10$$
$$-3n-35=10$$
$$-3n=45$$
$$n=-15$$

$$n=-15, m=-7$$

13. **(1)**

$$x+9=2y$$
$$x=2y-9$$

(2)

$$8y-13=4(2y-9)$$
$$8y-13=8y-36$$
$$-13=-36$$

No solution

14.

$$6x+10y-12=0$$
$$3x+2.5y-6=0$$

$$6x+10y=12$$
$$3x+2.5=6$$

$$6x+10y=12$$
$$-6x-5y=-12$$
$$5y=0$$
$$y=0$$
$$6x+10(0)-12=0$$
$$6x-12=0$$
$$6x=12$$
$$x=2$$
$$(2,0)$$

15. $6.5x+10y=17{,}650$
$x+y=1975$

$6.5x+10y=17{,}650$
$-10x-10y=-19{,}750$
$-3.5x=-2{,}100$
$x=600$

$600+y=1975$
$y=1375$

600 tickets were sold before 5:00 P.M. and 1375 tickets were sold after 5:00 P.M.

16. **(1)** $x=3y$

(2) $2x-50=5y$
$2(3y)-50=5y$
$6y-50=5y$
$y=50$

Fifty $5 tickets were printed.

17. $x+y=200{,}000$
$0.05x+0.06y=11{,}200$

$-0.05x-0.05y=10{,}000$
$0.05x+0.06y=11{,}200$
$0.01y=1200$
$y=120{,}000$

$x+120{,}000=200{,}000$
$x=80{,}000$

$80,000 was invested in the fund at 5% interest, and $120,000 was invested in the fund at 6%.

18. $2(b+c)=13$
$2(b-c)=11$

$2b+2c=13$
$2b-2c=11$
$4b=24$
$b=6$

$2(6+c)=13$
$12+2c=13$
$2c=1$
$c=0.5$

The speed of the boat was 6 mph, and the speed of the current was 0.5 mph.

19. $y<\frac{1}{2}x-1$

The boundary line is dashed

x	$y=\frac{1}{2}x-1$	(x,y)
0	$\frac{1}{2}(0)-1=-1$	$(0,-1)$
2	$\frac{1}{2}(2)-1=0$	$(2,0)$

The test point is (0, 0).

$0\overset{?}{<}\frac{1}{2}(0)-1\Rightarrow 0\not<-1$

Shade the half-plane not containing (0, 0).

$y\ge-3x+1$

The boundary line is solid.

x	$y=-3x+1$	(x,y)
0	$-3(0)+1=1$	$(0,1)$
1	$-3(1)+1=-2$	$(1,-2)$

Test point is $(0,0)$.

$0\overset{?}{\ge}-3(0)+1\Rightarrow 0\not\ge 1$; shade the half-plane not containing $(0,0)$.

$y\ge-4$

The boundary line is solid.

x	$y=-4$	(x,y)
0	-4	$(0,-4)$
2	-4	$(2,-4)$

The test point is $(0,0)$.

$0\ge-4$, shade the half-plane containing $(0,0)$.

The solutions to the system are all the points that lie in all three overlapping shaded regions.

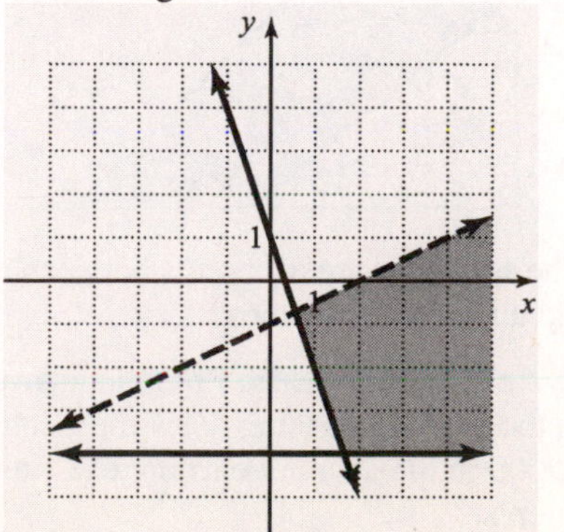

20. **a.** Let x = the amount saved at 5% and let y = the amount saved at 6%.

$x+y \le 10000$

$0.05x+0.06y \ge 300$

b. $x+y \le 10000$

$y \le -x+10000$

The boundary line is solid.

x	$y=-x+10,000$	(x,y)
0	$-(0)+10,000=10,000$	(0, 10,000)
10,000	$-(10,000)+10,000=0$	(10,000, 0)

Test point is $(0,0)$.

$(0)+(0)\overset{?}{\le}10,000$

$0 \le 10,000$, shade the half plane containing $(0,0)$.

$0.05x+0.06y \ge 300$

$0.06y \ge -0.05x+300$

$y \ge -\dfrac{5x}{6}+5000$

The boundary line is solid.

x	$y=-\dfrac{5x}{6}+5000$	(x,y)
0	$-\dfrac{5}{6}(0)+5000=5000$	(0,5000)
6000	$-\dfrac{5}{6}(6000)+5000=0$	(6000,0)

The test point is (0, 0).

$0.05(0)+0.06(0)\overset{?}{\ge}300 \Rightarrow 0 \not\ge 300$

The half plane not containing (0, 0) is shaded.

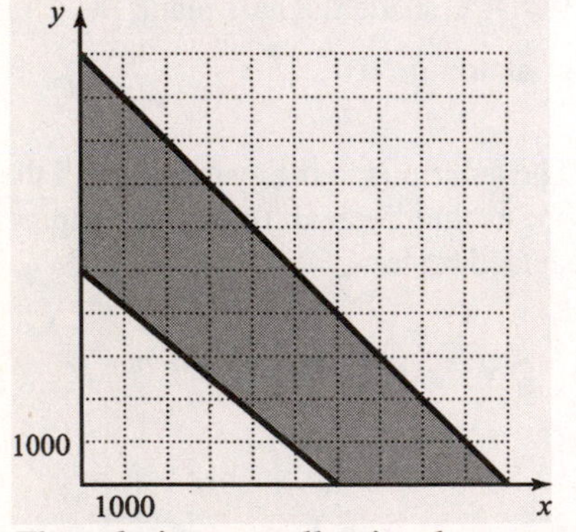

The solutions are all points between and including the two lines.

c. Answers may vary. Possible answer: $2500 in the account earning 5% simple interest and $5000 in the account earning 6% simple interest.

4.1 Solving Systems of Linear Equations by Graphing.

Practice

1. **a.** $3(0)+2(2.5)=5$

$5=5$ True

$4(0)-2(2.5)=-5$

$-5=-5$ True

Yes, it is a solution of the system.

b. $3(1)+2(-1)=5$

$3-2=5$

$1=5$ False

$4(1)-2(-1)=-5$

$4+2=-5$

$6=-5$ False

No, it is not a solution of the system.

2. **a.**

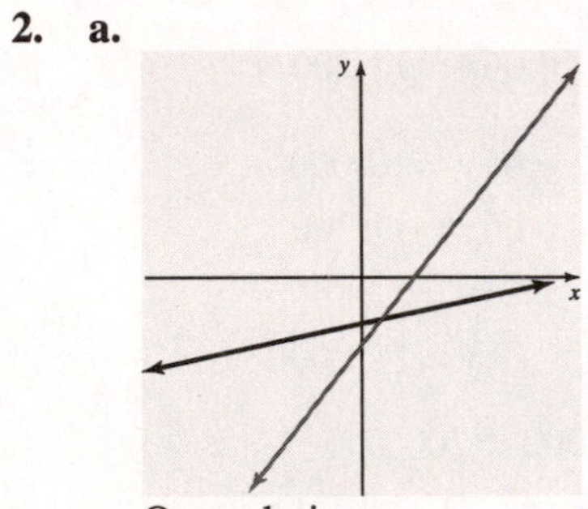

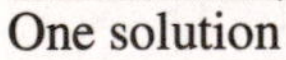

One solution

b.

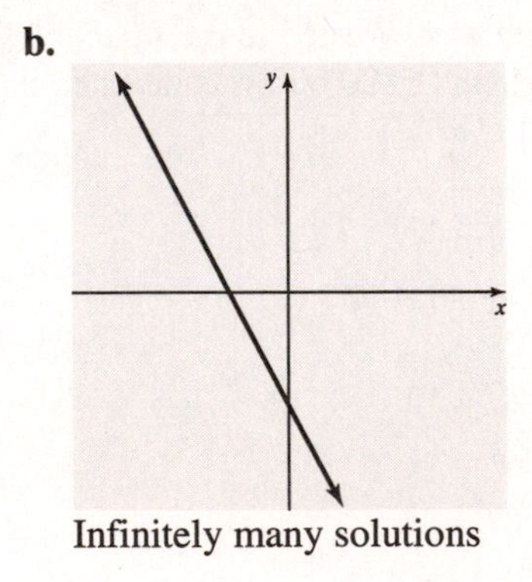

Infinitely many solutions

2. **c.**

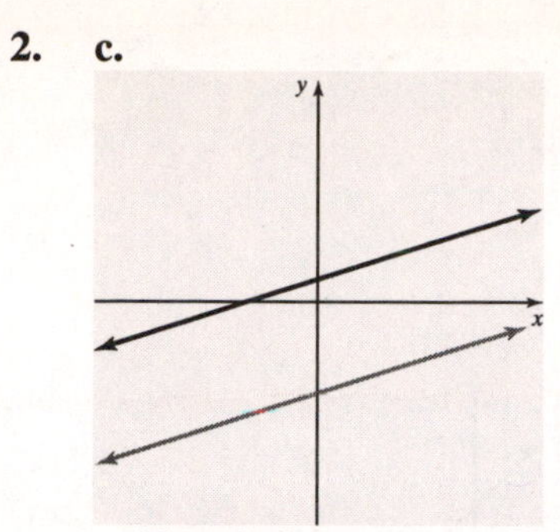

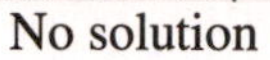

No solution

3.

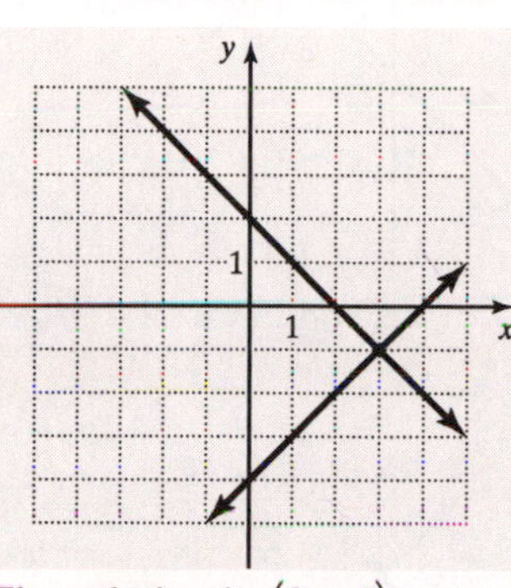

The solution is $(3,-1)$.

4.

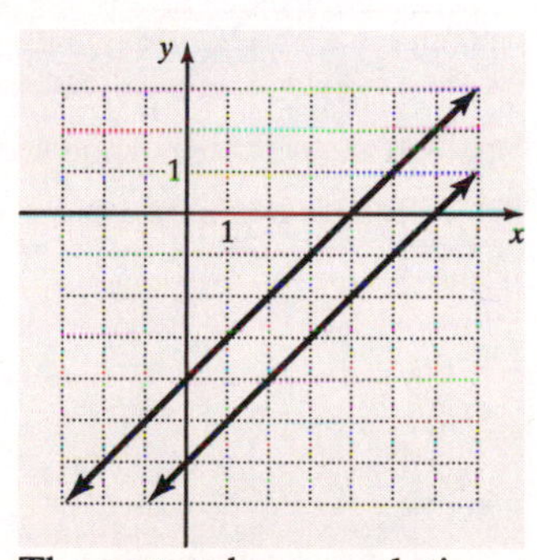

The system has no solution.

5.

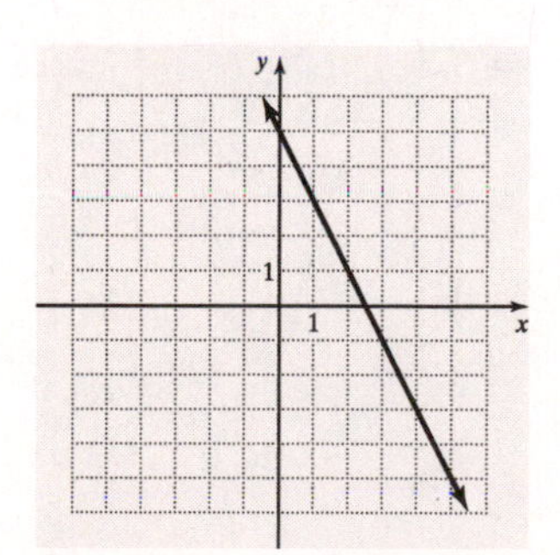

The system has infinitely many solutions.

6. **a.** $m+v=1150$

$v=m-100$

b.

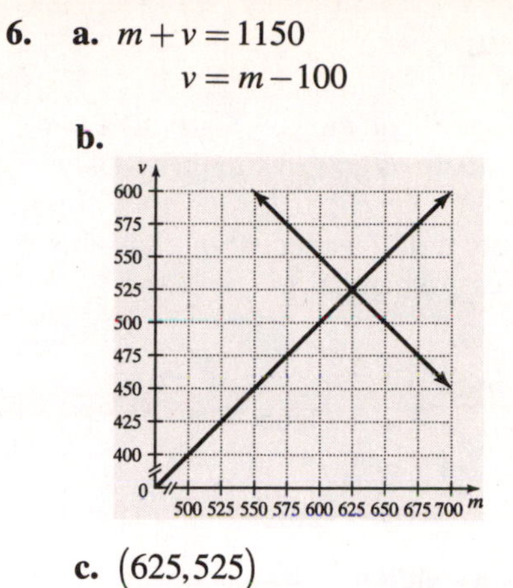

c. $(625,525)$

d. The point of intersection indicates that she got a score of 625 on her test of math skills and a score of 525 on her test of verbal skills.

7. **a.** $y=1.50+x$

b. $y=3x$

c.

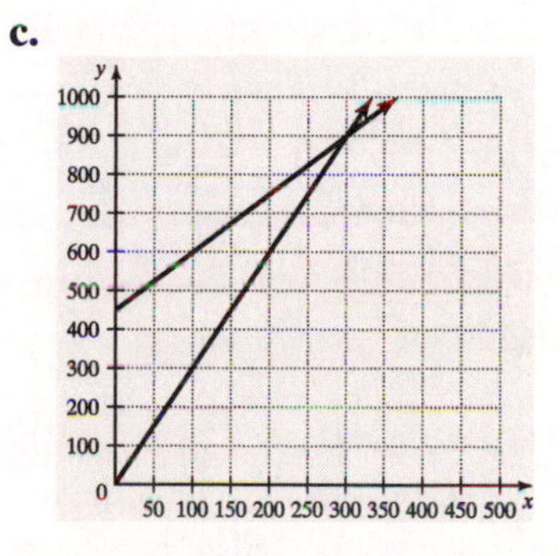

d. The break-even point is (300, 900). So when 300 newsletters are printed, the cost of printing the newsletter and the income from sales will be the same, \$900.

8.

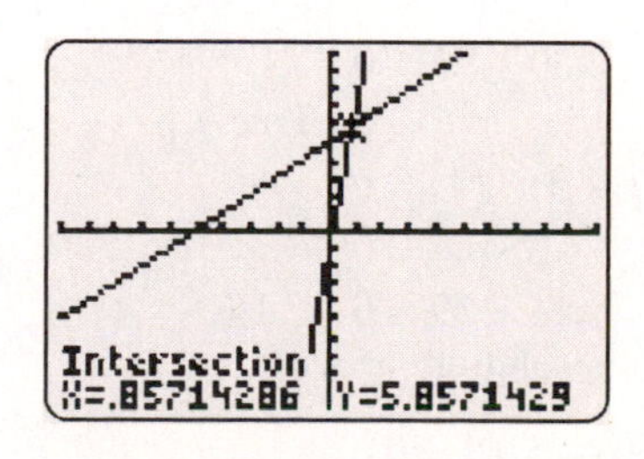

The approximate solution is (0.857, 5.857).

Exercises

1. A <u>system of equations</u> is a group of two or more equations solved simultaneously.

3. If a system of linear equations has no solution, the lines <u>are parallel</u>.

5. **a.**
$$(0)+(3)=3$$
$$3=3 \quad \text{True}$$
$$2(0)-(3)=6$$
$$-3=6 \quad \text{False}$$

Not a solution

b.
$$(3)+(3)=3$$
$$6=3 \quad \text{False}$$
$$2(3)-(3)=6$$
$$6-3=6$$
$$3=6 \quad \text{False}$$

Not a solution

c.
$$(3)+(0)=3$$
$$3=3 \quad \text{True}$$
$$2(3)-(0)=6$$
$$6=6$$
$$6=6 \quad \text{True}$$

A solution

7. **a.**
$$4(1)+5(7)=0$$
$$4+35=0$$
$$39=0 \quad \text{False}$$
$$7(1)-(7)=0$$
$$7-7=0$$
$$0=0 \quad \text{True}$$

Not a solution

b.
$$4(-5)+5(4)=0$$
$$-20+20=0$$
$$0=0 \quad \text{True}$$
$$7(-5)-(4)=0$$
$$-35-4=0$$
$$-39=0 \quad \text{False}$$

Not a solution

c.
$$4(0)+5(0)=0$$
$$0+0=0$$
$$0=0 \quad \text{True}$$
$$7(0)-(0)=0$$
$$0-0=0$$
$$0=0 \quad \text{True}$$

A solution

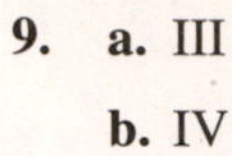

9. **a.** III

b. IV

c. II

d. I

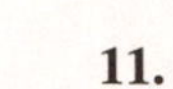

11.

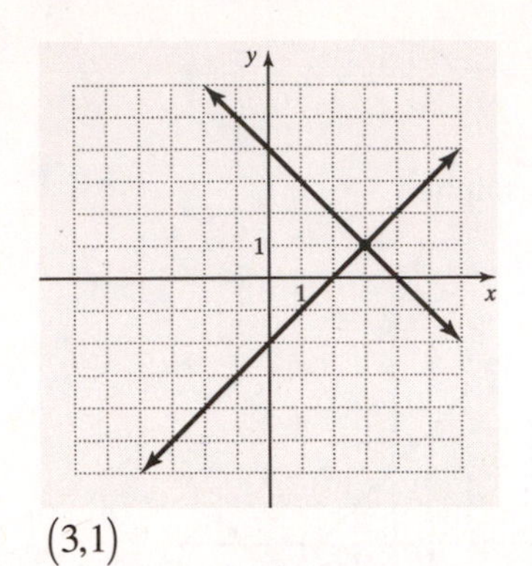

$(3,1)$

13.

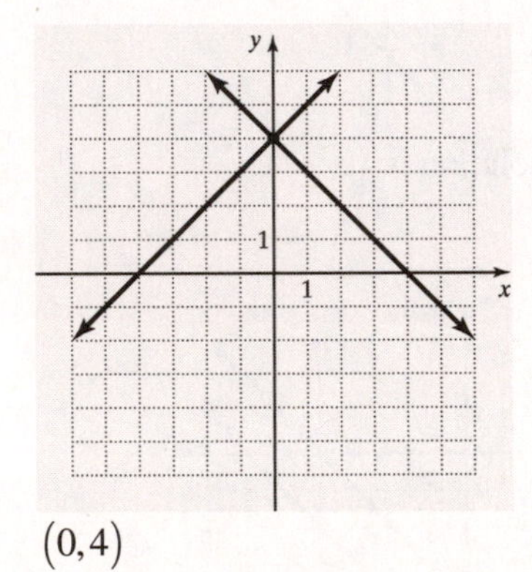

$(0,4)$

15.

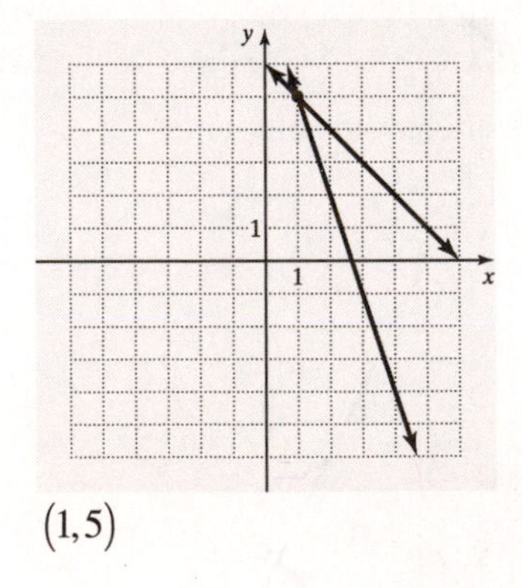

$(1,5)$

17.

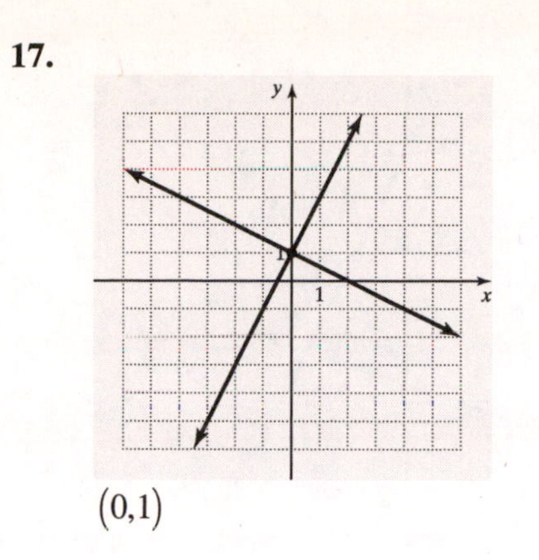

$(0,1)$

19.

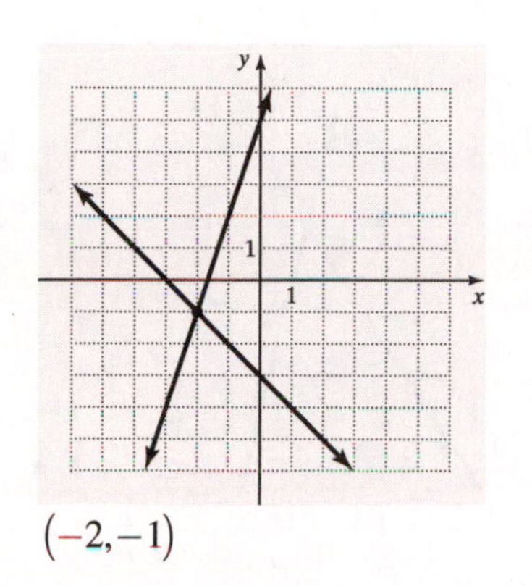

$(-2,-1)$

21.

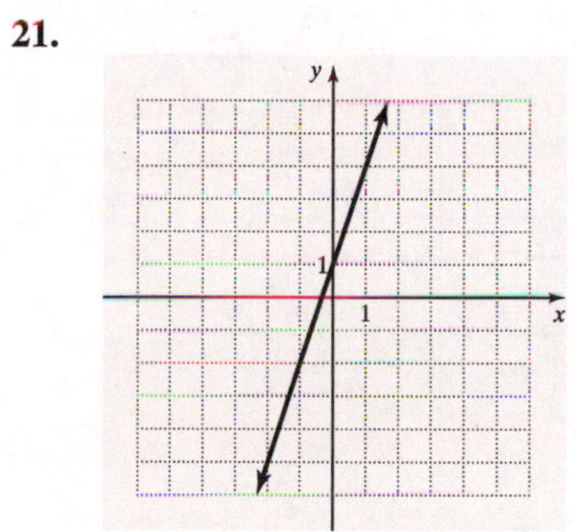

Infinitely many solutions

23.

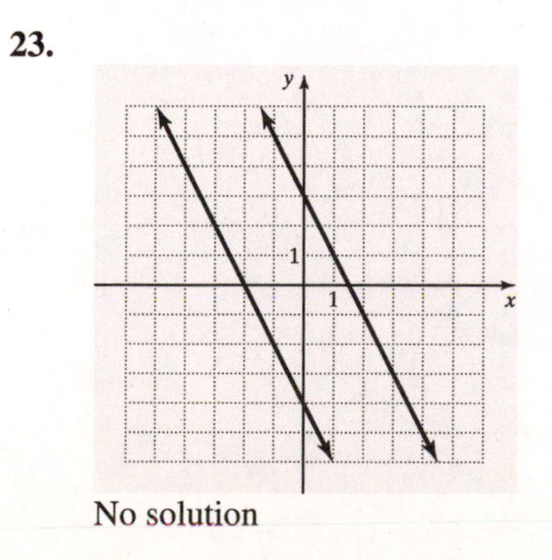

No solution

25.

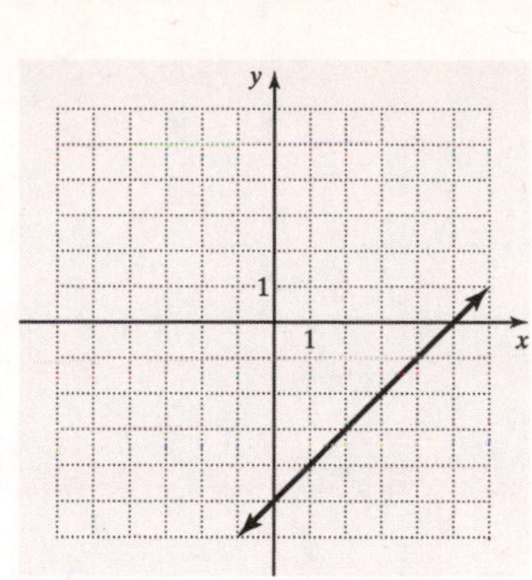

Infinitely many solutions

27.

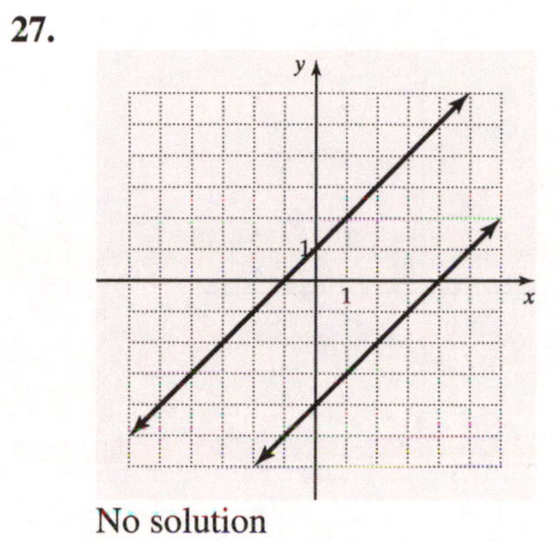

No solution

29.

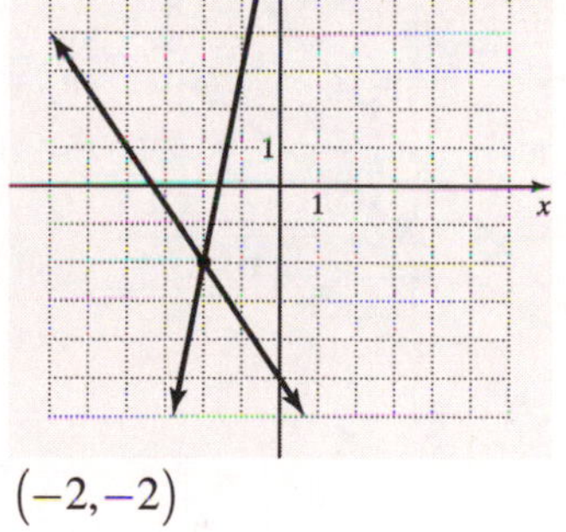

$(-2,-2)$

31.

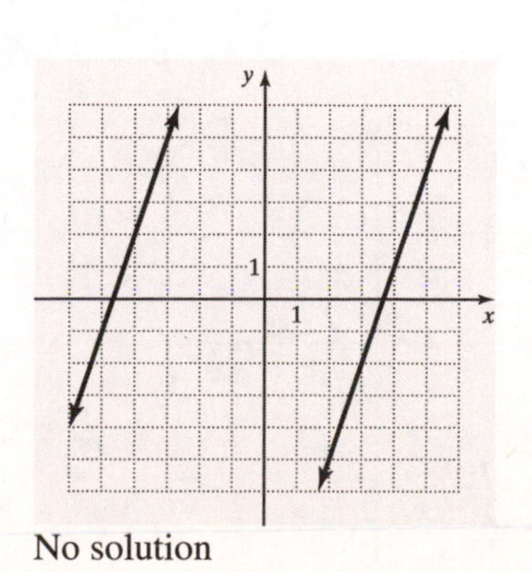

No solution

33.

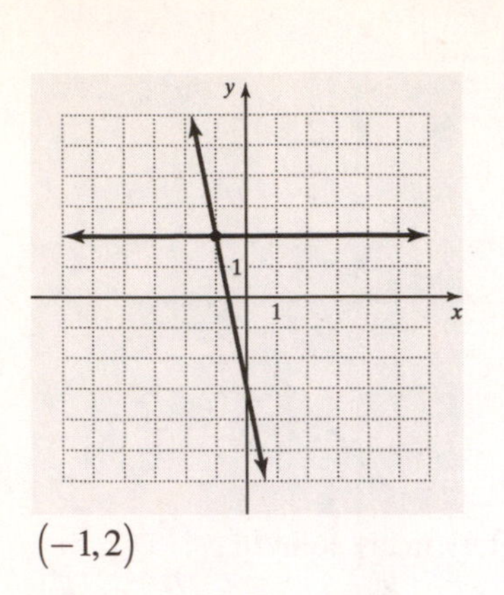

$(-1,2)$

35.

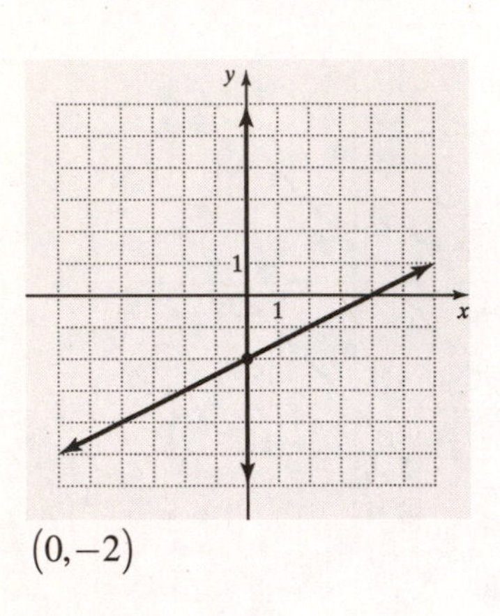

$(0,-2)$

37.

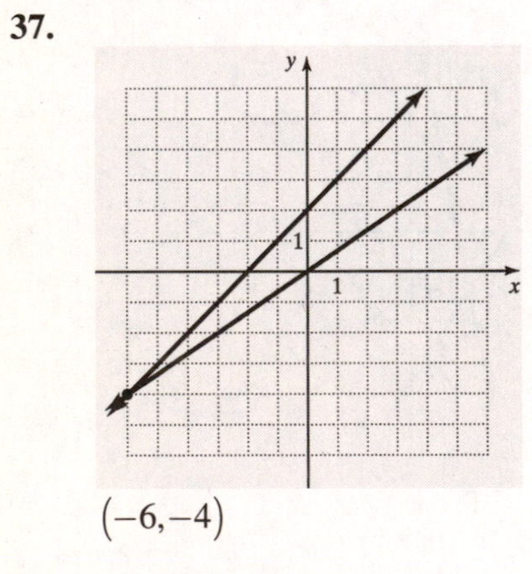

$(-6,-4)$

39.

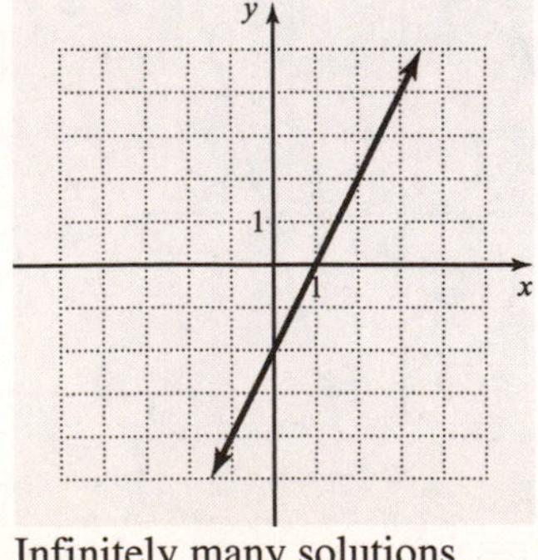

Infinitely many solutions

41.

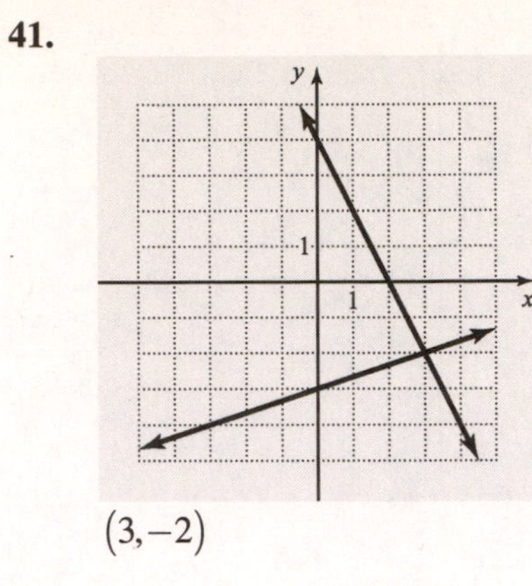

$(3,-2)$

43.

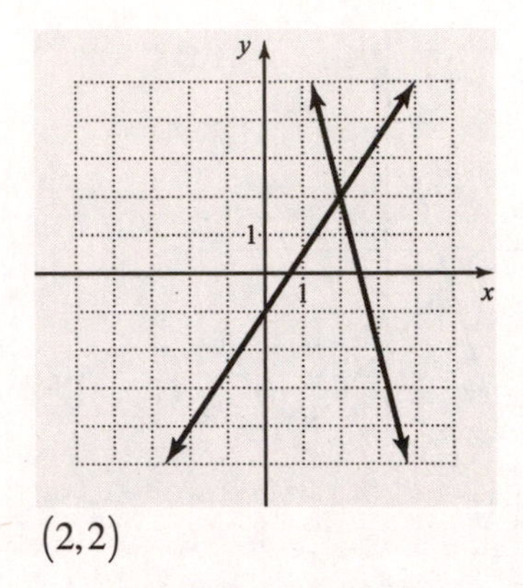

$(2,2)$

45.

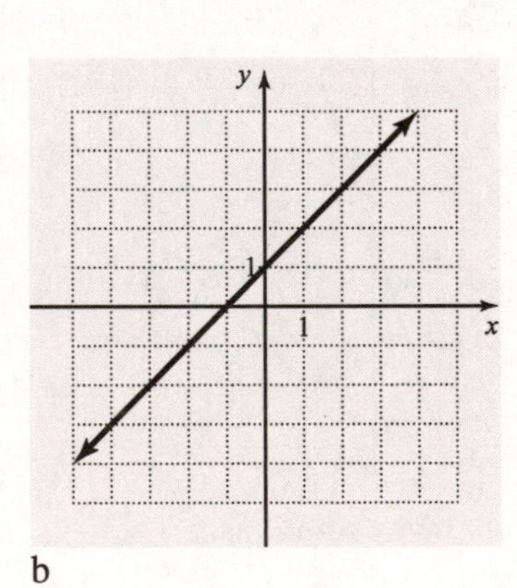

b

47. **a.** $x+y=57{,}000$
$y=x+3000$

b.

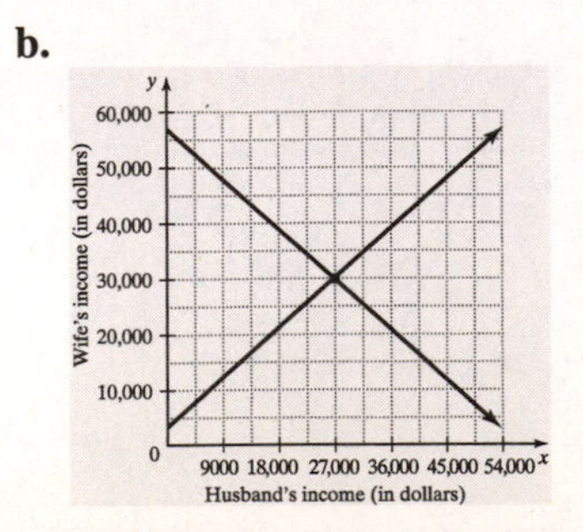

c. The husband made $27,000 and the wife made $30,000.

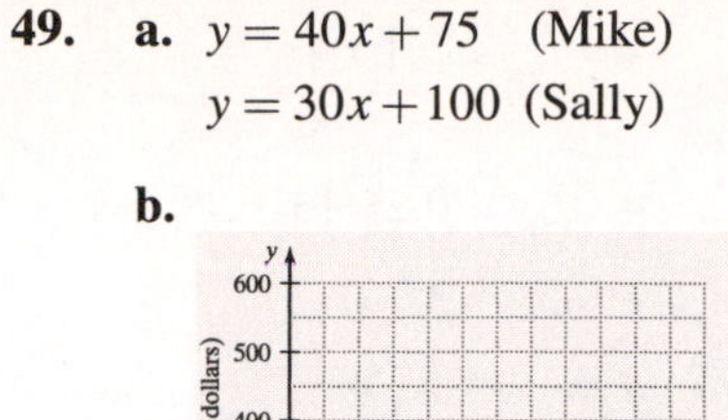

49. **a.** $y = 40x + 75$ (Mike)

$y = 30x + 100$ (Sally)

b.

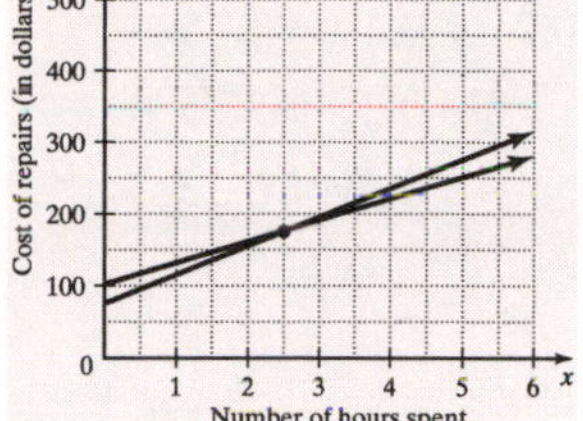

c. The plumbers would charge the same amount for 2.5 hr of work.

d. Sally charges less.

51.

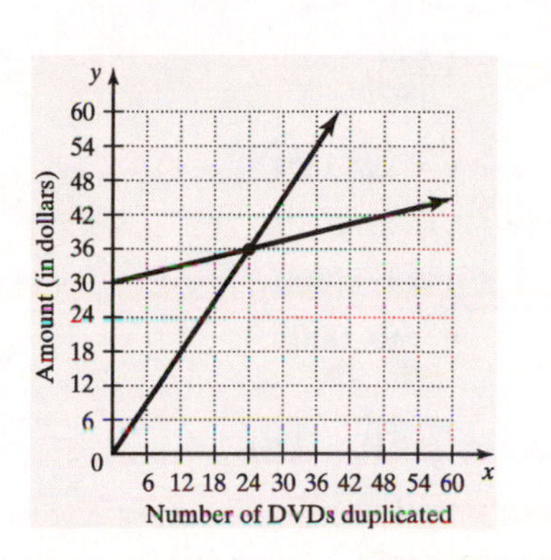

The break-even point for duplicating DVDs is (24, 36).

53.

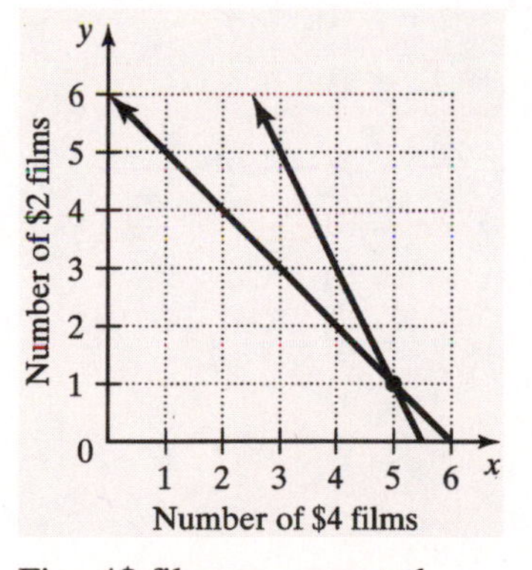

Five 4$ films were rented.

55.

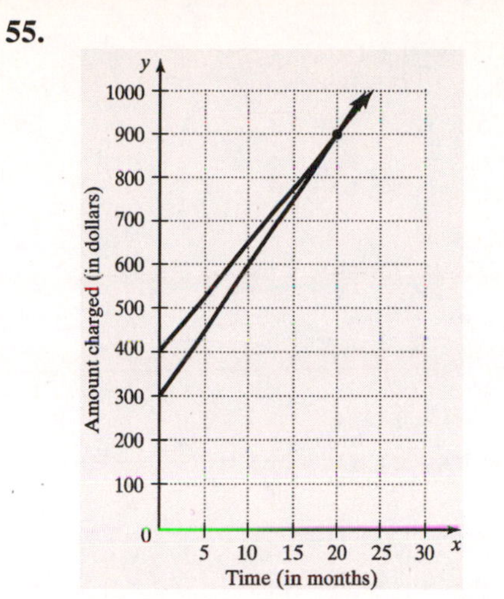

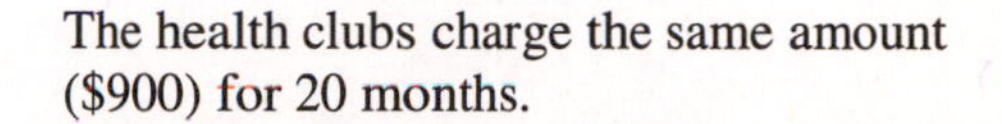

The health clubs charge the same amount ($900) for 20 months.

4.2 Solving Systems of Linear Equations by Substitution

Practice

1. **(1)**

$$\begin{aligned} x - y &= 7 \\ (-y+1) - y &= 7 \\ -y + 1 - y &= 7 \\ -2y + 1 &= 7 \\ -2y &= 6 \\ y &= -3 \end{aligned}$$

(2) $x = -(-3) + 1$

$x = 3 + 1$

$x = 4$

$(4, -3)$

2. **(2)**

$$\begin{aligned} 2m + 3n &= 7 \\ 2(-5n+1) + 3n &= 7 \\ -10n + 2 + 3n &= 7 \\ -7n + 2 &= 7 \\ -7n &= 5 \\ n &= -\frac{5}{7} \end{aligned}$$

2. **(1)** $m=-5\left(-\frac{5}{7}\right)+1$

$m=\frac{25}{7}+\frac{7}{7}$

$m=\frac{32}{7}$

$m=\frac{32}{7},\ n=-\frac{5}{7}$

3. **(2)** $6x-y=1$

$y=6x-1$

(1) $2x-7(6x-1)=7$

$2x-42x+7=7$

$-40x=0$

$x=0$

(2) $6(0)-y=1$

$y=-1$

$(0,-1)$

4. **(1)** $3x+y=10$

$y=10-3x$

(2) $-6x-2(10-3x)=1$

$-6x-20+6x=1$

$-20=1$

No solution

5. **(2)** $10x+5y=20$

$10x+5(-2x+4)=20$

$10x-10x+20=20$

$20=20$

Infinitely many solutions

6. **a.** $c=35n+20$ (TV Deal)

$c=25n+30$ (Movie Deal)

b. $35n+20=25n+30$

$10n=10$

$n=1$

$c=35(1)+20$

$c=55$

$n=1$ and $c=\$55$

c. The cost is the same ($55) for both cable deals if you sign up for one month.

7. **(1)** $x+y=15$

$x=15-y$

(2) $0.2x+0.5y=0.25(15)$

$0.2(15-y)+0.5y=3.75$

$3-0.2y+0.5y=3.75$

$3+0.3y=3.75$

$0.3y=0.75$

$y=2.5$

(1) $x+2.5=15$

$x=12.5$

12.5 oz of the 20% copper alloy and 2.5 oz of the 50% copper alloy are required.

8. **(1)** $0.04x=0.05y$

$x=1.25y$

(2) $x+y=198{,}000$

$1.25y+y=198{,}000$

$2.25y=198{,}000$

$y=88{,}0000$

(1) $0.04x=0.05(88{,}000)$

$0.04x=440$

$x=110{,}000$

The manager put $110,000 in the investment that pays 4% and $88,000 in the investment that pays 5%.

Exercises

1. **(1)** $x+y=10$

$x+(2x+1)=10$

$3x+1=10$

$3x=9$

$x=3$

(2) $y=2(3)+1$

$y=6+1$

$y=7$

Check:

(1) $3+7=10$

(2) $7=2(3)+1$

$7=6+1$

$7=7$

$(3,7)$

3. **(1)** $-x-7=-3x-15$
$2x-7=-15$
$2x=-8$
$x=-4$

(2) $y=-(-4)-7$
$y=4-7$
$y=-3$

Check:

(1) $-3=-3(-4)-15$
$-3=12-15$
$-3=-3$

(2) $-3=-(-4)-7$
$-3=4-7$
$-3=-3$

$(-4,-3)$

5. **(1)** $-(-3y)-y=8$
$3y-y=8$
$2y=8$
$y=4$

(2) $x=-3(4)$
$x=-12$

Check:

(1) $-(-12)-4=8$
$12-4=8$
$8=8$

(2) $-12=-3(4)$
$-12=-12$

$(-12,4)$

7. **(1)** $4(2)+2y=10$
$8+2y=10$
$2y=2$
$y=1$

Check:

(1) $4(2)+2(1)=10$
$8+2=10$
$10=10$

$(2,1)$

9. **(1)** $-x+20y=0$
$x=20y$

(2) $20y-y=0$
$19y=0$
$y=0$

(1) $-x+20(0)=0$
$-x=0$
$x=0$

Check:

(1) $-0+20(0)=0$
$0=0$

(2) $0-0=0$
$0=0$

$(0,0)$

11. **(2)** $2x+y=0$
$y=-2x$

(1) $6x+4(-2x)=2$
$6x-8x=2$
$-2x=2$
$x=-1$

(2) $2(-1)+y=0$
$-2+y=0$
$y=2$

Check:

(1) $6(-1)+4(2)=2$
$-6+8=2$
$2=2$

(2) $2(-1)+2=0$
$-2+2=0$
$0=0$

$(-1,2)$

13. **(2)** $x+2y=-6$

$x=-2y-6$

(1) $3(-2y-6)+5y=-12$

$-6y-18+5y=-12$

$-y-18=-12$

$-y=6$

$y=-6$

(2) $x+2(-6)=-6$

$x-12=-6$

$x=6$

Check:

(1) $3(6)+5(-6)=-12$

$18-30=-12$

$-12=-12$

(2) $6+2(-6)=-6$

$6-12=-6$

$-6=-6$

$(6,-6)$

15. **(3)** $x+3y=6$

$x=-3y+6$

Check:

(1) $2(-3y+6)+6y=12$

$-6y+12+6y=12$

$12=12$

Infinitely many solutions

17. **(2)** $3x-y=11$

$y=3x-11$

(1) $7x-3(3x-11)=26$

$7x-9x+33=26$

$-2x=-7$

$x=\frac{7}{2}$

17. **(2)** $3\left(\frac{7}{2}\right)-y=11$

$\frac{21}{2}-y=11$

$-y=11-\frac{21}{2}$

$-y=\frac{22}{2}-\frac{21}{2}$

$-y=\frac{1}{2}$

$y=-\frac{1}{2}$

Check:

(1) $7\left(\frac{7}{2}\right)-3\left(-\frac{1}{2}\right)=26$

$\frac{49}{2}+\frac{3}{2}=26$

$\frac{52}{2}=26$

$26=26$

(2) $3\left(\frac{7}{2}\right)-\left(-\frac{1}{2}\right)=11$

$\frac{21}{2}+\frac{1}{2}=11$

$\frac{22}{2}=11$

$11=11$

$\left(\frac{7}{2},-\frac{1}{2}\right)$

19. **(1)** $8x+2(-4x+1)=-1$

$8x-8x+2=-1$

$2=-1$

No solution

21. **(1)** $6x-2(3x-1)=2$

$6x-6x+2=2$

$2=2$

Infinitely many solutions

23. **(1)** $2(20-2n)+4n=-22$

$40-4n+4n=-22$

$40=-22$

No solution

25. (1) $p+2q=13$

$p=-2q+13$

(2) $q+7=4(-2q+13)$

$q+7=-8q+52$

$9q=45$

$q=5$

(1) $p+2(5)=13$

$p+10=13$

$p=3$

Check:

(1) $3+2(5)=13$

$3+10=13$

$13=13$

(2) $5+7=4(3)$

$12=12$

$p=3,\ q=5$

27. (1) $s-3t+5=0$

$s=3t-5$

(2) $-4(3t-5)+t-9=0$

$-12t+20+t-9=0$

$-11t+11=0$

$-11t=-11$

$t=1$

(1) $s-3(1)+5=0$

$s+2=0$

$s=-2$

Check:

(1) $-2-3(1)+5=0$

$-2-3+5=0$

$-5+5=0$

$0=0$

(2) $-4(-2)+1-9=0$

$8+1-9=0$

$9-9=0$

$0=0$

$s=-2,\ t=1$

29. (1) $-3x-3=2x+12$

$-5x=15$

$x=-3$

(2) $y=-3(-3)-3$

$y=9-3$

$y=6$

Check:

(1) $6=2(-3)+12$

$6=-6+12$

$6=6$

(2) $6=-3(-3)-3$

$6=9-3$

$6=6$

$(-3,6)$

31. (1) $6x+2(-3x+2)=2$

$6x-6x+4=2$

$4=2$

No solution

33. (2) $-3x+y=5$

$y=3x+5$

(1) $5x+3(3x+5)=-6$

$5x+9x+15=-6$

$14x+15=-6$

$14x=-21$

$x=\frac{-21}{14}$

$x=-\frac{3}{2}$

(2) $-3\left(-\frac{3}{2}\right)+y=5$

$\frac{9}{2}+y=5$

$y=5-\frac{9}{2}$

$y=\frac{10}{2}-\frac{9}{2}$

$y=\frac{1}{2}$

33. (continued)
Check:

(1) $5\left(-\frac{3}{2}\right)+3\left(\frac{1}{2}\right)=-6$

$-\frac{15}{2}+\frac{3}{2}=-6$

$-\frac{12}{2}=-6$

$-6=-6$

(2) $-3\left(-\frac{3}{2}\right)+\frac{1}{2}=5$

$\frac{9}{2}+\frac{1}{2}=5$

$\frac{10}{2}=5$

$5=5$

$\left(-\frac{3}{2},\frac{1}{2}\right)$

35. **a.** (1) $c=1.25m+3$

(2) $c=1.5m+2$

b. (1) $1.5m+2=1.25m+3$

$0.25m=1$

$m=4$

(2) $c=1.5(4)+2$

$c=6+2$

$c=8$

$m=4,\ c=8$

The solution indicates that both companies charge the same amount ($8) for a 4-mi taxi ride.

37. (1) $310f+210d=44{,}120$

(2) $f+d=172$

$d=172-f$

(1) $310f+210(172-f)=44{,}120$ 80

$310f+36{,}120-210f=44{,}120$

$100f=8000$

$f=80$

full-price tickets were sold.

39. (1) $x+y=10$

$x=10-y$

(2) $0.3x+0.7y=0.6(10)$

$0.3(10-y)+0.7y=6$

$3-0.3y+0.7y=6$

$3+0.4y=6$

$0.4y=3$

$y=7.5$

(1) $x+7.5=10$

$x=2.5$

She can combine 2.5 liters of the antiseptic that is 30% alcohol with 7.5 liters of the antiseptic that is 70% alcohol to get the desired concentration.

41. (1) $x+y=150$

$y=150-x$

(2) $0.05x+0.8y=0.5(150)$

$0.05x+0.8(150-x)=75$

$0.05x+120-0.8x=75$

$-0.75x=-45$

$x=60$

(1) $60+y=150$

$y=90$

$0.05(60)=3$

$0.8(90)=72$

There were 3 women in one department and 72 women in the other department.

43. (1) $x+y=5000$

$x=5000-y$

(2) $0.06x+0.07y=310$

$0.06(5000-y)+0.07y=310$

$300-0.06y+0.07y=310$

$300+0.01y=310$

$0.01y=10$

$y=1000$

(1) $x+1000=5000$

$x=4000$

The loan at 6% was $4000 and the loan at 7% was $1000.

45. **(1)** $x+y=40{,}000$

$y=40{,}000-x$

(2) $0.07x+0.09y=3140$

$0.07x+0.09(40{,}000-x)=3140$

$0.07x+3600-0.09x=3140$

$3600-0.02x=3140$

$-0.02x=-460$

$x=23{,}0000$

(1) $y=40{,}000-23{,}0000$

$y=17{,}000$

\$23,000 was invested at 7% and \$17,000 was invested at 9%.

4.3 Solving Systems of Linear Equations by Elimination

Practice

1. $x+y=6$

$x-y=-10$

$2x=-4$

$x=-2$

$-2+y=6$

$y=8$

$(-2,8)$

2. $4x+3y=-7$

$5x+3y=-5$

$-4x-3y=7$

$5x+3y=-5$

$x=2$

$4(2)+3y=-7$

$8+3y=-7$

$3y=-15$

$y=-5$

$(2,-5)$

3. $x-3y=-18$

$5x+2y=12$

$2x-6y=-36$

$15x+6y=36$

$17x=0$

$x=0$

$0-3y=-18$

$-3y=-18$

$y=6$

$(0,6)$

4. $5x-7y=24$

$3x-5y=16$

$-15x+21y=-72$

$15x-25y=80$

$-4y=8$

$y=-2$

$5x-7(-2)=24$

$5x+14=24$

$5x=10$

$x=2$

$(2,-2)$

5. $-2x+5y=20$

$3x=7y-26$

$-2x+5y=20$

$3x-7y=-26$

$-6x+15y=60$

$6x-14y=-52$

$y=8$

$-2x+5(8)=20$

$-2x+40=20$

$-2x=-20$

$x=10$

$(10,8)$

6.
$$3x=4+y$$
$$9x-3y=12$$

$$3x-y=4$$
$$9x-3y=12$$

$$-9x+3y=-12$$
$$9x-3y=12$$
$$0=0$$

Infinitely many solutions

7.
$$2(w+c)=80$$
$$2(w-c)=40$$

$$2w+2c=80$$
$$2w-2c=40$$
$$4w=120$$
$$w=30$$

$$2(30+c)=80$$
$$60+2c=80$$
$$2c=20$$
$$c=10$$

The whale's speed in calm water is 30 mph and the speed of the current is 10 mph.

8.
$$a+c=175$$
$$10a+6c=1450$$

$$-10a-10c=-1750$$
$$10a+6c=1450$$
$$-4c=-300$$
$$c=75$$

$$a+75=175$$
$$a=100$$

100 adults and 75 children attended the game.

Exercises

1.
$$x+y=3$$
$$x-y=7$$
$$2x=10$$
$$x=5$$

$$5+y=3$$
$$y=-2$$

$$(5,-2)$$

3.
$$x+y=-4$$
$$-x+3y=-6$$
$$4y=-10$$
$$y=-\frac{10}{4}=-\frac{5}{2}$$

$$x+\left(-\frac{5}{2}\right)=-4$$
$$x-\frac{5}{2}=-4$$
$$x=-4+\frac{5}{2}$$
$$x=-\frac{8}{2}+\frac{5}{2}$$
$$x=-\frac{3}{2}$$

$$\left(-\frac{3}{2},-\frac{5}{2}\right)$$

5.
$$10p-q=-14$$
$$-4p+q=-4$$
$$6p=-18$$
$$p=-3$$

$$10(-3)-q=-14$$
$$-30-q=-14$$
$$q=-16$$
$$p=-3,\ q=-16$$

7.
$$3x+y=-3$$
$$4x+y=-4$$

$$3x+y=-3$$
$$-4x-y=4$$
$$-x=1$$
$$x=-1$$

$$3(-1)+y=-3$$
$$-3+y=-3$$
$$y=0$$

$$(-1,0)$$

9. $3x+5y=10$
$3x+5y=11$

$-3x-5y=-10$
$3x+5y=11$
$0=1$

No solution

11. $9x+6y=-15$
$-3x-2y=5$

$9x+6y=-15$
$-9x-6y=15$
$0=0$

Infinitely many solutions

13. $5x+2y=-9$
$-5x+2y=11$
$4y=2$
$y=\frac{1}{2}$

$5x+2\left(\frac{1}{2}\right)=-9$
$5x+1=-9$
$5x=-10$
$x=-2$

$\left(-2,\frac{1}{2}\right)$

15. $2s+d=-2$
$5s+3d=-6$

$-10s-5d=10$
$10s+6d=-12$
$d=-2$

$2s+(-2)=-2$
$2s=0$
$s=0$

$s=0,\ d=-2$

17. $3x-5y=1$
$7x-8y=17$

$-21x+35y=-7$
$21x-24y=51$
$11y=44$
$y=4$

$3x-5(4)=1$
$3x-20=1$
$3x=21$
$x=7$

$(7,4)$

19. $5x+2y=-1$
$4x-5y=-14$

$25x+10y=-5$
$8x-10y=-28$
$33x=-33$
$x=-1$

$5(-1)+2y=-1$
$-5+2y=-1$
$2y=4$
$y=2$

$(-1,2)$

21. $7p+3q=15$
$-5p-7q=16$

$49p+21q=105$
$-15p-21q=48$
$34p=153$
$p=\frac{153}{34}=\frac{9}{2}$

21. (continued)

$$7\left(\frac{9}{2}\right)+3q=15$$
$$\frac{63}{2}+3q=15$$
$$3q=15-\frac{63}{2}$$
$$3q=\frac{30}{2}-\frac{63}{2}$$
$$3q=-\frac{33}{2}$$
$$q=-\frac{11}{2}$$
$$p=\frac{9}{2},\ q=-\frac{11}{2}$$

23. $6x+5y=-8.5$

$8x+10y=-3$

$$-12x-10y=17$$
$$8x+10y=-3$$
$$-4x=14$$
$$x=-3.5$$

$$6(-3.5)+5y=-8.5$$
$$-21+5y=-8.5$$
$$5y=12.5$$
$$y=2.5$$

$(-3.5, 2.5)$

25. $3.5x+5y=-3$

$2x=-2y$

$$3.5x+5y=-3$$
$$2x+2y=0$$

$$-7x-10y=6$$
$$10x+10y=0$$
$$3x=6$$
$$x=2$$

$$2(2)=-2y$$
$$4=-2y$$
$$-2=y$$

$(2,-2)$

27. $2x-4=-y$

$x+2y=0$

$$2x+y=4$$
$$x+2y=0$$

$$-4x-2y=-8$$
$$x+2y=0$$
$$-3x=-8$$
$$x=\frac{8}{3}$$

$$2\left(\frac{8}{3}\right)-4=-y$$
$$\frac{16}{3}-\frac{12}{3}=-y$$
$$\frac{4}{3}=-y$$
$$-\frac{4}{3}=y$$
$$\left(\frac{8}{3},-\frac{4}{3}\right)$$

29. $8x+10y=1$

$-4x-5y+6=0$

$$8x+10y=1$$
$$-4x-5y=-6$$

$$8x+10y=1$$
$$-8x-10y=-12$$
$$0=-11$$

No solution

31. $4a-b=-10$

$-3a+b=7$

$$a=-3$$

$$4(-3)-b=-10$$
$$-12-b=-10$$
$$-b=2$$
$$b=-2$$
$$a=-3,\ b=-2$$

33.
$$3x-5y=4$$
$$-6x+10y=-8$$

$$6x-10y=8$$
$$-6x+10y=-8$$
$$0=0$$

Infinitely many solutions

35.
$$5x+3y=-3$$
$$-7x-5y=4$$

$$25x+15y=-15$$
$$-21x-15y=12$$
$$4x=-3$$
$$x=-\frac{3}{4}$$

$$5\left(-\frac{3}{4}\right)+3y=-3$$
$$-\frac{15}{4}+3y=-3$$
$$3y=-3+\frac{15}{4}$$
$$3y=-\frac{12}{4}+\frac{15}{4}$$
$$3y=\frac{3}{4}$$
$$y=\frac{1}{4}$$

$$\left(-\frac{3}{4},\frac{1}{4}\right)$$

37.
$$2.5(p+w)=40$$
$$2(p-w)=20$$

$$p+w=16$$
$$p-w=10$$
$$2p=26$$
$$p=13$$

The speed of the pass if there were no wind would be 13 yd per second.

39.
$$f+h=223$$
$$5f+2.5h=765$$

$$-5f-5h=-1115$$
$$5f+2.5h=765$$
$$-2.5h=-350$$
$$h=140$$

$$f+140=223$$
$$f=83$$

The zoo collected 83 full-price admissions and 140 half-price admissions.

41.
$$s+c=227{,}000$$
$$s=c+23{,}000$$

$$s+c=227{,}000$$
$$s-c=23{,}200$$
$$2s=250{,}200$$
$$s=125{,}100$$

$$125{,}100+c=227{,}000$$
$$c=101{,}900$$

The salary of a senator is \$125,100 and the salary of a congressman is \$101,900.

43.
$$5s+7p=43$$
$$2s+9p=42$$

$$-10s-14p=-86$$
$$10s+45p=210$$
$$31p=124$$
$$p=4$$

$$5s+7(4)=43$$
$$5s+28=43$$
$$5s=15$$
$$s=3$$

It takes the computer 3 nanoseconds to carry out one sum and 4 nanoseconds to carry out one product.

45. $3x+5y=6075$

$4x+4y=6380$

$$-12x-20y=-24,300$$
$$12x+12y=19,140$$
$$-8y=-5160$$
$$y=645$$

$$3x+5(645)=6075$$
$$3x+3225=6075$$
$$3x=2850$$
$$x=950$$

The rate for full-page adds is \$950 and for half-page ads is \$645.

4.4 Solving Systems of Linear Inequalities

Practice

1. $y \leq 3x-6$

$y > -4x+2$

$y \leq 3x-6$ The border line will be solid.

x	$y=3x-6$	(x,y)
0	$3(0)-6=-6$	$(0,-6)$
1	$3(1)-6=-3$	$(1,-3)$

Test point is $(0,0)$.

$0 \stackrel{?}{\leq} 3(0)-6$

$0 \not\leq -6$, shade the half-plane not containing $(0,0)$.

$y > -4x+2$ The border line will be dashed.

x	$y=-4x+2$	(x,y)
0	$-4(0)+2=2$	$(0,2)$
1	$-4(1)+2=-2$	$(1,-2)$

Test point is $(0,0)$.

$0 \stackrel{?}{>} -4(0)+2$

$0 \not> -1$, shade the half-plane not containing $(0,0)$.

The solution of the system is the overlapping region.

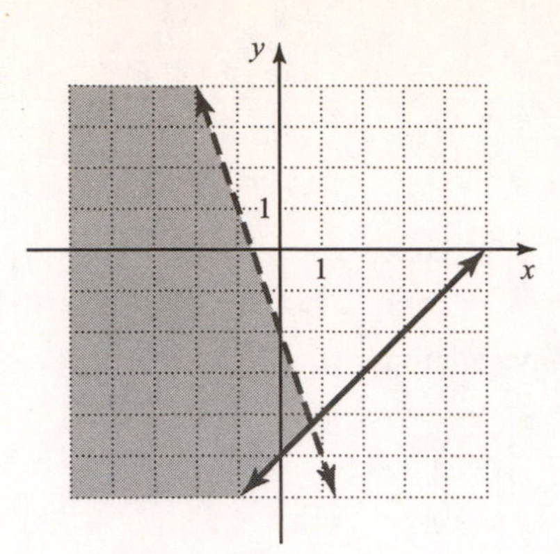

2. $2x+y<1$

$-y+3x<1$

$2x+y<1$

$\Rightarrow y<-2x+1$

The border line will be dashed.

x	$y=-2x+1$	(x,y)
0	$-2(0)+1=1$	$(0,1)$
1	$-2(1)+1=-1$	$(1,-1)$

Test point is $(0,0)$.

$0 \stackrel{?}{<} -2(0)+1$

$0<1$, shade the half-plane containing $(0,0)$.

$-y+3x<1$

$\Rightarrow -y<-3x+1$

$\Rightarrow y>3x-1$

The border line will be dashed.

x	$y=3x-1$	(x,y)
0	$3(0)-1=-1$	$(0,-1)$
1	$3(1)-1=2$	$(1,2)$

Test point is $(0,0)$.

$0 \stackrel{?}{>} 3(0)-1$

$0<-1$, shade the half-plane containing $(0,0)$.

The solution of the system is the overlapping region.

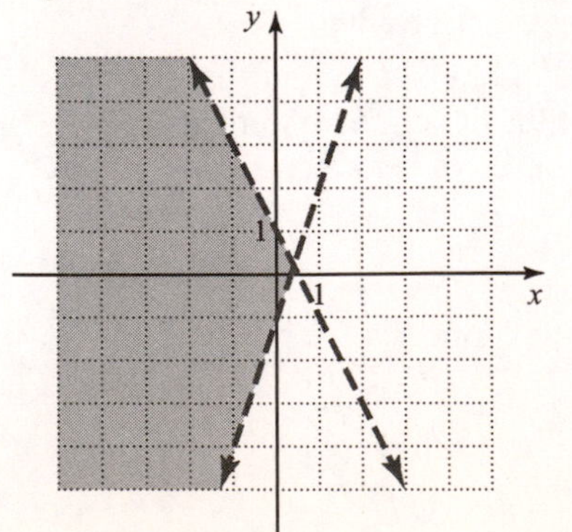

3. $y+2x \geq 4$
$x > -3$
$y \geq 1$

$y+2x \geq 4$
$\Rightarrow y \geq -2x+4$
The border line will be solid.

x	$y=-2x+4$	(x,y)
0	$-2(0)+4=4$	$(0,4)$
1	$-2(1)+4=2$	$(1,2)$

Test point is $(0,0)$.

$0 \overset{?}{\leq} -2(0)+4$

$0<4$, shade the half-plane containing $(0,0)$.

$x>-3$ Boundary line is dashed.

$x=-3$	y	(x,y)
-3	0	$(-3,0)$
-3	2	$(-3,2)$

Test point is $(0,0)$.

$0>-3$, so shade the half-plane containing $(0,0)$.

$y \geq 1$ Boundary line is solid.

x	$y=1$	(x,y)
-2	1	$(-2,1)$
2	1	$(2,1)$

Test point is $(0,5)$.

$5>1$, so shade the half-plane containing $(0,5)$.

The solution of the system is the overlapping region.

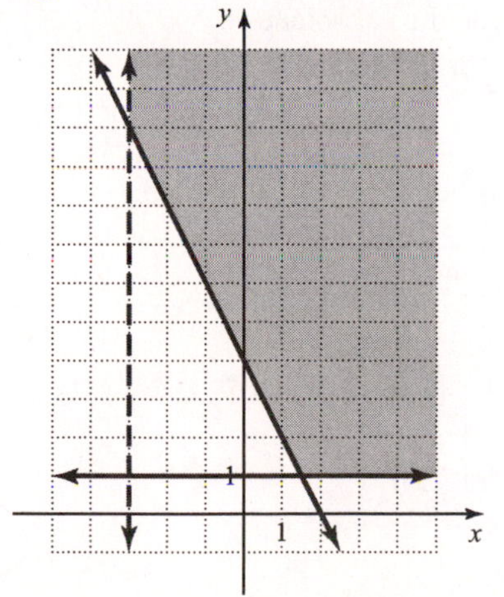

4. a. $x+y \geq 7$
$x+y \leq 10$
$x>y$

b.
$x+y \geq 7$
$\Rightarrow y \geq -x+7$
The border line will be solid.

x	$y=-x+7$	(x,y)
0	$-(0)+7=7$	$(0,7)$
1	$-(1)+7=6$	$(1,6)$

Test point is $(0,0)$.

$0 \overset{?}{\geq} -(0)+7$

$0 \not\geq 7$, shade the half-plane not containing $(0,0)$.

$x+y \leq 10$
$\Rightarrow y \leq -x+10$
The border line will be solid.

x	$y=-x+10$	(x,y)
0	$-(0)+10=10$	$(0,10)$
1	$-(1)+10=9$	$(1,9)$

Test point is $(0,0)$.

$0 \overset{?}{\geq} -(0)+10$

$0 \not\geq 10$, shade the half-plane not containing $(0,0)$.

$x>y$ Boundary line is dashed.

Shade below the line $y=x$.

The solution of the system is the overlapping region.

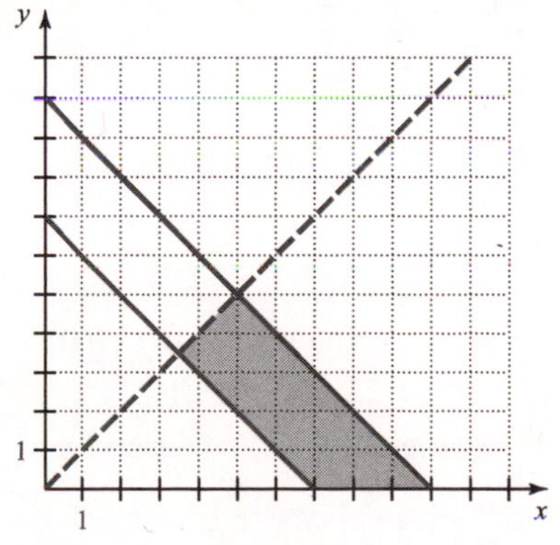

c.
(4,3); 4 freshmen and 3 returning students
(5,2); 5 freshmen and 2 returning students
(5,3); 5 freshmen and 3 returning students
(5,4); 5 freshmen and 4 returning students
(6,1); 6 freshmen and 1 returning students
(6,2); 6 freshmen and 2 returning students
(6,3); 6 freshmen and 3 returning students
(6,4); 6 freshmen and 4 returning students
(7,0); 7 freshmen and 0 returning students
(7,1); 7 freshmen and 1 returning students
(7,2): 7 freshmen and 2 returning students
(7,3); 7 freshmen and 3 returning students
(8,0); 8 freshmen and 0 returning students
(8,1); 8 freshmen and 1 returning students
(8,2); 8 freshmen and 2 returning students
(9,0); 9 freshmen and 0 returning students
(9,1); 9 freshmen and 1 returning students
(10,0); 10 freshmen and 0 returning students

Exercises

1. Two or more linear inequalities considered simultaneously, that is, together, are called a system of linear inequalities.

3. A system of linear inequalities in two variables can be solved by graphing each inequality on the same coordinate plane.

5. $y > 2x - 1$
$y < -x + 3$

$y > 2x - 1$ The border line will be dashed.

x	$y = 2x - 1$	(x, y)
0	$2(0) - 1 = -1$	$(0, -1)$
1	$2(1) - 1 = 1$	$(2,1)$

Test point is $(0,0)$.

$0 \overset{?}{>} 2(0) - 1$

$0 > -1$, shade the half-plane containing $(0,0)$.

$y < -x + 3$ The border line will be dashed.

x	$y = -x + 3$	(x, y)
0	$-(0) + 3 = 3$	$(0,3)$
1	$-(1) + 3 = 2$	$(1,2)$

Test point is $(0,0)$.

$0 \overset{?}{<} -(0) + 3$

$0 < 3$, shade the half-plane containing $(0,0)$.

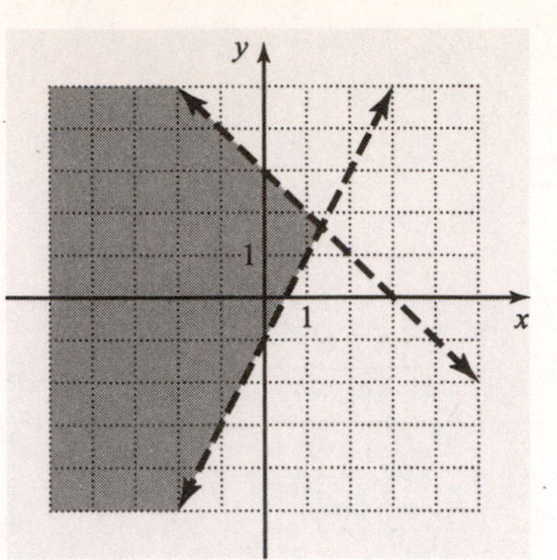

The solution of the system is the overlapping region.

7. $y \leq \frac{1}{3}x + 3$
$y < -\frac{1}{2}x + 1$

$y \leq \frac{1}{3}x + 3$

The border line is solid.

x	$y = \frac{1}{3}x + 3$	(x, y)
0	$\frac{1}{3}(0) + 3 = 3$	$(0,3)$
3	$\frac{1}{3}(3) + 3 = 4$	$(3,4)$

Test point is $(0,0)$.

$0 \overset{?}{\leq} \frac{1}{3}(0) + 3$

$0 \leq 3$, shade the half-plane containing $(0,0)$.

$y < -\frac{1}{2}x + 1$ The boundary line is dashed.

x	$y = -\frac{1}{2}x + 1$	(x, y)
0	$-\frac{1}{2}(0) + 1 = 1$	$(0,1)$
2	$-\frac{1}{2}(2) + 1 = 0$	$(2,0)$

Test point is $(0,0)$.

$0 \overset{?}{<} -\frac{1}{2}(0) + 1$

$0 < 1$

Shade the half-plane containing $(0,0)$.

The solution of the system is the overlapping region.

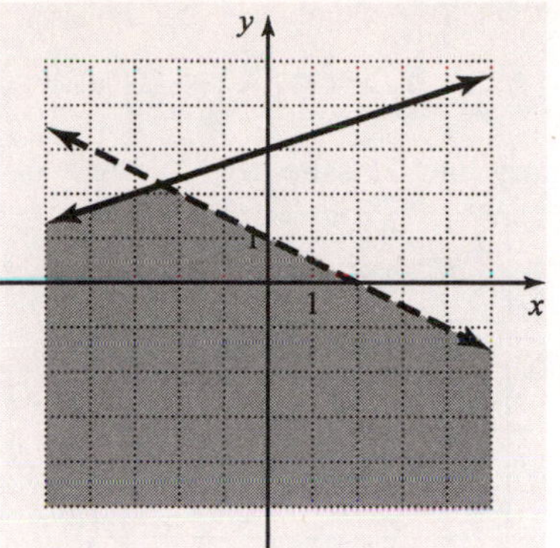

9. $4x+2y\geq -6$
$12x-3y\geq -6$

$4x+2y\geq -6$
$\Rightarrow 2y\geq -4x-6$
$\Rightarrow y\geq -2x-3$

The boundary line is solid.

x	$y=-2x-3$	(x,y)
-2	$-2(-2)-3=1$	$(-2,1)$
0	$-2(0)-3=-3$	$(0,-3)$

Test point is $(0,0)$.

$4(0)+2(0)\overset{?}{\geq}-6$
$0\geq -6$

Shade the half plane containing $(0,0)$.

$12x-3y\geq -6\Rightarrow -3y\geq -12x-6\Rightarrow y\leq 4x+2$

The boundary line is solid.

x	$y=4x+2$	(x,y)
-1	$4(-1)+2=-2$	$(-1,-2)$
0	$4(0)+2=2$	$(0,2)$

Test point is $(0,0)$.

$12(0)-3(0)\overset{?}{\geq}-6$
$0\geq -6$

Shade the half-plane containing $(0,0)$.

The solution of the system is the overlapping region.

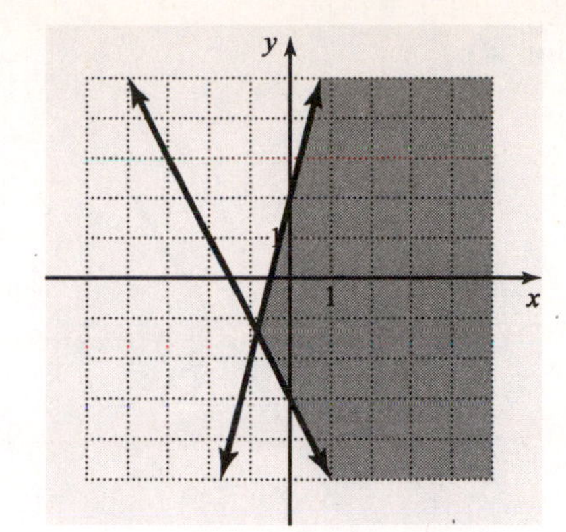

11. $3x-2y\leq 8$
$-x-3y>0$

$3x-2y\leq 8\Rightarrow -2y\leq -3x+8\Rightarrow y\geq \frac{3}{2}x-4$

The boundary line is solid.

x	$y=\frac{3}{2}x-4$	(x,y)
0	$\frac{3}{2}(0)-4=-4$	$(0,-4)$
2	$\frac{3}{2}(2)-4=-1$	$(2,-1)$

The test point is $(0,0)$.

$3(0)-2(0)\overset{?}{\leq}8$
$0\leq 8$

Shade the half-plane containing $(0,0)$.

$-x-3y>0\Rightarrow -3y>x\Rightarrow y<-\frac{1}{3}x$

The boundary line is dashed.

x	$y=-\frac{1}{3}x$	(x,y)
0	$-\frac{1}{3}(0)=0$	$(0,0)$
3	$-\frac{1}{3}(3)=-1$	$(3,-1)$

Test point is $(-5,-5)$.

$-(-5)-3(-5)\overset{?}{>}0$
$5+15=20>0$

Shade the half-plane containing $(-5,-5)$.

The solution of the system is the overlapping region.

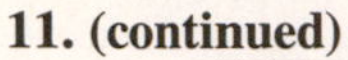

11. (continued)

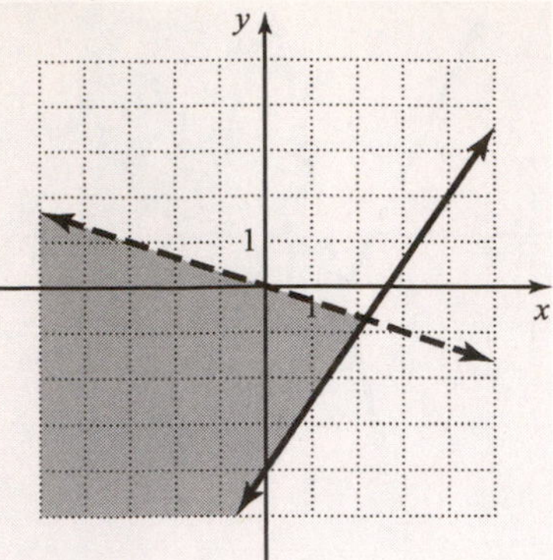

13. $4x+2y \le 4$
$2x+ \ y > -3$

$4x+2y \le 4 \Rightarrow 2y \le -4x+4 \Rightarrow y \le -2x+2$

The boundary line is solid.

x	$y=-2x+2$	(x,y)
0	$-2(0)+2=2$	$(0,2)$
1	$-2(1)+2=0$	$(1,0)$

Test point is $(0,0)$.

$4(0)+2(0)\overset{?}{\le}4$

$0 \le 4$

Shade the half plane containing $(0,0)$.

$2x+y>-3 \Rightarrow y>-2x-3$

The boundary line is dashed.

x	$y=-2x-3$	(x,y)
-1	$-2(-1)-3=-1$	$(-1,-1)$
0	$-2(0)-3=-3$	$(0,-3)$

Test point is $(0,0)$.

$2(0)+(0)\overset{?}{>}-3$

$0>-3$

Shade the half-plane containing $(0,0)$.

The solution of the system is the overlapping region.

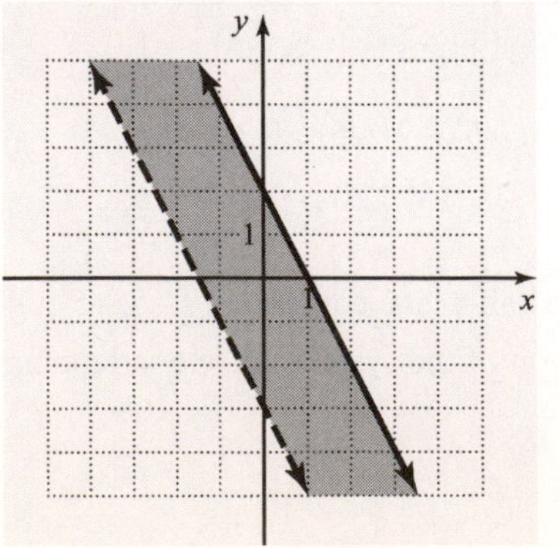

15. $3x- \ 9y<18$
$1.5x+0.5y>1.5$

$3x-9y<18 \Rightarrow -9y<-3x+18 \Rightarrow y>\frac{1}{3}x-2$

The boundary line is dashed.

x	$y=\frac{1}{3}x-2$	(x,y)
0	$\frac{1}{3}(0)-2=-2$	$(0,-2)$
3	$\frac{1}{3}(3)-2=-1$	$(3,-1)$

Test point is $(0,0)$,

$3(0)-9(0)\overset{?}{<}18$

$0<18$

Shade the half-plane containing $(0,0)$.

$1.5x+0.5y>1.5 \Rightarrow 0.5y>-1.5x+1.5 \Rightarrow$
$y>-3x+3$

The boundary line is dashed.

x	$y=-3x+3$	(x,y)
0	$-3(0)+3=3$	$(0,3)$
1	$-3(1)+3=0$	$(1,0)$

Test point is $(0,0)$.

$1.5(0)+0.5(0)\overset{?}{>}1.5$

$0 \not> 1.5$

Shade the half-plane not containing $(0,0)$.

The solution of the system is the overlapping region.

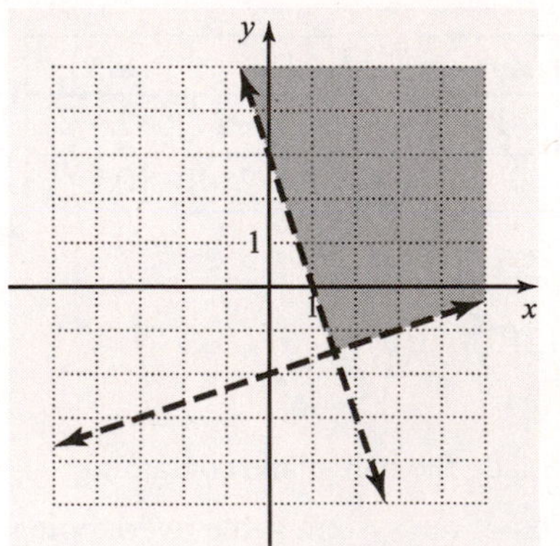

17. $y - x > 1$
$x \le 4$
$y > 0$

$y - x > 1 \Rightarrow y > x + 1$

The boundary line is dashed.

x	$y = x + 1$	(x, y)
0	$(0) + 1 = 1$	$(0,1)$
1	$(1) + 1 = 2$	$(1,2)$

Test point is $(0,0)$.

$0 - 0 \overset{?}{>} 1$

$0 \not> 1$

Shade the half-plane not containing $(0,0)$.

$x \le 4$ Boundary line is solid.

$x = 4$	y	(x, y)
4	0	$(4,0)$
4	2	$(4,2)$

Test point is $(0,0)$.

$0 \le 4$, so shade the half-plane containing $(0,0)$.

$y > 0$ Boundary line is dashed.

x	$y = 0$	(x, y)
-2	0	$(-2,0)$
2	0	$(2,0)$

Test point is $(0,5)$.

$5 > 0$, so shade the half-plane containing $(0,5)$.

The solution of the system is the overlapping region.

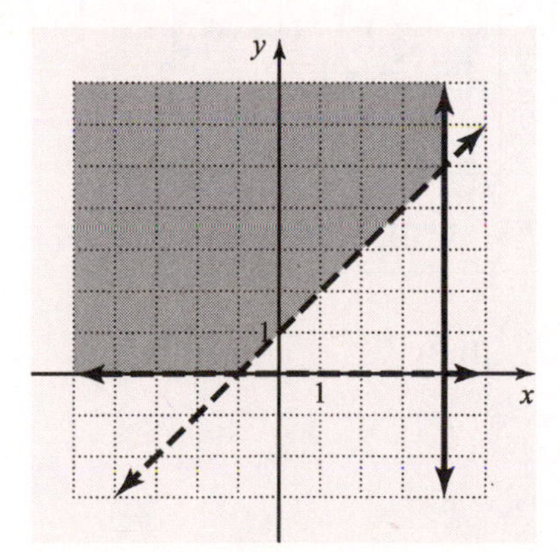

19. $-2x - y \ge 3$
$x \ge -4$
$y \ge -2$

$-2x - y \ge 3 \Rightarrow -y \ge 2x + 3 \Rightarrow y \le -2x - 3$

Boundary line is solid.

x	$y = -2x - 3$	(x, y)
-1	$-2(-1) - 3 = -1$	$(-1,-1)$
0	$-2(0) - 3 = -3$	$(0,-3)$

Test point is $(0,0)$.

$-2(0) - (0) \overset{?}{\ge} 3$

$0 \not\ge 3$

Shade the half-plane not containing $(0,0)$.

$x \ge -4$

The boundary line is solid

$x = -4$	y	(x, y)
-4	0	$(-4,0)$
-4	2	$(-4,2)$

Test point is $(0,0)$.

$0 \ge -4$, so shade the half-plane containing $(0,0)$.

$y \ge -2$

The boundary line is solid.

x	$y = -2$	(x, y)
0	-2	$(0,-2)$
2	-2	$(2,-2)$

Test point is $(0,0)$.

$0 \ge -2$, so shade the half-plane containing $(0,0)$.

The solution of the system is the overlapping region.

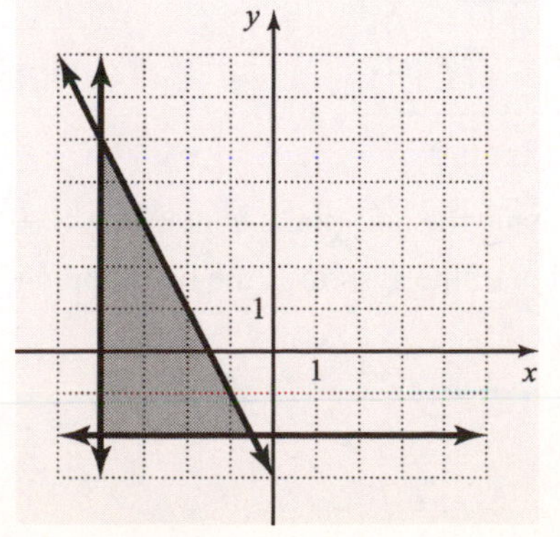

21. $3x-2y<2$
$x+3y<12$
$x>-2$

$3x-2y<2 \Rightarrow -2y<-3x+2 \Rightarrow y>\frac{3}{2}x-1$

Boundary line is dashed.

x	$y=\frac{3}{2}x-1$	(x,y)
0	$\frac{3}{2}(0)-1=-1$	$(0,-1)$
2	$\frac{3}{2}(2)-1=2$	$(2,2)$

Test point is $(0,0)$.

$3(0)-2(0)\overset{?}{<}2$
$0<2$

Shade the half-plane containing $(0,0)$.

$x+3y<12 \Rightarrow 3y<-x+12 \Rightarrow y<-\frac{1}{3}x+4$

The boundary line is dashed.

x	$y=-\frac{1}{3}x+4$	(x,y)
0	$-\frac{1}{3}(0)+4=4$	$(0,4)$
3	$-\frac{1}{3}(3)+4=3$	$(3,3)$

Test point is $(0,0)$.

$(0)+3(0)\overset{?}{<}12$
$0<12$

Shade the half-plane containing $(0,0)$.

$x>-2$ Boundary line is dashed.

$x=-2$	y	(x,y)
-2	0	$(-2,0)$
-2	2	$(-2,2)$

Test point is $(0,0)$.

$0>-2$, shade the half-plane containing $(0,0)$.

The solution of the system is the overlapping region.

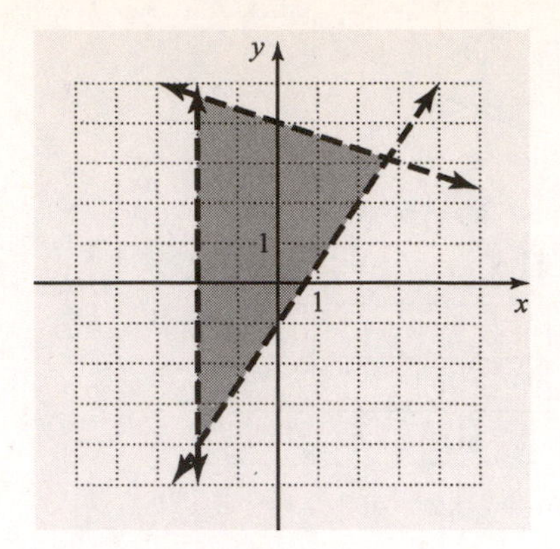

23. $12x-4y\geq 16$
$3x-6y<-6$
$y<4$

$12x-4y\geq 16 \Rightarrow -4y\geq -12x+16 \Rightarrow y\leq 3x-4$

The boundary line is solid.

x	$y=3x-4$	(x,y)
0	$3(0)-4=-4$	$(0,-4)$
1	$3(1)-4=-1$	$(1,-1)$

Test point is $(0,0)$.

$12(0)-4(0)\overset{?}{\geq}16$
$0\ngeq 16$

Shade the half-plane not containing $(0,0)$.

$3x-6y<-6 \Rightarrow -6y<-3x-6 \Rightarrow y>\frac{1}{2}x+1$

The boundary line is dashed.

x	$y=\frac{1}{2}x+1$	(x,y)
0	$\frac{1}{2}(0)+1=1$	$(0,1)$
2	$\frac{1}{2}(2)+1=2$	$(2,2)$

Test point is $(0,0)$.

$3(0)-6(0)\overset{?}{<}-6$
$0\not<-6$

Shade the half-plane not containing $(0,0)$.

23. (continued)

$y < 4$ The boundary line is dashed.

x	$y=4$	(x,y)
0	4	$(0,4)$
2	4	$(2,4)$

Test point is $(0,0)$.

$0 < 4$, shade the half-plane containing (0,0).

The solution of the system is the overlapping region.

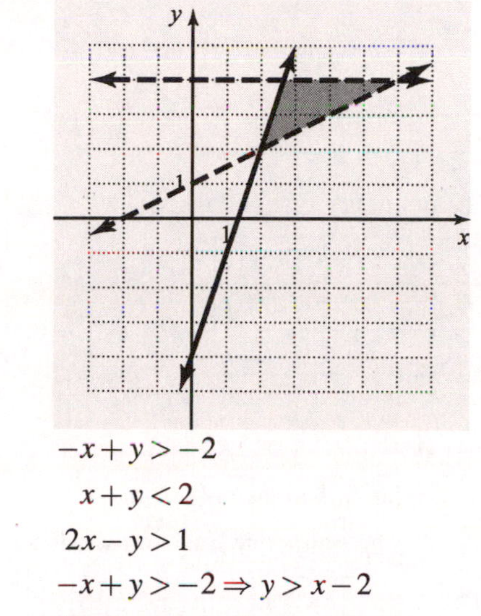

25. $-x+y>-2$

$x+y<2$

$2x-y>1$

$-x+y>-2 \Rightarrow y>x-2$

The boundary line is dashed.

x	$y=x-2$	(x,y)
0	$(0)-2=-2$	$(0,-2)$
2	$(2)-2=0$	$(2,0)$

Test point is $(0,0)$.

$-(0)+(0)\overset{?}{>}-2$

$0>-2$

Shade the half-plane containing $(0,0)$.

$y>-x+2$

The boundary line is dashed.

x	$y=-x+2$	(x,y)
0	$-(0)+2=2$	$(0,2)$
2	$-(2)+2=0$	$(2,0)$

Test point is $(0,0)$.

$(0)\overset{?}{<}-(0)+2$

$0<2$

So shade the half-plane containing $(0,0)$.

$2x-y>1 \Rightarrow -y>-2x+1 \Rightarrow y<2x-1$

The boundary line is dashed.

x	$y=2x-1$	(x,y)
0	$2(0)-1=-1$	$(0,-1)$
1	$2(1)-1=1$	$(1,1)$

Test point is $(0,0)$.

$2(0)-(0)\overset{?}{>}1$

$0 \not> 1$

Shade the half-plane not containing $(0,0)$.

The solution of the system is the overlapping region.

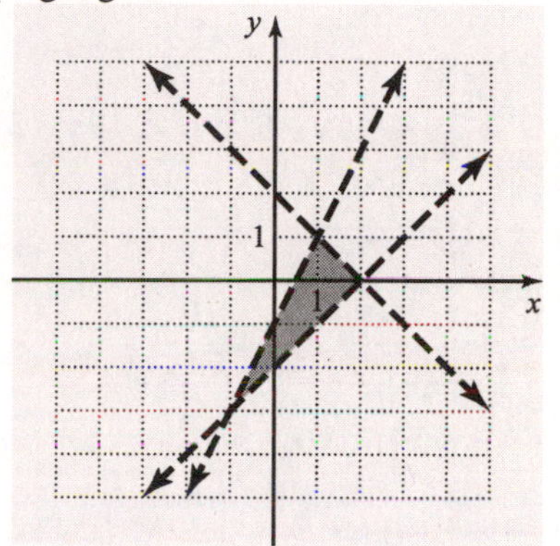

27. $y \le -3x-2$

$y<x+2$

$y \le -3x-2$ The boundary line is solid.

x	$y=-3x-2$	(x,y)
0	$-3(0)-2=-2$	$(0,-2)$
-1	$-3(-1)-2=1$	$(-1,1)$

Test point is $(0,0)$.

$0\overset{?}{=}-3(0)-2$

$0 \not< -2$

Shade the half-plane not containing $(0,0)$.

$y<x+2$ The boundary line is dashed.

x	$y=x+2$	(x,y)
0	$0+2=2$	$(0,2)$
1	$1+2=3$	$(1,3)$

Test point is $(0,0)$.

$0\overset{?}{<}0+2$

$0<2$

Shade the half-plane containing $(0,0)$.

27. (continued)
The solution of the system is the overlapping region.

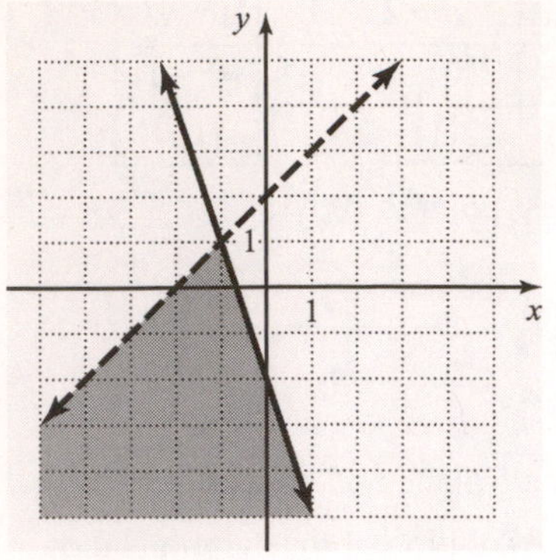

29. $$6x+3y<9$$
$$-4x-2y<4$$

$6x+3y<9 \Rightarrow 3y<-6x+9 \Rightarrow y<-2x+3$

The boundary line is dashed.

x	$y=-2x+3$	(x,y)
0	$-2(0)+3$	$(0,3)$
2	$-2(2)+3=0$	$(2,-1)$

Test point is $(0,0)$.

$$6(0)+3(0)\overset{?}{<}9$$
$$0<9$$

Shade the half-plane containing $(0,0)$.

$-2y<4x+4 \Rightarrow y>-2x-2$

The boundary line is dashed.

x	$y=-2x-2$	(x,y)
0	$-2(0)-2$	$(0,-2)$
-2	$-2(-2)-2=2$	$(-2,2)$

Test point is $(0,0)$.

$$-4(0)-2(0)\overset{?}{<}4$$
$$0<4$$

Shade the half-plane containing $(0,0)$.

The solution of the system is the overlapping region.

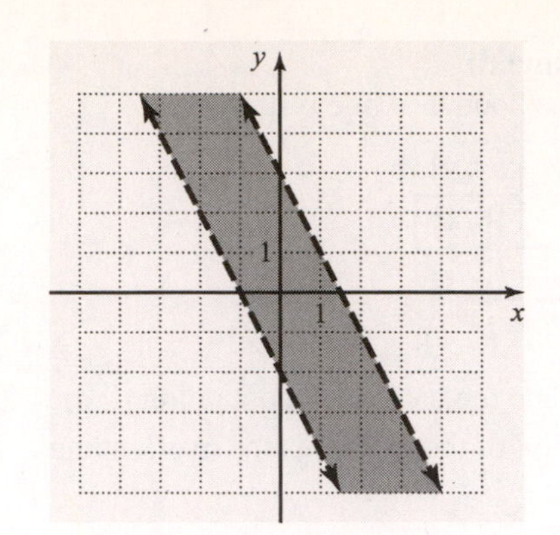

31. $$x>-3$$
$$y<4$$
$$-2x+y\geq 3$$

$x>-3$ The boundary line is dashed.

y	$x=-3$	(x,y)
0	-3	$(-3,0)$
2	-3	$(-3,2)$

Test point is $(0,0)$.

$0>-3$ Shade the half-plane containing $(0,0)$.

$y<4$ The boundary line is dashed.

x	$y=4$	(x,y)
0	4	$(0,4)$
2	4	$(2,4)$

Test point is $(0,0)$.

$0<4$ Shade the half-plane containing $(0,0)$.

$y\geq 2x+3$ The boundary line is solid.

x	$y=2x+3$	(x,y)
0	$2(0)+3=3$	$(0,3)$
-1	$2(-1)+3=1$	$(-1,1)$

Test point is $(0,0)$.

$$-2(0)+0\overset{?}{\geq}3$$
$$0\ngeq 3$$

Shade the half-plane not containing $(0,0)$.

The solution of the system is the overlapping region.

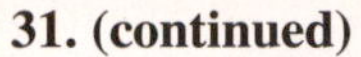

31. (continued)

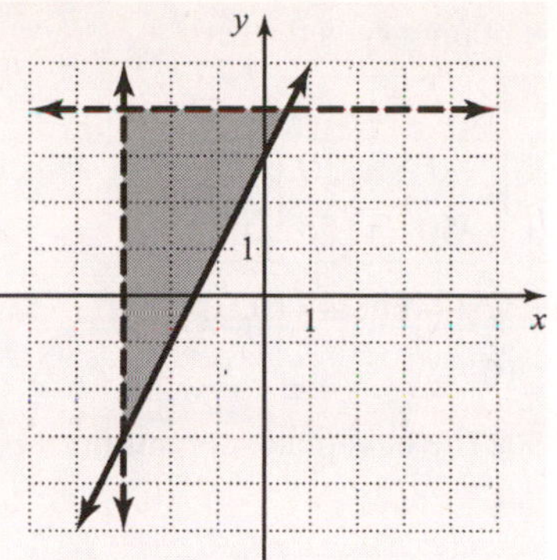

33. **a.** Let x = the hours she works at the retail job and let y = the hours she works at the office job.

$$x + y \geq 35$$
$$x \leq 20$$
$$y \leq 30$$

b.

$x + y \geq 35 \Rightarrow y \geq -x + 35$

The boundary line is solid.

x	$y = -x + 35$	(x, y)
0	$-(0) + 35 = 35$	$(0,35)$
35	$-(35) + 35 = 0$	$(35,0)$

Test point is $(0,0)$.

$$(0) + (0) \overset{?}{\geq} 35$$
$$0 \not\geq 35$$

Shade the half-plane not containing $(0,0)$.

$x \leq 20$ The boundary line is solid.

$x = 20$	y	(x, y)
20	0	$(20,0)$
20	20	$(20,20)$

Test point is $(0,0)$.

$0 \leq 20$, shade the half-plane containing $(0,0)$.

$y \leq 30$ The boundary line is solid.

x	$y = 30$	(x, y)
0	30	$(0,30)$
20	30	$(20,30)$

Test point is $(0,0)$.

$0 \leq 30$, shade the half-plane containing $(0,0)$.

The solution of the system is the overlapping region.

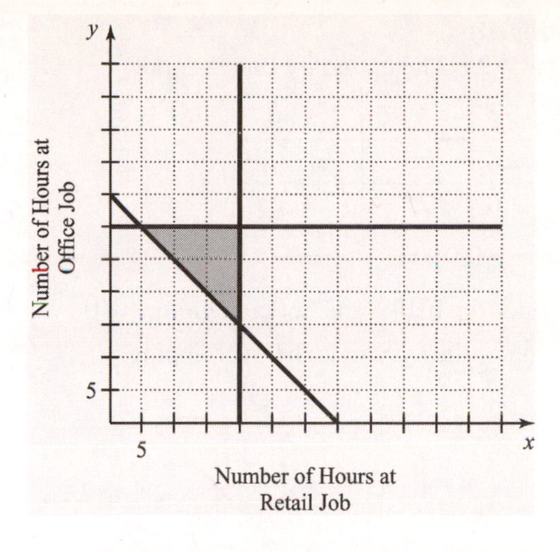

c. She must work between 20 and 30 hours at her office job.

35. **a.** Let l = the length of the lot and w = the width of the lot. The perimeter is twice the width plus twice the length.

$$2l + 2w \leq 400$$
$$l \geq w + 25$$
$$w \geq 25$$

b.

$2l + 2w \leq 400 \Rightarrow 2w \leq -2l + 400 \Rightarrow w \leq -l + 200$

The boundary line is solid.

l	$w = -l + 200$	(l, w)
0	$-(0) + 200 = 200$	$(0,200)$
200	$-(200) + 200 = 0$	$(200,0)$

Test point is $(0,0)$.

$$2(0) + 2(0) \overset{?}{\leq} 400$$
$$0 \leq 400$$

Shade the half-plane containing $(0,0)$.

$l \geq w + 25 \Rightarrow l - 25 \geq w \Rightarrow w \leq l - 25$

The boundary line is solid.

l	$w = l - 25$	(l, w)
25	$(25) - 25 = 0$	$(25,0)$
200	$(200) - 25 = 175$	$(200,175)$

Test point is $(0,0)$.

$$0 \overset{?}{\geq} 0 + 25$$
$$0 \not\geq 25$$

Shade the half-plane not containing $(0,0)$.

35. (continued)

$w \geq 25$ The boundary line is solid

l	$w=25$	(l,w)
25	25	$(25,25)$
200	25	$(200,25)$

Test point is $(0,0)$.

$0 \not\geq 25$, shade the half-plane not containing $(0,0)$.

The solution of the system is the overlapping region.

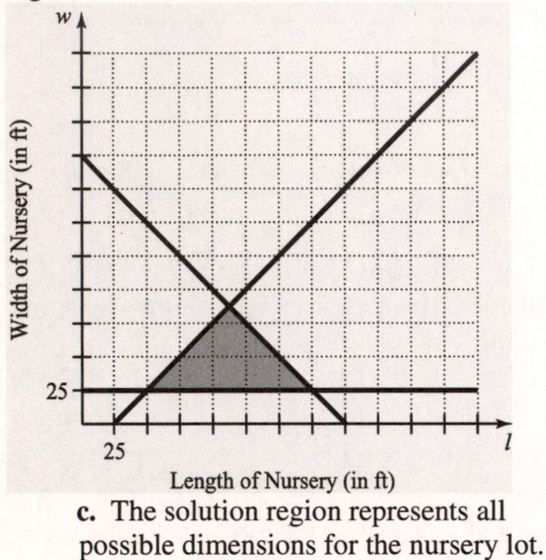

c. The solution region represents all possible dimensions for the nursery lot.

37. **a.** Let x = the number of half-page ads and let y = the number of full-page ads.

$$150x+225y<15000$$
$$x>y$$
$$x \geq 20$$

b. $150x+225y<15,000$

$$225y<-150x+15,000$$
$$y<-\frac{2}{3}x+\frac{200}{3}$$

The boundary line is dashed.

x	$y=-\frac{2}{3}x+\frac{200}{3}$	(x,y)
0	$-\frac{2}{3}(0)+\frac{200}{3}$	$\left(0,66\frac{2}{3}\right)$
100	$-\frac{2}{3}(100)+\frac{200}{3}=0$	$(100,0)$

Test point is $(0,0)$.

$$150(0)+225(0)\overset{?}{<}15,000$$
$$0<15,000$$

Shade the half-plane containing (0,0).

$x>y \Rightarrow y<x$

The boundary is dashed.

x	$y=x$	(x,y)
0	0	$(0,0)$
50	50	$(50,50)$

The test point is $(10,0)$.

$0<10$

Shade the half-plane containing $(10,0)$.

$x \geq 20$

The boundary is solid.

$x=20$	y	(x,y)
20	0	$(20,0)$
20	50	$(20,50)$

Test point is $(0,0)$.

$0 \not\geq 20$

Shade the half-plane not containing $(0,0)$.

The solution of the system is the overlapping region.

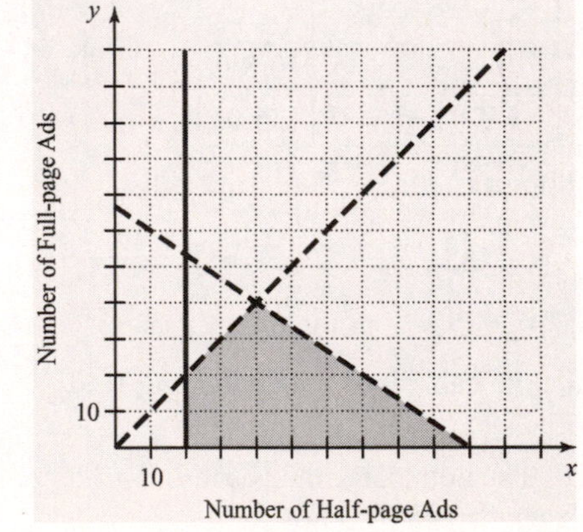

39. **a.** Let x = the number of tables and y = the chairs.

$2x+1.5y \le 360$ (assembly)

$3x+y \le 400$ (finishing and painting)

b.

$2x+1.5y \le 360 \Rightarrow 1.5y \le -2x+360 \Rightarrow$

$y \le -\frac{4}{3}x+240$

The boundary line is solid.

x	$y=-\frac{4}{3}x+240$	(x,y)
0	$-\frac{4}{3}(0)+240=240$	$(0,240)$
180	$-\frac{4}{3}(180)+240=0$	$(180,0)$

Test point is $(0,0)$.

$2(0)+1.5(0) \overset{?}{\le} 360$

$0 \le 360$

Shade the half-plane containing $(0,0)$.

$3x+y \le 400 \Rightarrow y \le -3x+400$

The boundary line is solid.

x	$y=-3x+400$	(x,y)
0	$-3(0)+400=400$	$(0,400)$
100	$-3(100)+400=100$	$(100,100)$

Test point is $(0,0)$.

$3(0)+(0) \overset{?}{\le} 400$

$0 \le 400$

Shade the half-plane containing $(0,0)$.

Chapter 4 Review Exercises

1. $(2)+2(-3)=-4$

$2-6=-4$

$-4=-4$ True

$3(2)-(-3)=3$

$6+3=3$

$9=3$ False

No, it is not a solution of the system.

2. **a.** No solution

b. One solution

c. Infinitely many solutions

3. **a.** III

b. IV

c. II

d. I

4.

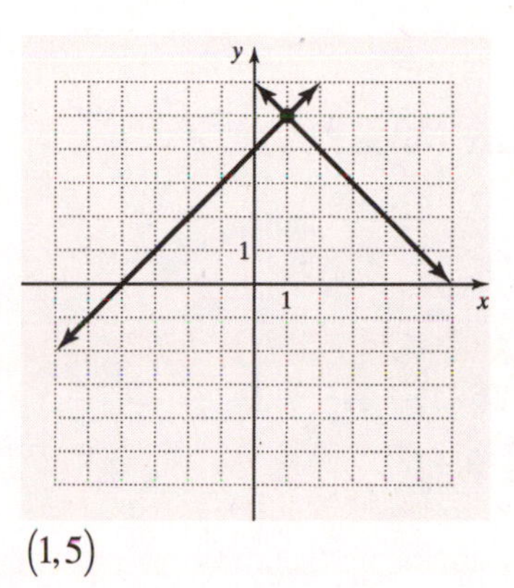

$(1,5)$

5.

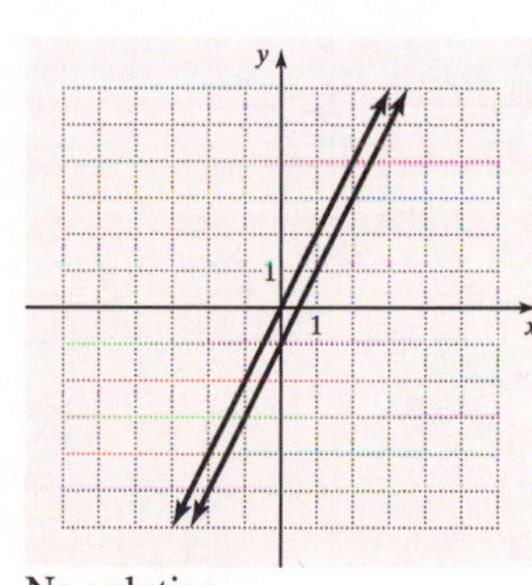

No solution

6.

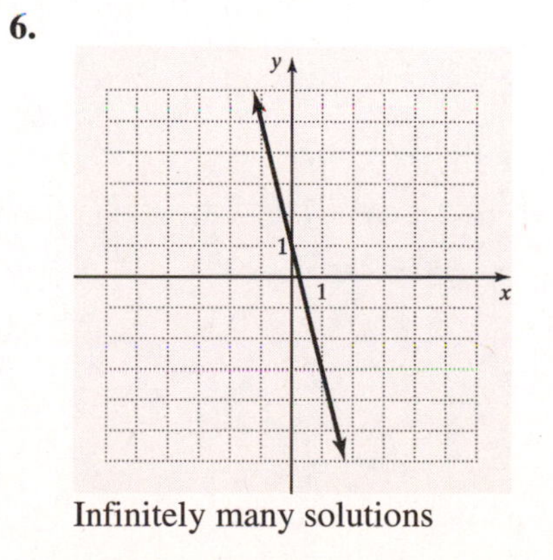

Infinitely many solutions

7.

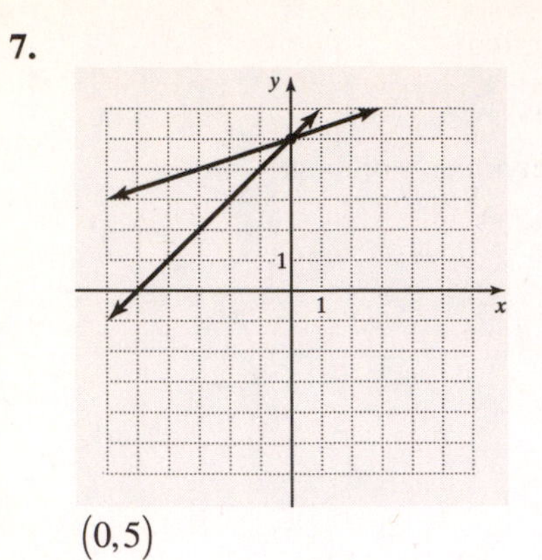

$(0,5)$

8. **(1)** $x+(2x+6)=3$
$3x+6=3$
$3x=-3$
$x=-1$

(2) $y=2(-1)+6$
$y=-2+6$
$y=4$

$(-1,4)$

9. **(2)** $(3b-4)+4b=10$
$7b-4=10$
$7b=14$
$b=2$

(1) $a=3(2)-4$
$a=6-4$
$a=2$

$a=2,\ b=2$

10. **(1)** $x-3y=1$
$x=3y+1$

(2) $-2(3y+1)+6y=7$
$-6y-2+6y=7$
$-2=7$

No solution

11. **(2)** $-y=5x-7$
$y=-5x+7$

(1) $10x+2(-5x+7)=14$
$10x-10x+14=14$
$14=14$

Infinitely many solutions

12. $x+y=1$
$x-y=7$
$2x=8$
$x=4$

$4+y=1$
$y=-3$

$(4,-3)$

13. $2x+3y=8$
$4x+6y=16$

$-4x-6y=-16$
$4x+6y=16$
$0=0$

Infinitely many solutions

14. $4x=9-5y$
$2x+3y=3$

$4x+5y=9$
$2x+3y=3$

$4x+5y=9$
$-4x-6y=-6$
$-y=3$
$y=-3$

$4x=9-5(-3)$
$4x=9+15$
$4x=24$
$x=6$

$(6,-3)$

15. $3x + 2y = -4$
$4x - 3y = 23$

$9x + 6y = -12$
$8x - 6y = 46$
$17x = 34$
$x = 2$

$3(2) + 2y = -4$
$6 + 2y = -4$
$2y = -10$
$y = -5$

$(2, -5)$

16. $6x - 4y \le 12$
$4x + 2y > 2$

$6x - 4y \le -12 \Rightarrow -4y \le -6x - 12 \Rightarrow$
$y \ge \frac{3}{2}x + 3$

The boundary line is solid

x	$y = \frac{3}{2}x + 3$	(x, y)
-2	$\frac{3}{2}(-2) + 3 = 0$	$(-2, 0)$
0	$\frac{3}{2}(0) + 3 = 3$	$(0, 3)$

Test point is $(0,0)$.

$6(0) - 4(0) \overset{?}{\le} -12$
$0 \not\le -12$

Shade the half-plane not containing (0, 0).

$4x + 2y > 2 \Rightarrow 2y > -4x + 2 \Rightarrow y > -2x + 1$

The boundary line is dashed.

x	$y = -2x + 1$	(x, y)
0	$-2(0) + 1 = 1$	$(0, 1)$
1	$-2(1) + 1 = -1$	$(1, -2)$

Test point is $(0,0)$.

$4(0) + 2(0) \overset{?}{>} 2$
$0 \not> 2$

Shade the half-plane not containing $(0,0)$.

The solution of the system is the overlapping region.

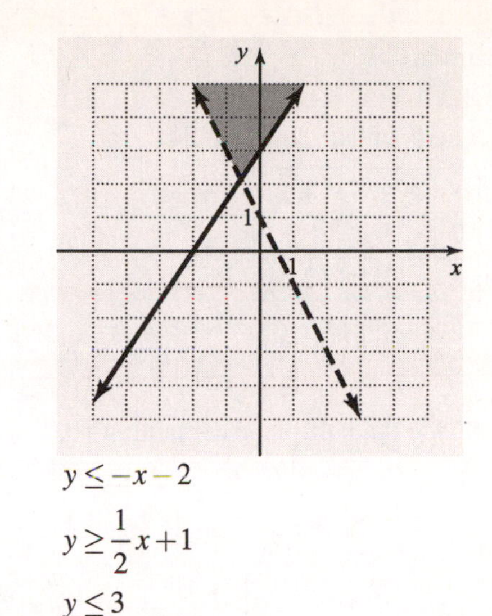

17. $y \le -x - 2$
$y \ge \frac{1}{2}x + 1$
$y \le 3$

$y \le -x - 2$

The boundary line is solid

x	$y = -x - 2$	(x, y)
0	$-(0) - 2 = -2$	$(0, -2)$
-2	$-(-2) - 2 = 0$	$(-2, 0)$

Test point is $(0,0)$.

$0 \overset{?}{\le} -(0) - 2$
$0 \not\le -2$

Shade the half-plane not containing $(0,0)$.

$y \ge \frac{1}{2}x + 1$

The boundary line is solid.

x	$y = \frac{1}{2}x + 1$	(x, y)
0	$\frac{1}{2}(0) + 1 = 1$	$(0, 1)$
2	$\frac{1}{2}(2) + 1 = 2$	$(2, 2)$

Test point is $(0,0)$.

$0 \overset{?}{\ge} \frac{1}{2}(0) + 1$
$0 \not\ge 1$

Shade the half-plane not containing $(0,0)$.

17. (continued)

$y \le 3$

The boundary line is solid.

x	$y=3$	(x,y)
0	3	$(0,3)$
2	3	$(2,3)$

Test point is $(0,0)$.

$0 \le 3$

Shade the half-plane containing $(0,0)$.

The solution of the system is the overlapping region.

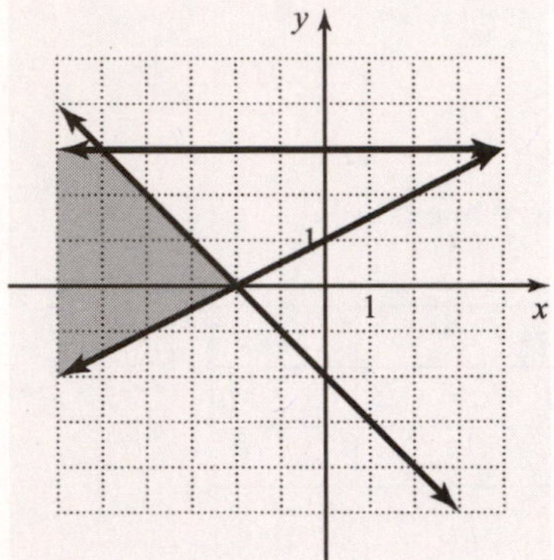

18. **a.** $y = 0.50x + 1750$

$y = 5.50x$

b. $5.50x = 0.50x + 1750$

$5x = 1750$

$x = 350$

The student must type 350 pages to break even.

19. **a.** $s = 10h$

$s = 8h + 50$

b. $10h = 8h + 50$

$2h = 50$

$h = 25$

25 hr

20. $2l + 2w = 332$

$l = w + 26$

$2l + 2w = 332$

$l - w = 26$

$2l + 2w = 332$

$2l - 2w = 52$

$4l = 384$

$l = 96$

$96 = w + 26$

$70 = w$

$lw = (70)(96) = 6720$

The area of the screen is 6720 sq ft.

21. $2l + 2w = 228$

$l = w + 51$

$2l + 2w = 228$

$l - w = 51$

$2l + 2w = 228$

$2l - 2w = 102$

$4l = 330$

$l = 82.5$

$82.5 = w + 51$

$31.5 = w$

The tennis court is 31.5 ft wide and 82.5 ft long.

22. $n + d = 350$

$0.05n + 0.1d = 25$

$-0.1n - 0.1d = -35$

$0.05n + 0.1d = 25$

$-0.05n = -10$

$n = 200$

$200 + d = 350$

$d = 150$

The coin box contained 200 nickels and 150 dimes.

23. $x - y = 5$
$4(x + y) = 500$

$x - y = 5$
$x + y = 125$
$2x = 130$
$x = 65$

$65 - y = 5$
$-y = -60$
$y = 60$

One train travels at a rate of 60 mph and the other travels at a rate of 65 mph.

24. $3x + 2y = 2437$
$x + y = 1085$

$3x + 2y = 2437$
$-3x - 3y = -3255$
$-y = -818$
$y = 818$

$x + 818 = 1085$
$x = 267$

The team made 818 two-point baskets and 267 three-point baskets.

25. $x + y = 200$
$0.1x + 0.3y = 0.25(200)$

$-0.1x + -0.1y = -20$
$0.1x + 0.3y = 50$
$0.2y = 30$
$y = 150$

$x + 150 = 200$
$x = 50$

The pharmacist should mix 150 ml of the 30% solution and 50 ml of the 10% solution.

26. $x + y = 2000$
$0.5x + y = 0.65(2000)$

$x + y = 2000$
$-0.5x - y = -1300$
$0.5x = 700$
$x = 1400$

$1400 + y = 2000$
$y = 600$

1400 L of the 50% solution and 600 L of water are needed to fill the tank.

27. $x + y = 50{,}000$
$0.06x + 0.08y = 3200$

$-0.08x - 0.08y = -4000$
$0.06x + 0.08y = 3200$
$-0.02x = -800$
$x = 40{,}000$

$40{,}000 + y = 50{,}000$
$y = 10{,}000$

The client put $40,000 in municipal bonds and $10,000 in corporate stocks.

28. $x + y = 10{,}000$
$0.12x + 0.02y = 900$

$-0.02x - 0.02y = -200$
$0.12x + 0.02y = 900$
$0.1x = 700$
$x = 7000$

$7000 + y = 10{,}000$
$y = 3000$

$7000 was invested in the high-risk fund and $3000 was invested in the low-risk fund.

29. $2(x + y) = 1800$
$x - y = 100$

$x + y = 900$
$x - y = 100$
$2x = 1000$
$x = 500$

$500 + y = 900$
$y = 400$

The speed of the slower plane was 400 mph.

30. $0.5(x+y)=13$
$0.5(x-y)=8$

$x+y=26$
$x-y=16$
$2x=42$
$x=21$

$21+y=26$
$y=5$

The bird flies at a speed of 21 mph in calm air and the speed of the wind is 5 mph.

31. **a.** $x+y=100$
$x-y=14$
$2x=114$
$x=57$
57 senators

b. $57+y=100$
$y=43$
43 senators

32. **a.** $y=2+3x$
$y=86-4x$

$2+3x=86-4x$
$2+7x=86$
$7x=84$
$x=12$

The two metals will be the same temperature after 12 min.

b. $y=2+3x$
$y-14=86-4x$

$y=2+3x$
$y=100-4x$
$2+3x=100-4x$
$2+7x=100$
$7x=98$
$x=14$

The iron will be 14° colder than the copper after 14 min.

33. **a.** $2t+3a\le 500$ (assembly)
$1.5t+2a\le 300$ (painting)

b. $3a\le -2t+500$
$a\le -\frac{2}{3}t+\frac{500}{3}$

The border line is solid.

t	$a=-\frac{2}{3}t+\frac{500}{3}$	(t,a)
0	$-\frac{2}{3}(0)+\frac{500}{3}=\frac{500}{3}$	$\left(0,166\frac{2}{3}\right)$
250	$-\frac{2}{3}(250)+\frac{500}{3}=0$	$(250,0)$

Test point is $(0,0)$.

$2(0)+3(0)\overset{?}{\le}500$
$0\le 500$

Shade the half-plane containing $(0,0)$.

$2a\le -1.5t+300$
$a\le -0.75t+150$
The boundary line is solid.

t	$a=-0.75t+150$	(t,a)
0	$-0.75(0)+150=150$	$(0,150)$
200	$-0.75(200)+150=0$	$(200,0)$

Test point is $(0,0)$.

$1.5(0)+2(0)\overset{?}{\le}300$
$0\le 300$

Shade the half-plane containing $(0,0)$.

The solution of the system is the overlapping region.

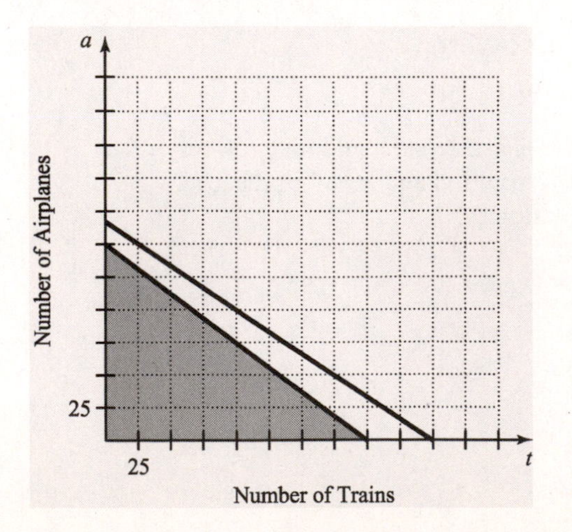

33. **c.** The solution region represents all the possible production units of trains and corresponding number of airplanes the company can assemble and paint each month.

34. **a.** Let x be the amount of money invested at 6% and y be the amount of money invested at 8%.

$$x + y \leq 12{,}000$$
$$0.08y > 0.06x$$
$$y \geq 4500$$

b. $y \leq -x + 12{,}000$

The boundary line is solid.

x	$y = -x + 12{,}000$	(x, y)
0	$-(0) + 12000 = 12{,}000$	$(0, 12{,}000)$
12,000	$-(12{,}000) + 12{,}000 = 0$	$(12{,}000, 0)$

The test point is $(0,0)$.

$$(0) + (0) \stackrel{?}{\leq} 12{,}000$$
$$0 \leq 12{,}000$$

Shade the half-plane containing $(0,0)$.

$$0.08y > 0.06x \Rightarrow y > \frac{3}{4}x$$

The boundary line is dashed.

x	$y = \frac{3}{4}x$	(x, y)
0	$\frac{3}{4}(0) = 0$	$(0,0)$
12,000	$\frac{3}{4}(12{,}000) = 9000$	$(12{,}000, 9000)$

Test point is $(0, 12{,}000)$.

$$0.08(12{,}000) \stackrel{?}{>} 0.06(0)$$
$$960 > 0$$

Shade the half-plane containing $(0, 12{,}000)$.

$y \geq 4500$

The boundary line is solid

x	$y = 4500$	(x, y)
0	4500	$(0, 4500)$
12,000	4500	$(12{,}000, 4500)$

Test point is $(0,0)$.

$0 \not\geq 4500$

Shade the half-plane not containing $(0,0)$.

The solution of the system is the overlapping region.

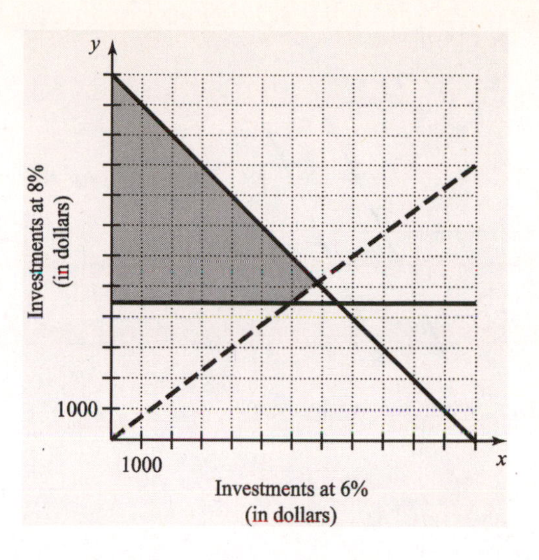

Chapter 4 Posttest

1. **a.** $(0) + (-1) = -1$

$-1 = -1$ True

$3(0) - (-1) = 1$

$1 = 1$ True

A solution

b. $(-1) + (0) = -1$

$-1 = -1$ True

$3(-1) - (0) = 1$

$-3 = 1$ False

Not a solution

c. $(2) + (-3) = -1$

$-1 = -1$ True

$3(2) - (-3) = 1$

$6 + 3 = 1$

$9 = 1$ False

Not a solution

2. The system has no solution.

3.

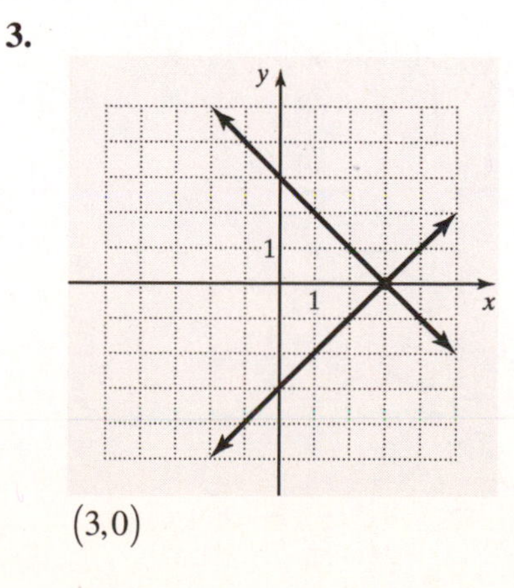

$(3,0)$

4.

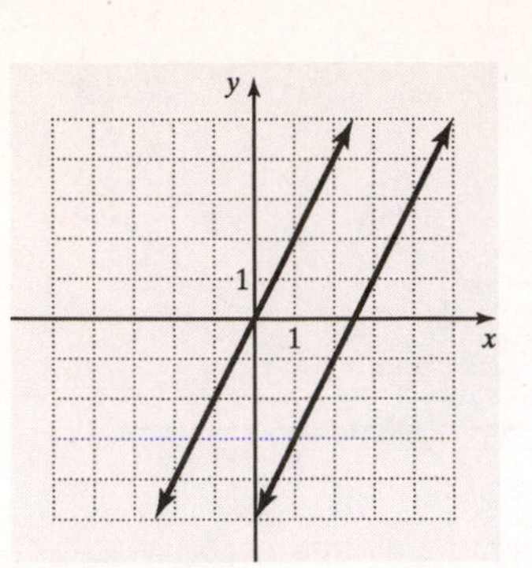

The system has no solution.

5.

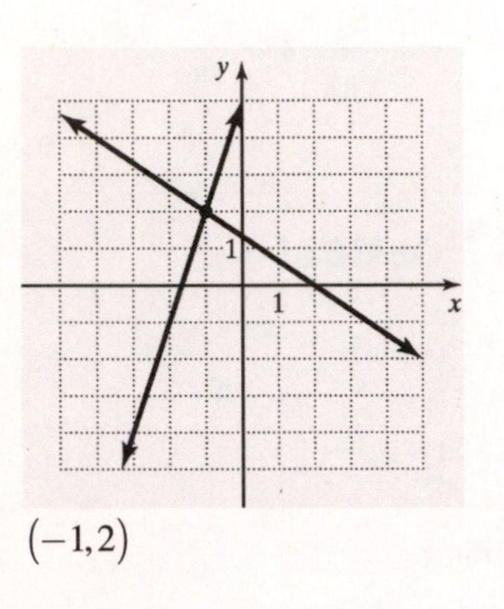

$(-1,2)$

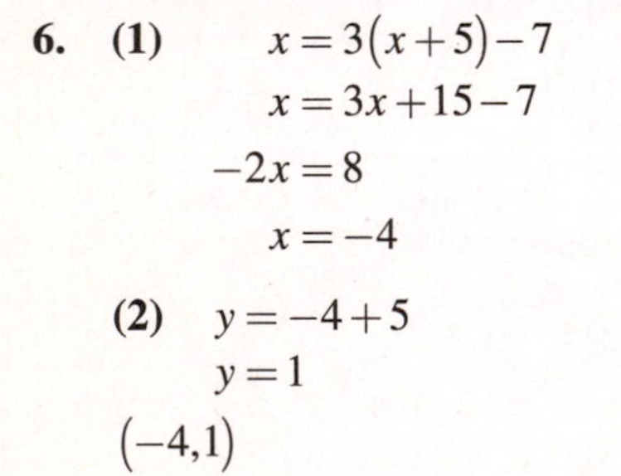

6. **(1)**

$$x=3(x+5)-7$$
$$x=3x+15-7$$
$$-2x=8$$
$$x=-4$$

(2)

$$y=-4+5$$
$$y=1$$

$(-4,1)$

7. **(1)**

$$u-3v=-12$$
$$u=3v-12$$

(2)

$$5(3v-12)+v=8$$
$$15v-60+v=8$$
$$16v=68$$
$$v=\frac{68}{16}$$
$$v=\frac{17}{4}$$

(1)

$$u-3\left(\frac{17}{4}\right)=-12$$
$$u-\frac{51}{4}=-\frac{48}{4}$$
$$u=-\frac{48}{4}+\frac{51}{4}$$
$$u=\frac{3}{4}$$

$u=\frac{3}{4},\ v=\frac{17}{4}$

8.

$$4x+y=3$$
$$7x-y=19$$
$$11x=22$$
$$x=2$$

$$4(2)+y=3$$
$$8+y=3$$
$$y=-5$$

$(2,-5)$

9.

$$x-y=5$$
$$2x-2y=5$$

$$-2x+2y=-10$$
$$2x-2y=5$$
$$0=-5$$

No solution

10

$$-5p+2q=1$$
$$4p+3q=1.5$$

$$15p-6q=-3$$
$$8p+6q=3$$
$$23p=0$$
$$p=0$$

$$-5(0)+2q=1$$
$$2q=1$$
$$q=0.5$$

$p=0,\ q=0.5$

11. (2) $-4x+y=1$

$$y=4x+1$$

(1) $7x=2(4x+1)$

$$7x=8x+2$$
$$-x=2$$
$$x=-2$$

(2) $-4(-2)+y=1$

$$8+y=1$$
$$y=-7$$

$(-2,-7)$

12.

$$4l=-(m+3)$$
$$8l+2m=-6$$

$$4l=-m-3$$
$$8l+2m=-6$$

$$4l+m=-3$$
$$8l+2m=-6$$

$$-8l-2m=6$$
$$8l+2m=-6$$
$$0=0$$

Infinitely many solutions

13.

$$5x+2y=-1$$
$$x-1=y$$

$$5x+2y=-1$$
$$x-y=1$$

$$5x+2y=-1$$
$$2x-2y=2$$
$$7x=1$$
$$x=\frac{1}{7}$$

$$\frac{1}{7}-1=y$$
$$-\frac{6}{7}=y$$

$\left(\frac{1}{7},-\frac{6}{7}\right)$

14. (1) $y=2x$

(2) $x+y=6306$

$$x+2x=6306$$
$$3x=6306$$
$$x=2102$$

(1) $y=2(2102)$

$$y=4204$$

The winning candidate got 4204 votes.

15.

$$3t+3s=9$$
$$135t+99s=333$$

$$-99t-99s=-297$$
$$135t+99s=333$$
$$36t=36$$
$$t=1$$

$$3(1)+3s=9$$
$$3s=6$$
$$s=2$$

One serving of turkey and two servings of salmon.

16. (1) $x=2y$

(2) $0.075x+0.06y=840$

$$0.075(2y)+0.06y=840$$
$$0.15y+0.06y=840$$
$$0.21y=840$$
$$y=4000$$

(1) $x=2(4000)$

$$x=8000$$

\$8000 was invested at 7.5% and \$4000 at 6%.

17.

$$x+y=4$$
$$0.2x+0.6y=0.5(4)$$

$$-0.2x-0.2y=-0.8$$
$$0.2x+0.6y=2$$
$$0.4y=1.2$$
$$y=3$$

$$x+3=4$$
$$x=1$$

1 gal of the 20% iodine solution and 3 gal of the 60% iodine solution

18. $a + w = 170$

$a - w = 130$

$2a = 300$

$a = 150$

$150 + w = 170$

$w = 20$

The speed of the wind was 20 mph and the speed of the plane in still air was 150 mph.

19. $6x - 4y \geq -16$

$x + 2y > -2$

$x \leq 3$

$6x - 4y \geq -16 \Rightarrow -4y \geq -6x - 16 \Rightarrow y \leq \frac{3}{2}x + 4$

The boundry line is solid

x	$y = \frac{3}{2}x + 4$	(x, y)
-2	$\frac{3}{2}(-2) + 4 = 1$	$(-2, 1)$
0	$\frac{3}{2}(0) + 4 = 4$	$(0, 4)$

Test point is $(0,0)$.

$6(0) - 4(0) \overset{?}{\geq} -16$

$0 \geq -16$

Shade the half-plane containing $(0,0)$.

$x + 2y > -2 \Rightarrow 2y > -x - 2 \Rightarrow y > -\frac{1}{2}x - 1$

The boundary line is dashed.

x	$y = -\frac{1}{2}x - 1$	(x, y)
0	$-\frac{1}{2}(0) - 1 = -1$	$(0, -1)$
2	$-\frac{1}{2}(2) - 1 = -2$	$(2, -2)$

Test point is $(0,0)$.

$(0) + 2(0) \overset{?}{>} -2$

$0 > -2$

Shade the half-plane containing $(0,0)$.

$x \leq 3$

The boundary line is solid.

$x = 3$	y	(x, y)
3	0	$(3, 0)$
3	2	$(3, 2)$

Test point is $(0,0)$.

$0 \leq 3$, shade the half-plane containing $(0,0)$.

The solution of the system is the overlapping region.

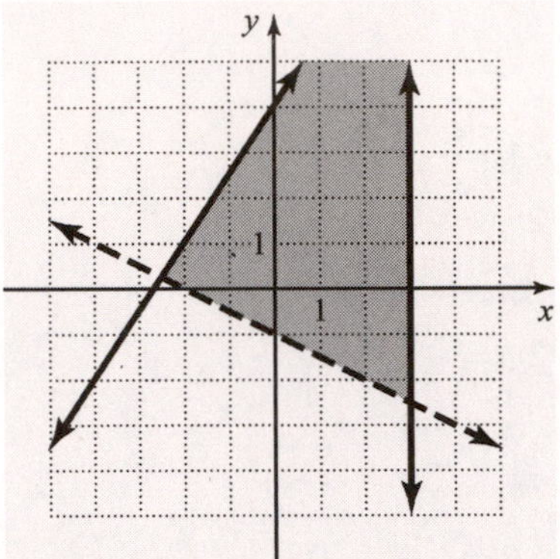

20. Let x be the hours the faster student drives per day and y be the hours the slower student drives per day.

a. $64x + 60y \geq 300$

$x + y \leq 12$

$x > y$

b.

$64x + 60y \geq 300 \Rightarrow 60y \geq -64x + 300 \Rightarrow$

$y \geq -\frac{16}{15}x + 5$

The boundary line is solid.

x	$y = -\frac{16}{15}x + 5$	(x, y)
0	$-\frac{16}{15}(0) + 5 = 5$	$(0, 5)$
3	$-\frac{16}{15}(3) + 5 = \frac{9}{5}$	$\left(3, \frac{9}{5}\right)$

Test point is $(0,0)$.

$64(0) + 60(0) \overset{?}{\geq} 300$

$0 \not\geq 300$

Shade the half plane not containing $(0,0)$.

20. (continued)

$x+y\leq 12 \Rightarrow y\leq -x+12$

The boundary line is solid

x	$y=-x+12$	(x,y)
0	$-(0)+12=12$	$(0,12)$
12	$-(12)+12=0$	$(12,0)$

Test point is $(0,0)$.

$$(0)+(0)\overset{?}{\leq}12$$

$$0\leq 12$$

Shade the half-plane containing $(0,0)$.

$x>y\Rightarrow y<x$

The boundary line is dashed.

x	$y=x$	(x,y)
0	0	$(0,0)$
10	10	$(10,10)$

Test point is $(10,0)$.

$0<10$

Shade the half-plane containing $(10,0)$.

The solution of the system is the overlapping region.

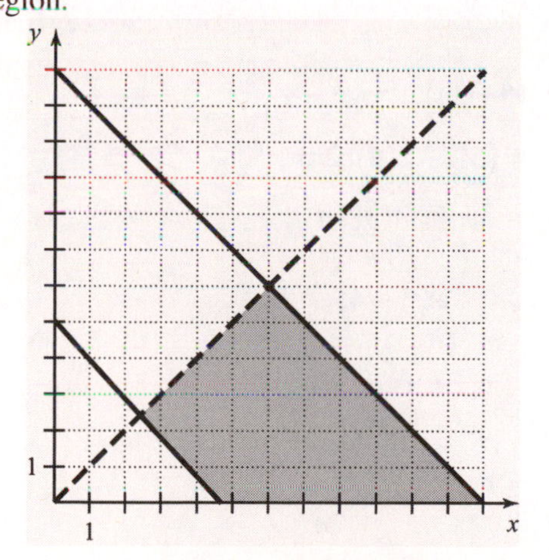

c. Answers may vary. Possible answers: $(6,4)$: The faster student drives 6 hr and the slower student drives 4 hr. $(5,3)$: The faster student drives 5 hr and the slower student drives 3 hr.

Chapter 5

EXPONENTS AND POLYNOMIALS

Pretest

1. $x^5 \cdot x^4 = x^{5+4} = x^9$

2. $y^7 \div y^3 = y^{7-3} = y^4$

3. $-3a^0 = -3(1) = -3$

4. $(4x^4y^3)^2 = 4^2(x^4)^2(y^3)^2 = 16x^8y^6$

5. $\left(\dfrac{a}{b^5}\right)^3 = \dfrac{a^3}{b^{5\cdot 3}} = \dfrac{a^3}{b^{15}}$

6. $(5x^{-1}y^4)^{-2} = 5^{-2}x^{-1\cdot -2}y^{4\cdot -2}$
$= \dfrac{x^2y^{-8}}{25}$
$= \dfrac{x^2}{25y^8}$

7. **a.** $6x^4, 5x^3, x^2, -7x,$ and 8
   b. $6, 5, 1, -7,$ and 8
   c. 4
   d. 8

8. $(2n^2+7n-10)+(n^2-6n+12)$
$= 2n^2+7n-10+n^2-6n+12$
$= 3n^2+n+2$

9. $(8x^2-9)-(7x^2-x-5)$
$= 8x^2-9-7x^2+x+5$
$= x^2+x-4$

10. $(6a^2b+ab-a^2)+(2a^2+3a^2b-5b^2)$
$\quad -(7a^2-2ab-b^2)$
$= 6a^2b+ab-a^2+2a^2+3a^2b-5b^2$
$\quad -7a^2+2ab+b^2$
$= -6a^2+9a^2b+3ab-4b^2$

11. $3x^2(x^2-4x+9)$
$= 3x^2(x^2)+3x^2(-4x)+3x^2(9)$
$= 3x^4-12x^3+27x^2$

12.
$$\begin{array}{r} 2n^2+n-6 \\ n+3 \\ \hline 6n^2+3n-18 \\ 2n^3+n^2-6n \\ \hline 2n^3+7n^2-3n-18 \end{array}$$

13. $(4x+9)(x-3)$
F: $(4x)(x) = 4x^2$
O: $(4x)(-3) = -12x$
I: $(9)(x) = 9x$
L: $(9)(-3) = -27$
$= 4x^2-12x+9x-27$
$= 4x^2-3x-27$

14. $(3y-7)(3y+7)$
$= 9y^2-49$

15. $(5-2n)^2$
$= (5)^2-2(5)(2n)+(2n)^2$
$= 25-20n+4n^2$

16. $\dfrac{9t^4-18t^3-45t^2}{9t^2}$
$= \dfrac{9t^4}{9t^2}-\dfrac{18t^3}{9t^2}-\dfrac{45t^2}{9t^2}$
$= t^2-2t-5$

17.
$$\begin{array}{r} 4x+5 \\ x-2\overline{)4x^2-3x-10} \\ \underline{4x^2-8x} \\ 5x-10 \\ \underline{5x-10} \\ 0 \end{array}$$

18. $(6\times10^{23})\cdot 200 = 1200\times10^{23} = 1.2\times10^{26}$
200 mol of hydrogen will contain 1.2×10^{26} molecules.

19. $A = LW$
$= (2x - 5 + 10)(2x - 5)$
$= (2x + 5)(2x - 5)$
$= (2x)^2 - (5)^2$
$= 4x^2 - 25$

The area of the sandbox is $(4x^2 - 25)$ sq ft.

20. $-0.535t^2 + 2.64t + 45.3$
$-0.535(6)^2 + 2.64(6) + 45.3$
$= -0.535(36) + 2.64(6) + 45.3$
$= -19.26 + 15.84 + 45.3$
$= -3.42 + 45.3$
$= 41.88$

The average monthly cell telephone bill was \$41.88 in 2006.

5.1 Laws of Exponents

Practice

1. **a.** $10^4 = (10)(10)(10)(10) = 10{,}000$

b. $\left(-\frac{1}{2}\right)^5 = \left(-\frac{1}{2}\right)\left(-\frac{1}{2}\right)\left(-\frac{1}{2}\right)\left(-\frac{1}{2}\right)\left(-\frac{1}{2}\right) = -\frac{1}{32}$

c. $(-y)^6 = (-y)(-y)(-y)(-y)(-y)(-y) = y^6$

d. $(-y)^1 = -y$

2. **a.** $8^0 = 1$

b. $\left(-\frac{2}{3}\right)^0 = 1$

c. $y^0 = 1$

d. $-y^0 = -(1) = -1$

e. $(4x)^0 = 1$

3. **a.** $10^8 \cdot 10^4 = 10^{8+4} = 10^{12}$

b. $(-4)^3 \cdot (-4)^3 = (-4)^{3+3} = (-4)^6$

c. $n^3 \cdot n^7 = n^{3+7} = n^{10}$

d. $y^5 \cdot y^0 = y^{5+0} = y^5$

e. $a \cdot b^4$

Cannot apply the product rule because the bases are not the same.

4. **a.** $\frac{7^7}{7^2} = 7^{7-2} = 7^5$

b. $(-9)^6 \div (-9)^5 = (-9)^{6-5} = (-9)^1$, or -9

c. $\frac{s^{10}}{s^{10}} = s^{10-10} = s^0$, or 1

d. $\frac{r^8}{r} = r^{8-1} = r^7$

e. $\frac{a^5}{b^3}$

Cannot apply the quotient rule because the bases are not the same.

5. **a.** $y^2 \cdot y^3 \cdot y^4 = y^{2+3+4} = y^9$

b. $(x^3y^3)(x^2y^3) = x^3 \cdot x^2 \cdot y^3 \cdot y^3 = x^5y^6$

c. $\frac{a^7}{a \cdot a^4} = \frac{a^7}{a^5} = a^2$

6. **a.** $9^{-2} = \frac{1}{9^2} = \frac{1}{81}$

b. $n^{-5} = \frac{1}{n^5}$

c. $-(3y)^{-1} = -\frac{1}{3y}$

d. $5^{-3} = \frac{1}{5^3} = \frac{1}{125}$

7. **a.** $8^{-1}s = \frac{s}{8}$

b. $3x^{-1} = \frac{3}{x}$

c. $\frac{r^3}{r^9} = r^{3-9}$
$= r^{-6}$
$= \frac{1}{r^6}$

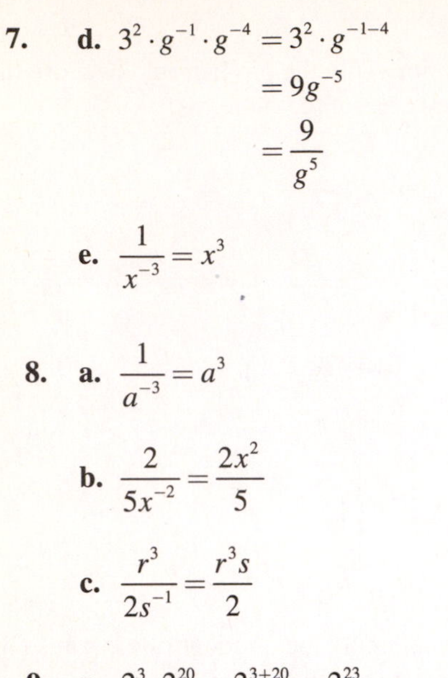

7. **d.** $3^2 \cdot g^{-1} \cdot g^{-4} = 3^2 \cdot g^{-1-4}$
$= 9g^{-5}$
$= \dfrac{9}{g^5}$

e. $\dfrac{1}{x^{-3}} = x^3$

8. **a.** $\dfrac{1}{a^{-3}} = a^3$

b. $\dfrac{2}{5x^{-2}} = \dfrac{2x^2}{5}$

c. $\dfrac{r^3}{2s^{-1}} = \dfrac{r^3 s}{2}$

9. **a.** $2^3 \cdot 2^{20} = 2^{3+20} = 2^{23}$

b. $2^{10} \cdot 2^{20} = 2^{10+20} = 2^{30}$

Exercises

1. The product rule of exponents states that when powers of the same base are multiplied, the exponents are <u>added</u> and the base is left the same.

3. Any nonzero real number raised to the <u>0 power</u> is 1.

5. $5^3 = (5)(5)(5) = 125$

7. $\left(\dfrac{3}{4}\right)^2 = \left(\dfrac{3}{4}\right)\left(\dfrac{3}{4}\right) = \dfrac{9}{16}$

9. $(0.4)^2 = (0.4)(0.4) = 0.16$

11. $-(0.5)^2 = -(0.5)(0.5) = -0.25$

13. $(-2)^3 = (-2)(-2)(-2) = -8$

15. $(-3)^4 = (-3)(-3)(-3)(-3) = 81$

17. $\left(-\dfrac{1}{2}\right)^3 = \left(-\dfrac{1}{2}\right)\left(-\dfrac{1}{2}\right)\left(-\dfrac{1}{2}\right) = -\dfrac{1}{8}$

19. $(-x)^4 = (-x)(-x)(-x)(-x) = x^4$

21. $(pq)^1 = pq$

23. $(-3)^0 = 1$

25. $-a^0 = -(1) = -1$

27. $10^9 \cdot 10^2 = 10^{9+2} = 10^{11}$

29. $4^3 \cdot 4^5 = 4^{3+5} = 4^8$

31. $a^4 \cdot a^2 = a^{4+2} = a^6$

33. $x^3 \cdot y^5$
Cannot be simplified

35. $n^6 \cdot n = n^{6+1} = n^7$

37. $x^2 y$
Cannot be simplified

39. $\dfrac{8^5}{8^3} = 8^{5-3} = 8^2$

41. $\dfrac{5^4}{5} = 5^{4-1} = 5^3$

43. $\dfrac{y^6}{y^5} = y^{6-5} = y^1$ or, y

45. $\dfrac{a^{10}}{a^4} = a^{10-4} = a^6$

47. $\dfrac{y^8}{x^4}$
Cannot be simplified

49. $\dfrac{x^6}{x^6} = x^{6-6} = x^0 = 1$

51. $y^2 \cdot y^3 \cdot y = y^{2+3+1} = y^6$

53. $(p^2 q^3)(p^5 q^2) = p^2 \cdot p^5 \cdot q^3 \cdot q^2 = p^7 q^5$

55. $(yx^2)(xz^2)(yz) = x^2 \cdot x \cdot y \cdot y \cdot z^2 \cdot z$
$= x^3y^2z^3$

57. $\dfrac{a^2 \cdot a^3}{a^4} = \dfrac{a^5}{a^4} = a^1$, or a

59. $\dfrac{x^2 \cdot x^4}{x^3 \cdot x} = \dfrac{x^6}{x^4} = x^2$

61. $5^{-1} = \dfrac{1}{5}$

63. $x^{-1} = \dfrac{1}{x}$

65. $(-3a)^{-1} = \dfrac{1}{-3a} = -\dfrac{1}{3a}$

67. $2^{-4} = \dfrac{1}{2^4}$

69. $-3^{-4} = -\dfrac{1}{3^4}$

71. $8n^{-3} = \dfrac{8}{n^3}$

73. $(-x)^{-2} = \dfrac{1}{(-x)^2} = \dfrac{1}{x^2}$

75. $-3^{-2}x = -\dfrac{x}{3^2}$

77. $x^{-2}y^3 = \dfrac{y^3}{x^2}$

79. $qr^{-1} = \dfrac{q}{r}$

81. $4x^{-1}y^2 = \dfrac{4y^2}{x}$

83. $p^{-2} \cdot p^{-3} = p^{-5} = \dfrac{1}{p^5}$

85. $p^{-1} \cdot p^4 = p^3$

87. $\dfrac{a^3}{a^4} = \dfrac{1}{a}$

89. $\dfrac{2}{n^{-4}} = 2n^4$

91. $\dfrac{p^4}{q^{-1}} = p^4q$

93. $\dfrac{t^{-2}}{t^3} = \dfrac{1}{t^5}$

95. $\dfrac{x^5}{x^{-2}} = x^7$

97. $\dfrac{a^{-4}}{a^{-5}} = a^1$, or a

99. $\dfrac{a^{-3}}{b^{-3}} = \dfrac{b^3}{a^3}$

101. $(s^2t^4)(st^2) = s^2 \cdot s \cdot t^4 \cdot t^2 = s^3t^6$

103. $\left(-\dfrac{2}{3}\right)^3 = \left(-\dfrac{2}{3}\right)\left(-\dfrac{2}{3}\right)\left(-\dfrac{2}{3}\right) = -\dfrac{8}{27}$

105. x^2y^3
Cannot be simplified

107. $(-y)^{-6} = \dfrac{1}{(-y)^6} = \dfrac{1}{y^6}$

109. $q^{-2} \cdot q^{-3} = q^{-5} = \dfrac{1}{q^5}$

111. $\dfrac{x^4}{y^{-3}} = x^4y^3$

113. **a.** $35 \cdot 2^5 = 1120$ people were ill on the sixth day of the epidemic.
$35 \cdot 2^9 = 17{,}290$ people were ill on the tenth day.

b. The number of people ill on the tenth day was 2^4, or 16, times as great as the number ill on the sixth day.

115. $60\times(0.95)^{11}$ ppm

117. **a.** Volume of small box: $(2x)^3=8x^3$

Volume of large box: $(5x)^3=125x^3$

b. $\dfrac{125x^3}{8x^3}=15\dfrac{5}{8}$

The volume of the larger box is $15\dfrac{5}{8}$ times the volume of the small box.

5.2 More Laws of Exponents and Scientific Notation

Practice

1. **a.** $(2^3)^2=2^{3\cdot 2}$
$=2^6$
$=64$

b. $(7^3)^{-1}=7^{3(-1)}$
$=7^{-3}$
$=\dfrac{1}{7^3}$
$=\dfrac{1}{343}$

c. $(q^2)^4=q^{2\cdot 4}=q^8$

d. $-(p^3)^{-5}=-p^{3(-5)}$
$=-p^{-15}$
$=-\dfrac{1}{p^{15}}$

2. **a.** $(7a)^2=7^2\cdot a^2=49a^2$

b. $(-4x)^3=(-4)^3\cdot x^3=-64x^3$

c. $-(4x)^3=-4^3\cdot x^3=-64x^3$

3. **a.** $(-6a^9)^2=(-6)^2(a^9)^2=36a^{18}$

b. $(q^8r^{10})^2=(q^8)^2(r^{10})^2=q^{16}r^{20}$

c. $-2(ab^7)^3=-2(a)^3(b^7)^3=-2a^3b^{21}$

d. $(7a^{-1}c^{-5})^2=7^2(a^{-1})^2(c^{-5})^2$
$=49a^{-2}c^{-10}$
$=\dfrac{49}{a^2c^{10}}$

4. **a.** $\left(\dfrac{y}{3}\right)^2=\dfrac{y^2}{3^2}=\dfrac{y^2}{9}$

b. $\left(\dfrac{-u}{v}\right)^{10}=\dfrac{(-u)^{10}}{v^{10}}=\dfrac{u^{10}}{v^{10}}$

c. $\left(\dfrac{3}{y}\right)^{-2}=\dfrac{3^{-2}}{y^{-2}}=\dfrac{y^2}{3^2}=\dfrac{y^2}{9}$

d. $\left(\dfrac{-10a^5}{3b^2c}\right)^2=\dfrac{(-10a^5)^2}{(3b^2c)^2}$
$=\dfrac{(-10)^2(a^5)^2}{3^2(b^2)^2c^2}$
$=\dfrac{100a^{10}}{9b^4c^2}$

e. $\left(\dfrac{5x}{y^{-2}}\right)^3=\dfrac{(5x)^3}{(y^{-2})^3}=\dfrac{5^3x^3}{y^{-6}}=125x^3y^6$

5. **a.** $\left(\dfrac{5}{a}\right)^{-2}=\left(\dfrac{a}{5}\right)^2=\dfrac{a^2}{5^2}=\dfrac{a^2}{25}$

b. $\left(\dfrac{4u}{v}\right)^{-1}=\left(\dfrac{v}{4u}\right)^1=\dfrac{v}{4u}$

c. $\left(\dfrac{a^5}{b^3}\right)^{-2}=\left(\dfrac{b^3}{a^5}\right)^2=\dfrac{(b^3)^2}{(a^5)^2}=\dfrac{b^6}{a^{10}}$

6. $2.539\times10^2=2.539\times100$
$=253.9$

7. $4.3\times10^{-9}=\dfrac{4.3}{1{,}000{,}000{,}000}$
$=0.0000000043$

8. $8,000,000,000,000 = 8\times10^{12}$

9. $0.000000000071 = 7.1\times10^{-11}$

10. **a.** $\left(7\times10^{-2}\right)\left(3.52\times10^{3}\right)$
$= \left(7\times3.52\right)\left(10^{-2}\times10^{3}\right)$
$= 24.64\times10^{-2+3}$
$= 24.64\times10$
$= 2.464\times10^{1}\times10$
$= 2.464\times10^{2}$

b. $\left(2.4\times10^{3}\right)\div\left(6\times10^{-9}\right)$
$= \dfrac{2.4\times10^{3}}{6\times10^{-9}}$
$= \dfrac{2.4}{6}\times\dfrac{10^{3}}{10^{-9}}$
$= 0.4\times10^{3-(-9)}$
$= 0.4\times10^{12}$
$= 4\times10^{-1}\times10^{12}$
$= 4\times10^{11}$

11. $\left(1.5\times10^{5}\right)\left(6\times10^{9}\right)$
$= \left(1.5\times6\right)\left(10^{5}\times10^{9}\right)$
$= 9\times10^{5+9}$
$= 9\times10^{14}$

12. $\left(2\times10^{-10}\right)\div\left(2\times10^{-7}\right)$
$= \dfrac{2\times10^{-10}}{2\times10^{-7}}$
$= \dfrac{2}{2}\times\dfrac{10^{-10}}{10^{-7}}$
$= 1\times10^{-10-(-7)}$
$= 1\times10^{-3}$
$= 0.001$

13. $2.5\text{E}-17$ (Answers may vary.)

14. $7.3\text{E}-10$ (Answers may vary.)

15. 4.6×10^{8}, or $460,000,000$

Exercises

1. The expression $\left(x^{3}\right)^{2}$ contains two factors of x^{3}.

3. To raise a product to a power, raise each factor to that power.

5. To raise a quotient to a negative power, raise the reciprocal of the quotient to the opposite of the given power.

7. To convert a number from scientific to standard notation, move the decimal point to the left if the power of 10 is negative.

9. $\left(2^{2}\right)^{4} = 2^{2\cdot4} = 2^{8} = 256$

11. $\left(5^{2}\right)^{2} = 5^{2\cdot2} = 5^{4} = 625$

13. $\left(10^{5}\right)^{2} = 10^{5\cdot2} = 10^{10} = 10,000,000,000$

15. $\left(4^{-2}\right)^{2} = 4^{-2\cdot2} = 4^{-4} = \dfrac{1}{4^{4}} = \dfrac{1}{256}$

17. $\left(x^{4}\right)^{6} = x^{4\cdot6} = x^{24}$

19. $\left(y^{4}\right)^{2} = y^{4\cdot2} = y^{8}$

21. $\left(x^{-2}\right)^{3} = x^{-2\cdot3} = x^{-6} = \dfrac{1}{x^{6}}$

23. $\left(n^{-2}\right)^{-2} = n^{-2\cdot(-2)} = n^{4}$

25. $(4x)^{3} = 4^{3}x^{3} = 64x^{3}$

27. $(-8y)^{2} = (-8)^{2}y^{2} = 64y^{2}$

29. $-\left(4n^{5}\right)^{3} = -(4)^{3}\left(n^{5}\right)^{3} = -64n^{15}$

31. $4\left(-2y^{2}\right)^{4} = 4(-2)^{4}\left(y^{2}\right)^{4} = 4(16)y^{8} = 64y^{8}$

33. $(3a)^{-2} = \dfrac{1}{3^{2}a^{2}} = \dfrac{1}{9a^{2}}$

35. $(pq)^{-7} = \frac{1}{(pq)7} = \frac{1}{p^7q^7}$

37. $(r^2t)^6 = r^{2\cdot6}t^6 = r^{12}t^6$

39. $(-2p^5q)^2 = (-2)^2 p^{5\cdot2}q^2 = 4p^{10}q^2$

41. $-2(m^4n^8)^3 = -2m^{4\cdot3}n^{8\cdot3} = -2m^{12}n^{24}$

43. $(-4m^5n^{-10})^3 = (-4)^3 m^{5\cdot3}n^{-10\cdot3} = -\frac{64m^{15}}{n^{30}}$

45. $(a^3b^2)^{-4} = a^{3(-4)}b^{2(-4)} = \frac{1}{a^{12}b^8}$

47. $(4x^{-2}y^3)^2 = (4)^2 x^{-2(2)}y^{3(2)} = \frac{16y^6}{x^4}$

49. $\left(\frac{5}{b}\right)^3 = \frac{5^3}{b^3} = \frac{125}{b^3}$

51. $\left(\frac{c}{b}\right)^2 = \frac{c^2}{b^2}$

53. $-\left(\frac{a}{b}\right)^7 = -\frac{a^7}{b^7}$

55. $\left(\frac{a^2}{3}\right)^3 = \frac{a^{2\cdot3}}{3^3} = \frac{a^6}{27}$

57. $\left(-\frac{p^3}{q^2}\right)^5 = \frac{(-1)^5 p^{3\cdot5}}{q^{2\cdot5}} = -\frac{p^{15}}{q^{10}}$

59. $\left(\frac{a}{4}\right)^{-1} = \left(\frac{4}{a}\right)^1 = \frac{4}{a}$

61. $\left(\frac{2x^5}{y^2}\right)^3 = \frac{2^3x^{5\cdot3}}{y^{2\cdot3}} = \frac{8x^{15}}{y^6}$

63. $\left(\frac{pq}{p^2q^2}\right)^5 = \frac{p^5q^5}{p^{2\cdot5}q^{2\cdot5}} = \frac{p^5q^5}{p^{10}q^{10}} = \frac{1}{p^5q^5}$

65. $\left(\frac{3x}{y^{-3}}\right)^4 = (3xy^3)^4 = 3^4x^4(y^3)^4 = 81x^4y^{12}$

67. $\left(\frac{-u^2v^3}{4vu^4}\right)^2 = \left(\frac{-v^2}{4u^2}\right)^2 = \frac{(-1)^2v^{2\cdot2}}{4^2u^{2\cdot2}} = \frac{v^4}{16u^4}$

69.
$$\left(-\frac{x^{-2}y}{2z^{-4}}\right)^4 = \left(-\frac{yz^4}{2x^2}\right)^4$$
$$= \frac{(-1)^4y^4z^{4\cdot4}}{2^4x^{2\cdot4}}$$
$$= \frac{y^4z^{16}}{16x^8}$$

71. $\left(\frac{r^5}{t^6}\right)^{-2} = \left(\frac{t^6}{r^5}\right)^2 = \frac{t^{6\cdot2}}{r^{5\cdot2}} = \frac{t^{12}}{r^{10}}$

73.
$$\left(\frac{-2a^4}{b^2}\right)^{-3} = \left(\frac{b^2}{-2a^4}\right)^3$$
$$= \frac{b^{2\cdot3}}{(-2)^3a^{4\cdot3}}$$
$$= \frac{b^6}{-8a^{12}} \text{ or } = -\frac{b^6}{8a^{12}}$$

75.
$$3.17\times10^8 = 3.17\times100{,}000{,}000$$
$$= 317{,}000{,}000$$

77.
$$1\times10^{-6} = \frac{1}{1{,}000{,}000}$$
$$= 0.000001$$

79.
$$6.2\times10^6 = 6.2\times1{,}000{,}000$$
$$= 6{,}200{,}000$$

81.
$$4.025\times10^{-5} = \frac{4.025}{100{,}000}$$
$$= 0.00004025$$

83. $420{,}000{,}000 = 4.2\times10^8$

85. $0.0000035 = 3.5\times10^{-6}$

87. $217{,}000{,}000{,}000 = 2.17\times10^{11}$

89. $0.00000000731 = 7.31\times10^{-9}$

91.

Standard Notation	Scientific Notation (written)	Scientific Notation (on a calculator)
975,000,000	9.75×10^{8}	9.75E8
487,000,000	4.87×10^{8}	4.87E8
0.0000000001652	1.652×10^{-10}	1.652E−10
0.000000067	6.7×10^{-8}	6.7E−8
0.0000000000001	1×10^{-13}	1E−13
3,281,000,000	3.281×10^{9}	3.281E9

93. $(3\times10^{2})(3\times10^{5})$
$=(3\times3)(10^{2}\times10^{5})$
$=9\times10^{2+5}$
$=9\times10^{7}$

95. $(2.5\times10^{-2})(8.3\times10^{-3})$
$=(2.5\times8.3)(10^{-2}\times10^{-3})$
$=20.75\times10^{-2+(-3)}$
$=20.75\times10^{-5}$
$=2.075\times10^{1}\times10^{-5}$
$=2.075\times10^{-4}$

97. $(8.6\times10^{9})(4.4\times10^{-12})$
$=(8.6\times4.4)(10^{9}\times10^{-12})$
$=37.84\times10^{(9+-12)}$
$=37.84\times10^{-3}$
$=3.784\times10^{1}\times10^{-3}$
$=3.784\times10^{-2}$

99. $(2.5\times10^{8})\div(2\times10^{-2})$
$=\frac{2.5}{2}\times\frac{10^{8}}{10^{-2}}$
$=1.25\times10^{8-(-2)}$
$=1.25\times10^{10}$

101. $(6\times10^{5})\div(2\times10^{3})$
$=\frac{6}{2}\times\frac{10^{5}}{10^{3}}$
$=3\times10^{5-3}$
$=3\times10^{2}$

103. $(9.6\times10^{20})\div(3.2\times10^{12})$
$=\frac{9.6}{3.2}\times\frac{10^{20}}{10^{12}}$
$=3\times10^{20-12}$
$=3\times10^{8}$

105. $3.067\times10^{-4}=\frac{3.067}{10,000}$
$=0.0003067$

107.

Standard Notation	Scientific Notation (written)	Scientific Notation (on a calculator)
428,000,000,000	4.28×10^{11}	4.28E11
3,240,000	3.24×10^{6}	3.24E6
0.000005224	5.224×10^{-6}	5.224E−6
0.000000057	5.7×10^{-8}	5.7E−8
0.000000682	6.82×10^{-7}	6.82E−7
48,360,000	4.836×10^{7}	4.836E7

109. $\left(\frac{a^{2}}{3b^{5}}\right)^{-2}=\left(\frac{3b^{5}}{a^{2}}\right)^{2}=\frac{3^{2}b^{5\cdot2}}{a^{2\cdot2}}=\frac{9b^{10}}{a^{4}}$

111. $-(4y)^{-3}=-\left(\frac{1}{4y}\right)^{3}=-\frac{1}{4^{3}y^{3}}=-\frac{1}{64y^{3}}$

113. $-\left(\frac{2x^{-2}}{y^{-3}z}\right)^{4}$
$=-\left(\frac{2y^{3}}{x^{2}z}\right)^{4}$
$=-\frac{2^{4}y^{3\cdot4}}{x^{2\cdot4}z^{4}}$
$=-\frac{16y^{12}}{x^{8}z^{4}}$

115. $(6.3\times10^{-4})\div(9\times10^{3})$
$=\frac{6.3}{9}\times\frac{10^{-4}}{10^{3}}$
$=0.7\times10^{-4-3}$
$=0.7\times10^{-7}$
$=7\times10^{-1}\times10^{-7}$
$=7\times10^{-8}$

117. Box 1: $(x)^3 = x^3$

Box 2: $(2x)^3 = 2^3 x^3 = 8x^3$

$\frac{8x^3}{x^3}$

The larger volume is 8 times the smaller volume.

119. 4,000,000,000 and 17,000,000,000 bytes

121. 7×10^{-7} m

123. 2×10^{11} cells

125. 0.0000000000000000000000017 g

127. 780,000,000

129. $100(3.2\times10^4)(5\times10^6)$

$=100(3.2\times5)(10^4\times10^6)$

$=1600\times10^{10}$

$=1.6\times10^3\times10^{10}$

$=1.6\times10^{13}$ red blood cells

131. **a.** 1.86×10^3 mi per sec

b. $(1.58\times10^{14})\div(1.86\times10^3)$

$=\frac{1.58}{1.86}\times\frac{10^{14}}{10^3}$

$=0.8495\times10^9$

$=8.495\times10^8$

About 8.495×10^8 seconds, or about 27 years.

5.3 Basic Concepts of Polynomials

Practice

1. **a.** $-10x^2, 4x,$ and 20

b. $-10, 4,$ and 20

2. $2x+9$; Binomial

$-4x^2$; Monomial

$12p-1$; Binomial

3. **a.** Degree 1

b. Degree 2

c. Degree 3

d. Degree 0

4.

Polynomial	Constant Term	Leading Term	Leading Coefficient
$-3x^7+9$	9	$-3x^7$	−3
x^5	0	x^5	1
x^4-7x-1	−1	x^4	1
$3x+5x^3+20$	20	$5x^3$	5

5. **a.** $9x^5-7x^4+9x^2-8x-6$

b. $7x^5+x^3-3x^2+8$

6. $2x^2+3x-x^2+5x^3+3x-5x^3+20$

$=x^2+6x+20$

7. **a.** $x^2-5x+5=(2)^2-5(2)+5$

$=4-5(2)+5$

$=4-10+5$

$=-1$

b. $x^2-5x+5=(-2)^2-5(-2)+5$

$=4-5(-2)+5$

$=4+10+5$

$=19$

8. $500-16t^2=500-16(3)^2$

$=500-16(9)$

$=500-144$

$=356$

356 ft

Exercises

1. A monomial is an expression that is the product of a real number and variables raised to nonnegative integer powers.

3. The degree of a monomial is the power of the monomial's variable.

5. The leading term of a polynomial is the term in the polynomial with the highest degree.

7. The term of degree 0 in a polynomial is called the constant term.

9. Polynomial

11. Not a polynomial

13. Polynomial

15. Not a polynomial

17.

$5x-1$;	Binomial
$-5a^2$;	Monomial
$-6a+3$;	Binomial
x^3+4x^2+2;	Trinomial

19. $-4x^3+3x^2-2x+8$; degree 3

21. $-3y+2$; degree 1

23. $-5x^2+7x$; degree 2

25. $-4y^5-y^3-2y+2$; degree 5

27. $5a^2-a$; degree 2

29. $3p^3+9p$; degree 3

31.

Polynomial	Constant Term	Leading Term	Leading Coefficient
$-x^7+2$	2	$-x^7$	-1
$2x-3$	-30	$2x$	2
$-5x+1+x^2$	1	x^2	1
$7x^3-2x-3$	-3	$7x^3$	7

33. $$9x^3-7x^2+1+x^3+10x+5$$
$$=10x^3-7x^2+10x+6$$

35. $$r^3+2r^2+15+r^2-8r-1$$
$$=r^3+3r^2-8r+14$$

37. $0x^2$

39. $0x$

41. $$\begin{aligned}7x-3&=7(2)-3\\&=14-3\\&=11\end{aligned}$$

$$\begin{aligned}7x-3&=7(-2)-3\\&=-14-3\\&=-17\end{aligned}$$

43. $$\begin{aligned}n^2-3n+9&=(7)^2-3(7)+9\\&=49-3(7)+9\\&=49-21+9\\&=37\end{aligned}$$

$$\begin{aligned}n^2-3n+9&=(-7)^2-3(-7)+9\\&=49-3(-7)+9\\&=49+21+9\\&=79\end{aligned}$$

45. $$\begin{aligned}&2.1x^2+3.9x-7.3\\&=2.1(2.37)^2+3.9(2.37)-7.3\\&=2.1(5.6169)+3.9(2.37)-7.3\\&=11.7955+9.243-7.3\\&=13.73849\end{aligned}$$

$$\begin{aligned}&2.1x^2+3.9x-7.3\\&=2.1(-2.37)^2+3.9(-2.37)-7.3\\&=2.1(5.6169)+3.9(-2.37)-7.3\\&=11.7955-9.243-7.3\\&=-4.74751\end{aligned}$$

47. $$4x^2-2x-x^2-10-3x+4$$
$$=3x^2-5x-6$$

49. $$6n^3+20n-n^2+2-4n^3+15n^2+8$$
$$=2n^3+14n^2+20n+10$$

51. $5a^4-a^3+2a-4$; degree 4

53.

Polynomial	Constant Term	Leading Term	Leading Coefficient
$4x-20$	-20	$4x$	4
$-7x^6+9$	9	$-7x^6$	-7
$3x+2+x^2$	2	x^2	1
$6x-x^5+11$	11	$-x^5$	-1

55. Not a polynomial

57.
$$\begin{aligned}&5.3x^2-2.7x-6.8\\&=5.3(1.25)^2-2.7(1.25)-6.8\\&=5.3(1.5625)-2.7(1.25)-6.8\\&=8.28125-3.375-6.8\\&=-1.89375\end{aligned}$$

$$\begin{aligned}&5.3x^2-2.7x-6.8\\&=5.3(-3.87)^2-2.7(-3.87)-6.8\\&=5.3(14.9769)-2.7(-3.87)-6.8\\&=79.3776+10.449-6.8\\&=83.02657\end{aligned}$$

59. $0x^3$ and $0x^0$ or 0

61. A polynomial in x; degree 16

63.
$$\begin{aligned}x+\frac{x^2}{20}&=(40)+\frac{(40)^2}{20}\\&=40+\frac{1600}{20}\\&=40+80\\&=120\end{aligned}$$
120 ft

65.
$$\begin{aligned}72x+2342&=72(50)+2342\\&=3600+2342\\&=5942\end{aligned}$$
5,942,000,000 people, or 5942 million people

67.
$$\begin{aligned}&1.68x^3+9.95x^2-11.6x+730\\&=1.68(4)^3+9.95(4)^2-11.6(4)+730\\&=1.68(64)+9.95(16)-11.6(4)+730\\&=107.52+159.2-46.4+730\\&=950.32\end{aligned}$$
About 950 U.S. radio stations

5.4 Addition and Subtraction of Polynomials

Practice

1.
$$\begin{aligned}&(6x-3)+(9x^2-3x-40)\\&=6x-3+9x^2-3x-40\\&=9x^2+3x-43\end{aligned}$$

2.
$$\begin{aligned}&(9p^2+4pq+2q^2)+(-p^2-5q^2)+\\&\qquad(2p^2-3pq-7q^2)\\&=9p^2+4pq+2q^2-p^2-5q^2+\\&\qquad 2p^2-3pq-7q^2\\&=10p^2+pq-10q^2\end{aligned}$$

3.
$$\begin{array}{rrr}8n^2 & +2n & -1\\ 3n^2 & & -2\\ \hline 11n^2 & +2n & -3\end{array}$$

4.
$$\begin{array}{rrrrr}7p^3 & -8p^2q & -3pq^2 & & +20\\ & 10p^2q & +pq^2 & -q^3 & +5\\ p^3 & & & -q^3 & \\ \hline 8p^3 & +2p^2q & -2pq^2 & -2q^3 & +25\end{array}$$

5. a.
$$\begin{aligned}&-(4r-3s)+7r\\&=-4r+3s+7r\\&=3r+3s\end{aligned}$$

b.
$$\begin{aligned}&(2p+5q)+(p-6q)-(3p+2q)\\&=2p+5q+p-6q-3p-2q\\&=-3q\end{aligned}$$

6.
$$\begin{aligned}&(2x-1)-(3x^2+15x-1)\\&=2x-1-3x^2-15x+1\\&=-3x^2-13x\end{aligned}$$

7.
$$\begin{array}{r} 20x-13 \\ \underline{-(5x^2-12x+13)} \end{array}$$

$$\begin{array}{r} 20x-13 \\ \underline{-5x^2+12x-13} \\ -5x^2+32x-26 \end{array}$$

8.
$$\begin{array}{r} 2p^2-7pq+\ 5q^2 \\ \underline{-(3p^2+4pq-12q^2)} \end{array}$$

$$\begin{array}{r} 2p^2-7pq+\ 5q^2 \\ \underline{-3p^2-4pq+12q^2} \\ -p^2-11pq+17q^2 \end{array}$$

9.
$$\begin{array}{r} 0.3x+74.8 \\ \underline{-(0.3x+67.2)} \end{array}$$

$$\begin{array}{r} 0.3x+74.8 \\ \underline{-0.3x-67.2} \\ 7.6 \end{array}$$

The polynomial 7.6 approximates how much greater the life expectancy is for females than for males.

9.
$$\left(2p^3-p^2q-5pq^2+1\right)+\left(3p^2q+2pq^2+4q^3\right)+\left(p^3+q^3\right)$$
$$=2p^3-p^2q-5pq^2+1+3p^2q+2pq^2+4q^3+p^3+q^3$$
$$=3p^3+2p^2q-3pq^2-3q^3+5$$

Exercises

1. $\left(3x^2+6x-5\right)+\left(-x^2+2x+7\right)$
$=3x^2+6x-5-x^2+2x+7$
$=2x^2+8x+2$

3. $\left(2n^3+n\right)+\left(3n^3+8n\right)$
$=2n^3+n+3n^3+8n$
$=5n^3+9n$

5. $\left(10p+3+p^2\right)+\left(p^2-7p-4\right)$
$=10p+3+p^2+p^2-7p-4$
$=2p^2+3p-1$

7. $\left(8x^2+7xy-y^2\right)+\left(3x^2-10xy+3y^2\right)$
$=8x^2+7xy-y^2+3x^2-10xy+3y^2$
$=11x^2-3xy+2y^2$

11.
$$\begin{array}{r} 10x^2-3x-8 \\ \underline{20x+3} \\ 10x^2+17x-5 \end{array}$$

13.
$$\begin{array}{r} 5x^3\qquad +7x-1 \\ \underline{x^2+2x+3} \\ 5x^3+x^2+9x+2 \end{array}$$

15.
$$\begin{array}{r} 5ab^2-3a^2\ +a^3 \\ \underline{2ab^2+9a^2-4a^3} \\ 7ab^2+6a^2-3a^3 \end{array}$$
$=-3a^3+6a^2+7ab^2$

17. $\left(x^2+x+4\right)-\left(2x^2+3x-7\right)$
$=x^2+x+4-2x^2-3x+7$
$=-x^2-2x+11$

19. $\left(x^3+10x^2-8x+3\right)-\left(3x^3+x^2+5x-8\right)$
$=x^3+10x^2-8x+3-3x^3-x^2-5x+8$
$=-2x^3+9x^2-13x+11$

21. $\left(x^2+3x\right)-\left(5x+9\right)$
$=x^2+3x-5x-9$
$=x^2-2x-9$

23. $\left(1-6xy+5x^2-y^2\right)-\left(4y^2-6xy-3\right)$
$=1-6xy+5x^2-y^2-4y^2+6xy+3$
$=5x^2-5y^2+4$

25.
$$\begin{array}{r} 2p^2-3p+5 \\ \underline{7p^2-10p-1} \end{array}$$

$$\begin{array}{r} 2p^2-3p+5 \\ \underline{-7p^2+10p+1} \\ -5p^2+7p+6 \end{array}$$

27.
$$\begin{array}{r} 9t^3 - 12t^2 + 3 \\ -(8t^3 \qquad\quad - 5) \\ \hline \end{array}$$

$$\begin{array}{r} 9t^3 - 12t^2 + 3 \\ -8t^3 \qquad\quad + 5 \\ \hline t^3 - 12t^2 + 8 \end{array}$$

29.
$$\begin{array}{r} 4r^3 - 20r^2s - 7 \\ -(r^3 - \; 3r^2s - 5) \\ \hline \end{array}$$

$$\begin{array}{r} 4r^3 - 20r^2s - 7 \\ -r^3 + \; 3r^2s + 5 \\ \hline 3r^3 - 17r^2s - 2 \end{array}$$

31. $7x - (8x + r)$
$= 7x - 8x - r$
$= -x - r$

33. $2p - (3q + r) = 2p - 3q - r$

35. $(4y - 1) + (3y^2 - y + 5)$
$= 4y - 1 + 3y^2 - y + 5$
$= 3y^2 + 3y + 4$

37. $(m^3 - 6m + 7) - (-9 + 6m)$
$= m^3 - 6m + 7 + 9 - 6m$
$= m^3 - 12m + 16$

39. $(2x^3 - 7x + 8) - (5x^2 + 3x - 1)$
$= 2x^3 - 7x + 8 - 5x^2 - 3x + 1$
$= 2x^3 - 5x^2 - 10x + 9$

41. $(8x^2 + 3x) + (x - 2) + (x^2 + 9)$
$= 8x^2 + 3x + x - 2 + x^2 + 9$
$= 9x^2 + 4x + 7$

43. $(3x - 7) + (2x + 9) - (7x - 10)$
$= 3x - 7 + 2x + 9 - 7x + 10$
$= -2x + 12$

45. $(7x^2y^2 - 10xy + 4) - (2xy + 8) + (x^2y^2 - 10)$
$= 7x^2y^2 - 10xy + 4 - 2xy - 8 + x^2y^2 - 10$
$= 8x^2y^2 - 12xy - 14$

47. $(5m^2 - 9m - 7) - (3m^2 - 4m + 2)$
$= 5m^2 - 9m - 7 - 3m^2 + 4m - 2$
$= 2m^2 - 5m - 9$

49. $(5x^3 - x - 8) - (-3x^2 - 3x + 4)$
$= 5x^3 - x - 8 + 3x^2 + 3x - 4$
$= 5x^3 + 3x^2 + 2x - 12$

51. $(3m - 1) - (4m + 5) + (6m - 2)$
$= 3m - 1 - 4m - 5 + 6m - 2$
$= 5m - 8$

53. $(3x^2 + 5x - 4) + (-5x^2 + x + 7)$
$= 3x^2 + 5x - 4 - 5x^2 + x + 7$
$= -2x^2 + 6x + 3$

55.
$$\begin{array}{r} 6p^2 + 7p - 4 \\ -(p^2 - 4p + 3) \\ \hline \end{array}$$

$$\begin{array}{r} 6p^2 + 7p - 4 \\ -\;p^2 + 4p - 3 \\ \hline 5p^2 + 11p - 7 \end{array}$$

57. **a.** $6.28r^2 + 6.28rh$

b. $6.28r^2 + 6.28rh$
$= 6.28r^2 + 6.28r(r)$
$= 6.28r^2 + 6.28r^2$
$= 12.56r^2$ or $12.56h^2$

59.
$$\begin{array}{r} 31x^3 - 522x^2 + 2083x + 6051 \\ -(-x^3 \quad + 16x^2 + \quad 22x + \; 189) \\ \hline \end{array}$$

$$\begin{array}{r} 31x^3 - 522x^2 + 2083x + 6051 \\ x^3 \quad - 16x^2 - \quad 22x - \; 189 \\ \hline 32x^3 - 538x^2 + 2061x + 5862 \end{array}$$

$(32x^3 - 538x^2 + 2061x + 5862)$ million

61.
$$\begin{array}{r} 2x^3-27x^2+172x+391 \\ -\left(2x^3-22x^2+\ 18x+360\right) \\ \hline \end{array}$$

$$\begin{array}{r} 2x^3-27x^2+172x+391 \\ -2x^3+22x^2-\ 18x-360 \\ \hline -5x^2+\ 154x+31 \end{array}$$

$\left(-5x^2+\ 154x+31\right)$ million

5.5 Multiplication of Polynomials

Practice

1. $\left(-10x^2\right)\left(-4x^3\right)=\left(-10\cdot -4\right)\left(x^{2+3}\right)$
$=40x^5$

2. $\left(7ab^2\right)\left(10a^2b^3\right)\left(-5a\right)$
$=\left(7\cdot 10\cdot -5\right)\left(a\cdot a^2\cdot a\right)\left(b^2b^3\right)$
$=-350\left(a^{1+2+1}\right)\left(b^{2+3}\right)$
$=-350a^4b^5$

3. $\left(-5xy^2\right)^2=\left(-5\right)^2\left(x\right)^2\left(y^2\right)^2$
$=25x^2y^4$

4. $\left(10s^2-3\right)\left(7s\right)=\left(7s\right)\left(10s^2\right)+\left(7s\right)\left(-3\right)$
$=70s^3-21s$

5. $-2m^3n^2\left(-6m^3n^5+2mn^2+n\right)$
$=\left(-2m^3n^2\right)\left(-6m^3n^5\right)+\left(-2m^3n^2\right)\left(2mn^2\right)$
$+\left(-2m^3n^2\right)\left(n\right)$
$=12m^6n^7-4m^4n^4-2m^3n^3$

6. $7s^3\left(-2s^2+5s+4\right)-s^2\left(s^2+6s-1\right)$
$=7s^3\left(-2s^3\right)+7s^3\left(5s\right)+7s^3\left(4\right)$
$-s^2\left(s^2\right)-s^2\left(6s\right)-s^2\left(-1\right)$
$=-14s^6+35s^4+28s^3-s^4-6s^3+s^2$
$=-14s^6+34s^4+22s^3+s^2$

7. $\left(a-1\right)\left(2a+3\right)$
$=a\left(2a+3\right)-1\left(2a+3\right)$
$=2a^2+3a-2a-3$
$=2a^2+a=3$
$=2a^2+3a-2a-3$
$=2a^2+a-3$

8. $\left(8x+3\right)\left(2x-1\right)$
F: $\left(8x\right)\left(2x\right)=16x^2$
O: $\left(8x\right)\left(-1\right)=-8x$
I: $\left(3\right)\left(2x\right)=6x$
L: $\left(3\right)\left(-1\right)=-3$
$=16x^2-8x+6x-3$
$=16x^2-2x-3$

9. $\left(7m-n\right)\left(2m+n\right)$
F: $\left(7m\right)\left(2m\right)=14m^2$
O: $\left(7m\right)\left(n\right)=7mn$
I: $\left(-n\right)\left(2m\right)=-2mn$
L: $\left(-n\right)\left(n\right)=-n^2$
$=14m^2+7mn-2mn-n^2$
$=14m^2+5mn-n^2$

10. $\left(n^2+n-2\right)\left(n-4\right)$
$=\left(n^2+n-2\right)\left(n\right)+\left(n^2+n-2\right)\left(-4\right)$
$=n^3+n^2-2n-4n^2-4n+8$
$=n^3-3n^2-6n+8$

11.
$$\begin{array}{r} 8n^2-n+3 \\ n+2 \\ \hline 16n^2-2n+6 \\ 8n^3-n^2+3n \\ \hline 8n^3+15n^2+n+6 \end{array}$$

12.
$$\begin{array}{r} 8x^3+0x^2+9x-1 \\ 3x+7 \\ \hline 56x^3+0x^2+63x-7 \\ 24x^4+0x^3+27x^2-3x \\ \hline 24x^4+56x^3+27x^2+60x-7 \end{array}$$

13.

$$\begin{array}{r} p^2-2pq+q^2 \\ p-q \\ \hline -p^2q+2pq^2-q^3 \\ p^3-2p^2q+pq^2 \quad \\ \hline p^3-3p^2q+3pq^2-q^3 \end{array}$$

14. $(P+Pr)+(P+Pr)r$

$=P+Pr+Pr+Pr^2$

$=P+2Pr+Pr^2$

Exercises

1. $(6x)(-4x)=(6\cdot-4)(x^{1+1})$

$=-24x^2$

3. $(9t^2)(-t^3)=(9\cdot-1)(t^{2+3})$

$=-9t^5$

5. $(-5x^2)(-4x^4)=(-5\cdot-4)(x^{2+4})$

$=20x^6$

7. $(10x^3)(-7x^5)=(10\cdot-7)(x^{3+5})$

$=-70x^8$

9. $(-4pq^2)(-4p^2qr^2)$

$=(-4\cdot-4)(p^{1+2}q^{2+1}r^2)$

$=16p^3q^3r^2$

11. $(-8x)^2=(-8)^2x^2$

$=64x^2$

13. $\left(\frac{1}{2}t^4\right)^3=\left(\frac{1}{2}\right)^3t^{4\cdot3}$

$=\frac{1}{8}t^{12}$

15. $(7a)(10a^2)(-5a)=(7\cdot10\cdot-5)(a\cdot a^2\cdot a)$

$=-350a^{1+2+1}$

$=-350a^4$

17. $(2ab^2)(-3abc)(4a^2)$

$=(2\cdot-3\cdot4)(a\cdot a\cdot a^2)(b^2\cdot b)\cdot c$

$=-24a^{1+1+2}b^{2+1}c$

$=-24a^4b^3c$

19. $(7x-5)x$

$=(7x)x+(-5)x$

$=7x^2-5x$

21. $(9t+t^2)(5t)$

$=(9t)(5t)+(t^2)(5t)$

$=45t^2+5t^3$

23. $6a^3(4a^2-7a)$

$=6a^3(4a^2)+6a^3(-7a)$

$=24a^5-42a^4$

25. $4x^2(3x-2)$

$=4x^2(3x)+4x^2(-2)$

$=12x^3-8x^2$

27. $x^3(x^2-2x+4)$

$=x^3(x^2)+x^3(-2x)+x^3(4)$

$=x^5-2x^4+4x^3$

29. $5x(3x^2+5x+6)$

$=5x(3x^2)+5x(5x)+5x(6)$

$=15x^3+25x^2+30x$

31. $(5x^2-3x-7)(-9x)$

$=(5x^2)(-9x)+(-3x)(-9x)+(-7)(-9x)$

$=-45x^3+27x^2+63x$

33. $6x^2(x^3+4x^2-x-1)$

$=6x^2(x^3)+6x^2(4x^2)+6x^2(-x)+6x^2(-1)$

$=6x^5+24x^4-6x^3-6x^2$

35. $4p(7q-p^2)$

$=4p(7q)+4p(-p^2)$

$=28pq-4p^3$

37. $(v+3w^2)(-7v)$
$=(v)(-7v)+(3w^2)(-7v)$
$=-7v^2-21vw^2$

39. $2a^2b^3(3a^4b^2+10ab^5)$
$=2a^2b^3(3a^4b^2)+2a^2b^3(10ab^5)$
$=6a^6b^5+20a^3b^8$

41. $10x+2x(-3x+8)$
$=10x-6x^2+16x$
$=-6x^2+26x$

43. $-x+8x(x^2-2x+1)$
$=-x+8x^3-16x^2+8x$
$=8x^3-16x^2+7x$

45. $9x(x^2+3x-5)+8x(-4x^2+x)$
$=9x^3+27x^2-45x-32x^3+8x^2$
$=-23x^3+35x^2-45x$

47. $-4xy(2x^2+4xy)+x^2y(7x^2-2y)$
$=-8x^3y-16x^2y^2+7x^4y-2x^2y^2$
$=7x^4y-8x^3y-18x^2y^2$

49. $5a^2b^2(3ab^4-a^3b^2)+4a^2b^2(9ab^4-10a^3b^2)$
$=15a^3b^6-5a^5b^4+36a^3b^6-40a^5b^4$
$=-45a^5b^4+51a^3b^6$

51. $(y+2)(y+3)$
F: $(y)(y)=y^2$
O: $(y)(3)=3y$
I: $(2)(y)=2y$
L: $(2)(3)=6$

$=y^2+3y+2y+6$
$=y^2+5y+6$

53. $(x-3)(x-5)$
F: $(x)(x)=x^2$
O: $(x)(-5)=-5x$
I: $(-3)(x)=-3x$
L: $(-3)(-5)=15$

$=x^2-5x-3x+15$
$=x^2-8x+15$

55. $(a-2)(a+2)$
F: $(a)(a)=a^2$
O: $(a)(2)=2a$
I: $(-2)(a)=-2a$
L: $(-2)(2)=-4$

$=a^2+2a-2a-4$
$=a^2-4$

57. $(w+3)(2w-7)$
F: $(w)(2w)=2w^2$
O: $(w)(-7)=-7w$
I: $(3)(2w)=6w$
L: $(3)(-7)=-21$

$=2w^2-7w+6w-21$
$=2w^2-w-21$

59. $(3-2y)(5y-1)$
F: $(3)(5y)=15y$
O: $(3)(-1)=-3$
I: $(-2y)(5y)=-10y^2$
L: $(-2y)(-1)=2y$

$=15y-3-10y^2+2y$
$=-10y^2+17y-3$

61. $(10p-4)(2p-1)$

F: $(10p)(2p)=20p^2$

O: $(10p)(-1)=-10p$

I: $(-4)(2p)=-8p$

L: $(-4)(-1)=4$

$=20p^2-10p-8p+4$

$=20p^2-18p+4$

63. $(u+v)(u-v)$

F: $(u)(u)=u^2$

O: $(u)(-v)=-uv$

I: $(v)(u)=vu$

L: $(v)(-v)=-v^2$

$=u^2-uv+vu-v^2$

$=u^2-v^2$

65. $(2p-q)(q-p)$

F: $(2p)(q)=2pq$

O: $(2p)(-p)=-2p^2$

I: $(-q)(q)=-q^2$

L: $(-q)(-p)=qp$

$=2pq-2p^2-q^2+qp$

$=-2p^2+3pq-q^2$

67. $(3a-b)(a-2b)$

F: $(3a)(a)=3a^2$

O: $(3a)(-2b)=-6ab$

I: $(-b)(a)=-ba$

L: $(-b)(-2b)=2b^2$

$=3a^2-6ab-ba+2b^2$

$=3a^2-7ab+2b^2$

69. $(p-8)(4q+3)$

F: $(p)(4q)=4pq$

O: $(p)(3)=3p$

I: $(-8)(4q)=-32q$

L: $(-8)(3)=-24$

$=4pq+3p-32q-24$

71.

$$\begin{array}{r} x^2-3x+1 \\ x-3 \\ \hline -3x^2+9x-3 \\ x^3-3x^2+x \quad \\ \hline x^3-6x^2+10x-3 \end{array}$$

73.

$$\begin{array}{r} x^2+3x-5 \\ 2x-1 \\ \hline -x^2-3x+5 \\ 2x^3+6x^2-10x \quad \\ \hline 2x^3+5x^2-13x+5 \end{array}$$

75.

$$\begin{array}{r} a^2+ab+b^2 \\ a-b \\ \hline -a^2b-ab^2-b^3 \\ a^3+a^2b+ab^2 \quad \\ \hline a^3-b^3 \end{array}$$

77. $(3x)(x+5)(x-7)$

$=(3x^2+15x)(x-7)$

F: $(3x^2)(x)=3x^3$

O: $(3x^2)(-7)=-21x^2$

I: $(15x)(x)=15x^2$

L: $(15x)(-7)=-105x$

$=3x^3-21x^2+15x^2-105x$

$=3x^3-6x^2-105x$

79. $(3n)(2n-1)(2n+1)$

$=(6n^2-3n)(2n+1)$

F: $(6n^2)(2n)=12n^3$

O: $(6n^2)(1)=6n^2$

I: $(-3n)(2n)=-6n^2$

L: $(-3n)(1)=-3n$

$=12n^3+6n^2-6n^2-3n$

$=12n^3-3n$

81. $m^2n(8m^2-6n)-2mn(3m^2-7mn)$
$=8m^4n-6m^2n^2-6m^3n+14m^2n^2$
$=8m^4n-6m^3n+8m^2n^2$

83. $(-8s^3)(-5s^4)=40s^7$

85. $(9y^2+7y-8)(-4y)=-36y^3-28y^2+32y$

87. $(3x^4)(-8x^5)(x^2)$
$=-24x^9(x^2)$
$=-24x^{11}$

89. $-a^2b(4a-2b^3)=-4a^3b+2a^2b^4$

91.
$$\begin{array}{r} x^2-3x+2 \\ x-1 \\ \hline -x^2+3x-2 \\ x^3-3x^2+2x \\ \hline x^3-4x^2+5x-2 \end{array}$$

93. $(x+30)x=x^2+30x$
$x^2+30x\,\text{mm}^2$

95. $(1500+100x)(1000-30x)$
F: $(1500)(1000)=1{,}500{,}000$
O: $(1500)(-30x)=-45{,}000x$
I: $(100x)(1000)=100{,}000x$
L: $(100x)(-30x)=-3000x^2$
$=1{,}500{,}000-45{,}000x$
$+100{,}000x-3000x^2$
$=-3000x^2+55{,}000x+1{,}500{,}000$
$(-3000x^2+55{,}000x+1{,}500{,}000)$ dollars

97. **a.** $P(1+r)^t$
$=5000(1+r)^3$
$=5000(1+r)(1+r)(1+r)$
$=5000(1+2r+r^2)(1+r)$
$=5000(1+3r+3r^2+r^3)$
$=5000+15000r+15000r^3+5000r^3$

$(5000+15{,}000r+15{,}000r^2+5000r^3)$ dollars

b.
$$\begin{array}{r} 5000+15{,}000r+15{,}000r^2+5000r^3 \\ -5000-10{,}000r\quad -5000r^2 \qquad\qquad \\ \hline 5000r+10{,}000r^2+5000r^3 \end{array}$$

$(5000r+10{,}000r^2+5000r^3)$ dollars

c. $5000(0.1)+10{,}000(0.1)^2+5000(0.1)^3$
$=5000(0.1)+10{,}000(0.01)+5000(0.001)$
$=500+100+5$
$=605$
$605

5.6 Special Products

Practice

1. $(p+10)^2=p^2+2(p)(10)+(10)^2$
$=p^2+20p+100$

2. $(s+t)^2=s^2+2(s)(t)+(t)^2$
$=s^2+2st+t^2$

3. $(4p+5q)^2=(4p)^2+2(4p)(5q)+(5q)^2$
$=16p^2+40pq+25q^2$

4. $(5x-2)^2=(5x)^2-2(5x)(2)+(-2)^2$
$=25x^2-20x+4$

5. $(u-v)^2=u^2-2(u)(v)+(v)^2$
$=u^2-2uv+v^2$

6. $(2x-9y)^2=(2x)^2-2(2x)(9y)+(9y)^2$
$=4x^2-36xy+81y^2$

7. $(t+10)(t-10)=t^2-10^2$
$=t^2-100$

8. **a.** $(r-s)(r+s)=r^2-s^2$

b. $(8s-3t)(8s+3t)=(8s)^2-(3t)^2$
$=64s^2-9t^2$

9. $(10-7k^2)(10+7k^2)=(10)^2-(7k^2)^2$
$=100-49k^4$

10. $(S+s)(S-s)=S^2-s^2$

Exercises

1. The square of the sum of two terms is equal to the square of the first term plus twice the product of the two terms plus the square of the second term.

3. In the formula for the square of a difference of two terms, the middle term of the trinomial is negative.

5. $(y+2)^2=y^2+2(y)(2)+2^2$
$=y^2+4y+4$

7. $(x+4)^2=x^2+2(x)(4)+4^2$
$=x^2+8x+16$

9. $(x-11)^2=x^2-2(x)(11)+11^2$
$=x^2-22x+121$

11. $(6-n)^2=6^2-2(n)(6)+(n)^2$
$=36-12n+n^2$

13. $(x+y)^2=x^2+2xy+y^2$

15. $(3x+1)^2=3x^2+2(3x)(1)+(1)^2$
$=9x^2+6x+1$

17. $(4n-5)^2=(4n)^2-2(4n)(5)+(5)^2$
$=16n^2-40n+25$

19. $(9x+2)^2=(9x)^2+2(9x)(2)+(2)^2$
$=81x^2+36x+4$

21. $\left(a+\frac{1}{2}\right)^2=a^2+2(a)\left(\frac{1}{2}\right)+\left(\frac{1}{2}\right)^2$
$=a^2+a+\frac{1}{4}$

23. $(8b+c)^2=(8b)^2+2(8b)(c)+(c)^2$
$=64b^2+16bc+c^2$

25. $(5x-2y)^2=(5x)^2-2(5x)(2y)+(2y)^2$
$=25x^2-20xy+4y^2$

27. $(-x+3y)^2=(x)^2-2(x)(3y)+(3y)^2$
$=x^2-6xy+9y^2$

29. $(4x^3+y^4)^2=(4x^3)^2+2(4x^3)(y^4)+(y^4)^2$
$=16x^6+8x^3y^4+y^8$

31. $(a+1)(a-1)=a^2-1^2$
$=a^2-1$

33. $(4x-3)(4x+3)=(4x^2)-(3)^2$
$=16x^2-9$

35. $(10+3y)(3y-10)=(3y+10)(3y-10)$
$=(3y)^2-(10)^2$
$=9y^2-100$

37. $\left(m-\frac{1}{2}\right)\left(m+\frac{1}{2}\right)=m^2-\left(\frac{1}{2}\right)^2$
$=m^2-\frac{1}{4}$

39. $(4a+b)(4a-b)=(4a)^2-b^2$
$=16a^2-b^2$

41. $(3x-2y)(3x+2y)=(3x)^2-(2y)^2$
$=9x^2-4y^2$

43. $(1-5n)(5n+1)=-(1-5n)(1+5n)$
$=(1)^2-(5n)^2$
$=1-25n^2$

45. $x(x+5)(x-5)=x(x^2-25)$
$=x^3-25x$

47. $5n^2(n+7)^2=5n^2(n^2+14n+49)$
$=5n^4+70n^3+245n^2$

49. $(n^2 - m^4)(n^2 + m^4) = (n^2)^2 - (m^4)^2$
$= n^4 - m^8$

51. $(a-b)(a+b)(a^2+b^2) = (a^2-b^2)(a^2+b^2)$
$= a^4 - b^4$

53. $(6n+4)^2 = (6n)^2 + 2(6n)(4) + (4)^2$
$= 36n^2 + 48n + 16$

55. $(4p-9)(4p+9) = (4p)^2 - (9)^2$
$= 16p^2 - 81$

57. $(8-a)^2 = 8^2 - 2(8)(a) + (a)^2$
$= 64 - 16a + a^2$

59.
$-2x^2(4x-3)^2 = -2x^2[(4x)^2 - 2(4x)(3) + (3)^2]$
$= -2x^2(16x^2 - 24x + 9)$
$= -32x^4 + 48x^3 - 18x^2$

61. $(x-5)^2 + (y-1)^2$
$= x^2 - 10x + 25 + y^2 - 2y + 1$
$= x^2 + y^2 - 10x - 2y + 26$

63. $A\left(1+\frac{P}{100}\right)\left(1+\frac{P}{100}\right) = A\left(1+\frac{P^2}{50}+\frac{P^2}{100,00}\right)$
$= A + \frac{AP}{50} + \frac{AP^2}{10,000}$

65.
$$\frac{(a-m)^2 + (b-m)^2 + (c-m)^2}{2}$$
$$= \frac{a^2 - 2am + m^2 + b^2 - 2bm + m^2 + c^2 - 2cm + m^2}{2}$$
$$= \frac{3m^2 - 2am - 2bm - 2cm + a^2 + b^2 + c^2}{2}$$

5.7 Division of Polynomials

Practice

1. $\frac{-12n^6}{3n} = \frac{-12}{3} \cdot \frac{n^6}{n}$
$= -4n^{6-1}$
$= -4n^5$

2. $\frac{20p^3q^2r^4}{-5p^2q^2r} = \frac{20}{-5} \cdot \frac{p^3q^2r^4}{p^2q^2r}$
$= -4 \cdot p^{3-2}q^{2-2}r^{4-1}$
$= -4pr^3$

3. $\frac{21x^3 - 14x^2}{7x} = \frac{21x^3}{7x} - \frac{14x^2}{7x}$
$= 3x^2 - 2x$

4. $\frac{14x^8 + 10x^5 - 8x^3}{-2x^3} = \frac{14x^8}{-2x^3} + \frac{10x^5}{-2x^3} - \frac{8x^3}{-2x^3}$
$= -7x^5 - 5x^2 + 4$

5. $\frac{-5a^7b^6 + a^2b^4 - 15ab^3}{5ab^3}$
$= \frac{-5a^7b^6}{5ab^3} + \frac{a^2b^4}{5ab^3} - \frac{15ab^3}{5ab^3}$
$= -a^6b^3 + \frac{ab}{5} - 3$

6.
$$\begin{array}{r} 2x+3 \\ 5x+1 \overline{)10x^2+17x+3} \\ \underline{10x^2+2x} \\ 15x+3 \\ \underline{15x+3} \\ 0 \end{array}$$

7.

$$
\begin{array}{r}
x^2+2x+3 \\
3x+1\overline{)3x^3+7x^2+11x+5} \\
\underline{3x^3+x^2} \\
6x^2+11x+5 \\
\underline{6x^2+2x} \\
9x+5 \\
\underline{9x+3} \\
2
\end{array}
$$

$$x^2+2x+3+\frac{2}{3x+1}$$

8.

$$
\begin{array}{r}
3s-5 \\
3s-2\overline{)9s^2-21s+10} \\
\underline{9s^2-6s} \\
-15s+10 \\
\underline{-15s+10} \\
0
\end{array}
$$

9.

$$
\begin{array}{r}
n^2-4n-3 \\
4n-3\overline{)4n^3-19n^2+0n-4} \\
\underline{4n^3-3n^2} \\
-16n^2+0n-4 \\
\underline{-16n^2+12n} \\
-12n-4 \\
\underline{-12n+9} \\
-13
\end{array}
$$

$$n^2-4n-3+\frac{-13}{4n-3}$$

10. a. The future valued of the investment after 1 year is $10(1+r)^1$, or $10(1+r)$. The future value of the investment after 2 years is $10(1+r)^2$.

b. $10r+10$ and $10r^2+20r+10$

c.

$$
\begin{array}{r}
r+1 \\
10r+10\overline{)10r^2+20r+10} \\
\underline{10r^2+10r} \\
10r+10 \\
\underline{10r+10} \\
0
\end{array}
$$

The future value of the investment after 2 years is $(r+1)$ times as great as the future value of the investment after 1 year.

Exercises

1. To divide a monomial by a monomial, divide the coefficients and use the quotient rule of exponents.

3. In dividing one polynomial by another, repeat the process until the degree of the remainder is less than the degree of the divisor.

5.
$$\frac{10x^4}{5x^2}=\frac{10}{5}\cdot\frac{x^4}{x^2}$$
$$=2x^{4-2}$$
$$=2x^2$$

7.
$$\frac{16a^8}{-4a}=\frac{16}{-4}\cdot\frac{a^8}{a}$$
$$=-4a^{8-1}$$
$$=-4a^7$$

9.
$$\frac{-8x^5}{-6x^4}=\frac{-8}{-6}\cdot\frac{x^5}{x^4}$$
$$=\frac{4}{3}x^{5-4}$$
$$=\frac{4}{3}x$$
$$=\frac{4x}{3}$$

11.
$$\frac{12p^2q^3}{3p^2q}=\frac{12}{3}\cdot\frac{p^2q^3}{p^2q}$$
$$=4\cdot p^{2-2}q^{3-1}$$
$$=4q^2$$

13.
$$\frac{-24u^6v^4}{-8u^4v^2}=\frac{-24}{-8}\cdot\frac{u^6v^4}{u^4v^2}$$
$$=3\cdot u^{6-4}v^{4-2}$$
$$=3u^2v^2$$

15. $$\frac{-15a^2b^5}{7ab^3} = \frac{-15}{7} \cdot \frac{a^2b^5}{ab^3} = -\frac{15}{7}a^{2-1}b^{5-3} = -\frac{15}{7}ab^2$$

17. $$\frac{-6u^5v^3w^3}{4u^2vw^3} = \frac{-6}{4} \cdot \frac{u^5v^3w^3}{u^2vw^3} = -\frac{3}{2}u^{5-2}v^{3-1}w^{3-3} = -\frac{3}{2}u^3v^2$$

19. $$\frac{6n^2 + 10n}{2n} = \frac{6n^2}{2n} + \frac{10n}{2n} = 3n + 5$$

21. $$\frac{20b^4 - 10b}{10b} = \frac{20b^4}{10b} - \frac{10b}{10b} = 2b^3 - 1$$

23. $$\frac{18a^2 + 12a}{-3a} = \frac{18a^2}{-3a} + \frac{12a}{-3a} = -6a - 4$$

25. $$\frac{9x^5 - 6x^7}{3x^5} = \frac{9x^5}{3x^5} - \frac{6x^7}{3x^5} = 3 - 2x^2$$

27. $$\frac{12a^4 - 18a^3 + 30a^2}{6a^2} = \frac{12a^4}{6a^2} - \frac{18a^3}{6a^2} + \frac{30a^2}{6a^2} = 2a^2 - 3a + 5$$

29. $$\frac{n^5 - 10n^4 - 5n^3}{-5n^3} = \frac{n^5}{-5n^3} - \frac{10n^4}{-5n^3} - \frac{5n^3}{-5n^3} = -\frac{n^2}{5} + 2n + 1$$

31. $$\frac{20a^2b + 4ab^3}{8ab} = \frac{20a^2b}{8ab} + \frac{4ab^3}{8ab} = \frac{5a}{2} + \frac{b^2}{2}$$

33. $$\frac{12x^2y^3 - 9xy - 3xy^2}{-3xy} = \frac{12x^2y^3}{-3xy} - \frac{9xy}{-3xy} - \frac{3xy^2}{-3xy} = -4xy^2 + 3 + y$$

35. $$\frac{8p^2q^3 - 4p^3q^3 + 6p^4q}{4p^2q} = \frac{8p^2q^3}{4p^2q} - \frac{4p^3q^3}{4p^2q} + \frac{6p^4q}{4p^2q} = 2q^2 - pq^2 + \frac{3p^2}{2}$$

37. $$\begin{array}{r} x - 7 \\ x + 3 \overline{)\, x^2 - 4x - 21} \\ \underline{x^2 + 3x} \\ -7x - 21 \\ \underline{-7x - 21} \\ 0 \end{array}$$

39. $$\begin{array}{r} 7x - 2 \\ 8x - 1 \overline{)\, 56x^2 - 23x + 2} \\ \underline{56x^2 - 7x} \\ -16x + 2 \\ \underline{-16x + 2} \\ 0 \end{array}$$

41. $$\begin{array}{r} 3x - 1 \\ 2x + 5 \overline{)\, 6x^2 + 13x - 5} \\ \underline{6x^2 + 15x} \\ -2x - 5 \\ \underline{-2x - 5} \\ 0 \end{array}$$

43. $$\begin{array}{r} 5x + 3 \\ x - 1 \overline{)\, 5x^2 - 2x - 3} \\ \underline{5x^2 - 5x} \\ 3x - 3 \\ \underline{3x - 3} \\ 0 \end{array}$$

45. $$\begin{array}{r} 7x + 2 \\ 3x + 2 \overline{)\, 21x^2 + 20x + 4} \\ \underline{21x^2 + 14x} \\ 6x + 4 \\ \underline{6x + 4} \\ 0 \end{array}$$

47. $$\begin{array}{r} x \\ x+2\overline{)x^2+2x+5} \\ \underline{x^2+2x} \\ 5 \end{array}$$

$$x+\frac{5}{x+2}$$

49. $$\begin{array}{r} 2x+1 \\ x-3\overline{)2x^2-5x-3} \\ \underline{2x^2-6x} \\ x-3 \\ \underline{x-1} \\ 0 \end{array}$$

51. $$\begin{array}{r} 2x-3 \\ 4x+3\overline{)8x^2-6x-11} \\ \underline{8x^2+6x} \\ -12x-11 \\ \underline{-12x-9} \\ -2 \end{array}$$

$$2x-3+\frac{-2}{4x+3}$$

53. $$\begin{array}{r} x^2-6x+5 \\ x+1\overline{)x^3-5x^2-x+5} \\ \underline{x^3+x^2} \\ -6x^2-x+5 \\ \underline{-6x^2-6x} \\ 5x+5 \\ \underline{5x+5} \\ 0 \end{array}$$

55. $$\begin{array}{r} 2x^2-x-3 \\ 3x-4\overline{)6x^3-11x^2-5x+19} \\ \underline{6x^3-8x^2} \\ -3x^2-5x+19 \\ \underline{-3x^2+4x} \\ -9x+19 \\ \underline{-9x+12} \\ 7 \end{array}$$

$$2x^2-x-3+\frac{7}{3x-4}$$

57. $$\begin{array}{r} 5x+20 \\ x-4\overline{)5x^2+0x-2} \\ \underline{5x^2-20x} \\ 20x-2 \\ \underline{20x-80} \\ 78 \end{array}$$

$$5x+20+\frac{78}{x-4}$$

59. $$\begin{array}{r} 2x^2+3x+4 \\ 2x-3\overline{)4x^3+0x^2-x+3} \\ \underline{4x^3-6x^2} \\ 6x^2-x+3 \\ \underline{6x^2-9x} \\ 8x+3 \\ \underline{8x-12} \\ 15 \end{array}$$

$$2x^2+3x+4+\frac{15}{2x-3}$$

61. $$\begin{array}{r} x^2-3x+9 \\ x+3\overline{)x^3+0x^2+0x+27} \\ \underline{x^3+3x^2} \\ -3x^2+0x+27 \\ \underline{-3x^2-9x} \\ 9x+27 \\ \underline{9x+27} \\ 0 \end{array}$$

63. $$\begin{aligned} \frac{56r^2}{-8r} &= \frac{56}{-8}\cdot\frac{r^2}{r} \\ &= -7r^{2-1} \\ &= -7r \end{aligned}$$

65. $$\begin{aligned} \frac{-18a^2b^3c}{27ab^2c} &= \frac{-18}{27}\cdot\frac{a^2b^3c}{ab^2c} \\ &= -\frac{2}{3}a^{2-1}b^{3-2}c^{1-1} \\ &= -\frac{2}{3}ab \end{aligned}$$

67. $$\frac{18n^6-48n^4-2n^3}{-6n^3}$$
$$=\frac{18n^6}{-6n^3}-\frac{48n^4}{-6n^3}-\frac{2n^3}{-6n^3}$$
$$=-3n^3+8n+\frac{1}{3}$$

69. $$\begin{array}{r} 2x+1 \\ 3x+5\overline{)6x^2+13x+5} \\ \underline{6x^2+10x} \\ 3x+5 \\ \underline{3x+5} \\ 0 \end{array}$$

71. $$\begin{array}{r} x^2-1 \\ 4x+1\overline{)4x^3+x^2-4x+2} \\ \underline{4x^3+x^2} \\ -4x+2 \\ \underline{-4x-1} \\ 3 \end{array}$$
$$x^2-1+\frac{3}{4x+1}$$

73. $$\frac{32m^3-72m}{-8m}=\frac{32m^3}{-8m}-\frac{72m}{-8m}$$
$$=-4m^2+9$$

75. $$A=2(1)^2+20(1)+48$$
$$=2+20+48$$
$$=70$$
$$w=(1)+6=7$$
$$l=\frac{A}{w}=\frac{70}{7}=10$$

77. **a.** $d=rt$

$\frac{d}{r}=t$, or $t=\frac{d}{r}$

b. $$\begin{array}{r} t^2-7t+14 \\ t+1\overline{)t^3-6t^2+7t+14} \\ \underline{t^3+t^2} \\ -7t^2+7t+14 \\ \underline{-7t^2-7t} \\ 14t+14 \\ \underline{14t+14} \\ 0 \end{array}$$

It takes $(t^2-7t+14)$ hr.

79. $$\begin{array}{r} 3x-14 \\ 200x-300\overline{)600x^2-3700x+4400} \\ \underline{600x^2-900x} \\ -2800x+4400 \\ \underline{-2800x+4200} \\ 200 \end{array}$$
$$3x-14+\frac{200}{200x-300}$$

There are $(3x-14)$ thousand subscribers per cell system.

Chapter 5 Review Exercises

1. $(-x)^3=(-x)(-x)(-x)=-x^3$

2. $-31^0=-(1)=-1$

3. $n^4\cdot n^7=n^{4+7}=n^{11}$

4. $x^6\cdot x=x^{6+1}=x^7$

5. $\frac{n^8}{n^5}=n^{8-5}=n^3$

6. $p^{10}\div p^7=p^{10-7}=p^3$

7. $y^4\cdot y^2\cdot y=y^{4+2+1}=y^7$

8. $(a^2b)(ab^2)=a^{2+1}b^{1+2}=a^3b^3$

9. $x^0y=(1)y=y$

10. $\dfrac{n^4 \cdot n^7}{n^9} = n^{4+7-9} = n^2$

11. $(5x)^{-1} = \dfrac{1}{(5x)^1} = \dfrac{1}{5x}$

12. $-3n^{-2} = \dfrac{-3}{n^2}$ or $-\dfrac{3}{n^2}$

13. $8^{-2}v^4 = \dfrac{v^4}{8^2} = \dfrac{v^4}{64}$

14. $\dfrac{1}{y^{-4}} = y^4$

15. $x^{-8} \cdot x^7 = x^{-8+7} = \dfrac{1}{x^1} = \dfrac{1}{x}$

16. $5^{-1} \cdot y^6 \cdot y^{-3} = \dfrac{y^{6+(-3)}}{5^1} = \dfrac{y^3}{5}$

17. $\dfrac{a^5}{a^{-5}} = a^{5-(-5)} = a^{10}$

18. $\dfrac{t^{-2}}{t^4} = t^{-2-4} = t^{-6} = \dfrac{1}{t^6}$

19. $\dfrac{x^{-2}}{y} = \dfrac{1}{x^2 y}$

20. $\dfrac{x^2}{y^{-1}} = x^2 y^1 = x^2 y$

21. $(10^2)^4 = 10^8 = 10{,}000{,}000$

22. $-(x^3)^3 = -x^{3\cdot 3} = -x^9$

23. $(2x^3)^2 = 2^2 x^{3\cdot 2} = 4x^6$

24. $(-4m^5n)^3 = (-4)^3 m^{5\cdot 3} n^3 = -64m^{15}n^3$

25. $3(x^{-2})^6 = 3x^{-2\cdot 6} = 3x^{-12} = \dfrac{3}{x^{12}}$

26.
$$\begin{aligned}(a^3b^{-4})^{-2} &= a^{3\cdot(-2)}b^{-4\cdot(-2)}\\ &= a^{-6}b^8\\ &= \frac{b^8}{a^6}\end{aligned}$$

27. $\left(\dfrac{x}{3}\right)^4 = \dfrac{x^4}{3^4} = \dfrac{x^4}{81}$

28. $\left(\dfrac{-a}{b^3}\right)^2 = \dfrac{(-a)^2}{b^{3\cdot 2}} = \dfrac{a^2}{b^6}$

29. $\left(\dfrac{x}{y}\right)^{-6} = \left(\dfrac{y}{x}\right)^6 = \dfrac{y^6}{x^6}$

30. $\left(\dfrac{x^2}{y^{-1}}\right)^5 = \dfrac{x^{2\cdot 5}}{y^{-1\cdot 5}} = \dfrac{x^{10}}{y^{-5}} = x^{10}y^5$

31. $\left(\dfrac{4a^3}{b^4c}\right)^2 = \dfrac{4^2a^{3\cdot 2}}{b^{4\cdot 2}c^2} = \dfrac{16a^6}{b^8c^2}$

32.
$$\begin{aligned}\left(\frac{-u^{-5}v^2}{7w}\right)^2 &= \frac{(-1)^2u^{-5\cdot 2}v^{2\cdot 2}}{7^2w^2}\\ &= \frac{u^{-10}v^4}{49w^2} = \frac{v^4}{49u^{10}w^2}\end{aligned}$$

33.
$$\begin{aligned}3.7\times 10^{10} &= 3.7\times 10{,}000{,}000{,}000\\ &= 37{,}000{,}000{,}000\end{aligned}$$

34.
$$\begin{aligned}1.63\times 10^{9} &= 1.63\times 1{,}000{,}000{,}000\\ &= 1{,}630{,}000{,}000\end{aligned}$$

35.
$$\begin{aligned}5.022\times 10^{-5} &= \frac{5.022}{10{,}000}\\ &= 0.00005022\end{aligned}$$

36.
$$\begin{aligned}6\times 10^{-11} &= \frac{6}{100{,}000{,}000{,}000}\\ &= 0.00000000006\end{aligned}$$

37. $1{,}200{,}000{,}000 = 1.2\times 10^{12}$

38. $427{,}000{,}000 = 4.27\times 10^{8}$

39. $0.00000000000004 = 4\times 10^{-14}$

40. $0.00000056 = 5.6\times10^{-7}$

41. $\left(1.4\times10^{6}\right)\left(4.2\times10^{3}\right) = (1.4\times4.2)\times\left(10^{6}\times10^{3}\right)$
$= 5.88\times10^{6+3}$
$= 5.88\times10^{9}$

42. $\left(3\times10^{-2}\right)\left(2.1\times10^{5}\right) = (3\times2.1)\times\left(10^{-2}\times10^{5}\right)$
$= 6.3\times10^{-2+5}$
$= 6.3\times10^{3}$

43. $\left(1.8\times10^{4}\right)\div\left(3\times10^{-3}\right) = \left(\frac{1.8}{3}\right)\times\left(\frac{10^{4}}{10^{-3}}\right)$
$= 0.6\times10^{4-(-3)}$
$= 0.6\times10^{7}$
$= 6\times10^{6}$

44. $\left(9.6\times10^{-4}\right)\div\left(1.6\times10^{6}\right) = \left(\frac{9.6}{1.6}\right)\times\left(\frac{10^{-4}}{10^{-6}}\right)$
$= 6\times10^{-4-6}$
$= 6\times10^{-10}$

45. Polynomial

46. Not a polynomial

47. Trinomial

48. Binomial

49. $-3y^3 + y^2 + 8y - 1$; degree 3
leading term: $-3y^3$
leading coefficient: -3

50. $n^4 - 7n^3 - 6n^2 + n$; degree 4
leading term: n^4
leading coefficient: 1

51. $10x - 8x^2 - 8x + 9x - x^3 + 13$
$= -x^3 + x^2 + 2x + 13$

52. $4n^3 - 7n + 9 - 3n^2 - n^3 + 7n^2 - 5 + n$
$= 3n^3 + 4n^2 - 6n + 4$

53. $2n^2 - 7n + 3 = 2(-1)^2 - 7(-1) + 3$
$2(1) - 7(-1) + 3$
$= 2 + 7 + 3$
$= 12$

$2n^2 - 7n + 3 = 2(3)^2 - 7(3) + 3$
$= 2(9) - 7(3) + 3$
$= 18 - 21 + 3$
$= 0$

54. $x^3 - 8 = (2)^3 - 8$
$= 8 - 8$
$= 0$

$x^3 - 8 = (-2)^3 - 8$
$= -8 - 8$
$= -16$

55. $\left(4x^2 - x + 4\right) + \left(-3x^2 + 9\right)$
$= 4x^2 + 4 - 3x^2 + 9$
$= x^2 - x + 13$

56. $\left(5y^4 - 2y^3 + 7y - 11\right) + \left(6 - 8y - y^2 - 5y^4\right)$
$= 5y^4 - 2y^3 + 7y - 11 + 6 - 8y - y^2 - 5y^4$
$= -2y^3 - y^2 - y - 5$

57. $\left(a^2 + 5ab + 6b^2\right) + \left(3a^2 - 9b^2\right) + \left(-7ab - 3a^2\right)$
$= a^2 + 5ab + 6b^2 + 3a^2 - 9b^2 - 7ab - 3a^2$
$= a^2 - 2ab - 3b^2$

58. $\left(5s^3t - 2st + t^2\right) + \left(s^2t - 5t^2\right) + \left(t^2 - 4st + 9s^2\right)$
$= 5s^3t - 2st + t^2 + s^2t - 5t^2 + t^2 - 4st + 9s^2$
$= 5s^3t + s^2t + 9s^2 - 6st - 3t^2$

59. $\left(x^2 - 5x + 2\right) - \left(-x^2 + 3x + 10\right)$
$= x^2 - 5x + 2 + x^2 - 3x - 10$
$= 2x^2 - 8x - 8$

60. $\left(10n^3 + n^2 - 4n + 1\right) - \left(11n^3 - 2n^2 - 5n + 1\right)$
$= 10n^3 + n^2 - 4n + 1 - 11n^3 + 2n^2 + 5n - 1$
$= -n^3 + 3n^2 + n$

61.
$$\begin{array}{r} 5y^4-4y^3 \qquad +y-6 \\ \underline{-(\quad y^3-2y^2+7y-3)} \end{array}$$

$$\begin{array}{r} 5y^4-4y^3 \qquad +y-6 \\ \underline{-y^3+2y^2-7y+3} \\ 5y^4-5y^3+2y^2-6y-3 \end{array}$$

62.
$$\begin{array}{r} -9x^3+8x^2-11x-12 \\ \underline{+(11x^3 \qquad -x+15)} \\ 2x^3+8x^2-12x+3 \end{array}$$

63. $14t^2-(10t^2-4t)$
$=14t^2-10t^2+4t$
$=4t^2+4t$

64. $-(5x-6y)+(3x-7y)$
$=-5x+6y+3x-7y$
$=-2x-y$

65. $(3y^2-1)-(y^2+3y+2)+(-2y+5)$
$=3y^2-1-y^2-3y-2-2y+5$
$=2y^2-5y+2$

66. $(1-4x-6x^2)-(7x-8)-(-11x-x^2)$
$=1-4x-6x^2-7x+8+11x+x^2$
$=-5x^2+9$

67. $-3x^4\cdot 2x=(-3)(2)x^{4+1}=-6x^5$

68. $(3ab)(8a^2b^3)(-6b)=(3)(8)(-6)a^{1+2}b^{1+3+1}$
$=-144a^3b^5$

69. $2xy^2(4x-5y)=-2xy^2(4x)-2xy^2(5y)$
$=8x^2y^2-10xy^3$

70. $(x^2-3x+1)(-5x^2)=-5x^{2+2}+15x^{1+2}-5x^2$
$=-5x^4+15x^3-5x^2$

71. $(n+3)(n+7)$
F: $(n)(n)=n^2$
O: $(n)(7)=7n$
I: $(3)(n)=3n$
L: $(3)(7)=21$

$n^2+7n+3n+21$
$=n^2+10n+21$

72. $(3x-9)(x+6)$
F: $(3x)(x)=3x^2$
O: $(3x)(6)=18x$
I: $(-9)(x)=-9x$
L: $(-9)(6)=-54$

$3x^2+18x-9x-54$
$=3x^2+9x-54$

73. $(2x-1)(4x-1)$
F: $(2x)(4x)=8x^2$
O: $(2x)(-1)=-2x$
I: $(-1)(4x)=-4x$
L: $(-1)(-1)=1$

$8x^2-2x-4x+1$
$=8x^2-6x+1$

74. $(3a-b)(3a+2b)$
F: $(3a)(3a)=9a^2$
O: $(3a)(2b)=6ab$
I: $(-b)(3a)=-3ab$
L: $(-b)(2b)=-2ab^2$

$9a^2+6ab-3ab-2b^2$
$=9a^2+3ab-2b^2$

75.
$$\begin{array}{r} 2x^3-5x+2 \\ \underline{x+3} \\ 6x^3 \qquad -15x+6 \\ \underline{2x^4 \qquad -5x^2+2x} \\ 2x^4+6x^3-5x^2-13x+6 \end{array}$$

76.

$$\begin{array}{r} y^2 - 7y + 1 \\ y - 2 \\ \hline -2y^2 + 14y - 2 \\ y^3 - 7y^2 + y \quad \\ \hline y^3 - 9y^2 + 15y - 2 \end{array}$$

77. $-y + 2y(-3y + 7)$
$= -y - 6y^2 + 14y$
$= -6y^2 + 13y$

78. $4x^2(2x - 6) - 3x(3x^2 - 10x + 2)$
$= 8x^3 - 24x^2 - 9x^3 + 30x^2 - 6x$
$= -x^3 + 6x^2 - 6x$

79. $(a - 1)^2 = -2(a)(1) + (-1)^2$
$= a^2 - 2a + 1$

80. $(s + 4)^2 = s^2 + 2(s)(4) + (4)^2$
$= s^2 + 8s + 16$

81. $(2x + 5)^2 = (2x)^2 + 2(2x)(5) + (5)^2$
$= 4x^2 + 20x + 25$

82. $(3 - 4t)^2 = 3^2 - (2)(3)(4t) + (4t)^2$
$= 9 - 24t + 16t^2$

83. $(5a - 2b)^2 = (5a)^2 - (2)(5a)(2b) + (2b)^2$
$= 25a^2 - 20ab + 4b^2$

84. $(u^2 + v^2)^2 = (u^2)^2 + 2(u^2)(v^2) + (v^2)^2$
$= u^4 + 2u^2v^2 + v^4$

85. $(m + 4)(m - 4) = m^2 - 4^2 = m^2 - 16$

86. $(6 - n)(6 + n) = 6^2 - n^2 = 36 - n^2$

87. $(7n - 1)(7n + 1) = (7n)^2 - 1^2 = 49n^2 - 1$

88. $(2x + y)(2x - y) = (2x)^2 - y^2 = 4x^2 - y^2$

89. $(4a - 3b)(4a + 3b) = 16a^2 - 9b^2$

90. $x(x + 10)(x - 10) = x(x^2 - 100)$
$= x^3 - 100x$

91. $-3t^2(4t - 5)^2 = -3t^2(16t^2 - 40t + 25)$
$= -48t^4 + 120t^3 - 75t^2$

92. $(p^2 - q^2)(p + q)(p - q) = (p^2 - q^2)(p^2 - q^2)$
$= (p^2 - q^2)^2$
$= p^4 - 2p^2q^2 + q^4$

93. $12x^4 \div 4x^2 = \frac{12}{4}x^{4-2}$
$= 3x^2$

94. $\frac{-20a^3b^5c}{10ab^2} = -2a^{3-1}b^{5-2}c$
$= -2a^2b^3c$

95. $(18x^3 - 6x) \div (3x) = \frac{18x^3}{3x} - \frac{6x}{3x}$
$= 6x^2 - 2$

96. $\frac{10x^5 + 6x^4 - 4x^3 - 2x^2}{2x^2}$
$= \frac{10x^5}{2x^2} + \frac{6x^4}{2x^2} - \frac{4x^3}{2x^2} - \frac{2x^2}{2x^2}$
$= 5x^3 + 3x^2 - 2x - 1$

97.

$$\begin{array}{r} 3x - 7 \\ x + 5 \overline{)\, 3x^2 + 8x - 35} \\ \underline{3x^2 + 15x \quad\;\;} \\ -7x - 35 \\ \underline{-7x - 35} \\ 0 \end{array}$$

98.

$$\begin{array}{r} x^2-2x-1 \\ 2x-1\overline{)2x^3-5x^2+0x+13} \\ \underline{2x^3-x^2} \\ -4x^2+0x+13 \\ \underline{-4x^2+2x} \\ -2x+13 \\ \underline{-2x+1} \\ 12 \end{array}$$

$$x^2-2x-1+\frac{12}{2x-1}$$

99. $13{,}900{,}000{,}000\text{ yr}=1.39\times10^{10}\text{ yr}$

100. $6.24\times10^{18}\text{ eV}$
$=6{,}240{,}000{,}000{,}000{,}000{,}000\text{ eV}$

101. $0.00003\text{ m}=3\times10^{-5}\text{ m}$

102. $1.1\times10^{-10}\text{ m}=0.00000000011\text{ m}$

103. $\frac{n^2}{2}-\frac{n}{2}=\frac{(9)^2}{2}-\frac{9}{2}$
$=\frac{81}{2}-\frac{9}{2}$
$=40.5-4.5$
$=36$
There will be 36 handshakes

104. $-4.9t^2+500=-4.9(2)^2+500$
$=-4.9(4)+500$
$=-19.6+500$
$=480.4$
The object is 480.4 m above the ground.

105. $-0.3x^2+36.7x+213$
$=-0.3(1)^2+36.7(1)+213$
$=-0.3+36.7+213$
$=249.4$
There were about 249,000 divorces in 1951.

106. $-0.8x^2+41.5x+898.6$
$-0.8(0)^2+41.5(0)+898.6$
$=898.6$
There were about 879 two-year colleges in 1970.

107. **a.** $(3w-10)w=3w^2-10w$
$(3w^2-10)$ sq. ft

b. $(3w-10+12)(w+12)-(3w^2-10w)$
$=(3w+2)(w+12)-(3w^2-10w)$
$=3w^2+38w+24-3w^2+10w$
$=48w+24$
$(48w+24)$ sq. ft

c. $48w+24$
$48(12)+24$
$=576+24$
$=600$
The area of the concrete wall is 600 sq. ft.

108. **a.**

$$\begin{array}{r} 2x+436 \\ -13x+2164\overline{)-26x^2-1340x+943{,}504} \\ \underline{-26x^2+4328x\phantom{+943{,}504}} \\ -5668x+943{,}504 \\ \underline{-5668x+943{,}504} \\ 0 \end{array}$$

b. $2x+436=2(2)+436$
$=4+436$
$=440$
440 acres

Chapter 5 Posttest

1. $x^6\cdot x=x^{6+1}=x^7$

2. $n^{10}\div n^4=n^{10-4}=n^6$

3. $7a^{-1}b^0=\frac{7}{a^1}=\frac{7}{a}$

4. $(-3x^2y)^3=(-3)^3x^{2\cdot3}y^3=-27x^6y^3$

5. $\left(\frac{x^2}{y^3}\right)^4=\frac{x^{2\cdot4}}{y^{3\cdot4}}=\frac{x^8}{y^{12}}$

6. $\left(\frac{3x^2}{y}\right)^{-3} = \left(\frac{y}{3x^2}\right)^3$

$= \frac{y^3}{3^3 x^{2\cdot 3}}$

$= \frac{y^3}{27x^6}$

7. **a.** $-x^3, 2x^2, 9x, -1$

b. $-1, 2, 9,$ and -1

c. 3

d. -1

8. $(y^2 - 1) + (y^2 - y + 6)$

$= y^2 - 1 + y^2 - y + 6$

$= 2y^2 - y + 5$

9. $(x^2 - 7x - 4) - (2x^2 - 8x + 5)$

$= x^2 - 7x - 4 - 2x^2 + 8x - 5$

$= -x^2 + x - 9$

10.

$(4x^2y^2 - 6xy - y^2) - (3x^2 + x^2y^2 - 2y^2) - (x^2 - 6xy + y^2)$

$= 4x^2y^2 - 6xy - y^2 - 3x^2 - x^2y^2 + 2y^2 - x^2 + 6xy + y^2$

$= 3x^2y^2 - 4x^2$

11.

$(2mn^2)(5m^2n - 10mn + mn^2)$

$= 2mn^2 \cdot (5m^2n) - 2mn^2 \cdot (10mn) + 2mn^2 \cdot (mn^2)$

$= 10m^3n^3 - 20m^2n^3 + 2m^2n^4$

12.

$$\begin{array}{r} y^3 - 2y^2 + 0y + 4 \\ y - 1 \\ \hline -y^3 + 2y^2 \qquad - 4 \\ y^4 - 2y^3 \qquad + 4y \\ \hline y^4 - 3y^3 + 2y^2 + 4y - 4 \end{array}$$

13. $(3x - 1)(2x + 7)$

F: $(3x)(2x) = 6x^2$

O: $(3x)(7) = 21x$

I: $(-1)(2x) = -2x$

L: $(-1)(7) = -7$

$= 6x^2 + 21x - 2x - 7$

$= 6x^2 + 19x - 7$

14. $(7 - 2n)(7 + 2n) = (7)^2 - (2n)^2$

$= 49 - 4n^2$

15. $(2m - 3)^2 = (2m)^2 - 2(2m)(3) + (3)^2$

$= 4m^2 - 12m + 9$

16. $\frac{12s^3 + 15s^2 - 27s}{-3s}$

$= \frac{12s^3}{-3s} + \frac{15s^2}{-3s} - \frac{27s}{-3s}$

$= -4s^2 - 5s + 9$

17.

$$\begin{array}{r} t^2 - t - 1 \\ 3t - 2 \overline{)3t^3 - 5t^2 - t + 6} \\ \underline{3t^3 - 2t^2} \\ -3t^2 \quad - t + 6 \\ \underline{-3t^2 - 2t} \\ -3t + 6 \\ \underline{-3t + 2} \\ 4 \end{array}$$

$t^2 - t - 1 + \frac{4}{3t - 2}$

18. $(10^{-10}) \cdot 1000$

$= 10^{-10} \cdot 10^3$

$= 10^{-7}$

$= 1 \times 10^{-7}$

1×10^{-7} m

19. **a.** First house: $(1500x + 140,000)$ dollars;

second house: $(800x + 90,000)$ dollars

b. $(1500x + 140,000) + (800x + 90,000)$

$= 1500x + 140,000 + 800x + 90,000$

$= 2300x + 230,000$

$(2300x + 230,000)$ dollars

20. $1000(1 + 0.03)^2$

$= 1000(1.03)^2$

$= 1000(1.0609)$

$= 1060.9$

The account balance is \$1060.90.

Chapter 6

FACTORING POLYNOMIALS

Pretest

1. $18ab = 2\cdot3\cdot3\cdot a\cdot b$
   $36a^4 = 2\cdot2\cdot3\cdot3\cdot a\cdot a\cdot a\cdot a$
   $GCF = 2\cdot3\cdot3\cdot a = 18a$

2. $4pq+16p = 4p(q+4)$

3. $10x^2y-5x^3y^3+5xy^2$
   $= 5xy(2x-x^2y^2+y)$

4. $3x^2+6x+2x+4$
   $= 3x(x+2)+2(x+2)$
   $= (x+2)(3x+2)$

5. $n^2-11n+24 = (n-3)(n-8)$

6. $4a+a^2-21 = a^2+4a-21$
   $= (a-3)(a+7)$

7. $9y-12y^2+3y^3$
   $= 3y^3-12y^2+9y$
   $= 3y(y^2-4y+3)$
   $= 3y(y-1)(y-3)$ or $3y(1-y)(3-y)$

8. $5a^2+6ab-8b^2 = (5a-4b)(a+2b)$

9. $-12n^2+38n+14 = -2(6n^2-19n-7)$
   $= -2(3n+1)(2n-7)$

10. $4x^2-28x+49 = (2x-7)^2$

11. $25n^2-9 = (5n+3)(5n-3)$

12. $x^2y-4y^3 = y(x^2-4y^2)$
    $= y(x+2y)(x-2y)$

13. $y^6-9y^3+20 = (y^3-4)(y^3-5)$

14. $n(n-6) = 0$
    $n = 0$ or $n-6 = 0$
    $n = 6$

15. $3x^2+x = 2$
    $3x^2+x-2 = 0$
    $(3x-2)(x+1) = 0$
    $3x-2 = 0$ or $x+1 = 0$
    $3x = 2$ $\quad x = -1$
    $x = \frac{2}{3}$

16. $(y+4)(y-2) = 7$
    $y^2+2y-8 = 7$
    $y^2+2y-15 = 0$
    $(y+5)(y-3) = 0$
    $y+5 = 0$ or $y-3 = 0$
    $y = -5$ $\quad y = 3$

17. $A = 2lw+2lh+2wh$
    $A-2lw = 2lh+2wh$
    $A-2lw = h(2l+2w)$
    $h = \frac{A-2lw}{2l+2w}$

18. $-16t^2+63t+4$
    $= -(16t^2-63t-4)$
    $= -(16t+1)(t-4)$
    $-(16t+1)(t-4)$ ft

19. $S^2-15^2 = (S+15)(S-15)$
    $(S+15)(S-15)$ sq ft

20. $x^2+(x+8)^2 = 40^2$
    $x^2+x^2+16x+64 = 1600$
    $2x^2+16x-1536 = 0$
    $2(x^2+8x-768) = 0$
    $2(x+32)(x-24) = 0$
    $x+32 = 0$ or $x-24 = 0$
    $x = -32$ $\quad x = 24$
    $x+8 = 24+8 = 32$
    The length of the screen is 32 in. and the width is 24 in.

6.1 Common Factoring and Factoring by Grouping

Practice

1. $24 = 2 \cdot 2 \cdot 2 \cdot 3 = 2^3 \cdot 3$

$72 = 2 \cdot 2 \cdot 2 \cdot 3 \cdot 3 = 2^3 \cdot 3^2$

$96 = 2 \cdot 2 \cdot 2 \cdot 2 \cdot 2 \cdot 3 = 2^5 \cdot 3$

$GCF = 2^3 \cdot 3 = 24$

2. $a^3 = a \cdot a \cdot a$

$a^2 = a \cdot a$

$a = a$

$GCF = a$

3. $-18x^3y^4 = -2 \cdot 3 \cdot 3 \cdot x \cdot x \cdot x \cdot y \cdot y \cdot y \cdot y$

$12xy^2 = 2 \cdot 2 \cdot 3 \cdot x \cdot y \cdot y$

$GCF = 2 \cdot 3 \cdot x \cdot y \cdot y = 6xy^2$

4. $$\begin{aligned} 10y^2 + 8y^5 &= 2y^2(5) + 2y^2(4y^3) \\ &= 2y^2(5 + 4y^3) \end{aligned}$$

5. $$\begin{aligned} 21a^2b - 14a &= 7a(3ab) - 7a(2) \\ &= 7a(3ab - 2) \end{aligned}$$

6. $$\begin{aligned} 8a^2b^2 - 6ab^3 &= 2ab^2(4a) - 2ab^2(3b) \\ &= 2ab^2(4a - 3b) \end{aligned}$$

7. $$\begin{aligned} &24a^2 - 48a + 12 \\ &= 12(2a^2) - 12(4a) + 12(1) \\ &= 12(2a^2 - 4a + 1) \end{aligned}$$

8. $$\begin{aligned} ab &= s^2 - ac \\ ab + ac &= s^2 - ac + ac \\ ab + ac &= s^2 \\ a(b+c) &= s^2 \\ \frac{a(b+c)}{b+c} &= \frac{s^2}{b+c} \\ a &= \frac{s^2}{b+c} \end{aligned}$$

9. $$\begin{aligned} 4(y-3) + y(y-3) &= (y-3)(4+y), \\ &\text{or } (y-3)(y+4) \end{aligned}$$

10. $$\begin{aligned} 3y(x-1) + 2(1-x) &= 3y(x-1) - 2(x-1) \\ &= (x-1)(3y-2) \end{aligned}$$

11. $$\begin{aligned} (4-3x) + 2x(4-3x) &= 1(4-3x) + 2x(4-3x) \\ &= (4-3x)(1+2x) \end{aligned}$$

12. $$\begin{aligned} 4b - 20 + ab - 5a &= (4b-20) + (ab-5a) \\ &= 4(b-5) + a(b-5) \\ &= (b-5)(4+a) \end{aligned}$$

13. $$\begin{aligned} 5y - 5z - y^2 + yz &= (5y-5z) + (-y^2 + yz) \\ &= (5y-5z) - (y^2 - yz) \\ &= 5(y-z) - y(y-z) \\ &= (y-z)(5-y) \end{aligned}$$

14. $$v_0t + \frac{1}{2}at^2 = t\left(v_0 + \frac{1}{2}at\right)$$

Exercises

1. Rewriting a polynomial as a product is called factoring the polynomial.

3. The greatest common factor (GCF) of two or more integers is the greatest integer that is a factor of each integer.

5. $27 = 3 \cdot 3 \cdot 3 = 3^3$

$54 = 3 \cdot 3 \cdot 3 \cdot 2 = 3^3 \cdot 2$

$81 = 3 \cdot 3 \cdot 3 \cdot 3 = 3^4$

$GCF = 3^3 = 27$

7. $x^4 = x \cdot x \cdot x \cdot x$

$x^6 = x \cdot x \cdot x \cdot x \cdot x \cdot x$

$x^3 = x \cdot x \cdot x$

$GCF = x^3$

9. $16b = 2 \cdot 2 \cdot 2 \cdot 2 \cdot b$

$8b^3 = 2 \cdot 2 \cdot 2 \cdot b \cdot b \cdot b$

$12b^2 = 2 \cdot 2 \cdot 3 \cdot b \cdot b$

$GCF = 2 \cdot 2 \cdot b = 4b$

11. $-12x^5y^7$
$= -2\cdot 2\cdot 3\cdot x\cdot x\cdot x\cdot x\cdot x\cdot y\cdot y\cdot y\cdot y\cdot y\cdot y\cdot y$
$4y^3 = 2\cdot 2\cdot y\cdot y\cdot y$
$GCF = 2\cdot 2\cdot y\cdot y\cdot y = 4y^3$

13. $18a^5b^4 = 2\cdot 3\cdot 3\cdot a\cdot a\cdot a\cdot a\cdot a\cdot b\cdot b\cdot b\cdot b$
$-6a^4b^3 = -2\cdot 3\cdot a\cdot a\cdot a\cdot a\cdot b\cdot b\cdot b$
$9a^2b^2 = 3\cdot 3\cdot a\cdot a\cdot b\cdot b$
$GCF = 3\cdot a\cdot a\cdot b\cdot b = 3a^2b^2$

15. $x(3x-1) = x\cdot(3x-1)$
$8(3x-1) = 2\cdot 2\cdot 2\cdot(3x-1)$
$GCF = 3x-1$

17. $4x(x+7) = 2\cdot 2\cdot x\cdot(x+7)$
$9x(x+7) = 3\cdot 3\cdot x\cdot(x+7)$
$GCF = x(x+7)$

19. $3x+6 = 3(x)+3(2)$
$= 3(x+2)$

21. $24x^2+8 = 8(3x^2)+8(1)$
$= 8(3x^2+1)$

23. $27m-9n = 9(3m)-9(n)$
$= 9(3m-n)$

25. $2x-7x^2 = x(2)-x(7x)$
$= x(2-7x)$

27. $5b^2-6b^3 = b^2(5)-b^2(6b)$
$= b^2(5-6b)$

29. $10x^3-15x = 5x(2x^2)-5x(3)$
$= 5x(2x^2-3)$

31. $a^2b^2-ab = ab(ab)-ab(1)$
$= ab(ab-1)$

33. $6xy^2+7x^2y = xy(6y)+xy(7x)$
$= xy(6y+7x)$

35. $27pq^2+18p^2q = 9pq(3q)+9pq(2p)$
$= 9pq(3q+2p)$

37. $2x^3y-12x^3y^4 = 2x^3y(1)-2x^3y(6y^3)$
$= 2x^3y(1-6y^3)$

39. $3c^3+6c^2+12 = 3(c^3)+3(2c^2)+3(4)$
$= 3(c^3+2c^2+4)$

41. $9b^4-3b^3+b^2 = b^2(9b^2)-b^2(3b)+b^2(1)$
$= b^2(9b^2-3b+1)$

43. $2m^4+10m^3-6m^2$
$= 2m^2(m^2)+2m^2(5m)-2m^2(3)$
$= 2m^2(m^2+5m-3)$

45. $5b^5-3b^3+2b^2$
$= b^2(5b^3)-b^2(3b)+b^2(2)$
$= b^2(5b^3-3b+2)$

47. $15x^4-10x^3-25x$
$= 5x(3x^3)-5x(2x^2)-5x(5)$
$= 5x(3x^3-2x^2-5)$

49. $4a^2b+8a^2b^2-12ab$
$= 4ab(a)+4ab(2ab)-4ab(3)$
$= 4ab(a+2ab-3)$

51. $9c^2d^2+12c^3d+3cd^3$
$= 3cd(3cd)+3cd(4c^2)+3cd(d^2)$
$= 3cd(3cd+4c^2+d^2)$

53. $x(x-1)+3(x-1) = (x-1)(x+3)$

55. $5a(a-1)-3(a-1) = (a-1)(5a-3)$

57. $r(s+7)-2(7+s) = r(s+7)-2(s+7)$
$= (s+7)(r-2)$

59. $a(x-y)-b(x-y) = (x-y)(a-b)$

61. $$\begin{aligned}3x(y+2)-(y+2)&=3x(y+2)-1(y+2)\\&=(y+2)(3x-1)\end{aligned}$$

63. $$\begin{aligned}b(b-1)+5(1-b)&=b(b-1)-5(b-1)\\&=(b-1)(b-5)\end{aligned}$$

65. $$\begin{aligned}y(y-1)-5(1-y)&=y(y-1)+5(y-1)\\&=(y-1)(y+5)\end{aligned}$$

67. $$\begin{aligned}(t-3)-t(3-t)&=1(t-3)+t(t-3)\\&=(t-3)(1+t)\end{aligned}$$

69.

$$\begin{aligned}9a(b-7)+2(7-b)&=9a(b-7)-2(b-7)\\&=(b-7)(9a-2)\end{aligned}$$

71. $$\begin{aligned}rs+3s+rt+3t&=s(r+3)+t(r+3)\\&=(r+3)(s+t)\end{aligned}$$

73. $$\begin{aligned}xy+6y-4x-24&=y(x+6)-4(x+6)\\&=(x+6)(y-4)\end{aligned}$$

75. $$\begin{aligned}&15xy-9yz+20xz-12z^2\\&=3y(5x-3z)-4z(5x-3z)\\&=(5x-3z)(3y+4z)\end{aligned}$$

77. $$\begin{aligned}&2xz+8x+5yz+20y\\&=2x(z+4)+5y(z+4)\\&=(z+4)(2x+5y)\end{aligned}$$

79. $$\begin{aligned}TM&=PC+PL\\TM&=P(C+L)\\\frac{TM}{C+L}&=\frac{P(C+L)}{C+L}\\P&=\frac{TM}{C+L}\end{aligned}$$

81. $$\begin{aligned}S&=2lw+2lh+2wh\\S&=2l(w+h)+2wh\\S-2wh&=2l(w+h)\\\frac{S-2wh}{2(w+h)}&=\frac{2l(w+h)}{2(w+h)}\\l&=\frac{S-2wh}{2w+2h}\end{aligned}$$

83. $$\begin{aligned}16p^3+24p&=8p(2p^2)+8p(3)\\&=8p(2p^2+3)\end{aligned}$$

85. $$\begin{aligned}48r^2s-60r^2s&=12rs(4s)-12rs(5r)\\&=12rs(4s-5r)\end{aligned}$$

87. $$\begin{aligned}42j^2-6&=6(7j^2)-6(1)\\&=6(7j^2-1)\end{aligned}$$

89. $$\begin{aligned}st-3t-7s+21&=t(s-3)-7(s-3)\\&=(s-3)(t-7)\end{aligned}$$

91. $$\begin{aligned}&3x(y-4)+5(4-y)\\&=3x(y-4)-5(y-4)\\&=(y-4)(3x-5)\end{aligned}$$

93. $$\begin{aligned}16a^5b^3&=2\cdot2\cdot2\cdot2\cdot a\cdot a\cdot a\cdot a\cdot a\cdot b\cdot b\cdot b\\-12a^2b^3&=-2\cdot2\cdot3\cdot a\cdot a\cdot b\cdot b\cdot b\\20a^3b&=2\cdot2\cdot5\cdot a\cdot a\cdot a\cdot b\\GCF&=2\cdot2\cdot a\cdot a\cdot b=4a^2b\end{aligned}$$

95. $mv_2-mv_1=m(v_2-v_1)$

97. $0.5n^2-0.5n=0.5n(n-1)$

99.

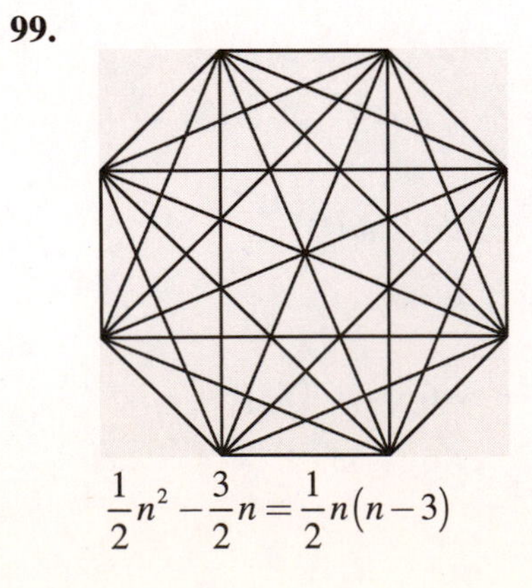

$$\frac{1}{2}n^2-\frac{3}{2}n=\frac{1}{2}n(n-3)$$

101.
$$\begin{aligned} P &= nC + nT + D \\ P &= n(C+T) + D \\ P - D &= n(C+T) \\ \frac{P-D}{C+T} &= \frac{n(C+T)}{C+T} \\ n &= \frac{P-D}{C+T} \end{aligned}$$

6.2 Factoring Trinomials Whose Leading Coefficient Is 1

Practice

1. $x^2 + 5x + 4 = (x+1)(x+4)$, or $(x+4)(x+1)$

2. $y^2 - 9y + 20 = (y-4)(y-5)$ or, $(y-5)(y-4)$

3. $x^2 + 3x + 5$
Prime polynomial; cannot be factored

4. $32 - 12y + y^2 = y^2 - 12y + 32 = (y-4)(y-8)$

5. $p^2 - 4pq + 3q^2 = (p-q)(p-3q)$, or $(p-3q)(p-q)$

6. $x^2 + x - 6 = (x-2)(x+3)$

7. $x^2 - 21x - 46 = (x+2)(x-23)$

8. $y^2 - 24 + 2y = y^2 + 2y - 24 = (y-4)(y+6)$

9. $a^2 - 5ab - 24b^2 = (a+3b)(a-8b)$

10. $y^3 - 9y^2 - 10y = y(y^2 - 9y - 10) = y(y+1)(y-10)$

11. $8x^3 - 24x^2 + 16x = 8x(x^2 - 3x + 2) = 8x(x-1)(x-2)$

12. $-x^2 - 10x + 11 = -(x + 10x - 11) = -(x-1)(x+11)$, or $(-x+1)(x+11)$, or $(x-1)(-x-11)$

13. $-16t^2 + 32t + 48 = -16(t^2 - 2t - 3) = -16(t+1)(t-3)$

Exercises

1. Polynomials that are not factorable are called prime polynomials.

3. If $c < 0$ in the trinomial $x^2 + bx + c$, then the constant terms of the binomial factors have opposite signs.

5. f

7. e

9. b

11. $(x-5)$

13. $(x+4)$

15. $(x-1)$

17. $x^2 + 6x + 8 = (x+2)(x+4)$

19. $x^2 + 5x - 6 = (x-1)(x+6)$

21. $x^2 + x + 2$
Prime polynomial

23. $x^2 + 5x + 4 = (x+1)(x+4)$

25. $x^2 - 4x + 3 = (x-1)(x-3)$

27. $y^2 - 12y + 32 = (y-4)(y-8)$

29. $t^2 - 4t - 45 = (t+1)(t-5)$

31. x^2+2x-1
Prime polynomial

33. $x^2+4x-45=(x-5)(x+9)$

35. $y^2-9y+20=(y-4)(y-5)$

37. $b^2+11b+28=(b+4)(b+7)$

39. y^2+4y+5
Prime polynomial

41. $$\begin{aligned}-y^2+5y+50&=-(y^2-5y-50)\\&=-(y+5)(y-10)\end{aligned}$$

43. $$\begin{aligned}x^2+64-16x&=x^2-16x+64\\&=(x-8)(x-8)\text{ or }(x-8)^2\end{aligned}$$

45. $$\begin{aligned}16-10x+x^2&=x^2-10x+16\\&=(x-2)(x-8)\end{aligned}$$

47. $$\begin{aligned}81-30w+w^2&=w^2-30w+81\\&=(w-3)(w-27)\end{aligned}$$

49. $p^2-8pq+7q^2=(p-q)(p-7q)$

51. $p^2-4pq-5q^2=(p+q)(p-5q)$

53. $m^2-12mn+35n^2=(m-5n)(m-7n)$

55. $x^2+9xy+8y^2=(x+y)(x+8y)$

57. $$\begin{aligned}5x^2-5x-30&=5(x^2-x-6)\\&=5(x+2)(x-3)\end{aligned}$$

59. $$\begin{aligned}2x^2+10x-28&=2(x^2+5x-14)\\&=2(x-2)(x+7)\end{aligned}$$

61. $$\begin{aligned}12-18t+6t^2&=6t^2-18t+12\\&=6(t^2-3t+2)\\&=6(t-1)(t-2)\end{aligned}$$

63. $$\begin{aligned}3x^2+24+18x&=3x^2+18x+24\\&=3(x^2+6x+8)\\&=3(x+2)(x+4)\end{aligned}$$

65. $$\begin{aligned}y^3+3y^2-10y&=y(y^2+3y-10)\\&=y(y-2)(y+5)\end{aligned}$$

67. $$\begin{aligned}a^3+8a^2+15a&=a(a^2+8x+15)\\&=a(a+3)(a+5)\end{aligned}$$

69. $$\begin{aligned}t^4-14t^3+24t^2&=t^2(t^2-14t+24)\\&=t^2(t-2)(t-12)\end{aligned}$$

71. $$\begin{aligned}4a^3-12a^2+8a&=4a(a^2-3a+2)\\&=4a(a-1)(a-2)\end{aligned}$$

73. $$\begin{aligned}2x^3+30x+16x^2&=2x^3+16x^2+30x\\&=2x(x^2+8x+15)\\&=2x(x+3)(x+5)\end{aligned}$$

75. $$\begin{aligned}4x^3+48x-28x^2&=4x^3-28x^2+48x\\&=4x(x^2-7x+12)\\&=4x(x-3)(x-4)\end{aligned}$$

77. $$\begin{aligned}-56s+6s^2+2s^3&=2s^3+6s^2-56s\\&=2s(s^2+3s-28)\\&=2s(s-4)(s+7)\end{aligned}$$

79. $$\begin{aligned}2c^4+4c^3-70c^2&=2c^2(c^2+2c-35)\\&=2c^2(c-5)(c+7)\end{aligned}$$

81. $$\begin{aligned}ax^3-18ax^2+32ax&=ax(x^2-18x+32)\\&=ax(x-2)(x-16)\end{aligned}$$

83. x^2+6x+3
Prime polynomial

85. $$\begin{aligned}5x^4-15x^3-50x^2&=5x^2(x^2-3x-10)\\&=5x^2(x-5)(x+2)\end{aligned}$$

87. $-w^2+6w+40=-(w^2-6w-40)$
$=-(w+4)(w-10)$

89. $6m^2-6m-36=6(m^2-m-6)$
$=6(m+2)(m-3)$

91. $t^2+60-17t=t^2-17t+60$
$=(t-5)(t-12)$

93. b

95. $p^2+2pq+q^2=1$
$(p+q)(p+q)=1$, or $(p+q)^2=1$

97. $16t^2+48t-160=16(t^2+3t-10)$
$=16(t-2)(t+5)$

99. **a.**

$$\begin{aligned}A&=2lw+2lh+2wh\\&=2(x+3)(x+3)+2x(x+3)+2x(x+3)\\&=2(x^2+6x+9)+4x(x+3)\\&=2x^2+12x+18+4x^2+12x\\&=6x^2+24x+18\end{aligned}$$

$(6x^2+24x+18)$ sq in.

b. $6x^2+24x+18=6(x^2+4x+3)$
$=6(x+1)(x+3)$

6.3 Factoring Trinomials Whose Leading Coefficient Is Not 1

Practice

1. $5x^2+14x+8=(5x+4)(x+2)$

2. $21-25x+6x^2=6x^2-25x+21$
$=(6x-7)(x-3)$

3. $7y^2+47y-14=(7y-2)(y+7)$

4. $2x^2-x-10=(2x-5)(x+2)$

5. $18x^3-21x^2-9x=3x(6x^2-7x-3)$
$=3x(3x+1)(2x-3)$

6. $36c^2-12cd-15d^2=3(12c^2-4cd-5d^2)$
$=3(6c-5d)(2c+d)$

7. $2x^2-7x-4=(2x+1)(x-4)$

8. $4x^3-24x^2+35x=x(4x^2-24x+35)$
$=x(2x-5)(2x-7)$

Exercises

1. e

3. b

5. c

7. $(3x+1)$

9. $(x-3)$

11. $(x-3)$

13. $3x^2+8x+5=(3x+5)(x+1)$

15. $2y^2-11y+5=(2y-1)(y-5)$

17. $3x^2+14x+8=(3x+2)(x+4)$

19. $5x^2+9x-6$
Cannot be factored

21. $6y^2-y-5=(6y+5)(y-1)$

23. $2y^2-11y+14=(2y-7)(y-2)$

25. $9a^2-18a-16=(3a+2)(3a-8)$

27. $4x^2-13x+3=(4x-1)(x-3)$

29. $6+17y+12y^2=12y^2+17y+6$
$=(3y+2)(4y+3)$

31. $-17m+21+2m^2 = 2m^2-17m+21$
$= (2m-3)(m-7)$

33. $-6a^2-7a+3 = -(6a^2+7a-3)$
$= -(3a-1)(2a+3)$

35. $8y^2+5y-22 = (8y-11)(y+2)$

37. $7y^2+36y-5$
Cannot be factored

39. $8a^2+65a+8 = (8a+1)(a+8)$

41. $6x^2+25x-9 = (3x-1)(2x+9)$

43. $8y^2-26y+15 = (4y-3)(2y-5)$

45. $14y^2-38y+20 = 2(7y^2-19y+10)$
$= 2(7y-5)(y-2)$

47. $28a^2+24a-4 = 4(7a^2+6a-1)$
$= 4(7a-1)(a+1)$

49. $-6b^2+40b+14 = -2(3b^2-20b-7)$
$= -2(3b+1)(b-7)$

51. $12y^3+50y^2+28y = 2y(6y^2+25y+14)$
$= 2y(3y+2)(2y+7)$

53.

$14a^4-38a^3+20a^2 = 2a^2(7a^2-19a+10)$
$= 2a^2(7a-5)(a-2)$

55.
$2x^3y+13x^2y+15xy = xy(2x^2+13x+15)$
$= xy(2x+3)(x+5)$

57.

$6ab^3-44ab^2+14ab = 2ab(3b^2-22b+7)$
$= 2ab(3b-1)(b-7)$

59. $20c^2-9cd+d^2 = (5c-d)(4c-d)$

61. $2x^2-5xy-3y^2 = (2x+y)(x-3y)$

63. $8a^2-6ab-b^2 = (4a-b)(2a-b)$

65. $18x^2+3xy-6y^2 = 3(6x^2+xy-2y^2)$
$= 3(3x+2y)(2x-y)$

67. $16c^2-44cd+30d^2$
$= 2(8c^2-22cd+15d^2)$
$= 2(4c-5d)(2c-3d)$

69. $27u^2+18uv+3v^2$
$= 3(9u^2+6uv+v^2)$
$= 3(3u+v)(3u+v)$ or $3(3u+v)^2$

71. $42x^3+45x^2y-27xy^2$
$= 3x(14x^2+15xy-9y^2)$
$= 3x(7x-3y)(2x+3y)$

73. $-30x^4y+35x^3y^2+15x^2y^3$
$= -5x^2y(6x^2-7xy-3y^2)$
$= -5x^2y(3x+y)(2x-3y)$

75. $5ax^2-28axy-12ay^2$
$= a(5x^2-28xy-12y^2)$
$= a(5x+2y)(x-6y)$

77. $-5m+2+3m^2 = 3m^2-5m+2$
$= (m-1)(3m-2)$

79. $8x^2-2x-3 = (4x-3)(2x+1)$

81. $14x^3+44x^2+6x = 2x(7x^2+22x+3)$
$= 2x(x+3)(7x+1)$

83. $8m^2-18mn+9n^2 = (2m-3n)(4m-3n)$

85. $7r^2-9r+2 = (r-1)(7r-2)$

87. d

89. $-5t^2 - 21t + 20 = -(5t^2 + 21 - 20)$
$= -(5t - 4)(t + 5)$

91.

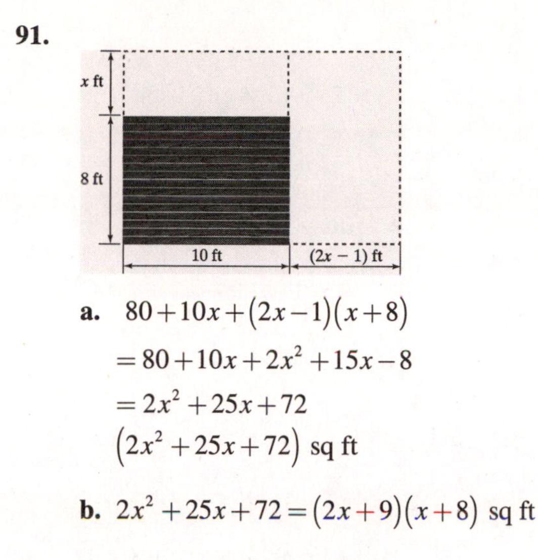

a. $80 + 10x + (2x - 1)(x + 8)$
$= 80 + 10x + 2x^2 + 15x - 8$
$= 2x^2 + 25x + 72$
$(2x^2 + 25x + 72)$ sq ft

b. $2x^2 + 25x + 72 = (2x + 9)(x + 8)$ sq ft

93. $4n^2 - 12n + 5 = (2n - 5)(2n - 1)$
Since the difference of the factors is $(2n - 1) - (2n - 5) = 2n - 1 - 2n + 5 = 4$, the factors represent two integers that differ by 4 no matter what integer n represents

6.4 Special Factoring

Practice

1. **a.** The trinomial is a perfect square.
b. The trinomial is not a perfect square.
c. The trinomial is a perfect square.
d. The trinomial is not a perfect square.
e. The trinomial is a perfect square.

2. $n^2 + 20n + 100 = (n + 10)^2$

3. $t^2 + 4 - 4t = t^2 - 4t + 4$
$= (t - 2)^2$

4. $25c^2 - 40cd + 16d^2 = (5c - 4d)^2$

5. $x^4 + 8x^2 + 16 = (x^2 + 4)^2$

6. **a.** The binomial is a difference of squares.
b. The binomial is not a difference of squares.
c. The binomial is not a difference of squares.
d. The binomial is a difference of squares.

7. $y^2 - 121 = (y + 11)(y - 11)$

8. $9x^2 - 25y^2 = (3x + 5y)(3x - 5y)$

9. $64x^8 - 81y^2 = (8x^4 + 9y)(8x^4 - 9y)$

10. $256 - 16t^2 = 16(16 - t^2)$
$= 16(4 + t)(4 - t)$

Exercises

1. A perfect square trinomial of the form $a^2 + 2ab + b^2$ can be factored as $(a + b)^2$.

3. The binomial $a^2 - b^2$, a difference of squares, can be factored as $(a + b)(a - b)$.

5. Perfect square trinomial

7. Neither

9. Perfect square trinomial

11. Difference of squares

13. Difference of squares

15. Perfect square trinomial

17. Neither

19. Neither

21. Neither

23. $x^2 - 12x + 36 = (x - 6)^2$

25. $y^2 + 20y + 100 = (y + 10)^2$

27. $a^2 - 4a + 4 = (a - 2)^2$

29. x^2-6x-9
Not factorable

31. $m^2-64=(m+8)(m-8)$

33. $y^2-81=(y+9)(y-9)$

35. $144-x^2=(12+x)(12-x)$

37. $4a^2-36a+81=(2a-9)^2$

39. $49x^2+28x+4=(7x+2)^2$

41. $36-60x+25x^2=(6-5x)^2$

43. $100m^2-81=(10m+9)(10m-9)$

45. $36x^2+121$
Not factorable

47. $1-9x^2=(1+3x)(1-3x)$

49. $m^2+26mn+169n^2=(m+13n)^2$

51. $4a^2+36ab+81b^2=(2a+9b)^2$

53. $x^2-4y^2=(x+2y)(x-2y)$

55. $100x^2-9y^2=(10x+3y)(10x-3y)$

57. $y^4+2y^2+1=(y^2+1)^2$

59. $6x^2+12x+6=6(x^2+2x+1)$
$=6(x+1)^2$

61.
$27m^3-36m^2+12m=3m(9m^2-12m+4)$
$=3m(3m-2)^2$

63. $4s^2t^3+80s^2t^2+400s^2t$
$=4s^2t(t^2+20t+100)$
$=4s^2t(t+10)^2$

65. $3k^3-147k=3k(k^2-49)$
$=3k(k+7)(k-7)$

67. $4y^4-36y^2=4y^2(y^2-9)$
$=4y^2(y+3)(y-3)$

69. $27x^2y-3x^2y^3=3x^2y(9-y^2)$
$=3x^2y(3+y)(3-y)$

71. $2a^2b^2-98=2(a^2b^2-49)$
$=2(ab+7)(ab-7)$

73. $256-r^4=(16+r^2)(16-r^2)$
$=(16+r^2)(4+r)(4-r)$

75. $5x^4-80y^4=5(x^4-16y^4)$
$=5(x^2+4y^2)(x^2-4y^2)$
$=5(x^2+4y^2)(x+2y)(x-2y)$

77. $x^2(c-d)-4(c-d)=(c-d)(x^2-4)$
$=(c-d)(x+2)(x-2)$

79. $16(x-y)-a^2(x-y)=(x-y)(16-a^2)$
$=(x-y)(4+a)(4-a)$

81. $9c^2+48cd+64d^2=(3c+8d)^2$

83. $a^2-225b^4=(a+15b^2)(a-15b^2)$

85. $54a^4b^2-36a^2b+6=6(9a^4b^2-6a^2b+1)$
$=6(3a^2b-1)^2$

87. $81-w^4=(9+w^2)(9-w^2)$
$=(9+w^2)(3-w)(3+w)$

89. $36u^2+60u+25=(6u+5)^2$

91. Difference of squares

93.

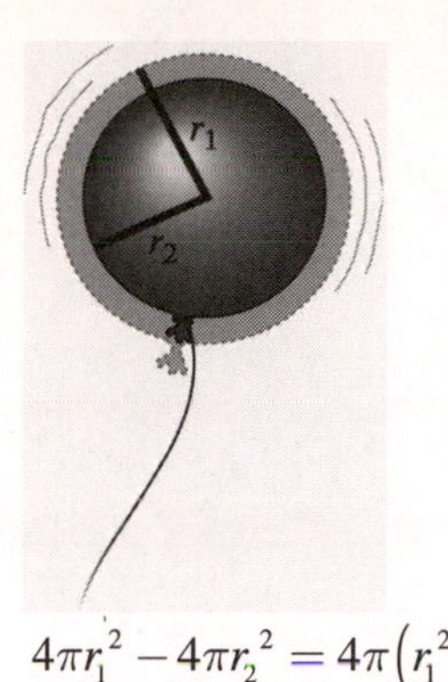

$$4\pi r_1^{\,2} - 4\pi r_2^{\,2} = 4\pi\left(r_1^{\,2} - r_2^{\,2}\right)$$
$$= 4\pi\left(r_1 + r_2\right)\left(r_1 - r_2\right)$$

95. $16{,}000 + 32{,}000r + 16{,}000r^2$
$$= 16{,}000\left(1 + 2r + r^2\right)$$
$$= 16{,}000\left(1 + r\right)^2 \text{ or } 16{,}000\left(r + 1\right)^2$$

97. $kv_2^{\,2} - kv_1^{\,2} = k\left(v_2^{\,2} - v_1^{\,2}\right)$
$$= k\left(v_2 + v_1\right)\left(v_2 - v_1\right)$$

6.5 More Factoring Strategies

Practice

1. **a.** $16 - (2x - y)^2$
$$= 4^2 - (2x - y)^2$$
$$= \left[4 + (2x - y)\right]\left[4 - (2x - y)\right]$$
$$= (4 + 2x - y)(4 - 2x + y)$$

b. $64x^8 - 25y^2$
$$= \left(8x^4\right)^2 - (5y)^2$$
$$= \left(8x^4 + 5y\right)\left(8x^4 - 5y\right)$$

2. **a.** Difference of cubes
b. Sum of cubes
c. Neither

3. **a.** $125 - y^3$
$$= 5^3 - y^3$$
$$= (5 - y)\left[(5)^2 + 5 \cdot y + (y)^2\right]$$
$$= (5 - y)\left(25 + 5y + y^2\right)$$

b. $27m^3n^6 + 1$
$$= 3^3m^3\left(n^2\right)^3 + 1^3$$
$$= \left(3mn^2\right)^3 + 1^3$$
$$= \left(3mn^2 + 1\right)\left[\left(3mn^2\right)^2 - \left(3mn^2\right) \cdot 1 + 1^2\right]$$
$$= \left(3mn^2 + 1\right)\left(9m^2n^4 - 3mn^2 + 1\right)$$

c. $(x + 1)^3 - (x - 1)^3$
$$= \left[(x + 1) - (x - 1)\right]\left[(x + 1)^2 + (x + 1)(x - 1) + (x - 1)^2\right]$$
$$= (x + 1 - x + 1)\left[\left(x^2 + 2x + 1\right) + \left(x^2 - 1\right) + \left(x^2 - 2x + 1\right)\right]$$
$$= (2)\left(3x^2 + 1\right)$$

4. $100 - 16t^2$
$$= 4\left(25 - 4t^2\right)$$
$$= 4\left[5^2 - (2t)^2\right]$$
$$= 4(5 + 2t)(5 - 2t) \text{ or } -4(2t + 5)(2t - 5)$$

Exercises

1. $(2u - v)^2 - 64 = (2u - v)^2 - (8)^2$
$$= (2u - v - 8)(2u - v + 8)$$

3. $49 - 4(x + y)^2 = (7)^2 - \left[2(x + y)\right]^2$
$$= \left(7 - 2(x + y)\right)\left(7 + 2(x + y)\right)$$
$$= (7 - 2x - 2y)(7 + 2x + 2y)$$

5. $p^6 - 22p^3 + 121 = \left(p^3\right)^2 - 2\left(p^3\right)(11) + (11)^2$
$$= \left(p^3 - 11\right)^2$$

7. $9a^8 + 48a^4b + 64b^2 = \left(3a^4\right)^2 + 2\left(3a^4\right)(8b) + (8b)^2$
$$= \left(3a^4 + 8b\right)^2$$

9.

$$4a^4 - 225 = (2a^2)^2 - (15)^2 = (2a^2 - 15)(2a^2 + 15)$$

11. $49x^6 - 144y^4 = (7x^3)^2 - (12y^2)^2$
$= (7x^3 - 12y^2)(7x^3 + 12y^2)$

13.
$100p^4q^2 - 9r^2 = (10p^2q)^2 - (3r)^2$
$= (10p^2q - 3r)(10p^2q + 3r)$

15.
$(3p+q)^2 - (2p-q)^2$
$= ((3p+q)-(2p-q))((3p+q)+(2p-q))$
$= (3p+q-2p+q)(3p+q+2p-q)$
$= (p+2q)(5p) = 5p(p+2q)$

17. $x^3 - 64 = (x)^3 - (4)^3$; difference of cubes

19. $x^6 - 3y^3 = (x^2)^3 - 3(y)^3$; neither

21. $y^{12} + 0.008x^3 = (y^4)^3 + (0.2x)^3$; sum of cubes.

23. $x^3 + 1$
$= (x)^2 + (1)^3$
$= (x+1)(x^2 - (x)(1) + 1)$
$= (x+1)(x^2 - x + 1)$

25. $p^3 - 8$
$= (p)^3 - (2)^2$
$= (p-2)(p^2 + (2)(p) + 2^2)$
$= (p-2)(p^2 + 2p + 4)$

27. $27t^2 - 4 = 27t^2 - 2^2$ is a prime polynomial.

29. $\frac{1}{8} - a^3 = \left(\frac{1}{2}\right)^3 - (a)^3$
$= \left(\frac{1}{2} - a\right)\left(\left(\frac{1}{2}\right)^2 + \left(\frac{1}{2}\right)a + a^2\right)$
$= \left(\frac{1}{2} - a\right)\left(\frac{1}{4} + \frac{1}{2}a + a^2\right)$

31. $125x^3 + y^3$
$= (5x)^3 + (y)^3$
$= (5x+y)((5x)^2 - (5x)(y) + y^2)$
$= (5x+y)(25x^2 - 5xy + y^2)$

33. $0.064b^3 - 0.027a^3$
$= (0.4b)^3 - (0.3a)^3$
$= (0.4b - 0.3a)((0.4b)^2 + (0.4b)(0.3a) + (0.3a)^2)$
$= (0.4b - 0.3a)(0.16b^2 + 0.12ab + 0.09a^2)$

35. $a^6 - 8 = (a^2)^3 - (2)^3$
$= (a^2 - 2)((a^2)^2 + (a^2)(2) + (2)^2)$
$= (a^2 - 2)(a^4 + 2a^2 + 4)$

37. $64x^9 + 27y^3 = (4x^3)^3 + (3y)^3$
$= (4x^3 + 3y)((4x^3)^2 - (4x^3)(3y) + (3y)^2)$
$= (4x^3 + 3y)(16x^6 - 12x^3y + 9y^2)$

39. $27 - (a+1)^3 = (3)^3 - (a+1)^3$
$= (3 - (a+1))((3)^2 + 3(a+1) + (a+1)^2)$
$= (3 - a - 1)(9 + 3a + 3 + a^2 + 2a + 1)$
$= (2 - a)(a^2 + 5a + 13)$

41. $(x-2)^3 + (x+2)^3$
$= ((x-2) + (x+2))$
$\cdot ((x-2)^2 - (x-2)(x+2) + (x+2)^2)$
$= (2x)(x^2 - 4x + 4 - (x^2 - 4) + x^2 + 4x + 4)$
$= 2x(x^2 + 12)$

43. Difference of cubes.

45.

$$\begin{aligned}25(2a-b)^2-9&=[5(2a-b)]^2-3^2\\&=[5(2a-b)+3][5(2a-b)-3]\\&=(10a-5b+3)(10a-5b-3)\end{aligned}$$

47. $32x^3y-18xy^3$

$2xy$ is a common factor

$$\begin{aligned}32x^3y-18xy^3&=2xy(16x^2-9y^2)\\&=2xy(4x+3y)(4x-3y)\end{aligned}$$

49. Volume of a cube is side^3.

$x^3+y^3=(x+y)(x^2-xy+y^2)$

6.6 Solving Quadratic Equations by Factoring

Practice

1. $(3x-1)(x+5)=0$

$3x-1=0$ or $x+5=0$

$3x=1 \qquad x=-5$

$x=\frac{1}{3}$

2. $y^2+6y=0$

$y(y+6)=0$

$y=0$ or $y+6=0$

$y=-6$

3.

$$\begin{aligned}4y^2-11y&=3\\4y^2-11y-3&=0\\(4y+1)(y-3)&=0\end{aligned}$$

$4y+1=0$ or $y-3=0$

$4y=-1 \qquad y=3$

$y=-\frac{1}{4}$

4.

$$\begin{aligned}3t(t+4)&=15\\3t^2+12t&=15\\3t^2+12t-15&=0\\3(t^2+4t-5)&=0\\3(t+5)(t-1)&=0\end{aligned}$$

$t+5=0$ or $t-1=0$

$t=-5 \qquad t=1$

5.

$$\begin{aligned}x(18-x)&=80\\18x-x^2&=80\\-x^2+18x-80&=0\\-(x-10)(x-8)&=0\end{aligned}$$

$x-10=0$ or $x-8=0$

$x=10 \qquad x=8$

The dimensions of the frame should be 8 in. by 10 in.

6.

$$\begin{aligned}x^2+5^2&=13^2\\x^2+25&=169\\x^2-144&=0\\(x-12)(x+12)&=0\end{aligned}$$

$x-12=0$ or $x+12=0$

$x=12 \qquad x=-12$

The scooter going north has traveled 12 mi.

Exercises

1. A second-degree or <u>quadratic equation</u> is an equation that can be written in the form $ax^2+bc+c=0$, where a, b, and c are real numbers and $a\neq 0$.

3. In using the zero-product property to solve a quadratic equation, the equation must be <u>in standard form</u>.

5. Quadratic

7. Linear

9. Quadratic

11. $(x+3)(x-4)=0$

$x+3=0$ or $x-4=0$

$x=-3 \qquad x=4$

13. $4(x-1)=0$
$x-1=0$
$x=1$

15. $y(3y+5)=0$
$y=0$ or $3y+5=0$
$3y=-5$
$y=-\frac{5}{3}$

17. $(2t+1)(t-5)=0$
$2t+1=0$ or $t-5=0$
$2t=-1$ $t=5$
$t=-\frac{1}{2}$

19. $(2x+3)(2x-3)=0$
$2x+3=0$ or $2x-3=0$
$2x=-3$ $2x=3$
$x=-\frac{3}{2}$ $x=\frac{3}{2}$

21. $t(2-3t)=0$
$t=0$ or $2-3t=0$
$-3t=-2$
$t=\frac{2}{3}$

23. $y^2-2y=0$
$y(y-2)=0$
$y=0$ or $y-2=0$
$y=2$

25. $5x-25x^2=0$
$5x(1-5x)=0$
$5x=0$ or $1-5x=0$
$x=0$ $-5x=-1$
$x=\frac{1}{5}$

27. $x^2+5x+6=0$
$(x+2)(x+3)=0$
$x+2=0$ or $x+3=0$
$x=-2$ $x=-3$

29. $x^2+x-56=0$
$(x-7)(x+8)=0$
$x-7=0$ or $x+8=0$
$x=7$ $x=-8$

31. $2x^2-5x-3=0$
$(2x+1)(x-3)=0$
$2x+1=0$ or $x-3=0$
$2x=-1$ $x=3$
$x=-\frac{1}{2}$

33. $6x^2-x-2=0$
$(3x-2)(2x+1)=0$
$3x-2=0$ or $2x+1=0$
$3x=2$ $2x=-1$
$y=\frac{2}{3}$ $x=-\frac{1}{2}$

35. $0=36x^2-12x+1$
$0=(6x-1)(6x-1)$
$6x-1=0$
$6x=1$
$x=\frac{1}{6}$

37. $r^2-121=0$
$(r+11)(r-11)=0$
$r+11=0$ or $r-11=0$
$r=-11$ $r=11$

39. $0=(2x-3)^2$
$0=2x-3$
$3=2x$
$\frac{3}{2}=x$

41. $16x^2-16x+4=0$
$4(4x^2-4x+1)=0$
$4(2x-1)(2x-1)=0$
$2x-1=0$
$2x=1$
$x=\frac{1}{2}$

43.
$$\begin{aligned} 9m^2+15m-6&=0 \\ 3(3m^2+5m-2)&=0 \\ 3(3m-1)(m+2)&=0 \\ 3m-1=0 \quad &\text{or} \quad m+2=0 \\ 3m=1 \quad &\quad\quad m=-2 \\ m=\frac{1}{3} \quad & \end{aligned}$$

45.
$$\begin{aligned} r^2-r&=6 \\ r^2-r-6&=0 \\ (r-3)(r+2)&=0 \\ r-3=0 \quad &\text{or} \quad r+2=0 \\ r=3 \quad &\quad\quad r=-2 \end{aligned}$$

47.
$$\begin{aligned} y^2-7y&=-12 \\ y^2-7y+12&=0 \\ (y-3)(y-4)&=0 \\ y-3=0 \quad &\text{or} \quad y-4=0 \\ y=3 \quad &\quad\quad y=4 \end{aligned}$$

49.
$$\begin{aligned} n^2+2n&=8 \\ n^2+2n-8&=0 \\ (n-2)(n+4)&=0 \\ n-2=0 \quad &\text{or} \quad n+4=0 \\ n=2 \quad &\quad\quad n=-4 \end{aligned}$$

51.
$$\begin{aligned} 3y^2+4y&=-1 \\ 3y^2+4y+1&=0 \\ (3y+1)(y+1)&=0 \\ 3y+1=0 \quad &\text{or} \quad y+1=0 \\ 3y=-1 \quad &\quad\quad y=-1 \\ y=-\frac{1}{3} \quad & \end{aligned}$$

53.
$$\begin{aligned} 4x^2+6x&=-2 \\ 4x^2+6x+2&=0 \\ 2(2x^2+3x+1)&=0 \\ (2x+1)(x+1)&=0 \\ 2x+1=0 \quad &\text{or} \quad x+1=0 \\ 2x=-1 \quad &\quad\quad x=-1 \\ x=-\frac{1}{2} \quad & \end{aligned}$$

55.
$$\begin{aligned} 2n^2&=-10n \\ 2n^2+10n&=0 \\ 2n(n+5)&=0 \\ 2n=0 \quad &\text{or} \quad n+5=0 \\ n=0 \quad &\quad\quad n=-5 \end{aligned}$$

57.
$$\begin{aligned} 4x^2&=1 \\ 4x^2-1&=0 \\ (2x+1)(2x-1)&=0 \\ 2x+1=0 \quad &\text{or} \quad 2x-1=0 \\ 2x=-1 \quad &\quad\quad 2x=1 \\ x=-\frac{1}{2} \quad &\quad\quad x=\frac{1}{2} \end{aligned}$$

59.
$$\begin{aligned} 8y^2&=2 \\ 8y^2-2&=0 \\ 2(4y^2-1)&=0 \\ 2(2y+1)(2y-1)&=0 \\ 2y+1=0 \quad &\text{or} \quad 2y-1=0 \\ 2y=-1 \quad &\quad\quad 2y=1 \\ y=-\frac{1}{2} \quad &\quad\quad y=\frac{1}{2} \end{aligned}$$

61.
$$\begin{aligned} 3r^2+6r&=2r^2-9 \\ r^2+6r+9&=0 \\ (r+3)(r+3)&=0 \\ r+3&=0 \\ r&=-3 \end{aligned}$$

63.
$$\begin{aligned} x(x-1)&=12 \\ x^2-x&=12 \\ x^2-x-12&=0 \\ (x-4)(x+3)&=0 \\ x-4=0 \quad &\text{or} \quad x+3=0 \\ x=4 \quad &\quad\quad x=-3 \end{aligned}$$

65.
$$\begin{aligned} 4t(t-1)&=24 \\ 4t^2-4t&=24 \\ 4t^2-4t-24&=0 \\ 4(t^2-t-6)&=0 \\ 4(t+2)(t-3)&=0 \\ t+2=0 \quad &\text{or} \quad t-3=0 \\ t=-2 \quad &\quad\quad t=3 \end{aligned}$$

67. $(y+3)(y-2)=14$

$y^2+y-6=14$

$y^2+y-20=0$

$(y-4)(y+5)=0$

$y-4=0$ or $y+5=0$

$y=4 \quad y=-5$

69. $(3n-2)(n+5)=-14$

$3n^2+13n-10=-14$

$3n^2+13n+4=0$

$(3n+1)(n+4)=0$

$3n+1=0$ or $n+4=0$

$3n=-1 \quad n=-4$

$n=-\frac{1}{3}$

71. $(n+2)(n+4)=12n$

$n^2+6n+8=12n$

$n^2-6n+8=0$

$(n-2)(n-4)=0$

$n-2=0$ or $n-4=0$

$n=2 \quad n=4$

73. $3x(2x-5)=x^2-10$

$6x^2-15x=x^2-10$

$5x^2-15x+10=0$

$5(x^2-3x+2)=0$

$5(x-1)(x-2)=0$

$x-1=0$ or $x-2=0$

$x=1 \quad x=2$

75. $4r^2+11r-3=0$

$(4r-1)(r+3)=0$

$4r-1=0$ or $r+3=0$

$4r=1 \quad r=-3$

$r=\frac{1}{4}$

77. $10x^2+25x-15=0$

$5(2x^2+5x-3)=0$

$5(2x-1)(x+3)=0$

$2x-1=0$ or $x+3=0$

$2x=1 \quad x=-3$

$x=\frac{1}{2}$

79. $5a^2+19a=4$

$5a^2+19a-4=0$

$(5a-1)(a+4)=0$

$5a-1=0$ or $a+4=0$

$5a=1 \quad a=-4$

$a=\frac{1}{5}$

81. $3s^2+64=16s+2s^2$

$s^2-16s+64=0$

$(s-8)(s-8)=0$

$s-8=0$

$s=8$

83. $6t(3t-5)=0$

$6t=0$ or $3t-5=0$

$t=0 \quad 3t=5$

$t=\frac{5}{3}$

85. $4j+8j^2=0$

$4j(1+2j)=0$

$4j=0$ or $1+2j=0$

$j=0 \quad 2j=-1$

$j=-\frac{1}{2}$

87. $n^2-n=210$

$n^2-n-210=0$

$(n+14)(n-15)=0$

$n+14=0$ or $n-15=0$

$n=-14 \quad n=15$

There were 15 teams in the league.

89.

$$x^2+(x+2)^2=10^2$$
$$x^2+x^2+4x+4=100$$
$$2x^2+4x-96=0$$
$$2(x^2+2x-48)=0$$
$$2(x-6)(x+8)=0$$
$$x-6=0 \quad \text{or} \quad x+8=0$$
$$x=6 \qquad x=-8$$
$$x+2=6+2=8$$

One car traveled 6 mi and the other car traveled 8 mi.

91.

$$x^2=40{,}000$$
$$x^2-40{,}000=0$$
$$(x-200)(x+200)=0$$
$$x-200=0 \quad \text{or} \quad x+200=0$$
$$x=200 \qquad x=-200$$

200 ft

93.

$$-16t^2+8t+24=0$$
$$-8(2t^2-t-3)=0$$
$$-8(2t-3)(t+1)=0$$
$$2t-3=0 \quad \text{or} \quad t+1=0$$
$$2t=3 \qquad t=-1$$
$$t=\frac{3}{2}$$

The diver will hit the water in $\frac{3}{2}$, or 1.5 sec.

Chapter 6 Review Exercises

1. $48=2\cdot2\cdot2\cdot2\cdot3=2^4\cdot3$

$36=2\cdot2\cdot3\cdot3=2^2\cdot3^2$

$60=2\cdot2\cdot3\cdot5=2^2\cdot3\cdot5$

$GCF=2^2\cdot3=12$

2. $9m^3n=3\cdot3\cdot m\cdot m\cdot m\cdot n$

$24m^4=2\cdot2\cdot2\cdot3\cdot m\cdot m\cdot m\cdot m$

$15m^2n^2=3\cdot5\cdot m\cdot m\cdot n\cdot n$

$GCF=3m^2$

3. $3x-6y=3(x-2y)$

4. $16p^3q^2+18p^2q-4pq^2$
$=2pq(8p^2q+9p-2q)$

5. $(n-1)+n(n-1)=1(n-1)+n(n-1)$
$=(n-1)(1+n)$

6. $xb-5b-2x+10=b(x-5)-2(x-5)$
$=(x-5)(b-2)$

7. $d=rt_1+rt_2$

$d=r(t_1+t_2)$

$r=\dfrac{d}{t_1+t_2}$

8. $ax+y=bx+c$

$ax-bx=c-y$

$x(a-b)=c-y$

$x=\dfrac{c-y}{a-b}$

9. x^2+x+1
Not factorable

10. m^2-m+3
Not factorable

11. $y^2+42+13y=y^2+13y+42$
$=(y+6)(y+7)$

12. $m^2-7mn+10n^2=(m-2n)(m-5n)$

13. $24-8x-2x^2=-2x^2-8x+24$
$=-2(x^2+4x-12)$
$=-2(x-2)(x+6)$

14. $-15xy^2+3x^3-12x^2y$
$=3x^3-12x^2y-15xy^2$
$=3x(x^2-4xy-5y^2)$
$=3x(x+y)(x-5y)$

15. $3x^2+5x-2=(3x-1)(x+2)$

16. $5n^2+13n+6=(5n+3)(n+2)$

17. $3n^2-n-1$
Not factorable

18. $6x^2-x-12=(3x+4)(2x-3)$

19. $2a^2+3ab-35b^2=(2a-7b)(a+5b)$

20. $16a-4a^2-15=-4a^2+16a-15$
$=-(4a^2-16a+15)$
$=-(2a-3)(2a-5)$

21. $9y^3-21y+60y^2=9y^3+60y^2-21y$
$=3y(3y^2+20y-7)$
$=3y(3y-1)(y+7)$

22. $2p^2q-3pq^2-2q^3$
$=q(2p^2-3pq-2q^2)$
$=q(2p+q)(p-2q)$

23. $b^2-6b+9=(b-3)^2$

24. $64-x^2=(8+x)(8-x)$

25. $25y^2-20y+4=(5y-2)^2$

26. $29a^2+24ab+16b^2=(3a+4b)^2$

27. $81p^2-100q^2=(9p+10q)(9p-10q)$

28. $4x^8-28x^4+49=(2x^4-7)^2$

29. $48x^4-3y^4$
$=3(16x^4-y^4)$
$=3(4x^2+y^2)(4x^2-y^2)$
$=3(4x^2+y^2)(2x+y)(2x-y)$

30. $x^2(x-1)-9(x-1)$
$=(x-1)(x^2-9)$
$=(x-1)(x+3)(x-3)$

31. $3u^3+81$
Common factor is 3.
$3u^3+81=3((u)^3+(3)^3)=3(u+3)(u^2-3u+9)$

32. $64c^3-27d^3$
$=(4c)^3-(3d)^3$
$=(4c-3d)((4c)^2+(4c)(3d)+(3d)^2)$
$=(4c-3d)(16c^2+12cd+9d^2)$

33. $(x+y)^2-z^2=((x+y)+z)((x+y)-z)$
$=(x+y+z)(x+y-z)$

34. $1+(3a+1)^3$
$=(1)^3+(3a+1)^3$
$=(1+(3a+1))(1^2-1(3a+1)+(3a+1)^2)$
$=(3a+2)(1-3a-1+9a^2+6a+1)$
$=(3a+2)(9a^2+3a+1)$

35. $32u^{2n}-2v^{2m}$
Common factor is 2.
$32u^{2n}-2v^{2m}=2(16u^{2n}-v^{2m})$
$=2((4u^n)^2-(v^m)^2)$
$=2(4u^n+v^m)(4u^n-v^m)$

36. $x^2(x-1)+9(1-x)$
$=x^2(x-1)+9(-1)(x-1)$
$=(x-1)(x^2-9)$
$=(x-1)(x+3)(x-3)$

37. $(x+2)(x-1)=0$

$x+2=0$ or $x-1=0$

$x=-2$ $\quad x=1$

38. $t(t-4)=0$

$t=0$ or $t-4=0$

$t=4$

39. $3x^2+18x=0$

$3x(x+6)=0$

$3x=0$ or $x+6=0$

$x=-6$

40. $4x^2+4x+1=0$

$(2x+1)(2x+1)=0$

$2x+1=0$

$2x=-1$

$x=-\frac{1}{2}$

41. $y^2-10y=-16$

$y^2-10y+16=0$

$(y-2)(y-8)=0$

$y-2=0$ or $y-8=0$

$y=2$ $\quad y=8$

42. $3k^2-k=2$

$3k^2-k-2=0$

$(3k+2)(k-1)=0$

$3k+2=0$ or $k-1=0$

$3k=-2$ $\quad k=1$

$k=-\frac{2}{3}$

43. $4n(2n+3)=20$

$8n^2+12n=20$

$8n^2+12n-20=0$

$4(2n^2+3n-5)=0$

$(2n+5)(n-1)=0$

$2n+5=0$ or $n-1=0$

$2n=-5$ $\quad n=1$

$n=-\frac{5}{2}$

44. $(y-1)(y+2)=10$

$y^2+y-2=10$

$y^2+y-12=0$

$(y-3)(y+4)=0$

$y-3=0$ or $y+4=0$

$y=3$ $\quad y=-4$

45.

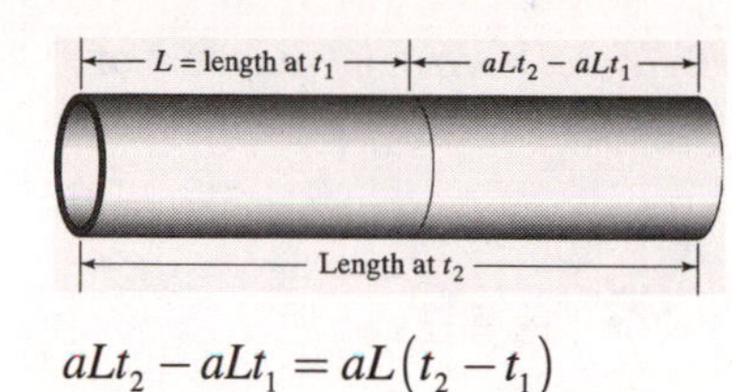

$aLt_2-aLt_1=aL(t_2-t_1)$

46. $at_2-16t_2^2-(at_1-16t_1^2)$

$=at_2-16t_2^2-at_1+16t_1^2$

$=at_2-at_1-16t_2^2+16t_1^2$

$=a(t_2-t_1)-16(t_2^2-t_1^2)$

$=a(t_2-t_1)-16(t_2+t_1)(t_2-t_1)$

$=(t_2-t_1)[a-16(t_2+t_1)]$

47.

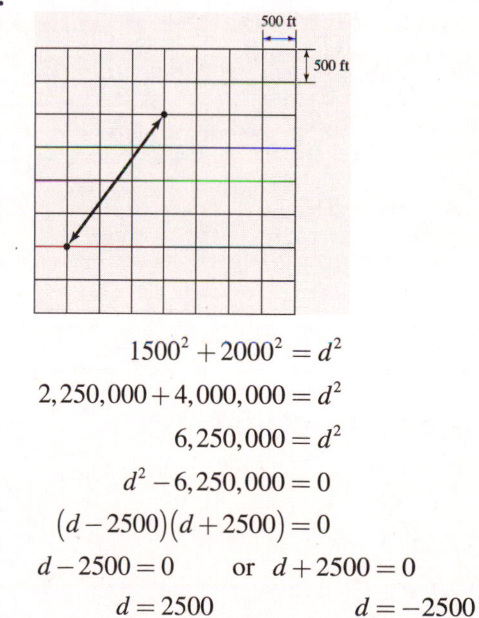

$1500^2+2000^2=d^2$

$2,250,000+4,000,000=d^2$

$6,250,000=d^2$

$d^2-6,250,000=0$

$(d-2500)(d+2500)=0$

$d-2500=0$ or $d+2500=0$

$d=2500$ $\quad d=-2500$

The distance between the two intersections is 2500 ft.

48. $52-40=12$

$$x^2+12^2=20^2$$
$$x^2+144=400$$
$$x^2-256=0$$
$$(x-16)(x+16)=0$$
$$x-16=0 \quad \text{or} \quad x+16=0$$
$$x=16 \qquad x=-16$$

The length of the horizontal diagonal of the kite is 32 in.

49.
$$76t-16t^2=18$$
$$-16t^2+76t-18=0$$
$$-2(8t^2-38t+9)=0$$
$$-2(4t-1)(2t-9)=0$$
$$4t-1=0 \quad \text{or} \quad 2t-9=0$$
$$4t=1 \qquad 2t=9$$
$$t=\frac{1}{4} \qquad t=\frac{9}{2}$$

The rocket will reach a height of 18 ft above the launch in $\frac{1}{4}$ sec and $\frac{9}{2}$ sec, or in 0.25 sec and 4.5 sec.

50.
$$A=\frac{1}{2}hb+\frac{1}{2}hB$$
$$A=\frac{1}{2}h(b+B)$$
$$2A=h(b+B)$$
$$h=\frac{2A}{b+B}$$

Chapter 6 Posttest

1.
$$12x^3=2\cdot 2\cdot 3\cdot x\cdot x\cdot x$$
$$15x^2=3\cdot 5\cdot x\cdot x$$
$$GCF=3x^2$$

2. $2xy-14y=2y(x-7)$

3.
$$6pq^2+8p^3-16p^2q$$
$$=8p^3-16p^2q+6pq^2$$
$$=2p(4p^2-8pq+3q^2)$$
$$=2p(2p-3q)(2p-q)$$

4.
$$ax-bx+by-ay=x(a-b)+y(b-a)$$
$$=x(a-b)-y(a-b)$$
$$=(a-b)(x-y)$$

5. $n^2-13n-48=(n+3)(n-16)$

6.
$$-8+x^2-2x=x^2-2x-8$$
$$=(x+2)(x-4)$$

7.
$$15x^2-5x^3+20x=-5x^3+15x^2+20x$$
$$=-5x(x^2-3x-4)$$
$$=-5x(x+1)(x-4)$$

8. $4x^2+13xy-12y^2=(4x-3y)(x+4y)$

9.
$$-12x^2+36x-27=-3(4x^2-12x+9)$$
$$=-3(2x-3)^2$$

10. $9x^2+30xy+25y^2=(3x+5y)^2$

11. $121-4x^2=(11+2x)(11-2x)$

12. $p^2q^2-1=(pq+1)(pq-1)$

13.
$$y^4-8y^2+16=(y^2-4)^2$$
$$=(y+2)^2(y-2)^2$$

14.
$$64-n^3$$
$$=(4)^3-(n)^3$$
$$=(4-n)(4^2+4n+n^2)$$
$$=(4-n)(16+4n+n^2)$$

15.
$$(n+8)(n-1)=0$$
$$n+8=0 \quad \text{or} \quad n-1=0$$
$$n=-8 \qquad n=1$$

16.

$$
\begin{aligned}
6x^2 + 10x &= 4 \\
6x^2 + 10x - 4 &= 0 \\
2(3x^2 + 5x - 2) &= 0 \\
2(3x - 1)(x + 2) &= 0
\end{aligned}
$$

$3x - 1 = 0$ or $x + 2 = 0$

$3x = 1 \qquad x = -2$

$x = \frac{1}{3}$

17. $(2n+1)(n-1) = 5$

$$
\begin{aligned}
2n^2 - n - 1 &= 5 \\
2n^2 - n - 6 &= 0 \\
(2n+3)(n-2) &= 0
\end{aligned}
$$

$2n + 3 = 0$ or $n - 2 = 0$

$2n = -3 \qquad n = 2$

$n = -\frac{3}{2}$

18.

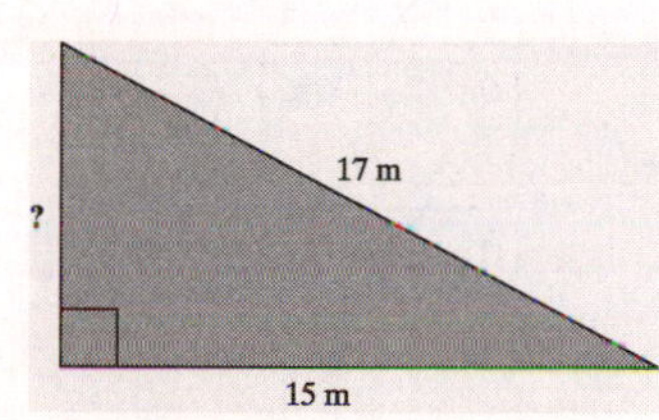

$$
\begin{aligned}
x^2 + 15^2 &= 17^2 \\
x^2 + 225 &= 289 \\
x^2 - 64 &= 0 \\
(x+8)(x-8) &= 0
\end{aligned}
$$

$x + 8 = 0$ or $x - 8 = 0$

$x = -8 \qquad x = 8$

8 m

19. $mgy_2 - mgy_1 = mg(y_2 - y_1)$

20.

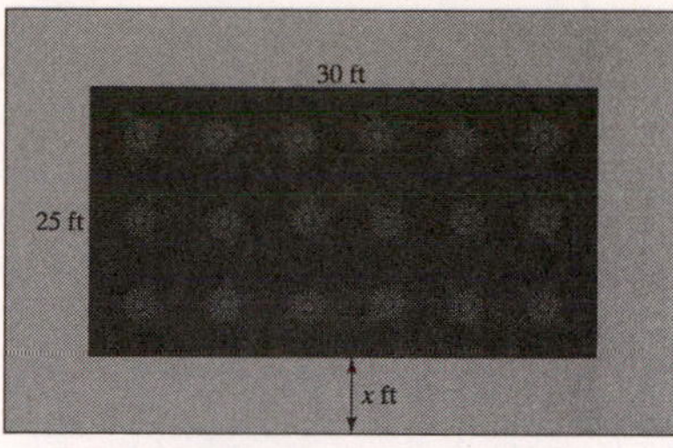

$$
\begin{aligned}
A &= (25 + 2x)(30 + 2x) - 25 \cdot 30 \\
&= 750 + 110x + 4x^2 - 750 \\
&= 4x^2 + 110x \\
&= 2x(2x + 55)
\end{aligned}
$$

$2x(2x + 55)$ sq ft

Chapter 7

RATIONAL EXPRESSIONS AND EQUATIONS

Pretest

1. $\dfrac{5}{x+6}$

$$x+6=0$$
$$x=-6$$

The expression is undefined for $x=-6$.

2. $\dfrac{n-2}{3}=\dfrac{(n-2)\cdot n}{3\cdot n}=\dfrac{n^2-2n}{3n}$

3. $\dfrac{24x^2y^3}{6xy^5}=\dfrac{6xy^3(4x)}{6xy^3(y^2)}=\dfrac{4x}{y^2}$

4. $\dfrac{4a^2-8a}{a-2}=\dfrac{4a(a-2)}{a-2}=4a$

5.
$$\frac{w^2-6w}{36-w^2}=\frac{w(w-6)}{(6-w)(6+w)}$$
$$=-\frac{w(w-6)}{(w-6)(w+6)}$$
$$=-\frac{w\cancel{(w-6)}}{\cancel{(w-6)}(w+6)}$$
$$=-\frac{w}{w+6}$$

6.
$$\frac{\frac{5n}{8}}{\frac{n^3}{16}}=\frac{5n}{8}\div\frac{n^3}{16}$$
$$=\frac{5n}{8}\cdot\frac{16}{n^3}$$
$$=\frac{5\cancel{n}}{\underset{1}{\cancel{8}}}\cdot\frac{\overset{2}{\cancel{16}}}{n^{\cancel{3}2}}$$
$$=\frac{10}{n^2}$$

7.
$$\frac{12y}{y+1}-\frac{7y+2}{y+1}=\frac{12y-(7y+2)}{y+1}$$
$$=\frac{12y-7y-2}{y+1}$$
$$=\frac{5y-2}{y+1}$$

8.
$$\frac{1}{6x^2}+\frac{5}{4x}=\frac{1}{6x^2}\cdot\frac{2}{2}+\frac{5}{4x}\cdot\frac{3x}{3x}$$
$$=\frac{2}{12x^2}+\frac{15x}{12x^2}$$
$$=\frac{15x+2}{12x^2}$$

9.
$$\frac{3}{c-3}-\frac{1}{c+3}=\frac{c+3}{c+3}\cdot\frac{3}{c-3}-\frac{c-3}{c-3}\cdot\frac{1}{c+3}$$
$$=\frac{3(c+3)-(c-3)}{(c+3)(c-3)}$$
$$=\frac{3c+9-c+3}{(c+3)(c-3)}$$
$$=\frac{2c+12}{(c-3)(c+3)}$$

10.
$$\frac{1}{x^2-2x+1}-\frac{2}{1-x^2}$$
$$=\frac{1}{x^2-2x+1}+\frac{2}{x^2-1}$$
$$=\frac{1}{(x-1)^2}+\frac{2}{(x-1)(x+1)}$$
$$=\frac{x+1}{x+1}\cdot\frac{1}{(x-1)^2}+\frac{x-1}{x-1}\cdot\frac{2}{(x-1)(x+1)}$$
$$=\frac{x+1}{(x+1)(x-1)^2}+\frac{2(x-1)}{(x-1)^2(x+1)}$$
$$=\frac{x+1+2(x-1)}{(x-1)^2(x+1)}$$
$$=\frac{x+1+2x-2}{(x-1)^2(x+1)}$$
$$=\frac{3x-1}{(x-1)^2(x+1)}$$

11.
$$\frac{15a^3b}{20n^4}\cdot\frac{16n^2}{9ab}=\frac{\overset{\cancel{5}}{\cancel{15}}\,a^{\cancel{3}2}\,\cancel{b}}{\underset{\cancel{5}}{\cancel{20}}\,n^{\cancel{4}2}}\cdot\frac{\overset{4}{\cancel{16}}\,\cancel{n^2}}{\underset{3}{\cancel{9}}\,\cancel{ab}}=\frac{4a^2}{3n^2}$$

12. $\dfrac{y-4}{5y^2+10y}\cdot\dfrac{y^2-2y-8}{y^2-16}$

$$=\frac{y-4}{5y(y+2)}\cdot\frac{(y+2)(y-4)}{(y-4)(y+4)}$$

$$=\frac{y-4}{5y\cancel{(y+2)}}\cdot\frac{\cancel{(y+2)}\cancel{(y-4)}}{\cancel{(y-4)}(y+4)}$$

$$=\frac{y-4}{5y(y+4)}$$

13. $\dfrac{x^2-x-2}{x^2+5x+4}\div\dfrac{x^2-7x+10}{x-5}$

$$=\frac{x^2-x-2}{x^2+5x+4}\cdot\frac{x-5}{x^2-7x+10}$$

$$=\frac{(x+1)(x-2)}{(x+1)(x+4)}\cdot\frac{x-5}{(x-2)(x-5)}$$

$$=\frac{\cancel{(x+1)}\cancel{(x-2)}}{\cancel{(x+1)}(x+4)}\cdot\frac{\cancel{x-5}}{\cancel{(x-2)}\cancel{(x-5)}}$$

$$=\frac{1}{x+4}$$

14. $\dfrac{3x-8}{x^2-4}+\dfrac{2}{x-2}=\dfrac{7}{x+2}$

$$\frac{3x-8}{(x-2)(x+2)}+\frac{2}{x-2}=\frac{7}{x+2}$$

$$(x-2)(x+2)\cdot\frac{3x-8}{(x-2)(x+2)}$$
$$+(x-2)(x+2)\cdot\frac{2}{x-2}$$
$$=(x-2)(x+2)\cdot\frac{7}{x+2}$$

$$\cancel{(x-2)(x+2)}\cdot\frac{3x-8}{\cancel{(x-2)(x+2)}}$$
$$+\cancel{(x-2)}(x+2)\cdot\frac{2}{\cancel{x-2}}$$
$$=(x-2)\cancel{(x+2)}\cdot\frac{7}{\cancel{x+2}}$$

$$3x-8+2(x+2)=7(x-2)$$
$$3x-8+2x+4=7x-14$$
$$-2x=-10$$
$$x=5$$

Check:

$$\frac{3(5)-8}{(5)^2-4}+\frac{2}{5-2}=\frac{7}{5+2}$$
$$\frac{15-8}{25-4}+\frac{2}{3}=\frac{7}{7}$$
$$\frac{7}{21}+\frac{2}{3}=1$$
$$\frac{1}{3}+\frac{2}{3}=1$$
$$1=1$$

15. $\dfrac{x}{x+4}-1=\dfrac{2}{x-1}$

$$(x-1)(x+4)\cdot\frac{x}{x+4}-(x-1)(x+4)\cdot 1$$
$$=(x-1)(x+4)\cdot\frac{2}{x-1}$$

$$(x-1)\cancel{(x+4)}\cdot\frac{x}{\cancel{x+4}}-(x-1)(x+4)$$
$$=\cancel{(x-1)}(x+4)\cdot\frac{2}{\cancel{x-1}}$$

$$x(x-1)-(x-1)(x+4)=2(x+4)$$
$$x^2-x-(x^2+3x-4)=2x+8$$
$$x^2-x-x^2-3x+4=2x+8$$
$$-6x=4$$
$$x=-\frac{4}{6}$$
$$x=-\frac{2}{3}$$

15. (continued)

Check:

$$\frac{-\frac{2}{3}}{-\frac{2}{3}+4}-1=\frac{2}{-\frac{2}{3}-1}$$

$$\frac{-\frac{2}{3}}{-\frac{2}{3}+\frac{12}{3}}-1=\frac{2}{-\frac{2}{3}-\frac{3}{3}}$$

$$-\frac{\frac{2}{3}}{\frac{10}{3}}-1=\frac{2}{-\frac{5}{3}}$$

$$-\frac{2}{10}-1=-\frac{6}{5}$$

$$-\frac{2}{10}-\frac{10}{10}=-\frac{6}{5}$$

$$-\frac{12}{10}=-\frac{6}{5}$$

$$-\frac{6}{5}=-\frac{6}{5}$$

16.

$$\frac{9}{2x}=\frac{x}{x-1}$$

$$9(x-1)=2x^2$$

$$9x-9=2x^2$$

$$2x^2-9x+9=0$$

$$(2x-3)(x-3)=0$$

$$2x-3=0 \quad \text{or} \quad x-3=0$$

$$2x=3 \qquad x=3$$

$$x=\frac{3}{2}$$

Check:

$$\frac{9}{2\left(\frac{3}{2}\right)}=\frac{\frac{3}{2}}{\frac{3}{2}-1}$$

$$\frac{9}{3}=\frac{\frac{3}{2}}{\frac{3}{2}-\frac{2}{2}}$$

$$3=3$$

$$\frac{9}{2(3)}=\frac{3}{3-1}$$

$$\frac{9}{6}=\frac{3}{2}$$

$$\frac{3}{2}=\frac{3}{2}$$

17.

$$2+\frac{1500}{x}=2\cdot\frac{x}{x}+\frac{1500}{x}$$

$$=\frac{2x}{x}+\frac{1500}{x}$$

$$=\frac{2x+1500}{x}$$

$\frac{2x+1500}{x}$ dollars

18.

$$\frac{195}{r}+\frac{165}{r-10}=6$$

$$r(r-10)\cdot\frac{195}{r}+r(r-10)\cdot\frac{165}{r-10}=r(r-10)\cdot 6$$

$$\cancel{r}(r-10)\cdot\frac{195}{\cancel{r}}+r\cancel{(r-10)}\cdot\frac{165}{\cancel{r-10}}=r(r-10)\cdot 6$$

$$195(r-10)+165r=6r(r-10)$$

$$195r-1950+165r=6r^2-60r$$

$$0=6r^2-420r+1950$$

$$0=6(r^2-70r+325)$$

$$0=6(r-5)(r-65)$$

$$r-5=0 \quad \text{or} \quad r-65=0$$

$$r=5 \qquad r=65$$

The average speed during the first part of the trip was 65 mph and the average speed during the second part was 55 mph.

19. $\frac{30}{t} = \frac{90}{t+5}$

$30(t+5) = 90t$

$30t + 150 = 90t$

$-60t = -150$

$t = 2.5$

It takes the photocopier $2\frac{1}{2}$ (2.5) min to make 30 copies.

20. $d = kF$

$3 = k(12)$

$k = \frac{3}{12} = \frac{1}{4}$

$d = \frac{1}{4}F$

$d = \frac{1}{4}(30) = \frac{15}{2} = 7\frac{1}{2}$

The spring will stretch 7½ in.

7.1 Rational Expressions and Functions

Practice

1. **a.** $\frac{n+1}{n-3}$

$n - 3 = 0$

$n = 3$

The expression is undefined when n is equal to three.

b. $\frac{6}{n^2 - 9}$

$n^2 - 9 = 0$

$(n+3)(n-3) = 0$

$n + 3 = 0$ or $n - 3 = 0$

$n = -3 \qquad n = 3$

The expression is undefined when n is equal to -3 or 3.

2. **a.** $\frac{b}{3} \cdot \frac{3b}{3b} = \frac{3b^2}{9b}$

The expressions are equivalent.

b. $\frac{5}{a+7} \cdot \frac{a}{a} = \frac{5a}{a^2 + 7a}$

The expressions are equivalent.

3. **a.** $\frac{-12n^3}{-6mn} = \frac{-6n(2n^2)}{-6n(m)}$

$= \frac{2n^2}{m}$

b. $\frac{-3x^4y}{2x^2y} = \frac{x^2y(-3x^2)}{x^2y(2)}$

$= -\frac{3x^2}{2}$

4. **a.** $\frac{2y-8}{4y+6} = \frac{2(y-4)}{2(2y+3)}$

$= \frac{y-4}{2y+3}$

b. $\frac{v-4}{v+3}$

The expression cannot be simplified.

c. $\frac{x+2}{3x+6} = \frac{x+2}{3(x+2)} = \frac{1}{3}$

5. **a.** $\frac{wt - wx}{wz - 3wg} = \frac{w(t-x)}{w(z-3g)}$

$= \frac{t-x}{z-3g}$

b. $\frac{4n^2 - 4}{n^2 + 3n + 2} = \frac{4(n^2 - 1)}{(n+1)(n+2)}$

$= \frac{4(n+1)(n-1)}{(n+1)(n+2)}$

$= \frac{4(n-1)}{n+2}$

c. $\frac{2y^2 - y - 1}{2y^2 + 7y + 3} = \frac{(2y+1)(y-1)}{(2y+1)(y+3)}$

$= \frac{y-1}{y+3}$

6. a. $\frac{-y-1}{1+y}=\frac{-(1+y)}{1+y}=-1$

b. $\frac{3x-12}{-x+4}=\frac{3(x-4)}{-(x-4)}=-3$

c. $\frac{3n-15}{25-n^2}=\frac{3(n-5)}{(5-n)(5+n)}$

$=-\frac{3(n-5)}{(n-5)(n+5)}$

$=-\frac{3}{n+5}$

d. $\frac{4-9s^2}{3s^2+s-2}=-\frac{9s^2-4}{3s^2+s-2}$

$=-\frac{(3s+2)(3s-2)}{(3s-2)(s+1)}$

$=-\frac{3s+2}{s+1}$

7. a. $\frac{\pi r^2}{2\pi r}=\frac{\pi r(r)}{\pi r(2)}=\frac{r}{2}$

b. $2\pi r=0$

$r=0$

The expression is undefined when $r=0$.

Exercises

1. A rational expression, $\frac{P}{Q}$, is an algebraic expression that can be written as the quotient of two polynomials, P and Q, where $Q\neq 0$.

3. The graph of $f(x)=\frac{3}{x-4}$ has an asymptote at $x = 4$.

5. Rational expressions are equivalent if they have the same value no matter what value replaces the variable.

7. $\frac{7}{x}$

$x=0$

9. $\frac{8}{y-2}$

$y-2=0$

$y=2$

11. $\frac{x-3}{x+5}$

$x+5=0$

$x=-5$

13. $\frac{n+11}{2n-1}$

$2n-1=0$

$2n=1$

$n=\frac{1}{2}$

15. $\frac{x^2+1}{x^2-1}$

$x^2-1=0$

$(x+1)(x-1)=0$

$x+1=0$ or $x-1=0$

$x=-1$ $\quad$ $x=1$

17. $\frac{x^2+x+1}{x^2-x-20}$

$x^2-x-20=0$

$(x+4)(x-5)=0$

$x+4=0$ or $x-5=0$

$x=-4$ $\quad$ $x=5$

19. The graph does not intersect the vertical line $x = 0$.

21. $\frac{p}{q}\cdot\frac{r}{r}=\frac{pr}{qr}$

The expressions are equivalent.

23. $\frac{3t+5}{t+1}\cdot\frac{t}{t}=\frac{3t^2+5t}{t^2+t}$

The expressions are equivalent.

25. $\dfrac{x-1}{x+3}$ and $-\dfrac{1}{3}$
The expressions are not equivalent.

27. $\dfrac{x-2}{2-x}\cdot\dfrac{-1}{-1}=\dfrac{2-x}{x-2}$
The expressions are equivalent.

29. $\dfrac{x^2+4}{x+2}$ and $x+2$
The expressions are not equivalent.

31. $\dfrac{10a^4}{12a}=\dfrac{2a\left(5a^3\right)}{2a(6)}=\dfrac{5a^3}{6}$

33. $\dfrac{3x^2}{12x^5}=\dfrac{3x^2(1)}{3x^2\left(4x^3\right)}=\dfrac{1}{4x^3}$

35. $\dfrac{9s^3t^2}{6s^5t}=\dfrac{3s^3t(3t)}{3s^3t\left(2s^2\right)}=\dfrac{3t}{2s^2}$

37. $\dfrac{24a^4b^5}{3ab^2}=\dfrac{3ab^2\left(8a^3b^3\right)}{3ab^2(1)}=8a^3b^3$

39. $\dfrac{-2p^2q^3}{-10pq^4}=\dfrac{-2pq^3(p)}{-2pq^3(5q)}=\dfrac{p}{5q}$

41. $\dfrac{5x(x+8)}{4x(x+8)}=\dfrac{5}{4}$

43. $\dfrac{8x^2(5-2x)}{3x(2x-5)}=\dfrac{-8x^2(2x-5)}{3x(2x-5)}=-\dfrac{8x}{3}$

45. $\dfrac{5x-10}{5}=\dfrac{5(x-2)}{5(1)}=x-2$

47. $\dfrac{2x^2+2x}{4x^2+6x}=\dfrac{2x(x+1)}{2x(2x+3)}=\dfrac{x+1}{2x+3}$

49. $\dfrac{a^2-4a}{ab-4b}=\dfrac{a(a-4)}{b(a-4)}=\dfrac{a}{b}$

51. $\dfrac{6x-4y}{9x-6y}=\dfrac{2(3x-2y)}{3(3x-2y)}=\dfrac{2}{3}$

53. $\dfrac{t^2-1}{t+1}=\dfrac{(t+1)(t-1)}{(t+1)}=t-1$

55. $\dfrac{p^2-q^2}{q^2-p^2}=\dfrac{(p+q)(p-q)}{(q+p)(q-p)}$
$=-\dfrac{(p+q)(p-q)}{(p+q)(p-q)}$
$=-1$

57. $\dfrac{n-1}{2-2n}=\dfrac{n-1}{-2(n-1)}=-\dfrac{1}{2}$

59. $\dfrac{(b-4)^2}{b^2-16}=\dfrac{(b-4)^2}{(b+4)(b-4)}=\dfrac{b-4}{b+4}$

61. $\dfrac{2x^2+5x-3}{10x-5}=\dfrac{(x+3)(2x-1)}{5(2x-1)}=\dfrac{x+3}{5}$

63. $\dfrac{a^3+9a^2+14a}{a^2-10a-24}=\dfrac{a\left(a^2+9a+14\right)}{(a-12)(a+2)}$
$=\dfrac{a(a+7)(a+2)}{(a-12)(a+2)}$
$=\dfrac{a(a+7)}{a-12}$

65. $\dfrac{t^2-4t-5}{t^2-3t-10}=\dfrac{(t-5)(t+1)}{(t-5)(t+2)}=\dfrac{t+1}{t+2}$

67. $\dfrac{9-16d^2}{16d^2-24d+9}=\dfrac{(3-4d)(3+4d)}{(4d-3)^2}$
$=-\dfrac{(4d-3)(4d+3)}{(4d-3)^2}$
$=-\dfrac{4d+3}{4d-3}$

69. $\dfrac{8s-2s^2}{2s^2-11s+12}=\dfrac{2s(4-s)}{(s-4)(2s-3)}$
$=-\dfrac{2s(s-4)}{(s-4)(2s-3)}$
$=-\dfrac{2s}{2s-3}$

71. $\frac{6x^2+5x+1}{6x^2-x-1}=\frac{(3x+1)(2x+1)}{(3x+1)(2x-1)}=\frac{2x+1}{2x-1}$

73. $\frac{6y^2-7y+2}{6y^3+5y^2-6y}=\frac{(3y-2)(2y-1)}{y(6y^2+5y-6)}$
$=\frac{(3y-2)(2y-1)}{(3y-2)(2y+3)}$
$=\frac{2y-1}{y(2y+3)}$

75. $\frac{2ab^2+4a^2b}{2b^2+5ab+2a^2}=\frac{2ab(b+2a)}{(2b+a)(b+2a)}$
$=\frac{2ab}{2b+a}$

77. $\frac{m^2+3mn-28n^2}{2m^2+4mn-48n^2}=\frac{(m+7n)(m-4n)}{2(m^2+2mn-24n^2)}$
$=\frac{(m+7n)(m-4n)}{2(m+6n)(m-4n)}$
$=\frac{m+7n}{2(m+6n)}$

79. $\frac{12x^2y^6}{18x^3y^4}=\frac{6x^2y^4(2y^2)}{6x^2y^4(3x)}=\frac{2y^2}{3x}$

81. $\frac{3x^3-9x^2+6x}{x^3+x^2-6x}=\frac{3x(x^2-3x+2)}{x(x^2+x-6)}$
$=\frac{3x(x-2)(x-1)}{x(x-2)(x+3)}$
$=\frac{3(x-1)}{x+3}$

83. $\frac{w^2-2y^2}{2y^2-w^2}=\frac{-(2y^2-w^2)}{2y^2-w^2}=-1$

85. $\frac{2x^2+x-3}{3x^2-3x}=\frac{(x-1)(2x+3)}{3x(x-1)}=\frac{2x+3}{3x}$

87. $\frac{p-5}{2p-3}$
$2p-3=0$
$2p=3$
$p=\frac{3}{2}$

89. $\frac{6y+7}{y+1}\cdot\frac{y}{y}=\frac{6y^2+7y}{y^2+y}$
The expressions are equivalent.

91. $\frac{mu^2-mv^2}{mu-mv}=\frac{m(u^2-v^2)}{m(u-v)}$
$=\frac{m(u-v)(u+v)}{m(u-v)}$
$=u+v$

93. **a.** $(x+2)(x+5)=x^2+7x+10$

b. $x(x+2)=x^2+2x$

c. $\frac{x^2+7x+10}{x^2+2x}=\frac{(x+2)(x+5)}{x(x+2)}=\frac{x+5}{x}$

d. $\frac{8+5}{8}=\frac{13}{8}$

95. **a.** $\frac{\pi r_1^2}{\pi r_2^2-\pi r_1^2}=\frac{\pi(r_1^2)}{\pi(r_2^2-r_1^2)}=\frac{r_1^2}{r_2^2-r_1^2}$

b. $\frac{\pi r_2^2-\pi r_1^2}{\pi r_3^2-\pi r_2^2}=\frac{\pi(r_2^2-r_1^2)}{\pi(r_3^2-r_2^2)}=\frac{r_2^2-r_1^2}{r_3^2-r_2^2}$

7.2 Multiplication and Division of Rational Expressions

Practice

1. a. $\frac{n}{7}\cdot\frac{2}{m}=\frac{2n}{7m}$

b. $\frac{p^2}{6q^2}\cdot\frac{2q}{5p}=\frac{p^2}{6q^2}\cdot\frac{2q}{5p}=\frac{p}{15q}$

2. a. $\frac{3t}{t^2+5t}\cdot\frac{3t+15}{6t}=\frac{3t}{t(t+5)}\cdot\frac{3(t+5)}{6t}$

$=\frac{3t}{t(t+5)}\cdot\frac{3(t+5)}{6t}$

$=\frac{3}{2t}$

b. $\frac{8}{36g^2-1}\cdot\frac{1+6g}{4}$

$=\frac{8}{(6g+1)(6g-1)}\cdot\frac{1+6g}{4}$

$=\frac{8}{(6g+1)(6g-1)}\cdot\frac{1+6g}{4}$

$=\frac{2}{6g-1}$

3. a. $\frac{y^2+4y-21}{3y-9}\cdot\frac{6y^2-24}{y+2}$

$=\frac{(y+7)(y-3)}{3(y-3)}\cdot\frac{6(y^2-4)}{y+2}$

$=\frac{(y+7)(y-3)}{3(y-3)}\cdot\frac{6(y-2)(y+2)}{y+2}$

$=\frac{(y+7)(y-3)}{3(y-3)}\cdot\frac{6(y-2)(y+2)}{y+2}$

$=2(y+7)(y-2)$

b. $\frac{x^2-x-30}{x^2+10x+9}\cdot\frac{18-7x-x^2}{x^2-8x+12}$

$=-\frac{x^2-x-30}{x^2+10x+9}\cdot\frac{x^2+7x-18}{x^2-8x+12}$

$=-\frac{(x+5)(x-6)}{(x+1)(x+9)}\cdot\frac{(x-2)(x+9)}{(x-2)(x-6)}$

$=-\frac{(x+5)(x-6)}{(x+1)(x+9)}\cdot\frac{(x-2)(x+9)}{(x-2)(x-6)}$

$=-\frac{x+5}{x+1}$

4. a. $\frac{a}{4}\div\frac{b}{6}=\frac{a}{4}\cdot\frac{6}{b}=\frac{a}{4}\cdot\frac{6}{b}=\frac{3a}{2b}$

b. $\frac{5pq}{7p^2q^4}\div\frac{p^3}{3q^2}=\frac{5pq}{7p^2q^4}\cdot\frac{3q^2}{p^3}=\frac{15}{7p^2q}$

5. a. $\frac{x+3}{x-10}\div\frac{x+1}{x+3}=\frac{x+3}{x-10}\cdot\frac{x+3}{x+1}$

$=\frac{(x+3)^2}{(x-10)(x+1)}$

b. $\frac{p+4q}{3p-6q}\div\frac{2p+8q}{2p-4q}$

$=\frac{p+4q}{3p-6q}\cdot\frac{2p-4q}{2p+8q}$

$=\frac{p+4q}{3(p-2q)}\cdot\frac{2(p-2q)}{2(p+4q)}$

$=\frac{p+4q}{3(p-2q)}\cdot\frac{2(p-2q)}{2(p+4q)}$

$=\frac{1}{3}$

5. **c.** $\dfrac{y^2+4y+3}{5y+10} \div \dfrac{y^2-1}{y}$

$= \dfrac{y^2+4y+3}{5y+10} \cdot \dfrac{y}{y^2-1}$

$= \dfrac{(y+1)(y+3)}{5(y+2)} \cdot \dfrac{y}{(y+1)(y-1)}$

$= \dfrac{\cancel{(y+1)}(y+3)}{5(y+2)} \cdot \dfrac{y}{\cancel{(y+1)}(y-1)}$

$= \dfrac{y(y+3)}{5(y+2)(y-1)}$

6. $\dfrac{W}{g} \cdot \dfrac{v^2}{r} = \dfrac{Wv^2}{gr}$

Exercises

1. $\dfrac{1}{t^2} \cdot \dfrac{t}{4} = \dfrac{1}{t^{\cancel{2}}} \cdot \dfrac{\cancel{t}}{4} = \dfrac{1}{4t}$

3. $\dfrac{2}{a} \cdot \dfrac{3}{b} = \dfrac{6}{ab}$

5. $\dfrac{2x^4}{3x^5} \cdot \dfrac{5}{x^8} = \dfrac{2\cancel{x^4}}{3x^5} \cdot \dfrac{5}{x^{\cancel{8}4}} = \dfrac{10}{3x^9}$

7. $-\dfrac{7x^2y}{3} \cdot \dfrac{6}{x^3y} = -\dfrac{7\cancel{x^2}\cancel{y}}{\underset{1}{\cancel{3}}} \cdot \dfrac{\overset{2}{\cancel{6}}}{x^{\cancel{3}}\cancel{y}} = -\dfrac{14}{x}$

9. $\dfrac{x}{x-2} \cdot \dfrac{5x-10}{x^4} = \dfrac{x}{x-2} \cdot \dfrac{5(x-2)}{x^4}$

$= \dfrac{\cancel{x}}{\cancel{x-2}} \cdot \dfrac{5\cancel{(x-2)}}{x^{\cancel{4}3}}$

$= \dfrac{5}{x^3}$

11. $\dfrac{8n-3}{n^2} \cdot n = \dfrac{8n-3}{n^{\cancel{2}}} \cdot \cancel{n} = \dfrac{8n-3}{n}$

13. $\dfrac{8x-6}{5x+20} \cdot \dfrac{2x+8}{4x-3} = \dfrac{2(4x-3)}{5(x+4)} \cdot \dfrac{2(x+4)}{4x-3}$

$= \dfrac{2\cancel{(4x-3)}}{5\cancel{(x+4)}} \cdot \dfrac{2\cancel{(x+4)}}{\cancel{4x-3}}$

$= \dfrac{4}{5}$

15. $\dfrac{5n-1}{6n+4} \cdot \dfrac{3n+2}{1-5n} = -\dfrac{5n-1}{2(3n+2)} \cdot \dfrac{3n+2}{5n-1}$

$= -\dfrac{\cancel{5n-1}}{2\cancel{(3n+2)}} \cdot \dfrac{\cancel{3n+2}}{\cancel{5n-1}}$

$= -\dfrac{1}{2}$

17. $\dfrac{x^2-4y^2}{x+y} \cdot \dfrac{3x+3y}{4x-8y}$

$= \dfrac{(x-2y)(x+2y)}{x+y} \cdot \dfrac{3(x+y)}{4(x-2y)}$

$= \dfrac{\cancel{(x-2y)}(x+2y)}{\cancel{x+y}} \cdot \dfrac{3\cancel{(x+y)}}{4\cancel{(x-2y)}}$

$= \dfrac{3(x+2y)}{4}$

19. $\dfrac{p^4-1}{p^4-16} \cdot \dfrac{p^2+4}{p^2+1}$

$= \dfrac{(p^2-1)\cancel{(p^2+1)}}{(p^2-4)\cancel{(p^2+4)}} \cdot \dfrac{\cancel{p^2+4}}{\cancel{p^2+1}}$

$= \dfrac{p^2-1}{p^2-4}$

21. $\dfrac{n^2-2n-24}{n^2+6n+8} \cdot \dfrac{n^2+5n+6}{n^2-5n-6}$

$= \dfrac{(n-6)(n+4)}{(n+2)(n+4)} \cdot \dfrac{(n+2)(n+3)}{(n-6)(n+1)}$

$= \dfrac{\cancel{(n-6)}\cancel{(n+4)}}{\cancel{(n+2)}\cancel{(n+4)}} \cdot \dfrac{\cancel{(n+2)}(n+3)}{\cancel{(n-6)}(n+1)}$

$= \dfrac{n+3}{n+1}$

23. $$\frac{2y^2-y-6}{2y^2+y-3}\cdot\frac{2y^2-3y+1}{2y^2-9y+10}$$
$$=\frac{(y-2)(2y+3)}{(2y+3)(y-1)}\cdot\frac{(2y-1)(y-1)}{(2y-5)(y-2)}$$
$$=\frac{\cancel{(y-2)}\cancel{(2y+3)}}{\cancel{(2y+3)}\cancel{(y-1)}}\cdot\frac{(2y-1)\cancel{(y-1)}}{(2y-5)\cancel{(y-2)}}$$
$$=\frac{2y-1}{2y-5}$$

25. $$\frac{2}{x^3}\cdot\frac{4x}{5}\cdot\frac{10}{x^2}=\frac{2}{x^3}\cdot\frac{4\cancel{x}}{\underset{1}{\cancel{5}}}\cdot\frac{\overset{2}{\cancel{10}}}{\cancel{x^2}}=\frac{16}{x^4}$$

27.
$$\frac{x^2-7x+10}{2x-2}\cdot\frac{6x}{x^2-2x-15}\cdot\frac{x^2+2x-3}{x-2}$$
$$=\frac{\cancel{(x-2)}\cancel{(x-5)}}{\underset{1}{\cancel{2}}\cancel{(x-1)}}\cdot\frac{\overset{3}{\cancel{6}}x}{\cancel{(x-5)}\cancel{(x+3)}}\cdot\frac{\cancel{(x-1)}\cancel{(x+3)}}{\cancel{x-2}}$$
$$=3x$$

29. $$\frac{7}{a}\div\frac{14}{a}=\frac{7}{a}\cdot\frac{a}{14}$$
$$=\frac{\overset{1}{\cancel{7}}}{\cancel{a}}\cdot\frac{\cancel{a}}{\underset{2}{\cancel{14}}}$$
$$=\frac{1}{2}$$

31. $$\frac{p^3}{10}\div\frac{p^3}{20}=\frac{\cancel{p^3}}{\underset{1}{\cancel{10}}}\cdot\frac{\overset{2}{\cancel{20}}}{\cancel{p^3}}=2$$

33. $$\frac{12}{x^3}\div\frac{6}{5x^2}=\frac{12}{x^3}\cdot\frac{5x^2}{6}$$
$$=\frac{\overset{2}{\cancel{12}}}{\cancel{x^3}}\cdot\frac{5\cancel{x^2}}{\underset{1}{\cancel{6}}}$$
$$=\frac{10}{x}$$

35. $$-\frac{3}{t}\div t=-\frac{3}{t}\cdot\frac{1}{t}=-\frac{3}{t^2}$$

37. $$\frac{9xy^2}{2x^3}\div\frac{3x^2y}{4y}=\frac{9xy^2}{2x^3}\cdot\frac{4y}{3x^2y}$$
$$=\frac{\overset{2}{\cancel{9}}\cancel{x}y^2}{\underset{1}{\cancel{2}}x^3}\cdot\frac{\overset{2}{\cancel{4}}\cancel{y}}{\underset{1}{\cancel{3}}x^{\cancel{2}}\cancel{y}}$$
$$=\frac{6y^2}{x^4}$$

39. $$\frac{c+3}{c-5}\div\frac{c+9}{c-7}=\frac{c+3}{c-5}\cdot\frac{c-7}{c+9}$$
$$=\frac{(c+3)(c-7)}{(c-5)(c+9)}$$

41. $$\frac{6a-12}{8a+32}\div\frac{9a-18}{5a+20}=\frac{6a-12}{8a+32}\cdot\frac{5a+20}{9a-18}$$
$$=\frac{6(a-2)}{8(a+4)}\cdot\frac{5(a+4)}{9(a-2)}$$
$$=\frac{\overset{1}{\cancel{6}}\cancel{(a-2)}}{\underset{4}{\cancel{8}}\cancel{(a+4)}}\cdot\frac{5\cancel{(a+4)}}{\underset{3}{\cancel{9}}\cancel{(a-2)}}$$
$$=\frac{5}{12}$$

43. $$\frac{x+1}{10}\div\frac{1-x^2}{5}=\frac{x+1}{10}\cdot\frac{5}{1-x^2}$$
$$=\frac{x+1}{10}\cdot\frac{5}{(1-x)(1+x)}$$
$$=\frac{\cancel{x+1}}{\underset{2}{\cancel{10}}}\cdot\frac{\overset{1}{\cancel{5}}}{(1-x)\cancel{(1+x)}}$$
$$=\frac{1}{2(1-x)}$$

45. $$\frac{p^2-1}{1-p}\div\frac{p+1}{p}=\frac{p^2-1}{1-p}\cdot\frac{p}{p+1}$$
$$=\frac{(p-1)(p+1)}{1-p}\cdot\frac{p}{p+1}$$
$$=-\frac{\cancel{(1-p)}\cancel{(p+1)}}{\cancel{1-p}}\cdot\frac{p}{\cancel{p+1}}$$
$$=-p$$

47. $\dfrac{x^2y+3xy^2}{x^2-9y^2}\div\dfrac{5x^2y}{x^2-2xy-3y^2}$

$$=\frac{x^2y+3xy^2}{x^2-9y^2}\cdot\frac{x^2-2xy-3y^2}{5x^2y}$$

$$=\frac{xy(x+3y)}{(x-3y)(x+3y)}\cdot\frac{(x-3y)(x+y)}{5x^2y}$$

$$=\frac{\cancel{x}\cancel{y}\cancel{(x+3y)}}{\cancel{(x-3y)}\cancel{(x+3y)}}\cdot\frac{\cancel{(x-3y)}(x+y)}{5x^{\cancel{2}}\cancel{y}}$$

$$=\frac{x+y}{5x}$$

49. $\dfrac{2t^2-3t-2}{2t+1}\div\left(4-t^2\right)$

$$=\frac{2t^2-3t-2}{2t+1}\cdot\frac{1}{\left(4-t^2\right)}$$

$$=-\frac{2t^2-3t-2}{2t+1}\cdot\frac{1}{\left(t^2-4\right)}$$

$$=-\frac{(2t+1)(t-2)}{2t+1}\cdot\frac{1}{(t-2)(t+2)}$$

$$=-\frac{\cancel{(2t+1)}\cancel{(t-2)}}{\cancel{2t+1}}\cdot\frac{1}{\cancel{(t-2)}(t+2)}$$

$$=-\frac{1}{t+2}$$

51. $\dfrac{x^2-11x+28}{x^2-x-42}\div\dfrac{x^2-2x-8}{x^2+7x+10}$

$$=\frac{x^2-11x+28}{x^2-x-42}\cdot\frac{x^2+7x+10}{x^2-2x-8}$$

$$=\frac{(x-4)(x-7)}{(x+6)(x-7)}\cdot\frac{(x+2)(x+5)}{(x-4)(x+2)}$$

$$=\frac{\cancel{(x-4)}\cancel{(x-7)}}{(x+6)\cancel{(x-7)}}\cdot\frac{\cancel{(x+2)}(x+5)}{\cancel{(x-4)}\cancel{(x+2)}}$$

$$=\frac{x+5}{x+6}$$

53. $\dfrac{3p^2-3p-18}{p^2+2p-15}\div\dfrac{2p^2+6p-20}{2p^2-12p+16}$

$$=\frac{3p^2-3p-18}{p^2+2p-15}\cdot\frac{2p^2-12p+16}{2p^2+6p-20}$$

$$=\frac{3\left(p^2-p-6\right)}{p^2+2p-15}\cdot\frac{2\left(p^2-6p+8\right)}{2\left(p^2+3p-10\right)}$$

$$=\frac{3(p+2)(p-3)}{(p-3)(p+5)}\cdot\frac{(p-2)(p-4)}{(p-2)(p+5)}$$

$$=\frac{3(p+2)\cancel{(p-3)}}{\cancel{(p-3)}(p+5)}\cdot\frac{\cancel{2}\cancel{(p-2)}(p-4)}{\cancel{2}\cancel{(p-2)}(p+5)}$$

$$=\frac{3(p+2)(p-4)}{(p+5)^2}$$

55. $\dfrac{r^2}{s}\div\left(-\dfrac{r^4}{s}\right)=\dfrac{\cancel{r^2}}{\cancel{s}}\cdot\left(-\dfrac{\cancel{s}}{r^{\cancel{4}2}}\right)=-\dfrac{1}{r^2}$

57. $\dfrac{3x+9}{12}\div\dfrac{9-x^2}{8}$

$$=\frac{3x+9}{12}\cdot\frac{8}{9-x^2}$$

$$=\frac{3(x+3)}{12}\cdot\frac{8}{(3+x)(3-x)}$$

$$=\frac{\overset{1}{\cancel{3}}\cancel{(x+3)}}{\underset{1}{\cancel{12}}}\cdot\frac{\overset{2}{\cancel{8}}}{\cancel{(3+x)}(3-x)}$$

$$=\frac{2}{3-x}$$

59. $-\dfrac{6q^2}{15p^2q}\div\dfrac{12p^2q^2}{5p^2q}$

$$=-\frac{6q^2}{15p^2q}\cdot\frac{5p^2q}{12p^2q^2}$$

$$=-\frac{\overset{1}{\cancel{6}}\cancel{q^2}}{\underset{3}{\cancel{15}}p^2\cancel{q}}\cdot\frac{\overset{1}{\cancel{5}}\cancel{p^2}\cancel{q}}{\underset{2}{\cancel{12}}\cancel{p^2}\cancel{q^2}}$$

$$=-\frac{1}{6p^2}$$

61. $$\frac{3y}{4}\cdot\frac{y+3}{2y^2}\cdot\frac{6y}{3}$$
$$=\frac{\cancel{3}\cancel{y}}{4}\cdot\frac{y+3}{\cancel{2}_1\cancel{y^2}}\cdot\frac{\cancel{6}^3\cancel{y}}{\cancel{3}}$$
$$=\frac{3(y+3)}{4}$$

63. $$\frac{x^2+x-2}{x+1}\cdot\frac{3x+3x^2}{3x^2+7x+2}$$
$$=\frac{(x+2)(x-1)}{x+1}\cdot\frac{3x(1+x)}{(3x+1)(x+2)}$$
$$=\frac{\cancel{(x+2)}(x-1)}{\cancel{x+1}}\cdot\frac{3x\cancel{(1+x)}}{(3x+1)\cancel{(x+2)}}$$
$$=\frac{3x(x-1)}{3x+1}$$

65. $$\left(\frac{A-p}{p}\right)\div r=\left(\frac{A-p}{p}\right)\cdot\frac{1}{r}$$
$$=\frac{A-p}{pr}$$

67. $$C=B\cdot\frac{p}{100}\cdot\frac{q}{100}$$
$$=\frac{pqB}{10{,}000}$$

69. $$P=\frac{V^2}{R+r}\cdot\frac{r}{R+r}$$
$$=\frac{V^2r}{(R+r)^2}$$

71. a. $$\frac{\frac{4}{3}\pi r^3}{\pi r^2h}=\frac{\frac{4}{3}\cancel{\pi}r^{\cancel{3}}}{\cancel{\pi}\cancel{r^2}h}=\frac{4r}{3h}$$

b. $$\frac{4\pi r^2}{2\pi rh+2\pi r^2}=\frac{4\pi r^2}{2\pi r(h+r)}$$
$$=\frac{\cancel{4}^2\cancel{\pi}r^{\cancel{2}}}{\cancel{2}_1\cancel{\pi}\cancel{r}(h+r)}$$
$$=\frac{2r}{h+r}$$

c. $$\frac{4r}{3h}\div\frac{2r}{h+r}=\frac{4r}{3h}\cdot\frac{h+r}{2r}$$
$$=\frac{\cancel{4}^2\cancel{r}}{3h}\cdot\frac{h+r}{\cancel{2}_1\cancel{r}}$$
$$=\frac{2(h+r)}{3h}$$

d. $$\frac{2(h+r)}{3h}=\frac{2(2r+r)}{3(2r)}$$
$$=\frac{2(3r)}{6r}$$
$$=\frac{6r}{6r}$$
$$=1$$

7.3 Addition and Subtraction of Rational Expressions

Practice

1. a. $$\frac{9}{y+2}+\frac{1}{y+2}=\frac{9+1}{y+2}$$
$$=\frac{10}{y+2}$$

b. $$\frac{10r}{3s}+\frac{5r}{3s}=\frac{10r+5r}{3s}$$
$$=\frac{15r}{3s}$$
$$=\frac{5r}{s}$$

c. $$\frac{6t-7}{t-1}+\frac{1}{t-1}=\frac{6t-7+1}{t-1}$$
$$=\frac{6t-6}{t-1}$$
$$=\frac{6\cancel{(t-1)}}{\cancel{t-1}}$$
$$=6$$

1. d. $\dfrac{n^2+10n+1}{n+5}+\dfrac{2n^2+4n-6}{n+5}$

$=\dfrac{n^2+10n+1+2n^2+4n-6}{n+5}$

$=\dfrac{3n^2+14n-5}{n+5}$

$=\dfrac{(3n-1)\cancel{(n+5)}}{\cancel{n+5}}$

$=3n-1$

2. a. $\dfrac{12}{v}-\dfrac{7}{v}=\dfrac{12-7}{v}$

$=\dfrac{5}{v}$

b. $\dfrac{7t}{10}-\dfrac{t}{10}=\dfrac{7t-t}{10}$

$=\dfrac{6t}{10}$

$=\dfrac{3t}{5}$

c. $\dfrac{7p}{3q}-\dfrac{8p}{3q}=\dfrac{7p-8p}{3q}$

$=-\dfrac{p}{3q}$

3. a. $\dfrac{7a-4b}{3a}-\dfrac{a-4b}{3a}=\dfrac{(7a-4b)-(a-4b)}{3a}$

$=\dfrac{7a-4b-a+4b}{3a}$

$=\dfrac{6a}{3a}$

$=2$

b. $\dfrac{9xy-5xz}{4y-z}-\dfrac{xy-3xz}{4y-z}$

$=\dfrac{(9xy-5xz)-(xy-3xz)}{4y-z}$

$=\dfrac{9xy-5xz-xy+3xz}{4y-z}$

$=\dfrac{8xy-2xz}{4y-z}$

$=\dfrac{2x\cancel{(4y-z)}}{\cancel{4y-z}}$

$=2x$

c. $\dfrac{2x+13}{x^2-7x+10}-\dfrac{5x+7}{x^2-7x+10}$

$=\dfrac{(2x+13)-(5x+7)}{x^2-7x+10}$

$=\dfrac{2x+13-5x-7}{x^2-7x+10}$

$=\dfrac{-3x+6}{x^2-7x+10}$

$=\dfrac{-3\cancel{(x-2)}}{\cancel{(x-2)}(x-5)}$

$=-\dfrac{3}{x-5}$

4. a. Factor $y: y$

Factor $20: 2\cdot2\cdot5$

$\text{LCD}=20y$

b. Factor $6t: 2\cdot3\cdot t$

Factor $3t^2: 3\cdot t\cdot t$

$\text{LCD}=6t^2$

c. Factor $5x: 5\cdot x$

Factor $15xy^3: 3\cdot5\cdot x\cdot y\cdot y\cdot y$

Factor $2x^2: 2\cdot x\cdot x$

$\text{LCD}=30x^2y^3$

5. a. Factor $2n^2+2n: 2n(n+1)$

Factor $n^2+2n+1: (n+1)(n+1)$

$\text{LCD}=2n(n+1)^2$

b. Factor $p+2: p+2$

Factor $p-1: p-1$

Factor $p+5: p+5$

$\text{LCD}=(p+2)(p-1)(p+5)$

c. Factor $s^2+4st+4t^2: (s+2t)(s+2t)$

Factor $s^2-4t^2: (s+2t)(s-2t)$

Factor $s^2-4st+4t^2: (s-2t)(s-2t)$

$\text{LCD}=(s+2t)^2(s-2t)^2$

6. **a.** Factor $7p^3 : 7 \cdot p \cdot p \cdot p$

Factor 2 : 2

$\text{LCD} = 14p^3$

$$\frac{2}{7p^3} \cdot \frac{2}{2} = \frac{4}{14p^3}$$

$$\frac{p+3}{2p} \cdot \frac{7p^2}{7p^2} = \frac{7p^2(p+3)}{14p^3}$$

b. Factor $y^2 - 9 : (y+3)(y-3)$

Factor $y^2 - 6y + 9 : (y-3)(y-3)$

$\text{LCD} = (y+3)(y-3)^2$

$$\frac{3y-2}{y^2-9} = \frac{(3y-2)(y-3)}{(y+3)(y-3)(y-3)} = \frac{(3y-2)(y-3)}{(y+3)(y-3)^2}$$

$$\frac{y}{y^2-6y+9} = \frac{y(y+3)}{(y-3)(y-3)(y+3)} = \frac{y(y+3)}{(y+3)(y-3)^2}$$

7. **a.** $$\frac{3}{4p} + \frac{1}{6p} = \frac{3}{4p} \cdot \frac{3}{3} + \frac{1}{6p} \cdot \frac{2}{2} = \frac{9}{12p} + \frac{2}{12p} = \frac{11}{12p}$$

b. $$\frac{1}{5y} - \frac{2}{15y^2} = \frac{1}{5y} \cdot \frac{3y}{3y} - \frac{2}{15y^2} = \frac{3y}{15y^2} - \frac{2}{15y^2} = \frac{3y-2}{15y^2}$$

8. **a.** $$\frac{x+2}{x} - \frac{x-4}{x+3}$$

$$= \frac{x+2}{x} \cdot \frac{x+3}{x+3} - \frac{x-4}{x+3} \cdot \frac{x}{x}$$

$$= \frac{(x+2)(x+3)}{x(x+3)} - \frac{x(x-4)}{x(x+3)}$$

$$= \frac{(x+2)(x+3) - x(x-4)}{x(x+3)}$$

$$= \frac{x^2 + 5x + 6 - x^2 + 4x}{x(x+3)}$$

$$= \frac{9x+6}{x(x+3)}$$

$$= \frac{3(3x+2)}{x(x+3)}$$

b. $$\frac{3x-4}{x-1} + \frac{x+1}{1-x}$$

$$= \frac{3x-4}{x-1} - \frac{x+1}{x-1}$$

$$= \frac{(3x-4)-(x+1)}{x-1}$$

$$= \frac{3x-4-x-1}{x-1}$$

$$= \frac{2x-5}{x-1}$$

9. **a.** $$\frac{1}{4x-16} + \frac{2x}{x^2-16}$$

$$= \frac{1}{4(x-4)} + \frac{2x}{(x-4)(x+4)}$$

$$= \frac{x+4}{4(x-4)(x+4)} + \frac{2x \cdot 4}{(x-4)(x+4) \cdot 4}$$

$$= \frac{x+4+8x}{4(x-4)(x+4)}$$

$$= \frac{9x+4}{4(x-4)(x+4)}$$

9. b.
$$\frac{3x-7}{x^2-1}+\frac{2}{1-x}$$
$$=\frac{3x-7}{(x-1)(x+1)}-\frac{2}{x-1}$$
$$=\frac{3x-7}{(x+1)(x-1)}-\frac{2(x+1)}{(x-1)(x+1)}$$
$$=\frac{(3x-7)-2(x+1)}{(x+1)(x-1)}$$
$$=\frac{3x-7-2x-2}{(x+1)(x-1)}$$
$$=\frac{x-9}{(x+1)(x-1)}$$

10. a.
$$\frac{y}{y^2+5y+6}-\frac{4y+1}{y^2+3y+2}$$
$$=\frac{y}{(y+2)(y+3)}-\frac{4y+1}{(y+1)(y+2)}$$
$$=\frac{y(y+1)}{(y+2)(y+3)(y+1)}-\frac{(4y+1)(y+3)}{(y+1)(y+2)(y+3)}$$
$$=\frac{y(y+1)-(4y+1)(y+3)}{(y+2)(y+3)(y+1)}$$
$$=\frac{(y^2+y)-(4y^2+13y+3)}{(y+2)(y+3)(y+1)}$$
$$=\frac{y^2+y-4y^2-13y-3}{(y+2)(y+3)(y+1)}$$
$$=\frac{-3y^2-12y-3}{(y+2)(y+3)(y+1)}$$
$$=\frac{-3(y^2+4y+1)}{(y+2)(y+3)(y+1)}$$

b.
$$\frac{2x}{5}-\frac{1}{x+1}-\frac{x-1}{4x}$$
$$=\frac{2x\cdot 4x(x+1)}{5\cdot 4x(x+1)}-\frac{1\cdot 20x}{20x(x+1)}-\frac{(x-1)\cdot 5(x+1)}{4x\cdot 5(x+1)}$$
$$=\frac{8x^2(x+1)}{20x(x+1)}-\frac{20x}{20x(x+1)}-\frac{5(x^2-1)}{20x(x+1)}$$
$$=\frac{8x^2(x+1)-20x-5(x^2-1)}{20x(x+1)}$$
$$=\frac{8x^3+8x^2-20x-5x^2+5}{20x(x+1)}$$
$$=\frac{8x^3+3x^2-20x+5}{20x(x+1)}$$

11.
$$100\left(\frac{C_1}{C_0}-1\right)=100\left(\frac{C_1}{C_0}-1\cdot\frac{C_0}{C_0}\right)$$
$$=100\left(\frac{C_1}{C_0}-\frac{C_0}{C_0}\right)$$
$$=100\left(\frac{C_1-C_0}{C_0}\right)$$
$$=\frac{100(C_1-C_0)}{C_0}$$

Exercises

1.
$$\frac{5a}{12}+\frac{11a}{12}=\frac{5a+11a}{12}$$
$$=\frac{16a}{12}$$
$$=\frac{4a}{3}$$

3.
$$\frac{5t}{3}-\frac{2t}{3}=\frac{5t-2t}{3}$$
$$=\frac{3t}{3}$$
$$=t$$

5. $\frac{10}{x}+\frac{1}{x}=\frac{10+1}{x}$
$=\frac{11}{x}$

7. $\frac{6}{7y}-\frac{1}{7y}=\frac{6-1}{7y}$
$=\frac{5}{7y}$

9. $\frac{5x}{2y}+\frac{x}{2y}=\frac{5x+x}{2y}$
$=\frac{6x}{2y}$
$=\frac{3x}{y}$

11. $\frac{2p}{5q}-\frac{3p}{5q}=\frac{2p-3p}{5q}$
$=-\frac{p}{5q}$

13. $\frac{2}{x+1}+\frac{7}{x+1}=\frac{2+7}{x+1}$
$=\frac{9}{x+1}$

15. $\frac{5}{x+2}-\frac{9}{2+x}=\frac{5-9}{x+2}$
$=-\frac{4}{x+2}$

17. $\frac{a}{a+3}+\frac{1}{a+3}=\frac{a+1}{a+3}$

19. $\frac{3x}{x-8}+\frac{2x+1}{x-8}=\frac{3x+2x+1}{x-8}$
$=\frac{5x+1}{x-8}$

21. $\frac{7x+1}{5x+2}-\frac{3x}{5x+2}=\frac{7x+1-3x}{5x+2}$
$=\frac{4x+1}{5x+2}$

23. $\frac{9x+17}{2x+5}-\frac{3x+2}{2x+5}=\frac{(9x+17)-(3x+2)}{2x+5}$
$=\frac{9x+17-3x-2}{2x+5}$
$=\frac{6x+15}{2x+5}$
$=\frac{3\cancel{(2x+5)}}{\cancel{2x+5}}$
$=3$

25. $\frac{-7+5n}{3n-1}+\frac{7n+3}{3n-1}=\frac{-7+5n+7n+3}{3n-1}$
$=\frac{12n-4}{3n-1}$
$=\frac{4\cancel{(3n-1)}}{\cancel{3n-1}}$
$=4$

27. $\frac{x^2-1}{x^2-4x-2}-\frac{x^2-x+3}{x^2-4x-2}$
$=\frac{(x^2-1)-(x^2-x+3)}{x^2-4x-2}$
$=\frac{x^2-1-x^2+x-3}{x^2-4x-2}$
$=\frac{x-4}{x^2-4x-2}$

29. $\frac{x}{x^2-3x+2}+\frac{2}{x^2-3x+2}+\frac{x^2-4x}{x^2-3x+2}$
$=\frac{x+2+x^2-4x}{x^2-3x+2}$
$=\frac{\cancel{x^2-3x+2}}{\cancel{x^2-3x+2}}$
$=1$

31. $\frac{2x-1}{3x^2-x+2}+\frac{8}{3x^2-x+2}-\frac{3x}{3x^2-x+2}$
$=\frac{2x-1+8-3x}{3x^2-x+2}$
$=\frac{-x+7}{3x^2-x+2}$

33. Factor $5(x+2): 5\cdot(x+2)$
Factor $3(x+2): 3\cdot(x+2)$
$\text{LCD}=15(x+2)$

35. Factor $(p-3)(p+8):(p-3)\cdot(p+8)$

Factor $(p-3)(p-8):(p-3)\cdot(p-8)$

$\text{LCD}=(p-3)(p+8)(p-8)$

37. Factor $t:t$

Factor $t+3:t+3$

Factor $t-3$

$\text{LCD}=t(t+3)(t-3)$

39. Factor $t^2+7t+10:(t+2)(t+5)$

Factor $t^2-25:(t+5)(t-5)$

$\text{LCD}=(t+2)(t+5)(t-5)$

41. Factor $3s^2-11s+6:(3s-2)(s-3)$

Factor $3s^2+4s-4:(3s-2)(s+2)$

$\text{LCD}=(3s-2)(s-3)(s+2)$

43. Factor $3x:3\cdot x$

Factor $4x:2\cdot 2\cdot x$

$\text{LCD}=12x^2$

$$\frac{1}{3x}\cdot\frac{4x}{4x}=\frac{4x}{12x^2}$$

$$\frac{5}{4x^2}\cdot\frac{3}{3}=\frac{15}{12x^2}$$

45. Factor $2a^2:2\cdot a\cdot a$

Factor $7ab:7\cdot a\cdot b$

$\text{LCD}=14a^2b$

$$\frac{5}{2a^2}\cdot\frac{7b}{7b}=\frac{35b}{14a^2b}$$

$$\frac{a-3}{7ab}\cdot\frac{2a}{2a}=\frac{2a(a-3)}{14a^2b}$$

47. Factor $n(n+1):n\cdot(n+1)$

Factor $(n+1)^2:(n+1)\cdot(n+1)$

$\text{LCD}=n(n+1)^2$

$$\frac{8}{n(n+1)}\cdot\frac{(n+1)}{(n+1)}=\frac{8(n+1)}{n(n+1)^2}$$

$$\frac{5}{(n+1)^2}\cdot\frac{n}{n}=\frac{5n}{n(n+1)^2}$$

49. Factor $4n+4:4(n+1)$

Factor $n^2-1:(n+1)(n-1)$

$\text{LCD}=4(n+1)(n-1)$

$$\frac{3n}{4n+4}=\frac{3n}{4(n+1)}\cdot\frac{n-1}{n-1}=\frac{3n(n-1)}{4(n+1)(n-1)}$$

$$\frac{2n}{n^2-1}=\frac{2n}{(n+1)(n-1)}\cdot\frac{4}{4}=\frac{8n}{4(n+1)(n-1)}$$

51. Factor $n^2+6n+5:(n+1)(n+5)$

Factor $n^2+2n-15:(n+5)(n-3)$

$\text{LCD}=(n+1)(n+5)(n-3)$

$$\frac{2n}{n^2+6n+5}=\frac{2n}{(n+1)(n+5)}\cdot\frac{(n-3)}{(n-3)}$$

$$=\frac{2n(n-3)}{(n+1)(n-3)(n+5)}$$

$$\frac{3n}{n^2+2n-15}=\frac{3n}{(n-3)(n+5)}\cdot\frac{(n+1)}{(n+1)}$$

$$=\frac{3n(n+1)}{(n+1)(n-3)(n+5)}$$

53.
$$\frac{5}{3x}+\frac{1}{2x}=\frac{5}{3x}\cdot\frac{2}{2}+\frac{1}{2x}\cdot\frac{3}{3}$$

$$=\frac{10}{6x}+\frac{3}{6x}$$

$$=\frac{13}{6x}$$

55.
$$\frac{2}{3x^2}-\frac{5}{6x}=\frac{2}{3x^2}\cdot\frac{2}{2}-\frac{5}{6x}\cdot\frac{x}{x}$$

$$=\frac{4}{6x^2}-\frac{5x}{6x^2}$$

$$=\frac{4-5x}{6x^2}$$

57.
$$\frac{-2}{3x^2y}+\frac{4}{3xy^2}=\frac{-2}{3x^2y}\cdot\frac{y}{y}+\frac{4}{3xy^2}\cdot\frac{x}{x}$$

$$=\frac{-2y}{3x^2y^2}+\frac{4x}{3x^2y^2}$$

$$=\frac{-2y+4x}{3x^2y^2}$$

$$=-\frac{2(y-2x)}{3x^2y^2}$$

59. $\dfrac{1}{x+1}+\dfrac{1}{x-1}$

$=\dfrac{1}{x+1}\cdot\dfrac{x-1}{x-1}+\dfrac{1}{x-1}\cdot\dfrac{x+1}{x+1}$

$=\dfrac{x-1}{(x-1)(x+1)}+\dfrac{x+1}{(x-1)(x+1)}$

$=\dfrac{x-1+x+1}{(x-1)(x+1)}$

$=\dfrac{2x}{(x+1)(x-1)}$

61. $\dfrac{p+6}{3}-\dfrac{2p+1}{7}$

$=\dfrac{p+6}{3}\cdot\dfrac{7}{7}-\dfrac{2p+1}{7}\cdot\dfrac{3}{3}$

$=\dfrac{7p+42}{21}-\dfrac{6p+3}{21}$

$=\dfrac{7p+42-(6p+3)}{21}$

$=\dfrac{7p+42-6p-3}{21}$

$=\dfrac{p+39}{21}$

63. $x-\dfrac{10-4x}{2}$

$=x\cdot\dfrac{2}{2}-\dfrac{10-4x}{2}$

$=\dfrac{2x}{2}-\dfrac{10-4x}{2}$

$=\dfrac{2x-(10-4x)}{2}$

$=\dfrac{2x-10+4x}{2}$

$=\dfrac{6x-10}{2}$

$=\dfrac{\not{2}(3x-5)}{\not{2}}$

$=3x-5$

65. $\dfrac{3a+1}{6a}-\dfrac{a^2-2}{2a^2}$

$=\dfrac{3a+1}{6a}\cdot\dfrac{a}{a}-\dfrac{a^2-2}{2a^2}\cdot\dfrac{3}{3}$

$=\dfrac{3a^2+a}{6a^2}-\dfrac{3a^2-6}{6a^2}$

$=\dfrac{3a^2+a-(3a^2-6)}{6a^2}$

$=\dfrac{3a^2+a-3a^2+6}{6a^2}$

$=\dfrac{a+6}{6a^2}$

67. $\dfrac{a^2}{a-1}-\dfrac{1}{1-a}$

$=\dfrac{a^2}{a-1}-\dfrac{1}{1-a}\cdot\dfrac{-1}{-1}$

$=\dfrac{a^2}{a-1}+\dfrac{1}{a-1}$

$=\dfrac{a^2+1}{a-1}$

69. $\dfrac{4}{c-4}+\dfrac{c}{4-c}$

$=\dfrac{4}{c-4}-\dfrac{c}{c-4}$

$=\dfrac{4-c}{c-4}$

$=\dfrac{-(c-4)}{c-4}$

$=-1$

71. $\dfrac{x-5}{x+1}-\dfrac{x+2}{x}$

$=\dfrac{x-5}{x+1}\cdot\dfrac{x}{x}-\dfrac{x+2}{x}\cdot\dfrac{x+1}{x+1}$

$=\dfrac{x(x-5)}{x(x+1)}-\dfrac{(x+1)(x+2)}{x(x+1)}$

$=\dfrac{x(x-5)-(x+1)(x+2)}{x(x+1)}$

$=\dfrac{(x^2-5x)-(x^2+3x+2)}{x(x+1)}$

$=\dfrac{x^2-5x-x^2-3x-2}{x(x+1)}$

$=\dfrac{-8x-2}{x(x+1)}$

$=-\dfrac{2(4x+1)}{x(x+1)}$

73. $\dfrac{4x-5}{x-4}+\dfrac{1-3x}{4-x}$

$=\dfrac{4x-5}{x-4}+\dfrac{1-3x}{4-x}\cdot\dfrac{-1}{-1}$

$=\dfrac{4x-5}{x-4}+\dfrac{3x-1}{x-4}$

$=\dfrac{4x-5+3x-1}{x-4}$

$=\dfrac{7x-6}{x-4}$

75. $\dfrac{5x}{x^2+x-2}+\dfrac{6}{x+2}$

$=\dfrac{5x}{(x-1)(x+2)}+\dfrac{6}{x+2}$

$=\dfrac{5x}{(x-1)(x+2)}+\dfrac{6}{x+2}\cdot\dfrac{x-1}{x-1}$

$=\dfrac{5x+6x-6}{(x-1)(x+2)}$

$=\dfrac{11x-6}{(x-1)(x+2)}$

77. $\dfrac{4}{3n-9}-\dfrac{n}{n^2+2n-15}$

$=\dfrac{4}{3(n-3)}-\dfrac{n}{(n-3)(n+5)}$

$=\dfrac{4}{3(n-3)}\cdot\dfrac{n+5}{n+5}-\dfrac{n}{(n-3)(n+5)}\cdot\dfrac{3}{3}$

$=\dfrac{4n+20-3n}{3(n-3)(n+5)}$

$=\dfrac{n+20}{3(n-3)(n+5)}$

79. $\dfrac{2}{t+5}-\dfrac{t+6}{25-t^2}$

$=\dfrac{2}{t+5}-\dfrac{t+6}{(5-t)(5+t)}$

$=\dfrac{2}{t+5}+\dfrac{t+6}{(t-5)(t+5)}$

$=\dfrac{2}{t+5}\cdot\dfrac{t-5}{t-5}+\dfrac{t+6}{(t+5)(t-5)}$

$=\dfrac{2t-10+t+6}{(t+5)(t-5)}$

$=\dfrac{3t-4}{(t+5)(t-5)}$

81. $\dfrac{4x}{x^2+2x+1}-\dfrac{2x+5}{x^2+4x+3}$

$=\dfrac{4x}{(x+1)^2}-\dfrac{2x+5}{(x+1)(x+3)}$

$=\dfrac{4x}{(x+1)^2}\cdot\dfrac{x+3}{x+3}-\dfrac{2x+5}{(x+1)(x+3)}\cdot\dfrac{x+1}{x+1}$

$=\dfrac{4x(x+3)-(2x+5)(x+1)}{(x+1)^2(x+3)}$

$=\dfrac{(4x^2+12x)-(2x^2+7x+5)}{(x+1)^2(x+3)}$

$=\dfrac{4x^2+12x-2x^2-7x-5}{(x+1)^2(x+3)}$

$=\dfrac{2x^2+5x-5}{(x+1)^2(x+3)}$

83. $$\frac{2t-1}{2t^2+t-3}+\frac{2}{t-1}$$
$$=\frac{2t-1}{(2t+3)(t-1)}+\frac{2}{t-1}$$
$$=\frac{2t-1}{(2t+3)(t-1)}+\frac{2}{t-1}\cdot\frac{2t+3}{2t+3}$$
$$=\frac{2t-1+4t+6}{(2t+3)(t-1)}$$
$$=\frac{6t+5}{(2t+3)(t-1)}$$

85. $$\frac{4x}{x-1}+\frac{2}{3x}+\frac{x}{x^2-1}$$
$$=\frac{4x}{x-1}+\frac{2}{3x}+\frac{x}{(x+1)(x-1)}$$
$$=\frac{4x}{x-1}\cdot\frac{3x(x+1)}{3x(x+1)}+\frac{2}{3x}\cdot\frac{(x-1)(x+1)}{(x-1)(x+1)}$$
$$+\frac{x}{(x+1)(x-1)}\cdot\frac{3x}{3x}$$
$$=\frac{12x^2(x+1)+2(x-1)(x+1)+3x^2}{3x(x+1)(x-1)}$$
$$=\frac{12x^3+12x^2+2(x^2-1)+3x^2}{3x(x+1)(x-1)}$$
$$=\frac{12x^3+12x^2+2x^2-2+3x^2}{3x(x+1)(x-1)}$$
$$=\frac{12x^3+17x^2-2}{3x(x+1)(x-1)}$$

87. $$\frac{5y}{3y-1}-\frac{3}{y-4}+\frac{y+1}{3y^2-13y+4}$$
$$=\frac{5y}{3y-1}-\frac{3}{y-4}+\frac{y+1}{(3y-1)(y-4)}$$
$$=\frac{5y}{3y-1}\cdot\frac{y-4}{y-4}-\frac{3}{y-4}\cdot\frac{3y-1}{3y-1}$$
$$+\frac{y+1}{(3y-1)(y-4)}$$
$$=\frac{5y(y-4)-3(3y-1)+y+1}{(3y-1)(y-4)}$$
$$=\frac{5y^2-20y-9y+3+y+1}{(3y-1)(y-4)}$$
$$=\frac{5y^2-28y+4}{(3y-1)(y-4)}$$

89. $$\frac{a-1}{(a+3)^2}-\frac{2a-3}{a+3}-\frac{a}{4a+12}$$
$$=\frac{a-1}{(a+3)^2}-\frac{2a-3}{a+3}-\frac{a}{4(a+3)}$$
$$=\frac{a-1}{(a+3)^2}\cdot\frac{4}{4}-\frac{2a-3}{a+3}\cdot\frac{4(a+3)}{4(a+3)}$$
$$-\frac{a}{4(a+3)}\cdot\frac{a+3}{a+3}$$
$$=\frac{4(a-1)-4(2a-3)(a+3)-a(a+3)}{4(a+3)^2}$$
$$=\frac{4a-4-4(2a^2+3a-9)-a^2-3a}{4(a+3)^2}$$
$$=\frac{4a-4-8a^2-12a+36-a^2-3a}{4(a+3)^2}$$
$$=\frac{-9a^2-11a+32}{4(a+3)^2}$$

91. Factor $6p^2: 2\cdot3\cdot p\cdot p$

Factor $8pq: 2\cdot2\cdot2\cdot p\cdot q$

$\text{LCD}=2\cdot2\cdot2\cdot3\cdot p\cdot p\cdot q=24p^2q$

$$\frac{p+3}{6p^2}\cdot\frac{4q}{4q}=\frac{4q(p+3)}{24p^2q}$$
$$\frac{5}{8pq}\cdot\frac{3p}{3p}=\frac{15p}{24p^2q}$$

93. $$\frac{9}{xy^2}+\frac{6}{x^2y}=\frac{9}{xy^2}\cdot\frac{x}{x}+\frac{6}{x^2y}\cdot\frac{y}{y}$$
$$=\frac{9x+6y}{x^2y^2}$$
$$=\frac{3(3x+2y)}{x^2y^2}$$

95. $$\frac{c-3}{c}-\frac{c-2}{c+1}=\frac{c-3}{c}\cdot\frac{c+1}{c+1}-\frac{c-2}{c+1}\cdot\frac{c}{c}$$
$$=\frac{(c-3)(c+1)-c(c-2)}{c(c+1)}$$
$$=\frac{c^2-2c-3-c^2+2c}{c(c+1)}$$
$$=-\frac{3}{c(c+1)}$$

97. $\dfrac{b+3}{b^2-2b-3}-\dfrac{4}{b^2-6b+9}$

$$=\frac{b+3}{(b+1)(b-3)}-\frac{4}{(b-3)^2}$$

$$=\frac{b+3}{(b+1)(b-3)}\cdot\frac{b-3}{b-3}-\frac{4}{(b-3)^2}\cdot\frac{b+1}{b+1}$$

$$=\frac{(b+3)(b-3)-4(b+1)}{(b-3)^2(b+1)}$$

$$=\frac{b^2-9-4b-4}{(b-3)^2(b+1)}$$

$$=\frac{b^2-4b-13}{(b-3)^2(b+1)}$$

99. $\dfrac{6}{m+n}-\dfrac{11}{n+m}=\dfrac{6}{m+n}-\dfrac{11}{m+n}$

$$=\frac{6-11}{m+n}$$

$$=-\frac{5}{m+n}$$

101. $\dfrac{r}{r^2-r-6}+\dfrac{3}{r^2-r-6}+\dfrac{r^2-5r}{r^2-r-6}$

$$=\frac{r+3+r^2-5r}{r^2-r-6}$$

$$=\frac{r^2-4r+3}{r^2-r-6}$$

$$=\frac{(r-1)\cancel{(r-3)}}{(r+2)\cancel{(r-3)}}$$

$$=\frac{r-1}{r+2}$$

103. $vt+\dfrac{at^2}{2}=vt\cdot\dfrac{2}{2}+\dfrac{at^2}{2}$

$$=\frac{2vt}{2}+\frac{at^2}{2}$$

$$=\frac{2vt+at^2}{2}$$

105. $\dfrac{1000}{1+r}-\dfrac{1000}{(1+r)^2}=\dfrac{1000}{1+r}\cdot\dfrac{1+r}{1+r}-\dfrac{1000}{(1+r)^2}$

$$=\frac{1000+1000r-1000}{(1+r)^2}$$

$$=\frac{1000r}{(1+r)^2}$$

$\dfrac{1000r}{(1+r)^2}$ dollars

107. $\dfrac{20}{r}+\dfrac{20}{2r}=\dfrac{20}{r}\cdot\dfrac{2}{2}+\dfrac{20}{2r}$

$$=\frac{40}{2r}+\frac{20}{2r}$$

$$=\frac{60}{2r}$$

$$=\frac{30}{r}$$

The trip took $\dfrac{30}{r}$ hr.

109. $\dfrac{3}{x}+0.1=\dfrac{3}{x}+0.1\cdot\dfrac{x}{x}$

$$=\frac{3}{x}+\frac{0.1x}{x}$$

$$=\frac{3+0.1x}{x}$$

$\dfrac{3+0.1x}{x}$ dollars

7.4 Complex Rational Expressions

Practice

1. a. $\dfrac{\frac{3}{x^4}}{\frac{5}{x}}=\dfrac{3}{x^4}\div\dfrac{5}{x}$

$$=\frac{3}{x^4}\cdot\frac{x}{5}$$

$$=\frac{3}{x^{\cancel{4}3}}\cdot\frac{\cancel{x}}{5}$$

$$=\frac{3}{5x^3}$$

1. b. $$\frac{2x}{\frac{x^2}{4}+\frac{2x}{4}}=\frac{2x}{\frac{x^2+2x}{4}}$$
$$=2x\div\frac{x^2+2x}{4}$$
$$=2x\cdot\frac{4}{x^2+2x}$$
$$=2\cancel{x}\cdot\frac{4}{\cancel{x}(x+2)}$$
$$=\frac{8}{x+2}$$

2. $$\frac{2-\frac{1}{n}}{2+\frac{1}{n}}=\frac{\frac{2n-1}{n}}{\frac{2n+1}{n}}$$
$$=\frac{2n-1}{n}\div\frac{2n+1}{n}$$
$$=\frac{2n-1}{\cancel{n}}\cdot\frac{\cancel{n}}{2n+1}$$
$$=\frac{2n-1}{2n+1}$$

3. a. $$\frac{\frac{4}{x}}{\frac{2}{x^3}}=\frac{\frac{4}{x}\cdot x^3}{\frac{2}{x^3}\cdot x^3}$$
$$=\frac{4x^2}{2}$$
$$=2x^2$$

b. $$\frac{\frac{1}{y}+\frac{3}{y^2}}{2y}=\frac{\left(\frac{1}{y}+\frac{3}{y^2}\right)\cdot y^2}{2y\cdot y^2}$$
$$=\frac{\frac{1}{y}\cdot y^2+\frac{3}{y^2}\cdot y^2}{2y\cdot y^2}$$
$$=\frac{y+3}{2y^3}$$

4. a. $$\frac{4+\frac{1}{y}}{4-\frac{1}{y^2}}=\frac{\left(4+\frac{1}{y}\right)\cdot y^2}{\left(4-\frac{1}{y^2}\right)\cdot y^2}$$
$$=\frac{4\cdot y^2+\frac{1}{y}\cdot y^2}{4\cdot y^2-\frac{1}{y^2}\cdot y^2}$$
$$=\frac{4y^2+y}{4y^2-1}$$

b. $$\frac{\frac{1}{2a^2}-\frac{1}{2b^2}}{\frac{5}{a}+\frac{5}{b}}=\frac{\left(\frac{1}{2a^2}-\frac{1}{2b^2}\right)2a^2b^2}{\left(\frac{5}{a}+\frac{5}{b}\right)2a^2b^2}$$
$$=\frac{\frac{1}{2a^2}\cdot 2a^2b^2-\frac{1}{2b^2}\cdot 2a^2b^2}{\frac{5}{a}\cdot 2a^2b^2+\frac{5}{b}\cdot 2a^2b^2}$$
$$=\frac{b^2-a^2}{10ab^2+10a^2b}$$
$$=\frac{\cancel{(b+a)}(b-a)}{10ab\cancel{(b+a)}}$$
$$=\frac{b-a}{10ab}$$

5. $$\frac{3}{\frac{1}{a}+\frac{1}{b}+\frac{1}{c}}=\frac{3\cdot abc}{\left(\frac{1}{a}+\frac{1}{b}+\frac{1}{c}\right)\cdot abc}$$
$$=\frac{3\cdot abc}{\frac{1}{a}\cdot abc+\frac{1}{b}\cdot abc+\frac{1}{c}\cdot abc}$$
$$=\frac{3abc}{bc+ac+ab}$$

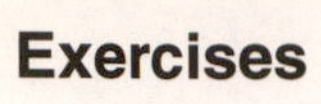

Exercises

1. $$\frac{\frac{x}{5}}{\frac{x^2}{10}}=\frac{x}{5}\div\frac{x^2}{10}$$
$$=\frac{x}{5}\cdot\frac{10}{x^2}$$
$$=\frac{\cancel{x}}{\cancel{5}_{1}}\cdot\frac{\cancel{10}^{2}}{x^{\cancel{2}}}$$
$$=\frac{2}{x}$$

3. $$\frac{\frac{a+1}{2}}{\frac{a-1}{2}}=\frac{a+1}{2}\div\frac{a-1}{2}$$
$$=\frac{a+1}{2}\cdot\frac{2}{a-1}$$
$$=\frac{a+1}{\cancel{2}}\cdot\frac{\cancel{2}}{a-1}$$
$$=\frac{a+1}{a-1}$$

5. $$\frac{3+\frac{1}{x}}{3-\frac{1}{x^2}}=\frac{\frac{3x+1}{x}}{\frac{3x^2-1}{x^2}}$$
$$=\frac{3x+1}{x}\div\frac{3x^2-1}{x^2}$$
$$=\frac{3x+1}{\cancel{x}}\cdot\frac{x^{\cancel{2}}}{3x^2-1}$$
$$=\frac{x(3x+1)}{3x^2-1}$$

7. $$\frac{\frac{1}{3d}-\frac{1}{d^2}}{d-\frac{9}{d}}=\frac{\frac{d-3}{3d^2}}{\frac{d^2-9}{d}}$$
$$=\frac{d-3}{3d^2}\div\frac{d^2-9}{d}$$
$$=\frac{d-3}{3d^2}\cdot\frac{d}{d^2-9}$$
$$=\frac{\cancel{d-3}}{3d^{\cancel{2}}}\cdot\frac{\cancel{d}}{\cancel{(d-3)}(d+3)}$$
$$=\frac{1}{3d(d+3)}$$

9. $$\frac{1-\frac{4y^2}{x^2}}{3+\frac{6y}{x}}=\frac{\frac{x^2-4y^2}{x^2}}{\frac{3x+6y}{x}}$$
$$=\frac{x^2-4y^2}{x^2}\div\frac{3x+6y}{x}$$
$$=\frac{x^2-4y^2}{x^2}\cdot\frac{x}{3x+6y}$$
$$=\frac{(x-2y)\cancel{(x+2y)}}{x^{\cancel{2}}}\cdot\frac{\cancel{x}}{3\cancel{(x+2y)}}$$
$$=\frac{x-2y}{3x}$$

11. $$\frac{\frac{2}{y}-\frac{1}{5}}{\frac{5}{y}-1}==\frac{\frac{10-y}{5y}}{\frac{5-y}{y}}$$
$$=\frac{10-y}{5y}\div\frac{5-y}{y}$$
$$=\frac{10-y}{5\cancel{y}}\cdot\frac{\cancel{y}}{5-y}$$
$$=\frac{10-y}{5(5-y)}$$

13. $$\frac{1+\frac{4}{x}+\frac{4}{x^2}}{1+\frac{5}{x}+\frac{6}{x^2}}=\frac{\frac{x^2+4x+4}{x^2}}{\frac{x^2+5x+6}{x^2}}$$

$$=\frac{x^2+4x+4}{x^2}\div\frac{x^2+5x+6}{x^2}$$

$$=\frac{x^2+4x+4}{x^2}\cdot\frac{x^2}{x^2+5x+6}$$

$$=\frac{(x+2)^{\cancel{2}}}{\cancel{x^2}}\cdot\frac{\cancel{x^2}}{\cancel{(x+2)}(x+3)}$$

$$=\frac{x+2}{x+3}$$

15. $$\frac{3+\frac{1}{y+1}}{5-\frac{1}{y+1}}=\frac{\frac{3(y+1)+1}{y+1}}{\frac{5(y+1)-1}{y+1}}$$

$$=\frac{\frac{3y+3+1}{y+1}}{\frac{5y+5-1}{y+1}}$$

$$=\frac{\frac{3y+4}{y+1}}{\frac{5y+4}{y+1}}$$

$$=\frac{3y+4}{y+1}\div\frac{5y+4}{y+1}$$

$$=\frac{3y+4}{\cancel{y+1}}\cdot\frac{\cancel{y+1}}{5y+4}$$

$$=\frac{3y+4}{5y+4}$$

17. $$\frac{\frac{x}{4}-\frac{x}{8}}{\frac{2}{y^2}+\frac{2}{y}}=\frac{\left(\frac{x}{4}-\frac{x}{8}\right)\cdot 8y^2}{\left(\frac{2}{y^2}+\frac{2}{y}\right)\cdot 8y^2}$$

$$=\frac{\frac{x}{4}\cdot 8y^2-\frac{x}{8}\cdot 8y^2}{\frac{2}{y^2}\cdot 8y^2+\frac{2}{y}\cdot 8y^2}$$

$$=\frac{2xy^2-xy^2}{16+16y}$$

$$=\frac{xy^2}{16(y+1)}$$

19. $$\frac{\frac{x}{x+1}-\frac{2}{x}}{\frac{x}{3}}=\frac{\left(\frac{x}{x+1}-\frac{2}{x}\right)\cdot 3x(x+1)}{\left(\frac{x}{3}\right)\cdot 3x(x+1)}$$

$$=\frac{\frac{x}{x+1}\cdot 3x(x+1)-\frac{2}{x}\cdot 3x(x+1)}{\frac{x}{3}\cdot 3x(x+1)}$$

$$=\frac{3x^2-6(x+1)}{x^2(x+1)}$$

$$=\frac{3x^2-6x-6}{x^2(x+1)}$$

$$=\frac{3(x^2-2x-2)}{x^2(x+1)}$$

21. $$\frac{\frac{m+2}{3m}}{\frac{m-1}{m}}=\frac{m+2}{3m}\div\frac{m-1}{m}$$

$$=\frac{m+2}{3m}\cdot\frac{m}{m-1}$$

$$=\frac{m+2}{3\cancel{m}}\cdot\frac{\cancel{m}}{m-1}$$

$$=\frac{m+2}{3(m-1)}$$

23. $$\frac{\frac{4}{u}-\frac{2}{u+1}}{\frac{u}{2}}=\frac{\frac{4(u+1)-2u}{u(u+1)}}{\frac{u}{2}}$$

$$=\frac{\frac{4u+4-2u}{u(u+1)}}{\frac{u}{2}}$$

$$=\frac{\frac{2u+4}{u(u+1)}}{\frac{u}{2}}$$

$$=\frac{2u+4}{u(u+1)}\div\frac{u}{2}$$

$$=\frac{2(u+2)}{u(u+1)}\cdot\frac{2}{u}$$

$$=\frac{4(u+2)}{u^2(u+1)}$$

25. $$\frac{\frac{3}{y^2}+\frac{4}{y}}{6+\frac{3}{y}}=\frac{\left(\frac{3}{y^2}+\frac{4}{y}\right)\cdot y^2}{\left(6+\frac{3}{y}\right)\cdot y^2}$$
$$=\frac{\frac{3}{y^2}\cdot y^2+\frac{4}{y}\cdot y^2}{6\cdot y^2+\frac{3}{y}\cdot y^2}$$
$$=\frac{3+4y}{6y^2+3y}$$
$$=\frac{4y+3}{3y(2y+1)}$$

27. $$\frac{V}{\frac{1}{2R}+\frac{1}{2R+2}}$$
$$=\frac{V\cdot 2R(2R+2)}{\left(\frac{1}{2R}+\frac{1}{2R+2}\right)\cdot 2R(2R+2)}$$
$$=\frac{2VR(2R+2)}{\frac{1}{\cancel{2R}}\cdot\cancel{2R}\,(2R+2)+\frac{1}{\cancel{2R+2}}2R(\cancel{2R+2})}$$
$$=\frac{2VR\cdot 2(R+1)}{2R+2+2R}$$
$$=\frac{4VR(R+1)}{4R+2}$$
$$=\frac{4VR(R+1)}{2(2R+1)}$$
$$=\frac{2VR(R+1)}{2R+1}$$

29. $$\frac{E}{\frac{I}{9}}=E\div\frac{I}{9}=E\cdot\frac{9}{I}=\frac{9E}{I}$$

31. $$\frac{2}{\frac{1}{a}+\frac{1}{b}}=\frac{2\cdot ab}{\left(\frac{1}{a}+\frac{1}{b}\right)\cdot ab}$$
$$=\frac{2ab}{\frac{1}{a}\cdot ab+\frac{1}{b}\cdot ab}$$
$$=\frac{2ab}{b+a}$$

$\frac{2ab}{b+a}$ mph

33. $$\frac{w}{\left(1+\frac{h}{6400}\right)^2}=\frac{w}{\left(\frac{6400+h}{6400}\right)^2}$$
$$=\frac{w}{\frac{(6400+h)^2}{6400^2}}$$
$$=w\div\frac{(6400+h)^2}{6400^2}$$
$$=w\cdot\frac{6400^2}{(6400+h)^2}$$
$$=\frac{6400^2 w}{(6400+h)^2}$$

7.5 Solving Rational Equations

Practice

1. $$\frac{y}{2}-\frac{y}{3}=\frac{1}{12}$$
$$12\cdot\left(\frac{y}{2}-\frac{y}{3}\right)=12\cdot\frac{1}{12}$$
$$12\cdot\frac{y}{2}-12\cdot\frac{y}{3}=12\cdot\frac{1}{12}$$
$$6y-4y=1$$
$$2y=1$$
$$y=\frac{1}{2}$$

Check: $$\frac{\frac{1}{2}}{2}-\frac{\frac{1}{2}}{3}=\frac{1}{12}$$
$$\frac{1}{4}-\frac{1}{6}=\frac{1}{12}$$
$$\frac{3}{12}-\frac{2}{12}=\frac{1}{12}$$
$$\frac{1}{12}=\frac{1}{12}$$

2.

$$\frac{x-2}{5}-1=-\frac{2}{x}$$
$$5x\cdot\left(\frac{x-2}{5}-1\right)=5x\cdot\left(-\frac{2}{x}\right)$$
$$5x\cdot\frac{x-2}{5}-5x\cdot 1=5x\cdot\left(-\frac{2}{x}\right)$$
$$x(x-2)-5x=-10$$
$$x^2-2x-5x=-10$$
$$x^2-7x+10=0$$
$$(x-2)(x-5)=0$$
$$x-2=0 \text{ or } x-5=0$$
$$x=2 \qquad x=5$$

Check:

$$\left(\frac{2-2}{5}\right)-1=-\frac{2}{2} \qquad \left(\frac{5-2}{5}\right)-1=-\frac{2}{5}$$
$$\frac{0}{5}-1=-1 \qquad \frac{3}{5}-1=-\frac{2}{5}$$
$$-1=-1 \qquad \frac{3}{5}-\frac{5}{5}=-\frac{2}{5}$$
$$-\frac{2}{5}=-\frac{2}{5}$$

3.

$$\frac{4}{y+2}+\frac{2}{y-1}=\frac{12}{y^2+y-2}$$
$$\frac{4}{y+2}+\frac{2}{y-1}=\frac{12}{(y-1)(y+2)}$$
$$(y+2)(y-1)\cdot\frac{4}{y+2}+(y+2)(y-1)\cdot\frac{2}{y-1}$$
$$=(y+2)(y-1)\cdot\frac{12}{(y-1)(y+2)}$$
$$\cancel{(y+2)}(y-1)\cdot\frac{4}{\cancel{y+2}}+(y+2)\cancel{(y-1)}\cdot\frac{2}{\cancel{y-1}}$$
$$=\cancel{(y+2)}\cancel{(y-1)}\cdot\frac{12}{\cancel{(y-1)}\cancel{(y+2)}}$$
$$4(y-1)+2(y+2)=12$$
$$4y-4+2y+4=12$$
$$6y=12$$
$$y=2$$

Check: $$\frac{4}{2+2}+\frac{2}{2-1}=\frac{12}{(2)^2+2-2}$$
$$\frac{4}{4}+\frac{2}{1}=\frac{12}{4}$$
$$1+2=3$$
$$3=3$$

4.

$$x=\frac{9}{x+3}+\frac{3x}{x+3}$$
$$x=\frac{9+3x}{x+3}$$
$$(x+3)\cdot x=(x+3)\cdot\frac{9+3x}{x+3}$$
$$(x+3)\cdot x=\cancel{(x+3)}\cdot\frac{9+3x}{\cancel{x+3}}$$
$$x(x+3)=9+3x$$
$$x^2+3x=9+3x$$
$$x^2-9=0$$
$$(x-3)(x+3)=0$$
$$x-3=0 \text{ or } x+3=0$$
$$x=3 \qquad x=-3$$

Check:

$$3=\frac{9}{3+3}+\frac{3(3)}{3+3} \qquad -3=\frac{9}{-3+3}+\frac{3(-3)}{-3+3}$$
$$3=\frac{9}{6}+\frac{9}{6} \qquad -3=\frac{9}{0}+\frac{-9}{0}$$
$$3=\frac{18}{6} \qquad \text{Undefined}$$
$$3=3$$

5.

$$\frac{1}{10}\cdot t+\frac{1}{6}\cdot t=1$$
$$\frac{t}{10}+\frac{t}{6}=1$$
$$30\cdot\frac{t}{10}+30\cdot\frac{t}{6}=30\cdot 1$$
$$3t+5t=30$$
$$8t=30$$
$$t=\frac{30}{8}=3\frac{3}{4}$$

Working together, it will take both pumps $3\frac{3}{4}$ hr (or 3 hr and 45 min) to fill the tank.

6.

$$\frac{1800}{3r}+\frac{300}{r}=6$$
$$r\cdot\frac{600}{r}+r\cdot\frac{300}{r}=r\cdot 6$$
$$\cancel{r}\cdot\frac{600}{\cancel{r}}+\cancel{r}\cdot\frac{300}{\cancel{r}}=6r$$
$$600+300=6r$$
$$900=6r$$
$$r=\frac{900}{6}$$
$$r=150$$

The speed of the propeller plane was 150 mph.

7. a.

$$D=\frac{500}{x}+50$$
$$x\cdot D=x\cdot\left(\frac{500}{x}+50\right)$$
$$x\cdot D=x\cdot\frac{500}{x}+x\cdot 50$$
$$x\cdot D=\cancel{x}\cdot\frac{500}{\cancel{x}}+50x$$
$$Dx=500+50x$$
$$Dx-50x=500$$
$$(D-50)x=500$$
$$x=\frac{500}{D-50}$$

b.

$$x=\frac{500}{450-50}$$
$$x=\frac{500}{400}$$
$$x=\frac{5}{4}$$
$$x=1.25$$

\$1.25 per unit

Exercises

1.

$$\frac{y}{2}+\frac{7}{10}=-\frac{4}{5}$$
$$10\cdot\left(\frac{y}{2}+\frac{7}{10}\right)=10\cdot\left(-\frac{4}{5}\right)$$
$$10\cdot\frac{y}{2}+10\cdot\frac{7}{10}=10\cdot\left(-\frac{4}{5}\right)$$
$$5y+7=-8$$
$$5y=-15$$
$$y=-3$$

Check :
$$\frac{-3}{2}+\frac{7}{10}=-\frac{4}{5}$$
$$-\frac{15}{10}+\frac{7}{10}=-\frac{4}{5}$$
$$-\frac{8}{10}=-\frac{4}{5}$$
$$-\frac{4}{5}=-\frac{4}{5}$$

3.

$$\frac{1}{t}-\frac{7}{3}=-\frac{1}{3}$$
$$3t\cdot\left(\frac{1}{t}-\frac{7}{3}\right)=3t\cdot\left(-\frac{1}{3}\right)$$
$$3t\cdot\frac{1}{t}-3t\cdot\frac{7}{3}=3t\cdot\left(-\frac{1}{3}\right)$$
$$3-7t=-t$$
$$-6t=-3$$
$$t=\frac{1}{2}$$

Check:
$$\frac{1}{\frac{1}{2}}-\frac{7}{3}=-\frac{1}{3}$$
$$2-\frac{7}{3}=-\frac{1}{3}$$
$$\frac{6}{3}-\frac{7}{3}=-\frac{1}{3}$$
$$-\frac{1}{3}=-\frac{1}{3}$$

5.
$$x+\frac{1}{x}=2$$
$$x+\frac{1}{x}=2$$
$$x\cdot\left(x+\frac{1}{x}\right)=x\cdot 2$$
$$x\cdot x+x\cdot\frac{1}{x}=x\cdot 2$$
$$x^2+1=2x$$
$$x^2-2x+1=0$$
$$(x-1)^2=0$$
$$x-1=0$$
$$x=1$$

Check: $1+\frac{1}{1}=2$
$$1+1=2$$
$$2=2$$

7.
$$\frac{t+1}{2t-1}-\frac{5}{7}=0$$
$$\frac{t+1}{2t-1}=\frac{5}{7}$$
$$7(2t-1)\cdot\frac{t+1}{2t-1}=7(2t-1)\cdot\frac{5}{7}$$
$$7\cancel{(2t-1)}\cdot\frac{t+1}{\cancel{2t-1}}=\cancel{7}(2t-1)\cdot\frac{5}{\cancel{7}}$$
$$7(t+1)=5(2t-1)$$
$$7t+7=10t-5$$
$$-3t=-12$$
$$t=4$$

Check: $\frac{4+1}{2(4)-1}-\frac{5}{7}=0$
$$\frac{5}{8-1}-\frac{5}{7}=0$$
$$\frac{5}{7}-\frac{5}{7}=0$$
$$0=0$$

9.
$$\frac{t-2}{3}=4$$
$$3\cdot\frac{t-2}{3}=3\cdot 4$$
$$t-2=12$$
$$t=14$$

Check: $\frac{14-2}{3}=4$
$$\frac{12}{3}=4$$
$$4=4$$

11.
$$\frac{2n}{n+1}=\frac{-2}{n+1}+1$$
$$(n+1)\cdot\frac{2n}{n+1}=(n+1)\cdot\frac{-2}{n+1}+(n+1)\cdot 1$$
$$\cancel{(n+1)}\cdot\frac{2n}{\cancel{n+1}}=\cancel{(n+1)}\cdot\frac{-2}{\cancel{n+1}}+(n+1)\cdot 1$$
$$2n=-2+n+1$$
$$n=-1$$

Check: $\frac{2(-1)}{-1+1}=\frac{-2}{-1+1}$
$$\frac{-2}{0}=\frac{-2}{0}\quad\text{Undefined}$$

No solution

13.
$$\frac{5x}{x+1}=\frac{x^2}{x+1}+2$$
$$(x+1)\cdot\frac{5x}{x+1}=(x+1)\cdot\frac{x^2}{x+1}+(x+1)\cdot 2$$
$$\cancel{(x+1)}\cdot\frac{5x}{\cancel{x+1}}=\cancel{(x+1)}\cdot\frac{x^2}{\cancel{x+1}}+2(x+1)$$
$$5x=x^2+2x+2$$
$$0=x^2-3x+2$$
$$0=(x-1)(x-2)$$
$$x-1=0 \quad\text{or}\quad x-2=0$$
$$x=1 \qquad\qquad x=2$$

Check:
$$\frac{5(1)}{1+1}=\frac{(1)^2}{1+1}+2 \qquad \frac{5(2)}{2+1}=\frac{(2)^2}{2+1}+2$$
$$\frac{5}{2}=\frac{1}{2}+\frac{4}{2} \qquad \frac{10}{3}=\frac{4}{3}+\frac{6}{3}$$
$$\frac{5}{2}=\frac{5}{2} \qquad \frac{10}{3}=\frac{10}{3}$$

15.

$$\frac{x}{x-3}-\frac{6}{x}=1$$

$$x(x-3)\cdot\frac{x}{x-3}-x(x-3)\cdot\frac{6}{x}=x(x-3)\cdot 1$$

$$x\cancel{(x-3)}\cdot\frac{x}{\cancel{x-3}}-\cancel{x}(x-3)\cdot\frac{6}{\cancel{x}}=x(x-3)$$

$$x^2-6(x-3)=x(x-3)$$

$$x^2-6x+18=x^2-3x$$

$$-3x=-18$$

$$x=6$$

Check: $\frac{6}{6-3}-\frac{6}{6}=1$

$$\frac{6}{3}-1=1$$

$$2-1=1$$

$$1=1$$

17.

$$1+\frac{4}{x^2}=\frac{4}{x}$$

$$x^2\cdot\left(1+\frac{4}{x^2}\right)=x^2\cdot\frac{4}{x}$$

$$x^2\cdot 1+x^2\cdot\frac{4}{x^2}=x^2\cdot\frac{4}{x}$$

$$x^2+4=4x$$

$$x^2-4x+4=0$$

$$(x-2)^2=0$$

$$x-2=0$$

$$x=2$$

Check: $1+\frac{4}{2^2}=\frac{4}{2}$

$$1+\frac{4}{4}=2$$

$$1+1=2$$

$$2=2$$

19.

$$\frac{2}{p+1}-\frac{1}{p-1}=\frac{2p}{p^2-1}$$

$$\frac{2}{p+1}-\frac{1}{p-1}=\frac{2p}{(p+1)(p-1)}$$

$$(p+1)(p-1)\cdot\frac{2}{p+1}-(p+1)(p-1)\cdot\frac{1}{p-1}$$

$$=(p+1)(p-1)\cdot\frac{2p}{(p+1)(p-1)}$$

$$\cancel{(p+1)}(p-1)\cdot\frac{2}{\cancel{p+1}}-(p+1)\cancel{(p-1)}\cdot\frac{1}{\cancel{p-1}}$$

$$=\cancel{(p+1)}\cancel{(p-1)}\cdot\frac{2p}{\cancel{(p+1)}\cancel{(p-1)}}$$

$$2(p-1)-(p+1)=2p$$

$$2p-2-p-1=2p$$

$$p-3=2p$$

$$-p=3$$

$$p=-3$$

Check: $\frac{2}{-3+1}-\frac{1}{-3-1}=\frac{2(-3)}{(-3)^2-1}$

$$\frac{2}{-2}-\frac{1}{-4}=\frac{-6}{9-1}$$

$$-1+\frac{1}{4}=-\frac{6}{8}$$

$$-\frac{4}{4}+\frac{1}{4}=-\frac{3}{4}$$

$$-\frac{3}{4}=-\frac{3}{4}$$

21. $\frac{3}{x}-\frac{1}{x+4}=\frac{5}{x^2+4x}$

$$\frac{3}{x}-\frac{1}{x+4}=\frac{5}{x(x+4)}$$

$$x(x+4)\cdot\frac{3}{x}-x(x+4)\cdot\frac{1}{x+4}=x(x+4)\cdot\frac{5}{x(x+4)}$$

$$\cancel{x}(x+4)\cdot\frac{3}{\cancel{x}}-x\cancel{(x+4)}\cdot\frac{1}{\cancel{x+4}}=\cancel{x(x+4)}\cdot\frac{5}{\cancel{x(x+4)}}$$

$$3(x+4)-x=5$$
$$3x+12-x=5$$
$$2x+12=5$$
$$2x=-7$$
$$x=-\frac{7}{2}$$

Check:

$$\frac{3}{-\frac{7}{2}}-\frac{1}{-\frac{7}{2}+4}=\frac{5}{\left(-\frac{7}{2}\right)^2+4\left(-\frac{7}{2}\right)}$$

$$\frac{3}{-\frac{7}{2}}-\frac{1}{-\frac{7}{2}+\frac{8}{2}}=\frac{5}{\frac{49}{4}-\frac{56}{4}}$$

$$\frac{3}{-\frac{7}{2}}-\frac{1}{\frac{1}{2}}=\frac{5}{-\frac{7}{4}}$$

$$-\frac{6}{7}-2=-\frac{20}{7}$$

$$-\frac{6}{7}-\frac{14}{7}=-\frac{20}{7}$$

$$-\frac{20}{7}=-\frac{20}{7}$$

23. $1-\frac{6x}{(x-4)^2}=\frac{2x}{x-4}$

$$(x-4)^2\cdot 1-(x-4)^2\cdot\frac{6x}{(x-4)^2}=(x-4)^2\cdot\frac{2x}{x-4}$$

$$(x-4)^2\cdot 1-\cancel{(x-4)^2}\cdot\frac{6x}{\cancel{(x-4)^2}}=(x-4)^{\cancel{2}}\cdot\frac{2x}{\cancel{x-4}}$$

$$(x-4)^2-6x=2x(x-4)$$
$$x^2-8x+16-6x=2x^2-8x$$
$$x^2+6x-16=0$$
$$(x-2)(x+8)=0$$
$$x-2=0 \text{ or } x+8=0$$
$$x=2 \qquad x=-8$$

Check:

$$1-\frac{6(2)}{(2-4)^2}=\frac{2(2)}{2-4}$$

$$1-\frac{12}{(-2)^2}=\frac{4}{-2}$$

$$1-\frac{12}{4}=-2$$

$$1-3=-2$$

$$-2=-2$$

$$1-\frac{6(-8)}{(-8-4)^2}=\frac{2(-8)}{-8-4}$$

$$1-\frac{-48}{(-12)^2}=\frac{-16}{-12}$$

$$1+\frac{48}{144}=\frac{-16}{-12}$$

$$1+\frac{1}{3}=\frac{4}{3}$$

$$\frac{3}{3}+\frac{1}{3}=\frac{4}{3}$$

$$\frac{4}{3}=\frac{4}{3}$$

25.
$$\frac{n+1}{n^2+2n-3}=\frac{n}{n+3}-\frac{1}{n-1}$$
$$\frac{n+1}{(n-1)(n+3)}=\frac{n}{n+3}-\frac{1}{n-1}$$
$$(n-1)(n+3)\cdot\frac{n+1}{(n-1)(n+3)}$$
$$=(n-1)(n+3)\cdot\frac{n}{n+3}$$
$$-(n-1)(n+3)\cdot\frac{1}{n-1}$$
$$\cancel{(n-1)}\,\cancel{(n+3)}\cdot\frac{n+1}{\cancel{(n-1)}\,\cancel{(n+3)}}$$
$$=(n-1)\,\cancel{(n+3)}\cdot\frac{n}{\cancel{n+3}}$$
$$-\cancel{(n-1)}\,(n+3)\cdot\frac{1}{\cancel{n-1}}$$
$$n+1=n(n-1)-(n+3)$$
$$n+1=n^2-n-n-3$$
$$0=n^2-3n-4$$
$$0=(n+1)(n-4)$$
$$n+1=0 \quad\text{or}\quad n-4=0$$
$$n=-1 \qquad n=4$$

Check:
$$\frac{-1+1}{(-1)^2+2(-1)-3}=\frac{(-1)}{(-1)+3}-\frac{1}{(-1)-1}$$
$$\frac{0}{1-2-3}=\frac{-1}{2}-\frac{1}{-2}$$
$$\frac{0}{-4}=-\frac{1}{2}+\frac{1}{2}$$
$$0=0$$
$$\frac{4+1}{(4)^2+2(4)-3}=\frac{4}{4+3}-\frac{1}{4-1}$$
$$\frac{5}{16+8-3}=\frac{4}{7}-\frac{1}{3}$$
$$\frac{5}{21}=\frac{12}{21}-\frac{7}{21}$$
$$\frac{5}{21}=\frac{5}{21}$$

27.
$$\frac{2}{b+3}=\frac{5b}{b^2-9}-\frac{3}{3-b}$$
$$\frac{2}{b+3}=\frac{5b}{(b+3)(b-3)}+\frac{3}{b-3}$$
$$(b+3)(b-3)\cdot\frac{2}{b+3}$$
$$=(b+3)(b-3)\cdot\frac{5b}{(b+3)(b-3)}$$
$$+(b+3)(b-3)\cdot\frac{3}{b-3}$$
$$\cancel{(b+3)}\,(b-3)\cdot\frac{2}{\cancel{b+3}}$$
$$=\cancel{(b+3)}\,\cancel{(b-3)}\cdot\frac{5b}{\cancel{(b+3)}\,\cancel{(b-3)}}$$
$$+(b+3)\,\cancel{(b-3)}\cdot\frac{3}{\cancel{b-3}}$$
$$2(b-3)=5b+3(b+3)$$
$$2b-6=5b+3b+9$$
$$-6b=15$$
$$b=\frac{15}{-6}$$
$$b=-\frac{5}{2}$$

Check:
$$\frac{2}{-\frac{5}{2}+3}=\frac{5\left(-\frac{5}{2}\right)}{\left(-\frac{5}{2}\right)^2-9}-\frac{3}{3-\left(-\frac{5}{2}\right)}$$
$$\frac{2}{-\frac{5}{2}+\frac{6}{2}}=\frac{-\frac{25}{2}}{\frac{25}{4}-9}-\frac{3}{\frac{6}{2}+\frac{5}{2}}$$
$$\frac{2}{\frac{1}{2}}=\frac{-\frac{25}{2}}{\frac{25}{4}-\frac{36}{4}}-\frac{3}{\frac{11}{2}}$$
$$4=\frac{-\frac{25}{2}}{-\frac{11}{4}}-\frac{6}{11}$$
$$4=\frac{50}{11}-\frac{6}{11}$$
$$4=\frac{44}{11}$$
$$4=4$$

29. $\frac{6m}{m+1}-3=\frac{8m}{(m+1)^2}$

$(m+1)^2\cdot\frac{6m}{m+1}-(m+1)^2\cdot 3$

$=(m+1)^2\cdot\frac{8m}{(m+1)^2}$

$(m+1)^{\not 2}\cdot\frac{6m}{\cancel{m+1}}-3(m+1)^2$

$=\cancel{(m+1)^2}\cdot\frac{8m}{\cancel{(m+1)^2}}$

$$6m(m+1)-3(m+1)^2=8m$$
$$6m^2+6m-3(m^2+2m+1)=8m$$
$$6m^2+6m-3m^2-6m-3=8m$$
$$3m^2-8m-3=0$$
$$(3m+1)(m-3)=0$$
$$3m+1=0 \quad\text{or}\quad m-3=0$$
$$3m=-1 \qquad m=3$$
$$m=-\frac{1}{3}$$

Check:

$$\frac{6\left(-\frac{1}{3}\right)}{-\frac{1}{3}+1}-3=\frac{8\left(-\frac{1}{3}\right)}{\left(-\frac{1}{3}+1\right)^2}$$
$$\frac{-2}{-\frac{1}{3}+\frac{3}{3}}-3=\frac{-\frac{8}{3}}{\left(-\frac{1}{3}+\frac{3}{3}\right)^2}$$
$$\frac{-2}{\frac{2}{3}}-3=\frac{-\frac{8}{3}}{\left(\frac{2}{3}\right)^2}$$
$$-3-3=\frac{-\frac{8}{3}}{\frac{4}{9}}$$
$$-6=-\frac{24}{4}$$
$$-6=-6$$

$$\frac{6(3)}{3+1}-3=\frac{8(3)}{(3+1)^2}$$
$$\frac{18}{4}-\frac{12}{4}=\frac{24}{16}$$
$$\frac{6}{4}=\frac{24}{16}$$
$$\frac{3}{2}=\frac{3}{2}$$

31. $2-\frac{2x}{x+3}=\frac{9}{x+1}$

$(x+3)(x+1)\cdot 2-(x+3)(x+1)\cdot\frac{2x}{x+3}$

$=(x+3)(x+1)\cdot\frac{9}{x+1}$

$2(x+3)(x+1)-\cancel{(x+3)}(x+1)\cdot\frac{2x}{\cancel{x+3}}$

$=(x+3)\cancel{(x+1)}\cdot\frac{9}{\cancel{x+1}}$

$$2(x+3)(x+1)-2x(x+1)=9(x+3)$$
$$2(x^2+4x+3)-2x^2-2x=9x+27$$
$$2x^2+8x+6-2x^2-2x=9x+27$$
$$-3x=21$$
$$x=-7$$

Check: $2-\frac{2(-7)}{-7+3}=\frac{9}{-7+1}$

$$2-\frac{-14}{-4}=\frac{9}{-6}$$
$$\frac{4}{2}-\frac{7}{2}=-\frac{3}{2}$$
$$-\frac{3}{2}=-\frac{3}{2}$$

33. $\frac{t}{30}+\frac{t}{45}=1$

$$90\cdot\left(\frac{t}{30}+\frac{t}{45}\right)=90\cdot 1$$
$$90\cdot\frac{t}{30}+90\cdot\frac{t}{45}=90\cdot 1$$
$$3t+2t=90$$
$$5t=90$$
$$t=18$$

It will take them 18 min to clean the attic.

35. $\frac{3}{x}+\frac{3}{\frac{x}{4}}=1$

$\frac{3}{x}+\frac{12}{x}=1$

$\frac{15}{x}=1$

$15=x$

It would take the clerical worker 15 hr to finish the job working alone.

37. $\frac{60}{2x}+\frac{60}{x}=3$

$\frac{30}{x}+\frac{60}{x}=3$

$\frac{90}{x}=3$

$90=3x$

$30=x$

$2x=2\cdot 30=60$

The speed on the dry road was 60 mph.

39. $p=\frac{P}{LD}$

$LD\cdot p=LD\cdot\frac{P}{LD}$

$LDp=P$

$\frac{LDp}{Lp}=\frac{P}{Lp}$

$D=\frac{P}{Lp}$

7.6 Ratio and Proportion; Variation

Practice

1. $\frac{4}{5}=\frac{p}{10}$

$4\cdot 10=5p$

$40=5p$

$\frac{40}{5}=\frac{5p}{5}$

$8=p$

Check: $\frac{4}{5}=\frac{8}{10}$

$\frac{4}{5}=\frac{4}{5}$

2. $\frac{x}{49}=\frac{80}{98}$

$98x=(49)(90)$

$\frac{98x}{98}=\frac{3920}{98}$

$x=40$

40 lb of sodium hydroxide are needed to neutralize 49 lb of sulfuric acid.

3. $\frac{900-x}{75{,}000}=\frac{900}{100{,}000}$

$100{,}000(900-x)=900\cdot 75{,}000$

$90{,}000{,}000-100{,}000x=67{,}500{,}000$

$-100{,}000x=-22{,}500{,}000$

$x=\frac{-22{,}500{,}000}{-100{,}000}$

$x=225$

She would be paying \$225 less if she had a \$75,000 mortgage at the same interest rate.

4. $\frac{8}{5+5}=\frac{x}{5}$

$\frac{8}{10}=\frac{x}{5}$

$10x=40$

$x=4$

The length of $\overline{DE}$ is 4 in.

5. $\frac{1000}{x+300}=\frac{250}{x-300}$

$1000(x-300)=250(x+300)$

$1000x-300{,}000=250x+75{,}000$

$750x=375{,}000$

$x=500$

The speed of the plane in still air is 500 mph.

6. $y = kx$

$16 = k(20)$

$\frac{16}{20} = k$

$k = \frac{4}{5};\ y = \frac{4}{5}x$

7. $A = kM$

$160 = k(40)$

$\frac{160}{40} = k$

$k = 4$

$A = 4M$

$A = 4(55)$

$A = 220$

220 mg of the drug should be administered.

8. $y = \frac{k}{x}$

$1.6 = \frac{k}{30}$

$k = 48;\ y = \frac{48}{x}$

9. $f = \frac{k}{a}$

$2 = \frac{k}{25}$

$k = 50$

$f = \frac{50}{a}$

$f = \frac{50}{12.5}$

$f = 4$

The *f*-stop of the lens is 4.

10. $y = kxz$

$150 = k(15)(20)$

$150 = 300k$

$\frac{150}{300} = k$

$k = \frac{1}{2};\ y = \frac{1}{2}xz$

11. $I = prt$

$140 = 2000(0.035)t$

$140 = 70t$

$\frac{140}{70} = t$

$t = 2$

$I = 2pr$

$180 = 2(p)(0.04)$

$180 = 0.08p$

$\frac{180}{0.08} = p$

$p = 2250$

The other employee must invest \$2250.

12. **a.** $w = \frac{kxy}{z^2}$

b. $8 = \frac{k(2)(6)}{(6)^2}$

$8 = \frac{12k}{36}$

$288 = 12k$

$\frac{288}{12} = k$

$k = 24$

c. $w = \frac{24xy}{z^2}$

$w = \frac{24(2)(3)}{(4)^2}$

$w = \frac{144}{16}$

$w = 9$

13. $BMI = \frac{kw}{h^2}$

$$23 = \frac{k(165)}{(70)^2}$$
$$112,700 = 165k$$
$$\frac{112,700}{165} = k$$
$$k \approx 683$$
$$BMI = \frac{683w}{h^2}$$
$$BMI = \frac{683(120)}{(65)^2}$$
$$BMI \approx 19$$

The BMI is approximately 19.

Exercises

1. A(n) ratio is a comparison of two numbers, written as a quotient.

3. A(n) proportion is a statement that two ratios $\frac{a}{b}$ and $\frac{c}{d}$ are equal, and is written $\frac{a}{b} = \frac{c}{d}$, where $b \neq 0$ and $d \neq 0$.

5. The number k is called the constant of variation.

7. The relationship is a(n) inverse variation.

9. $\frac{x}{10} = \frac{4}{5}$

$$5x = 10 \cdot 4$$
$$5x = 40$$
$$\frac{5x}{5} = \frac{40}{5}$$
$$x = 8$$

Check: $\frac{8}{10} = \frac{4}{5}$

$$\frac{4}{5} = \frac{4}{5}$$

11. $\frac{n}{100} = \frac{4}{5}$

$$5n = 100 \cdot 4$$
$$5n = 400$$
$$\frac{5n}{5} = \frac{400}{5}$$
$$n = 80$$

Check: $\frac{80}{100} = \frac{4}{5}$

$$\frac{4}{5} = \frac{4}{5}$$

13. $\frac{8}{7} = \frac{s}{21}$

$$8 \cdot 21 = 7s$$
$$168 = 7s$$
$$\frac{168}{7} = \frac{7s}{7}$$
$$\frac{168}{7} = s$$
$$24 = s$$

Check: $\frac{8}{7} = \frac{24}{21}$

$$\frac{8}{7} = \frac{8}{7}$$

15. $\frac{8+x}{12} = \frac{22}{36}$

$$36(8+x) = 12 \cdot 22$$
$$288 + 36x = 264$$
$$36x = -24$$
$$\frac{36x}{36} = -\frac{24}{36}$$
$$x = -\frac{2}{3}$$

Check: $\frac{8 + \left(-\frac{2}{3}\right)}{12} = \frac{22}{36}$

$$\frac{\frac{24}{3} - \frac{2}{3}}{12} = \frac{11}{18}$$
$$\frac{\frac{22}{3}}{12} = \frac{11}{18}$$
$$\frac{22}{36} = \frac{11}{18}$$

17.
$$\frac{y+3}{14}=\frac{y}{7}$$
$$7(y+3)=14y$$
$$7y+21=14y$$
$$-7y=-21$$
$$\frac{-7y}{-7}=\frac{-21}{-7}$$
$$y=3$$

Check: $\frac{3+3}{14}=\frac{3}{7}$
$$\frac{6}{14}=\frac{3}{7}$$
$$\frac{3}{7}=\frac{3}{7}$$

19.
$$\frac{x-1}{8}=\frac{x+1}{12}$$
$$12(x-1)=8(x+1)$$
$$12x-12=8x+8$$
$$4x=20$$
$$x=5$$

Check: $\frac{5-1}{8}=\frac{5+1}{12}$
$$\frac{4}{8}=\frac{6}{12}$$
$$\frac{1}{2}=\frac{1}{2}$$

21.
$$\frac{x}{8}=\frac{2}{x}$$
$$x^2=16$$
$$x^2-16=0$$
$$(x+4)(x-4)=0$$
$$x+4=0 \text{ or } x-4=0$$
$$x=-4 \qquad x=4$$

Check:
$$\frac{-4}{8}=\frac{2}{-4} \qquad \frac{4}{8}=\frac{2}{4}$$
$$-\frac{1}{2}=-\frac{1}{2} \qquad \frac{1}{2}=\frac{1}{2}$$

23.
$$\frac{2}{y}=\frac{y-4}{16}$$
$$32=y(y-4)$$
$$32=y^2-4y$$
$$y^2-4y-32=0$$
$$(y+4)(y-8)=0$$
$$y+4=0 \text{ or } y-8=0$$
$$y=-4 \qquad y=8$$

Check:
$$\frac{2}{-4}=\frac{-4-4}{16} \qquad \frac{2}{8}=\frac{8-4}{16}$$
$$-\frac{1}{2}=\frac{-8}{16} \qquad \frac{1}{4}=\frac{4}{16}$$
$$-\frac{1}{2}=-\frac{1}{2} \qquad \frac{1}{4}=\frac{1}{4}$$

25.
$$\frac{a}{a+3}=\frac{4}{5a}$$
$$5a^2=4(a+3)$$
$$5a^2=4a+12$$
$$5a^2-4a-12=0$$
$$(5a+6)(a-2)=0$$
$$5a+6=0 \text{ or } a-2=0$$
$$5a=-6 \qquad a=2$$
$$a=-\frac{6}{5}$$

Check:
$$\frac{-\frac{6}{5}}{-\frac{6}{5}+3}=\frac{4}{5\left(-\frac{6}{5}\right)}$$
$$\frac{-\frac{6}{5}}{-\frac{6}{5}+\frac{15}{5}}=\frac{4}{-6}$$
$$\frac{-\frac{6}{5}}{\frac{9}{5}}=-\frac{2}{3}$$
$$-\frac{2}{3}=-\frac{2}{3}$$

25. (continued)
$$\frac{2}{2+3}=\frac{4}{5(2)}$$
$$\frac{2}{5}=\frac{4}{10}$$
$$\frac{2}{5}=\frac{2}{5}$$

27.
$$\frac{y+1}{y+6}=\frac{y}{y+6}$$
$$(y+1)(y+6)=y(y+6)$$
$$y^2+7y+6=y^2+6y$$
$$y+6=0$$
$$y=-6$$

Check: $\dfrac{-6+1}{-6+6}=\dfrac{-6}{-6+6}$

$\dfrac{-5}{0}=\dfrac{-6}{0}$ Undefined

No solution

29. Decreases; inverse variation.

31. Increases; direct variation.

33. Increases; direct variation.

35.
$$y=kx$$
$$48=k(16)$$
$$k=\frac{48}{16}=3$$
$$y=3x$$

37.
$$y=kx$$
$$6=k(36)$$
$$k=\frac{6}{36}=\frac{1}{6}$$
$$y=\frac{1}{6}x$$

39.
$$y=kx$$
$$3=k\left(\frac{1}{3}\right)$$
$$k=3\left(\frac{3}{1}\right)=9$$
$$y=9x$$

41.
$$y=kx$$
$$0.9=k(0.6)$$
$$k=\frac{0.9}{0.6}=\frac{3}{2}$$
$$y=\frac{3}{2}x$$

43.
$$y=\frac{k}{x}$$
$$13=\frac{k}{3}$$
$$k=13(3)=39$$
$$y=\frac{39}{x}$$

45.
$$y=\frac{k}{x}$$
$$1.8=\frac{k}{15}$$
$$k=1.8(15)=27$$
$$y=\frac{27}{x}$$

47.
$$y=\frac{k}{x}$$
$$0.7=\frac{k}{0.4}$$
$$k=0.7(0.4)=0.28=\frac{28}{100}=\frac{7}{25}$$
$$y=\frac{\frac{7}{25}}{x}=\frac{7}{25x}$$

49.
$$y=\frac{k}{x}$$
$$27=\frac{k}{\frac{2}{3}}$$
$$k=27\left(\frac{2}{3}\right)=18$$
$$y=\frac{18}{x}$$

51.
$$\begin{aligned} y &= kxz \\ 160 &= k(10)(4) \\ 160 &= k(40) \\ k &= \frac{160}{40} = 4 \\ y &= 4xz \end{aligned}$$

53.
$$\begin{aligned} y &= kxz \\ 360 &= k(25)(12) \\ 360 &= k(300) \\ k &= \frac{360}{300} = \frac{6}{5} \\ y &= \frac{6}{5}xz \end{aligned}$$

55.
$$\begin{aligned} y &= kxz \\ 63 &= k(4.2)(5) \\ 63 &= k(21) \\ k &= \frac{63}{21} = 3 \\ y &= 3xz \end{aligned}$$

57.
$$\begin{aligned} y &= kxz \\ 4.5 &= k(0.6)(0.3) \\ 4.5 &= k(0.18) \\ k &= \frac{4.5}{0.18} = 25 \\ y &= 25xz \end{aligned}$$

59.
$$\begin{aligned} y &= \frac{kx}{z^2} \\ 20 &= \frac{k(4)}{(5)^2} \\ 20 &= \frac{k(4)}{25} \\ k &= 20\left(\frac{25}{4}\right) = 125 \\ y &= \frac{125x}{z^2} \end{aligned}$$

61.
$$\begin{aligned} y &= \frac{k}{xz^2} \\ 100 &= \frac{k}{(20)(0.5)^2} \\ 100 &= \frac{k}{(20)(0.25)} = \frac{k}{5} \\ 100 &= \frac{k}{5} \\ k &= 100(5) = 500 \\ y &= \frac{500}{xz^2} \end{aligned}$$

63.
$$\begin{aligned} y &= \frac{kxw}{z^2} \\ 130 &= \frac{k(13)(16)}{(0.4)^2} \\ 130 &= \frac{k(208)}{0.16} = k(1300) \\ 130 &= k(1300) \\ k &= \frac{130}{1300} = \frac{1}{10} \\ y &= \frac{\frac{1}{10}xw}{z^2} \\ y &= \frac{xw}{10z^2} \end{aligned}$$

65.
$$\begin{aligned} \frac{3}{r} &= \frac{r-4}{7} \\ 21 &= r(r-4) \\ 21 &= r^2 - 4r \\ r^2 - 4r - 21 &= 0 \\ (r-7)(r+3) &= 0 \end{aligned}$$

$r - 7 = 0$ or $r + 3 = 0$

$r = 7 \qquad r = -3$

Check:

$$\frac{3}{7} = \frac{7-4}{7} \qquad \frac{3}{7} = \frac{3}{7}$$

$$\frac{3}{-3} = \frac{-3-4}{7} \qquad -1 = \frac{-7}{7} \qquad -1 = -1$$

67.
$$\frac{4}{w-3}=\frac{7}{w+3}$$
$$4(w+3)=7(w-3)$$
$$4w+12=7w-21$$
$$-3w=-33$$
$$w=11$$

Check: $\frac{4}{11-3}=\frac{7}{11+3}$
$$\frac{4}{8}=\frac{7}{14}$$
$$\frac{1}{2}=\frac{1}{2}$$

69.
$$\frac{7}{m}=\frac{63}{6}$$
$$7\cdot 6=63m$$
$$42=63m$$
$$\frac{42}{63}=\frac{63m}{63}$$
$$\frac{2}{3}=m$$

Check: $\frac{7}{\frac{2}{3}}=\frac{63}{6}$
$$\frac{21}{2}=\frac{21}{2}$$

71.
$$y=k\frac{xw}{z^3}$$
$$15=k\left(\frac{\frac{1}{2}\bullet 8}{2^3}\right)$$
$$15=k\left(\frac{4}{8}\right)$$
$$15=k\left(\frac{1}{2}\right)$$
$$30=k$$
$$y=\frac{30xw}{z^3}$$

73.
$$y=kx$$
$$\frac{3}{7}=k\left(\frac{9}{14}\right)$$
$$\frac{3}{7}\bullet\frac{14}{9}=k$$
$$k=\frac{2}{3}$$
$$y=\frac{2}{3}x$$

75. $y=\frac{k}{x}$
$$36=\frac{k}{\frac{2}{3}}$$
$$\frac{2}{3}\cdot 36=k$$
$$k=24;\ y=\frac{24}{x}$$

77. Decreases; inverse variation

79.
$$\frac{6}{2}=\frac{25}{x}$$
$$6x=50$$
$$x=\frac{50}{6}=8\frac{1}{3}$$

It would take $8\frac{1}{3}$ min (or 8 min and 20 sec) to print a 25-page report.

81.
$$\frac{60}{8}=\frac{120}{x}$$
$$60x=8\cdot 120$$
$$60x=960$$
$$x=16$$

It will take 16 gal of gas to drive 120 mi.

83.
$$\frac{10}{s}=\frac{3}{s-14}$$
$$10(s-14)=3s$$
$$10s-140=3s$$
$$7s=140$$
$$s=20$$

The cyclist's speed was 20 mph.

85.
$$\frac{400}{s+30}=\frac{250}{s}$$
$$400s=250(s+30)$$
$$400s=250s+7500$$
$$150s=7500$$
$$s=50$$
$$s+30=50+30=80$$

The speed of the bus is 50 mph and the speed of the train is 80 mph.

87.
$$\frac{y}{20}=\frac{10}{25}$$
$$25y=200$$
$$y=8$$
$$AB=8\text{ ft}$$

89.
$$\frac{6}{4}=\frac{h}{4+8}$$
$$\frac{3}{2}=\frac{h}{12}$$
$$36=2h$$
$$18=h$$

The height of the tree is 18 ft.

91.
$$\frac{w+4}{20+4}=\frac{1}{2}$$
$$\frac{w+4}{24}=\frac{1}{2}$$
$$2(w+4)=24$$
$$2w+8=24$$
$$2w=16$$
$$w=8$$

There are 8 women at the party.

93. **a.**
$$A=ki$$
$$1080=k(36000)$$
$$k=\frac{1080}{36000}=\frac{3}{100}=0.03$$
$$A=\frac{3}{100}i\text{ or }A=0.03i$$

The constant of variation represents the income tax rate, which is 3%.

b.
$$A=0.03i$$
$$A=0.03(26500)=795$$

A person will pay $795 in income tax.

95.
$$w=\frac{k}{f}$$
$$5.1=\frac{k}{300}$$
$$k=5.1(300)=1530$$
$$w=\frac{1530}{f}$$
$$w=\frac{1530}{500}=3.06$$

The wavelength of the 500 Hz sound is 3.06m.

97.
$$E=kmv^2$$
$$113.6=k(0.142)(40)^2$$
$$113.6=k(227.2)$$
$$k=\frac{113.6}{227.2}=\frac{1}{2}$$
$$E=\frac{1}{2}mv^2$$
$$E=\frac{1}{2}(0.142)(20)^2=28.4$$

The baseball has an energy of 28.4 J.

99.
$$I=\frac{k}{d^2}$$
$$21.6=\frac{k}{2^2}$$
$$k=21.6(2^2)=86.4$$
$$I=\frac{86.4}{d^2}$$
$$I=\frac{86.4}{4^2}=\frac{86.4}{16}=5.4$$

The illumination 4 m above the table is 5.4 lumens per sq m.

Chapter 7 Review Exercises

1. **a.** $\dfrac{4}{x+1}$

$$x+1=0$$
$$x=-1$$

b. $\dfrac{6x+12}{x^2-x-6}$

$$x^2-x-6=0$$
$$(x+2)(x-3)=0$$
$$x+2=0 \quad \text{or} \quad x-3=0$$
$$x=-2 \qquad\qquad x=3$$

2. **a.** $\dfrac{2x}{y}\cdot\dfrac{5xy}{5xy}=\dfrac{10x^2y}{5xy^2}$

Equivalent

b.

$$\frac{x-3}{x+3}\cdot\frac{x+3}{x+3}=\frac{(x-3)(x+3)}{(x+3)^2}=\frac{x^2-9}{x^2+6x+9}$$

Equivalent

3. $\dfrac{12m}{20m^2}=\dfrac{4m(3)}{4m(5m)}=\dfrac{3}{5m}$

4. $\dfrac{15n-18}{9n+6}=\dfrac{3(5n-6)}{3(3n+2)}=\dfrac{5n-6}{3n+2}$

5. $\dfrac{x^2+2x-8}{4-x^2}=\dfrac{(x-2)(x+4)}{(2-x)(2+x)}$

$$=-\frac{(x-2)(x+4)}{(x-2)(x+2)}$$
$$=-\frac{x+4}{x+2}$$

6. $\dfrac{2x^2-3x-20}{3x^2-13x+4}=\dfrac{(2x+5)(x-4)}{(3x-1)(x-4)}=\dfrac{2x+5}{3x-1}$

7. $\dfrac{10mn}{3p^2}\cdot\dfrac{9np}{5m^2}=\dfrac{\overset{2}{\cancel{10}}\,\cancel{m}n}{\underset{1}{\cancel{3}}\,p^{\cancel{2}}}\cdot\dfrac{\overset{3}{\cancel{9}}\,n\cancel{p}}{\underset{1}{\cancel{5}}\,m^{\cancel{2}}}=\dfrac{6n^2}{pm}$

8. $\dfrac{y-5}{4y+6}\cdot\dfrac{6y+9}{3y-15}=\dfrac{y-5}{2(2y+3)}\cdot\dfrac{3(2y+3)}{3(y-5)}$

$$=\frac{\cancel{y-5}}{2\cancel{(2y+3)}}\cdot\frac{\cancel{3}\cancel{(2y+3)}}{\cancel{3}\cancel{(y-5)}}$$
$$=\frac{1}{2}$$

9. $\dfrac{x+6}{x^2+x-30}\cdot\dfrac{x^2-10x+25}{2x+5}$

$$=\frac{x+6}{(x-5)(x+6)}\cdot\frac{(x-5)^2}{2x+5}$$
$$=\frac{\cancel{x+6}}{\cancel{(x-5)}\cancel{(x+6)}}\cdot\frac{(x-5)^{\cancel{2}}}{2x+5}$$
$$=\frac{x-5}{2x+5}$$

10. $\dfrac{2a^2-2a-4}{4-a^2}\cdot\dfrac{2a^2+a-6}{4a^2-2a-6}$

$$=\frac{2(a^2-a-2)}{(2-a)(2+a)}\cdot\frac{2a^2+a-6}{2(2a^2-a-3)}$$
$$=-\frac{2(a-2)(a+1)}{(a-2)(a+2)}\cdot\frac{(2a-3)(a+2)}{2(2a-3)(a+1)}$$
$$=-\frac{\cancel{2}\cancel{(a-2)}\cancel{(a+1)}}{\cancel{(a-2)}\cancel{(a+2)}}\cdot\frac{\cancel{(2a-3)}\cancel{(a+2)}}{\cancel{2}\cancel{(2a-3)}\cancel{(a+1)}}$$
$$=-1$$

11. $\dfrac{x^2y}{2x}\div xy^2=\dfrac{x^2y}{2x}\cdot\dfrac{1}{xy^2}=\dfrac{\cancel{x^2}\cancel{y}}{2\cancel{x}}\cdot\dfrac{1}{\cancel{x}y^{\cancel{2}}}=\dfrac{1}{2y}$

12. $\dfrac{5m+10}{2m-20}\div\dfrac{7m+14}{14m-20}$

$$=\frac{5m+10}{2m-20}\cdot\frac{14m-20}{7m+14}$$
$$=\frac{5(m+2)}{2(m-10)}\cdot\frac{2(7m-10)}{7(m+2)}$$
$$=\frac{5\cancel{(m+2)}}{\cancel{2}(m-10)}\cdot\frac{\cancel{2}(7m-10)}{7\cancel{(m+2)}}$$
$$=\frac{5(7m-10)}{7(m-10)}$$

13. $\frac{5y^2}{x^2-36} \div \frac{25xy-25y}{x^2-7x+6}$

$= \frac{5y^2}{x^2-36} \cdot \frac{x^2-7x+6}{25xy-25y}$

$= \frac{5y^2}{(x-6)(x+6)} \cdot \frac{(x-1)(x-6)}{25y(x-1)}$

$= \frac{\overset{1}{\cancel{5}} y^{\cancel{2}}}{\cancel{(x-6)}(x+6)} \cdot \frac{\cancel{(x-1)}\cancel{(x-6)}}{\underset{5}{\cancel{25}}\cancel{y}\cancel{(x-1)}}$

$= \frac{y}{5(x+6)}$

14. $\frac{2x^2+x-1}{x^2+8x+7} \div \frac{6x^2+x-2}{x^2+14x+49}$

$= \frac{2x^2+x-1}{x^2+8x+7} \cdot \frac{x^2+14x+49}{6x^2+x-2}$

$= \frac{(2x-1)(x+1)}{(x+1)(x+7)} \cdot \frac{(x+7)^2}{(2x-1)(3x+2)}$

$= \frac{\cancel{(2x-1)}\cancel{(x+1)}}{\cancel{(x+1)}\cancel{(x+7)}} \cdot \frac{(x+7)^{\cancel{2}}}{\cancel{(2x-1)}(3x+2)}$

$= \frac{x+7}{3x+2}$

15. Factor $5x: 5 \cdot x$

Factor $20x^2: 2 \cdot 2 \cdot 5 \cdot x \cdot x$

$\text{LCD} = 20x^2$

$\frac{1}{5x} \cdot \frac{4x}{4x} = \frac{4x}{20x^2}$

$\frac{3}{20x^2} = \frac{3}{20x^2}$

16. Factor $n-1: n-1$

Factor $n+4: n+4$

$\text{LCD} = (n-1)(n+4)$

$\frac{4}{n-1} \cdot \frac{n+4}{n+4} = \frac{4(n+4)}{(n-1)(n+4)}$

$\frac{n}{n+4} \cdot \frac{n-1}{n-1} = \frac{n(n-1)}{(n-1)(n+4)}$

17. Factor $3x+9: 3(x+3)$

Factor $x^2+4x+3: (x+1)(x+3)$

$\text{LCD} = 3(x+3)(x+1)$

$\frac{1}{3x+9} \cdot \frac{(x+1)}{(x+1)} = \frac{x+1}{3(x+3)(x+1)}$

$\frac{x}{x^2+4x+3} \cdot \frac{3}{3} = \frac{3x}{3(x+3)(x+1)}$

18. Factor $3x^2-5x-2: (3x+1)(x-2)$

Factor $4-x^2: (2-x)(2+x) = -(x-2)(x+2)$

$\text{LCD} = -(3x+1)(x-2)(x+2)$

$\frac{2}{3x^2-5x-2} = \frac{2}{(3x+1)(x-2)} \cdot \frac{x+2}{x+2}$

$= \frac{2(x+2)}{(3x+1)(x-2)(x+2)}$

$\frac{1}{4-x^2} = -\frac{1}{x^2-4}$

$= -\frac{1}{(x-2)(x+2)} \cdot \frac{3x+1}{3x+1}$

$= -\frac{3x+1}{(3x+1)(x-2)(x+2)}$

19. $\frac{3t+1}{2t} + \frac{t-1}{2t} = \frac{(3t+1)+(t-1)}{2t}$

$= \frac{3t+1+t-1}{2t}$

$= \frac{4t}{2t}$

$= 2$

20. $\frac{5y}{y+7} - \frac{y-28}{y+7} = \frac{5y-(y-28)}{y+7}$

$= \frac{5y-y+28}{y+7}$

$= \frac{4y+28}{y+7}$

$= \frac{4(y+7)}{y+7}$

$= 4$

21.
$$\frac{5y+4}{4y^2-2y}-\frac{2}{2y-1}$$
$$=\frac{5y+4}{2y(2y-1)}-\frac{2}{2y-1}$$
$$=\frac{5y+4}{2y(2y-1)}-\frac{2}{2y-1}\cdot\frac{2y}{2y}$$
$$=\frac{5y+4-4y}{2y(2y-1)}$$
$$=\frac{y+4}{2y(2y-1)}$$

22.
$$\frac{n}{3n+15}+\frac{n-2}{n^2+5n}$$
$$=\frac{n}{3(n+5)}+\frac{n-2}{n(n+5)}$$
$$=\frac{n}{3(n+5)}\cdot\frac{n}{n}+\frac{n-2}{n(n+5)}\cdot\frac{3}{3}$$
$$=\frac{n^2+3(n-2)}{3n(n+5)}$$
$$=\frac{n^2+3n-6}{3n(n+5)}$$

23.
$$\frac{4}{x-3}-\frac{4x+1}{9-x^2}$$
$$=\frac{4}{x-3}-\frac{4x+1}{(3-x)(3+x)}$$
$$=\frac{4}{x-3}+\frac{4x+1}{(x-3)(x+3)}$$
$$=\frac{4}{x-3}\cdot\frac{x+3}{x+3}+\frac{4x+1}{(x-3)(x+3)}$$
$$=\frac{4(x+3)+4x+1}{(x-3)(x+3)}$$
$$=\frac{4x+12+4x+1}{(x-3)(x+3)}$$
$$=\frac{8x+13}{(x-3)(x+3)}$$

24.
$$\frac{y+3}{4-y^2}+\frac{1}{2-y}$$
$$=\frac{y+3}{(2-y)(2+y)}+\frac{1}{2-y}$$
$$=\frac{y+3}{(2-y)(2+y)}+\frac{1}{2-y}\cdot\frac{2+y}{2+y}$$
$$=\frac{y+3+2+y}{(2-y)(2+y)}$$
$$=\frac{2y+5}{(2-y)(2+y)}$$

25.
$$\frac{2}{m+1}+\frac{6m-2}{m^2-2m-3}$$
$$=\frac{2}{m+1}+\frac{6m-2}{(m+1)(m-3)}$$
$$=\frac{2}{m+1}\cdot\frac{m-3}{m-3}+\frac{6m-2}{(m+1)(m-3)}$$
$$=\frac{2(m-3)+6m-2}{(m+1)(m-3)}$$
$$=\frac{2m-6+6m-2}{(m+1)(m-3)}$$
$$=\frac{8m-8}{(m+1)(m-2)}$$
$$=\frac{8(m-1)}{(m+1)(m-2)}$$

26.
$$\frac{3x-2}{x^2-x-12}-\frac{x+3}{x-4}$$
$$=\frac{3x-2}{(x+3)(x-4)}-\frac{x+3}{x-4}$$
$$=\frac{3x-2}{(x+3)(x-4)}-\frac{x+3}{x-4}\cdot\frac{x+3}{x+3}$$
$$=\frac{3x-2-(x+3)^2}{(x+3)(x-4)}$$
$$=\frac{(3x-2)-(x^2+6x+9)}{(x+3)(x-4)}$$
$$=\frac{3x-2-x^2-6x-9}{(x+3)(x-4)}$$
$$=\frac{-x^2-3x-11}{(x+3)(x-4)}$$

27. $$\frac{2x}{x^2+4x+4}-\frac{x-1}{x^2-2x-8}$$
$$=\frac{2x}{x^2+4x+4}-\frac{x-1}{x^2-2x-8}$$
$$=\frac{2x}{(x+2)^2}-\frac{x-1}{(x+2)(x-4)}$$
$$=\frac{2x}{(x+2)^2}\cdot\frac{x-4}{x-4}-\frac{x-1}{(x+2)(x-4)}\cdot\frac{x+2}{x+2}$$
$$=\frac{2x(x-4)-(x-1)(x+2)}{(x+2)^2(x-4)}$$
$$=\frac{2x^2-8x-(x^2+x-2)}{(x+2)^2(x-4)}$$
$$=\frac{2x^2-8x-x^2-x+2}{(x+2)^2(x-4)}$$
$$=\frac{x^2-9x+2}{(x+2)^2(x-4)}$$

28. $$\frac{n+4}{2n^2-3n+1}+\frac{n+1}{2n^2+5n-3}$$
$$=\frac{n+4}{(2n-1)(n-1)}+\frac{n+1}{(2n-1)(n+3)}$$
$$=\frac{n+4}{(2n-1)(n-1)}\cdot\frac{n+3}{n+3}+\frac{n+1}{(2n-1)(n+3)}\cdot\frac{n-1}{n-1}$$
$$=\frac{(n+4)(n+3)+(n+1)(n-1)}{(2n-1)(n-1)(n+3)}$$
$$=\frac{n^2+7n+12+n^2-1}{(2n-1)(n-1)(n+3)}$$
$$=\frac{2n^2+7n+11}{(2n-1)(n-1)(n+3)}$$

29. $$\frac{\frac{x}{2}}{\frac{3x^2}{7}}=\frac{x}{2}\div\frac{3x^2}{7}$$
$$=\frac{x}{2}\cdot\frac{7}{3x^2}$$
$$=\frac{\cancel{x}}{2}\cdot\frac{7}{3x^{\cancel{2}}}$$
$$=\frac{7}{6x}$$

30. $$\frac{1-\frac{9}{y}}{1-\frac{81}{y^2}}=\frac{\frac{y-9}{y}}{\frac{y^2-81}{y^2}}$$
$$=\frac{y-9}{y}\div\frac{y^2-81}{y^2}$$
$$=\frac{y-9}{y}\cdot\frac{y^2}{y^2-81}$$
$$=\frac{\cancel{y-9}}{\cancel{y}}\cdot\frac{y^{\cancel{2}}}{\cancel{(y-9)}(y+9)}$$
$$=\frac{y}{y+9}$$

31. $$\frac{\frac{1}{x}+\frac{1}{y}}{\frac{1}{2x}+\frac{1}{2y}}=\frac{\left(\frac{1}{x}+\frac{1}{y}\right)\cdot 2xy}{\left(\frac{1}{2x}+\frac{1}{2y}\right)\cdot 2xy}$$
$$=\frac{\frac{1}{x}\cdot 2xy+\frac{1}{y}\cdot 2xy}{\frac{1}{2x}\cdot 2xy+\frac{1}{2y}\cdot 2xy}$$
$$=\frac{2y+2x}{y+x}$$
$$=\frac{2(y+x)}{y+x}$$
$$=\frac{2\cancel{(y+x)}}{\cancel{y+x}}$$
$$=2$$

32.
$$\frac{4-\frac{3}{x}-\frac{1}{x^2}}{2-\frac{5}{x}+\frac{3}{x^2}}=\frac{\left(4-\frac{3}{x}-\frac{1}{x^2}\right)\cdot x^2}{\left(2-\frac{5}{x}+\frac{3}{x^2}\right)\cdot x^2}$$
$$=\frac{4\cdot x^2-\frac{3}{x}\cdot x^2-\frac{1}{x^2}\cdot x^2}{2\cdot x^2-\frac{5}{x}\cdot x^2+\frac{3}{x^2}\cdot x^2}$$
$$=\frac{4x^2-3x-1}{2x^2-5x+3}$$
$$=\frac{(4x+1)(x-1)}{(2x-3)(x-1)}$$
$$=\frac{(4x+1)\cancel{(x-1)}}{(2x-3)\cancel{(x-1)}}$$
$$=\frac{4x+1}{2x-3}$$

33.
$$\frac{2x}{x-4}=5-\frac{1}{x-4}$$
$$(x-4)\cdot\frac{2x}{x-4}=(x-4)\cdot 5-(x-4)\cdot\frac{1}{x-4}$$
$$\cancel{(x-4)}\cdot\frac{2x}{\cancel{x-4}}=(x-4)\cdot 5-\cancel{(x-4)}\cdot\frac{1}{\cancel{x-4}}$$
$$2x=5(x-4)-1$$
$$2x=5x-20-1$$
$$-3x=-21$$
$$x=7$$

Check: $\frac{2(7)}{7-4}=5-\frac{1}{7-4}$
$$\frac{14}{3}=5-\frac{1}{3}$$
$$\frac{14}{3}=\frac{15}{3}-\frac{1}{3}$$
$$\frac{14}{3}=\frac{14}{3}$$

34.
$$\frac{y+1}{y}+\frac{1}{2y}=4$$
$$2y\cdot\left(\frac{y+1}{y}+\frac{1}{2y}\right)=2y\cdot 4$$
$$2y\cdot\frac{y+1}{y}+2y\cdot\frac{1}{2y}=2y\cdot 4$$
$$2(y+1)+1=8y$$
$$2y+2+1=8y$$
$$-6y=-3$$
$$y=\frac{1}{2}$$

Check: $\frac{\frac{1}{2}+1}{\frac{1}{2}}+\frac{1}{2\left(\frac{1}{2}\right)}=4$
$$\frac{\frac{3}{2}}{\frac{1}{2}}+\frac{1}{1}=4$$
$$3+1=4$$
$$4=4$$

35.
$$\frac{5}{2x}+\frac{3}{x+1}=\frac{7}{x}$$
$$2x(x+1)\cdot\frac{5}{2x}+2x(x+1)\cdot\frac{3}{x+1}=2x(x+1)\cdot\frac{7}{x}$$
$$\cancel{2x}(x+1)\cdot\frac{5}{\cancel{2x}}+2x\cancel{(x+1)}\cdot\frac{3}{\cancel{x+1}}=2\cancel{x}(x+1)\cdot\frac{7}{\cancel{x}}$$
$$5(x+1)+6x=14(x+1)$$
$$5x+5+6x=14x+14$$
$$-3x=9$$
$$x=-3$$

Check: $\frac{5}{2(-3)}+\frac{3}{-3+1}=\frac{7}{-3}$
$$\frac{5}{-6}+\frac{3}{-2}=-\frac{7}{3}$$
$$-\frac{5}{6}-\frac{9}{6}=-\frac{7}{3}$$
$$-\frac{14}{6}=-\frac{7}{3}$$
$$-\frac{7}{3}=-\frac{7}{3}$$

36. $\frac{y-2}{y-4}=\frac{1}{y+2}+\frac{y+3}{y^2-2y-8}$

$\frac{y-2}{y-4}=\frac{1}{y+2}+\frac{y+3}{(y+2)(y-4)}$

$(y+2)(y-4)\cdot\frac{y-2}{y-4}$

$=(y+2)(y-4)\cdot\frac{1}{y+2}$

$+(y+2)(y-4)\cdot\frac{y+3}{(y+2)(y-4)}$

$(y+2)\cancel{(y-4)}\cdot\frac{y-2}{\cancel{y-4}}$

$=\cancel{(y+2)}(y-4)\cdot\frac{1}{\cancel{y+2}}$

$+\cancel{(y+2)}\cancel{(y-4)}\cdot\frac{y+3}{\cancel{(y+2)}\cancel{(y-4)}}$

$(y+2)(y-2)=(y-4)+y+3$

$y^2-4=y-4+y+3$

$y^2-2y-3=0$

$(y-3)(y+1)=0$

$y-3=0$ or $y+1=0$

$y=3$ $\quad$ $y=-1$

Check:

$\frac{-1-2}{-1-4}=\frac{1}{-1+2}+\frac{-1+3}{(-1)^2-2(-1)-8}$

$\frac{-3}{-5}=\frac{1}{1}+\frac{2}{1+2-8}$

$\frac{3}{5}=1+\frac{2}{-5}$

$\frac{3}{5}=\frac{5}{5}-\frac{2}{5}$

$\frac{3}{5}=\frac{3}{5}$

$\frac{3-2}{3-4}=\frac{1}{3+2}+\frac{3+3}{(3)^2-2(3)-8}$

$\frac{1}{-1}=\frac{1}{5}+\frac{6}{9-6-8}$

$-1=\frac{1}{5}+\frac{6}{-5}$

$-1=\frac{1}{5}-\frac{6}{5}$

$-1=-\frac{5}{5}$

$-1=-1$

37. $\frac{x}{x+2}-\frac{2}{2-x}=\frac{x+6}{x^2-4}$

$\frac{x}{x+2}-\frac{2}{2-x}=\frac{x+6}{(x-2)(x+2)}$

$\frac{x}{x+2}+\frac{2}{x-2}=\frac{x+6}{(x-2)(x+2)}$

$(x-2)(x+2)\cdot\frac{x}{x+2}$

$+(x-2)(x+2)\cdot\frac{2}{x-2}$

$=(x-2)(x+2)\frac{x+6}{(x-2)(x+2)}$

$(x-2)\cancel{(x+2)}\cdot\frac{x}{\cancel{x+2}}$

$+\cancel{(x-2)}(x+2)\cdot\frac{2}{\cancel{x-2}}$

$=\cancel{(x-2)}\cancel{(x+2)}\frac{x+6}{\cancel{(x-2)}\cancel{(x+2)}}$

$x(x-2)+2(x+2)=x+6$

$x^2-2x+2x+4=x+6$

$x^2-x-2=0$

$(x+1)(x-2)=0$

$x+1=0$ or $x-2=0$

$x=-1$ $\quad$ $x=2$

37. (continued)

Check:

$$\frac{-1}{-1+2}-\frac{2}{2-(-1)}=\frac{-1+6}{(-1)^2-4}$$
$$\frac{-1}{1}-\frac{2}{2+1}=\frac{5}{1-4}$$
$$-1-\frac{2}{3}=\frac{5}{-3}$$
$$-\frac{3}{3}-\frac{2}{3}=-\frac{5}{3}$$
$$-\frac{5}{3}=-\frac{5}{3} \quad \text{True}$$

$$\frac{2}{2+2}-\frac{2}{2-2}=\frac{2+6}{(2)^2-4}$$
$$\frac{2}{4}-\frac{2}{0}=\frac{8}{4-4}$$
$$\frac{1}{2}-\frac{2}{0}=\frac{8}{0} \quad \text{Undefined}$$

38.

$$\frac{3}{n^2-5n+4}-\frac{1}{n^2-4n+3}=\frac{n-3}{n^2-7n+12}$$
$$\frac{3}{(n-1)(n-4)}-\frac{1}{(n-1)(n-3)}=\frac{n-3}{(n-3)(n-4)}$$
$$(n-1)(n-4)(n-3)\cdot\frac{3}{(n-1)(n-4)}-(n-1)(n-4)(n-3)\cdot\frac{1}{(n-1)(n-3)}=(n-1)(n-4)(n-3)\cdot\frac{n-3}{(n-3)(n-4)}$$
$$\cancel{(n-1)}\cancel{(n-4)}(n-3)\cdot\frac{3}{\cancel{(n-1)}\cancel{(n-4)}}-\cancel{(n-1)}(n-4)\cancel{(n-3)}\cdot\frac{1}{\cancel{(n-1)}\cancel{(n-3)}}=(n-1)\cancel{(n-4)}\cancel{(n-3)}\cdot\frac{n-3}{\cancel{(n-3)}\cancel{(n-4)}}$$
$$3(n-3)-(n-4)=(n-1)(n-3)$$
$$3n-9-n+4=n^2-4n+3$$
$$n^2-6n+8=0$$
$$(n-2)(n-4)=0$$
$$n-2=0 \quad \text{or} \quad n-4=0$$
$$n=2 \qquad\qquad n=4$$

Check:

$$\frac{3}{(2)^2-5(2)+4}-\frac{1}{(2)^2-4(2)+3}=\frac{2-3}{(2)^2-7(2)+12}$$
$$\frac{3}{4-10+4}-\frac{1}{4-8+3}=\frac{2-3}{4-14+12}$$
$$\frac{3}{-2}-\frac{1}{-1}=\frac{-1}{2}$$
$$-\frac{3}{2}+\frac{2}{2}=-\frac{1}{2}$$
$$-\frac{1}{2}=-\frac{1}{2} \quad \text{True}$$

$$\frac{3}{(4)^2-5(4)+4}-\frac{1}{(4)^2-4(4)+3}=\frac{4-3}{(4)^2-7(4)+12}$$
$$\frac{3}{16-20+4}-\frac{1}{16-16+3}=\frac{4-3}{16-28+12}$$
$$\frac{3}{0}-\frac{1}{3}=\frac{1}{0} \quad \text{Undefined}$$

39.

$$\frac{8}{5}=\frac{72}{x}$$
$$8x=5\cdot 72$$
$$8x=360$$
$$x=45$$

Check:
$$\frac{8}{5}=\frac{72}{x}$$
$$\frac{8}{5}=\frac{72}{45}$$
$$\frac{8}{5}=\frac{8}{5}$$

40.

$$\frac{28}{x+3}=\frac{7}{9}$$
$$\frac{28}{x+3}=\frac{7}{9}$$
$$28\cdot 9=7(x+3)$$
$$252=7x+21$$
$$231=7x$$
$$33=x$$

40. (continued)

Check: $\frac{28}{33+3}=\frac{7}{9}$

$\frac{28}{36}=\frac{7}{9}$

$\frac{7}{9}=\frac{7}{9}$

41. $\frac{5}{3+y}=\frac{3}{7y+1}$

$5(7y+1)=3(3+y)$

$35y+5=9+3y$

$32y=4$

$y=\frac{1}{8}$

Check: $\frac{5}{3+\frac{1}{8}}=\frac{3}{7\left(\frac{1}{8}\right)+1}$

$\frac{5}{\frac{24}{8}+\frac{1}{8}}=\frac{3}{\frac{7}{8}+\frac{8}{8}}$

$\frac{5}{\frac{25}{8}}=\frac{3}{\frac{15}{8}}$

$\frac{8}{5}=\frac{8}{5}$

42. $\frac{11}{x-2}=\frac{x+7}{2}$

$11\cdot 2=(x-2)(x+7)$

$22=x^2+5x-14$

$0=x^2+5x-36$

$0=(x-4)(x+9)$

$x-4=0$ or $x+9=0$

$x=4$ $\quad$ $x=-9$

Check:

$\frac{11}{4-2}=\frac{4+7}{2}$ $\qquad$ $\frac{11}{-9-2}=\frac{-9+7}{2}$

$\frac{11}{2}=\frac{11}{2}$ $\qquad$ $\frac{11}{-11}=\frac{-2}{2}$

$-1=-1$

43. $y=kx$

$1.6=k(4)$

$\frac{1.6}{4}=k$

$k=0.4$

$y=0.4x$

44. $y=\frac{k}{x}$

$\frac{1}{2}=\frac{k}{3}$

$2k=3$

$k=\frac{3}{2}$

$y=\frac{\frac{3}{2}}{x}=\frac{3}{2}\bullet\left(\frac{1}{x}\right)\Rightarrow y=\frac{3}{2x}$

45. $y=kxz$

$144=k(4)(6)$

$144=24k$

$k=\frac{144}{24}=6$

$y=6xz$

46. $y=\frac{kx^2}{z^2}$

$2=\frac{k(5)^2}{(10)^2}$

$2=\frac{25k}{100}\Rightarrow 2=\frac{k}{4}\Rightarrow k=8$

$y=\frac{8x^2}{z^2}$

47. $0.72+\frac{200}{x}=0.72\cdot\frac{x}{x}+\frac{200}{x}$

$=\frac{0.72x}{x}+\frac{200}{x}$

$=\frac{0.72x+200}{x}$

$\frac{0.72x+200}{x}$ dollars

48.
$$\frac{c}{4}-10=\frac{c}{5}$$
$$20\cdot\frac{c}{4}-20\cdot10=20\cdot\frac{c}{5}$$
$$5c-200=4c$$
$$c=200$$
The cost of the car rental is \$200.

49.
$$\frac{2d}{\frac{d}{r}+\frac{d}{s}}=\frac{2d\cdot rs}{\left(\frac{d}{r}+\frac{d}{s}\right)\cdot rs}$$
$$=\frac{2drs}{\frac{d}{r}\cdot rs+\frac{d}{s}\cdot rs}$$
$$=\frac{2drs}{ds+dr}$$
$$=\frac{2drs}{d(s+r)}$$
$$=\frac{2rs}{s+r}$$

50.
$$\frac{t}{10}-\frac{t}{15}=1$$
$$30\cdot\frac{t}{10}-30\cdot\frac{t}{15}=30\cdot1$$
$$3t-2t=30$$
$$t=30$$
It will take 30 min to fill the tub.

51.
$$\frac{x}{50}+\frac{(400-x)}{60}=7$$
$$300\cdot\frac{x}{50}+300\cdot\frac{(400-x)}{60}=300\cdot7$$
$$6x+5(400-x)=2100$$
$$6x+2000-5x=2100$$
$$x=100$$
The family drove 100 mi at 50 mph.

52.
$$\frac{1}{x}+\frac{1}{x+1}$$
$$=\frac{x+1}{x+1}\cdot\frac{1}{x}+\frac{x}{x}\cdot\frac{1}{x+1}$$
$$=\frac{x+1+x}{x(x+1)}$$
$$=\frac{2x+1}{x(x+1)}$$
$\frac{2x+1}{x(x+1)}$ of the job will be in an hour.

53.
$$\frac{31}{15}=\frac{x}{20}$$
$$31\cdot20=15x$$
$$620=15x$$
$$x=\frac{620}{15}$$
$$x\approx41.333$$
She should expect to spend about \$41,333.

54.
$$\frac{1}{n}+\frac{1}{n+1}+\frac{1}{n+2}$$
$$=\frac{(n+1)(n+2)}{(n+1)(n+2)}\cdot\frac{1}{n}+\frac{n(n+2)}{n(n+2)}\cdot\frac{1}{n+1}+\frac{n(n+1)}{n(n+1)}\cdot\frac{1}{n+2}$$
$$=\frac{(n+1)(n+2)+n(n+2)+n(n+1)}{n(n+1)(n+2)}$$
$$=\frac{n^2+3n+2+n^2+2n+n^2+n}{n(n+1)(n+2)}$$
$$=\frac{3n^2+6n+2}{n(n+1)(n+2)}$$

55.
$$v=kt$$
$$19.6=k(2)$$
$$k=\frac{19.6}{2}=9.8$$
$$v=9.8t$$
$$v=9.8(5)=49$$
Its velocity after 5 sec is 49 m per sec.

56.
$$A=\frac{k}{d}$$
$$8=\frac{k}{12.5}$$
$$8(12.5)=k$$
$$k=100$$
$$A=\frac{100}{d}$$
$$A=\frac{100}{10}=10$$
The accommodation is 10 diopters.

Chapter 7 Posttest

1. $\dfrac{3x}{x-8}$

$x-8=0$

$x=8$

The expression is undefined when $x=8$.

2. $$-\frac{3y-y^2}{y^2}=-\frac{y(3-y)}{y^2}=\frac{y(y-3)}{y^2}=\frac{y-3}{y}$$

3. $$\frac{15a^3b}{12ab^2}=\frac{3ab(5a^2)}{3ab(4b)}=\frac{5a^2}{4b}$$

4. $$\frac{x^2-4x}{xy-4y}=\frac{x(x-4)}{y(x-4)}=\frac{x}{y}$$

5. $$\frac{3b^2-27}{b^2-4b-21}=\frac{3(b^2-9)}{(b+3)(b-7)}=\frac{3(b+3)(b-3)}{(b+3)(b-7)}=\frac{3(b+3)(b-3)}{(b+3)(b-7)}=\frac{3(b-3)}{(b-7)}$$

6. $$\frac{\frac{3}{x^2}-\frac{1}{x}}{\frac{9}{x^2}-1}=\frac{\left(\frac{3}{x^2}-\frac{1}{x}\right)\cdot x^2}{\left(\frac{9}{x^2}-1\right)\cdot x^2}=\frac{\frac{3}{x^2}\cdot x^2-\frac{1}{x}\cdot x^2}{\frac{9}{x^2}\cdot x^2-1\cdot x^2}=\frac{3-x}{9-x^2}=\frac{3-x}{(3-x)(3+x)}=\frac{3-x}{(3-x)(3+x)}=\frac{1}{x+3}$$

7. Factor $n^2+6n-16$: $(n+8)(n-2)$

Factor $n+8$: $(n+8)$

Factor $4n-8$: $4\cdot(n-2)$

LCD: $4(n+8)(n-2)$

$$\frac{4n-1}{n^2+6n-16}=\frac{4n-1}{(n+8)(n-2)}\cdot\frac{4}{4}=\frac{4(4n-1)}{4(n+8)(n-2)}$$

$$\frac{2}{n+8}=\frac{2}{n+8}\cdot\frac{4(n-2)}{4(n-2)}=\frac{8(n-2)}{4(n+8)(n-2)}$$

$$\frac{n}{4n-8}=\frac{n}{4(n-2)}\cdot\frac{(n+8)}{(n+8)}=\frac{n(n+8)}{4(n+8)(n-2)}$$

8. $\dfrac{7x-10}{x+6}-\dfrac{5x-22}{x+6}$

$=\dfrac{(7x-10)-(5x-22)}{x+6}$

$=\dfrac{7x-10-5x+22}{x+6}$

$=\dfrac{2x+12}{x+6}$

$=\dfrac{2(x+6)}{x+6}$

$=\dfrac{2\cancel{(x+6)}}{\cancel{x+6}}$

$=2$

9. $\dfrac{3}{2y-8}+\dfrac{2}{4y^2-16y}$

$=\dfrac{3}{2(y-4)}+\dfrac{2}{4y(y-4)}$

$=\dfrac{3}{2(y-4)}\cdot\dfrac{2y}{2y}+\dfrac{2}{4y(y-4)}$

$=\dfrac{6y+2}{4y(y-4)}$

$=\dfrac{2(3y+1)}{4y(y-4)}$

$=\dfrac{3y+1}{2y(y-4)}$

10. $\dfrac{5}{d-3}-\dfrac{d-4}{d^2-d-6}$

$=\dfrac{5}{d-3}-\dfrac{d-4}{(d-3)(d+2)}$

$=\dfrac{5}{d-3}\cdot\dfrac{d+2}{d+2}-\dfrac{d-4}{(d-3)(d+2)}$

$=\dfrac{5(d+2)-(d-4)}{(d-3)(d+2)}$

$=\dfrac{5d+10-d+4}{(d-3)(d+2)}$

$=\dfrac{4d+14}{(d-3)(d+2)}$

$=\dfrac{2(2d+7)}{(d-3)(d+2)}$

11. $\dfrac{5}{2x^2-3x-2}-\dfrac{x}{4-x^2}$

$=\dfrac{5}{(2x+1)(x-2)}-\dfrac{x}{(2-x)(2+x)}$

$=\dfrac{5}{(2x+1)(x-2)}+\dfrac{x}{(x-2)(x+2)}$

$=\dfrac{x+2}{x+2}\cdot\dfrac{5}{(2x+1)(x-2)}$

$+\dfrac{2x+1}{2x+1}\cdot\dfrac{x}{(x-2)(x+2)}$

$=\dfrac{5(x+2)+x(2x+1)}{(2x+1)(x+2)(x-2)}$

$=\dfrac{5x+10+2x^2+x}{(2x+1)(x+2)(x-2)}$

$=\dfrac{2x^2+6x+10}{(2x+1)(x+2)(x-2)}$

$=\dfrac{2(x^2+3x+5)}{(2x+1)(x+2)(x-2)}$

12. $\dfrac{n+1}{3n-18}\cdot\dfrac{n-6}{6n^3-6n}$

$=\dfrac{n+1}{3(n-6)}\cdot\dfrac{n-6}{6n(n^2-1)}$

$=\dfrac{n+1}{3(n-6)}\cdot\dfrac{n-6}{6n(n+1)(n-1)}$

$=\dfrac{\cancel{n+1}}{3\cancel{(n-6)}}\cdot\dfrac{\cancel{n-6}}{6n\cancel{(n+1)}(n-1)}$

$=\dfrac{1}{18n(n-1)}$

13. $\dfrac{a^2-25}{a^2-2a-24}\div\dfrac{a^2+a-30}{a^2-36}$

$=\dfrac{a^2-25}{a^2-2a-24}\cdot\dfrac{a^2-36}{a^2+a-30}$

$=\dfrac{(a-5)(a+5)}{(a+4)(a-6)}\cdot\dfrac{(a-6)(a+6)}{(a-5)(a+6)}$

$=\dfrac{\cancel{(a-5)}(a+5)}{(a+4)\cancel{(a-6)}}\cdot\dfrac{\cancel{(a-6)}\cancel{(a+6)}}{\cancel{(a-5)}\cancel{(a+6)}}$

$=\dfrac{a+5}{a+4}$

14. $\dfrac{x^2+6x+8}{x^2+x-2} \div \dfrac{x+4}{2x^2+12x+16}$

$$= \frac{x^2+6x+8}{x^2+x-2} \cdot \frac{2x^2+12x+16}{x+4}$$

$$= \frac{x^2+6x+8}{x^2+x-2} \cdot \frac{2(x^2+6x+8)}{x+4}$$

$$= \frac{(x+2)(x+4)}{(x-1)(x+2)} \cdot \frac{2(x+4)(x+2)}{x+4}$$

$$= \frac{\cancel{(x+2)}\,\cancel{(x+4)}}{(x-1)\,\cancel{(x+2)}} \cdot \frac{2(x+4)(x+2)}{\cancel{x+4}}$$

$$= \frac{2(x+4)(x+2)}{x-1}$$

15. $\dfrac{1}{y-5} + \dfrac{y+4}{25-y^2} = \dfrac{1}{y+5}$

$$\frac{1}{y-5} + \frac{y+4}{(5-y)(5+y)} = \frac{1}{y+5}$$

$$\frac{1}{y-5} - \frac{y+4}{(y-5)(y+5)} = \frac{1}{y+5}$$

$$(y-5)(y+5)\cdot\frac{1}{y-5} - (y-5)(y+5)\cdot\frac{y+4}{(y-5)(y+5)} = (y-5)(y+5)\cdot\frac{1}{y+5}$$

$$\cancel{(y-5)}(y+5)\cdot\frac{1}{\cancel{y-5}} - \cancel{(y-5)}\,\cancel{(y+5)}\cdot\frac{y+4}{\cancel{(y-5)}\,\cancel{(y+5)}} = (y-5)\,\cancel{(y+5)}\cdot\frac{1}{\cancel{y+5}}$$

$$y+5-(y+4) = y-5$$

$$y+5-y-4 = y-5$$

$$-y = -6$$

$$y = 6$$

Check: $\dfrac{1}{6-5} + \dfrac{6+4}{25-(6)^2} = \dfrac{1}{6+5}$

$$\frac{1}{1} + \frac{10}{25-36} = \frac{1}{11}$$

$$\frac{1}{1} + \frac{10}{-11} = \frac{1}{11}$$

$$\frac{11}{11} - \frac{10}{11} = \frac{1}{11}$$

$$\frac{1}{11} = \frac{1}{11}$$

16. $\dfrac{2y}{y-4} - 2 = \dfrac{4}{y+5}$

$$(y-4)(y+5)\cdot\frac{2y}{y-4} - (y-4)(y+5)\cdot 2 = (y-4)(y+5)\cdot\frac{4}{y+5}$$

$$\cancel{(y-4)}(y+5)\cdot\frac{2y}{\cancel{y-4}} - (y-4)(y+5)\cdot 2 = (y-4)\,\cancel{(y+5)}\cdot\frac{4}{\cancel{y+5}}$$

$$2y(y+5) - 2(y-4)(y+5) = 4(y-4)$$

$$2y^2+10y-2(y^2+y-20) = 4y-16$$

$$2y^2+10y-2y^2-2y+40 = 4y-16$$

$$4y = -56$$

$$y = -14$$

Check: $\dfrac{2(-14)}{-14-4} - 2 = \dfrac{4}{-14+5}$

$$\frac{-28}{-18} - 2 = \frac{4}{-9}$$

$$\frac{14}{9} - \frac{18}{9} = -\frac{4}{9}$$

$$-\frac{4}{9} = -\frac{4}{9}$$

17.

$$\frac{x}{x+6}=\frac{1}{x+2}$$
$$x(x+2)=x+6$$
$$x^2+2x=x+6$$
$$x^2+x-6=0$$
$$(x-2)(x+3)=0$$
$$x-2=0 \quad \text{or} \quad x+3=0$$
$$x=2 \qquad\qquad x=-3$$

Check:

$$\frac{2}{2+6}=\frac{1}{2+2} \qquad \frac{-3}{-3+6}=\frac{1}{-3+2}$$
$$\frac{2}{8}=\frac{1}{4} \qquad \frac{-3}{3}=\frac{1}{-1}$$
$$\frac{1}{4}=\frac{1}{4} \qquad -1=-1$$

18.

$$\frac{1}{R}=\frac{1}{R_1}+\frac{1}{R_2}$$
$$RR_1R_2\cdot\frac{1}{R}=RR_1R_2\cdot\frac{1}{R_1}+RR_1R_2\cdot\frac{1}{R_2}$$
$$R_1R_2=RR_2+RR_1$$
$$R_1R_2-RR_1=RR_2$$
$$R_1(R_2-R)=RR_2$$
$$R_1=\frac{RR_2}{R_2-R}$$

19.

$$\frac{1}{t}+\frac{1}{2t}=\frac{1}{20}$$
$$20t\cdot\frac{1}{t}+20t\cdot\frac{1}{2t}=20t\cdot\frac{1}{20}$$
$$20+10=t$$
$$30=t$$

$$2t=2\cdot 30=60$$

Working alone, the newer machine can process 1000 pieces of mail in 30 min and the older machine can process 1000 pieces of mail in 60 min.

20.

$$\frac{1.5}{3}=\frac{h}{120}$$
$$15\cdot 120=3h$$
$$180=3h$$
$$60=h$$

The height of the tree is 60 m.

Chapter 8

RADICAL EXPRESSIONS AND EQUATIONS

Pretest

1. a. $2\sqrt{36} = 2\sqrt{(6)^2} = 2(6) = 12$

   b. $\sqrt[3]{-64} = \sqrt[3]{(-4)^3} = -4$

2. $\sqrt{100u^2v^4} = \sqrt{10^2u^2\left(v^2\right)^2} = 10uv^2$

3. $\left(81x^4\right)^{3/4} = \left(\sqrt[4]{81x^4}\right)^3 = \left(\sqrt[4]{3^4x^4}\right)^3$
   $= (3x)^3 = 3^3x^3 = 27x^3$

4. a.

$$\frac{8p^{2/3}}{\left(36p^{4/3}\right)^{1/2}} = \frac{8p^{2/3}}{36^{1/2}p^{(4/3)(1/2)}} = \frac{8p^{2/3}}{6p^{2/3}} = \frac{8}{6} = \frac{4}{3}$$

   b. $\sqrt[6]{x^2y^4} = \left(x^2y^4\right)^{1/6} = x^{2/6}y^{4/6} = x^{1/3}y^{2/3}$
   $= \left(xy^2\right)^{1/3} = \sqrt[3]{xy^2}$

5. $\sqrt{6a} \bullet \sqrt{7b} = \sqrt{(6a)(7b)} = \sqrt{42ab}$

6. $\dfrac{\sqrt[3]{18r^2}}{\sqrt[3]{3r}} = \sqrt[3]{\dfrac{18r^2}{3r}} = \sqrt[3]{6r}$

7. $\sqrt{147x^5y^4} = \sqrt{49 \bullet 3 \bullet x^4 \bullet x \bullet y^4}$
   $= \sqrt{49x^4y^4}\sqrt{3x} = 7x^2y^2\sqrt{3x}$

8. $\sqrt{\dfrac{2p}{25q^8}} = \dfrac{\sqrt{2p}}{\sqrt{25q^8}} = \dfrac{\sqrt{2p}}{5q^4}$

9.

$$3\sqrt{12} - 5\sqrt{3} + \sqrt{108} = 3\sqrt{4 \bullet 3} - 5\sqrt{3} + \sqrt{36 \bullet 3}$$
$$= 3 \bullet 2\sqrt{3} - 5\sqrt{3} + 6\sqrt{3}$$
$$= (6 - 5 + 6)\sqrt{3} = 7\sqrt{3}$$

10. $\left(\sqrt{x} - 8\right)\left(2\sqrt{x} + 1\right)$

    F O I L

    $2\left(\sqrt{x}\right)^2 + \sqrt{x} - 16\sqrt{x} - 8 = 2x - 15\sqrt{x} - 8$

11. $\dfrac{\sqrt{2x}}{\sqrt{27y}} = \dfrac{\sqrt{2x}}{\sqrt{27y}} \bullet \dfrac{\sqrt{3y}}{\sqrt{3y}} = \dfrac{\sqrt{6xy}}{\sqrt{81y^2}} = \dfrac{\sqrt{6xy}}{9y}$

12. $\dfrac{\sqrt{n}}{\sqrt{n} + \sqrt{3}} = \dfrac{\sqrt{n}}{\sqrt{n} + \sqrt{3}} \bullet \dfrac{\left(\sqrt{n} - \sqrt{3}\right)}{\left(\sqrt{n} - \sqrt{3}\right)}$
    $= \dfrac{\left(\sqrt{n}\right)^2 - \sqrt{3}\sqrt{n}}{\left(\sqrt{n}\right)^2 - \left(\sqrt{3}\right)^2} = \dfrac{n - \sqrt{3n}}{n - 3}$

13. $\sqrt{x + 4} - 9 = -3$
    $\sqrt{x + 4} = 6$
    $\left(\sqrt{x + 4}\right)^2 = 6^2$
    $x + 4 = 36$
    $x = 32$

    Check $x = 32$
    $\sqrt{32 + 4} - 9 \stackrel{?}{=} -3$
    $\sqrt{36} - 9 \stackrel{?}{=} -3$
    $6 - 9 \stackrel{?}{=} -3$
    $-3 = -3$ True

14. $\sqrt{x^2 - 2} = \sqrt{9x - 10}$
    $\left(\sqrt{x^2 - 2}\right)^2 = \left(\sqrt{9x - 10}\right)^2$
    $x^2 - 2 = 9x - 10$
    $x^2 - 9x + 8 = 0$
    $(x - 8)(x - 1) = 0$
    $x - 8 = 0 \quad x - 1 = 0$
    $x = 8 \quad x = 1$

    Check $x = 1$
    $\sqrt{1^2 - 2} = \sqrt{1 - 2} = \sqrt{-1}$
    $\sqrt{-1}$ is not a real number.
    $x = 1$ is not a solution
    Check $x = 8$
    $\sqrt{8^2 - 2} = \sqrt{9(8) - 10}$
    $\sqrt{64 - 2} = \sqrt{72 - 10}$
    $\sqrt{62} = \sqrt{62}$ True

15. $\sqrt{2x}-\sqrt{x+7}=-1$

$\sqrt{2x}=\sqrt{x+7}-1$

$\left(\sqrt{2x}\right)^2=\left(\sqrt{x+7}-1\right)^2$

$2x=(x+7)-2\sqrt{x+7}+1$

$x-8=-2\sqrt{x+7}$

$(x-8)^2=\left(-2\sqrt{x+7}\right)^2$

$x^2-16x+64=4(x+7)$

$x^2-16x+64=4x+28$

$x^2-20x+36=0$

$(x-18)(x-2)=0$

$x-18=0 \quad x-2=0$

$x=18 \quad\quad x=2$

Check $x=18$

$\sqrt{2(18)}-\sqrt{18+7}\stackrel{?}{=}-1$

$\sqrt{36}-\sqrt{25}\stackrel{?}{=}-1$

$6-5\stackrel{?}{=}-1$

$1=-1$ False

$x=18$ is not a solution

Check $x=2$

$\sqrt{2(2)}-\sqrt{2+7}\stackrel{?}{=}-1$

$\sqrt{4}-\sqrt{9}\stackrel{?}{=}-1$

$2-3\stackrel{?}{=}-1$

$-1=-1$ True

16.

$\left(5-\sqrt{-9}\right)\left(4-\sqrt{-25}\right)$

$=\left(5-\left(\sqrt{9}\right)\left(\sqrt{-1}\right)\right)\left(4-\left(\sqrt{25}\right)\left(\sqrt{-1}\right)\right)$

$=(5-3i)(4-5i)$

F O I L

$=20-25i-12i+15i^2=20-37i+15(-1)$

$=20-37i-15=5-37i$

17.

$\frac{1+2i}{1-2i}=\frac{1+2i}{1-2i}\bullet\frac{1+2i}{1+2i}=\frac{1+4i+4i^2}{1-4i^2}=\frac{1+4i-4}{1+4}$

$=\frac{-3+4i}{5}=-\frac{3}{5}+\frac{4i}{5}$

18. **a.** $\sqrt{\frac{h-d}{16}}=\frac{\sqrt{h-d}}{\sqrt{16}}=\frac{\sqrt{h-d}}{4}$

b. $\frac{\sqrt{h-d}}{4}=\frac{\sqrt{200-125}}{4}=\frac{\sqrt{75}}{4}$

$=\frac{\sqrt{25\bullet 3}}{4}=\frac{5\sqrt{3}}{4}=2.16506...$

It takes the stone $\frac{5\sqrt{3}}{4}$ sec, or approximately 2.2 sec to be 125 ft above the ground.

19. Use the Pythagorean theorem. Let x be the length of the diagonal.

$x^2=50^2+94^2=2500+8836=11336$

$x=\sqrt{11336}=\sqrt{4\bullet 2834}=2\sqrt{2834}$

$x=106.4706...$

The diagonal of the court is $2\sqrt{2834}$ ft or about
106.5 ft.

20. $f=120\sqrt{p}$

$\frac{f}{120}=\frac{120\sqrt{p}}{120}$

$\frac{f}{120}=\sqrt{p}$

$\left(\frac{f}{120}\right)^2=\left(\sqrt{p}\right)^2$

$p=\frac{f^2}{14400}$

8.1 Radical Expressions and Functions

Practice

1. **a.** $\sqrt{25}=5$ since $5^2=25$.

b. $\sqrt{\frac{9}{100}}=\frac{3}{10}$ since $\left(\frac{3}{10}\right)^2=\frac{9}{100}$.

c. $-4\sqrt{81}=-4\cdot 9=-36$

d. $\sqrt{-1}$ is not a real number.

2. $\sqrt{6}=2.449489743...\approx 2.449$

3. **a.** $-\sqrt{4y^2}=-\sqrt{(2y)^2}=-2y$

b. $\sqrt{36x^6y^6}=\sqrt{\left(6x^3y^3\right)^2}=6x^3y^3$

4. **a.** $\sqrt[3]{216}=\sqrt[3]{6^3}=6$

b. $\sqrt[3]{-27}=\sqrt[3]{(-3)^3}=-3$

c. $\sqrt[3]{\frac{1}{125}}=\sqrt[3]{\left(\frac{1}{5}\right)^3}=\frac{1}{5}$

d. $\sqrt[3]{-64x^6}=\sqrt[3]{(-4x^2)^3}=-4x^2$

5. **a.** $\sqrt[6]{64}=\sqrt[6]{2^6}=2$

b. $\sqrt[5]{-32}=\sqrt[5]{(-2)^5}=-2$

c. $\sqrt[4]{256y^8}=\sqrt[4]{(4y^2)^4}=4y^2$

6. **a.** $f(85)=\sqrt{85-4}=\sqrt{81}=9$

b. $g(121)=3+\sqrt{121}=3+11=14$

c. $f(49)=\sqrt{49-4}=\sqrt{45}=3\sqrt{5}\approx 6.71$

d. $g(48)=3+\sqrt{48}=3+4\sqrt{3}\approx 9.93$

7. $h(x)=\sqrt{x-3}$

x	$h(x)=\sqrt{x-3}$	(x,y)
3	$h(3)=\sqrt{3-3}=\sqrt{0}=0$	$(3,0)$
4	$h(4)=\sqrt{4-3}=\sqrt{1}=1$	$(4,1)$
5	$h(5)=\sqrt{5-3}=\sqrt{2}\approx 1.41$	$(5,1.41)$
6	$h(6)=\sqrt{6-3}=\sqrt{3}\approx 1.73$	$(6,1.73)$
7	$h(7)=\sqrt{7-3}=\sqrt{4}=2$	$(7,2)$

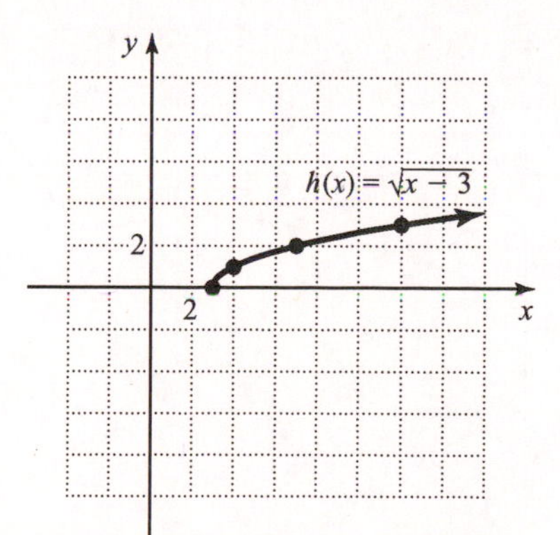

Exercises

1. The number b is a(n) square root of a if $b^2=a$, for any real numbers a and b and for a nonnegative.

3. The number under the radical sign is called the radicand.

5. The square root of a number that is not a perfect square is a(n) irrational number.

7. The cube root of a is written $\sqrt[3]{a}$, where 3 is called the index of the radical.

9. The function $f(x)=\sqrt{x}$ is called the square root function.

11. $\sqrt{64}=\sqrt{8^2}=8$

13. $-\sqrt{100}=-\sqrt{10^2}=-10$

15. $\sqrt{-36}$ is not a real number

17. $2\sqrt{16}=2\sqrt{4^2}=2(4)=8$

19. $\sqrt[3]{27}=\sqrt[3]{3^3}=3$

21. $5\sqrt[3]{-8}=5\sqrt[3]{(-2)^3}=5(-2)=-10$

23. $\sqrt[4]{256}=\sqrt[4]{4^4}=4$

25. $8\sqrt[5]{-1}=8\sqrt[5]{(-1)^5}=8(-1)=-8$

27. $\sqrt{\frac{9}{16}}=\frac{\sqrt{9}}{\sqrt{16}}=\frac{\sqrt{3^2}}{\sqrt{4^2}}=\frac{3}{4}$

29. $\sqrt[3]{-\frac{8}{125}}=\frac{\sqrt[3]{-8}}{\sqrt[3]{125}}=\frac{\sqrt[3]{(-2)^3}}{\sqrt[3]{5^3}}=\frac{-2}{5}=-\frac{2}{5}$

31. $\sqrt{0.04}=\sqrt{(0.2)^2}=0.2$

33. $\sqrt{21}=4.58257...\approx 4.583$

35. $\sqrt{46}=6.78232...\approx 6.782$

37. $\sqrt{14.25}=3.77491...\approx 3.775$

39. $\sqrt[3]{112}=4.82028...\approx 4.820$

41. $\sqrt[5]{150}=2.72406...\approx 2.724$

43. $\sqrt{x^8}=\sqrt{(x^4)^2}=x^4$

45. $\sqrt{16a^6}=\sqrt{(4a^3)^2}=4a^3$

47. $9\sqrt{p^8q^4}=9\sqrt{(p^4q^2)^2}=9p^4q^2$

49. $\frac{1}{3}\sqrt{36x^{10}y^2}=\frac{1}{3}\sqrt{(6x^5y)^2}=\frac{1}{3}(6x^5y)=2x^5y$

51. $\sqrt[3]{-125u^9}=\sqrt[3]{(-5u^3)^3}=-5u^3$

53. $2\sqrt[3]{216u^3v^{12}}=2\sqrt[3]{(6uv^4)^3}=2(6uv^4)=12uv^4$

55. $\sqrt[4]{16t^{12}} = \sqrt[4]{\left(2t^3\right)^4} = 2t^3$

57. $\sqrt[5]{p^5q^{15}} = \sqrt[5]{\left(pq^3\right)^5} = pq^3$

59. $f(10) = \sqrt{2\cdot 10+5} = \sqrt{25} = 5$

61. $g(-12) = \sqrt[3]{-12+4} = \sqrt[3]{-8} = -2$

63. $f(0) = \sqrt{2\cdot 0+5} = \sqrt{5}$

65. $g(36) = \sqrt[3]{36+4} = \sqrt[3]{40} = 2\sqrt[3]{5}$

67. $f(x) = \sqrt{x}+3$

x	$f(x)=\sqrt{x}+3$	(x,y)
0	$f(0)=\sqrt{0}+3=0+3=3$	$(0,3)$
1	$f(1)=\sqrt{1}+3=1+3=4$	$(1,4)$
2	$f(2)=\sqrt{2}+3\approx 1.41+3=4.41$	$(2,4.41)$
3	$f(3)=\sqrt{3}+3\approx 1.73+3=4.73$	$(3,4.73)$
4	$f(4)=\sqrt{4}+3=2+3=5$	$(4,5)$

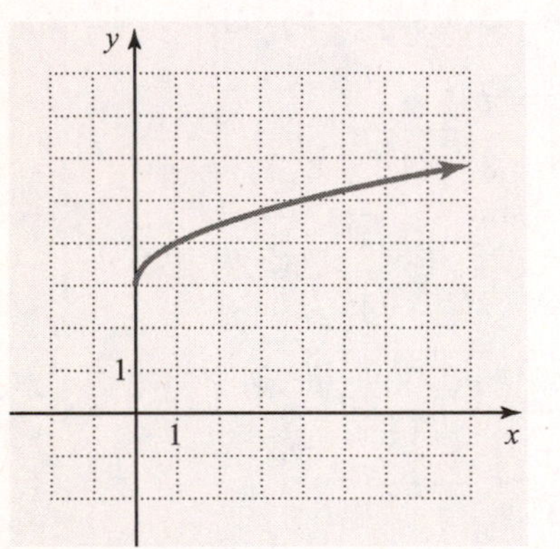

69. $h(x) = \sqrt{x+4}$

x	$h(x)=\sqrt{x+4}$	(x,y)
−4	$h(-4)=\sqrt{-4+4}=\sqrt{0}=0$	$(-4,0)$
−3	$h(-3)=\sqrt{-3+4}=\sqrt{1}=1$	$(-3,1)$
0	$h(0)=\sqrt{0+4}=\sqrt{4}=2$	$(0,2)$
1	$h(1)=\sqrt{1+4}=\sqrt{5}\approx 2.24$	$(1,2.24)$
5	$h(5)=\sqrt{5+4}=\sqrt{9}=3$	$(5,3)$

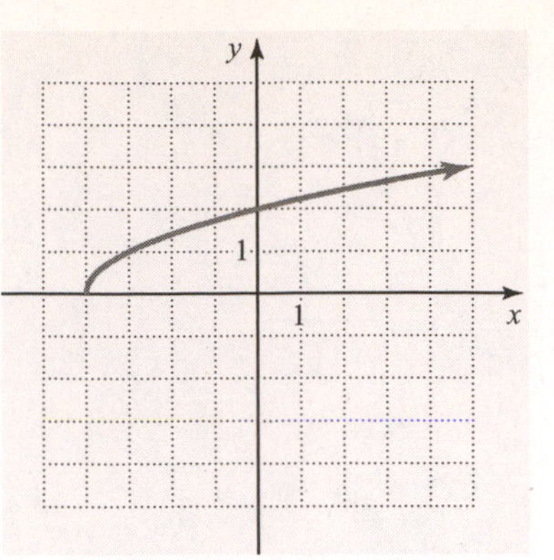

71. $\sqrt{\dfrac{25}{121}} = \dfrac{\sqrt{25}}{\sqrt{121}} = \dfrac{\sqrt{5^2}}{\sqrt{11^2}} = \dfrac{5}{11}$

73. $2\sqrt[3]{-125} = 2\sqrt[3]{(-5)^3} = 2(-5) = -10$

75. $\sqrt{0.009} = 0.094868... = 0.095$

77. $\dfrac{1}{4}\sqrt{(8)^2\left(a^3\right)^2\left(b^2\right)^2} = \dfrac{1}{4}(8)a^3b^2 = 2a^3b^2$

79. $f(1) = \sqrt[3]{2\cdot 1-10} = \sqrt[3]{-8} = \sqrt[3]{(-2)^3} = -2$

81. $h(x) = \sqrt{x}-1$

x	$h(x)=\sqrt{x}-1$	(x,y)
0	$h(0)=\sqrt{0}-1=0-1=-1$	$(0,-1)$
1	$h(1)=\sqrt{1}-1=1-1=0$	$(1,0)$
2	$h(2)=\sqrt{2}-1\approx 1.41-1=0.41$	$(2,0.41)$
3	$h(3)=\sqrt{3}-1\approx 1.73-1=0.73$	$(3,0.73)$
4	$h(4)=\sqrt{4}-1=2-1=1$	$(4,1)$

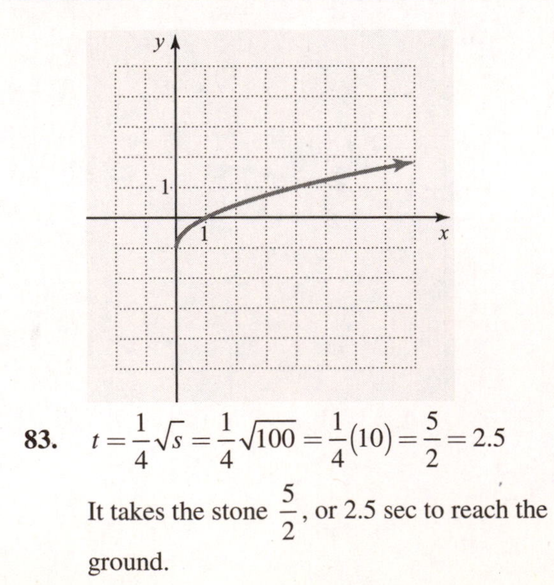

83. $t = \dfrac{1}{4}\sqrt{s} = \dfrac{1}{4}\sqrt{100} = \dfrac{1}{4}(10) = \dfrac{5}{2} = 2.5$

It takes the stone $\dfrac{5}{2}$, or 2.5 sec to reach the ground.

85. **a.** $s = \sqrt{A} = \sqrt{25} = 5$. The length of the side of the frame is 5 in.
b. A 1 in. border on top, bottom and each side will leave 5 – 2(1) = 3 in. A 3-in. by 3-in. photograph fits in the frame.

8.2 Rational Exponents

Practice

1. **a.** $36^{1/2} = \sqrt{36} = \sqrt{6^2} = 6$

b. $27^{1/3} = \sqrt[3]{27} = \sqrt[3]{(3)^3} = 3$

c. $-n^{1/4} = -\sqrt[4]{n}$

d.

$$\left(-125y^9\right)^{1/3} = \sqrt[3]{-125y^9} = \sqrt[3]{\left(-5y^3\right)^3} = -5y^3$$

2. **a.** $81^{3/4} = \left(\sqrt[4]{81}\right)^3 = \left(\sqrt[4]{3^4}\right)^3 = 3^3 = 27$

b.

$$(-64)^{2/3} = \left(\sqrt[3]{-64}\right)^2 = \left(\sqrt[3]{(-4)^3}\right)^2 = (-4)^2 = 16$$

c.

$$\left(\frac{4}{9}y^4\right)^{5/2} = \left(\sqrt{\frac{4}{9}y^4}\right)^5$$

$$= \left(\sqrt{\left(\frac{2}{3}y^2\right)^2}\right)^5 = \left(\frac{2}{3}y^2\right)^5 = \frac{32}{243}y^{10}$$

3. **a.**

$$-81^{-1/4} = -\frac{1}{81^{1/4}} = -\frac{1}{\sqrt[4]{81}} = -\frac{1}{\sqrt[4]{3^4}} = -\frac{1}{3}$$

b.

$$\left(27a^6\right)^{-4/3} = \frac{1}{\left(27a^6\right)^{4/3}} = \frac{1}{\left(\sqrt[3]{27a^6}\right)^4}$$

$$= \frac{1}{\left(\sqrt[3]{\left(3a^2\right)^3}\right)^4} = \frac{1}{\left(3a^2\right)^4} = \frac{1}{81a^8}$$

4. **a.**

$$81^{1/4}81^{1/2} = 81^{1/4+1/2} = 81^{3/4} = \left(\sqrt[4]{81}\right)^3 = 3^3 = 27$$

b. $\frac{n^{3/5}}{n^{2/5}} = n^{3/5-2/5} = n^{1/5} = \sqrt[5]{n}$

c. $\left(r^{1/6}\right)^3 = r^{(1/6)\cdot 3} = r^{1/2} = \sqrt{r}$

d. $\left(\frac{x^4}{4y^2}\right)^{1/2} = \frac{\left(x^4\right)^{1/2}}{\left(4y^2\right)^{1/2}} = \frac{\sqrt{x^4}}{\sqrt{4y^2}} = \frac{x^2}{2y}$

5.

$$2\pi\sqrt{\frac{d^3}{Gm}} = 2\pi\left(\frac{d^3}{Gm}\right)^{1/2} = 2\pi\frac{\left(d^3\right)^{1/2}}{(Gm)^{1/2}}$$

$$= \frac{2\pi d^{3/2}}{(Gm)^{1/2}} = \frac{2\pi d^{3/2}}{G^{1/2}m^{1/2}}$$

Exercises

1. A number raised to the power $\frac{1}{n}$ is the nth root of that number.

3. When $a^{m/n}$ is rewritten as a radical expression, n is the index.

5. $16^{1/2} = \sqrt{16} = \sqrt{4^2} = 4$

7. $-16^{1/2} = -\sqrt{16} = -\sqrt{4^2} = -4$

9. $(-64)^{1/3} = \sqrt[3]{-64} = \sqrt[3]{(-4)^3} = -4$

11. $6x^{1/4} = 6\sqrt[4]{x}$

13. $\left(36a^2\right)^{1/2} = \sqrt{36a^2} = \sqrt{(6a)^2} = 6a$

15. $\left(-216u^6\right)^{1/3} = \sqrt[3]{-216u^6} = \sqrt[3]{\left(-6u^2\right)^3} = -6u^2$

17. $27^{4/3} = \left(\sqrt[3]{27}\right)^4 = \left(\sqrt[3]{3^3}\right)^4 = \left(3^4\right) = 81$

19. $-16^{3/2} = -\left(\sqrt{16}\right)^3 = -\left(\sqrt{4^2}\right)^3 = -(4)^3 = -64$

21. $\left(-27y^3\right)^{2/3} = \left(\sqrt[3]{-27y^3}\right)^2 = \left(\sqrt[3]{(-3y)^3}\right)^2$

$$= (-3y)^2 = 9y^2$$

23. $-81^{-3/4} = -\frac{1}{81^{3/4}} = -\frac{1}{\left(\sqrt[4]{81}\right)^3} = -\frac{1}{\left(\sqrt[4]{3^4}\right)^3}$

$$= -\frac{1}{3^3} = -\frac{1}{27}$$

25. $\left(\frac{x^{10}}{4}\right)^{-1/2} = \left(\frac{4}{x^{10}}\right)^{1/2} = \sqrt{\frac{4}{x^{10}}} = \frac{\sqrt{4}}{\sqrt{x^{10}}}$

$$= \frac{\sqrt{2^2}}{\sqrt{\left(x^5\right)^2}} = \frac{2}{x^5}$$

27. $16 \cdot 16^{1/2} = 16^1 \cdot 16^{1/2} = 16^{3/2} = \left(\sqrt{16}\right)^3$
$= \left(\sqrt{4^2}\right)^3 = 4^3 = 64$

29. $\frac{6^{3/5}}{6^{2/5}} = 6^{3/5-2/5} = 6^{1/5} = \sqrt[5]{6}$

31. $\left(\frac{1}{2}\right)^{3/2}\left(\frac{1}{2}\right)^{-1/2} = \left(\frac{1}{2}\right)^{3/2+(-1/2)} = \left(\frac{1}{2}\right)^{2/2}$
$= \left(\frac{1}{2}\right)^1 = \frac{1}{2}$

33. $\left(9^{3/4}\right)^{2/3} = 9^{(3/4)(2/3)} = 9^{1/2} = \sqrt{9} = \sqrt{3^2} = 3$

35. $4n^{2/5}n^{1/5} = 4n^{2/5+1/5} = 4n^{3/5} = 4\sqrt[5]{n^3}$

37. $\left(y^{-4}\right)^{-1/8} = y^{(-4)\cdot(-1/8)} = y^{4/8} = y^{1/2} = \sqrt{y}$

39.

$$\left(4x^2\right)^{-1/2} = \frac{1}{\left(4x^2\right)^{1/2}} = \frac{1}{\sqrt{4x^2}} = \frac{1}{\sqrt{(2x)^2}} = \frac{1}{2x}$$

41. $\frac{5r^{3/4}}{r^{1/2}} = 5r^{3/4-1/2} = 5r^{1/4} = 5\sqrt[4]{r}$

43. $\left(\frac{a^2}{b^6}\right)^{1/3} = \sqrt[3]{\frac{a^2}{b^6}} = \frac{\sqrt[3]{a^2}}{\sqrt[3]{\left(b^2\right)^3}} = \frac{\sqrt[3]{a^2}}{b^2}$

45. $3x\left(16x^8\right)^{1/2} = 3x\sqrt{16x^8} = 3x\sqrt{\left(4x^4\right)^2}$
$= 3x\left(4x^4\right) = 12x^5$

47. $\frac{\left(2p^{1/6}\right)^6}{16p^3} = \frac{2^6 p^{6/6}}{16p^3} = \frac{64p}{16p^3} = \frac{4}{p^2}$

49. $9^{3/2} = \left(\sqrt{9}\right)^3 = 3^3 = 27$

51. $\left(-125u^3\right)^{2/3} = \left(\sqrt[3]{-125u^3}\right)^2 = \left(\sqrt[3]{(-5u)^3}\right)^2$
$= (-5u)^2 = 25u^2$

53. $n^{-1/6} \cdot 3n^{2/3} = 3n^{-1/6+2/3} = 3n^{-1/6+4/6}$
$= 3n^{3/6} = 3n^{1/2} = 3\sqrt{n}$

55.
$2x\left(64x^6\right)^{1/3} = 2x(64)^{1/3}\left(x^6\right)^{1/3} = 2x\sqrt[3]{64}\,(x)^{6\cdot(1/3)}$
$= 2x(4)x^2 = 8x^{1+2} = 8x^3$

57. **a.** $8000(0.5)^{t/3} = 8000\sqrt[3]{0.5^t}$

b.
$8000(0.5)^{6/3} = 8000(0.5)^2 = 8000(0.25) = 2000$
The value of the equipment 6 yr after it was purchased is \$2000.

8.3 Simplifying Radical Expressions

Practice

1. **a.** $\sqrt[8]{y^2} = y^{2/8} = y^{1/4} = \sqrt[4]{y}$
b. $\sqrt[6]{25} = 25^{1/6} = \left(5^2\right)^{1/6} = 5^{2/6} = 5^{1/3} = \sqrt[3]{5}$
c. $\sqrt{n}\cdot\sqrt[5]{n} = n^{1/2}\cdot n^{1/5} = n^{1/2+1/5} = n^{7/10} = \sqrt[10]{n^7}$
d. $\frac{\sqrt{t}}{\sqrt[4]{t}} = \frac{t^{1/2}}{t^{1/4}} = t^{1/2-1/4} = t^{1/4} = \sqrt[4]{t}$

2. **a.** $\sqrt[3]{\sqrt[4]{n}} = \left(n^{1/4}\right)^{1/3} = n^{1/12} = \sqrt[12]{n}$
b. $\sqrt[8]{a^2b^6} = \left(a^2b^6\right)^{1/8}$
$= \left(a^2\right)^{1/8}\left(b^6\right)^{1/8} = a^{2/8}b^{6/8}$
$= a^{1/4}b^{3/4} = \left(ab^3\right)^{1/4} = \sqrt[4]{ab^3}$
c. $\sqrt[5]{5}\cdot\sqrt{7} = 5^{1/5}\cdot 7^{1/2}$ cannot be simplified because the bases are not the same.

3. **a.** $\sqrt{7}\cdot\sqrt{3} = \sqrt{7\cdot 3} = \sqrt{21}$
b. $\sqrt[5]{2}\cdot\sqrt[5]{y^3} = \sqrt[5]{2\cdot y^3} = \sqrt[5]{2y^3}$
c. $\sqrt[3]{4p}\cdot\sqrt[3]{7p} = \sqrt[3]{4p\cdot 7p} = \sqrt[3]{28p^2}$
d. $\sqrt{\frac{3}{v}}\cdot\sqrt{\frac{5}{u}} = \sqrt{\frac{3}{v}\cdot\frac{5}{u}} = \sqrt{\frac{15}{uv}}$

4. **a.** $\sqrt{72} = \sqrt{36\cdot 2} = \sqrt{36}\cdot\sqrt{2} = 6\sqrt{2}$
b. $-7\sqrt{18} = -7\sqrt{9\cdot 2} = -7\sqrt{9}\cdot\sqrt{2}$
$= -7\cdot 3\sqrt{2} = -21\sqrt{2}$
c. $\sqrt[3]{56} = \sqrt[3]{8\cdot 7} = \sqrt[3]{8}\cdot\sqrt[3]{7} = 2\sqrt[3]{7}$
d. $\sqrt[4]{48} = \sqrt[4]{16\cdot 3} = \sqrt[4]{16}\cdot\sqrt[4]{3} = 2\sqrt[4]{3}$

5. **a.** $\sqrt{y^7} = \sqrt{y^6\cdot y} = \sqrt{y^6}\cdot\sqrt{y} = y^3\sqrt{y}$
b.
$\sqrt{18x^5} = \sqrt{9x^4\cdot 2x} = \sqrt{9x^4}\cdot\sqrt{2x} = 3x^2\sqrt{2x}$
c. $\sqrt[3]{81a^4b^6} = \sqrt[3]{27a^3b^6\cdot 3a} = \sqrt[3]{27a^3b^6}\cdot\sqrt[3]{3a}$
$= 3ab^2\sqrt[3]{3a}$

6. **a.** $\dfrac{\sqrt{42n}}{\sqrt{6}}=\sqrt{\dfrac{42n}{6}}=\sqrt{7n}$

b. $\dfrac{\sqrt{72}}{\sqrt{2}}=\sqrt{\dfrac{72}{2}}=\sqrt{36}=6$

c. $\dfrac{\sqrt[3]{10x^2y^2}}{\sqrt[3]{5x^2y}}=\sqrt[3]{\dfrac{10x^2y^2}{5x^2y}}=\sqrt[3]{2y}$

7. **a.** $\sqrt{\dfrac{16}{25}}=\dfrac{\sqrt{16}}{\sqrt{25}}=\dfrac{4}{5}$

b. $\sqrt{\dfrac{n}{64}}=\dfrac{\sqrt{n}}{\sqrt{64}}=\dfrac{\sqrt{n}}{8}$

c. $\sqrt[3]{\dfrac{64x}{125y^9}}=\dfrac{\sqrt[3]{64x}}{\sqrt[3]{125y^9}}=\dfrac{4\sqrt[3]{x}}{5y^3}$

d. $\sqrt{\dfrac{7p^6q^2}{36r^4}}=\dfrac{\sqrt{7p^6q^2}}{\sqrt{36r^4}}=\dfrac{p^3q\sqrt{7}}{6r^2}$

8.
$$\begin{aligned}d&=\sqrt{(x_2-x_1)^2+(y_2-y_1)^2}\\&=\sqrt{(3-5)^2+(-1-(-7))^2}\\&=\sqrt{(-2)^2+(6)^2}\\&=\sqrt{4+36}\\&=\sqrt{40}\\&=2\sqrt{10}\text{ units}\end{aligned}$$

9. **a.** $(-20,50)$ and $(10,60)$

b.
$$\begin{aligned}d&=\sqrt{(x_2-x_1)^2+(y_2-y_1)^2}\\&=\sqrt{(10-(-20))^2+(60-50)^2}\\&=\sqrt{(30)^2+(10)^2}\\&=\sqrt{900+100}\\&=\sqrt{1000}\\&=10\sqrt{10}\approx 31.6\end{aligned}$$

The drivers were $10\sqrt{10}$ mi (or about 31.6 mi) apart.

Exercises

1. The product rule of radicals states that to multiply radicals with the same index, multiply the radicands.

3. The quotient rule of radicals states that to divide radicals with the same index, divide the radicands.

5. $\sqrt[6]{n^2}=n^{2/6}=n^{1/3}=\sqrt[3]{n}$

7. $\sqrt[8]{16}=\sqrt[8]{2^4}=2^{4/8}=2^{1/2}=\sqrt{2}$

9. $\sqrt[3]{x}\bullet\sqrt[6]{x}=x^{1/3}\bullet x^{1/6}=x^{1/3+1/6}=x^{1/2}=\sqrt{x}$

11. $\sqrt[3]{p}\bullet\sqrt[4]{q}$ cannot be simplified.

13. $\dfrac{\sqrt[3]{x}}{\sqrt[4]{x}}=\dfrac{x^{1/3}}{x^{1/4}}=x^{1/3-1/4}=x^{1/12}=\sqrt[12]{x}$

15. $\sqrt{\sqrt{y}}=\left(y^{1/2}\right)^{1/2}=y^{(1/2)(1/2)}=y^{1/4}=\sqrt[4]{y}$

17. $\sqrt[4]{x^8y^2}=x^{8/4}y^{2/4}=x^2y^{1/2}=x^2\sqrt{y}$

19. $\sqrt[6]{x^4}\bullet\sqrt[3]{x}=x^{4/6}\bullet x^{1/3}=x^{2/3+1/3}=x^1=x$

21. $\dfrac{\sqrt[3]{y^2}}{\sqrt[9]{y^3}}=\dfrac{y^{2/3}}{y^{3/9}}=y^{2/3-1/3}=y^{1/3}=\sqrt[3]{y}$

23.
$$\begin{aligned}\sqrt[4]{p^2}\bullet\sqrt{q}&=p^{2/4}\bullet q^{1/2}=p^{1/2}\bullet q^{1/2}\\&=(pq)^{1/2}=\sqrt{pq}\end{aligned}$$

25. $\sqrt{6}\bullet\sqrt{5}=\sqrt{6\bullet5}=\sqrt{30}$

27. $\sqrt{3x}\bullet\sqrt{2y}=\sqrt{3x\bullet2y}=\sqrt{6xy}$

29. $\sqrt[3]{4a^2}\bullet\sqrt[3]{9b}=\sqrt[3]{4a^2\bullet9b}=\sqrt[3]{36a^2b}$

31. $\sqrt[4]{3n}\bullet\sqrt[4]{7n^2}=\sqrt[4]{3n\bullet7n^2}=\sqrt[4]{21n^3}$

33. $\sqrt{\dfrac{x}{2}}\bullet\sqrt{\dfrac{6}{y}}=\sqrt{\dfrac{x}{2}\bullet\dfrac{6}{y}}=\sqrt{\dfrac{3x}{y}}$

35. $\sqrt{24}=\sqrt{4\bullet6}=\sqrt{4}\bullet\sqrt{6}=2\sqrt{6}$

37.
$$\begin{aligned}-3\sqrt{80}&=-3\sqrt{16\bullet5}=-3\sqrt{16}\bullet\sqrt{5}\\&=-3\bullet4\sqrt{5}=-12\sqrt{5}\end{aligned}$$

39. $\sqrt[3]{81}=\sqrt[3]{27\bullet3}=\sqrt[3]{27}\bullet\sqrt[3]{3}=3\sqrt[3]{3}$

41. $\sqrt[4]{96}=\sqrt[4]{16\bullet6}=\sqrt[4]{16}\bullet\sqrt[4]{6}=2\sqrt[4]{6}$

43. $\sqrt[5]{64}=\sqrt[5]{32\bullet2}=\sqrt[5]{32}\bullet\sqrt[5]{2}=2\sqrt[5]{2}$

45. $6\sqrt{x^7}=6\sqrt{x^6\bullet x}=6\sqrt{x^6}\bullet\sqrt{x}=6x^3\sqrt{x}$

47. $\sqrt{20y}=\sqrt{4\bullet5y}=\sqrt{4}\bullet\sqrt{5y}=2\sqrt{5y}$

49.
$$\sqrt{200r^4}=\sqrt{100r^4\bullet2}=\sqrt{100r^4}\bullet\sqrt{2}=10r^2\sqrt{2}$$

51.
$$\begin{aligned}\sqrt{54x^5y^7}&=\sqrt{9x^4y^6\bullet6xy}\\&=\sqrt{9x^4y^6}\bullet\sqrt{6xy}=3x^2y^3\sqrt{6xy}\end{aligned}$$

53.
$$\sqrt[3]{32n^5}=\sqrt[3]{8n^3\bullet4n^2}=\sqrt[3]{8n^3}\bullet\sqrt[3]{4n^2}=2n\sqrt[3]{4n^2}$$

55.
$$\sqrt[5]{64n^8}=\sqrt[5]{32n^5\bullet2n^3}=\sqrt[5]{32n^5}\bullet\sqrt[5]{2n^3}=2n\sqrt[5]{2n^3}$$

57. $\sqrt[3]{72x^7y^9} = \sqrt[3]{8x^6y^9 \cdot 9x} = \sqrt[3]{8x^6y^9} \cdot \sqrt[3]{9x}$
$= 2x^2y^3\sqrt[3]{9x}$

59.

$\sqrt[4]{64x^5y^{10}} = \sqrt[4]{16x^4y^8 \cdot 4xy^2} = \sqrt[4]{16x^4y^8} \cdot \sqrt[4]{4xy^2}$
$= 2xy^2\sqrt[4]{4xy^2}$

61. $\frac{\sqrt{90}}{\sqrt{10}} = \sqrt{\frac{90}{10}} = \sqrt{9} = 3$

63. $\frac{\sqrt{30n}}{\sqrt{6}} = \sqrt{\frac{30n}{6}} = \sqrt{5n}$

65. $\frac{\sqrt{12x^3y}}{\sqrt{3x}} = \sqrt{\frac{12x^3y}{3x}} = \sqrt{4x^2y} = 2x\sqrt{y}$

67. $\frac{\sqrt[3]{16u^2}}{\sqrt[3]{4u}} = \sqrt[3]{\frac{16u^2}{4u}} = \sqrt[3]{4u}$

69. $\frac{\sqrt[4]{24a^3b^2}}{\sqrt[4]{4a^2b}} = \sqrt[4]{\frac{24a^3b^2}{4a^2b}} = \sqrt[4]{6ab}$

71. $\sqrt{\frac{25}{16}} = \frac{\sqrt{25}}{\sqrt{16}} = \frac{5}{4}$

73. $\sqrt{\frac{7}{81}} = \frac{\sqrt{7}}{\sqrt{81}} = \frac{\sqrt{7}}{9}$

75. $\sqrt{\frac{2}{n^6}} = \frac{\sqrt{2}}{\sqrt{n^6}} = \frac{\sqrt{2}}{n^3}$

77. $\sqrt{\frac{7a}{121}} = \frac{\sqrt{7a}}{\sqrt{121}} = \frac{\sqrt{7a}}{11}$

79. $\sqrt{\frac{a}{9b^4}} = \frac{\sqrt{a}}{\sqrt{9b^4}} = \frac{\sqrt{a}}{3b^2}$

81. $\sqrt{\frac{9u}{25v^2}} = \frac{\sqrt{9u}}{\sqrt{25v^2}} = \frac{3\sqrt{u}}{5v}$

83. $\sqrt[3]{\frac{27a^2}{64}} = \frac{\sqrt[3]{27a^2}}{\sqrt[3]{64}} = \frac{3\sqrt[3]{a^2}}{4}$

85. $\sqrt[3]{\frac{9a}{8b^6c^9}} = \frac{\sqrt[3]{9a}}{\sqrt[3]{8b^6c^9}} = \frac{\sqrt[3]{9a}}{2b^2c^3}$

87. $\sqrt[4]{\frac{2a^4b^3}{81c^8}} = \frac{\sqrt[4]{2a^4b^3}}{\sqrt[4]{81c^8}} = \frac{a\sqrt[4]{2b^3}}{3c^2}$

89. $\sqrt{\frac{13a^4b}{9c^6d^2}} = \frac{\sqrt{13a^4b}}{\sqrt{9c^6d^2}} = \frac{a^2\sqrt{13b}}{3c^3d}$

91. $d = \sqrt{(x_2 - x_1)^2 + (y_2 - y_1)^2}$
$= \sqrt{(4-10)^2 + (-3-3)^2} = \sqrt{(-6)^2 + (-6)^2}$
$= \sqrt{72} = \sqrt{36 \cdot 2} = 6\sqrt{2}$

93. $d = \sqrt{(x_2 - x_1)^2 + (y_2 - y_1)^2}$
$= \sqrt{(6-12)^2 + (12-15)^2} = \sqrt{(-6)^2 + (-3)^2}$
$= \sqrt{36+9} = \sqrt{45} = \sqrt{9 \cdot 5} = 3\sqrt{5}$

95. $d = \sqrt{(x_2 - x_1)^2 + (y_2 - y_1)^2}$
$= \sqrt{(-8-(-4))^2 + (0-(-3))^2}$
$= \sqrt{(-4)^2 + (3)^2} = \sqrt{16+9} = \sqrt{25} = 5$

97. $\sqrt[4]{5x^2} \cdot \sqrt[4]{6xy^3} = \sqrt[4]{(5x^2)(6xy^3)} = \sqrt[4]{30x^3y^3}$

99. $\sqrt[3]{x^2} \cdot \sqrt[3]{x^4} = \sqrt[3]{(x^2)(x^4)} = \sqrt[3]{x^6} = \sqrt[3]{(x^2)^3} = x^2$

101. $\sqrt{50r^6s^7} = \sqrt{2 \cdot 25(r^3)^2(s^3)^2 s} = 5r^3s^3\sqrt{2s}$

103. $\sqrt[3]{\frac{25x}{64y^3z^9}} = \frac{\sqrt[3]{25x}}{\sqrt[3]{4^3y^3(z^3)^3}} = \frac{\sqrt[3]{25x}}{4yz^3}$

105. $\sqrt[3]{128u^4v^6} = \sqrt[3]{64 \cdot 2(u^3)u(v^2)^3} = 4uv^2\sqrt[3]{2u}$

107. $\sqrt{64d} = \sqrt{64} \cdot \sqrt{d} = 8\sqrt{d}$

109. $w = \sqrt{d^2 - l^2} = \sqrt{42^2 - 30^2} = \sqrt{1764 - 900}$
$= \sqrt{864} = \sqrt{144 \cdot 6} = 12\sqrt{6} = 29.3938....$
The width of the screen is $12\sqrt{6}$ or approximately 29 in.

111. $d = \sqrt{68^2 + 24^2} = \sqrt{4624 + 576} = \sqrt{5200}$
$= \sqrt{400 \cdot 13} = 20\sqrt{13} = 72.1110...$
The length of the string let out is $20\sqrt{13}$ ft, or about 72.1 ft.

113. **a.** The second college is located at (48, 20).

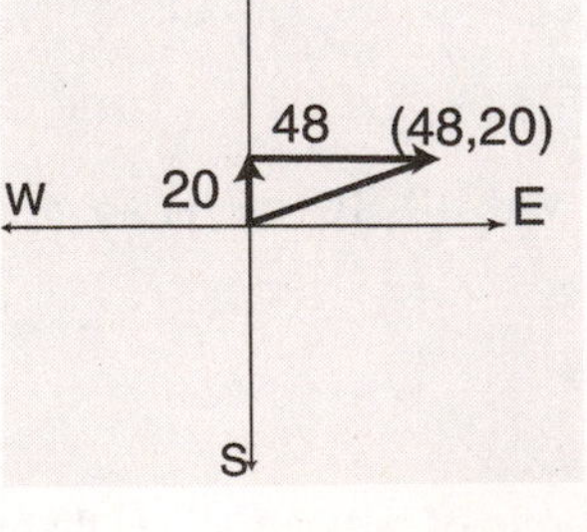

b.
$d = \sqrt{48^2 + 20^2} = \sqrt{2304 + 400} = \sqrt{2704} = 52$
The second college is 52 mi from her home.

8.4 Addition and Subtraction of Radical Expressions

Practice

1. a. $-\sqrt[4]{9}+3\sqrt[4]{9}=(-1+3)\sqrt[4]{9}=2\sqrt[4]{9}$

   b. $6\sqrt{p}+\sqrt{p}-2\sqrt{p}=(6+1-2)\sqrt{p}$
   $=5\sqrt{p}$

   c. $-5\sqrt{t^2-4}+\sqrt{t^2-4}$
   $=(-5+1)\sqrt{t^2-4}=-4\sqrt{t^2-4}$

   d. $5\sqrt[3]{a}-2\sqrt[3]{b}$ cannot be simplified because the terms are not like terms.

2. a. $\sqrt{75}+\sqrt{27}=\sqrt{25\cdot 3}+\sqrt{9\cdot 3}$
   $=\sqrt{25}\sqrt{3}+\sqrt{9}\sqrt{3}=5\sqrt{3}+3\sqrt{3}$
   $=(5+3)\sqrt{3}=8\sqrt{3}$

   b. $\sqrt{96}-3\sqrt{54}-7\sqrt{24}$
   $=\sqrt{16\cdot 6}-3\sqrt{9\cdot 6}-7\sqrt{4\cdot 6}$
   $=\sqrt{16}\sqrt{6}-3\sqrt{9}\sqrt{6}-7\sqrt{4}\sqrt{6}$
   $=4\sqrt{6}-3\cdot 3\sqrt{6}-7\cdot 2\sqrt{6}$
   $=(4-9-14)\sqrt{6}$
   $=-19\sqrt{6}$

   c. $9\sqrt[4]{2}-4\sqrt[4]{162}-\sqrt[4]{32}$
   $=9\sqrt[4]{2}-4\sqrt[4]{81\cdot 2}-\sqrt[4]{16\cdot 2}$
   $=9\sqrt[4]{2}-4\sqrt[4]{81}\sqrt[4]{2}-\sqrt[4]{16}\sqrt[4]{2}$
   $=9\sqrt[4]{2}-4\cdot 3\sqrt[4]{2}-2\sqrt[4]{2}$
   $=(9-12-2)\sqrt[4]{2}$
   $=-5\sqrt[4]{2}$

   d. $\sqrt{64}-\sqrt{28}+\sqrt{63}$
   $=\sqrt{64}-\sqrt{4\cdot 7}+\sqrt{9\cdot 7}$
   $=8-2\sqrt{7}+3\sqrt{7}$
   $=8+(-2+3)\sqrt{7}=8+\sqrt{7}$

3. a. $4\sqrt{25n}+\sqrt{n}=4\cdot 5\sqrt{n}+\sqrt{n}$
   $=(20+1)\sqrt{n}=21\sqrt{n}$

   b. $\sqrt[3]{27ab^3}-11\sqrt[3]{a}=\sqrt[3]{27b^3\cdot a}-11\sqrt[3]{a}$
   $=3b\sqrt[3]{a}-11\sqrt[3]{a}=(3b-11)\sqrt[3]{a}$

   c.
   $\sqrt{50x}-\sqrt{32x^3}=\sqrt{25\cdot 2x}-\sqrt{16x^2\cdot 2x}$
   $=5\sqrt{2x}-4x\sqrt{2x}=(5-4x)\sqrt{2x}$

4. a. $p(x)+q(x)=x\sqrt[3]{81x}+\sqrt[3]{24x^4}$
   $=x\sqrt[3]{27\cdot 3x}+\sqrt[3]{8x^3\cdot 3x}$
   $=3x\sqrt[3]{3x}+2x\sqrt[3]{3x}$
   $=(3x+2x)\sqrt[3]{3x}$
   $=5x\sqrt[3]{3x}$

   b. $p(x)-q(x)=x\sqrt[3]{81x}-\sqrt[3]{24x^4}$
   $=x\sqrt[3]{27\cdot 3x}-\sqrt[3]{8x^3\cdot 3x}$
   $=3x\sqrt[3]{3x}-2x\sqrt[3]{3x}$
   $=(3x-2x)\sqrt[3]{3x}$
   $=x\sqrt[3]{3x}$

5. a. $\sqrt{16^2+9^2}=\sqrt{256+81}=\sqrt{337}$ in.
   $\sqrt{16^2+12^2}=\sqrt{256+144}=\sqrt{400}=20$ in.

   b. $\left(20-\sqrt{337}\right)\approx 1.6$ in.

Exercises

1. Radicals that have the same index and the same radicand are called like radicals.
3. Some unlike radicals can become like radicals when they are simplified.
5. $4\sqrt{3}+\sqrt{3}=(4+1)\sqrt{3}=5\sqrt{3}$
7. $3\sqrt[3]{2}-6\sqrt[3]{2}=(3-6)\sqrt[3]{2}=-3\sqrt[3]{2}$
9. $4\sqrt{2}-2\sqrt{7}$ cannot be simplified.
11. $-8\sqrt{y}-12\sqrt{y}=(-8-12)\sqrt{y}=-20\sqrt{y}$
13. $7\sqrt{x}+4\sqrt[3]{x}$ cannot be simplified.
15. $-11\sqrt{n}+2\sqrt{n}+9\sqrt{n}=(-11+2+9)\sqrt{n}=0$
17. $12\sqrt[4]{p}-5\sqrt[4]{p}+\sqrt[4]{p}=(12-5+1)\sqrt[4]{p}=8\sqrt[4]{p}$
19. $9y\sqrt{3x}+4y\sqrt{3x}=(9y+4y)\sqrt{3x}=13y\sqrt{3x}$
21. $-7\sqrt{2r-1}+3\sqrt{2r-1}=(-7+3)\sqrt{2r-1}$
    $=-4\sqrt{2r-1}$
23. $\sqrt{x^2-9}+\sqrt{x^2-9}=(1+1)\sqrt{x^2-9}$
    $=2\sqrt{x^2-9}$
25. $\sqrt{6}+\sqrt{24}=\sqrt{6}+\sqrt{4\bullet 6}=\sqrt{6}+2\sqrt{6}$
    $=(1+2)\sqrt{6}=3\sqrt{6}$

27. $\sqrt{72}-\sqrt{8}=\sqrt{36\bullet 2}-\sqrt{4\bullet 2}$
$=6\sqrt{2}-2\sqrt{2}$
$=(6-2)\sqrt{2}=4\sqrt{2}$

29. $3\sqrt{18}-4\sqrt{2}+\sqrt{50}$
$=3\sqrt{9\bullet 2}-4\sqrt{2}+\sqrt{25\bullet 2}$
$=9\sqrt{2}-4\sqrt{2}+5\sqrt{2}$
$=(9-4+5)\sqrt{2}=10\sqrt{2}$

31. $-5\sqrt[3]{24}+\sqrt[3]{-81}+9\sqrt[3]{3}$
$=-5\sqrt[3]{8\bullet 3}+\sqrt[3]{(-27)\bullet 3}+9\sqrt[3]{3}$
$=-10\sqrt[3]{3}-3\sqrt[3]{3}+9\sqrt[3]{3}$
$=(-10-3+9)\sqrt[3]{3}=-4\sqrt[3]{3}$

33. $\sqrt{54x^3}-x\sqrt{150x}$
$=\sqrt{9x^2\bullet 6x}-x\sqrt{25\bullet 6x}$
$=3x\sqrt{6x}-5x\sqrt{6x}$
$=(3x-5x)\sqrt{6x}=-2x\sqrt{6x}$

35. $\frac{1}{4}\sqrt{128n^5}+\sqrt{242n}$
$=\frac{1}{4}\sqrt{64n^4\bullet 2n}+\sqrt{121\bullet 2n}$
$=\frac{8n^2}{4}\sqrt{2n}+11\sqrt{2n}$
$=(2n^2+11)\sqrt{2n}$

37. $-4y\sqrt{xy^3}+7x\sqrt{x^3y}$
$=-4y\sqrt{y^2\bullet xy}+7x\sqrt{x^2\bullet xy}$
$=-4y^2\sqrt{xy}+7x^2\sqrt{xy}$
$=(-4y^2+7x^2)\sqrt{xy}$

39. $\sqrt[3]{-125p^9}+p\sqrt[3]{-8p^6}$
$=-5p^3+p(-2p^2)$
$=-5p^3-2p^3=-7p^3$

41. $5\sqrt[4]{48ab^3}-2\sqrt[4]{3a^5b^3}$
$=5\sqrt[4]{16\bullet 3ab^3}-2\sqrt[4]{a^4\bullet 3ab^3}$
$=10\sqrt[4]{3ab^3}-2a\sqrt[4]{3ab^3}$
$=(10-2a)\sqrt[4]{3ab^3}$

43. $2\sqrt{125}+\frac{1}{6}\sqrt{144}-\frac{3}{4}\sqrt{80}$
$=2\sqrt{25\bullet 5}+\frac{1}{6}(12)-\frac{3}{4}\sqrt{16\bullet 5}$
$=10\sqrt{5}+2-\frac{3}{4}(4)\sqrt{5}$
$=10\sqrt{5}+2-3\sqrt{5}$
$=(10-3)\sqrt{5}+2=7\sqrt{5}+2$

45. $\frac{3}{5}\sqrt[3]{-125}-\frac{1}{4}\sqrt[3]{48}+\frac{1}{2}\sqrt[3]{162}$
$=\frac{3}{5}(-5)-\frac{1}{4}\sqrt[3]{8\bullet 6}+\frac{1}{2}\sqrt[3]{27\bullet 6}$
$=-3-\frac{1}{4}(2)\sqrt[3]{6}+\frac{1}{2}(3)\sqrt[3]{6}$
$=-3+\left(-\frac{1}{2}+\frac{3}{2}\right)\sqrt[3]{6}=-3+\frac{2}{2}\sqrt[3]{6}=\sqrt[3]{6}-3$

47. $f(x)+g(x)=5x\sqrt{20x}+3\sqrt{5x^3}$
$=5x\sqrt{4\bullet 5x}+3\sqrt{x^2\bullet 5x}$
$=10x\sqrt{5x}+3x\sqrt{5x}$
$=(10x+3x)\sqrt{5x}=13x\sqrt{5x}$

$f(x)-g(x)=5x\sqrt{20x}-3\sqrt{5x^3}$
$=5x\sqrt{4\bullet 5x}-3\sqrt{x^2\bullet 5x}$
$=10x\sqrt{5x}-3x\sqrt{5x}$
$=(10x-3x)\sqrt{5x}=7x\sqrt{5x}$

49. $f(x)+g(x)=2x\sqrt[4]{64x}+\left(-\sqrt[4]{4x^5}\right)$
$=2x\sqrt[4]{16\bullet 4x}-\sqrt[4]{x^4\bullet 4x}$
$=4x\sqrt[4]{4x}-x\sqrt[4]{4x}$
$=(4x-x)\sqrt[4]{4x}=3x\sqrt[4]{4x}$

$f(x)-g(x)=2x\sqrt[4]{64x}-\left(-\sqrt[4]{4x^5}\right)$
$=2x\sqrt[4]{16\bullet 4x}+\sqrt[4]{x^4\bullet 4x}$
$=4x\sqrt[4]{4x}+x\sqrt[4]{4x}$
$=(4x+x)\sqrt[4]{4x}=5x\sqrt[4]{4x}$

51. $-8\sqrt{24}-3\sqrt{6}+2\sqrt{150}$
$=-8\sqrt{4\bullet 6}-3\sqrt{6}+2\sqrt{25\bullet 6}$
$=-8(2)\sqrt{6}-3\sqrt{6}+2(5)\sqrt{6}$
$=-16\sqrt{6}-3\sqrt{6}+10\sqrt{6}$
$=(-16-3+10)\sqrt{6}-9\sqrt{6}$

53. $-9\sqrt{3b}-\sqrt{3b}+5\sqrt{3b}$
$=(-9-1+5)\sqrt{3b}=-5\sqrt{3b}$

55. $\frac{1}{3}\sqrt{405k^3}-\sqrt{180k}$
$=\frac{1}{3}\sqrt{81\bullet 5k\bullet k^2}-\sqrt{36\bullet 5k}$
$=\frac{9k}{3}\sqrt{5k}-6\sqrt{5k}$
$=3k\sqrt{5k}-6\sqrt{5k}$
$=(3k-6)\sqrt{5k}$

57. $\sqrt{64(80)}-\sqrt{64(20)}$
$=\sqrt{5120}-\sqrt{1280}$
$=\sqrt{1024\bullet 5}-\sqrt{256\bullet 5}$
$=32\sqrt{5}-16\sqrt{5}$
$=(32-16)\sqrt{5}=16\sqrt{5}$
$=35.7770...$
The difference in velocity is $16\sqrt{5}$ ft per sec or about 35.8 ft per sec.

59. $\sqrt{128}-\sqrt{72}=\sqrt{64\bullet 2}-\sqrt{36\bullet 2}$
$=8\sqrt{2}-6\sqrt{2}=(8-6)\sqrt{2}$
$=2\sqrt{2}=2.82842....$
The side of the larger tile is $2\sqrt{2}$ in. or about 2.8 in. longer than the smaller tile.

61. **a.** $l=\sqrt{b^2+h^2}=\sqrt{4^2+2^2}=\sqrt{16+4}$
$=\sqrt{20}=\sqrt{4\bullet 5}=2\sqrt{5}$
The length of the ramp is $2\sqrt{5}$ ft.

b. $\sqrt{(4+2)^2+2^2}-2\sqrt{5}$
$=\sqrt{6^2+2^2}-2\sqrt{5}$
$=\sqrt{36+4}-2\sqrt{5}$
$=\sqrt{40}-2\sqrt{5}$
$=6.32455...\ -4.47213...$
$=1.8524...$
The length of the ramp increases by approximately 1.9 ft.

8.5 Multiplication and Division of Radical Expressions

Practice

1. **a.** $\sqrt{6y}\cdot\sqrt{12y^7}=\sqrt{6y\cdot 12y^7}$
$=\sqrt{72y^8}=\sqrt{36y^2\cdot 2}$
$=6y^4\sqrt{2}$

b. $\left(-2\sqrt[3]{3}\right)\left(7\sqrt[3]{5}\right)=-2\cdot 7\cdot\sqrt[3]{3\cdot 5}=-14\sqrt[3]{15}$

2. **a.** $\sqrt[4]{5}\left(\sqrt[4]{7}+6\sqrt[4]{2}\right)$
$=\sqrt[4]{5}\cdot\sqrt[4]{7}+\sqrt[4]{5}\cdot 6\sqrt[4]{2}$
$=\sqrt[4]{35}+6\sqrt[4]{10}$

b. $\left(3\sqrt{x}-1\right)\left(2\sqrt{x}+1\right)$
$=\left(3\sqrt{x}\right)\left(2\sqrt{x}\right)+\left(3\sqrt{x}\right)(1)$
$+(-1)\left(2\sqrt{x}\right)+(-1)(1)$
$=3\cdot 2\cdot\sqrt{x^2}+3\cdot 1\sqrt{x}-1\cdot 2\sqrt{x}-1$
$=6x+3\sqrt{x}-2\sqrt{x}-1$
$=6x+\sqrt{x}-1$

3. **a.** $\left(\sqrt{2}-\sqrt{5}\right)\left(\sqrt{2}+\sqrt{5}\right)$
$=\left(\sqrt{2}\right)^2-\left(\sqrt{5}\right)^2$
$=2-5=-3$

b. $\left(\sqrt{2x}+3\right)\left(\sqrt{2x}-3\right)$
$=\left(\sqrt{2x}\right)^2-(3)$
$=2x-9$

c. $\left(\sqrt{p}-4\sqrt{q}\right)\left(\sqrt{p}+4\sqrt{q}\right)$
$=\left(\sqrt{p}\right)^2-\left(4\sqrt{q}\right)^2$
$=p-16q$

4. **a.** $\left(\sqrt{3}+2b\right)^2$
$=\left(\sqrt{3}\right)^2+2\left(\sqrt{3}\right)(2b)+(2b)^2$
$=3+4\sqrt{3}b+4b^2$ or $4b^2+4\sqrt{3}b+3$

4. **b.** $\left(\sqrt{n+1}-3\right)^2$

$=\left(\sqrt{n+1}\right)^2-2\left(\sqrt{n+1}\right)(3)+(3)^2$

$=n+1-6\sqrt{n+1}+9$

$=n-6\sqrt{n+1}+10$

5. Area $=\left(\sqrt{6}\right)\left(\sqrt{20}\right)=\sqrt{120}=2\sqrt{30}\text{ cm}^2$

6. **a.** $\dfrac{5\sqrt{12}}{\sqrt{3}}=5\sqrt{\dfrac{12}{3}}=5\sqrt{4}=5\cdot 2=10$

b.

$\dfrac{\sqrt[3]{18p^7}}{-5p\sqrt[3]{2p^2}}=-\dfrac{1}{5p}\cdot\sqrt[3]{\dfrac{18p^7}{2p^2}}=-\dfrac{1}{5p}\cdot\sqrt[3]{9p^5}$

$=-\dfrac{p\sqrt[3]{9p^2}}{5p}=-\dfrac{\sqrt[3]{9p^2}}{5}$

7. **a.** $\sqrt{\dfrac{y^2}{9x^8}}=\dfrac{\sqrt{y^2}}{\sqrt{9x^8}}=\dfrac{y}{3x^4}$

b. $\sqrt[3]{\dfrac{2n^5}{27}}=\dfrac{\sqrt[3]{2n^5}}{\sqrt[3]{27}}=\dfrac{n\sqrt[3]{2n^2}}{3}$

8. **a.** $\dfrac{4}{\sqrt{7}}=\dfrac{4}{\sqrt{7}}\cdot\dfrac{\sqrt{7}}{\sqrt{7}}=\dfrac{4\sqrt{7}}{\sqrt{7^2}}=\dfrac{4\sqrt{7}}{7}$

b.

$\dfrac{\sqrt{y}}{\sqrt{32}}=\dfrac{\sqrt{y}}{\sqrt{32}}\cdot\dfrac{\sqrt{32}}{\sqrt{32}}=\dfrac{\sqrt{32y}}{\sqrt{32^2}}=\dfrac{4\sqrt{2y}}{32}=\dfrac{\sqrt{2y}}{8}$

c. $\dfrac{\sqrt{4p^4}}{\sqrt{18}}=\dfrac{2p^2}{\sqrt{18}}=\dfrac{2p^2}{\sqrt{18}}\cdot\dfrac{\sqrt{18}}{\sqrt{18}}$

$=\dfrac{2p^2\sqrt{18}}{\sqrt{18^2}}=\dfrac{2p^2\cdot 3\sqrt{2}}{18}=\dfrac{p^2\sqrt{2}}{3}$

d.

$\dfrac{\sqrt[3]{5pq}}{\sqrt[3]{2r^2}}=\dfrac{\sqrt[3]{5pq}}{\sqrt[3]{2r^2}}\cdot\dfrac{\sqrt[3]{4r}}{\sqrt[3]{4r}}=\dfrac{\sqrt[3]{20pqr}}{\sqrt[3]{8r^3}}=\dfrac{\sqrt[3]{20pqr}}{2r}$

9. **a.** $\sqrt{\dfrac{x}{3y}}=\dfrac{\sqrt{x}}{\sqrt{3y}}=\dfrac{\sqrt{x}}{\sqrt{3y}}\cdot\dfrac{\sqrt{3y}}{\sqrt{3y}}=\dfrac{\sqrt{3xy}}{3y}$

b. $\sqrt[3]{\dfrac{5}{a}}=\dfrac{\sqrt[3]{5}}{\sqrt[3]{a}}=\dfrac{\sqrt[3]{5}}{\sqrt[3]{a}}\cdot\dfrac{\sqrt[3]{a^2}}{\sqrt[3]{a^2}}=\dfrac{\sqrt[3]{5a^2}}{a}$

10. **a.** $\dfrac{\sqrt{6}-9}{\sqrt{3}}=\dfrac{\sqrt{6}-9}{\sqrt{3}}\cdot\dfrac{\sqrt{3}}{\sqrt{3}}$

$=\dfrac{\left(\sqrt{6}-9\right)\sqrt{3}}{3}=\dfrac{\sqrt{6}\cdot\sqrt{3}-9\sqrt{3}}{3}$

$=\dfrac{\sqrt{18}-9\sqrt{3}}{3}=\dfrac{3\sqrt{2}-9\sqrt{3}}{3}$

$=\sqrt{2}-3\sqrt{3}$

b.

$\dfrac{5+\sqrt{b}}{\sqrt{b}}=\dfrac{5+\sqrt{b}}{\sqrt{b}}\cdot\dfrac{\sqrt{b}}{\sqrt{b}}=\dfrac{\left(5+\sqrt{b}\right)\sqrt{b}}{b}=\dfrac{5\sqrt{b}+b}{b}$

11. **a.** $\dfrac{4}{2+\sqrt{2}}=\dfrac{4}{2+\sqrt{2}}\cdot\dfrac{2-\sqrt{2}}{2-\sqrt{2}}$

$=\dfrac{4\left(2-\sqrt{2}\right)}{\left(2+\sqrt{2}\right)\left(2-\sqrt{2}\right)}=\dfrac{8-4\sqrt{2}}{2^2-\left(\sqrt{2}\right)^2}$

$=\dfrac{8-4\sqrt{2}}{4-2}=\dfrac{8-4\sqrt{2}}{2}=4-2\sqrt{2}$

b. $\dfrac{a}{\sqrt{b}-\sqrt{c}}=\dfrac{a}{\sqrt{b}-\sqrt{c}}\cdot\dfrac{\sqrt{b}+\sqrt{c}}{\sqrt{b}+\sqrt{c}}$

$=\dfrac{a\left(\sqrt{b}+\sqrt{c}\right)}{\left(\sqrt{b}\right)^2-\left(\sqrt{c}\right)^2}=\dfrac{a\sqrt{b}+a\sqrt{c}}{b-c}$

12. **a.** $p(x)\cdot q(x)=x\sqrt[3]{81x}\cdot\sqrt[3]{24x^4}$

$=x\sqrt[3]{81\cdot 24\cdot x^5}=x\sqrt[3]{1944x^5}$

$=x\sqrt[3]{6^3x^3\cdot 9x^2}=6x\cdot x\sqrt[3]{9x^2}$

$=6x^2\sqrt[3]{9x^2}$

b. $\dfrac{p(x)}{q(x)}=\dfrac{x\sqrt[3]{81x}}{\sqrt[3]{24x^4}}=x\sqrt[3]{\dfrac{81x}{24x^4}}$

$=x\sqrt[3]{\dfrac{27}{8x^3}}=x\cdot\dfrac{3}{2x}=\dfrac{3}{2}$

13. **a.** $\dfrac{\sqrt[3]{3V}}{\sqrt[3]{4\pi}}=\dfrac{\sqrt[3]{3V}}{\sqrt[3]{4\pi}}\cdot\dfrac{\sqrt[3]{2\pi^2}}{\sqrt[3]{2\pi^2}}=\dfrac{\sqrt[3]{6\pi^2V}}{2\pi}$

b. No, $\dfrac{\sqrt[3]{6\pi^2(2V)}}{2\pi}=\dfrac{\sqrt[3]{12\pi^2V}}{2\pi}\neq 2\left(\dfrac{\sqrt[3]{6\pi^2V}}{2\pi}\right)$

Exercises

1. To multiply radical expressions containing more than one term, we use the distributive property.

3. To simplify a radical that has its radicand in the form of a fraction, we can use the quotient rule of radicals "in reverse."

5. A denominator can be rationalized by multiplying the numerator and denominator by a factor that makes the radicand in the denominator a(n) perfect power.

7. $\sqrt{12}\cdot\sqrt{8}=\sqrt{12\cdot 8}=\sqrt{96}$
$=\sqrt{16\cdot 6}=4\sqrt{6}$

9. $\left(-4\sqrt{3}\right)\left(\sqrt{7}\right)=-4\sqrt{3\cdot 7}=-4\sqrt{21}$

11. $5\sqrt[3]{6}\left(3\sqrt[3]{9}\right)=15\sqrt[3]{6\cdot 9}=15\sqrt[3]{54}$
$=15\sqrt[3]{27\cdot 2}=15\cdot 3\sqrt[3]{2}$
$=45\sqrt[3]{2}$

13. $\left(2\sqrt{10x}\right)\left(7\sqrt{5x^5}\right)=14\sqrt{10x\cdot 5x^5}$
$=14\sqrt{50x^6}$
$=14\sqrt{25x^6\cdot 2}$
$=14\cdot 5x^3\sqrt{2}$
$=70x^3\sqrt{2}$

15. $\left(8\sqrt{ab^3}\right)\left(-2\sqrt{a^3b}\right)=-16\sqrt{ab^3\cdot a^3b}$
$=-16\sqrt{a^4b^4}$
$=-16a^2b^2$

17. $\sqrt[3]{12x^2y}\cdot\sqrt[3]{-16xy^4}=\sqrt[3]{12x^2y\cdot\left(-16xy^4\right)}$
$=\sqrt[3]{-192x^3y^5}$
$=\sqrt[3]{-64x^3y^3\left(3y^2\right)}$
$=-4xy\sqrt[3]{3y^2}$

19. $\sqrt{2}\left(\sqrt{8}-4\right)=\sqrt{2}\cdot\sqrt{8}-4\sqrt{2}$
$=\sqrt{2\cdot 8}-4\sqrt{2}=\sqrt{16}-4\sqrt{2}$
$=4-4\sqrt{2}$

21. $\sqrt{6}\left(2\sqrt{3}+\sqrt{12}\right)=2\sqrt{6}\cdot\sqrt{3}+\sqrt{6}\sqrt{12}$
$=2\sqrt{18}+\sqrt{72}$
$=2\sqrt{9\cdot 2}+\sqrt{36\cdot 2}$
$=6\sqrt{2}+6\sqrt{2}=12\sqrt{2}$

23. $-2\sqrt{3}\left(2\sqrt{5}-6\sqrt{3}\right)=-4\sqrt{3}\cdot\sqrt{5}+12\sqrt{3}\sqrt{3}$
$=-4\sqrt{15}+12\sqrt{9}$
$=-4\sqrt{15}+12\cdot 3$
$=-4\sqrt{15}+36$

25. $\sqrt[3]{4}\left(5\sqrt[3]{12}+2\sqrt[3]{3}\right)=5\sqrt[3]{4}\cdot\sqrt[3]{12}+2\sqrt[3]{4}\cdot\sqrt[3]{3}$
$=5\sqrt[3]{48}+2\sqrt[3]{12}$
$=5\sqrt[3]{8\cdot 6}+2\sqrt[3]{12}$
$=5\cdot 2\sqrt[3]{6}+2\sqrt[3]{12}$
$=10\sqrt[3]{6}+2\sqrt[3]{12}$

27. $\sqrt{x}\left(\sqrt{x^3}+\sqrt{2x}\right)=\sqrt{x}\cdot\sqrt{x^3}+\sqrt{x}\cdot\sqrt{2x}$
$=\sqrt{x^4}+\sqrt{2x^2}=x^2+x\sqrt{2}$

29. $\sqrt[4]{a^2}\left(3\sqrt[4]{2a^3}-\sqrt[4]{10a^2}\right)$
$=3\sqrt[4]{a^2}\cdot\sqrt[4]{2a^3}-\sqrt[4]{a^2}\cdot\sqrt[4]{10a^2}$
$=3\sqrt[4]{2a^5}-\sqrt[4]{10a^4}$
$=3\sqrt[4]{a^4\cdot 2a}-a\sqrt[4]{10}$
$=3a\sqrt[4]{2a}-a\sqrt[4]{10}$

31. $\left(\sqrt{2}-3\right)\left(\sqrt{2}+4\right)$
F O I L
$=\sqrt{2}\cdot\sqrt{2}+4\sqrt{2}-3\sqrt{2}-12$
$=\left(\sqrt{2}\right)^2+(4-3)\sqrt{2}-12$
$=2+\sqrt{2}-12=-10+\sqrt{2}$

33. $\left(2-4\sqrt{3}\right)\left(4+3\sqrt{3}\right)$
F O I L
$=8+6\sqrt{3}-16\sqrt{3}-12\left(\sqrt{3}\right)^2$
$=8-10\sqrt{3}-36=-28-10\sqrt{3}$

35. $\left(\sqrt{8}+\sqrt{3}\right)\left(\sqrt{2}+\sqrt{12}\right)$
F O I L
$=\sqrt{8}\cdot\sqrt{2}+\sqrt{8}\cdot\sqrt{12}+\sqrt{3}\cdot\sqrt{2}+\sqrt{3}\cdot\sqrt{12}$
$=\sqrt{16}+\sqrt{96}+\sqrt{6}+\sqrt{36}$
$=4+\sqrt{16\cdot 6}+\sqrt{6}+6=10+4\sqrt{6}+\sqrt{6}$
$=10+5\sqrt{6}$

37. $\left(2\sqrt{r}-4\right)\left(8\sqrt{r}+6\right)$
F O I L
$=16\left(\sqrt{r}\right)^2+12\sqrt{r}-32\sqrt{r}-24$
$=16r-20\sqrt{r}-24$

39. $\left(\sqrt{6}+\sqrt{2}\right)\left(\sqrt{6}-\sqrt{2}\right)=\left(\sqrt{6}\right)^2-\left(\sqrt{2}\right)^2$
$=6-2=4$

41. $\left(\sqrt{x}-8\right)\left(\sqrt{x}+8\right)=\left(\sqrt{x}\right)^2-8^2=x-64$

43. $\left(\sqrt{5x}+\sqrt{y}\right)\left(\sqrt{5x}-\sqrt{y}\right)$
$=\left(\sqrt{5x}\right)^2-\left(\sqrt{y}\right)^2=5x-y$

45. $\left(\sqrt{x-1}-5\right)\left(\sqrt{x-1}+5\right)=\left(\sqrt{x-1}\right)^2-5^2$
$=x-1-25$
$=x-26$

47. $\left(3\sqrt{x}-2\right)^2=\left(3\sqrt{x}\right)^2-2\left(3\sqrt{x}\right)(2)+2^2$
$=9x-12\sqrt{x}+4$

49. $\left(\sqrt{6a}+4\right)^2=\left(\sqrt{6a}\right)^2+2\left(\sqrt{6a}\right)(4)+4^2$
$=6a+8\sqrt{6a}+16$

51. $\left(1-\sqrt{n+7}\right)^2$
$=1^2-2(1)\left(\sqrt{n+7}\right)+\left(\sqrt{n+7}\right)^2$
$=1-2\sqrt{n+7}+n+7$
$=n-2\sqrt{n+7}+8$

53. $-\frac{\sqrt{32}}{4\sqrt{2}}=-\frac{1}{4}\sqrt{\frac{32}{2}}=-\frac{1}{4}\sqrt{16}=-\frac{4}{4}=-1$

55. $\frac{\sqrt{84x^3}}{\sqrt{7x}}=\sqrt{\frac{84x^3}{7x}}=\sqrt{12x^2}$
$=\sqrt{4x^2\cdot 3}=2x\sqrt{3}$

57. $-\frac{n\sqrt{60n^7}}{2\sqrt{5n^3}}=-\frac{n}{2}\sqrt{\frac{60n^7}{5n^3}}=-\frac{n}{2}\sqrt{12n^4}$
$=-\frac{n}{2}\sqrt{4n^4\cdot 3}=-\frac{n}{2}\left(2n^2\right)\sqrt{3}$
$=-n^3\sqrt{3}$

59. $\frac{\sqrt[4]{128r^{10}}}{6r\sqrt[4]{4r^3}}=\frac{1}{6r}\sqrt[4]{\frac{128r^{10}}{4r^3}}=\frac{1}{6r}\sqrt[4]{32r^7}$
$=\frac{1}{6r}\sqrt[4]{16r^4\cdot 2r^3}=\frac{2r}{6r}\sqrt[4]{2r^3}$
$=\frac{\sqrt[4]{2r^3}}{3}$

61. $\sqrt{\frac{5a}{16}}=\frac{\sqrt{5a}}{\sqrt{16}}=\frac{\sqrt{5a}}{4}$

63. $\sqrt{\frac{12x}{y^6}}=\frac{\sqrt{12x}}{\sqrt{y^6}}=\frac{\sqrt{4\cdot 3x}}{y^3}=\frac{2\sqrt{3x}}{y^3}$

65. $\sqrt[3]{\frac{54}{n^9}}=\frac{\sqrt[3]{54}}{\sqrt[3]{n^9}}=\frac{\sqrt[3]{27\cdot 2}}{n^3}=\frac{3\sqrt[3]{2}}{n^3}$

67. $\sqrt{\frac{32a^5}{9b^8}}=\frac{\sqrt{32a^5}}{\sqrt{9b^8}}=\frac{\sqrt{16a^4\cdot 2a}}{3b^4}=\frac{4a^2\sqrt{2a}}{3b^4}$

69. $\frac{4}{\sqrt{5}}=\frac{4}{\sqrt{5}}\cdot\frac{\sqrt{5}}{\sqrt{5}}=\frac{4\sqrt{5}}{5}$

71. $\frac{\sqrt{y}}{\sqrt{40}}=\frac{\sqrt{y}}{2\sqrt{10}}\cdot\frac{\sqrt{10}}{\sqrt{10}}=\frac{\sqrt{10y}}{2\cdot 10}=\frac{\sqrt{10y}}{20}$

73. $\frac{\sqrt{3a}}{\sqrt{18}}=\frac{\sqrt{3a}}{3\sqrt{2}}\cdot\frac{\sqrt{2}}{\sqrt{2}}=\frac{\sqrt{6a}}{3\cdot 2}=\frac{\sqrt{6a}}{6}$

75. $\frac{\sqrt{9u}}{6\sqrt{v}}=\frac{3\sqrt{u}}{6\sqrt{v}}=\frac{\sqrt{u}}{2\sqrt{v}}\cdot\frac{\sqrt{v}}{\sqrt{v}}=\frac{\sqrt{uv}}{2v}$

77. $\frac{\sqrt{25x^4}}{\sqrt{3y}}=\frac{5x^2}{\sqrt{3y}}\cdot\frac{\sqrt{3y}}{\sqrt{3y}}=\frac{5x^2\sqrt{3y}}{3y}$

79. $\frac{\sqrt{10xy}}{\sqrt{9z}}=\frac{\sqrt{10xy}}{3\sqrt{z}}\cdot\frac{\sqrt{z}}{\sqrt{z}}=\frac{\sqrt{10xyz}}{3z}$

81. $\frac{\sqrt[3]{9x}}{\sqrt[3]{2y^2}}=\frac{\sqrt[3]{9x}}{\sqrt[3]{2y^2}}\cdot\frac{\sqrt[3]{4y}}{\sqrt[3]{4y}}=\frac{\sqrt[3]{36xy}}{\sqrt[3]{8y^3}}=\frac{\sqrt[3]{36xy}}{2y}$

83. $\sqrt{\frac{11}{x}}=\frac{\sqrt{11}}{\sqrt{x}}\cdot\frac{\sqrt{x}}{\sqrt{x}}=\frac{\sqrt{11x}}{x}$

85. $\sqrt{\frac{7x}{3y}}=\frac{\sqrt{7x}}{\sqrt{3y}}\cdot\frac{\sqrt{3y}}{\sqrt{3y}}=\frac{\sqrt{21xy}}{3y}$

87. $\sqrt{\frac{5x^3}{48y^3}}=\frac{\sqrt{5x^3}}{\sqrt{48y^3}}=\frac{x\sqrt{5x}}{4y\sqrt{3y}}\cdot\frac{\sqrt{3y}}{\sqrt{3y}}$
$=\frac{x\sqrt{15xy}}{4y\cdot 3y}=\frac{x\sqrt{15xy}}{12y^2}$

89. $\sqrt[3]{\frac{a^2b}{32c^4}}=\frac{\sqrt[3]{a^2b}}{\sqrt[3]{32c^4}}=\frac{\sqrt[3]{a^2b}}{\sqrt[3]{8c^3\cdot 4c}}$
$=\frac{\sqrt[3]{a^2b}}{2c\sqrt[3]{4c}}\cdot\frac{\sqrt[3]{2c^2}}{\sqrt[3]{2c^2}}=\frac{\sqrt[3]{2a^2bc^2}}{2c\sqrt[3]{8c^3}}$
$=\frac{\sqrt[3]{2a^2bc^2}}{2c\cdot 2c}=\frac{\sqrt[3]{2a^2bc^2}}{4c^2}$

91. $\frac{2-\sqrt{3}}{\sqrt{6}}=\frac{2-\sqrt{3}}{\sqrt{6}}\cdot\frac{\sqrt{6}}{\sqrt{6}}=\frac{2\sqrt{6}-\sqrt{18}}{6}$
$=\frac{2\sqrt{6}-\sqrt{9\cdot 2}}{6}=\frac{2\sqrt{6}-3\sqrt{2}}{6}$

93. $\frac{\sqrt{a}-\sqrt{b}}{\sqrt{b}}=\frac{\sqrt{a}-\sqrt{b}}{\sqrt{b}}\cdot\frac{\sqrt{b}}{\sqrt{b}}$

$=\frac{\sqrt{a}\cdot\sqrt{b}-\left(\sqrt{b}\right)^2}{\left(\sqrt{b}\right)^2}$

$=\frac{\sqrt{ab}-b}{b}$

95. $\frac{\sqrt{5}+10\sqrt{t}}{\sqrt{15t}}\cdot\frac{\sqrt{15t}}{\sqrt{15t}}$

$=\frac{\sqrt{5}\cdot\sqrt{15t}+10\sqrt{t}\cdot\sqrt{15t}}{\left(\sqrt{15t}\right)^2}$

$=\frac{\sqrt{75t}+10\sqrt{15t^2}}{15t}$

$=\frac{\sqrt{25\cdot 3t}+10\sqrt{t^2\cdot 15}}{15t}$

$=\frac{5\sqrt{3t}+10t\sqrt{15}}{15t}$

$=\frac{5\left(\sqrt{3t}+2t\sqrt{15}\right)}{15t}$

$=\frac{\sqrt{3t}+2t\sqrt{15}}{3t}$

97. $\frac{\sqrt[3]{x}-4}{\sqrt[3]{x^2}}=\frac{\sqrt[3]{x}-4}{\sqrt[3]{x^2}}\cdot\frac{\sqrt[3]{x}}{\sqrt[3]{x}}$

$=\frac{\sqrt[3]{x\cdot x}-4\sqrt[3]{x}}{\sqrt[3]{x^3}}=\frac{\sqrt[3]{x^2}-4\sqrt[3]{x}}{x}$

99. $\frac{1}{2+\sqrt{2}}=\frac{1}{2+\sqrt{2}}\cdot\frac{2-\sqrt{2}}{2-\sqrt{2}}=\frac{2-\sqrt{2}}{2^2-\left(\sqrt{2}\right)^2}$

$=\frac{2-\sqrt{2}}{4-2}=\frac{2-\sqrt{2}}{2}$

101. $\frac{6}{\sqrt{2}-\sqrt{5}}=\frac{6}{\sqrt{2}-\sqrt{5}}\cdot\frac{\sqrt{2}+\sqrt{5}}{\sqrt{2}+\sqrt{5}}$

$=\frac{6\left(\sqrt{2}+\sqrt{5}\right)}{\left(\sqrt{2}\right)^2-\left(\sqrt{5}\right)^2}$

$=\frac{6\left(\sqrt{2}+\sqrt{5}\right)}{2-5}=\frac{6\left(\sqrt{2}+\sqrt{5}\right)}{-3}$

$=-2\left(\sqrt{2}+\sqrt{5}\right)=-2\sqrt{2}-2\sqrt{5}$

103. $\frac{8}{2+\sqrt{2x}}=\frac{8}{2+\sqrt{2x}}\cdot\frac{2-\sqrt{2x}}{2-\sqrt{2x}}$

$=\frac{8\left(2-\sqrt{2x}\right)}{(2)^2-\left(\sqrt{2x}\right)^2}$

$=\frac{8\left(2-\sqrt{2x}\right)}{4-2x}=\frac{8\left(2-\sqrt{2x}\right)}{2(2-x)}$

$=\frac{4\left(2-\sqrt{2x}\right)}{2-x}=\frac{8-4\sqrt{2x}}{2-x}$

105. $\frac{\sqrt{x}}{\sqrt{x}+y}=\frac{\sqrt{x}}{\sqrt{x}+y}\cdot\frac{\sqrt{x}-y}{\sqrt{x}-y}=\frac{\left(\sqrt{x}\right)^2-y\sqrt{x}}{\left(\sqrt{x}\right)^2-y^2}$

$=\frac{x-y\sqrt{x}}{x-y^2}$

107. $\frac{\sqrt{a}+3}{\sqrt{a}-\sqrt{2}}=\frac{\sqrt{a}+3}{\sqrt{a}-\sqrt{2}}\cdot\frac{\sqrt{a}+\sqrt{2}}{\sqrt{a}+\sqrt{2}}$

$=\frac{\left(\sqrt{a}\right)^2+\sqrt{a}\cdot\sqrt{2}+3\cdot\sqrt{a}+3\cdot\sqrt{2}}{\left(\sqrt{a}\right)^2-\left(\sqrt{2}\right)^2}$

$=\frac{a+\sqrt{2a}+3\sqrt{a}+3\sqrt{2}}{a-2}$

109. $\frac{\sqrt{x}-\sqrt{y}}{\sqrt{x}+\sqrt{y}}=\frac{\sqrt{x}-\sqrt{y}}{\sqrt{x}+\sqrt{y}}\cdot\frac{\sqrt{x}-\sqrt{y}}{\sqrt{x}-\sqrt{y}}$

$=\frac{\left(\sqrt{x}\right)^2-2\sqrt{x}\cdot\sqrt{y}+\left(\sqrt{y}\right)^2}{\left(\sqrt{x}\right)^2-\left(\sqrt{y}\right)^2}$

$=\frac{x-2\sqrt{xy}+y}{x-y}$

111. $\frac{2\sqrt{a}+3\sqrt{b}}{3\sqrt{a}-2\sqrt{b}}$

$=\frac{2\sqrt{a}+3\sqrt{b}}{3\sqrt{a}-2\sqrt{b}}\cdot\frac{3\sqrt{a}+2\sqrt{b}}{3\sqrt{a}+2\sqrt{b}}$

$=\frac{6\left(\sqrt{a}\right)^2+4\sqrt{ab}+9\sqrt{ab}+6\left(\sqrt{b}\right)^2}{\left(3\sqrt{a}\right)^2-\left(2\sqrt{b}\right)^2}$

$=\frac{6a+13\sqrt{ab}+6b}{9a-4b}$

113. $f(x)\bullet g(x)=\left(3x\sqrt{2x}\right)\bullet\left(\frac{1}{3}\sqrt{6x}\right)$

$$=\frac{3x}{3}\sqrt{12x^2}=x\sqrt{4x^2\bullet 3}$$
$$=2x^2\sqrt{3}$$

$$\frac{f(x)}{g(x)}=\frac{3x\sqrt{2x}}{\frac{1}{3}\sqrt{6x}}=9x\sqrt{\frac{2x}{6x}}=9x\sqrt{\frac{1}{3}}$$
$$=9x\frac{1}{\sqrt{3}}\bullet\frac{\sqrt{3}}{\sqrt{3}}=9x\frac{\sqrt{3}}{3}=3x\sqrt{3}$$

115. $f(x)\bullet g(x)=\left(\sqrt{x}+1\right)\bullet\left(\sqrt{x}-1\right)$

$$=\left(\sqrt{x}\right)^2-1^2=x-1$$

$$\frac{f(x)}{g(x)}=\frac{\sqrt{x}+1}{\sqrt{x}-1}=\frac{\sqrt{x}+1}{\sqrt{x}-1}\bullet\frac{\sqrt{x}+1}{\sqrt{x}+1}$$
$$=\frac{\left(\sqrt{x}\right)^2+2\sqrt{x}+1^2}{\left(\sqrt{x}\right)^2-1^2}=\frac{x+2\sqrt{x}+1}{x-1}$$

117. $\left(\sqrt{12}+\sqrt{3}\right)\left(\sqrt{3}+\sqrt{6}\right)$

F O I L

$$=\sqrt{12}\sqrt{3}+\sqrt{12}\sqrt{6}+\sqrt{3}\sqrt{3}+\sqrt{3}\sqrt{6}$$
$$=\sqrt{36}+\sqrt{72}+\sqrt{9}+\sqrt{18}$$
$$=6+\sqrt{36\bullet 2}+3+\sqrt{9\bullet 2}$$
$$=9+6\sqrt{2}+3\sqrt{2}=9+(6+3)\sqrt{2}$$
$$=9+9\sqrt{2}$$

119. $\left(-8\sqrt{ab^3}\right)\left(7\sqrt{a^5b}\right)=(-8)(7)\sqrt{aa^5b^3b}$

$$=-56\sqrt{a^6b^4}$$
$$=-56\sqrt{\left(a^3\right)^2\left(b^2\right)^2}$$
$$=-56a^3b^2$$

121. $\sqrt{\frac{75y}{z^4}}=\frac{\sqrt{75y}}{\sqrt{z^4}}=\frac{\sqrt{25\bullet 3y}}{\sqrt{\left(z^2\right)^2}}=\frac{5\sqrt{3y}}{z^2}$

123. $\sqrt{\frac{5x}{7y}}=\frac{\sqrt{5x}}{\sqrt{7y}}\bullet\frac{\sqrt{7y}}{\sqrt{7y}}=\frac{\sqrt{35xy}}{\sqrt{(7y)^2}}=\frac{\sqrt{35xy}}{7y}$

125. $\frac{\sqrt{p}-4}{\sqrt{p}-\sqrt{3}}$

$$=\frac{\sqrt{p}-4}{\sqrt{p}-\sqrt{3}}\bullet\frac{\sqrt{p}+\sqrt{3}}{\sqrt{p}+\sqrt{3}}$$
$$=\frac{\sqrt{p}\sqrt{p}+\sqrt{p}\sqrt{3}-4\sqrt{p}-4\sqrt{3}}{\left(\sqrt{p}\right)^2-\left(\sqrt{3}\right)^2}$$
$$=\frac{p+\sqrt{3p}-4\sqrt{p}-4\sqrt{3}}{p-3}$$

127. **a.** Area of a triangle is half the base times the height. The base is a.

$$A=\left(\frac{1}{2}a\right)\left(\frac{\sqrt{3}}{2}a\right)=\frac{\sqrt{3}}{4}a^2$$

b. $A=\frac{\sqrt{3}}{4}(8)^2=\frac{64\sqrt{3}}{4}=16\sqrt{3}$ sq in.

129. $2\pi\sqrt{\frac{L}{32}}=2\pi\frac{\sqrt{L}}{\sqrt{32}}\bullet\frac{\sqrt{2}}{\sqrt{2}}=2\pi\frac{\sqrt{2L}}{\sqrt{64}}$

$$=\frac{2\pi\sqrt{2L}}{8}=\frac{\pi\sqrt{2L}}{4}$$

131. $\sqrt{\frac{A}{4\pi}}=\frac{\sqrt{A}}{\sqrt{4\pi}}\bullet\frac{\sqrt{\pi}}{\sqrt{\pi}}=\frac{\sqrt{\pi A}}{\sqrt{4\pi^2}}=\frac{\sqrt{\pi A}}{2\pi}$

8.6 Solving Radical Equations

Practice

1.

$$\sqrt{y}+2=10$$
$$\sqrt{y}+2-2=10-2$$
$$\sqrt{y}=8$$
$$\left(\sqrt{y}\right)^2=(8)^2$$
$$y=64$$

Check:

$$\sqrt{y}+2=10$$
$$\sqrt{64}+2\overset{?}{=}10$$
$$8+2\overset{?}{=}10$$
$$10=10$$

2.

$$\sqrt[3]{2x+7}=3$$
$$\left(\sqrt[3]{2x+7}\right)^3=(3)^3$$
$$2x+7=27$$
$$2x+7-7=27-7$$
$$2x=20$$
$$x=10$$

Check:

$$\sqrt[3]{2x+7}=3$$
$$\sqrt[3]{2\cdot 10+7}\overset{?}{=}3$$
$$\sqrt[3]{27}\overset{?}{=}3$$
$$3=3$$

3.

$$\sqrt{3t+1}+4=0$$
$$\sqrt{3t+1}=-4$$
$$\left(\sqrt{3t+1}\right)=(-4)^2$$
$$3t+1=16$$
$$3t=15$$
$$t=5$$

Check:

$$\sqrt{3t+1}+4=0$$
$$\sqrt{3\cdot5+1}+4\overset{?}{=}0$$
$$\sqrt{16}+4\overset{?}{=}0$$
$$4+4\overset{?}{=}0$$
$$8\neq0$$

There is no solution.

4.

$$\sqrt{4n+5}-\sqrt{7n-4}=0$$
$$\sqrt{4n+5}=\sqrt{7n-4}$$
$$\left(\sqrt{4n+5}\right)^2=\left(\sqrt{7n-4}\right)^2$$
$$4n+5=7n-4$$
$$3n=9$$
$$n=3$$

Check:

$$\sqrt{4n+5}-\sqrt{7n-4}=0$$
$$\sqrt{4\cdot3+5}-\sqrt{7\cdot3-4}\overset{?}{=}0$$
$$\sqrt{17}-\sqrt{17}\overset{?}{=}0$$
$$0=0$$

5.

$$\sqrt{2y+3}=\sqrt{y-2}+2$$
$$\left(\sqrt{2y+3}\right)^2=\left(\sqrt{y-2}+2\right)^2$$
$$2y+3=y-2+2\cdot2\sqrt{y-2}+2\cdot2$$
$$2y+3=y-2+4\sqrt{y-2}+4$$
$$y+1=-4\sqrt{y-2}$$
$$(y+1)^2=\left(-4\sqrt{y-2}\right)^2$$
$$y^2+2y+1=16(y-2)$$
$$y^2+2y+1=16y-32$$
$$y^2-14y+33=0$$
$$(y-3)(y-11)=0$$
$$y-3=0 \text{ or } y-11=0$$
$$y=3 \text{ or } y=11$$

Check $y=3$:

$$\sqrt{2y+3}=\sqrt{y-2}+2$$
$$\sqrt{2\cdot3+3}\overset{?}{=}\sqrt{3-2}+2$$
$$\sqrt{9}\overset{?}{=}\sqrt{1}+2$$
$$3\overset{?}{=}1+2$$
$$3=3$$

Check $y=11$:

$$\sqrt{2y+3}=\sqrt{y-2}+2$$
$$\sqrt{2\cdot11+3}\overset{?}{=}\sqrt{11-2}+2$$
$$\sqrt{25}\overset{?}{=}\sqrt{9}+2$$
$$5\overset{?}{=}3+2$$
$$5=5$$

6.

$$\sqrt[3]{n^2+8}+\sqrt[3]{4-7n}=0$$
$$\sqrt[3]{n^2+8}=-\sqrt[3]{4-7n}$$
$$\left(\sqrt[3]{n^2+8}\right)^3=\left(-\sqrt[3]{4-7n}\right)^3$$
$$n^2+8=-(4-7n)$$
$$n^2+8=-4+7n$$
$$n^2-7n+12=0$$
$$(n-4)(n-3)=0$$
$$n-4=0 \text{ or } n-3=0$$
$$n=4 \text{ or } n=3$$

Check $n=4$:

$$\sqrt[3]{n^2+8}+\sqrt[3]{4-7n}=0$$
$$\sqrt[3]{4^2+8}+\sqrt[3]{4-7\cdot4}\overset{?}{=}0$$
$$\sqrt[3]{24}+\sqrt[3]{-24}\overset{?}{=}0$$
$$\sqrt[3]{24}+\sqrt[3]{(-1)24}\overset{?}{=}0$$
$$0=0$$

Check $n=3$:

$$\sqrt[3]{n^2+8}+\sqrt[3]{4-7n}=0$$
$$\sqrt[3]{3^2+8}+\sqrt[3]{4-7\cdot3}\overset{?}{=}0$$
$$\sqrt[3]{17}+\sqrt[3]{-17}\overset{?}{=}0$$
$$\sqrt[3]{17}+\sqrt[3]{(-1)17}\overset{?}{=}0$$
$$0=0$$

7.

$$\sqrt{y+4}-y=2$$
$$\sqrt{y+4}=y+2$$
$$\left(\sqrt{y+4}\right)^2=(y+2)^2$$
$$y+4=y^2+4y+4$$
$$y^2+3y=0$$
$$y(y+3)=0$$
$$y+3=0 \text{ or } y=0$$
$$y=-3$$

7. (continued)

Check $y=-3$:

$$\sqrt{y+4}-y=2$$
$$\sqrt{-3+4}-(-3)\stackrel{?}{=}2$$
$$\sqrt{1}+3\stackrel{?}{=}2$$
$$1+3\stackrel{?}{=}2$$
$$4\neq 2$$

Check $y=0$:

$$\sqrt{y+4}-y=2$$
$$\sqrt{0+4}-(0)\stackrel{?}{=}2$$
$$\sqrt{4}-0\stackrel{?}{=}2$$
$$2-0\stackrel{?}{=}2$$
$$2=2$$

So the solution is 0.

8.
$$I=\sqrt{\frac{P}{R}}$$
$$I^2=\left(\sqrt{\frac{P}{R}}\right)^2$$
$$I^2=\frac{P}{R}$$
$$I^2R=P$$
$$P=I^2R$$

9. **a.**
$$r=\sqrt[3]{\frac{A}{P}}-1$$
$$r+1=\sqrt[3]{\frac{A}{P}}$$
$$(r+1)^3=\left(\sqrt[3]{\frac{A}{P}}\right)^3$$
$$(r+1)^3=\frac{A}{P}$$
$$P(r+1)^3=A$$
$$A=P(r+1)^3$$

b.
$$A=P(r+1)^3$$
$$=10{,}000(0.03+1)^3$$
$$=10{,}000(1.03)^3$$
$$=10{,}927.27$$

The value after 3 yr is about \$10,927.27.

Exercises

Note: Solutions should be checked in the original equation.

1.
$$\sqrt{3n}=6$$
$$\left(\sqrt{3n}\right)^2=6^2$$
$$3n=36$$
$$n=12$$

3.
$$\sqrt{x+6}=3$$
$$\left(\sqrt{x+6}\right)^2=3^2$$
$$x+6=9$$
$$x=3$$

5.
$$\sqrt{5x-6}=2$$
$$\left(\sqrt{5x-6}\right)^2=2^2$$
$$5x-6=4$$
$$5x=10$$
$$x=2$$

7.
$$\sqrt[3]{3y+10}=-2$$
$$\left(\sqrt[3]{3y+10}\right)^3=(-2)^3$$
$$3y+10=-8$$
$$3y=-18$$
$$y=-6$$

9.
$$\sqrt{x}+9=8$$
$$\sqrt{x}=-1$$
$\sqrt{x}\geq 0$, no solution.

11.
$$\sqrt{x}-20=-9$$
$$\sqrt{x}=11$$
$$\left(\sqrt{x}\right)^2=11^2$$
$$x=121$$

13.
$$14-\sqrt[3]{x}=11$$
$$3=\sqrt[3]{x}$$
$$(3)^3=\left(\sqrt[3]{x}\right)^3$$
$$x=27$$

15.
$$\sqrt{6x}+17=29$$
$$\sqrt{6x}=12$$
$$\left(\sqrt{6x}\right)^2=12^2$$
$$6x=144$$
$$x=24$$

17. $\sqrt{2y-1}-8=5$

$\sqrt{2y-1}=13$

$\left(\sqrt{2y-1}\right)^2=13^2$

$2y-1=169$

$2y=170$

$y=85$

19. $14-\sqrt{4a+9}=13$

$-\sqrt{4a+9}=-1$

$\left(-\sqrt{4a+9}\right)^2=(-1)^2$

$4a+9=1$

$4a=-8$

$a=-2$

21. $\sqrt{5x-1}-\sqrt{3x+11}=0$

$\sqrt{5x-1}=\sqrt{3x+11}$

$\left(\sqrt{5x-1}\right)^2=\left(\sqrt{3x+11}\right)^2$

$5x-1=3x+11$

$2x=12$

$x=6$

23. $\left(2\sqrt{x-3}\right)^2=\left(\sqrt{7x+15}\right)^2$

$4(x-3)=7x+15$

$4x-12=7x+15$

$-3x=27$

$x=-9$

$\sqrt{x-3}=\sqrt{-12}$, not a real number

No solution

25. $\left(\sqrt[3]{3y-19}\right)^3=\left(\sqrt[3]{6y+26}\right)^3$

$3y-19=6y+26$

$-3y=45$

$y=-15$

27. $\left(\sqrt{a^2+7}\right)^2=\left(\sqrt{5a+1}\right)^2$

$a^2+7=5a+1$

$a^2-5a+6=0$

$(a-2)(a-3)=0$

$a-2=0 \quad a-3=0$

$a=2 \quad a=3$

Check $a=2$

$\sqrt{2^2+7}\stackrel{?}{=}\sqrt{5(2)+1}$

$\sqrt{11}=\sqrt{11}$ True

Check $a=3$

$\sqrt{3^2+7}\stackrel{?}{=}\sqrt{5(3)+1}$

$\sqrt{16}=\sqrt{16}$ True

29. $\sqrt[3]{a^2-6}+\sqrt[3]{1-4a}=0$

$\sqrt[3]{a^2-6}=-\sqrt[3]{1-4a}$

$\left(\sqrt[3]{a^2-6}\right)^3=\left(-\sqrt[3]{1-4a}\right)^3$

$a^2-6=-(1-4a)$

$a^2-6=-1+4a$

$a^2-4a-5=0$

$(a-5)(a+1)=0$

$a-5=0 \quad a+1=0$

$a=5 \quad a=-1$

Check $a=5$

$\sqrt[3]{5^2-6}+\sqrt[3]{1-4(5)}\stackrel{?}{=}0$

$\sqrt[3]{19}+\sqrt[3]{-19}\stackrel{?}{=}0$

$\sqrt[3]{19}-\sqrt[3]{19}\stackrel{?}{=}0$

$0=0$ True

Check $a=-1$

$\sqrt[3]{(-1)^2-6}+\sqrt[3]{1-4(-1)}\stackrel{?}{=}0$

$\sqrt[3]{-5}+\sqrt[3]{5}\stackrel{?}{=}0$

$-\sqrt[3]{5}+\sqrt[3]{5}\stackrel{?}{=}0$

$0=0$ True

31. $\sqrt{2x+8}=-x$

$\left(\sqrt{2x+8}\right)^2=(-x)^2$

$2x+8=x^2$

$x^2-2x-8=0$

$(x-4)(x+2)=0$

$x-4=0 \quad x+2=0$

$x=4 \quad x=-2$

31. (continued)

Check $x = 4$

$\sqrt{2(4)+8} \stackrel{?}{=} -4$

$\sqrt{16} \stackrel{?}{=} -4$

$4 = -4$ False

$x = 4$ is not a solution.

Check $x = -2$

$\sqrt{2(-2)+8} \stackrel{?}{=} -(-2)$

$\sqrt{4} \stackrel{?}{=} 2$

$2 = 2$ True

33.

$$2n - \sqrt{6-5n} = 0$$
$$2n = \sqrt{6-5n}$$
$$(2n)^2 = \left(\sqrt{6-5n}\right)^2$$
$$4n^2 = 6-5n$$
$$4n^2 + 5n - 6 = 0$$
$$(4n-3)(n+2) = 0$$
$$4n-3=0 \quad n+2=0$$
$$4n = 3 \quad n = -2$$
$$n = \frac{3}{4}$$

Check $n = \frac{3}{4}$

$2\left(\frac{3}{4}\right) - \sqrt{6-5\left(\frac{3}{4}\right)} \stackrel{?}{=} 0$

$\frac{3}{2} - \sqrt{\frac{24}{4} - \frac{15}{4}} \stackrel{?}{=} 0$

$\frac{3}{2} - \sqrt{\frac{9}{4}} \stackrel{?}{=} 0$

$\frac{3}{2} - \frac{3}{2} \stackrel{?}{=} 0$

$0 = 0$ True

Check $n = -2$

$2(-2) - \sqrt{6-5(-2)} \stackrel{?}{=} 0$

$-4 - \sqrt{16} \stackrel{?}{=} 0$

$-4 - 4 \stackrel{?}{=} 0$

$-8 = 0$ False

$n = -2$ is not a solution.

35.

$$x-2 = \sqrt{4x-11}$$
$$(x-2)^2 = \left(\sqrt{4x-11}\right)^2$$
$$x^2 - 4x + 4 = 4x - 11$$
$$x^2 - 8x + 15 = 0$$
$$(x-5)(x-3) = 0$$
$$x-5=0 \quad x-3=0$$
$$x=5 \quad x=3$$

Check $x = 3$

$3-2 \stackrel{?}{=} \sqrt{4(3)-11}$

$1 \stackrel{?}{=} \sqrt{1}$

$1 = 1$ True

Check $x = 5$

$5-2 \stackrel{?}{=} \sqrt{4(5)-11}$

$3 \stackrel{?}{=} \sqrt{9}$

$3 = 3$ True

37.

$$\sqrt{3t+1} - 1 = 2t$$
$$\sqrt{3t+1} = 2t+1$$
$$\left(\sqrt{3t+1}\right)^2 = (2t+1)^2$$
$$3t+1 = 4t^2 + 4t + 1$$
$$4t^2 + t = 0$$
$$t(4t+1) = 0$$
$$t = 0 \quad 4t+1 = 0$$
$$4t = -1$$
$$t = -\frac{1}{4}$$

Check $t = 0$

$\sqrt{3(0)+1} - 1 \stackrel{?}{=} 2(0)$

$\sqrt{1} - 1 \stackrel{?}{=} 0$

$1 - 1 \stackrel{?}{=} 0$

$0 = 0$ True

Check $t = -\frac{1}{4}$

$\sqrt{3\left(-\frac{1}{4}\right)+1} - 1 \stackrel{?}{=} 2\left(-\frac{1}{4}\right)$

$\sqrt{\frac{1}{4}} - 1 \stackrel{?}{=} -\frac{1}{2}$

$\frac{1}{2} - 1 \stackrel{?}{=} -\frac{1}{2}$

$-\frac{1}{2} = -\frac{1}{2}$ True

39.
$$\sqrt{3n}+\sqrt{n-2}=4$$
$$\sqrt{n-2}=4-\sqrt{3n}$$
$$\left(\sqrt{n-2}\right)^2=\left(4-\sqrt{3n}\right)^2$$
$$n-2=16-8\sqrt{3n}+3n$$
$$8\sqrt{3n}=18+2n$$
$$4\sqrt{3n}=9+n$$
$$\left(4\sqrt{3n}\right)^2=(9+n)^2$$
$$16(3n)=81+18n+n^2$$
$$n^2-30n+81=0$$
$$(n-27)(n-3)=0$$
$$n-27=0 \qquad n-3=0$$
$$n=27 \qquad n=3$$

Check $n=3$

$\sqrt{3(3)}+\sqrt{3-2}\stackrel{?}{=}4$

$\sqrt{9}+\sqrt{1}\stackrel{?}{=}4$

$3+1\stackrel{?}{=}4$

$4=4$ True

Check $n=27$

$\sqrt{3(27)}+\sqrt{27-2}\stackrel{?}{=}4$

$\sqrt{81}+\sqrt{25}\stackrel{?}{=}4$

$9+5\stackrel{?}{=}4$

$14=4$ False,

$n=27$ is not a solution

41.
$$\sqrt{x-2}+1=-\sqrt{x+3}$$
$$\left(\sqrt{x-2}+1\right)^2=\left(-\sqrt{x+3}\right)^2$$
$$x-2+2\sqrt{x-2}+1=x+3$$
$$2\sqrt{x-2}=4$$
$$\sqrt{x-2}=2$$
$$\left(\sqrt{x-2}\right)^2=2^2$$
$$x-2=4$$
$$x=6$$

Check $x=6$

$\sqrt{6-2}+1\stackrel{?}{=}-\sqrt{6+3}$

$\sqrt{4}+1\stackrel{?}{=}-\sqrt{9}$

$2+1\stackrel{?}{=}-3$

$3=-3$ False, no solution

43.
$$\sqrt{x+5}-2=\sqrt{x-1}$$
$$\left(\sqrt{x+5}-2\right)^2=\left(\sqrt{x-1}\right)^2$$
$$x+5-4\sqrt{x+5}+4=x-1$$
$$10=4\sqrt{x+5}$$
$$5=2\sqrt{x+5}$$
$$5^2=\left(2\sqrt{x+5}\right)^2$$
$$25=4(x+5)$$
$$25=4x+20$$
$$-4x=-5$$
$$x=\frac{5}{4}$$

Check $x=\frac{5}{4}$

$\sqrt{\frac{5}{4}+5}-2\stackrel{?}{=}\sqrt{\frac{5}{4}-1}$

$\sqrt{\frac{25}{4}}-2\stackrel{?}{=}\sqrt{\frac{1}{4}}$

$\frac{5}{2}-\frac{4}{2}\stackrel{?}{=}\frac{1}{2}$

$\frac{1}{2}=\frac{1}{2}$ True

45.
$$\sqrt{2y+3}-\sqrt{3y+7}=-1$$
$$\sqrt{2y+3}=\sqrt{3y+7}-1$$
$$\left(\sqrt{2y+3}\right)^2=\left(\sqrt{3y+7}-1\right)^2$$
$$2y+3=3y+7-2\sqrt{3y+7}+1$$
$$2\sqrt{3y+7}=y+5$$
$$\left(2\sqrt{3y+7}\right)^2=(y+5)^2$$
$$4(3y+7)=y^2+10y+25$$
$$12y+28=y^2+10y+25$$
$$y^2-2y-3=0$$
$$(y-3)(y+1)=0$$
$$y-3=0 \qquad y+1=0$$
$$y=3 \qquad y=-1$$

Check $y=3$

$\sqrt{2(3)+3}-\sqrt{3(3)+7}\stackrel{?}{=}-1$

$\sqrt{9}-\sqrt{16}\stackrel{?}{=}-1$

$3-4\stackrel{?}{=}-1$

$-1=-1$ True

45. (continued)

Check $y=-1$

$\sqrt{2(-1)+3}-\sqrt{3(-1)+7}\stackrel{?}{=}-1$

$\sqrt{1}-\sqrt{4}\stackrel{?}{=}-1$

$1-2\stackrel{?}{=}-1$

$-1=-1$ True

47. $\sqrt[3]{x^3+8}=x+2$

$\left(\sqrt[3]{x^3+8}\right)^3=(x+2)^3$

$x^3+8=x^3+6x^2+12x+8$

$6x^2+12x=0$

$6x(x+2)=0$

$6x=0 \quad x+2=0$

$x=0 \quad x=-2$

Check $x=0$

$\sqrt[3]{(0)^3+8}\stackrel{?}{=}0+2$

$\sqrt[3]{8}\stackrel{?}{=}2$

$2=2$ True

Check $x=-2$

$\sqrt[3]{(-2)^3+8}\stackrel{?}{=}-2+2$

$\sqrt[3]{0}\stackrel{?}{=}0$

$0=0$ True

49. $13-\sqrt{4x-7}=18$

$-\sqrt{4x-7}=5$

$\sqrt{4x-7}=-5$

$\sqrt{4x-7}\geq 0$ No solution

51. $\sqrt{6x+1}=7$

$\left(\sqrt{6x+1}\right)^2=7^2$

$6x+1=49$

$6x=48$

$x=8$

Check

$\sqrt{6(8)+1}\stackrel{?}{=}7$

$\sqrt{49}\stackrel{?}{=}7$

$7=7$ True

53.

$\sqrt[3]{n^2+16}+\sqrt[3]{10n+8}=0$

$\sqrt[3]{n^2+16}=-\sqrt[3]{10n+8}$

$\left(\sqrt[3]{n^2+16}\right)^3=\left(-\sqrt[3]{10n+8}\right)^3$

$n^2+16=-(10n+8)$

$n^2+16=-10n-8$

$n^2+10n+24=0$

$(n+6)(n+4)=0$

$n+6=0$ or $n+4=0$

$n=-6 \quad n=-4$

Check $n=-6$

$\sqrt[3]{(-6)^2+16}+\sqrt[3]{10(-6)+8}\stackrel{?}{=}0$

$\sqrt[3]{36+16}+\sqrt[3]{-60+8}\stackrel{?}{=}0$

$\sqrt[3]{52}+\sqrt[3]{-52}\stackrel{?}{=}0$

$\sqrt[3]{52}-\sqrt[3]{52}=0$ True

Check $n=-4$

$\sqrt[3]{(-4)^2+16}+\sqrt[3]{10(-4)+8}\stackrel{?}{=}0$

$\sqrt[3]{16+16}+\sqrt[3]{-40+8}\stackrel{?}{=}0$

$\sqrt[3]{32}+\sqrt[3]{-32}\stackrel{?}{=}0$

$\sqrt[3]{32}-\sqrt[3]{32}=0$ True

55.

$900=\sqrt{8000h+h^2}$

$900^2=\left(\sqrt{8000h+h^2}\right)^2$

$810000=8000h+h^2$

$h^2+8000h-810000=0$

$(h-100)(h+8100)=0$

$h-100=0 \quad h+8100=0$

$h=100 \quad h=-8100$

The negative answer makes no sense in this application. The satellite is 100 mi above the Earth's surface.

57. **a.** $S=\sqrt{30fL}$

$S^2=\left(\sqrt{30fL}\right)^2$

$S^2=30fL$

$L=\dfrac{S^2}{30f}$

b. $L=\dfrac{30^2}{30(0.5)}=\dfrac{900}{15}=60$

The length of the skid marks is 60 ft.

59. $15^2=12^2+d^2$

$d^2=15^2-12^2$

$d=\sqrt{15^2-12^2}=\sqrt{225-144}=\sqrt{81}=9$

The painter must place the ladder 9 ft from the side of the house.

8.7 Complex Numbers

Practice

1. **a.** $\sqrt{-36}=i\sqrt{36}=i\cdot 6=6i$

b. $\sqrt{-2}=i\sqrt{2}$

c.

$-10\sqrt{-\dfrac{1}{4}}=-10i\sqrt{\dfrac{1}{4}}=-10i\left(\dfrac{1}{2}\right)=-5i$

2. **a.** $(6+5i)+(-2-6i)$

$=(6+(-2))+(5+(-6))i$

$=4+(-1)i$

$=4-i$

b. $(-4+3i)-(8-i)=-4+3i-8+i$

$=(-4-8)+(3+1)i$

$=-12+4i$

c. $\left(1+5\sqrt{-9}\right)+\left(4+\sqrt{-16}\right)$

$=(1+5\cdot 3i)+(4+4i)$

$=(1+15i)+(4+4i)$

$=(1+4)+(15+4)i$

$=5+19i$

3. **a.** $\sqrt{-36}\cdot\sqrt{-4}=(6i)(2i)=12i^2=-12$

b. $\sqrt{-5}\cdot\sqrt{-3}=\left(i\sqrt{5}\right)\left(i\sqrt{3}\right)$

$=i^2\sqrt{15}=-1\cdot\sqrt{15}=-\sqrt{15}$

c. $-4i\cdot 10i=(-40)i^2=(-40)(-1)=40$

d. $-2i(5+6i)=-10i-12i^2=-10i-12(-1)$

$=-10i+12=12-10i$

4. **a.** $(4+7i)(2+i)=8+4i+14i+7i^2$

$=8+4i+14i-7=1+18i$

b. $(3-2i)(1-3i)=3-9i-2i+6i^2$

$=3-9i-2i-6=-3-11i$

5. **a.** $-1+7i$

b. $8-9i$

c. $3i$

6. $(2-5i)(2+5i)=2^2-(5i)^2=4+25=29$

7. **a.** $\dfrac{7}{1-2i}=\dfrac{7}{1-2i}\cdot\dfrac{1+2i}{1+2i}=\dfrac{7(1+2i)}{(1-2i)(1+2i)}$

$=\dfrac{7+14i}{1+4}=\dfrac{7+14i}{5}=\dfrac{7}{5}+\dfrac{14i}{5}$

b. $-\dfrac{2}{5i}=\dfrac{2}{-5i}\cdot\dfrac{5i}{5i}=\dfrac{10i}{-25i^2}=\dfrac{10i}{25}=\dfrac{2}{5}i$

8. **a.** $\dfrac{2+i}{1-3i}=\dfrac{2+i}{1-3i}\cdot\dfrac{1+3i}{1+3i}=\dfrac{(2+i)(1+3i)}{(1-3i)(1+3i)}$

$=\dfrac{2+6i+i+3i^2}{1+9}=\dfrac{2+6i+i-3}{10}$

$=\dfrac{-1+7i}{10}=-\dfrac{1}{10}+\dfrac{7}{10}i$

b. $\dfrac{1-\sqrt{-4}}{4-\sqrt{-64}}=\dfrac{1-2i}{4-8i}=\dfrac{1-2i}{4(1-2i)}=\dfrac{1}{4}$

9. $i^{30}=\left(i^4\right)^7i^2=1^7i^2=i^2=-1$

10. **a.** $I=\dfrac{V}{Z}$

b. $I=\dfrac{V}{Z}=\dfrac{3+5i}{1-i}\cdot\dfrac{1+i}{1+i}=\dfrac{3+3i+5i+5i^2}{1-i^2}$

$=\dfrac{3+8i-5}{1+1}=\dfrac{-2+8i}{2}=-1+4i$

The current is $(-1+4i)$ amps.

c. $V=(-1+4i)(1-i)=-1+i+4i-4i^2$

$=-1+5i+4=3+5i$

Exercises

7. $\sqrt{-4}=\sqrt{(-1)(4)}=\sqrt{4}\sqrt{-1}=2i$

9. $\sqrt{-\frac{1}{16}}=\sqrt{\left(\frac{1}{16}\right)(-1)}=\frac{\sqrt{1}}{\sqrt{16}}\sqrt{-1}=\frac{1}{4}i$

11. $\sqrt{-3}=\sqrt{(-1)3}=\sqrt{-1}\sqrt{3}=i\sqrt{3}$

13. $\sqrt{-18}=\sqrt{(9)(-1)(2)}$
$=\sqrt{9}\sqrt{-1}\sqrt{2}=3i\sqrt{2}$

15. $\sqrt{-500}=\sqrt{(100)(-1)(5)}$
$=\sqrt{100}\sqrt{-1}\sqrt{5}=10i\sqrt{5}$

17. $-\sqrt{-9}=-\sqrt{(9)(-1)}=-\sqrt{9}\sqrt{-1}=-3i$

19. $6\sqrt{\frac{-5}{16}}=6\frac{\sqrt{(-1)(5)}}{\sqrt{16}}=6\frac{i\sqrt{5}}{4}=\frac{3i\sqrt{5}}{2}$

21. $-\frac{1}{4}\sqrt{-12}=-\frac{1}{4}\sqrt{(4)(-1)(3)}$
$=-\frac{1}{4}\sqrt{4}\sqrt{-1}\sqrt{3}$
$=-\frac{1}{4}2i\sqrt{3}=-\frac{i\sqrt{3}}{2}$

23. $(1+12i)+8i=1+12i+8i$
$=1+(12+8)i=1+20i$

25. $(3-15i)+(2+9i)=3-15i+2+9i$
$=(3+2)+(-15+9)i$
$=5-6i$

27. $(7-i)-(7+5i)=7-i-7-5i$
$=(7-7)+(-1-5)i$
$=-6i$

29. $(-8-6i)-(-1-3i)$
$=-8-6i+1+3i$
$=(-8+1)+(-6+3)i$
$=-7-3i$

31. $16-\left(18+\sqrt{-4}\right)=16-18-2i=-2-2i$

33. $\left(10-3\sqrt{-16}\right)+\left(2-\sqrt{-25}\right)$
$=(10-3(4i))+(2-5i)$
$=10-12i+2-5i$
$=(10+2)+(-12-5)i$
$=12-17i$

35. $\sqrt{-25}\bullet\sqrt{-4}=(5i)(2i)=10i^2=-10$

37. $\sqrt{-3}\left(-\sqrt{-27}\right)=\left(i\sqrt{3}\right)\left(-3i\sqrt{3}\right)$
$=-3i^2\left(\sqrt{3}\right)^2=3(3)=9$

39. $7i\bullet 9i=7\bullet 9i^2=-63$

41. $-2i(14i)=-28i^2=28$

43. $3i(1-i)=3i-3i^2=3i+3=3+3i$

45. $-i(12+7i)=-12i-7i^2=-12i+7=7-12i$

47. $\sqrt{-9}\left(7+\sqrt{-16}\right)=3i(7+4i)=21i+12i^2$
$=21i-12=-12+21i$

49. $-\sqrt{2}\left(\sqrt{8}-\sqrt{-18}\right)=-\sqrt{2}\left(2\sqrt{2}-3i\sqrt{2}\right)$
$=-2\left(\sqrt{2}\right)^2+3i\left(\sqrt{2}\right)^2$
$=-4+6i$

51. $(4+2i)(2+3i)$
F O I L
$=8+12i+4i+6i^2=8+16i-6=2+16i$

53. $(10-i)(4+6i)$
F O I L
$=40+60i-4i-6i^2$
$=40+56i+6=46+56i$

55. $(7i-7)(3-5i)$
F O I L
$=21i-35i^2-21+35i=56i+35-21$
$=14+56i$

57. $(-4-2i)(2-4i)$
F O I L
$=-8+16i-4i+8i^2$
$=-8+12i-8=-16+12i$

59. $(6+5i)(6-5i)=(6)^2-(5i)^2=36-25i^2$
$=36+25=61$

61. $(3+2i)^2=(3)^2+2(3)(2i)+(2i)^2$
$=9+12i+4i^2=9+12i-4$
$=5+12i$

63. $(2-3i)^2=(2)^2-(2)(2)(3i)+(3i)^2$
$=4-12i+9i^2=4-12i-9$
$=-5-12i$

65. $\left(\sqrt{-1}+\sqrt{2}\right)\left(\sqrt{-9}-\sqrt{8}\right)$
$=\left(i+\sqrt{2}\right)\left(3i-2\sqrt{2}\right)$
$=3i^2-2i\sqrt{2}+3i\sqrt{2}-2\left(\sqrt{2}\right)^2$
$=-3+i\sqrt{2}-4=-7+i\sqrt{2}$

67.

Complex Number	Conjugate	Product
$1+10i$	$1-10i$	$1^2-(10i)^2=1-100i^2$ $=1+100$ $=101$

69.

Complex Number	Conjugate	Product
$4-3i$	$4+3i$	$4^2-(3i)^2=16-9i^2$ $=16+9=25$

71.

Complex Number	Conjugate	Product
$-9+6i$	$-9-6i$	$(-9)^2-(6i)^2=81-36i^2$ $=81+36$ $=117$

73.

Complex Number	Conjugate	Product
$8i$	$-8i$	$-(8i)^2=-64i^2=64$

75.

Complex Number	Conjugate	Product
$-11i$	$+11i$	$-(11i)^2=-121i^2=121$

77. $$\frac{7}{4+i}=\frac{7}{4+i}\bullet\frac{4-i}{4-i}=\frac{7(4-i)}{(4+i)(4-i)}$$
$$=\frac{28-7i}{16-i^2}=\frac{28-7i}{17}=\frac{28}{17}-\frac{7}{17}i$$

79.
$$\frac{-3}{1-5i}=\frac{-3}{1-5i}\bullet\frac{1+5i}{1+5i}=\frac{-3(1+5i)}{(1-5i)(1+5i)}$$
$$=\frac{-3-15i}{1^2-25i^2}=\frac{-3-15i}{26}=-\frac{3}{26}-\frac{15}{26}i$$

81. $$\frac{5}{4i}=\frac{5}{4i}\bullet\frac{-4i}{-4i}=\frac{-20i}{-16i^2}=\frac{-20i}{16}=-\frac{5}{4}i$$

83. $$-\frac{2}{\sqrt{-49}}=\frac{-2}{7i}\bullet\frac{-7i}{-7i}=\frac{14i}{-49i^2}=\frac{14i}{49}=\frac{2}{7}i$$

85. $$\frac{4-3i}{i}=\frac{4-3i}{i}\bullet\frac{-i}{-i}=\frac{-4i+3i^2}{-i^2}$$
$$=\frac{-3-4i}{1}=-3-4i$$

87. $$\frac{3+5i}{1+i}=\frac{3+5i}{1+i}\bullet\frac{1-i}{1-i}$$
$$=\frac{3+2i-5i^2}{1-i^2}=\frac{3+2i+5}{1+1}$$
$$=\frac{8+2i}{2}=4+i$$

89. $$\frac{6+3i}{2-2i}=\frac{6+3i}{2-2i}\bullet\frac{2+2i}{2+2i}=\frac{12+18i+6i^2}{4-4i^2}$$
$$=\frac{12+18i-6}{4+4}=\frac{6+18i}{8}=\frac{6}{8}+\frac{18}{8}i$$
$$=\frac{3}{4}+\frac{9}{4}i$$

91. $$\frac{2-\sqrt{-16}}{5-\sqrt{-100}}=\frac{2-4i}{5-10i}\bullet\frac{5+10i}{5+10i}$$
$$=\frac{10+20i-20i-40i^2}{25-100i^2}$$
$$=\frac{10+40}{25+100}=\frac{50}{125}=\frac{2}{5}$$

93. $$\frac{8-\sqrt{-36}}{6+\sqrt{-64}}=\frac{8-6i}{6+8i}\bullet\frac{6-8i}{6-8i}=\frac{48-100i+48i^2}{36-64i^2}$$
$$=\frac{48-100i-48}{100}=\frac{-100i}{100}=-i$$

95. $i^{18}=i^{16}\bullet i^2=(i^4)^4 i^2=(1)^4 i^2=i^2=-1$

97. $i^{35}=i^{32}\bullet i^3=(i^4)^8\bullet i^2\bullet i=(1)^8(-1)i=-i$

99. $i^{12}\bullet i^9=i^{21}=i^{20}\bullet i=(i^4)^5 i=(1)^5 i=i$

101.
$$\frac{i^{38}}{i^{19}}=i^{19}=i^{16}\bullet i^3=(i^4)^4\bullet i^2\bullet i=(1)^4(-1)i=-i$$

103. The complex conjugate of $-8+5i$ is $-8-5i$.
$$(-8+5i)(-8-5i)=(-8)^2-(5i)^2$$
$$=64-25i^2=64-25(-1)$$
$$=64+25=89$$

105. $$4\frac{\sqrt{-3}}{\sqrt{25}}=4\frac{i\sqrt{3}}{5}=\frac{4i}{5}\sqrt{3}$$

107. $$\frac{2-7i}{1-i}=\frac{2-7i}{1-i}\bullet\frac{1+i}{1+i}=\frac{2-5i-7i^2}{1-i^2}$$
$$=\frac{2-5i+7}{1+1}=\frac{9-5i}{2}=\frac{9}{2}-\frac{5i}{2}$$

109.

$$-\sqrt{2}\left(\sqrt{32}-\sqrt{-50}\right)=-\sqrt{2}\sqrt{32}-\left(-\sqrt{2}\right)\sqrt{-50}$$
$$=-\sqrt{64}+\sqrt{-100}$$
$$=-8+10i$$

111. $18-20-\sqrt{-16}=-2-4i$

113.

$$(3+9i)+(5-8i)=(3+5)+(9-8)i=8+i$$

The total impedance is $(8+i)$ ohms.

115. $V=IZ=(8+5i)(9+3i)$

$$=72+69i+15i^2$$
$$=72+69i-15=57+69i$$

The voltage is $(57+69i)$ volts.

Chapter 8 Review Exercises

1. $-6\sqrt{121}=-6\sqrt{11^2}=-6(11)=-66$

2. $2\sqrt[3]{-125}=2\sqrt[3]{(-5)^3}=2(-5)=-10$

3. $\sqrt{\frac{1}{9}}=\frac{\sqrt{1}}{\sqrt{9}}=\frac{1}{3}$

4. $\sqrt{0.36}=\sqrt{(0.6)^2}=0.6$

5. $\sqrt{81y^8}=\sqrt{\left(9y^4\right)^2}=9y^4$

6. $-\sqrt{49a^6b^2}=-\sqrt{\left(7a^3b\right)^2}=-7a^3b$

7. $\sqrt[3]{-216x^9}=\sqrt[3]{\left(-6x^3\right)^3}=-6x^3$

8. $\sqrt[5]{243p^{15}}=\sqrt[5]{\left(3p^3\right)^5}=3p^3$

9. $f(9)=\sqrt{4\cdot 9}=\sqrt{36}=6$

10. $f(10)=\sqrt{4\cdot 10}=\sqrt{4}\sqrt{10}=2\sqrt{10}$

11. $f(x)=-\sqrt{x}$

x	$f(x)=-\sqrt{x}$	(x,y)
0	$f(0)=-\sqrt{0}=0$	$(0,0)$
1	$f(1)=-\sqrt{1}=-1$	$(1,-1)$
2	$f(2)=-\sqrt{2}\approx-1.41$	$(2,-1.41)$
3	$f(3)=-\sqrt{3}\approx-1.73$	$(3,-1.73)$
4	$f(4)=-\sqrt{4}=-2$	$(4,-2)$

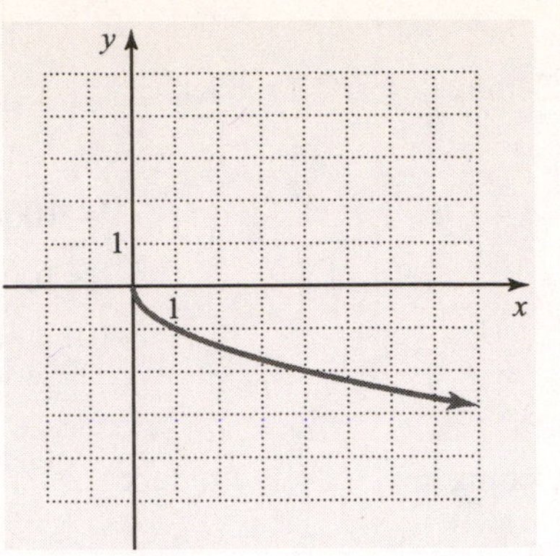

12. $g(x)=\sqrt{x}+1$

x	$g(x)=\sqrt{x}+1$	(x,y)
0	$g(0)=\sqrt{0}+1=0+1=1$	$(0,1)$
1	$g(1)=\sqrt{1}+1=1+1=2$	$(1,2)$
2	$g(2)=\sqrt{2}+1\approx1.41+1=2.41$	$(2,2.41)$
3	$g(3)=\sqrt{3}+1\approx1.73+1=2.73$	$(3,2.73)$
4	$g(4)=\sqrt{4}+1=2+1=3$	$(4,3)$

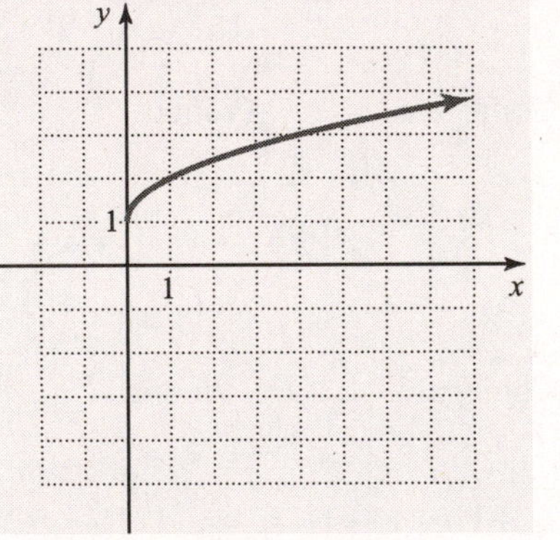

13. $-64^{1/2}=-\sqrt{64}=-\sqrt{8^2}=-8$

14. $7x^{1/3}=7\sqrt[3]{x}$

15. $-\left(16n^4\right)^{3/4}=-\left(\sqrt[4]{16n^4}\right)^3=-\left(\sqrt[4]{(2n)^4}\right)^3$

$$=-(2n)^3=-2^3n^3=-8n^3$$

16. $8^{-2/3}=\frac{1}{8^{2/3}}=\frac{1}{\left(\sqrt[3]{8}\right)^2}=\frac{1}{\left(\sqrt[3]{2^3}\right)^2}=\frac{1}{2^2}=\frac{1}{4}$

17. $x^{1/4}\bullet x^{1/2}=x^{1/4+1/2}=x^{3/4}=\sqrt[4]{x^3}$

18. $\frac{r^{2/3}}{6r^{1/6}}=\frac{1}{6}r^{2/3-1/6}=\frac{1}{6}r^{1/2}=\frac{1}{6}\sqrt{r}=\frac{\sqrt{r}}{6}$

19.

$$\left(25y^2\right)^{-1/2}=\frac{1}{\left(25y^2\right)^{1/2}}=\frac{1}{\sqrt{25y^2}}=\frac{1}{\sqrt{(5y)^2}}=\frac{1}{5y}$$

20. $\dfrac{3a^{2/3}}{\left(6a^{1/6}\right)^2} = \dfrac{3a^{2/3}}{36a^{1/3}} = \dfrac{a^{2/3-1/3}}{12} = \dfrac{a^{1/3}}{12} = \dfrac{\sqrt[3]{a}}{12}$

21. $\sqrt[8]{x^2} = x^{2/8} = x^{1/4} = \sqrt[4]{x}$

22. $\sqrt[6]{n^4} \bullet \sqrt[3]{n} = n^{4/6} \bullet n^{1/3} = n^{2/3+1/3} = n^1 = n$

23. $\sqrt{\sqrt[4]{y^2}} = \left(y^{2/4}\right)^{1/2} = y^{(2/4)(1/2)} = y^{1/4} = \sqrt[4]{y}$

24.
$\sqrt[6]{p^3q^6} = \left(p^3q^6\right)^{1/6} = p^{3/6}q^{6/6} = p^{1/2}q = q\sqrt{p}$

25. $\dfrac{\sqrt[3]{a^2}}{\sqrt{a}} = \dfrac{a^{2/3}}{a^{1/2}} = a^{2/3-1/2} = a^{1/6} = \sqrt[6]{a}$

26.
$\sqrt[4]{x^2} \bullet \sqrt[10]{y^5} = x^{2/4}y^{5/10} = x^{1/2}y^{1/2} = (xy)^{1/2} = \sqrt{xy}$

27. $\sqrt{10r} \bullet \sqrt{3s} = \sqrt{10r \bullet 3s} = \sqrt{30rs}$

28. $\sqrt[3]{4p} \bullet \sqrt[3]{7pq^2} = \sqrt[3]{4p \bullet 7pq^2} = \sqrt[3]{28p^2q^2}$

29.
$\sqrt{300n^3} = \sqrt{100n^2 \bullet 3n} = \sqrt{100n^2} \bullet \sqrt{3n} = 10n\sqrt{3n}$

30. $\sqrt{45x^5y^4} = \sqrt{9x^4y^4 \bullet 5x} = \sqrt{9x^4y^4} \bullet \sqrt{5x}$
$= 3x^2y^2\sqrt{5x}$

31.
$\sqrt[3]{128t^7} = \sqrt[3]{64t^6 \bullet 2t} = \sqrt[3]{64t^6} \bullet \sqrt[3]{2t} = 4t^2\sqrt[3]{2t}$

32. $\sqrt[4]{96a^5b^{10}} = \sqrt[4]{16a^4b^8 \bullet 6ab^2}$
$= \sqrt[4]{16a^4b^8} \bullet \sqrt[4]{6ab^2}$
$= 2ab^2\sqrt[4]{6ab^2}$

33. $\dfrac{\sqrt{35a}}{\sqrt{7}} = \sqrt{\dfrac{35a}{7}} = \sqrt{5a}$

34. $\dfrac{\sqrt[3]{12p^2q^2}}{\sqrt[3]{6pq^2}} = \sqrt[3]{\dfrac{12p^2q^2}{6pq^2}} = \sqrt[3]{2p}$

35. $\sqrt{\dfrac{n}{25}} = \dfrac{\sqrt{n}}{\sqrt{25}} = \dfrac{\sqrt{n}}{5}$

36. $\sqrt{\dfrac{6}{49y^4}} = \dfrac{\sqrt{6}}{\sqrt{49y^4}} = \dfrac{\sqrt{6}}{7y^2}$

37. $\sqrt[3]{\dfrac{64u^2}{125v^9}} = \dfrac{\sqrt[3]{64u^2}}{\sqrt[3]{125v^9}} = \dfrac{4\sqrt[3]{u^2}}{5v^3}$

38. $\sqrt[4]{\dfrac{4p^4q^3}{81r^4s^8}} = \dfrac{\sqrt[4]{4p^4q^3}}{\sqrt[4]{81r^4s^8}} = \dfrac{p\sqrt[4]{4q^3}}{3rs^2}$

39. $9\sqrt{x} - 5\sqrt{x} = (9-5)\left(\sqrt{x}\right) = 4\sqrt{x}$

40. $3\sqrt[3]{q^2} + 8\sqrt[3]{q^2} = (3+8)\sqrt[3]{q^2} = 11\sqrt[3]{q^2}$

41. $\sqrt{48} + \sqrt{27} = \sqrt{16 \bullet 3} + \sqrt{9 \bullet 3}$
$= 4\sqrt{3} + 3\sqrt{3} = (4+3)\sqrt{3}$
$= 7\sqrt{3}$

42. $-\sqrt{96} - 5\sqrt{6} + 3\sqrt{54}$
$= -\sqrt{16 \bullet 6} - 5\sqrt{6} + 3\sqrt{9 \bullet 6}$
$= -4\sqrt{6} - 5\sqrt{6} + 9\sqrt{6}$
$= (-4-5+9)\sqrt{6} = 0$

43. $6\sqrt[3]{56a^4} - \sqrt[3]{189a} = 6\sqrt[3]{8a^3 \bullet 7a} - \sqrt[3]{27 \bullet 7a}$
$= 12a\sqrt[3]{7a} - 3\sqrt[3]{7a}$
$= (12a-3)\sqrt[3]{7a}$

44. $\dfrac{1}{3}\sqrt[3]{27p^5q} + 2p\sqrt[3]{p^2q}$
$= \dfrac{1}{3}\sqrt[3]{27p^3 \bullet p^2q} + 2p\sqrt[3]{p^2q}$
$= \dfrac{3p}{3}\sqrt[3]{p^2q} + 2p\sqrt[3]{p^2q}$
$= (p+2p)\sqrt[3]{p^2q}$
$= 3p\sqrt[3]{p^2q}$

45. $\left(-2\sqrt{3a}\right)\left(3\sqrt{6a}\right) = -6\sqrt{3a \bullet 6a} = -6\sqrt{18a^2}$
$= -6\sqrt{9a^2 \bullet 2} = -6(3a)\sqrt{2}$
$= -18a\sqrt{2}$

46. $\sqrt{5}\left(4\sqrt{10} - 2\sqrt{5}\right) = 4\sqrt{10 \bullet 5} - 2\sqrt{5 \bullet 5}$
$= 4\sqrt{50} - 2\sqrt{25}$
$= 4\sqrt{25 \bullet 2} - 2 \bullet 5$
$= 4 \bullet 5\sqrt{2} - 10 = 20\sqrt{2} - 10$

47. $\sqrt[3]{2n}\left(\sqrt[3]{n^2} - \sqrt[3]{4}\right) = \sqrt[3]{2n^3} - \sqrt[3]{8n} = n\sqrt[3]{2} - 2\sqrt[3]{n}$

48. $\left(4\sqrt{t} - 5\right)\left(\sqrt{t} - 3\right) = 4\left(\sqrt{t}\right)^2 - 12\sqrt{t} - 5\sqrt{t} + 15$
$= 4t - 17\sqrt{t} + 15$

49. $\left(\sqrt{6} - \sqrt{x}\right)\left(\sqrt{6} + \sqrt{x}\right) = \left(\sqrt{6}\right)^2 - \left(\sqrt{x}\right)^2 = 6 - x$

50. $\left(\sqrt{2y} - 1\right)^2 = \left(\sqrt{2y}\right)^2 - 2\sqrt{2y} + 1$
$= 2y - 2\sqrt{2y} + 1$

51. $\dfrac{\sqrt{72n^3}}{4\sqrt{6}} = \dfrac{1}{4}\sqrt{\dfrac{72n^3}{6}} = \dfrac{1}{4}\sqrt{12n^3} = \dfrac{1}{4}\sqrt{4n^2 \bullet 3n}$
$= \dfrac{2n}{4}\sqrt{3n} = \dfrac{n\sqrt{3n}}{2}$

52. $\sqrt{\dfrac{32a}{9b^4}} = \dfrac{\sqrt{16 \bullet 2a}}{\sqrt{9b^4}} = \dfrac{4\sqrt{2a}}{3b^2}$

53. $\frac{1}{\sqrt{8}}\cdot\frac{\sqrt{2}}{\sqrt{2}}=\frac{\sqrt{2}}{\sqrt{16}}=\frac{\sqrt{2}}{4}$

54. $\frac{\sqrt{16x}}{2\sqrt{y}}=\frac{4\sqrt{x}}{2\sqrt{y}}\cdot\frac{\sqrt{y}}{\sqrt{y}}=\frac{2\sqrt{xy}}{y}$

55. $\sqrt{\frac{14p^2}{3q}}=\frac{\sqrt{14p^2}}{\sqrt{3q}}\cdot\frac{\sqrt{3q}}{\sqrt{3q}}$

$=\frac{\sqrt{42p^2q}}{\sqrt{9q^2}}=\frac{p\sqrt{42q}}{3q}$

56. $\sqrt[3]{\frac{5v}{54u^5}}=\frac{\sqrt[3]{5v}}{\sqrt[3]{27u^3\cdot 2u^2}}=\frac{\sqrt[3]{5v}}{3u\sqrt[3]{2u^2}}\cdot\frac{\sqrt[3]{4u}}{\sqrt[3]{4u}}$

$=\frac{\sqrt[3]{20uv}}{3u\sqrt[3]{8u^3}}=\frac{\sqrt[3]{20uv}}{6u^2}$

57. $\frac{\sqrt{10}-3}{\sqrt{5}}=\frac{\sqrt{10}-3}{\sqrt{5}}\cdot\frac{\sqrt{5}}{\sqrt{5}}$

$=\frac{\sqrt{50}-3\sqrt{5}}{5}=\frac{5\sqrt{2}-3\sqrt{5}}{5}$

58. $\frac{2\sqrt{6a}+\sqrt{2}}{\sqrt{2a}}=\frac{2\sqrt{6a}+\sqrt{2}}{\sqrt{2a}}\cdot\frac{\sqrt{2a}}{\sqrt{2a}}$

$=\frac{2\sqrt{12a^2}+\sqrt{4a}}{2a}$

$=\frac{2(2a)\sqrt{3}+2\sqrt{a}}{2a}$

$=\frac{2\left(2a\sqrt{3}+\sqrt{a}\right)}{2a}$

$=\frac{2a\sqrt{3}+\sqrt{a}}{a}$

59. $\frac{4}{\sqrt{3}-1}=\frac{4}{\sqrt{3}-1}\cdot\frac{\sqrt{3}+1}{\sqrt{3}+1}$

$=\frac{4\left(\sqrt{3}+1\right)}{\left(\sqrt{3}\right)^2-1^2}=\frac{4\left(\sqrt{3}+1\right)}{2}$

$=2\left(\sqrt{3}+1\right)=2\sqrt{3}+2$

60. $\frac{\sqrt{x}+2}{\sqrt{x}-\sqrt{5}}=\frac{\sqrt{x}+2}{\sqrt{x}-\sqrt{5}}\cdot\frac{\sqrt{x}+\sqrt{5}}{\sqrt{x}+\sqrt{5}}$

$=\frac{\left(\sqrt{x}\right)^2+\sqrt{5x}+2\sqrt{x}+2\sqrt{5}}{\left(\sqrt{x}\right)^2-\left(\sqrt{5}\right)^2}$

$=\frac{x+\sqrt{5x}+2\sqrt{x}+2\sqrt{5}}{x-5}$

61. $\left(\sqrt{x+8}\right)^2=4^2$

$x+8=16$

$x=8$

62. $\sqrt{n}-2=3$

$\sqrt{n}=5$

$\left(\sqrt{n}\right)^2=5^2$

$n=25$

63. $\sqrt{3n-4}+1=-2$

$\sqrt{3n-4}=-3$

$\left(\sqrt{3n-4}\right)^2=(-3)^2$

$3n-4=9$

$3n=13$

$n=\frac{13}{3}$

Check $n=\frac{13}{3}$

$\sqrt{3\frac{13}{3}-4}+1\stackrel{?}{=}-2$

$\sqrt{9}+1\stackrel{?}{=}-2$

$3+1\stackrel{?}{=}-2$

$4=-2$ False, No solution.

Note: The solution could have been halted with the second step since $\sqrt{3n-4}\geq 0$ and cannot be a negative number.

64. $\sqrt{x^2-7}=\sqrt{5x+7}$

$\left(\sqrt{x^2-7}\right)^2=\left(\sqrt{5x+7}\right)^2$

$x^2-7=5x+7$

$x^2-5x-14=0$

$(x+2)(x-7)=0$

$x+2=0 \quad x-7=0$

$x=-2 \qquad x=7$

Check $x=-2$

$\sqrt{(-2)^2-7}=\sqrt{-3}$, not a real number.

$x=-2$ is not a solution.

Check $x=7$

$\sqrt{7^2-7}\stackrel{?}{=}\sqrt{5(7)+7}$

$\sqrt{42}=\sqrt{42}$ True

65. $\sqrt[3]{2x-3}=-2$

$\left(\sqrt[3]{2x-3}\right)^3=(-2)^3$

$2x-3=-8$

$2x=-5$

$x=-\dfrac{5}{2}$

66. $\sqrt{x+5}+1=\sqrt{3x+4}$

$\left(\sqrt{x+5}+1\right)^2=\left(\sqrt{3x+4}\right)^2$

$x+5+2\sqrt{x+5}+1=3x+4$

$2\sqrt{x+5}=2x-2$

$\sqrt{x+5}=x-1$

$\left(\sqrt{x+5}\right)^2=(x-1)^2$

$x+5=x^2-2x+1$

$x^2-3x-4=0$

$(x-4)(x+1)=0$

$x-4=0 \quad x+1=0$

$x=4 \quad x=-1$

Check $x=4$

$\sqrt{4+5}+1\stackrel{?}{=}\sqrt{3(4)+4}$

$\sqrt{9}+1\stackrel{?}{=}\sqrt{16}$

$3+1\stackrel{?}{=}4$

$4=4$ True

Check $x=-1$

$\sqrt{-1+5}+1\stackrel{?}{=}\sqrt{3(-1)+4}$

$\sqrt{4}+1\stackrel{?}{=}\sqrt{1}$

$2+1\stackrel{?}{=}1$

$3=1$ False, $x=-1$, is not a solution

67. $\sqrt{-36}=\sqrt{36(-1)}=\sqrt{36}\bullet\sqrt{-1}=6i$

68. $\sqrt{-125}=\sqrt{(25)(-1)(5)}$

$=\sqrt{25}\bullet\sqrt{-1}\bullet\sqrt{5}=5i\sqrt{5}$

69.

$(6-4i)+(2+9i)=(6+2)+(-4+9)i=8+5i$

70. $\left(\sqrt{-4}-3\right)-\left(\sqrt{-16}-7\right)$

$=2i-3-4i+7$

$=(-3+7)+(2-4)i$

$=4-2i$

71. $\sqrt{-81}\bullet\sqrt{-1}=9i\bullet i=9i^2=-9$

84. $d^2=8^2+4^2$

72.

$-2i(5i+1)=-10i^2-2i=-10(-1)-2i=10-2i$

73. $(3-3i)(8+3i)=24+9i-24i-9i^2$

$=24-15i+9=33-15i$

74.

$(5-i)^2=5^2-2(5)i+i^2=25-10i-1=24-10i$

75. $\dfrac{-1}{4-4i}=\dfrac{-1}{4-4i}\bullet\dfrac{4+4i}{4+4i}=\dfrac{-(4+4i)}{4^2-(4i)^2}$

$=\dfrac{-4(1+i)}{16-16i^2}=\dfrac{-4(1+i)}{16+16}$

$=\dfrac{-4(1+i)}{32}=\dfrac{-(1+i)}{8}=-\dfrac{1}{8}-\dfrac{1}{8}i$

76. $\dfrac{3-4i}{6-2i}\bullet\dfrac{6+2i}{6+2i}=\dfrac{18-18i-8i^2}{36-4i^2}=\dfrac{18-18i+8}{36+4}$

$=\dfrac{26-18i}{40}=\dfrac{26}{40}-\dfrac{18}{40}i$

$=\dfrac{13}{20}-\dfrac{9}{20}i$

77. $i^{38}=i^{36}\bullet i^2=\left(i^4\right)^9(-1)=(1)^9(-1)=-1$

78. $i^{53}=i^{52}\bullet i=\left(i^4\right)^{13}\bullet i=(1)^{13}\bullet i=i$

79. **a.** $N(2)^{t/20}=N\sqrt[20]{2^t}$

b. $N=10,\ t=60$

$10(2)^{60/20}=10(2)^3=10(8)=80$

80 bacteria are present after 1 hr.

80. $\sqrt{64R}=\sqrt{8^2R}=8\sqrt{R}$

81. **a.** $v=\sqrt{2ad},\ a=2,\ d=50$

$v=\sqrt{2(2)(50)}=\sqrt{2(100)}=10\sqrt{2}=14.14213...$

The velocity of the car is $10\sqrt{2}$ m per sec or about 14.1 m per sec.

b.

$\sqrt{2(4)(50)}-\sqrt{2(2)(50)}$

$=\sqrt{400}-\sqrt{200}$

$=20-\sqrt{100\bullet 2}$

$=20-10\sqrt{2}=5.85786...$

The velocity is $\left(20-10\sqrt{2}\right)$ m per sec or about 5.9 m per sec faster.

82. $\sqrt{\dfrac{k}{I}}=\dfrac{\sqrt{k}}{\sqrt{I}}\bullet\dfrac{\sqrt{I}}{\sqrt{I}}=\dfrac{\sqrt{kI}}{I}$

83. $2\pi\sqrt{\dfrac{m}{k}}=2\pi\dfrac{\sqrt{m}}{\sqrt{k}}\bullet\dfrac{\sqrt{k}}{\sqrt{k}}=2\pi\dfrac{\sqrt{mk}}{k}=\dfrac{2\pi\sqrt{mk}}{k}$

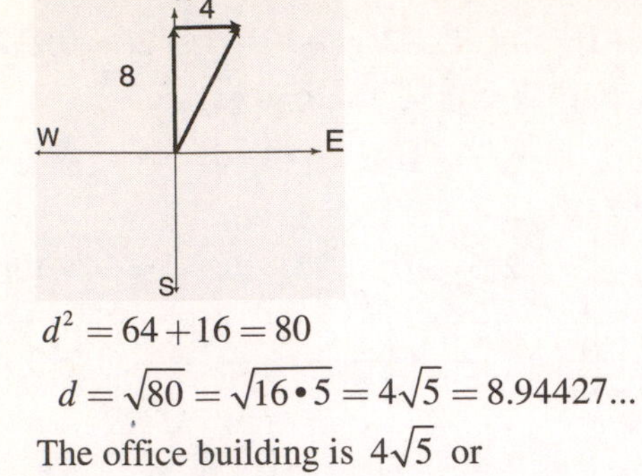

$d^2 = 64 + 16 = 80$

$d = \sqrt{80} = \sqrt{16 \bullet 5} = 4\sqrt{5} = 8.94427...$

The office building is $4\sqrt{5}$ or approximately 8.9 mi from her home.

85. $24^2 = s^2 + 20^2$

24
20
S

$576 = s^2 + 400$

$s^2 = 176$

$s = \sqrt{176} = \sqrt{16 \bullet 11} = 4\sqrt{11} = 13.2664...$

The wire needs to be anchored to the ground at $4\sqrt{11}$ ft or approximately 13.3 ft from the pole.

86. $13.5 = 18 - 0.5\sqrt{x-4}$

$-4.5 = -0.5\sqrt{x-4}$

$\sqrt{x-4} = 9$

$\left(\sqrt{x-4}\right)^2 = 9^2$

$x - 4 = 81$

$x = 85$

85 bottles per week are demanded.

Chapter 8 Posttest

1. **a.** $-3\sqrt{81} = -3\sqrt{9^2} = -3(9) = -27$

b. $\sqrt[3]{-216} = \sqrt[3]{(-6)^3} = -6$

2. $\sqrt{144a^6b^2} = \sqrt{\left(12a^3b\right)^2} = 12a^3b$

3. $\left(32x^{10}\right)^{2/5} = \left(\sqrt[5]{32x^{10}}\right)^2 = \left(\sqrt[5]{\left(2x^2\right)^5}\right)^2$

$= \left(2x^2\right)^2 = 4x^4$

4. **a.** $\dfrac{\left(16p^{1/3}\right)^{3/2}}{8p^{1/3}} = \dfrac{16^{3/2}p^{1/2}}{8p^{1/3}} = \dfrac{\left(\sqrt{16}\right)^3 p^{1/2-1/3}}{8}$

$= \dfrac{4^3 p^{1/6}}{8} = \dfrac{64\sqrt[6]{p}}{8} = 8\sqrt[6]{p}$

b. $\sqrt[8]{x^6y^2} = \left(x^6y^2\right)^{1/8} = x^{6/8}y^{2/8} = x^{3/4}y^{1/4}$

$= \left(x^3y\right)^{1/4} = \sqrt[4]{x^3y}$

5. $\sqrt[3]{5p^2} \bullet \sqrt[3]{4q} = \sqrt[3]{5p^2 \bullet 4q} = \sqrt[3]{20p^2q}$

6. $\dfrac{\sqrt{56n}}{\sqrt{7n}} = \sqrt{\dfrac{56n}{7n}} = \sqrt{8} = \sqrt{4 \bullet 2} = 2\sqrt{2}$

7. $\sqrt{117x^3y^7} = \sqrt{9x^2y^6 \bullet 13xy} = \sqrt{\left(3xy^3\right)^2 \bullet 13xy}$

$= 3xy^3\sqrt{13xy}$

8. $\sqrt{\dfrac{6u}{49v^6}} = \dfrac{\sqrt{6u}}{\sqrt{49v^6}} = \dfrac{\sqrt{6u}}{\sqrt{\left(7v^3\right)^2}} = \dfrac{\sqrt{6u}}{7v^3}$

9.

$-4\sqrt{24} + 2\sqrt{54} - 7\sqrt{6} = -4\sqrt{4 \bullet 6} + 2\sqrt{9 \bullet 6} - 7\sqrt{6}$

$= -8\sqrt{6} + 6\sqrt{6} - 7\sqrt{6}$

$= (-8 + 6 - 7)\sqrt{6} = -9\sqrt{6}$

10. $\left(4\sqrt{2} + 3\right)\left(2\sqrt{2} - 5\right)$

$= 8\left(\sqrt{2}\right)^2 - 20\sqrt{2} + 6\sqrt{2} - 15$

$= 16 - 14\sqrt{2} - 15 = 1 - 14\sqrt{2}$

11.

$\dfrac{\sqrt{3a}}{\sqrt{50b}} = \dfrac{\sqrt{3a}}{\sqrt{50b}} \bullet \dfrac{\sqrt{2b}}{\sqrt{2b}} = \dfrac{\sqrt{6ab}}{\sqrt{100b^2}} = \dfrac{\sqrt{6ab}}{\sqrt{(10b)^2}} = \dfrac{\sqrt{6ab}}{10b}$

12. $\dfrac{\sqrt{x}}{\sqrt{x} - \sqrt{y}} = \dfrac{\sqrt{x}}{\sqrt{x} - \sqrt{y}} \bullet \dfrac{\sqrt{x} + \sqrt{y}}{\sqrt{x} + \sqrt{y}}$

$= \dfrac{\left(\sqrt{x}\right)^2 + \sqrt{xy}}{\left(\sqrt{x}\right)^2 - \left(\sqrt{y}\right)^2}$

$= \dfrac{x + \sqrt{xy}}{x - y}$

13. $\sqrt{2x-1}+9=5$

$\sqrt{2x-1}=-4$

$\left(\sqrt{2x-1}\right)^2=(-4)^2$

$2x-1=16$

$2x=17$

$x=\frac{17}{2}$

Check $x=\frac{17}{2}$

$\sqrt{2\left(\frac{17}{2}\right)-1}+9\stackrel{?}{=}5$

$\sqrt{16}+9\stackrel{?}{=}5$

$4+9\stackrel{?}{=}5$

$13=5$ False

No solution.

The same conclusion could have been reached at the end of the second step since $\sqrt{2x-1}\geq 0$ and cannot be negative.

14. $\sqrt{8-3x}=\sqrt{6-x^2}$

$\left(\sqrt{8-3x}\right)^2=\left(\sqrt{6-x^2}\right)^2$

$8-3x=6-x^2$

$x^2-3x+2=0$

$(x-1)(x-2)=0$

$x-1=0 \quad x-2=0$

$x=1 \qquad x=2$

Check $x=1$

$\sqrt{8-3(1)}\stackrel{?}{=}\sqrt{6-1^2}$

$\sqrt{5}=\sqrt{5}$ True

Check $x=2$

$\sqrt{8-3(2)}\stackrel{?}{=}\sqrt{6-2^2}$

$\sqrt{2}=\sqrt{2}$ True

15.

$\sqrt{x+3}+\sqrt{2x+5}=2$

$\sqrt{x+3}=2-\sqrt{2x+5}$

$\left(\sqrt{x+3}\right)^2=\left(2-\sqrt{2x+5}\right)^2$

$x+3=4-4\sqrt{2x+5}+2x+5$

$4\sqrt{2x+5}=x+6$

$\left(4\sqrt{2x+5}\right)^2=(x+6)^2$

$16(2x+5)=x^2+12x+36$

$32x+80=x^2+12x+36$

$x^2-20x-44=0$

$(x-22)(x+2)=0$

$x-22=0 \quad x+2=0$

$x=22 \qquad x=-2$

Check $x=22$

$\sqrt{22+3}+\sqrt{2\bullet 22+5}\stackrel{?}{=}2$

$\sqrt{25}+\sqrt{49}\stackrel{?}{=}2$

$5+7\stackrel{?}{=}2$

$12=2$ False

$x=22$ is not a solution.

Check $x=-2$

$\sqrt{-2+3}+\sqrt{2(-2)+5}\stackrel{?}{=}2$

$\sqrt{1}+\sqrt{1}\stackrel{?}{=}2$

$2=2$ True

16. $\left(3+\sqrt{-49}\right)\left(1-\sqrt{-16}\right)=(3+7i)(1-4i)$

$=3-12i+7i-28i^2$

$=3-5i+28=31-5i$

17. $\frac{3-5i}{2+3i}=\frac{3-5i}{2+3i}\bullet\frac{2-3i}{2-3i}=\frac{6-19i+15i^2}{4-9i^2}$

$=\frac{6-19i-15}{4+9}=\frac{-9-19i}{13}=-\frac{9}{13}-\frac{19}{13}i$

18. **a.** $\sqrt{\frac{S}{4\pi}}=\frac{\sqrt{S}}{\sqrt{4\pi}}\bullet\frac{\sqrt{\pi}}{\sqrt{\pi}}=\frac{\sqrt{\pi S}}{\sqrt{4\pi^2}}=\frac{\sqrt{\pi S}}{2\pi}$

b. $\frac{\sqrt{\pi(512)}}{2\pi}=\frac{\sqrt{256\bullet 2\pi}}{2\pi}=\frac{16\sqrt{2\pi}}{2\pi}=\frac{8\sqrt{2\pi}}{\pi}$

The radius of the beach ball is $\frac{8\sqrt{2\pi}}{\pi}$ in.

19. $\sqrt{8000(200)+(200)^2}=\sqrt{1{,}600{,}000+40{,}000}$

$=\sqrt{1{,}640{,}000}$

$=\sqrt{40{,}000\bullet 41}$

$=200\sqrt{41}=1280.624...$

The distance to the horizon is $200\sqrt{41}$ mi or about 1280.6 mi.

20. $20=32-\sqrt{x-5}$

$\sqrt{x-5}=12$

$\left(\sqrt{x-5}\right)^2=12^2$

$x-5=144$

$x=149$

The daily demand is 149 units.

Chapter 9

QUADRATIC EQUATIONS, FUNCTIONS, AND INEQUALITIES

Pretest

1. $3x^2+17=5$

$3x^2=-12$

$x^2=-4$

$x=\pm\sqrt{-4}=\pm\sqrt{4(-1)}=\pm 2i$

$x=2i \quad x=-2i$

2. $4a^2+12a=7$

$4\left(a^2+3a\right)=7$

$4\left(a^2+3a+\left(\frac{3}{2}\right)^2\right)=7+4\left(\frac{3}{2}\right)^2$

$4\left(a+\frac{3}{2}\right)^2=7+4\left(\frac{9}{4}\right)=7+9=16$

$4\left(a+\frac{3}{2}\right)^2=16$

$\left(a+\frac{3}{2}\right)^2=4$

$a+\frac{3}{2}=\pm\sqrt{4}=\pm 2$

$a=2-\frac{3}{2}=\frac{4}{2}-\frac{3}{2}=\frac{1}{2}$

$a=-2-\frac{3}{2}=-\frac{4}{2}-\frac{3}{2}=-\frac{7}{2}$

3. $2x^2-6x+3=0$

$a=2 \quad b=-6 \quad c=3$

$$x=\frac{-(-6)\pm\sqrt{(-6)^2-4(2)(3)}}{2(2)}$$

$$x=\frac{6\pm\sqrt{36-24}}{4}=\frac{6\pm\sqrt{12}}{4}=\frac{6\pm 2\sqrt{3}}{4}$$

$$x=\frac{2(3\pm\sqrt{3})}{4} \quad x=\frac{3+\sqrt{3}}{2},\ x=\frac{3-\sqrt{3}}{2}$$

4. $3x^2+5x-1=0$

$a=3 \quad b=5 \quad c=-1$

$b^2-4ac=(5)^2-4(3)(-1)=25+12=37.$

37 is positive, not a perfect square.

Two real solutions.

5. $6(2n-3)^2=48$

$(2n-3)^2=8$

$2n-3=\pm\sqrt{8}=\pm 2\sqrt{2}$

$2n=3\pm 2\sqrt{2}$

$n=\frac{3+2\sqrt{2}}{2},\ n=\frac{3-2\sqrt{2}}{2}$

6. $x^2-4x+12=0$

$a=1 \quad b=-4 \quad c=12$

$$x=\frac{-(-4)\pm\sqrt{(-4)^2-4(1)(12)}}{2(1)}$$

$$x=\frac{4\pm\sqrt{16-48}}{2}=\frac{4\pm\sqrt{-32}}{2}=\frac{4\pm 4i\sqrt{2}}{2}$$

$x=2+2i\sqrt{2},\ x=2-2i\sqrt{2}$

7. $3x^2+15x+16=x$

$3x^2+14x+16=0$

$(x+2)(3x+8)=0$

$x+2=0 \qquad 3x+8=0$

$x=-2 \qquad 3x=-8$

$x=-\frac{8}{3}$

8. $2x^2+2x=1-6x$

$2x^2+8x-1=0$

$a=2 \quad b=8 \quad c=-1$

$$x=\frac{-(8)\pm\sqrt{(8)^2-4(2)(-1)}}{2(2)}$$

$$x=\frac{-8\pm\sqrt{64+8}}{4}=\frac{-8\pm\sqrt{72}}{4}$$

$$x=\frac{-8\pm 6\sqrt{2}}{4}=\frac{-4\pm 3\sqrt{2}}{2}$$

$$x=\frac{-4+3\sqrt{2}}{2},x=\frac{-4-3\sqrt{2}}{2}$$

9.
$$5x^2-10x+9=3$$
$$5x^2-10x+6=0$$
$$a=5 \qquad b=-10 \qquad c=6$$
$$x=\frac{-(-10)\pm\sqrt{(-10)^2-4(5)(6)}}{2(5)}$$
$$x=\frac{10\pm\sqrt{100-120}}{10}=\frac{10\pm\sqrt{-20}}{10}$$
$$x=\frac{10\pm\sqrt{4(5)(-1)}}{10}=\frac{10\pm 2i\sqrt{5}}{10}$$
$$x=\frac{5+i\sqrt{5}}{5},\quad x=\frac{5-i\sqrt{5}}{5}$$

10.
$$0.04x^2-0.12x+0.09=0$$
$$100\left(0.04x^2-0.12x+0.09\right)=100(0)$$
$$4x^2-12x+9=0$$
$$(2x-3)^2=0$$
$$2x-3=\pm\sqrt{0}$$
$$=0\Rightarrow 2x=3$$
$$x=\frac{3}{2}$$

11.
$$x^4-x^2-72=0$$
Let $u=x^2$
$$\left(x^2\right)^2-\left(x^2\right)-72=0$$
$$u^2-u-72=0$$
$$(u-9)(u+8)=0$$
$$u-9=0 \qquad u+8=0$$
$$u=9 \qquad u=-8$$
$$x^2=9 \qquad x^2=-8$$
$$x=\pm\sqrt{9} \qquad x=\pm\sqrt{-8}=\pm\sqrt{4(-1)(2)}$$
$$x=\pm 3 \qquad x=\pm 2i\sqrt{2}$$

12.
$$x=-\frac{3}{2}, \qquad x=4$$
$$2x=-3 \qquad x-4=0$$
$$2x+3=0$$
$$(2x+3)(x-4)=0$$
$$2x^2-5x-12=0$$

13.
$$f(x)=x^2-4x+3$$
$$a=1 \qquad b=-4$$
$$x=-\frac{b}{2a}=-\frac{-4}{2(1)}=2$$
$$f(2)=(2)^2-4(2)+3=4-8+3=-1$$
Vertex: $(2,-1)$

Axis of symmetry: $x=2$
$$f(0)=(0)^2-4(0)+3=0$$
y-intercept: $(0,3)$
$$x^2-4x+3=0$$
$$(x-3)(x-1)=0$$
$$x-3=0 \qquad x-1=0$$
$$x=3 \qquad x=1$$
x-intercepts: $(3,0)$ and $(1,0)$

14.
$$f(x)=12-x-x^2$$
$$a=-1 \qquad b=-1$$
$$x=-\frac{b}{2a}=-\frac{-1}{2(-1)}=-\frac{1}{2}$$
$$f\left(-\frac{1}{2}\right)=12-\left(-\frac{1}{2}\right)-\left(-\frac{1}{2}\right)^2$$
$$=\frac{48}{4}+\frac{2}{4}-\frac{1}{4}=\frac{49}{4}$$
Vertex: $\left(-\frac{1}{2},\frac{49}{4}\right)$

Axis of symmetry: $x=-\frac{1}{2}$
$$f(0)=12-(0)-(0)^2=12$$
y-intercept: $(0,12)$
$$12-x-x^2=0$$
$$(4+x)(3-x)=0$$
$$4+x=0 \qquad 3-x=0$$
$$x=-4 \qquad x=3$$
x-intercepts: $(-4,0)$ and $(3,0)$

14. (continued)

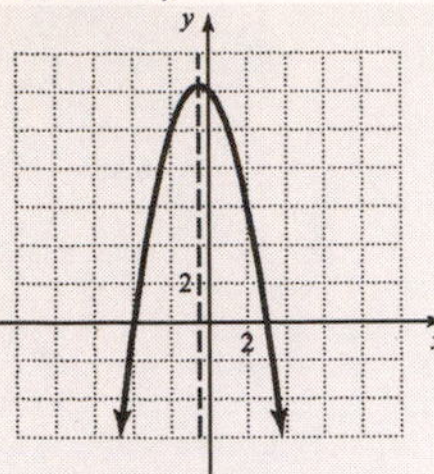

15.

x	$f(x)=\frac{1}{2}x^2+x-4$	(x,y)
-4	$\frac{1}{2}(-4)^2+(-4)-4=0$	$(-4,0)$
-2	$\frac{1}{2}(-2)^2+(-2)-4=-4$	$(-2,-4)$
0	$\frac{1}{2}(0)^2+(0)-4=-4$	$(0,-4)$
2	$\frac{1}{2}(2)^2+(2)-4=0$	$(2,0)$
4	$\frac{1}{2}(4)^2+(4)-4=8$	$(4,8)$

$$x=-\frac{b}{2a}=-\frac{1}{2\left(\frac{1}{2}\right)}=-\frac{1}{1}=-1$$

$$f(-1)=\frac{1}{2}(-1)^2+(-1)-4=-\frac{9}{2}$$

Vertex: $\left(-1,-\frac{9}{2}\right)$

Domain: $(-\infty,\infty)$ Range: $\left[-\frac{9}{2},\infty\right)$

16. $x^2-6x+8>0$

$$x^2-6x+8=0$$
$$(x-4)(x-2)=0$$
$$x-4=0 \quad x-2=0$$
$$x=4 \quad x=2$$

Interval	Test Value	$x^2-6x+8>0$	Conclusion
$x\le 2$	-6	$(-6)^2-6(-6)+8=80$	$80\overset{?}{>}0$ True
$2<x<4$	3	$(3)^2-6(3)+8=-1$	$-1\overset{?}{>}0$ False
$4\le x$	6	$(6)^2-6(6)+8=8$	$8\overset{?}{>}0$ True

$(-\infty,2)\cup(4,\infty)$

17. $h=-16t^2+20t=-50$

$$-16t^2+20t+50=0$$
$$8t^2-10t-25=0$$
$$(4t+5)(2t-5)=0$$
$$4t+5=0 \quad 2t-5=0$$
$$4t=-5 \quad 2t=5$$
$$t=-\frac{5}{4} \quad t=\frac{5}{2}$$

The negative answer makes no sense in this problem. The object is 50 ft below the point of release $\frac{5}{2}$ sec, or 2.5 sec, after it is tossed upward.

18. $\text{time}=\frac{\text{distance}}{\text{rate}}$

Let x be the speed of the current.

$$\frac{6}{9+x}=\frac{6}{9-x}-0.5$$
$$\frac{6}{9+x}\bullet(9-x)(9+x)=\frac{6}{9-x}\bullet(9-x)(9+x)-0.5\bullet(9-x)(9+x)$$
$$54-6x=54+6x-0.5\left(81-x^2\right)$$
$$54-6x=54+6x-40.5+0.5x^2$$
$$0.5x^2+12x-40.5=0$$
$$x^2+24x-81=0$$
$$(x+27)(x-3)=0$$

18. (continued)

$x+27=0 \quad x-3=0$

$x=-27 \quad x=3$

The negative answer makes no sense in this problem. The speed of the current is 3 mph.

19.

Let $r =$ the radius of the total planting area.

$\pi r^2 = 157$

$3.14r^2 = 157$

$r^2 = 50$

$r = \pm\sqrt{50} = 7.071...$

The radius of the planting area is approximately 6.1 ft. The maximum planting area is $(3.14)(6.1)^2 = 116.8$ sq ft.

20. $R(x) = 55x - 0.5x^2 \geq 1200$

$-0.5x^2 + 55x - 1200 \geq 0$

$0.5x^2 - 55x + 1200 \leq 0$

$x^2 - 110x + 2400 = 0$

$(x-80)(x-30) = 0$

$x-80=0 \quad x-30=0$

$x=80 \quad x=30$

Interval	Test	$55x-0.5x^2 \geq 1200$	Conclusion
$x \leq 30$	0	$55(0)-0.5(0)^2 = 0$	$0 \overset{?}{\geq} 1200$ False
$30 < x < 80$	50	$55(50)-0.5(50)^2 = 1500$	$1500 \overset{?}{\geq} 1200$ True
$80 \leq x$	90	$55(90)-0.5(90)^2 = 900$	$900 \overset{?}{\geq} 1200$ False

The company must sell between and including 30 and 80 bottles for revenue of at least $1200.

9.1 Solving Quadratic Equations: Square Root Property and Completing the Square

Practice

1. **a.** $n^2 = 49$

$n = \pm\sqrt{49}$

$n = \pm 7$

b. $y^2 + 9 = 0$

$y^2 = -9$

$y = \pm\sqrt{-9}$

$y = \pm 3i$

2. **a.** $2y^2 = 16$

$y^2 = 8$

$y = \pm\sqrt{8}$

$y = \pm 2\sqrt{2}$

b. $3x^2 + 54 = 0$

$3x^2 = -54$

$x^2 = -18$

$x = \pm\sqrt{-18}$

$x = \pm 3i\sqrt{2}$

3. **a.** $(2x+1)^2 = 12$

$2x+1 = \pm\sqrt{12}$

$2x = -1 \pm 2\sqrt{3}$

$x = \dfrac{-1 \pm 2\sqrt{3}}{2}$

b. $(n+3)^2 + 2 = 0$

$(n+3)^2 = -2$

$n+3 = \pm\sqrt{-2}$

$n = -3 \pm i\sqrt{2}$

c. $9(y-1)^2 = 49$

$(y-1)^2 = \dfrac{49}{9}$

$y-1 = \pm\sqrt{\dfrac{49}{9}}$

$y = 1 \pm \dfrac{7}{3}$

$y = \dfrac{10}{3}, -\dfrac{4}{3}$

4. $y^2 - 6y + 9 = 24$

$(y-3)^2 = 24$

$y-3 = \pm\sqrt{24}$

$y = 3 \pm 2\sqrt{6}$

5. **a.** $x^2+6x+\underline{9}$

b. $t^2-t+\underline{\frac{1}{4}}$

6.
$$\begin{aligned} t^2+3t&=1\\ t^2+3t+\frac{9}{4}&=1+\frac{9}{4}\\ t^2+3t+\frac{9}{4}&=\frac{13}{4}\\ \left(t+\frac{3}{2}\right)^2&=\frac{13}{4}\\ t+\frac{3}{2}&=\pm\sqrt{\frac{13}{4}}\\ t&=-\frac{3}{2}\pm\frac{\sqrt{13}}{2}\\ t&=\frac{-3\pm\sqrt{13}}{2} \end{aligned}$$

7.
$$\begin{aligned} 2y^2-y+1&=0\\ y^2-\frac{1}{2}y+\frac{1}{2}&=0\\ y^2-\frac{1}{2}y&=-\frac{1}{2}\\ y^2-\frac{1}{2}y+\frac{1}{16}&=-\frac{1}{2}+\frac{1}{16}\\ y^2-\frac{1}{2}y+\frac{1}{16}&=-\frac{7}{16}\\ \left(y-\frac{1}{4}\right)^2&=-\frac{7}{16}\\ y-\frac{1}{4}&=\pm\sqrt{-\frac{7}{16}}\\ y&=\frac{1}{4}\pm\frac{i\sqrt{7}}{4}\\ y&=\frac{1\pm i\sqrt{7}}{4} \end{aligned}$$

8.
$$\begin{aligned} g(x)&=h(x)\\ x^2+6x&=2\\ x^2+6x+9&=2+9\\ x^2+6x+9&=11\\ (x+3)^2&=11\\ x+3&=\pm\sqrt{11}\\ x&=-3\pm\sqrt{11} \end{aligned}$$

9.
$$\begin{aligned} 6000&=-x^2+200x-4000\\ x^2-200x&=-10{,}000\\ x^2-200x+10{,}000&=-10{,}000+10{,}000\\ x^2-200x+10{,}000&=0\\ (x-100)^2&=0\\ x-100&=\pm\sqrt{0}\\ x&=100+0\\ x&=100 \end{aligned}$$

Exercises

1.
$$\begin{aligned} x^2&=16\\ x&=\pm4\\ x&=4,\ x=-4 \end{aligned}$$

3.
$$\begin{aligned} y^2&=24\\ y&=\pm\sqrt{24}=\pm2\sqrt{6}\\ y&=2\sqrt{6},\ y=-2\sqrt{6} \end{aligned}$$

5.
$$\begin{aligned} a^2+25&=0\\ a^2&=-25\\ a&=\pm\sqrt{-25}=\pm5i\\ a&=5i,\ a=-5i \end{aligned}$$

7.
$$\begin{aligned} 4n^2-8&=0\\ 4n^2&=8\\ n^2&=2\\ n&=\pm\sqrt{2}\\ n&=\sqrt{2},\ n=-\sqrt{2} \end{aligned}$$

9.
$$\begin{aligned} \frac{1}{6}y^2&=12\\ y^2&=72\\ y&=\pm\sqrt{72}=\pm6\sqrt{2}\\ y&=6\sqrt{2},\ y=-6\sqrt{2} \end{aligned}$$

11.
$$\begin{aligned} 3x^2+6&=21\\ 3x^2&=15\\ x^2&=5\\ x&=\pm\sqrt{5}\\ x&=\sqrt{5},\ x=-\sqrt{5} \end{aligned}$$

13. $8-9n^2=14$

$$-9n^2=6$$

$$n^2=-\frac{6}{9}$$

$$n=\pm\sqrt{-\frac{6}{9}}=\pm\frac{\sqrt{-6}}{\sqrt{9}}=\pm\frac{i\sqrt{6}}{3}$$

$$n=\frac{i\sqrt{6}}{3},\ n=-\frac{i\sqrt{6}}{3}$$

15. $(x-1)^2=48$

$$x-1=\pm\sqrt{48}=\pm4\sqrt{3}$$

$$x=1\pm4\sqrt{3}$$

$$x=1+4\sqrt{3},\ x=1-4\sqrt{3}$$

17. $(2n+5)^2=75$

$$2n+5=\pm\sqrt{75}=\pm5\sqrt{3}$$

$$2n=-5\pm5\sqrt{3}$$

$$n=\frac{-5\pm5\sqrt{3}}{2}$$

$$n=\frac{-5+5\sqrt{3}}{2},\ n=\frac{-5-5\sqrt{3}}{2}$$

19. $(4a-3)^2+9=1$

$$(4a-3)^2=-8$$

$$4a-3=\pm\sqrt{-8}=\pm2i\sqrt{2}$$

$$4a=3\pm2i\sqrt{2}$$

$$a=\frac{3\pm2i\sqrt{2}}{4}$$

$$a=\frac{3+2i\sqrt{2}}{4},\ a=\frac{3-2i\sqrt{2}}{4}$$

21. $16(x+4)^2=81$

$$(x+4)^2=\frac{81}{16}$$

$$x+4=\pm\sqrt{\frac{81}{16}}=\pm\frac{\sqrt{81}}{\sqrt{16}}=\pm\frac{9}{4}$$

$$x=-4\pm\frac{9}{4}=-\frac{16}{4}\pm\frac{9}{4}$$

$$x=-\frac{7}{4},\ x=-\frac{25}{4}$$

23. $x^2-6x+9=80$

$$(x-3)^2=80$$

$$x-3=\pm\sqrt{80}=\pm4\sqrt{5}$$

$$x=3\pm4\sqrt{5}$$

$$x=3+4\sqrt{5},\ x=3-4\sqrt{5}$$

25. $4p^2+12p+9=32$

$$(2p+3)^2=32$$

$$2p+3=\pm\sqrt{32}=\pm4\sqrt{2}$$

$$2p=-3\pm4\sqrt{2}$$

$$p=\frac{-3\pm4\sqrt{2}}{2}$$

$$p=\frac{-3+4\sqrt{2}}{2},\ p=\frac{-3-4\sqrt{2}}{2}$$

27.

$$(3n-2)(3n+2)=-52$$

$$9n^2-4=-52$$

$$9n^2=-48$$

$$n^2=\frac{-48}{9}$$

$$n=\pm\sqrt{\frac{-48}{9}}=\pm\frac{\sqrt{-48}}{\sqrt{9}}=\pm\frac{4i\sqrt{3}}{3}$$

$$n=\frac{4i\sqrt{3}}{3},\ n=-\frac{4i\sqrt{3}}{3}$$

29. $2x-1=\frac{18}{2x-1}$

$$(2x-1)^2=18$$

$$2x-1=\pm\sqrt{18}=\pm3\sqrt{2}$$

$$2x=1\pm3\sqrt{2}$$

$$x=\frac{1\pm3\sqrt{2}}{2}$$

$$x=\frac{1+3\sqrt{2}}{2},\ x=\frac{1-3\sqrt{2}}{2}$$

31. $V = \pi r^2 h$

$$\frac{V}{\pi h} = \frac{\pi r^2 h}{\pi h}$$

$$r^2 = \frac{V}{\pi h}$$

$$r = \pm\sqrt{\frac{V}{\pi h}}$$

r is a radius and always positive.

$$r = \sqrt{\frac{V}{\pi h}} = \frac{\sqrt{V}}{\sqrt{\pi h}} \bullet \frac{\sqrt{\pi h}}{\sqrt{\pi h}} = \frac{\sqrt{\pi V h}}{\pi h}$$

33. $F = \dfrac{mv^2}{r}$

$$Fr = mv^2$$

$$\frac{Fr}{m} = v^2$$

$$v = \pm\sqrt{\frac{Fr}{m}} = \pm\frac{\sqrt{Fr}}{\sqrt{m}} \bullet \frac{\sqrt{m}}{\sqrt{m}} = \pm\frac{\sqrt{Frm}}{m}$$

35. $x^2 - 12x + \left[\left(\dfrac{-12}{2}\right)^2\right] = x^2 - 12x + 36$

37. $n^2 + 7n + \left[\left(\dfrac{7}{2}\right)^2\right] = n^2 + 7n + \dfrac{49}{4}$

39. $t^2 - \dfrac{4}{3}t + \left[\left(\dfrac{1}{2} \bullet \dfrac{-4}{3}\right)^2\right] = t^2 - \dfrac{4}{3}t + \left[\left(\dfrac{-2}{3}\right)^2\right]$

$$= t^2 - \frac{4}{3}t + \frac{4}{9}$$

41. $x^2 - 8x = 0$

$$x^2 - 8x + \left(\frac{1}{2} \bullet (-8)\right)^2 = 0 + \left(\frac{1}{2} \bullet (-8)\right)^2$$

$$x^2 - 8x + 16 = 16$$

$$(x-4)^2 = 16$$

$$x - 4 = \pm\sqrt{16}$$

$$x - 4 = \pm 4$$

$$x = 4 \pm 4$$

$$x = 8, \ x = 0$$

43. $n^2 - 3n = 4$

$$n^2 - 3n + \left(\frac{-3}{2}\right)^2 = 4 + \left(\frac{-3}{2}\right)^2$$

$$n^2 - 3n + \frac{9}{4} = \frac{16}{4} + \frac{9}{4}$$

$$\left(n - \frac{3}{2}\right)^2 = \frac{25}{4}$$

$$n - \frac{3}{2} = \pm\sqrt{\frac{25}{4}}$$

$$n - \frac{3}{2} = \pm\frac{5}{2}$$

$$n = \frac{3}{2} \pm \frac{5}{2}$$

$$n = 4, \ n = -1$$

45. $x^2 + 4x - 2 = 0$

$$x^2 + 4x = 2$$

$$x^2 + 4x + \left(\frac{4}{2}\right)^2 = 2 + \left(\frac{4}{2}\right)^2$$

$$x^2 + 4x + 4 = 2 + 4$$

$$(x+2)^2 = 6$$

$$x + 2 = \pm\sqrt{6}$$

$$x = -2 \pm \sqrt{6}$$

$$x = -2 + \sqrt{6}, \ x = -2 - \sqrt{6}$$

47. $a^2 + 7a = 3a - 4$

$$a^2 + 4a = -4$$

$$a^2 + 4a + \left(\frac{4}{2}\right)^2 = -4 + \left(\frac{4}{2}\right)^2$$

$$a^2 + 4a + 4 = -4 + 4$$

$$(a+2)^2 = 0$$

$$a + 2 = \pm\sqrt{0}$$

$$a = -2$$

49. $x^2 - 9x + 4 = x - 25$

$$x^2 - 10x = -29$$

$$x^2 - 10x + \left(\frac{-10}{2}\right)^2 = -29 + \left(\frac{-10}{2}\right)^2$$

$$x^2 - 10x + 25 = -29 + 25$$

$$(x-5)^2 = -4$$

$$x - 5 = \pm\sqrt{-4}$$

$$x = 5 \pm 2i$$

$$x = 5 + 2i, \ x = 5 - 2i$$

51.
$$2n^2-8n=-24$$
$$2(n^2-4n)=2(-12)$$
$$n^2-4n=-12$$
$$n^2-4n+\left(\frac{-4}{2}\right)^2=-12+\left(\frac{-4}{2}\right)^2$$
$$n^2-4n+4=-12+4$$
$$(n-2)^2=-8$$
$$n-2=\pm\sqrt{-8}$$
$$n=2\pm 2i\sqrt{2}$$
$$n=2+2i\sqrt{2},\ n=2-2i\sqrt{2}$$

53.
$$3x^2-12x-84=0$$
$$3(x^2-4x-28)=0$$
$$x^2-4x-28=0$$
$$x^2-4x=28$$
$$x^2-4x+\left(\frac{-4}{2}\right)^2=28+\left(\frac{-4}{2}\right)^2$$
$$x^2-4x+4=28+4$$
$$(x-2)^2=32$$
$$x-2=\pm\sqrt{32}$$
$$x=2\pm 4\sqrt{2}$$
$$x=2+4\sqrt{2},\ x=2-4\sqrt{2}$$

55.
$$4a^2+20a-12=0$$
$$4(a^2+5a-3)=0$$
$$a^2+5a-3=0$$
$$a^2+5a=3$$
$$a^2+5a+\left(\frac{5}{2}\right)^2=3+\left(\frac{5}{2}\right)^2$$
$$a^2+5a+\frac{25}{4}=\frac{12}{4}+\frac{25}{4}$$
$$\left(a+\frac{5}{2}\right)^2=\frac{37}{4}$$
$$a+\frac{5}{2}=\pm\sqrt{\frac{37}{4}}$$
$$a=-\frac{5}{2}\pm\frac{\sqrt{37}}{2}$$
$$a=\frac{-5+\sqrt{37}}{2},\ a=\frac{-5-\sqrt{37}}{2}$$

57.
$$3y^2-9y+15=0$$
$$3(y^2-3y+5)=0$$
$$y^2-3y+5=0$$
$$y^2-3y=-5$$
$$y^2-3y+\left(\frac{-3}{2}\right)^2=-5+\left(\frac{-3}{2}\right)^2$$
$$y^2-3y+\frac{9}{4}=\frac{-20}{4}+\frac{9}{4}$$
$$\left(y-\frac{3}{2}\right)^2=\frac{-11}{4}$$
$$y-\frac{3}{2}=\pm\sqrt{\frac{-11}{4}}$$
$$y-\frac{3}{2}=\pm\frac{i\sqrt{11}}{2}$$
$$y=\frac{3+i\sqrt{11}}{2},\ y=\frac{3-i\sqrt{11}}{2}$$

59.
$$x^2-\frac{4}{3}x-4=0$$
$$x^2-\frac{4}{3}x=4$$
$$x^2-\frac{4}{3}x+\left(\frac{1}{2}\bullet\frac{-4}{3}\right)^2=4+\left(\frac{1}{2}\bullet\frac{-4}{3}\right)^2$$
$$x^2-\frac{4}{3}x+\left(\frac{-2}{3}\right)^2=4+\left(\frac{-2}{3}\right)^2$$
$$x^2-\frac{4}{3}x+\frac{4}{9}=\frac{36}{9}+\frac{4}{9}$$
$$\left(x-\frac{2}{3}\right)^2=\frac{40}{9}$$
$$x-\frac{2}{3}=\pm\sqrt{\frac{40}{9}}=\pm\frac{2\sqrt{10}}{3}$$
$$x=\frac{2}{3}\pm\frac{2\sqrt{10}}{3}$$
$$x=\frac{2+2\sqrt{10}}{3},\ x=\frac{2-2\sqrt{10}}{3}$$

61.

$$4y^2+11y+6=0$$
$$4y^2+11y=-6$$
$$\frac{4y^2+11y}{4}=\frac{-6}{4}$$
$$y^2+\frac{11}{4}y=\frac{-3}{2}$$
$$y^2+\frac{11}{4}y+\left(\frac{1}{2}\bullet\frac{11}{4}\right)^2=\frac{-3}{2}+\left(\frac{1}{2}\bullet\frac{11}{4}\right)^2$$
$$y^2+\frac{11}{4}y+\frac{121}{64}=\frac{-3}{2}+\frac{121}{64}$$
$$\left(y+\frac{11}{8}\right)^2=\frac{-96}{64}+\frac{121}{64}$$
$$\left(y+\frac{11}{8}\right)^2=\frac{25}{64}$$
$$y+\frac{11}{8}=\pm\sqrt{\frac{25}{64}}$$
$$y+\frac{11}{8}=\pm\frac{5}{8}$$
$$y=-\frac{11}{8}\pm\frac{5}{8}$$
$$y=-2,\ y=-\frac{6}{8}=-\frac{3}{4}$$

63.

$$2p^2+7p=6p-8$$
$$2p^2+p=-8$$
$$\frac{2p^2+p}{2}=\frac{-8}{2}$$
$$p^2+\frac{1}{2}p=-4$$
$$p^2+\frac{1}{2}p+\left(\frac{1}{2}\bullet\frac{1}{2}\right)^2=-4+\left(\frac{1}{2}\bullet\frac{1}{2}\right)^2$$
$$p^2+\frac{1}{2}p+\frac{1}{16}=-\frac{64}{16}+\frac{1}{16}$$
$$\left(p+\frac{1}{4}\right)^2=\frac{-63}{16}$$
$$p+\frac{1}{4}=\pm\sqrt{\frac{-63}{16}}$$
$$p+\frac{1}{4}=\pm\frac{3i\sqrt{7}}{4}$$
$$p=\frac{-1+3i\sqrt{7}}{4},\ p=\frac{-1-3i\sqrt{7}}{4}$$

65.

$$x^2-9=4x-6$$
$$x^2-4x=3$$
$$x^2-4x+\left(\frac{-4}{2}\right)^2=3+\left(\frac{-4}{2}\right)^2$$
$$x^2-4x+4=3+4$$
$$(x-2)^2=7$$
$$x-2=\pm\sqrt{7}$$
$$x=2+\sqrt{7},\ x=2-\sqrt{7}$$

67.

$$4x^2=x^2-6x+6$$
$$3x^2+6x-6=0$$
$$3(x^2+2x-2)=0$$
$$x^2+2x-2=0$$
$$x^2+2x=2$$
$$x^2+2x+\left(\frac{2}{2}\right)^2=2+\left(\frac{2}{2}\right)^2$$
$$x^2+2x+1=2+1$$
$$(x+1)^2=3$$
$$x+1=\pm\sqrt{3}$$
$$x=-1+\sqrt{3},\ x=-1-\sqrt{3}$$

69.

$$V=\frac{1}{3}\pi r^2h$$
$$3V=\pi r^2h$$
$$\frac{3V}{\pi h}=r^2$$
$$\pm\sqrt{\frac{3V}{\pi h}}=r$$

The negative solution makes no sense in this context, so

$$r=\sqrt{\frac{3V}{\pi h}}=\frac{\sqrt{3V}}{\sqrt{\pi h}}\bullet\frac{\sqrt{\pi h}}{\sqrt{\pi h}}=\frac{\sqrt{3V\pi h}}{\pi h}.$$

71.

$$f(x)=g(x)$$
$$x^2-4x=2x-2$$
$$x^2-6x=-2$$

Solve by completing the square.

$$x^2-6x+\left(\frac{-6}{2}\right)^2=-2+\left(\frac{-6}{2}\right)^2$$
$$x^2-6x+9=7$$
$$(x-3)^2=7$$
$$x-3=\pm\sqrt{7}$$
$$x=3+\sqrt{7},\ x=3-\sqrt{7}$$

73. $-3x^2 - 24x + 6 = 0$

$-3\left(x^2 + 8x - 2\right) = 0$

$x^2 + 8x - 2 = 0$

$x^2 + 8x = 2$

Solve by completing the square.

$x^2 + 8x + \left(\frac{8}{2}\right)^2 = 2 + \left(\frac{8}{2}\right)^2$

$x^2 + 8x + 16 = 18$

$(x+4)^2 = 18$

$x + 4 = \pm\sqrt{18} = \pm 3\sqrt{2}$

$x = -4 + 3\sqrt{2},\quad x = -4 - 3\sqrt{2}$

75. $x^2 - 10x + 25 = 45$

$(x-5)^2 = 45$

$x - 5 = \pm\sqrt{45} = \pm 3\sqrt{5}$

$x = 5 - 3\sqrt{5},\quad x = 5 + 3\sqrt{5}$

77. $A = \pi r^2$

$78.5 = 3.14r^2$

$r^2 = \frac{78.5}{3.14} = 25$

$r = \pm\sqrt{25} \Rightarrow r = 5,\ r = -5$

The negative answer makes no sense in this problem. The team is searching 5 mi from the last known location of the hikers.

79. $1102.50 = 1000(1+r)^2$

$(1+r)^2 = \frac{1102.50}{1000}$

$(1+r)^2 = 1.1025$

$1 + r = \pm\sqrt{1.1025}$

$1 + r = \pm 1.05$

$r = -1 \pm 1.05 \Rightarrow r = -2.05,\ r = 0.05$

The negative answer makes no sense in this problem. The interest rate is 5%.

81. **a.** $1000 = -16t^2 + 2000$

$-16t^2 + 2000 = 1000$

$-16t^2 = -1000$

$t^2 = \frac{1000}{16}$

$t = \sqrt{\frac{1000}{16}} = \frac{10\sqrt{10}}{4} = \frac{5\sqrt{10}}{2} = 7.9056...$

The sandbag will be 1000 ft above the ground in approximately 7.9 sec after it is dropped.

b. $h = -16(2 \cdot 7.9)^2 + 2000 = -1994.24 \neq 0$

$0 = -16t^2 + 2000$

$16t^2 = 2000$

$t^2 = \frac{2000}{16}$

$t = \sqrt{\frac{2000}{16}} = \frac{20\sqrt{5}}{4} = 5\sqrt{5} = 11.1803...$

No, the sandbag will reach the ground in approximately 11.2 sec which is not equal to 2(7.9) or about 15.8 sec.

83. **a.** $A = lw,\ l = w + 10$

$A = (w+10)w$

$144 = w^2 + 10w$

$w^2 + 10w = 144$

$w^2 + 10w + \left(\frac{10}{2}\right)^2 = 144 + \left(\frac{10}{2}\right)^2$

$w^2 + 10w + 25 = 144 + 25$

$(w+5)^2 = 169$

$w + 5 = \pm\sqrt{169}$

$w = -5 \pm 13$

$w = 8,\ w = -18$

The negative answer makes no sense in this problem. The width is 8 ft. and the length is 8 + 10 or 18 ft.

b. $\text{Fencing} = 2w + l = 2(8) + 18 = 34$ ft

$\text{cost} = 34(14.95) = 508.30$

It will cost \$508.30 to enclose the patio.

85. Let x = the average speed of the truck driving east in miles per hour. Then $x - 16$ = the average speed of the truck driving north in miles per hour. The trucks travel for 15 minutes or $\frac{1}{4}$ hour.

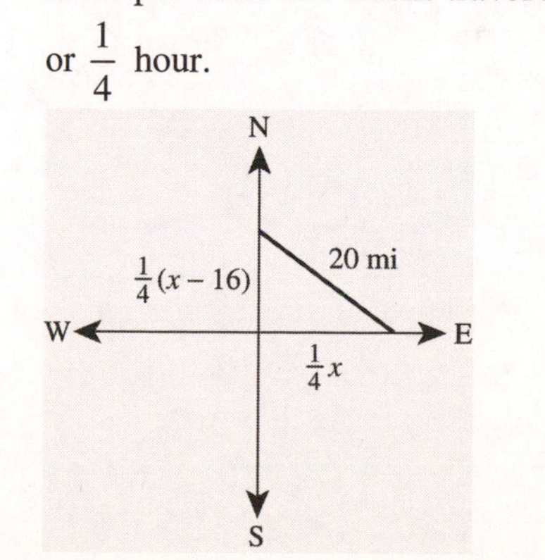

85. (continued)
Using the formula $rt = d$ and the Pythagorean theorem, we have

$$\left(\frac{1}{4}x\right)^2 + \left(\frac{1}{4}(x-16)\right)^2 = 20^2$$
$$\frac{x^2}{16} + \frac{1}{16}\left(x^2 - 32x + 256\right) = 400$$
$$\frac{x^2}{16} + \frac{x^2}{16} - 2x + 16 = 400$$
$$\frac{x^2}{8} - 2x + 16 = 400$$
$$x^2 - 16x + 128 = 3200$$
$$x^2 - 16x - 3072 = 0$$
$$(x+48)(x-64) = 0$$
$$x = -48, \quad x = 64$$

The negative answer makes no sense in this problem. The truck traveling east averages 64 miles per hour, and the truck traveling north averages 48 miles per hour.

9.2 Solving Quadratic Equations: The Quadratic Formula

Practice

1. $y^2 + 3y - 5 = 0$

$$y = \frac{-b \pm \sqrt{b^2 - 4ac}}{2a}$$
$$y = \frac{-3 \pm \sqrt{3^2 - 4(1)(-5)}}{2(1)}$$
$$y = \frac{-3 \pm \sqrt{9+20}}{2}$$
$$y = \frac{-3 \pm \sqrt{29}}{2}$$

2. $6p^2 - 2p = -1$

$$6p^2 - 2p + 1 = 0$$
$$p = \frac{-b \pm \sqrt{b^2 - 4ac}}{2a}$$
$$p = \frac{-(-2) \pm \sqrt{(-2)^2 - 4(6)(1)}}{2(6)}$$
$$p = \frac{2 \pm \sqrt{4-24}}{12}$$
$$p = \frac{2 \pm \sqrt{-20}}{12}$$
$$p = \frac{2 \pm 2i\sqrt{5}}{12}$$
$$p = \frac{1 \pm i\sqrt{5}}{6}$$

3. $\frac{n^2}{5} - \frac{n}{2} = 3$

$$10 \cdot \frac{n^2}{5} - 10 \cdot \frac{n}{2} = 10 \cdot 3$$
$$2n^2 - 5n = 30$$
$$2n^2 - 5n - 30 = 0$$
$$n = \frac{-b \pm \sqrt{b^2 - 4ac}}{2a}$$
$$n = \frac{-(-5) \pm \sqrt{(-5)^2 - 4(2)(-30)}}{2(2)}$$
$$n = \frac{5 \pm \sqrt{25+240}}{4}$$
$$n = \frac{5 \pm \sqrt{265}}{4}$$
$$n \approx -2.820, \ 5.320$$

4. $0.01v^2 = -0.18v - 0.81$

$$0.01v^2 + 0.18v + 0.81 = 0$$
$$100 \cdot 0.01v^2 + 100 \cdot 0.18v + 100 \cdot 0.81 = 100 \cdot 0$$
$$v^2 + 18v + 81 = 0$$
$$v = \frac{-b \pm \sqrt{b^2 - 4ac}}{2a}$$
$$v = \frac{-18 \pm \sqrt{18^2 - 4(1)(81)}}{2(1)}$$
$$v = \frac{-18 \pm \sqrt{324-324}}{2}$$
$$v = \frac{-18 \pm 0}{2}$$
$$v = -9$$

5. **a.** $a = 1, b = 6, c = 9$

$$b^2 - 4ac = 6^2 - 4(1)(9) = 36 - 36 = 0$$

There is 1 solution.
The solution is a real number.

b. $a = 1, b = 6, c = -9$

$$b^2 - 4ac = 6^2 - 4(1)(-9) = 36 + 36 = 72$$

There are 2 solutions.
The solutions are real numbers.

c. $4n^2 + 2n + 3 = 0$

$a = 4, b = 2, c = 3$

$$b^2 - 4ac = 2^2 - 4(4)(3) = 4 - 48 = -44$$

There are 2 solutions.
The solutions are complex numbers (containing i).

6. $1000 = 11.8n^2 + 83.2n + 432.9$

$11.8n^2 + 83.2n - 567.1 = 0$

$$n = \frac{-b \pm \sqrt{b^2 - 4ac}}{2a}$$

$$n = \frac{-83.2 \pm \sqrt{(83.2)^2 - 4(11.8)(-567.1)}}{2(11.8)}$$

$$n = \frac{-83.2 \pm \sqrt{6922.24 + 26,767.12}}{23.6}$$

$$n = \frac{-83.2 \pm \sqrt{33,689.36}}{23.6}$$

$n \approx 4.25, \ -11.30$

The negative solutions does not make sense in this problem. The value of the index was approximately 1000 in 1998.

Exercises

1. The quadratic formula states that if $ax^2 + bx + c = 0$, where a, b, and c are real numbers and $a \neq 0$, then

$$x = \frac{-b \pm \sqrt{b^2 - 4ac}}{2a}.$$

3. If the discriminant of a quadratic equation equals zero, the equation has one real solution(s).

5. $x^2 + 3x + 2 = 0$

$a = 1 \quad b = 3 \quad c = 2$

$$x = \frac{-(3) \pm \sqrt{(3)^2 - 4(1)(2)}}{2(1)} = \frac{-3 \pm \sqrt{9 - 8}}{2}$$

$$x = \frac{-3 \pm \sqrt{1}}{2} = \frac{-3 \pm 1}{2}$$

$x = -2, \ x = -1$

7. $x^2 - 6x - 1 = 0$

$a = 1 \quad b = -6 \quad c = -1$

$$x = \frac{-(-6) \pm \sqrt{(-6)^2 - 4(1)(-1)}}{2(1)}$$

$$x = \frac{6 \pm \sqrt{36 + 4}}{2} = \frac{6 \pm \sqrt{40}}{2} = \frac{6 \pm 2\sqrt{10}}{2}$$

$x = 3 \pm \sqrt{10}$

$x = 3 + \sqrt{10}, \ x = 3 - \sqrt{10}$

9. $x^2 = x + 11$

$x^2 - x - 11 = 0$

$a = 1 \quad b = -1 \quad c = -11$

$$x = \frac{-(-1) \pm \sqrt{(-1)^2 - 4(1)(-11)}}{2(1)}$$

$$x = \frac{1 \pm \sqrt{1 + 44}}{2} = \frac{1 \pm \sqrt{45}}{2} = \frac{1 \pm 3\sqrt{5}}{2}$$

$$x = \frac{1 + 3\sqrt{5}}{2}, \ x = \frac{1 - 3\sqrt{5}}{2}$$

11. $x^2 - 4x + 13 = 8$

$x^2 - 4x + 5 = 0$

$a = 1 \quad b = -4 \quad c = 5$

$$x = \frac{-(-4) \pm \sqrt{(-4)^2 - 4(1)(5)}}{2(1)}$$

$$x = \frac{4 \pm \sqrt{16 - 20}}{2} = \frac{4 \pm \sqrt{-4}}{2} = \frac{4 \pm 2i}{2}$$

$x = 2 + i, \ x = 2 - i$

13. $3x^2+6x=7$

$3x^2+6x-7=0$

$a=3 \qquad b=6 \qquad c=-7$

$$x=\frac{-(6)\pm\sqrt{(6)^2-4(3)(-7)}}{2(3)}$$

$$x=\frac{-6\pm\sqrt{36+84}}{6}=\frac{-6\pm\sqrt{120}}{6}=\frac{-6\pm2\sqrt{30}}{6}$$

$$x=\frac{2\left(-3\pm\sqrt{30}\right)}{6}$$

$$x=\frac{-3+\sqrt{30}}{3},\ x=\frac{-3-\sqrt{30}}{3}$$

15. $6t^2-8t=3t-4$

$6t^2-11t+4=0$

$a=6 \qquad b=-11 \qquad c=4$

$$t=\frac{-(-11)\pm\sqrt{(-11)^2-4(6)(4)}}{2(6)}$$

$$t=\frac{11\pm\sqrt{121-96}}{12}=\frac{11\pm\sqrt{25}}{12}=\frac{11\pm5}{12}$$

$$t=\frac{16}{12}=\frac{4}{3},\ t=\frac{6}{12}=\frac{1}{2}$$

17. $2x^2+8x+9=0$

$a=2 \qquad b=8 \qquad c=9$

$$x=\frac{-(8)\pm\sqrt{(8)^2-4(2)(9)}}{2(2)}$$

$$x=\frac{-8\pm\sqrt{64-72}}{4}=\frac{-8\pm\sqrt{-8}}{4}$$

$$x=\frac{-8\pm2i\sqrt{2}}{4}=\frac{2\left(-4\pm i\sqrt{2}\right)}{4}$$

$$x=\frac{-4+i\sqrt{2}}{2},\ x=\frac{-4-i\sqrt{2}}{2}$$

19. $1-5x^2=4x^2+6x$

$-9x^2-6x+1=0$

$a=-9 \qquad b=-6 \qquad c=1$

$$x=\frac{-(-6)\pm\sqrt{(-6)^2-4(-9)(1)}}{2(-9)}$$

$$x=\frac{6\pm\sqrt{36+36}}{-18}=\frac{6\pm\sqrt{72}}{-18}=\frac{6\pm6\sqrt{2}}{-18}$$

$$x=\frac{6\left(1\pm\sqrt{2}\right)}{-18}=\frac{-\left(1\pm\sqrt{2}\right)}{3}$$

$$x=\frac{-1-\sqrt{2}}{3},\ x=\frac{-1+\sqrt{2}}{3}$$

21. $2y^2-9y+10=1+3y-2y^2$

$4y^2-12y+9=0$

$a=4 \qquad b=-12 \qquad c=9$

$$y=\frac{-(-12)\pm\sqrt{(-12)^2-4(4)(9)}}{2(4)}$$

$$y=\frac{12\pm\sqrt{144-144}}{8}=\frac{12\pm\sqrt{0}}{8}$$

$$y=\frac{12}{8}=\frac{3}{2}$$

23. $\dfrac{x^2}{4}-\dfrac{x}{2}=-3$

$$\frac{x^2}{4}-\frac{x}{2}+3=0$$

$$4\left(\frac{x^2}{4}-\frac{x}{2}+3\right)=4(0)$$

$x^2-2x+12=0$

$a=1 \qquad b=-2 \qquad c=12$

$$x=\frac{-(-2)\pm\sqrt{(-2)^2-4(1)(12)}}{2(1)}$$

$$x=\frac{2\pm\sqrt{4-48}}{2}=\frac{2\pm\sqrt{-44}}{2}$$

$$x=\frac{2\pm2i\sqrt{11}}{2}$$

$x=1+i\sqrt{11} \quad x=1-i\sqrt{11}$

25. $\frac{1}{2}x^2+\frac{2}{3}x-\frac{5}{6}=0$

$6\left(\frac{1}{2}x^2+\frac{2}{3}x-\frac{5}{6}\right)=6(0)$

$3x^2+4x-5=0$

$a=3 \quad b=4 \quad c=-5$

$x=\frac{-(4)\pm\sqrt{(4)^2-4(3)(-5)}}{2(3)}$

$x=\frac{-4\pm\sqrt{16+60}}{6}=\frac{-4\pm\sqrt{76}}{6}$

$x=\frac{-4\pm2\sqrt{19}}{6}=\frac{2\left(-2\pm\sqrt{19}\right)}{6}$

$x=\frac{-2+\sqrt{19}}{3},\ x=\frac{-2-\sqrt{19}}{3}$

27. $0.2x^2+x+0.8=0$

$5\left(0.2x^2+x+0.8\right)=5(0)$

$x^2+5x+4=0$

$a=1 \quad b=5 \quad c=4$

$x=\frac{-(5)\pm\sqrt{(5)^2-4(1)(4)}}{2(1)}$

$x=\frac{-5\pm\sqrt{25-16}}{2}=\frac{-5\pm\sqrt{9}}{2}=\frac{-5\pm3}{2}$

$x=\frac{-8}{2}=-4,\ x=\frac{-2}{2}=-1$

29. $0.03x^2-0.12x+0.24=0$

$100\left(0.03x^2-0.12x+0.24\right)=100(0)$

$3x^2-12x+24=0$

$3\left(x^2-4x+8\right)=0$

$x^2-4x+8=0$

$a=1 \quad b=-4 \quad c=8$

$x=\frac{-(-4)\pm\sqrt{(-4)^2-4(1)(8)}}{2(1)}$

$x=\frac{4\pm\sqrt{16-32}}{2}=\frac{4\pm\sqrt{-16}}{2}=\frac{4\pm4i}{2}$

$x=2+2i,\ x=2-2i$

31. $(x+6)(x+2)=8$

$x^2+8x+12=8$

$x^2+8x+4=0$

$a=1 \quad b=8 \quad c=4$

$x=\frac{-(8)\pm\sqrt{(8)^2-4(1)(4)}}{2(1)}$

$x=\frac{-8\pm\sqrt{64-16}}{2}=\frac{-8\pm\sqrt{48}}{2}=\frac{-8\pm4\sqrt{3}}{2}$

$x=\frac{2\left(-4\pm2\sqrt{3}\right)}{2}=-4\pm2\sqrt{3}$

$x=-4+2\sqrt{3},\ x=-4-2\sqrt{3}$

33. $(2x-3)^2=8(x+1)$

$4x^2-12x+9=8x+8$

$4x^2-20x+1=0$

$a=4 \quad b=-20 \quad c=1$

$x=\frac{-(-20)\pm\sqrt{(-20)^2-4(4)(1)}}{2(4)}$

$x=\frac{20\pm\sqrt{400-16}}{8}=\frac{20\pm\sqrt{384}}{8}$

$x=\frac{20\pm8\sqrt{6}}{8}=\frac{4\left(5\pm2\sqrt{6}\right)}{8}$

$x=\frac{5+2\sqrt{6}}{2},\ x=\frac{5-2\sqrt{6}}{2}$

35. $1.4x^2-2.7x-0.1=0$

$a=1.4 \quad b=-2.7 \quad c=-0.1$

$x=\frac{-(-2.7)\pm\sqrt{(-2.7)^2-4(1.4)(-0.1)}}{2(1.4)}$

$x=\frac{2.7\pm\sqrt{7.29+0.56}}{2.8}=\frac{2.7\pm\sqrt{7.85}}{2.8}$

$x=\frac{2.7\pm2.8017...}{2.8}$

$x=1.9649...\approx1.965$

$x=-0.03635...\approx-0.036$

37. $0.003x^2+0.23x+1.124=0$

$a=0.003 \qquad b=0.23 \qquad c=1.124$

$$x=\frac{-(0.23)\pm\sqrt{(0.23)^2-4(0.003)(1.124)}}{2(0.003)}$$

$$x=\frac{-0.23\pm\sqrt{0.0529-0.013488}}{0.006}$$

$$x=\frac{-0.23\pm\sqrt{0.039412}}{0.006}$$

$$x=\frac{-0.23\pm 0.198524....}{0.006}$$

$x=-5.2459...\approx -5.246$

$x=-71.4207...\approx -71.421$

39. $x^2+2x+4=0$

$a=1 \qquad b=2 \qquad c=4$

$(2)^2-4(1)(4)=4-16=-12$

Two complex solutions (containing i).

41. $4x^2-12x=-9$

$4x^2-12x+9=0$

$a=4 \qquad b=-12 \qquad c=9$

$(-12)^2-4(4)(9)=144-144=0$

One real solution.

43. $6x^2=2-5x$

$6x^2+5x-2=0$

$a=6 \qquad b=5 \qquad c=-2$

$(5)^2-4(6)(-2)=25+48=73$

Two real solutions.

45. $7x^2-x+3=0$

$a=7 \qquad b=-1 \qquad c=3$

$(-1)^2-4(7)(3)=1-84=-83$

Two complex solutions (containing i).

47. $3x^2+10=0$

$3x^2+0x+10=0$

$a=3 \qquad b=0 \qquad c=10$

$(0)^2-4(3)(10)=-120$

Two complex solutions (containing i).

49. $2x^2+8=7x$

$2x^2-7x+8=0$

$a=2 \quad b=-7 \quad c=8$

$(-7)^2-4(2)(8)=-15$

Two complex solutions (containing i).

51. $2x^2-4x+2=-3$

$2x^2-4x+5=0$

$a=2 \qquad b=-4 \qquad c=5$

$$x=\frac{-(-4)\pm\sqrt{(-4)^2-4(2)(5)}}{2(2)}$$

$$x=\frac{4\pm\sqrt{16-40}}{4}=\frac{4\pm\sqrt{-24}}{4}=\frac{4\pm 2i\sqrt{6}}{4}$$

$$x=\frac{2+i\sqrt{6}}{2}, \quad x=\frac{2-i\sqrt{6}}{2}$$

53. $\frac{x^2}{4}-\frac{2x}{3}-\frac{1}{6}=0$

$$12\left(\frac{x^2}{4}-\frac{2x}{3}-\frac{1}{6}\right)=12(0)$$

$3x^2-8x-2=0$

$a=3 \qquad b=-8 \qquad c=-2$

$$x=\frac{-(-8)\pm\sqrt{(-8)^2-4(3)(-2)}}{2(3)}$$

$$x=\frac{8\pm\sqrt{64+24}}{6}=\frac{8\pm\sqrt{88}}{6}=\frac{8\pm 2\sqrt{22}}{6}$$

$$x=\frac{4+\sqrt{22}}{3}, \quad x=\frac{4-\sqrt{22}}{3}$$

55. $300=-16t^2-20t+800$

$16t^2+20t-500=0$

$4(4t^2+5t-125)=0$

$4t^2+5t-125=0$

$a=4 \qquad b=5 \qquad c=-125$

$$t=\frac{-(5)\pm\sqrt{(5)^2-4(4)(-125)}}{2(4)}$$

$$t=\frac{-5\pm\sqrt{25+2000}}{8}=\frac{-5\pm\sqrt{2025}}{8}$$

$$t=\frac{-5\pm 45}{8}$$

$t=5, \ t=-6.25$

The negative answer makes no sense in this problem. The stone is 300 ft above the ground 5 sec after it is thrown downward.

57. length $= 36 - 2x$

width $= 20 - 2x$

Area $= (36 - 2x)(20 - 2x)$

$465 = 720 - 112x + 4x^2$

$4x^2 - 112x + 255 = 0$

$a = 4 \qquad b = -112 \qquad c = 255$

$$x = \frac{-(-112) \pm \sqrt{(-112)^2 - 4(4)(255)}}{2(4)} =$$

$$x = \frac{112 \pm \sqrt{12544 - 4080}}{8} = \frac{112 \pm \sqrt{8464}}{8}$$

$$x = \frac{112 \pm 92}{8}$$

$x = 25.5, \ x = 2.5$

25.5 in. cannot be cut from 20 in., so that answer is discarded. Squares measuring 2.5 in. by 2.5 in. should be cut from each corner.

59. $x(300 - 5x) = 3520$

$300x - 5x^2 = 3520$

$-5x^2 + 300x - 3520 = 0$

$-5(x^2 - 60x + 704) = 0$

$x^2 - 60x + 704 = 0$

$a = 1 \qquad b = -60 \qquad c = 704$

$$x = \frac{-(-60) \pm \sqrt{(-60)^2 - 4(1)(704)}}{2(1)}$$

$$x = \frac{60 \pm \sqrt{3600 - 2816}}{2}$$

$$x = \frac{60 \pm \sqrt{784}}{2} = \frac{60 \pm 28}{2}$$

$x = 16$

$300 - 5x = 300 - 5(16) = 220$

$x = 44$

$300 - 5x = 300 - 5(44) = 80$

44 tickets will have a ticket price of only \$80.

16 tickets have a ticket price of \$220. The company must sell 16 tickets to generate revenue of \$3520, and maintain a ticket price greater than \$200.

61. $215 = -\frac{1}{2}t^2 + 64t + 29$

$-\frac{1}{2}t^2 + 64t - 186 = 0$

$a = -\frac{1}{2} \qquad b = 64 \qquad c = -186$

$$t = \frac{-(64) \pm \sqrt{(64)^2 - 4\left(-\frac{1}{2}\right)(-186)}}{2\left(-\frac{1}{2}\right)}$$

$$t = \frac{-64 \pm \sqrt{4096 - 372}}{-1} = -\left(-64 \pm \sqrt{3724}\right)$$

$t = 64 \pm 61.0245...$

$t = 2.97541..., \ t = 125.0245...$

The solution $x = 125$ is discarded. There were approximately 215,000 reports of identity theft in the year 2003.

9.3 More on Quadratic Equations

Practice

1. $1 = \frac{2}{x+2} + x$

$(x+2)(1) = (x+2)\frac{2}{x+2} + (x+2)x$

$x + 2 = 2 + x^2 + 2x$

$x^2 + x = 0$

$x(x+1) = 0$

$x + 1 = 0$ or $x = 0$

$x = -1$

Check $x = -1$	Check $x = 0$
$1 \stackrel{?}{=} \frac{2}{-1+2} + (-1)$	$1 \stackrel{?}{=} \frac{2}{0+2} + 0$
$1 \stackrel{?}{=} \frac{2}{1} - 1$	$1 \stackrel{?}{=} \frac{2}{2} + 0$
$1 \stackrel{?}{=} 2 - 1$	$1 \stackrel{?}{=} 1 + 0$
$1 = 1$	$1 = 1$

2. $n^4 - 9n^2 + 8 = 0$

Let $u = n^2$.

$$(n^2)^2 - 9(n^2) + 8 = 0$$
$$u^2 - 9u + 8 = 0$$
$$(u-1)(u-8) = 0$$
$$u - 1 = 0 \text{ or } u - 8 = 0$$
$$u = 1 \qquad u = 8$$
$$n^2 = 1 \qquad n^2 = 8$$
$$n = \pm\sqrt{1} \qquad n = \pm\sqrt{8}$$
$$n = \pm 1 \qquad n = \pm 2\sqrt{2}$$

Check $n = 1$

$$(1)^4 - 9(1)^2 + 8 \stackrel{?}{=} 0$$
$$1 - 9(1) + 8 \stackrel{?}{=} 0$$
$$1 - 9 + 8 \stackrel{?}{=} 0$$
$$0 = 0$$

Check $n = -1$

$$(-1)^4 - 9(-1)^2 + 8 \stackrel{?}{=} 0$$
$$1 - 9(1) + 8 \stackrel{?}{=} 0$$
$$1 - 9 + 8 \stackrel{?}{=} 0$$
$$0 = 0$$

Check $n = 2\sqrt{2}$

$$(2\sqrt{2})^4 - 9(2\sqrt{2})^2 + 8 \stackrel{?}{=} 0$$
$$64 - 9(8) + 8 \stackrel{?}{=} 0$$
$$64 - 72 + 8 \stackrel{?}{=} 0$$
$$0 = 0$$

Check $n = -2\sqrt{2}$

$$(-2\sqrt{2})^4 - 9(-2\sqrt{2})^2 + 8 \stackrel{?}{=} 0$$
$$64 - 9(8) + 8 \stackrel{?}{=} 0$$
$$64 - 72 + 8 \stackrel{?}{=} 0$$
$$0 = 0$$

3. $x - 3\sqrt{x} = 10$

Let $u = \sqrt{x}$.

$$u^2 - 3u = 10$$
$$u^2 - 3u - 10 = 0$$
$$(u-5)(u+2) = 0$$
$$u - 5 = 0 \text{ or } u + 2 = 0$$
$$u = 5 \qquad u = -2$$
$$\sqrt{x} = 5 \qquad \sqrt{x} = -2$$
$$(\sqrt{x})^2 = 5^2 \qquad (\sqrt{x})^2 = (-2)^2$$
$$x = 25 \qquad x = 4$$

Check $x = 25$

$$25 - 3\sqrt{25} \stackrel{?}{=} 10$$
$$25 - 3(5) \stackrel{?}{=} 10$$
$$25 - 15 \stackrel{?}{=} 10$$
$$10 = 10$$

Check $x = 4$

$$4 - 3\sqrt{4} \stackrel{?}{=} 10$$
$$4 - 3(2) \stackrel{?}{=} 10$$
$$4 - 6 \stackrel{?}{=} 10$$
$$-2 \neq 10$$

The solution is $x = 25$.

4. $y^{2/3} - y^{1/3} - 20 = 0$

Let $u = y^{1/3}$.

$$u^2 - u - 20 = 0$$
$$(u-5)(u+4) = 0$$
$$u - 5 = 0 \quad \text{or} \quad u + 4 = 0$$
$$u = 5 \qquad u = -4$$
$$y^{1/3} = 5 \qquad y^{1/3} = -4$$
$$(y^{1/3})^3 = (5)^3 \qquad (y^{1/3})^3 = (-4)^3$$
$$y = 125 \qquad y = -64$$

Check $y = 125$

$$125^{2/3} - 125^{1/3} - 20 \stackrel{?}{=} 0$$
$$25 - 5 - 20 \stackrel{?}{=} 0$$
$$0 = 0$$

Check $y = -64$

$$(-64)^{2/3} - (-64)^{1/3} - 20 \stackrel{?}{=} 0$$
$$16 - (-4) - 20 \stackrel{?}{=} 0$$
$$20 - 20 \stackrel{?}{=} 0$$
$$0 = 0$$

5. Let P equal the time the first painter takes to paint the apartment alone.

$$\text{time} = \frac{\text{work}}{\text{rate}}$$

Work done by the first painter in one hour + work done by the second painter in one hour = the work done by both painters in one hour.

	Rate of work	Time Worked	Part of Task
First painter	$\frac{1}{P}$	1	$\frac{1}{P}$
Second painter	$\frac{1}{P-1}$	1	$\frac{1}{P-1}$

$$\frac{1}{P}+\frac{1}{P-1}=\frac{1}{5}$$

$$\frac{1}{P}\bullet 5P(P-1)+\frac{1}{P-1}\bullet 5P(P-1)$$

$$=\frac{1}{5}\bullet 5P(P-1)$$

$$5(P-1)+5P=P^2-P$$

$$5P-5+5P=P^2-P$$

$$P^2-11P+5=0$$

$a=1 \quad b=-11 \quad c=5$

$$P=\frac{-(-11)\pm\sqrt{(-11)^2-4(1)(5)}}{2(1)}$$

$$P=\frac{11\pm\sqrt{121-20}}{2}=\frac{11\pm\sqrt{101}}{2}$$

$$P\approx\frac{11\pm 10.04987...}{2}$$

$P\approx 10.5249...,\ P\approx 0.4750...$

$P-1\approx 9.5249...,\ P-1\approx -0.5249...$

The negative answer makes no sense in this problem. If only one painter works, it takes the first painter about 10.5 hours and the second painter 9.5 hours to paint the apartment.

6. $y=0 \qquad y=-3$

$$y+3=0$$

$$y(y+3)=0$$

$$y^2+3y=0$$

7. $n=\frac{3}{4} \qquad n=\frac{1}{6}$

$$n-\frac{3}{4}=0 \qquad n-\frac{1}{6}=0$$

$$4n-3=0 \qquad 6n-1=0$$

$$(4n-3)(6n-1)=0$$

$$24n^2-22n+3=0$$

Exercises

1. The equation $\frac{4}{x-4}+\frac{3}{x-1}=1$ leads to a quadratic equation.

3. To solve an equation that is quadratic in form, rewrite it in terms of a new variable.

5. $2-\frac{3}{x}+\frac{1}{x^2}=0$

$$x^2\left(2-\frac{3}{x}+\frac{1}{x^2}\right)=x^2(0)$$

$$2x^2-3x+1=0$$

$$(2x-1)(x-1)=0$$

$$2x-1=0 \qquad x-1=0$$

$$2x=1 \qquad x=1$$

$$x=\frac{1}{2}$$

Check $x=\frac{1}{2}$

$$2-\frac{3}{\frac{1}{2}}+\frac{1}{\left(\frac{1}{2}\right)^2}\stackrel{?}{=}0$$

$$2-3\left(\frac{2}{1}\right)+1\left(\frac{4}{1}\right)\stackrel{?}{=}0$$

$$2-6+4\stackrel{?}{=}0$$

$0=0$ True

Check $x=1$

$$2-\frac{3}{1}+\frac{1}{(1)^2}\stackrel{?}{=}0$$

$$2-3+1\stackrel{?}{=}0$$

$0=0$ True

The solutions of the original equation are $\frac{1}{2}$ and 1.

7. $\frac{3}{p-1}+\frac{4}{p-4}=1$

$$\frac{3(p-1)(p-4)}{p-1}+\frac{4(p-1)(p-4)}{p-4}$$

$$=1(p-1)(p-4)$$

$$3(p-4)+4(p-1)=(p-1)(p-4)$$

$$3p-12+4p-4=p^2-5p+4$$

$$p^2-12p+20=0$$

$$(p-2)(p-10)=0$$

$$p-2=0 \qquad p-10=0$$

$$p=2 \qquad p=10$$

Check $p=2$

$$\frac{3}{2-1}+\frac{4}{2-4}\stackrel{?}{=}1$$

$$\frac{3}{1}+\frac{4}{-2}\stackrel{?}{=}1$$

$$3-2\stackrel{?}{=}1$$

$1=1$ True

Check $p=10$

$$\frac{3}{10-1}+\frac{4}{10-4}\stackrel{?}{=}1$$

$$\frac{3}{9}+\frac{4}{6}\stackrel{?}{=}1$$

$$\frac{1}{3}+\frac{2}{3}\stackrel{?}{=}1$$

$1=1$ True

The solutions of the original equation are 2 and 10.

9. $n^4 - 8n^2 + 12 = 0$

Let $u = n^2$

$(n^2)^2 - 8(n^2) + 12 = 0$

$u^2 - 8u + 12 = 0$

$(u-6)(u-2) = 0$

$u - 6 = 0 \quad u - 2 = 0$

$u = 6 \quad u = 2$

$n^2 = 6 \quad n^2 = 2$

$n = \pm\sqrt{6} \quad n = \pm\sqrt{2}$

Check $n = \sqrt{6}$

$(\sqrt{6})^4 - 8(\sqrt{6})^2 + 12 \stackrel{?}{=} 0$

$36 - 8(6) + 12 \stackrel{?}{=} 0$

$36 - 48 + 12 \stackrel{?}{=} 0$

$0 = 0$ True

Check $n = -\sqrt{6}$

$(-\sqrt{6})^4 - 8(-\sqrt{6})^2 + 12 \stackrel{?}{=} 0$

$36 - 8(6) + 12 \stackrel{?}{=} 0$

$36 - 48 + 12 \stackrel{?}{=} 0$

$0 = 0$ True

Check $n = \sqrt{2}$

$(\sqrt{2})^4 - 8(\sqrt{2})^2 + 12 \stackrel{?}{=} 0$

$4 - 8(2) + 12 \stackrel{?}{=} 0$

$4 - 16 + 12 \stackrel{?}{=} 0$

$0 = 0$ True

Check $n = -\sqrt{2}$

$(-\sqrt{2})^4 - 8(-\sqrt{2})^2 + 12 \stackrel{?}{=} 0$

$4 - 8(2) + 12 \stackrel{?}{=} 0$

$4 - 16 + 12 \stackrel{?}{=} 0$

$0 = 0$ True

The solutions of the original equation are $\pm\sqrt{6}$ and $\pm\sqrt{2}$.

11. $3x^4 + 11x^2 - 3 = -x^4$

$4x^4 + 11x^2 - 3 = 0$

Let $u = x^2$

$4(x^2)^2 + 11(x^2) - 3 = 0$

$4u^2 + 11u - 3 = 0$

$(4u-1)(u+3) = 0$

$4u - 1 = 0 \quad u + 3 = 0$

$4u = 1 \quad u = -3$

$u = \frac{1}{4}$

$x^2 = \frac{1}{4} \quad x^2 = -3$

$x = \pm\sqrt{\frac{1}{4}} \quad x = \pm\sqrt{-3}$

$x = \pm\frac{1}{2} \quad x = \pm i\sqrt{3}$

Check $x = \frac{1}{2}$

$3\left(\frac{1}{2}\right)^4 + 11\left(\frac{1}{2}\right)^2 - 3 \stackrel{?}{=} -\left(\frac{1}{2}\right)^4$

$\frac{3}{16} + \frac{11}{4} - 3 \stackrel{?}{=} -\frac{1}{16}$

$\frac{3}{16} + \frac{44}{16} - \frac{48}{16} \stackrel{?}{=} -\frac{1}{16}$

$-\frac{1}{16} = -\frac{1}{16}$ True

Check $x = -\frac{1}{2}$

$3\left(-\frac{1}{2}\right)^4 + 11\left(-\frac{1}{2}\right)^2 - 3 \stackrel{?}{=} -\left(-\frac{1}{2}\right)^4$

$\frac{3}{16} + \frac{11}{4} - 3 \stackrel{?}{=} -\frac{1}{16}$

$\frac{3}{16} + \frac{44}{16} - \frac{48}{16} \stackrel{?}{=} -\frac{1}{16}$

$-\frac{1}{16} = -\frac{1}{16}$ True

Check $x = i\sqrt{3}$

$3(i\sqrt{3})^4 + 11(i\sqrt{3})^2 - 3 \stackrel{?}{=} -(i\sqrt{3})^4$

$3(9i^4) + 11(3i^2) - 3 \stackrel{?}{=} -(9i^4)$

$27 - 33 - 3 \stackrel{?}{=} -9$

$-9 = -9$ True

Check $x = -i\sqrt{3}$

$3(-i\sqrt{3})^4 + 11(-i\sqrt{3})^2 - 3 \stackrel{?}{=} -(-i\sqrt{3})^4$

$3(9i^4) + 11(3i^2) - 3 \stackrel{?}{=} -(9i^4)$

$27 - 33 - 3 \stackrel{?}{=} -9$

$-9 = -9$ True

The solutions of the original equation are $\pm\frac{1}{2}$ and $\pm i\sqrt{3}$.

13. $h-4(\sqrt{h})+4=0$

Let $u=\sqrt{h}$

$(\sqrt{h})^2-4(\sqrt{h})+4=0$

$u^2-4u+4=0$

$(u-2)^2=0 \Rightarrow u-2=0$

$u=2$

$2=\sqrt{h}$

$2^2=(\sqrt{h})^2$

$h=4$

Check $h=4$

$4-4\sqrt{4}+4\overset{?}{=}0$

$4-8+4\overset{?}{=}0$

$0=0$ True

The solution of the original equation is 4.

15. $x^{1/2}+4x^{1/4}-32=0$

Let $u=x^{1/4}$.

$(x^{1/4})^2+4(x^{1/4})-32=0$

$u^2+4u-32=0$

$(u-4)(u+8)=0$

$u-4=0 \quad u+8=0$

$u=4 \quad u=-8$

$x^{1/4}=4 \quad x^{1/4}=-8$

$x=(4)^4 \quad x=(-8)^4$

$x=256 \quad x=4096$

Check $x=256$

$256^{1/2}+4(256)^{1/4}-32\overset{?}{=}0$

$16+4(4)-32\overset{?}{=}0$

$16+16-32\overset{?}{=}0$

$0=0$ True

Check $x=4096$

$4096^{1/2}+4(4096)^{1/4}-32\overset{?}{=}0$

$64+4(8)-32\overset{?}{=}0$

$64+32-32\overset{?}{=}0$

$64=0$ False

$x=4096$ is not a solution.

The solution of the original equation is 256.

17. $t^{2/3}+3t^{1/3}+2=0$

Let $u=t^{1/3}$

$(t^{1/3})^2+3(t^{1/3})+2=0$

$u^2+3u+2=0$

$(u+2)(u+1)=0$

$u+2=0 \quad u+1=0$

$u=-2 \quad u=-1$

$t^{1/3}=-2 \quad t^{1/3}=-1$

$t=(-2)^3 \quad t=(-1)^3$

$t=-8 \quad t=-1$

Check $t=-8$

$(-8)^{2/3}+3(-8)^{1/3}+2\overset{?}{=}0$

$(-2)^2+3(-2)+2\overset{?}{=}0$

$4-6+2\overset{?}{=}0$

$0=0$ True

Check $t=-1$

$(-1)^{2/3}+3(-1)^{1/3}+2\overset{?}{=}0$

$(-1)^2+3(-1)+2\overset{?}{=}0$

$1-3+2\overset{?}{=}0$

$0=0$ True

The solutions of the original equation are –8 and –1.

19. $(n-3)^2-5(n-3)+6=0$

Let $u=n-3$

$u^2-5u+6=0$

$(u-3)(u-2)=0$

$u-3=0 \quad u-2=0$

$u=3 \quad u=2$

$n-3=3 \quad n-3=2$

$n=6 \quad n=5$

Check $n=6$

$(6-3)^2-5(6-3)+6\overset{?}{=}0$

$3^2-5(3)+6\overset{?}{=}0$

$9-15+6\overset{?}{=}0$

$0=0$ True

19. (continued)

Check $n=5$

$(5-3)^2-5(5-3)+6\stackrel{?}{=}0$

$2^2-5(2)+6\stackrel{?}{=}0$

$4-10+6\stackrel{?}{=}0$

$0=0$ True

The solutions of the original equation are 5 and 6.

21. $(2x+4)+2\sqrt{2x+4}=24$

Let $u=\sqrt{2x+4}$

$\left(\sqrt{2x+4}\right)^2+2\left(\sqrt{2x+4}\right)=24$

$u^2+2u=24$

$u^2+2u-24=0$

$(u+6)(u-4)=0$

$u+6=0 \quad u-4=0$

$u=-6 \quad u=4$

$\sqrt{2x+4}=-6 \qquad \sqrt{2x+4}=4$

$\left(\sqrt{2x+4}\right)^2=(-6)^2 \qquad \left(\sqrt{2x+4}\right)^2=4^2$

$2x+4=36 \qquad 2x+4=16$

$2x=32 \qquad 2x=12$

$x=16 \qquad x=6$

Check $x=16$

$(2\cdot16+4)+2\sqrt{2\cdot16+4}\stackrel{?}{=}24$

$(36)+2\sqrt{36}\stackrel{?}{=}24$

$36+12\stackrel{?}{=}24$

$48=24$ False

$x=16$ is not a solution

Check $x=6$

$(2\cdot6+4)+2\sqrt{2\cdot6+4}\stackrel{?}{=}24$

$(16)+2\sqrt{16}\stackrel{?}{=}24$

$16+8\stackrel{?}{=}24$

$24=24$ True

The solution of the original equation is 6.

23. $x=7 \qquad x=2$

$x-7=0 \quad x-2=0$

$(x-7)(x-2)=0$

$x^2-9x+14=0$

25. $y=-4 \qquad y=9$

$y+4=0 \quad y-9=0$

$(y+4)(y-9)=0$

$y^2-5y-36=0$

27. $t=\dfrac{2}{3} \qquad t=-2$

$3t=2 \qquad t+2=0$

$3t-2=0$

$(3t-2)(t+2)=0$

$3t^2+4t-4=0$

29. $n=-\dfrac{3}{2} \quad n=-\dfrac{4}{3}$

$2n=-3 \qquad 3n=-4$

$2n+3=0 \quad 3n+4=0$

$(2n+3)(3n+4)=0$

$6n^2+17n+12=0$

31. $x=4$

$x-4=0$

$(x-4)^2=0$

$x^2-8x+16=0$

33. $t=\sqrt{2} \qquad t=-\sqrt{2}$

$t-\sqrt{2}=0 \quad t+\sqrt{2}=0$

$\left(t-\sqrt{2}\right)\left(t+\sqrt{2}\right)=0$

$t^2-2=0$

35. $x=3i \qquad x=-3i$

$x-3i=0 \quad x+3i=0$

$(x-3i)(x+3i)=0$

$x^2-9i^2=0$

$x^2+9=0$

37. $y^{2/3} - y^{1/3} = 2$

Let $u = y^{1/3}$

$$u^2 - u = 2$$
$$u^2 - u - 2 = 0$$
$$(u-2)(u+1) = 0$$
$$u - 2 = 0 \qquad u + 1 = 0$$
$$u = 2 \qquad u = -1$$

$$u = 2 \qquad u = -1$$
$$y^{1/3} = 2 \qquad y^{1/3} = -1$$
$$y = 2^3 = 8 \qquad y = (-1)^3 = -1$$

Check $y = 8$

$$8^{2/3} - 8^{1/3} \stackrel{?}{=} 2$$
$$4 - 2 \stackrel{?}{=} 2$$
$$2 = 2 \text{ True}$$

Check $y = -1$

$$(-1)^{2/3} - (-1)^{1/3} \stackrel{?}{=} 2$$
$$1 - (-1) \stackrel{?}{=} 2$$
$$2 = 2 \text{ True}$$

The solutions of the original equation are –1 and 8.

39. $a^4 + 7a^2 - 12 = 6$

$$a^4 + 7a^2 - 18 = 0$$

Let $u = a^2$

$$\left(a^2\right)^2 + 7a^2 - 18 = 0$$
$$u^2 + 7u - 18 = 0$$
$$(u+9)(u-2) = 0$$
$$u + 9 = 0 \qquad u - 2 = 0$$
$$u = -9 \qquad u = 2$$

$$u = -9 \qquad u = 2$$
$$a^2 = -9 \qquad a^2 = 2$$
$$a = \sqrt{-9} = \pm 3i \qquad a = \pm\sqrt{2}$$

Check $a = 3i$

$$(3i)^4 + 7(3i)^2 - 12 \stackrel{?}{=} 6$$
$$3^4 i^4 + 7\left(3^2 i^2\right) - 12 \stackrel{?}{=} 6$$
$$81(1) + 7(9)(-1) - 12 \stackrel{?}{=} 6$$
$$81 - 63 - 12 \stackrel{?}{=} 6$$
$$6 = 6 \text{ True}$$

Check $a = -3i$

$$(-3i)^4 + 7(-3i)^2 - 12 \stackrel{?}{=} 6$$
$$(-3)^4 i^4 + 7\left((-3)^2 i^2\right) - 12 \stackrel{?}{=} 6$$
$$81(1) + 7(9)(-1) - 12 \stackrel{?}{=} 6$$
$$81 - 63 - 12 \stackrel{?}{=} 6$$
$$6 = 6 \text{ True}$$

Check $a = \sqrt{2}$

$$\left(\sqrt{2}\right)^4 + 7\left(\sqrt{2}\right)^2 - 12 \stackrel{?}{=} 6$$
$$4 + 7(2) - 12 \stackrel{?}{=} 6$$
$$4 + 14 - 12 \stackrel{?}{=} 6$$
$$6 = 6 \text{ True}$$

Check $a = -\sqrt{2}$

$$\left(-\sqrt{2}\right)^4 + 7\left(-\sqrt{2}\right)^2 - 12 \stackrel{?}{=} 6$$
$$4 + 7(2) - 12 \stackrel{?}{=} 6$$
$$4 + 14 - 12 \stackrel{?}{=} 6$$
$$6 = 6 \text{ True}$$

The solutions of the original equation are $\pm 3i$ and $\pm\sqrt{2}$.

41. $a = -\dfrac{2}{5} \qquad a = 3$

$$a + \frac{2}{5} = 0 \qquad a - 3 = 0$$
$$5a + 2 = 0$$
$$(5a+2)(a-3) = 0$$
$$5a^2 - 13a - 6 = 0$$

43. $BMI = \dfrac{W}{H^2}$

$$20 = \frac{52}{H^2}$$
$$20H^2 = 52$$
$$H^2 = \frac{52}{20} = 2.6$$
$$H = \sqrt{2.6} = 1.61245...$$

The person's height is approximately 1.6 m.

45. $\text{time} = \dfrac{\text{distance}}{\text{rate}}$

Time going to work + time going home = 1

$\dfrac{15}{r} + \dfrac{15}{r-8} = 1$

$\dfrac{15}{r} \bullet (r)(r-8) + \dfrac{15}{r-8} \bullet (r)(r-8) = 1 \bullet (r)(r-8)$

$15(r-8) + 15(r) = r^2 - 8r$

$15r - 120 + 15r = r^2 - 8r$

$r^2 - 38r + 120 = 0$

$a = 1 \qquad b = -38 \qquad c = 120$

$r = \dfrac{-(-38) \pm \sqrt{(-38)^2 - 4(1)(120)}}{2(1)}$

$r = \dfrac{38 \pm \sqrt{1444 - 480}}{2} = \dfrac{38 \pm \sqrt{964}}{2}$

$r = \dfrac{38 \pm 2\sqrt{241}}{2} = 19 \pm \sqrt{241} = 19 \pm 15.5241...$

$r = 34.5241...,\ r = 3.4758...$

$r - 8 = 26.5241...,\ \ r - 8 = -4.5241...$

The negative answer makes no sense in this problem. She drives to work at an average speed of about 35 mph and she drives home at an average speed of about 27 mph.

47. Let L equal the time the large pipe takes to empty the pool alone.

$\text{time} = \dfrac{\text{work}}{\text{rate}}$

Work done by the large pipe in one hour + work done by the small pipe in one hour = the work done by both pipes in one hour.

	Rate of work	Time Worked	Part of Task
Large pipe	$\dfrac{1}{L}$	1	$\dfrac{1}{L}$
small pipe	$\dfrac{1}{L+1.5}$	1	$\dfrac{1}{L+1.5}$

$\dfrac{1}{L} + \dfrac{1}{L+1.5} = \dfrac{1}{2}$

$\dfrac{1}{L} \bullet 2L(L+1.5) + \dfrac{1}{L+1.5} \bullet 2L(L+1.5)$

$= \dfrac{1}{2} \bullet 2L(L+1.5)$

$2(L+1.5) + 2L = L^2 + 1.5L$

$2L + 3 + 2L = L^2 + 1.5L$

$L^2 - 2.5L - 3 = 0$

$a = 1 \qquad b = -2.5 \qquad c = -3$

$L = \dfrac{-(-2.5) \pm \sqrt{(-2.5)^2 - 4(1)(-3)}}{2(1)}$

$L = \dfrac{2.5 \pm \sqrt{6.25 + 12}}{2} = \dfrac{2.5 \pm \sqrt{18.25}}{2}$

$L = \dfrac{2.5 \pm 4.27200...}{2}$

$L = 3.38600...,\ L = -0.88600...$

$L + 1.5 = 3.38600... + 1.5 = 4.88600...$

The negative answer makes no sense in this problem. If only one drain is open, it takes the large drain 3.4 hour and the small drain 4.9 hr to empty the pool.

9.4 Graphing Quadratic Functions

Practice

1.

x	$y = -3x^2$	(x, y)
-2	$-3(-2)^2 = -12$	$(-2, -12)$
-1	$-3(-1)^2 = -3$	$(-1, -3)$
0	$-3(0)^2 = 0$	$(0, 0)$
1	$-3(1)^2 = -3$	$(1, -3)$
2	$-3(2)^2 = -12$	$(2, -12)$

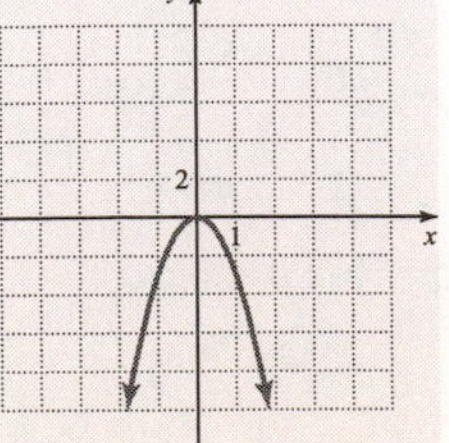

2.

x	$y=-4x^2+1$	(x,y)
-2	$-4(-2)^2+1=-15$	$(-2,-15)$
-1	$-4(-1)^2+1=-3$	$(-1,-3)$
0	$-4(0)^2+1=1$	$(0,1)$
1	$-4(1)^2+1=-3$	$(1,-3)$
2	$-4(2)^2+1=-15$	$(2,-15)$

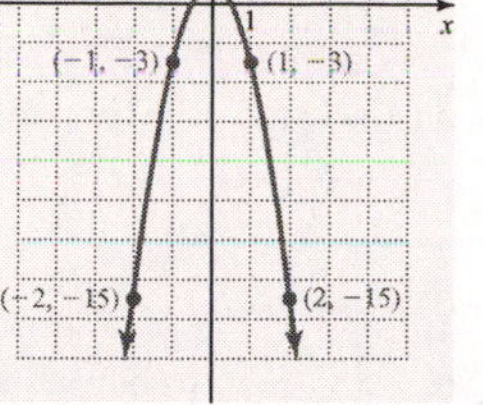

3. **a.** $y=-x^2+4x+5$

$a=-1 \qquad b=4$

$-\dfrac{b}{2a}=-\dfrac{4}{2(-1)}=-\dfrac{4}{-2}=2$

$y=-(2)^2+4(2)+5$

$=-4+8+5=9$

Vertex: $(2,9)$

Axis of symmetry: $x=2$

b. The graph opens downward, since the leading coefficient is negative.

c. $0=-x^2+4x+5$

$0=x^2-4x-5$

$0=(x-5)(x+1)$

$x-5=0 \quad \text{or} \quad x+1=0$

$x=5 \qquad x=-1$

The x-intercepts are $(5,0)$ and $(-1,0)$.

$y=-(0)^2+4(0)+5$

$=0+0+5$

$=5$

The y-intercept is $(0,5)$.

d.

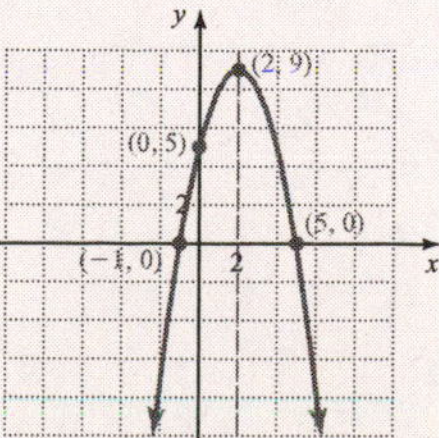

4. **a.** The leading coefficient is positive, so the graph opens upward and has a minimum point.

$b^2-4ac=(-9)^2-4(4)(2)=81-32=49$

Since the discriminant is positive, there are two x-intercepts and one y-intercept.

b. The leading coefficient is negative, so the graph opens downward and has a maximum point.

$b^2-4ac=(5)^2-4(-3)(-7)=25-84=-59$

Since the discriminant is negative, there is no x-intercept and there is one y-intercept.

c. The leading coefficient is negative, so the graph opens downward and has a maximum point.

$b^2-4ac=(4)^2-4(-4)(-1)=16-16=0$

Since the discriminant is zero, there is one x-intercept and one y-intercept.

5. **a.** $-\dfrac{b}{2a}=-\dfrac{(-3)}{2(-1)}=-\dfrac{3}{2}$

$f\left(-\dfrac{3}{2}\right)=-\left(\dfrac{3}{2}\right)^2-3\left(-\dfrac{3}{2}\right)-4$

$=-\dfrac{9}{4}+\dfrac{9}{2}-4$

$=-\dfrac{9}{4}+\dfrac{18}{4}-\dfrac{16}{4}=-\dfrac{7}{4}$

The vertex is $\left(-\dfrac{3}{2},-\dfrac{7}{4}\right)$.

b. Since the leading coefficient is negative, the parabola opens downward.

c.

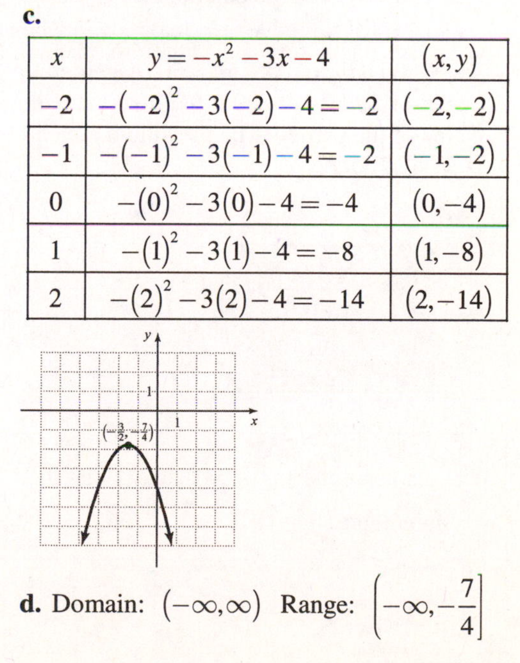

x	$y=-x^2-3x-4$	(x,y)
-2	$-(-2)^2-3(-2)-4=-2$	$(-2,-2)$
-1	$-(-1)^2-3(-1)-4=-2$	$(-1,-2)$
0	$-(0)^2-3(0)-4=-4$	$(0,-4)$
1	$-(1)^2-3(1)-4=-8$	$(1,-8)$
2	$-(2)^2-3(2)-4=-14$	$(2,-14)$

d. Domain: $(-\infty,\infty)$ Range: $\left(-\infty,-\dfrac{7}{4}\right]$

6. **a.**

$$200 = 2l + 2w$$
$$200 - 2w = 2l$$
$$l = \frac{200-2w}{2}$$
$$l = 100 - w$$

b.

$$A(w) = lw$$
$$= (100 - w)w$$
$$= 100w - w^2$$

c. $-\frac{b}{2a} = -\frac{100}{2(-1)} = \frac{100}{2} = 50$

$$A(50) = 100(50) - (50)^2$$
$$= 5000 - 2500 = 2500$$

The vertex is $(50, 2500)$.

$$0 = 100w - w^2$$
$$0 = (100 - w)w$$
$$100 - w = 0 \quad \text{or} \quad w = 0$$
$$w = 100$$

The w-intercepts are $(0,0)$ and $(100,0)$.

d. The vertex of the parabola is $(50, 2500)$, so the area is maximized when $w = 50$ and $l = 100 - 50 = 50$. A garden measuring 50 ft by 50 ft will maximize the area of the garden. The maximum area is $A(50) = 2500$ ft^2.

7. **a.**

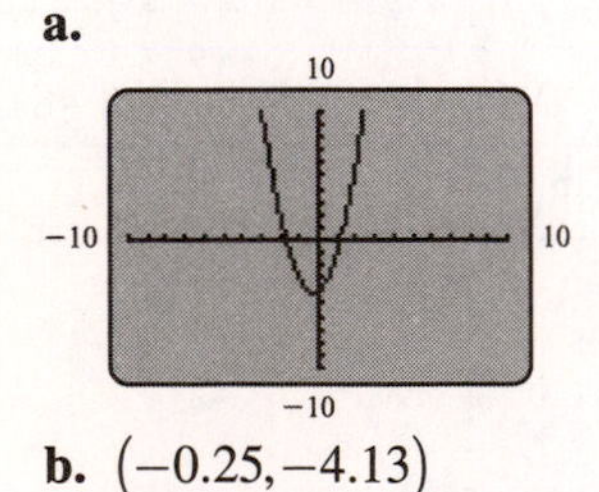

b. $(-0.25, -4.13)$

c. x-intercepts: $(-1.69, 0)$ and $(1.19, 0)$; y-intercept: $(0, -4)$

Exercises

1. Functions of the form $f(x) = ax^2 + bx + c$ are called quadratic functions.

3. The highest or lowest point of a parabola is called a(n) vertex.

5. When a is positive, the parabola given by the function $f(x) = ax^2 + bx + c$ opens upward.

7. If the discriminant of the quadratic equation given by the function $f(x) = ax^2 + bx + c$ is positive, the parabola has two x-intercept(s).

9.

x	$y = f(x) = 2x^2$	(x, y)
-2	$2(-2)^2 = 8$	$(-2, 8)$
-1	$2(-1)^2 = 2$	$(-1, 2)$
0	$2(0)^2 = 0$	$(0, 0)$
1	$2(1)^2 = 2$	$(1, 2)$
2	$2(2)^2 = 8$	$(2, 8)$

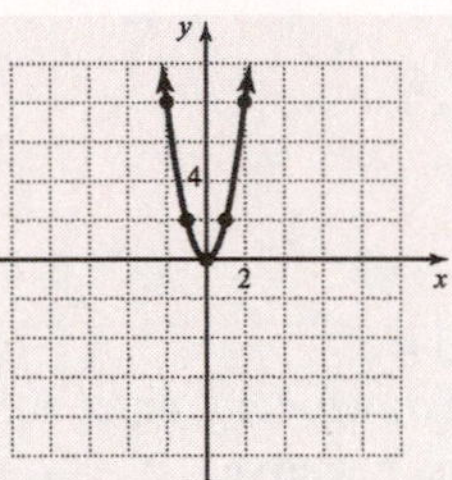

11.

x	$y = f(x) = \frac{1}{2}x^2$	(x, y)
-4	$\frac{1}{2}(-4)^2 = 8$	$(-4, 8)$
-2	$\frac{1}{2}(-2)^2 = 2$	$(-2, 2)$
0	$\frac{1}{2}(0)^2 = 0$	$(0, 0)$
2	$\frac{1}{2}(2)^2 = 2$	$(2, 2)$
4	$\frac{1}{2}(4)^2 = 8$	$(4, 8)$

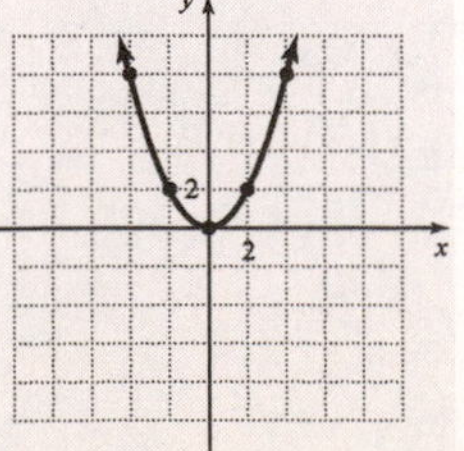

13.

x	$f(x)=2-x^2$	(x,y)
-3	$2-(-3)^2=-7$	$(-3,-7)$
-2	$2-(-2)^2=-2$	$(-2,-2)$
-1	$2-(-1)^2=1$	$(-1,1)$
0	$2-(0)^2=2$	$(0,2)$
1	$2-(1)^2=1$	$(1,1)$
2	$2-(2)^2=-2$	$(2,-2)$
3	$2-(3)^2=-7$	$(3,-7)$

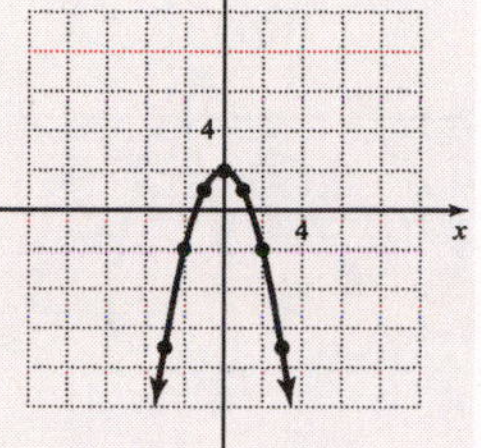

15. $f(x)=x^2-8x$

$a=1 \qquad b=-8$

$$-\frac{b}{2a}=-\frac{-8}{2(1)}=4$$

$f(4)=(4)^2-8(4)=-16$

Vertex: $(4,-16)$

Axis of symmetry: $x=4$

$f(0)=(0)^2-8(0)=0$

y-intercept: $(0,0)$

$x^2-8x=0$

$x(x-8)=0$

$x=0 \quad x-8=0$

$x=8$

x-intercepts: $(0,0)$ and $(8,0)$

17. $f(x)=x^2-2x-3$

$a=1 \qquad b=-2$

$$-\frac{b}{2a}=-\frac{-2}{2(1)}=1$$

$f(1)=(1)^2-2(1)-3=-4$

Vertex: $(1,-4)$

Axis of symmetry: $x=1$

$x^2-2x-3=0$

$(x-3)(x+1)=0$

$x-3=0 \quad x+1=0$

$x=3 \qquad x=-1$

x-intercepts: $(-1,0)$ and $(3,0)$

$f(0)=(0)^2-2(0)-3=-3$

y-intercept: $(0,-3)$

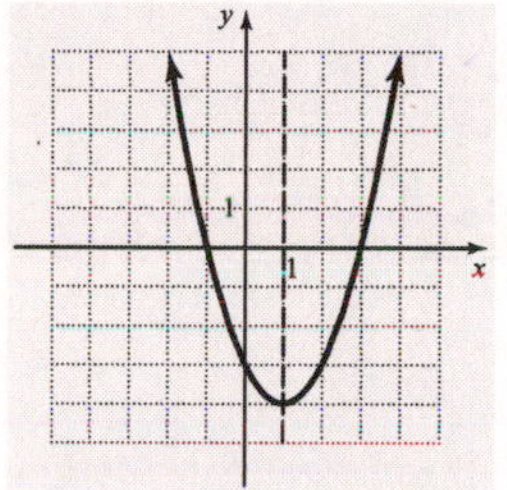

19. $f(x)=-x^2+4x+12$

$a=-1 \qquad b=4$

$$-\frac{b}{2a}=-\frac{4}{2(-1)}=2$$

$f(2)=-(2)^2+4(2)+12=16$

Vertex: $(2,16)$

Axis of symmetry: $x=2$

$f(0)=-(0)^2+4(0)+12=12$

y-intercept: $(0,12)$

$-x^2+4x+12=0$

$(x+2)(-x+6)=0$

$x+2=0 \quad -x+6=0$

$x=-2 \qquad x=6$

x-intercepts: $(-2,0)$ and $(6,0)$

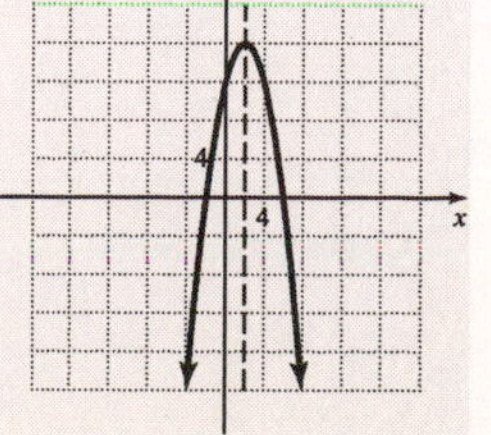

21. $f(x)=x^2-1$

$a=1 \qquad b=0$

$-\frac{b}{2a}=-\frac{0}{2(1)}=0$

$f(0)=(0)^2-1=-1$

Vertex: $(0,-1)$

Axis of symmetry: $x=0$

$f(0)=(0)^2-1=-1$

y-intercept: $(0,-1)$

$x^2-1=0$

$(x+1)(x-1)=0$

$x+1=0 \quad x-1=0$

$x=-1 \qquad x=1$

x-intercepts: $(-1,0)$ and $(1,0)$

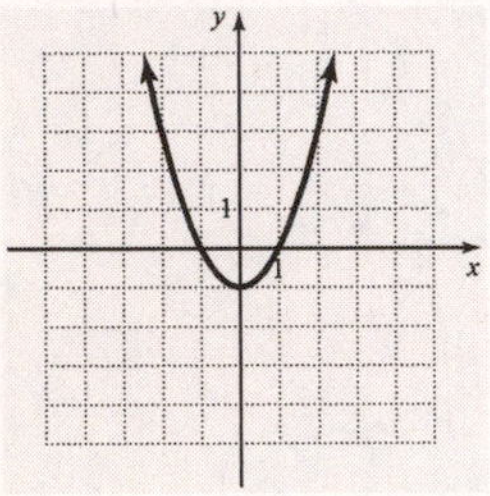

23. $f(x)=(x+1)^2$

$f(x)=x^2+2x+1$

$a=1 \ b=2$

$-\frac{b}{2a}=-\frac{2}{2(1)}=-1$

$f(-1)=(-1+1)^2=0$

Vertex: $(-1,0)$

Axis of symmetry: $x=-1$

$(x+1)^2=0$

$x+1=\sqrt{0}=0$

$x=-1$

x-intercept: $(-1,0)$

$f(0)=(0+1)^2=1$

y-intercept: (0, 1)

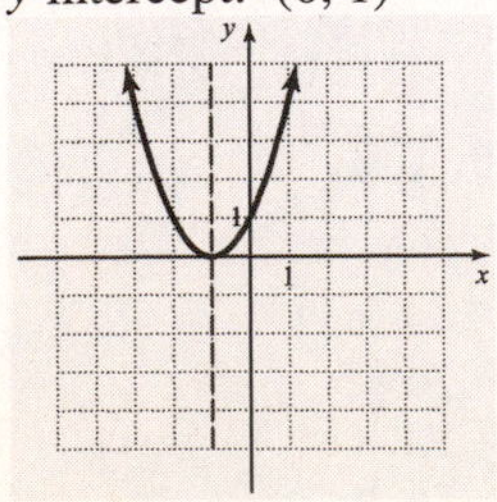

25. $f(x)=-x^2+6x-9$

$a=-1 \ b=6$

$-\frac{b}{2a}=-\frac{6}{2(-1)}=3$

$f(3)=-(3)^2+6(3)-9=0$

Vertex: $(3,0)$

Axis of symmetry: $x=3$

$f(0)=-(0)^2+6(0)-9=-9$

y-intercept: $(0,-9)$

$-x^2+6x-9=0$

$x^2-6x+9=0$

$(x-3)^2=0$

$x-3=\pm\sqrt{0}=0$

$x=3$

x-intercept: $(3,0)$

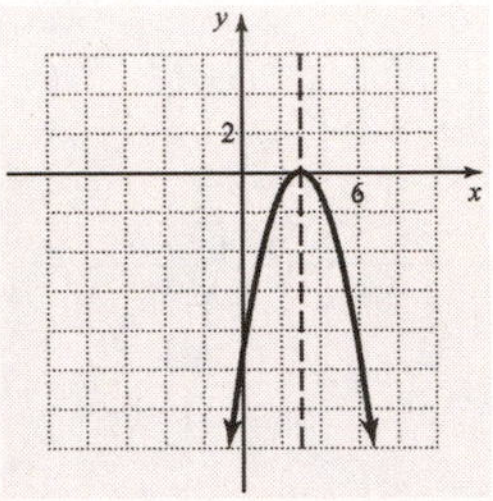

27. $f(x)=x^2-3x-10$

$a=1 \ b=-3$

$-\frac{b}{2a}=-\frac{-3}{2(1)}=\frac{3}{2}$

$f\left(\frac{3}{2}\right)=\left(\frac{3}{2}\right)^2-3\left(\frac{3}{2}\right)-10=-\frac{49}{4}$

Vertex: $\left(\frac{3}{2},-\frac{49}{4}\right)$

Axis of symmetry: $x=\frac{3}{2}$

$f(0)=(0)^2-3(0)-10=-10$

y-intercept: $(0,-10)$

$x^2-3x-10=0$

$(x-5)(x+2)=0$

$x-5=0 \quad x+2=0$

$x=5 \qquad x=-2$

x-intercepts: $(5,0)$ and $(-2,0)$

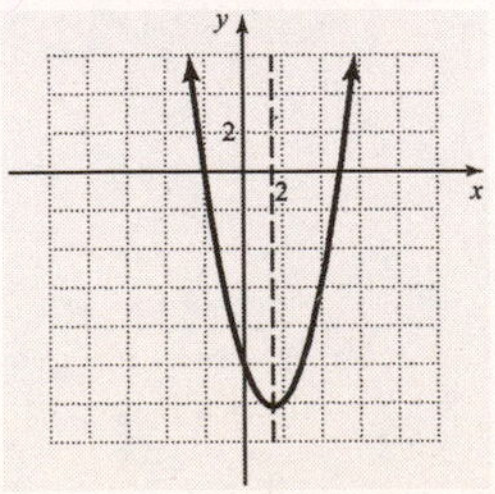

29.

Function	Opens	Maximum Minimum	Number of x-intercepts	Number of y-intercepts
$f(x)=x^2+5$	$a>0$ Upward	$a>0$ Minimum	$x^2+5=0$ $x^2=-5$ No real solutions 0	$f(0)=5$ 1
$f(x)=1-4x+4x^2$	$a>0$ Upward	$a>0$ Minimum	b^2-4ac $(-4)^2-4(4)(1)=0$ 1	$f(0)=1-4(0)+4(0)^2=1$ 1
$f(x)=2-3x^2$	$a<0$ Downward	$a<0$ Maximum	b^2-4ac $0^2-4(-3)(2)=24$ Two real solutions 2	$f(0)=2-3(0)^2=2$ 1
$f(x)=-2x^2-5x-8$	$a<0$ Downward	$a<0$ Maximum	b^2-4ac $(-5)^2-4(-2)(-8)=-39$ No real solutions 0	$f(0)=-2(0)^2-5(0)-8=-8$ 1
$f(x)=4x^2-4x-1$	$a>0$ Upward	$a>0$ Minimum	$b^2-4ac=$ $(-4)^2-4(4)(-1)=32$ Two real solutions 2	$f(0)=4(0)^2-4(0)-1=-1$ 1

31.

x	$f(x)=2x^2-1$	(x,y)
-2	$2(-2)^2-1=7$	$(-2,7)$
-1	$2(-1)^2-1=1$	$(-1,1)$
0	$2(0)^2-1=-1$	$(0,-1)$
1	$2(1)^2-1=1$	$(1,1)$
2	$2(2)^2-1=7$	$(2,7)$

$-\dfrac{b}{2a}=-\dfrac{0}{2(2)}=0 \quad f(0)=2(0)^2-1=-1$

Vertex: $(0,-1)$

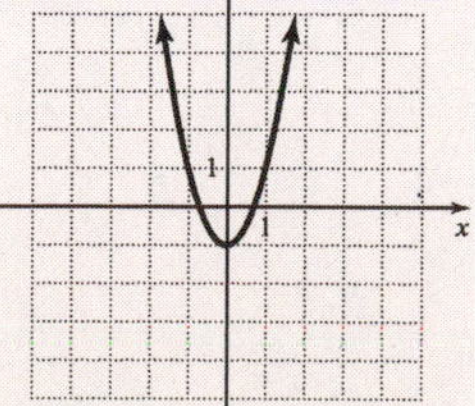

Domain: $(-\infty,\infty)$ Range: $[-1,\infty)$

33.

x	$g(x)=-3x^2+6x-2$	(x,y)
-1	$-3(-1)^2+6(-1)-2=-11$	$(-1,-11)$
0	$-3(0)^2+6(0)-2=-2$	$(0,-2)$
1	$-3(1)^2+6(1)-2=1$	$(1,1)$
2	$-3(2)^2+6(2)-2=-2$	$(2,-2)$
3	$-3(3)^2+6(3)-2=-11$	$(3,-11)$

$-\dfrac{b}{2a}=-\dfrac{6}{2(-3)}=1 \quad f(1)=-3(1)^2+6(1)-2=1$

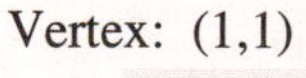

Vertex: $(1,1)$

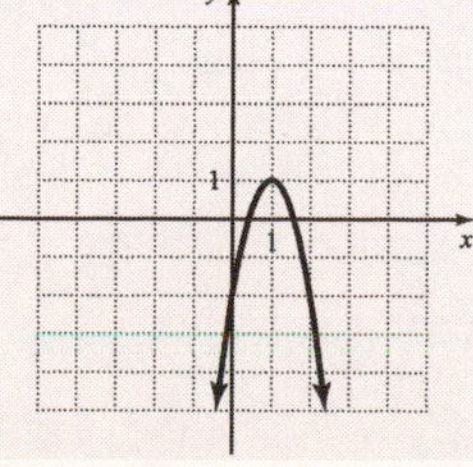

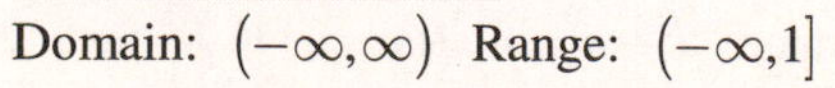

Domain: $(-\infty,\infty)$ Range: $(-\infty,1]$

35.

x	$f(x)=0.5x^2+2$	(x,y)
-2	$0.5(-2)^2+2=4$	$(-2,4)$
-1	$0.5(-1)^2+2=2.5$	$(-1,2.5)$
0	$0.5(0)^2+2=2$	$(0,2)$
1	$0.5(1)^2+2=2.5$	$(1,2.5)$
2	$0.5(2)^2+2=4$	$(2,4)$

$$-\frac{b}{2a}=-\frac{0}{2(0.5)}=0$$

$$f(0)=0.5(0)^2+2=2$$

Vertex: (0, 2)

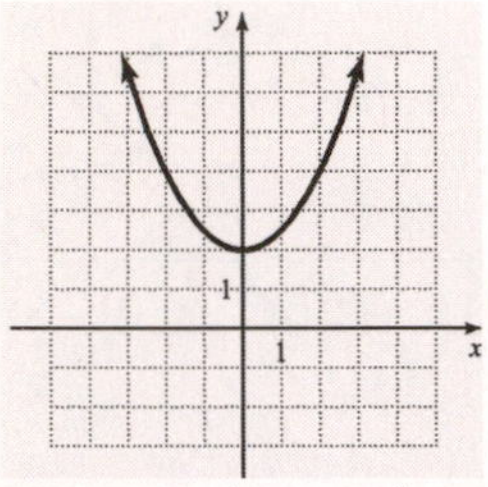

Domain: $(-\infty,\infty)$ Range: $[2,\infty)$

37.

x	$f(x)=x^2-3x-4$	(x,y)
-2	$(-2)^2-3(-2)-4=6$	$(-2,6)$
-1	$(-1)^2-3(-1)-4=0$	$(-1,0)$
0	$(0)^2-3(0)-4=-4$	$(0,-4)$
1	$(1)^2-3(1)-4=-6$	$(1,-6)$
2	$(2)^2-3(2)-4=-6$	$(2,-6)$
3	$(3)^2-3(3)-4=-4$	$(3,-4)$
4	$(4)^2-3(4)-4=0$	$(4,0)$

$$-\frac{b}{2a}=-\frac{-3}{2(1)}=\frac{3}{2}$$

$$f\left(\frac{3}{2}\right)=\left(\frac{3}{2}\right)^2-3\left(\frac{3}{2}\right)-4=-\frac{25}{4}$$

Vertex: $\left(\frac{3}{2},-\frac{25}{4}\right)$

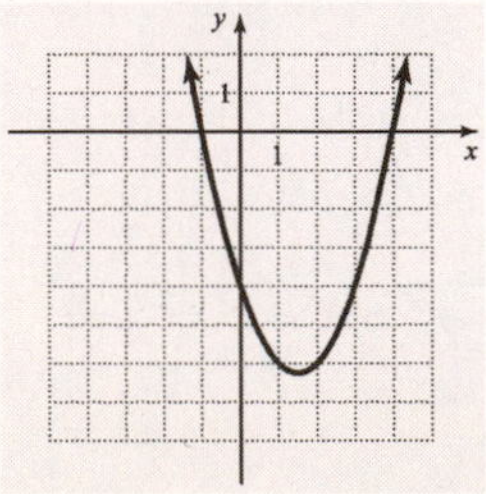

Domain: $(-\infty,\infty)$ Range: $\left[-\frac{25}{4},\infty\right)$

39.

x	$f(x)=-2x^2+2x-3$	(x,y)
-2	$-2(-2)^2+2(-2)-3=-15$	$(-2,-15)$
-1	$-2(-1)^2+2(-1)-3=-7$	$(-1,-7)$
0	$-2(0)^2+2(0)-3=-3$	$(0,-3)$
1	$-2(1)^2+2(1)-3=-3$	$(1,-3)$
2	$-2(2)^2+2(2)-3=-7$	$(2,-7)$

$$-\frac{b}{2a}=-\frac{2}{2(-2)}=\frac{1}{2}$$

$$f\left(\frac{1}{2}\right)=-2\left(\frac{1}{2}\right)^2+2\left(\frac{1}{2}\right)-3=-\frac{5}{2}$$

Vertex: $\left(\frac{1}{2},-\frac{5}{2}\right)$

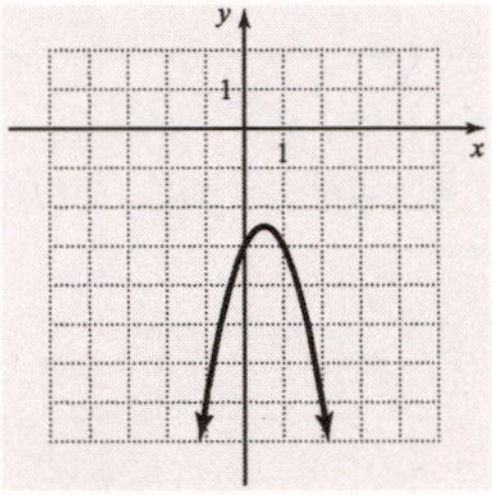

Domain: $(-\infty,\infty)$ Range: $\left(-\infty,-\frac{5}{2}\right]$

41. $f(x)=x^2+0.2x-1$

Vertex: $(-0.1,1.01)$

x-intercepts: $(-1.10,0)$ and $(0.90,0)$

y-intercept: $(0,-1)$

43. $f(x)=-0.15x^2-x+0.5$

Vertex: $(-3.33,2.17)$

x-intercepts: $(-7.13,0)$ and $(0.47,0)$

y-intercept: $(0,0.5)$

45. $f(x)=5x^2+3x+7$

Vertex: $(-0.30,6.55)$

x-intercepts: none

y-intercept: $(0,7)$

47.

x	$y=f(x)=-3+x^2$	(x,y)
-3	$-3+(-3)^2=6$	$(-3,6)$
-2	$-3+(-2)^2=1$	$(-2,1)$
-1	$-3+(-1)^2=-2$	$(-1,-2)$
0	$-3+(0)^2=-3$	$(0,-3)$
1	$-3+(1)^2=-2$	$(1,-2)$
2	$-3+(2)^2=1$	$(2,1)$
3	$-3+(3)^2=6$	$(3,6)$

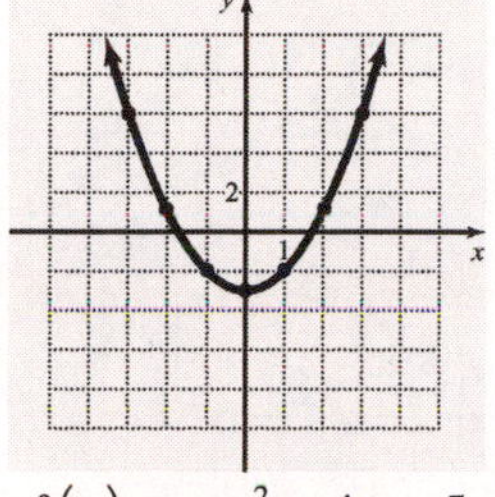

49. $f(x)=-x^2-4x+5$

$a=-1 \quad b=-4$

$$-\frac{b}{2a}=-\frac{-4}{2(-1)}=-2$$

$f(-2)=-(-2)^2-4(-2)+5=9$

Vertex: $(-2,9)$

Axis of symmetry: $x=-2$

$f(0)=-(0)^2-4(0)+5=5$

y-intercept: $(0,5)$

$f(x)=0$

$-x^2-4x+5=0$

$x^2+4x-5=0$

$(x+5)(x-1)=0$

$x+5=0 \qquad x-1=0$

$x=-5 \qquad x=1$

x-intercepts: $(-5, 0)$, $(1, 0)$

51.

x	$g(x)=-x^2+2x+1$	(x,y)
-2	$-(-2)^2+2(-2)+1=-7$	$(-2,-7)$
-1	$-(-1)^2+2(-1)+1=-2$	$(-1,-2)$
0	$-(0)^2+2(0)+1=1$	$(0,1)$
1	$-(1)^2+2(1)+1=2$	$(1,2)$
2	$-(2)^2+2(2)+1=1$	$(2,1)$
3	$-(3)^2+2(3)+1=-2$	$(3,-2)$
4	$-(4)^2+2(4)+1=-7$	$(4,-7)$

$$-\frac{b}{2a}=-\frac{2}{2(-1)}=1 \quad f(1)=-(1)^2+2(1)+1=2$$

Vertex: (1, 2)

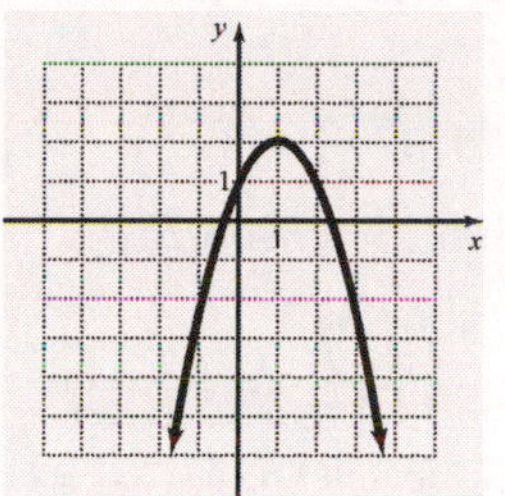

Domain: $(-\infty,\infty)$ Range: $(-\infty,2]$

53. **a.** $s=-16t^2+48t+280$

$$-\frac{b}{2a}=-\frac{48}{2(-16)}=\frac{3}{2}$$

$$f\left(\frac{3}{2}\right)=-16\left(-\frac{3}{2}\right)^2+48\left(\frac{3}{2}\right)+280$$

$$-16\left(\frac{9}{4}\right)+72+280=-36+352=316$$

Vertex: $\left(\frac{3}{2},316\right)$

b.

c. At $\frac{3}{2}$ sec (or 1.5 sec) after it is thrown, the stone reaches its maximum height of 316 ft.

55. **a.** $2l + 2w = 150$

$2l = 150 - 2w$

$l = \dfrac{150 - 2w}{2} = 75 - w$

b. $A = lw; \; l = 75 - w$

$A(w) = (75 - w)w = 75w - w^2$

c.

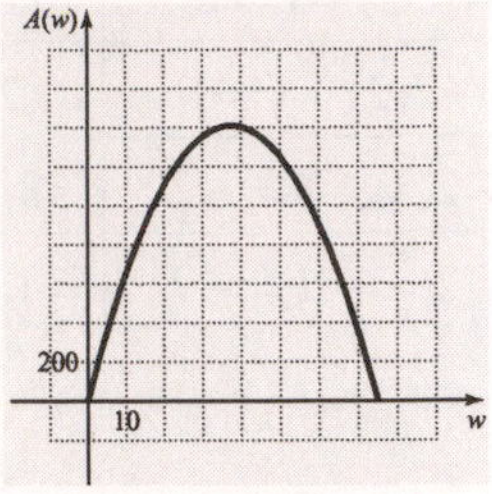

d. $-\dfrac{b}{2a} = -\dfrac{75}{2(-1)} = 37.5$

$w = A(37.5) = 75(37.5) - (37.5)^2$

$= 2812.5 - 1406.25 = 1406.25$

Vertex: $(37.5, 1406.25)$

$l = 75 - w = 75 - 37.5 = 37.5$

An exercise yard measuring 37.5 ft by 37.5 ft will produce a maximum area of 1406.25 sq ft.

57. **a.** $C(x) = 0.005x^2 - x + 100$

$-\dfrac{b}{2a} = -\dfrac{-1}{2(0.005)} = \dfrac{1}{0.010} = 100$

100 units must be produced in order to minimize the cost.

b. $C(100) = 0.005(100)^2 - (100) + 100$

$C(100) = 0.005(10{,}000) = 50$

The minimum daily cost is $50.

9.5 Solving Quadratic and Rational Inequalities

Practice

1. $x^2 - 3x - 4 \geq 0$

$x^2 - 3x - 4 = 0$

$(x-4)(x+1) = 0$

$x - 4 = 0$ or $x + 1 = 0$

$x = 4 \qquad x = -1$

Interval	Test Value	$x^2 - 3x - 4 \geq 0$	Conclusion
$x < -1$	-2	$(-2)^2 - 3(-2) - 4 = 6$	$6 \geq 0$
$-1 < x < 4$	0	$(0)^2 - 3(0) - 4 = -4$	$-4 < 0$
$4 < x$	5	$(5)^2 - 3(5) - 4 = 6$	$6 \geq 0$

$(-\infty, -1] \cup [4, \infty)$

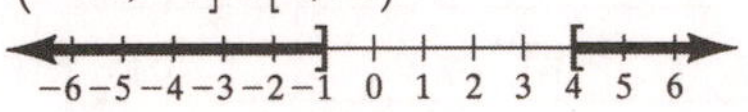

2. $x^2 - 6x - 2 < -7$

$x^2 - 6x - 2 = -7$

$x^2 - 6x + 5 = 0$

$(x-5)(x-1) = 0$

$x - 5 = 0$ or $x - 1 = 0$

$x = 5 \qquad x = 1$

Interval	Test Value	$x^2 - 6x - 2 < -7$	Conclusion
$x < 1$	0	$(0)^2 - 6(0) - 2 = -2$	$-2 > -7$
$1 < x < 5$	2	$(2)^2 - 6(2) - 2 = -10$	$-10 < -7$
$5 < x$	6	$(6)^2 - 6(6) - 2 = -2$	$-2 > -7$

$(1, 5)$

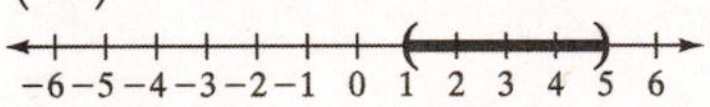

3. $\dfrac{x-5}{x+5} < 0$

$x + 5 = 0$

$x = -5$

$\dfrac{x-5}{x+5} = 0$

$\dfrac{x-5}{x+5} \cdot (x+5) = 0 \cdot (x+5)$

$x - 5 = 0$

$x = 5$

3. (continued)

Interval	Test Value	$\frac{x-5}{x+5}<0$	Conclusion
$x<-5$	-6	$\frac{-6-5}{-6+5}=11$	$11>0$
$-5<x<5$	0	$\frac{0-5}{0+5}=-1$	$-1<0$
$5<x$	6	$\frac{6-5}{6+5}=\frac{1}{11}$	$\frac{1}{11}>0$

$(-5,5)$

−6 −5 −4 −3 −2 −1 0 1 2 3 4 5 6

4. $\frac{x-2}{x+3}\geq -1$

$$x+3=0$$
$$x=-3$$
$$\frac{x-2}{x+3}=-1$$
$$\frac{x-2}{x+3}\cdot(x+3)=-1\cdot(x+3)$$
$$x-2=-x-3$$
$$2x=-1$$
$$x=-\frac{1}{2}$$

Interval	Test Value	$\frac{x-2}{x+3}\geq -1$	Conclusion
$x<-3$	-4	$\frac{-4-2}{-4+3}=6$	$6\geq -1$
$-3<x<-\frac{1}{2}$	-1	$\frac{-1-2}{-1+3}=-\frac{3}{2}$	$-\frac{3}{2}<-1$
$-\frac{1}{2}<x$	0	$\frac{0-2}{0+3}=-\frac{2}{3}$	$-\frac{2}{3}\geq -1$

$(-\infty,-3]\cup\left[-\frac{1}{2},\infty\right)$

−6 −5 −4 −3 −2 −1 0 1 2 3 4 5 6

5. $C(x)=5x^2+25x+70$

Solve $C(x)\leq 400$.

$$5x^2+25x+70=400$$
$$x^2+5x+14=80$$
$$x^2+5x-66=0$$
$$(x+11)(x-6)=0$$
$$x+11=0 \quad\text{or}\quad x-6=0$$
$$x=-11 \qquad x=6$$

Interval	Test Value	$5x^2+25x+70\leq 400$	Conclusion
$x<-11$	-12	$5(-12)^2+25(-12)+70=490$	$490>400$
$-11<x<6$	0	$5(0)^2+25(0)+70=70$	$70\leq 400$
$6<x$	7	$5(7)^2+25(7)+70=490$	$490>400$

Since the number of weeks cannot be negative, we do not include solutions less than 0. So the stand can run for six weeks or less.

Exercises

1. To determine if the value of an expression is greater than or less than 0 in a particular interval, a(n) test value is used.

3. A(n) rational inequality is an inequality that contains a rational expression.

5. $x^2-x>0$

$$x^2-x=0$$
$$x(x-1)=0$$
$$x=0 \quad x-1=0;\quad x=1$$

Interval	Test Value	$x^2-x>0$	Conclusion
$x<0$	-5	$(-5)^2-(-5)=30$	$30>0$
$0<x<1$	$.5$	$(.5)^2-(.5)=-.25$	$-.25<0$
$1<x$	2	$(2)^2-(2)=2$	$2>0$

$(-\infty,0)\cup(1,\infty)$

−6 −5 −4 −3 −2 −1 0 1 2 3 4 5 6

7. $x^2 < 4$

$x^2 - 4 < 0$

$x^2 - 4 = 0$

$(x+2)(x-2) = 0$

$x+2=0 \quad x-2=0$

$x=-2 \quad x=2$

Interval	Test Value	$x^2 < 4$	Conclusion
$x \le -2$	-5	$(-5)^2 = 25$	$25 \ge 4$
$-2 < x < 2$	0	$(0)^2 = 0$	$0 < 4$
$2 \le x$	4	$(4)^2 = 16$	$16 \ge 4$

$(-2,2)$

−6 −5 −4 −3 −2 −1 0 1 2 3 4 5 6

9. $x^2 - x - 2 \le 10$

$x^2 - x - 2 = 10$

$x^2 - x - 12 = 0$

$(x+3)(x-4) = 0$

$x+3=0 \quad x-4=0$

$x=-3 \quad x=4$

Interval	Test Value	$x^2 - x - 2 \le 10$	Conclusion
$x \le -3$	-5	$(-5)^2 - (-5) - 2 = 28$	$28 > 10$
$-3 < x < 4$	0	$(0)^2 - (0) - 2 = -2$	$-2 \le 10$
$4 \le x$	5	$(5)^2 - (5) - 2 = 18$	$18 > 10$

$[-3,4]$

−6 −5 −4 −3 −2 −1 0 1 2 3 4 5 6

11. $x^2 + 6x + 9 \ge 0$

$x^2 + 6x + 9 = 0$

$(x+3)^2 = 0$

$x+3=0$

$x=-3$

Interval	Test Value	$x^2 + 6x + 9 \ge 0$	Conclusion
$x < -3$	-5	$(-5)^2 + 6(-5) + 9 = 4$	$4 \ge 0$
$x = -3$	-3	$(-3)^2 + 6(-3) + 9 = 0$	$0 \ge 0$
$-3 < x$	5	$(5)^2 + 6(5) + 9 = 64$	$64 \ge 0$

$(-\infty, \infty)$

−6 −5 −4 −3 −2 −1 0 1 2 3 4 5 6

13. $6 + x - x^2 \le 0$

$x^2 - x - 6 \ge 0$

$x^2 - x - 6 = 0$

$(x-3)(x+2) = 0$

$x-3=0 \quad x+2=0$

$x=3 \quad x=-2$

Interval	Test Value	$6 + x - x^2 \le 0$	Conclusion
$x \le -2$	-3	$6 + (-3) - (-3)^2 = -6$	$-6 \le 0$
$-2 < x < 3$	0	$6 + (0) - (0)^2 = 6$	$6 > 0$
$3 \le x$	4	$6 + (4) - (4)^2 = -6$	$-6 \le 0$

$(-\infty, -2] \cup [3, \infty)$

−6 −5 −4 −3 −2 −1 0 1 2 3 4 5 6

15. $x^2+5x+4<0$

$x^2+5x+4=0$

$(x+4)(x+1)=0$

$x+4=0 \quad x+1=0$

$x=-4 \quad x=-1$

Interval	Test Value	$x^2+5x+4<0$	Conclusion
$x<-4$	-5	$(-5)^2+5(-5)+4=10$	$10>0$
$-4<x<-1$	-2	$(-2)^2+5(-2)+4=-2$	$-2<0$
$-1<x$	0	$(0)^2+5(0)+4=4$	$4>0$

$(-4,-1)$

−6 −5 −4 −3 −2 −1 0 1 2 3 4 5 6

17. $2x^2-3x+1\geq 0$

$2x^2-3x+1=0$

$(2x-1)(x-1)=0$

$2x-1=0 \quad x-1=0$

$2x=1 \quad x=1$

$x=\frac{1}{2}$

Interval	Test Value	$2x^2-3x+1\geq 0$	Conclusion
$x\leq\frac{1}{2}$	0	$2(0)^2-3(0)+1=1$	$1\geq 0$
$\frac{1}{2}<x<1$	$\frac{3}{4}$	$2\left(\frac{3}{4}\right)^2-3\left(\frac{3}{4}\right)+1=-\frac{1}{8}$	$-\frac{1}{8}<0$
$1\leq x$	2	$2(2)^2-3(2)+1=3$	$3\geq 0$

$\left(-\infty,\frac{1}{2}\right]\cup[1,\infty)$

−6 −5 −4 −3 −2 −1 0 1 2 3 4 5 6

19. $1-4x^2\leq 0$

$1-4x^2=0$

$(1-2x)(1+2x)=0$

$1-2x=0 \quad 1+2x=0$

$-2x=-1 \quad 2x=-1$

$x=\frac{1}{2} \quad x=-\frac{1}{2}$

Interval	Test Value	$1-4x^2\leq 0$	Conclusion
$x\leq-\frac{1}{2}$	-1	$1-4(-1)^2=-3$	$-3\leq 0$
$-\frac{1}{2}<x<\frac{1}{2}$	0	$1-4(0)^2=1$	$1>0$
$\frac{1}{2}\leq x$	1	$1-4(1)^2=-3$	$-3\leq 0$

$\left(-\infty,-\frac{1}{2}\right]\cup\left[\frac{1}{2},\infty\right)$

−6 −5 −4 −3 −2 −1 0 1 2 3 4 5 6

21. $3>6x^2-7x$

$6x^2-7x-3=0$

$(3x+1)(2x-3)=0$

$3x+1=0 \quad 2x-3=0$

$3x=-1 \quad 2x=3$

$x=-\frac{1}{3} \quad x=\frac{3}{2}$

Interval	Test Value	$3>6x^2-7x$	Conclusion
$x<-\frac{1}{3}$	-1	$6(-1)^2-7(-1)=13$	$3\leq 13$
$-\frac{1}{3}<x<\frac{3}{2}$	0	$6(0)^2-7(0)=0$	$3>0$
$\frac{3}{2}<x$	2	$6(2)^2-7(2)=10$	$3\leq 10$

$\left(-\frac{1}{3},\frac{3}{2}\right)$

−6 −5 −4 −3 −2 −1 0 1 2 3 4 5 6

23. $\frac{x}{x+2} \le 0$

$\frac{x}{x+2}$ is undefined at $x = -2$

$\frac{x}{x+2} = 0$ when $x = 0$

Interval	Test Value	$\frac{x}{x+2} \le 0$	Conclusion
$x < -2$	-3	$\frac{-3}{-3+2} = 3$	$3 > 0$
$-2 < x \le 0$	-1	$\frac{-1}{-1+2} = -\frac{1}{1}$	$-1 \le 0$
$0 < x$	1	$\frac{1}{1+2} = \frac{1}{3}$	$\frac{1}{3} > 0$

$(-2, 0]$

−6 −5 −4 −3 −2 −1 0 1 2 3 4 5 6

25. $\frac{1}{6-x} < 0$

$\frac{1}{6-x}$ is undefined when $x = 6$

Interval	Test value	$\frac{1}{6-x} < 0$	Conclusion
$x < 6$	0	$\frac{1}{6-0} = \frac{1}{6}$	$\frac{1}{6} \ge 0$
$x > 6$	7	$\frac{1}{6-7} = -1$	$-1 < 0$

$(6, \infty)$

0 1 2 3 4 5 6 7 8 9 10 11 12

27.

$\frac{x+3}{x-3} > 2 \qquad \frac{x+3}{x-3}$ is undefined when $x = 3$

$$\frac{x+3}{x-3} = 2$$
$$x+3 = 2(x-3)$$
$$x+3 = 2x-6$$
$$x = 9$$

Interval	Test value	$\frac{x+3}{x-2} > 2$	Conclusion
$x < 3$	0	$\frac{0+3}{0-2} = -\frac{3}{2}$	$-\frac{3}{2} \le 2$
$3 < x < 9$	5	$\frac{5+3}{5-2} = \frac{8}{3}$	$\frac{8}{3} > 2$
$x > 9$	10	$\frac{10+3}{10-2} = \frac{13}{8}$	$\frac{13}{8} \le 2$

$(3, 9)$

0 1 2 3 4 5 6 7 8 9 10 11 12

29. $\frac{2x-1}{x+4} \ge 0$

$\frac{2x-1}{x+4}$ is undefined when $x = -4$

$\frac{2x-1}{x+4} = 0$ when $2x - 1 = 0$ or $x = \frac{1}{2}$

Interval	Test value	$\frac{2x-1}{x+4} \ge 0$	Conclusion
$x < -4$	-5	$\frac{2(-5)-1}{-5+4} = 11$	$11 \ge 0$
$-4 < x < \frac{1}{2}$	0	$\frac{2(0)-1}{0+4} = -\frac{1}{4}$	$-\frac{1}{4} < 0$
$\frac{1}{2} \le x$	1	$\frac{2(1)-1}{1+4} = \frac{1}{5}$	$\frac{1}{5} \ge 0$

$(-\infty, -4) \cup \left[\frac{1}{2}, \infty\right)$

−6 −5 −4 −3 −2 −1 0 1 2 3 4 5 6

31. $\frac{2x-1}{2x+5} \le 2$

$\frac{2x-1}{2x+5}$ is undefined when $2x + 5 = 0$

$2x = -5, \; x = -\frac{5}{2}$

$$\frac{2x-1}{2x+5} = 2$$
$$2x - 1 = 2(2x+5)$$
$$2x - 1 = 4x + 10 \Rightarrow -2x = 11 \Rightarrow x = -\frac{11}{2}$$

31. (continued)

Interval	Test value	$\frac{2x-1}{2x+5} \le 2$	Conclusion
$x \le -\frac{11}{2}$	-6	$\frac{2(-6)-1}{2(-6)+5} = -\frac{13}{7}$	$-\frac{13}{7} \le 2$
$-\frac{11}{2} < x < -\frac{5}{2}$	-4	$\frac{2(-4)-1}{2(-4)+5} = 3$	$3 > 2$
$-\frac{5}{2} < x$	0	$\frac{2(0)-1}{2(0)+5} = -\frac{1}{5}$	$-\frac{1}{5} < 2$

$$\left(-\infty, -\frac{11}{2}\right] \cup \left(-\frac{5}{2}, \infty\right)$$

33.

$$x^2 + x - 5 < 7$$
$$x^2 + x - 12 < 0$$
$$x^2 + x - 12 = 0$$
$$(x+4)(x-3) = 0$$
$$x+4=0 \qquad x-3=0$$
$$x=-4 \qquad x=3$$

Interval	Test Value	$x^2 + x - 5 < 7$	Conclusion
$x < -4$	-5	$(-5)^2 + (-5) - 5 = 15$	$15 \ge 7$
$-4 < x < 3$	0	$(0)^2 + 0 - 5 = -5$	$-5 < 7$
$3 < x$	4	$(4)^2 + 4 - 5 = 15$	$15 \ge 7$

$$(-4, 3)$$

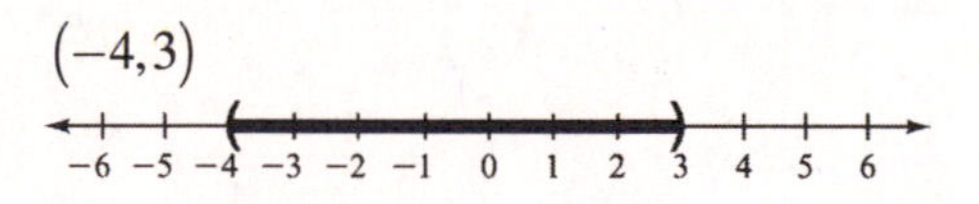

35.

$$\frac{x+2}{x-4} < 2$$

$\frac{x+2}{x-4}$ is undefined when $x = 4$.

$$\frac{x+2}{x-4} = 2$$
$$x+2 = 2(x-4)$$
$$x+2 = 2x-8$$
$$10 = x$$

Interval	Test value	$\frac{x+2}{x-4} < 2$	Conclusion
$x < 4$	0	$\frac{0+2}{0-4} = -\frac{1}{2}$	$-\frac{1}{2} < 2$
$4 < x < 10$	5	$\frac{5+2}{5-4} = 7$	$7 \ge 2$
$10 < x$	14	$\frac{14+2}{14-4} = \frac{8}{5}$	$\frac{8}{5} < 2$

$$(-\infty, 4) \cup (10, \infty)$$

37.

$$2x^2 + 5x - 3 \ge 0$$
$$(2x-1)(x+3) = 0$$
$$2x-1=0 \qquad x+3=0$$
$$2x=1 \qquad x=-3$$
$$x=\frac{1}{2}$$

Interval	Test value	$2x^2 + 5x - 3 \ge 0$	Conclusion
$x \le -3$	-5	$2(-5)^2 + 5(-5) - 3 = 22$	$22 \ge 0$
$-3 < x < \frac{1}{2}$	0	$2(0)^2 + 5(0) - 3 = -3$	$-3 < 0$
$\frac{1}{2} \le x$	1	$2(1)^2 + 5(1) - 3 = 4$	$4 \ge 0$

$$(-\infty, 3] \cup \left[\frac{1}{2}, \infty\right)$$

39. $h(t) = -16t^2 + 48t > 0$

$-16t^2 + 48t = 0$

$-16t(t-3) = 0$

$-16t = 0 \quad t-3=0$

$t = 0 \qquad t = 3$

Interval	Test value	$-16t^2 + 48t > 0$	Conclusion
$x<0$	-1	$-16(-1)^2 + 48(-1) = -64$	$-64 \le 0$
$0<x<3$	1	$-16(1)^2 + 48(1) = 32$	$32 > 0$
$3<x$	4	$-16(4)^2 + 48(4) = -64$	$-64 \le 0$

The penny is above the point of release between, but not including, 0 sec and 3 sec.

41. $C(x) = 2x^2 - 60x + 900 < 500$

$2x^2 - 60x + 900 = 500$

$2x^2 - 60x + 400 = 0$

$x^2 - 30x + 200 = 0$

$(x-10)(x-20) = 0$

$x - 10 = 0 \quad x - 20 = 0$

$x = 10 \qquad x = 20$

Interval	Test value	$2x^2 - 60x + 900 < 500$	Conclusion
$x<10$	9	$2(9)^2 - 60(9) + 900 = 522$	$522 \ge 500$
$10<x<20$	15	$2(15)^2 - 60(15) + 900 = 450$	$450 < 500$
$20<x$	21	$2(21)^2 - 60(21) + 900 = 522$	$522 \ge 500$

Producing between, but not including, 10 and 20 end tables per week will keep the cost under \$500.

43. Let $l =$ the length and $w =$ the width.

$2l + 2w = 90$

$2w = 90 - 2l$

$w = \frac{90-2l}{2} = 45 - l$

Area $A = lw = l(45 - l)$

$A = 45l - l^2$

$45l - l^2 > 450$

$l^2 - 45l + 450 = 0$

$(l-15)(l-30) = 0$

$l - 15 = 0 \quad l - 30 = 0$

$l = 15 \qquad l = 30$

Interval	Test value	$45l - l^2 > 450$	Conclusion
$x<15$	10	$45(10) - (10)^2 = 350$	$350 \le 450$
$15<x<30$	20	$45(20) - (20)^2 = 500$	$500 > 450$
$30<x$	40	$45(40) - (40)^2 = 200$	$200 \le 450$

The area will exceed 450 sq ft for any length between, but not including, 15 ft and 30 ft.

45. $C(x) = \frac{864+2x}{x} < 8$

$\frac{864+2x}{x}$ is undefined when $x = 0$

$\frac{864+2x}{x} = 8$

$864 + 2x = 8x$

$864 = 6x$

$x = 144$

x is units sold so $x > 0$

Interval	Test value	$\frac{864+2x}{x} < 8$	Conclusion
$0<x<144$	90	$\frac{864+2(90)}{90} = \frac{58}{5}$	$\frac{58}{5} > 8$
$144<x$	288	$\frac{864+2(288)}{288} = 5$	$5 < 8$

The average cost will be less than \$8 if more than 144 units are sold.

Chapter 9 Review Exercises

1.
$$x^2-81=0$$
$$x^2=81$$
$$\sqrt{x^2}=\pm\sqrt{81}$$
$$x=\pm 9$$
$$x=9,\ x=-9$$

2.
$$3n^2-7=8$$
$$3n^2=15$$
$$n^2=5$$
$$\sqrt{n^2}=\pm\sqrt{5}$$
$$n=\pm\sqrt{5}$$
$$n=\sqrt{5},\ n=-\sqrt{5}$$

3.
$$4(a-5)^2=1$$
$$(a-5)^2=\frac{1}{4}$$
$$\sqrt{(a-5)^2}=\pm\sqrt{\frac{1}{4}}$$
$$a-5=\pm\frac{1}{2}$$
$$a=5\pm\frac{1}{2}$$
$$a=\frac{11}{2},\quad a=\frac{9}{2}$$

4.
$$(2x+1)^2+10=6$$
$$(2x+1)^2=-4$$
$$\sqrt{(2x+1)^2}=\pm\sqrt{-4}$$
$$2x+1=\pm 2i$$
$$2x=-1\pm 2i$$
$$x=\frac{-1\pm 2i}{2}$$
$$x=\frac{-1+2i}{2},\quad x=\frac{-1-2i}{2}$$

5.
$$x^2+8x+16=2$$
$$(x+4)^2=2$$
$$\sqrt{(x+4)^2}=\pm\sqrt{2}$$
$$x+4=\pm\sqrt{2}$$
$$x=-4\pm\sqrt{2}$$
$$x=-4+\sqrt{2},\quad x=-4-\sqrt{2}$$

6.
$$(n-5)(n+5)=-33$$
$$n^2-25=-33$$
$$n^2=-8$$
$$\sqrt{n^2}=\pm\sqrt{-8}$$
$$n=\pm 2i\sqrt{2}$$
$$n=2i\sqrt{2},\quad n=-2i\sqrt{2}$$

7.
$$x^2+10x+\left[\left(\frac{10}{2}\right)^2\right]=x^2+10x+[25]$$

8.
$$n^2-9n+\left[\left(\frac{-9}{2}\right)^2\right]=n^2-9n+\left[\frac{81}{4}\right]$$

9.
$$x^2-6x+2=0$$
$$x^2-6x=-2$$
$$x^2-6x+\left(\frac{6}{2}\right)^2=-2+\left(\frac{6}{2}\right)^2$$
$$x^2-6x+9=-2+9$$
$$(x-3)^2=7$$
$$\sqrt{(x-3)^2}=\pm 7$$
$$x-3=\pm\sqrt{7}$$
$$x=3\pm\sqrt{7}$$
$$x=3+\sqrt{7},\quad x=3-\sqrt{7}$$

10.
$$a^2+a-3=0$$
$$a^2+a=3$$
$$a^2+a+\left(\frac{1}{2}\right)^2=3+\left(\frac{1}{2}\right)^2$$
$$a^2+a+\frac{1}{4}=3+\frac{1}{4}$$
$$\left(a+\frac{1}{2}\right)^2=\frac{13}{4}$$
$$\sqrt{\left(a+\frac{1}{2}\right)^2}=\pm\sqrt{\frac{13}{4}}$$
$$a+\frac{1}{2}=\pm\frac{\sqrt{13}}{2}$$
$$a=-\frac{1}{2}\pm\frac{\sqrt{13}}{2}$$
$$a=\frac{-1\pm\sqrt{13}}{2}$$

11.
$$2n^2+2n+9=3-4n$$
$$2n^2+6n=-6$$
$$n^2+3n=-3$$
$$n^2+3n+\left(\frac{3}{2}\right)^2=-3+\left(\frac{3}{2}\right)^2$$
$$n^2+3n+\frac{9}{4}=-3+\frac{9}{4}$$
$$\left(n+\frac{3}{2}\right)^2=-\frac{3}{4}$$
$$\sqrt{\left(n+\frac{3}{2}\right)^2}=\pm\sqrt{-\frac{3}{4}}$$
$$n+\frac{3}{2}=\pm\frac{i\sqrt{3}}{2}$$
$$n=-\frac{3}{2}\pm\frac{i\sqrt{3}}{2}$$
$$n=\frac{-3\pm i\sqrt{3}}{2}$$

12.
$$3x^2-2x-9=0$$
$$3x^2-2x=9$$
$$3\left(x^2-\frac{2}{3}x\right)=9$$
$$x^2-\frac{2}{3}x=3$$
$$x^2-\frac{2}{3}x+\left(\frac{1}{2}\bullet\frac{-2}{3}\right)^2=3+\left(\frac{1}{2}\bullet\frac{-2}{3}\right)^2$$
$$x^2-\frac{2}{3}x+\left(-\frac{1}{3}\right)^2=3+\left(-\frac{1}{3}\right)^2$$
$$x^2-\frac{2}{3}x+\frac{1}{9}=3+\frac{1}{9}$$
$$\left(x-\frac{1}{3}\right)^2=\frac{28}{9}$$
$$\sqrt{\left(x-\frac{1}{3}\right)^2}=\pm\sqrt{\frac{28}{9}}$$
$$x-\frac{1}{3}=\pm\frac{2\sqrt{7}}{3}$$
$$x=\frac{1}{3}\pm\frac{2\sqrt{7}}{3}$$
$$x=\frac{1\pm2\sqrt{7}}{3}$$

13.
$$x^2+7x+6=0$$
$a=1 \qquad b=7 \qquad c=6$
$$x=\frac{-(7)\pm\sqrt{(7)^2-4(1)(6)}}{2(1)}$$
$$x=\frac{-7\pm\sqrt{49-24}}{2}=\frac{-7\pm\sqrt{25}}{2}=\frac{-7\pm5}{2}$$
$$x=\frac{-2}{2}=-1,\quad x=\frac{-12}{2}=-6$$

14.
$$x^2-4x+5=0$$
$a=1 \qquad b=-4 \qquad c=5$
$$x=\frac{-(-4)\pm\sqrt{(-4)^2-4(1)(5)}}{2(1)}$$
$$x=\frac{4\pm\sqrt{16-20}}{2}=\frac{4\pm\sqrt{-4}}{2}=\frac{4\pm2i}{2}$$
$$x=\frac{4+2i}{2}=2+i,\quad x=\frac{4-2i}{2}=2-i$$

15. $3x^2-13x=5-7x$

$3x^2-6x-5=0$

$a=3 \quad b=-6 \quad c=-5$

$$x=\frac{-(-6)\pm\sqrt{(-6)^2-4(3)(-5)}}{2(3)}$$

$$x=\frac{6\pm\sqrt{36+60}}{6}=\frac{6\pm\sqrt{96}}{6}=\frac{6\pm4\sqrt{6}}{6}$$

$$x=\frac{6+4\sqrt{6}}{6}=\frac{3+2\sqrt{6}}{3}$$

$$x=\frac{6-4\sqrt{6}}{6}=\frac{3-2\sqrt{6}}{3}$$

16. $4x^2+12x=-9$

$4x^2+12x+9=0$

$a=4 \quad b=12 \quad c=9$

$$x=\frac{-(12)\pm\sqrt{(12)^2-4(4)(9)}}{2(4)}$$

$$x=\frac{-12\pm\sqrt{144-144}}{8}=\frac{-12\pm\sqrt{0}}{8}$$

$$x=-\frac{12}{8}=-\frac{3}{2}$$

17. $\frac{1}{3}x^2+\frac{3}{2}x+1=0$

$$6\left(\frac{1}{3}x^2+\frac{3}{2}x+1\right)=6(0)$$

$2x^2+9x+6=0$

$a=2 \quad b=9 \quad c=6$

$$x=\frac{-(9)\pm\sqrt{(9)^2-4(2)(6)}}{2(2)}$$

$$x=\frac{-9\pm\sqrt{81-48}}{4}=\frac{-9\pm\sqrt{33}}{4}$$

$$x=\frac{-9+\sqrt{33}}{4},\quad x=\frac{-9-\sqrt{33}}{4}$$

18. $0.01x^2+0.1x+0.34=0$

$100(0.01x^2+0.1x+0.34)=100(0)$

$x^2+10x+34=0$

$a=1 \quad b=10 \quad c=34$

$$x=\frac{-(10)\pm\sqrt{(10)^2-4(1)(34)}}{2(1)}$$

$$x=\frac{-10\pm\sqrt{100-136}}{2}=\frac{-10\pm\sqrt{-36}}{2}$$

$$x=\frac{-10\pm6i}{2}$$

$$x=\frac{-10+6i}{2}=-5+3i$$

$$x=\frac{-10-6i}{2}=-5-3i$$

19. $2x^2-x=2x-5$

$2x^2-3x+5=0$

$a=2 \quad b=-3 \quad c=5$

$b^2-4ac=(-3)^2-4(2)(5)=9-40=-31$

Two complex solutions (containing i).

20. $4x^2+9x-3=0$

$a=4 \quad b=9 \quad c=-3$

$b^2-4ac=(9)^2-4(4)(-3)=81+48=129$

Two real solutions.

21. $$\frac{2}{n-3}+\frac{1}{n-1}=-1$$

$$\frac{2}{n-3}\bullet(n-3)(n-1)+\frac{1}{n-1}\bullet(n-3)(n-1)=-1\bullet(n-3)(n-1)$$

$2(n-1)+1(n-3)=-1(n^2-4n+3)$

$2n-2+n-3=-n^2+4n-3$

$n^2-n-2=0$

$(n-2)(n+1)=0$

$n-2=0 \quad n+1=0$

$n=2 \quad n=-1$

Check $n=2$

$$\frac{2}{2-3}+\frac{1}{2-1}\overset{?}{=}-1$$

$-2+1\overset{?}{=}-1$

$-1=-1$ True

Check $n=-1$

$$\frac{2}{-1-3}+\frac{1}{-1-1}\overset{?}{=}-1$$

$$-\frac{1}{2}-\frac{1}{2}\overset{?}{=}-1$$

$-1=-1$ True

The solutions of the original equation are -1 and 2.

22. $x^4 - 2x^2 - 24 = 0$

Let $u = x^2$

$(x^2)^2 - 2(x^2) - 24 = 0$

$u^2 - 2u - 24 = 0$

$(u-6)(u+4) = 0$

$u - 6 = 0 \quad u + 4 = 0$

$u = 6 \quad u = -4$

$x^2 = 6 \quad x^2 = -4$

$x = \pm\sqrt{6} \quad x = \pm\sqrt{-4}$

$x = \pm 2i$

Check $x = \sqrt{6}$

$(\sqrt{6})^4 - 2(\sqrt{6})^2 - 24 \stackrel{?}{=} 0$

$36 - 12 - 24 \stackrel{?}{=} 0$

$0 = 0$ True

Check $x = -\sqrt{6}$

$(-\sqrt{6})^4 - 2(-\sqrt{6})^2 - 24 \stackrel{?}{=} 0$

$36 - 12 - 24 \stackrel{?}{=} 0$

$0 = 0$ True

Check $x = 2i$

$(2i)^4 - 2(2i)^2 - 24 \stackrel{?}{=} 0$

$16 + 8 - 24 \stackrel{?}{=} 0$

$0 = 0$ True

Check $x = -2i$

$(-2i)^4 - 2(-2i)^2 - 24 \stackrel{?}{=} 0$

$16 + 8 - 24 \stackrel{?}{=} 0$

$0 = 0$ True

The solutions of the original equation are $\pm 2i$ and $\pm\sqrt{6}$.

23. $x - 7\sqrt{x} + 12 = 0$

Let $u = \sqrt{x}$

$(\sqrt{x})^2 - 7\sqrt{x} + 12 = 0$

$u^2 - 7u + 12 = 0$

$(u-4)(u-3) = 0$

$u - 4 = 0 \quad u - 3 = 0$

$u = 4 \quad u = 3$

$\sqrt{x} = 4 \quad \sqrt{x} = 3$

$(\sqrt{x})^2 = 4^2 \quad (\sqrt{x})^2 = (3)^2$

$x = 16 \quad x = 9$

Check $x = 16$

$(16) - 7\sqrt{16} + 12 \stackrel{?}{=} 0$

$16 - 28 + 12 \stackrel{?}{=} 0$

$0 = 0$ True

Check $x = 9$

$(9) - 7\sqrt{9} + 12 \stackrel{?}{=} 0$

$9 - 21 + 12 \stackrel{?}{=} 0$

$0 = 0$ True

The solutions of the original equation are 9 and 16.

24. $(p-1)^2 + 3(p-1) + 2 = 0$

Let $u = (p-1)$

$u^2 + 3u + 2 = 0$

$(u+1)(u+2) = 0$

$u + 1 = 0 \quad u + 2 = 0$

$u = -1 \quad u = -2$

$p - 1 = -1 \quad p - 1 = -2$

$p = 0 \quad p = -1$

Check $p = 0$

$((0)-1)^2 + 3((0)-1) + 2 \stackrel{?}{=} 0$

$1 - 3 + 2 \stackrel{?}{=} 0$

$0 = 0$ True

Check $p = -1$

$((-1)-1)^2 + 3((-1)-1) + 2 \stackrel{?}{=} 0$

$4 - 6 + 2 \stackrel{?}{=} 0$

$0 = 0$ True

The solutions of the original equation are -1 and 0.

25. $x = \dfrac{5}{2} \quad x = -3$

$2x = 5 \quad x + 3 = 0$

$2x - 5 = 0$

$(2x-5)(x+3) = 0$

$2x^2 + x - 15 = 0$

26. $n = -7$

$n + 7 = 0$

$(n+7)^2 = 0^2$

$n^2 + 14n + 49 = 0$

27. $f(x) = x^2 - 6x + 5$

$-\frac{b}{2a} = -\frac{-6}{2(1)} = 3$

$f(3) = (3)^2 - 6(3) + 5 = -4$

Vertex: $(3, -4)$

Axis of symmetry: $x = 3$

$x^2 - 6x + 5 = 0$

$(x-5)(x-1) = 0$

$x - 5 = 0 \quad x - 1 = 0$

$x = 5 \qquad x = 1$

x-intercepts: $(1, 0)$ and $(5, 0)$

$f(0) = (0)^2 - 6(0) + 5 = 5$

y-intercept: $(0, 5)$

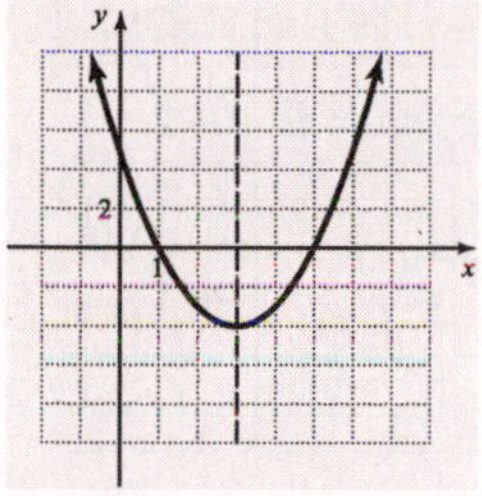

28. $f(x) = 3 + 2x - x^2$

$-\frac{b}{2a} = -\frac{2}{2(-1)} = 1$

$f(1) = 3 + 2(1) - (1)^2 = 4$

Vertex: $(1, 4)$

Axis of symmetry: $x = 1$

$3 + 2x - x^2 = 0$

$(3 - x)(1 + x) = 0$

$3 - x = 0 \quad 1 + x = 0$

$x = 3 \qquad x = -1$

x-intercepts: $(-1, 0)$ and $(3, 0)$

$f(0) = 3 + 2(0) - (0)^2 = 3$

y-intercept: $(0, 3)$

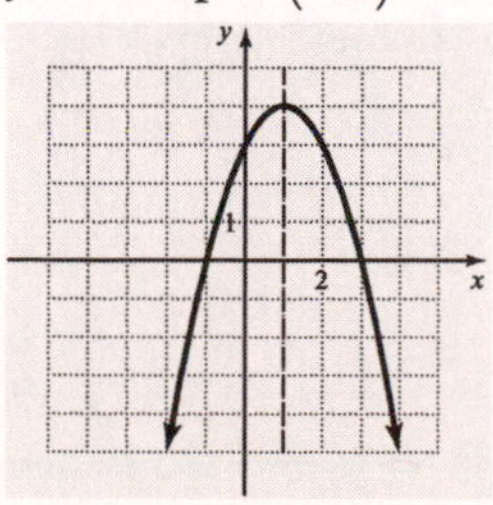

29. $f(x) = x^2 + 8x + 16$

$-\frac{b}{2a} = -\frac{8}{2(1)} = -4$

$f(-4) = (-4)^2 + 8(-4) + 16 = 0$

Vertex: $(-4, 0)$

Axis of symmetry: $x = -4$

$x^2 + 8x + 16 = 0$

$(x + 4)^2 = 0$

$x + 4 = 0$

$x = -4$

x-intercept: $(-4, 0)$

$f(0) = (0)^2 + 8(0) + 16 = 16$

y-intercept: $(0, 16)$

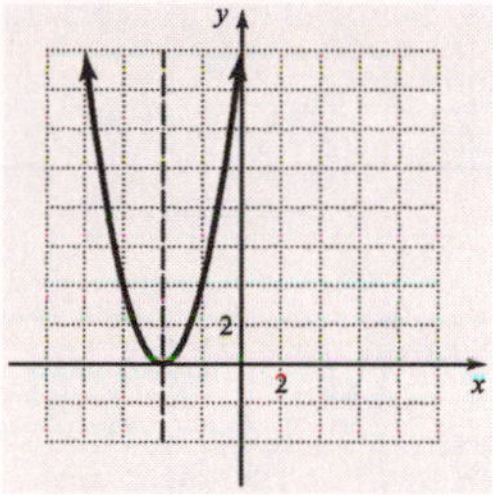

30. $f(x) = x^2 - 5x - 6$

$-\frac{b}{2a} = -\frac{-5}{2(1)} = \frac{5}{2}$

$f\left(\frac{5}{2}\right) = \left(\frac{5}{2}\right)^2 - 5\left(\frac{5}{2}\right) - 6 = -\frac{49}{4}$

Vertex: $\left(\frac{5}{2}, -\frac{49}{4}\right)$

Axis of symmetry: $x = \frac{5}{2}$

$x^2 - 5x - 6 = 0$

$(x + 1)(x - 6) = 0$

$x + 1 = 0 \quad x - 6 = 0$

$x = -1 \qquad x = 6$

x-intercepts: $(-1, 0)$ and $(6, 0)$

$f(0) = (0)^2 - 5(0) - 6 = -6$

y-intercept: $(0, -6)$

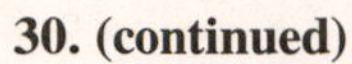

30. (continued)

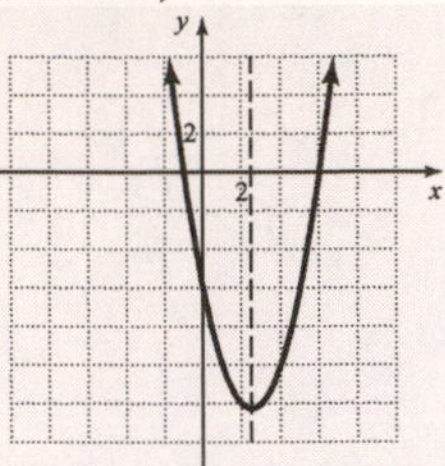

31.

Function	Opens	Max/Min
$f(x)=x^2+9$	$a>0$ Upward	$a>0$ Minimum

Number of x-intercepts	Number of y-intercepts
$x^2+9=0$ $x^2=-9$ No real solutions 0	$f(0)=9$ 1

32.

Function	Opens	Max/Min
$f(x)=1+3x-2x^2$	$a<0$ Downward	$a<0$ Maximum

Number of x-intercepts	Number of y-intercepts
$b^2-4ac=$ $3^2-4(-2)(1)=17$ Two real solutions 2	$f(0)=$ $1+3(0)-2(0)^2=1$ 1

33.

x	$f(x)=3x-4x^2$	(x,y)
-1	$3(-1)-4(-1)^2=-7$	$(-1,-7)$
-0.5	$3(-0.5)-4(-0.5)^2$ $=-2.5$	$(-0.5,-2.5)$
-0.25	$3(-0.25)-4(-0.25)^2$ $=-1$	$(-0.25,-1)$
0	$3(0)-4(0)^2=0$	$(0,0)$
0.25	$3(0.25)-4(0.25)^2=0.5$	$(0.25,0.5)$
0.5	$3(0.5)-4(0.5)^2=0.5$	$(0.5,0.5)$
1	$3(1)-4(1)^2=-1$	$(1,-1)$

$$-\frac{b}{2a}=-\frac{3}{2(-4)}=\frac{3}{8}$$

$$f\left(\frac{3}{8}\right)=3\left(\frac{3}{8}\right)-4\left(\frac{3}{8}\right)^2=\frac{9}{16}$$

Vertex: $\left(\frac{3}{8},\frac{9}{16}\right)$

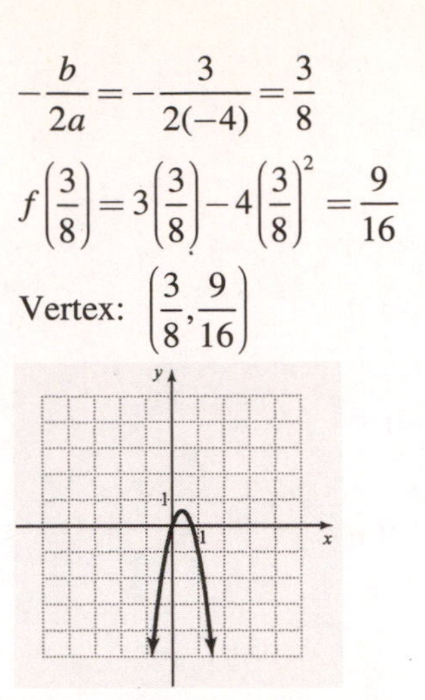

Domain: $(-\infty,\infty)$ Range: $\left(-\infty.\frac{9}{16}\right]$

34.

x	$f(x)=2x^2-x-1$	(x,y)
-2	$2(-2)^2-(-2)-1=9$	$(-2,9)$
-1	$2(-1)^2-(-1)-1=2$	$(-1,2)$
0	$2(0)^2-(0)-1=-1$	$(0,-1)$
1	$2(1)^2-(1)-1=0$	$(1,0)$
2	$2(2)^2-(2)-1=5$	$(2,5)$

$$-\frac{b}{2a}=-\frac{-1}{2(2)}=\frac{1}{4}$$

$$f\left(\frac{1}{4}\right)=2\left(\frac{1}{4}\right)^2-\left(\frac{1}{4}\right)-1=-\frac{9}{8}$$

Vertex: $\left(\frac{1}{4},-\frac{9}{8}\right)$

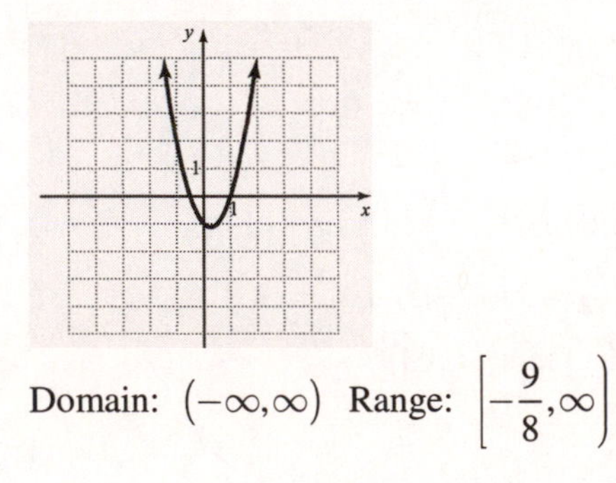

Domain: $(-\infty,\infty)$ Range: $\left[-\frac{9}{8},\infty\right)$

35. $x^2 - 4x < 12$

$x^2 - 4x = 12$

$x^2 - 4x - 12 = 0$

$(x+2)(x-6) = 0$

$x+2=0 \quad x-6=0$

$x=-2 \qquad x=6$

Interval	Test Value	$x^2 - 4x < 12$	Conclusion
$x<-2$	-3	$(-3)^2 - 4(-3) = 21$	$21 \geq 12$
$-2<x<6$	0	$(0)^2 - 4(0) = 0$	$0 < 12$
$6<x$	7	$(7)^2 - 4(7) = 21$	$21 \geq 12$

$(-2, 6)$

−6 −5 −4 −3 −2 −1 0 1 2 3 4 5 6

36. $10 + 3x - x^2 \leq 0$

$10 + 3x - x^2 = 0$

$(5-x)(2+x) = 0$

$5-x=0 \quad 2+x=0$

$x=5 \qquad x=-2$

Interval	Test Value	$10 + 3x - x^2 \leq 0$	Conclusion
$x \leq -2$	-3	$10+3(-3)-(-3)^2 = -8$	$-8 \leq 0$
$-2<x<5$	0	$10+3(0)-(0)^2 = 10$	$0<10$
$5 \leq x$	6	$10+3(6)-(6)^2 = -8$	$-8 \leq 0$

$(-\infty, -2] \cup [5, \infty)$

−6 −5 −4 −3 −2 −1 0 1 2 3 4 5 6

37. $2x^2 - 9x - 4 \geq -8$

$2x^2 - 9x - 4 = -8$

$2x^2 - 9x + 4 = 0$

$(2x-1)(x-4) = 0$

$2x-1=0 \quad x-4=0$

$2x=1 \qquad x=4$

$x = \frac{1}{2}$

Interval	Test Value	$2x^2 - 9x - 4 \geq -8$	Conclusion
$x < \frac{1}{2}$	0	$2(0)^2 - 9(0) - 4 = -4$	$-4 \geq -8$
$\frac{1}{2} < x < 4$	2	$2(2)^2 - 9(2) - 4 = -14$	$-14 < -8$
$4 < x$	5	$2(5)^2 - 9(5) - 4 = 1$	$1 \geq -8$

$\left(-\infty, \frac{1}{2}\right] \cup [4, \infty)$

−2 −1 0 1 2 3 4 5 6 7 8 9 10

38. $3 > 4x^2 - 4x$

$4x^2 - 4x - 3 = 0$

$(2x+1)(2x-3) = 0$

$2x+1=0 \quad 2x-3=0$

$2x=-1 \qquad 2x=3$

$x=-\frac{1}{2} \qquad x=\frac{3}{2}$

Interval	Test Value	$3 > 4x^2 - 4x$	Conclusion
$x \leq -\frac{1}{2}$	-1	$4(-1)^2 - 4(-1) = 8$	$3 \leq 8$
$-\frac{1}{2} < x < \frac{3}{2}$	0	$4(0)^2 - 4(0) = 0$	$3 > 0$
$\frac{3}{2} \leq x$	3	$4(3)^2 - 4(3) = 24$	$3 \leq 24$

$\left(-\frac{1}{2}, \frac{3}{2}\right)$

−6 −5 −4 −3 −2 −1 0 1 2 3 4 5 6

39. $\frac{x+3}{x-5} > -3$

$\frac{x+3}{x-5}$ is undefined when $x = 5$

$\frac{x+3}{x-5} = -3$

$x+3 = -3(x-5)$

$x+3 = -3x+15$

$4x = 12 \Rightarrow x = 3$

Interval	Test Value	$\frac{x+3}{x-5} > -3$	Conclusion
$x < 3$	0	$\frac{0+3}{0-5} = -\frac{3}{5}$	$-\frac{3}{5} > -3$
$3 < x < 5$	4	$\frac{4+3}{4-5} = -7$	$-7 \le -3$
$5 < x$	6	$\frac{6+3}{6-5} = 9$	$9 > -3$

$(-\infty, 3) \cup (5, \infty)$

−1 0 1 2 3 4 5 6 7 8 9 10 11

40. $\frac{4x-12}{3x} \le 0$

$\frac{4x-12}{3x}$ is undefined when $x = 0$

$\frac{4x-12}{3x} = 0$

$4x-12 = 0$

$4x = 12$

$x = 3$

Interval	Test Value	$\frac{4x-12}{3x} \le 0$	Conclusion
$x < 0$	−1	$\frac{4(-1)-12}{3(-1)} = \frac{16}{3}$	$\frac{16}{3} > 0$
$0 < x < 3$	2	$\frac{4(2)-12}{3(2)} = -\frac{2}{3}$	$-\frac{2}{3} \le 0$
$3 < x$	6	$\frac{4(6)-12}{3(6)} = \frac{2}{3}$	$\frac{2}{3} > 0$

$(0, 3]$

−6 −5 −4 −3 −2 −1 0 1 2 3 4 5 6

41. $s = 16t^2$

$400 = 16t^2$

$25 = t^2$

$\sqrt{t^2} = \pm\sqrt{25}$

$t = \pm 5$

$t = 5 \quad t = -5$

The negative answer makes no sense in this problem. It will take 5 sec for the object to fall 400 ft.

42. $(3d)^2 + d^2 = 10^2$

$9d^2 + d^2 = 100$

$10d^2 = 100$

$d^2 = 10$

$\sqrt{d^2} = \pm\sqrt{10}$

$d = \sqrt{10}$ $\quad$ d is a distance and must be non-negative

$3d = 3\sqrt{10} = 9.4868...$

The ladder reaches approximately 9.5 ft up the side of the house.

43. Let r be the rate of the train traveling west in miles per hour. Then $r + 8$ = the rate of the train traveling south in miles per hour. The trains travel for 30 minutes or $\frac{1}{2}$ hour.

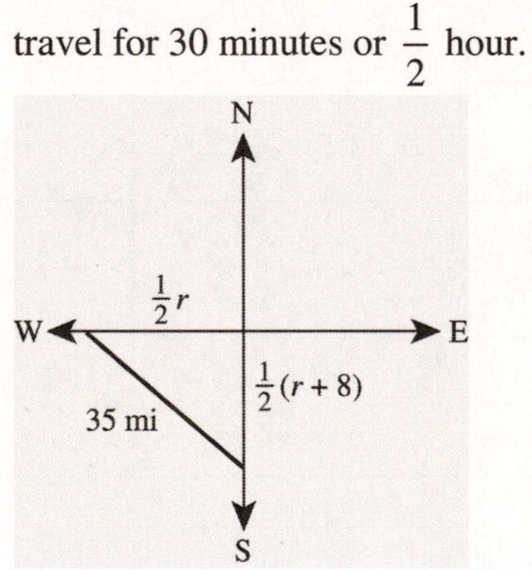

Using the formula $rt = d$ and the Pythagorean theorem, we have

$$\left(\frac{1}{2}r\right)^2 + \left(\frac{1}{2}(r+8)\right)^2 = 35^2$$

$$\frac{1}{4}r^2 + \left(\frac{1}{2}r + 4\right)^2 = 1225$$

$$\frac{1}{4}r^2 + \frac{1}{4}r^2 + 4r + 16 = 1225$$

$$\frac{1}{2}r^2 + 4r - 1209 = 0$$

43. (continued)

$a = \frac{1}{2} \qquad b = 4 \qquad c = -1209$

$$r = \frac{-(4) \pm \sqrt{(4)^2 - 4(0.5)(-1209)}}{2\left(\frac{1}{2}\right)}$$

$$r = \frac{-4 \pm \sqrt{16 + 2418}}{1} = -4 \pm \sqrt{2434}$$

$r = -4 \pm 49.3355...$

$r = 45.3355... \qquad r = -53.3355...$

$r + 8 = 53.3355... \qquad r + 8 = -45.3355...$

The negative answer makes no sense in this problem. The speed of the train traveling south is about 53 mph, and the speed of the train traveling west is about 45 mph.

44. $A = 5000\left[(1 + r)^2 - 1\right]$

$1100 = 5000\left[(1 + r)^2 - 1\right]$

$\frac{1100}{5000} = (1 + r)^2 - 1$

$\frac{11}{50} = 1 + 2r + r^2 - 1$

$r^2 + 2r - \frac{11}{50} = 0$

$50\left(r^2 + 2r - \frac{11}{50}\right) = 50(0)$

$50r^2 + 100r - 11 = 0$

$a = 50 \qquad b = 100 \qquad c = -11$

$$r = \frac{-(100) \pm \sqrt{(100)^2 - 4(50)(-11)}}{2(50)}$$

$$r = \frac{-100 \pm \sqrt{10000 + 2200}}{100}$$

$$r = \frac{-100 \pm \sqrt{12200}}{100} = \frac{-100 \pm 110.4536...}{100}$$

$r = -1 \pm 1.1045...$

$r = -2.1045... \qquad r = 0.1045...$

The negative answer makes no sense in this problem. The average rate of return was about 10.5%.

45. The length of the base will be $40 - 2x$ and the width will be $25 - 2x$. Area is length $\times$ width.

$A = (40 - 2x)(25 - 2x) = 700$

$1000 - 130x + 4x^2 = 700$

$4x^2 - 130x + 300 = 0$

$2(2x^2 - 65x + 150) = 0$

$2x^2 - 65x + 150 = 0$

$a = 2 \qquad b = -65 \qquad c = 150$

$$x = \frac{-(-65) \pm \sqrt{(-65)^2 - 4(2)(150)}}{2(2)}$$

$$x = \frac{65 \pm \sqrt{4225 - 1200}}{4} = \frac{65 \pm \sqrt{3025}}{4}$$

$$x = \frac{65 \pm 55}{4}$$

$x = \frac{120}{4} = 30, \quad x = \frac{10}{4} = \frac{5}{2}$

Two thirty inch squares cannot be cut from the original piece of cardboard. $\frac{5}{2}$ in. by $\frac{5}{2}$ in. squares should be cut from each corner.

46. $R(x) = 100x - 2x^2 = 1250$

$2x^2 - 100x + 1250 = 0$

$2(x^2 - 50x + 625) = 0$

$x^2 - 50x + 625 = 0$

$(x - 25)^2 = 0$

$x - 25 = 0$

$x = 25$

He needs to sell 25 cases per week.

47. $S = 7n^2 + 8n + 131 = 218$

$7n^2 + 8n - 87 = 0$

$a = 7 \qquad b = 8 \qquad c = -87$

$$n = \frac{-(8) \pm \sqrt{(8)^2 - 4(7)(-87)}}{2(7)}$$

$$n = \frac{-8 \pm \sqrt{64 + 2436}}{14} = \frac{-8 \pm \sqrt{2500}}{14}$$

$$n = \frac{-8 \pm 50}{14}$$

$n = \frac{-58}{14} = -\frac{29}{7}, \quad n = \frac{42}{14} = 3$

The negative answer makes no sense in this problem. 1999 + 3 = 2002. There were 218 stores in 2002.

48. $A = 0.009x^2 + 0.05x + 0.6 = 1.2$

$0.009x^2 + 0.05x - 0.6 = 0$

$a = 0.009 \qquad b = 0.05 \qquad c = -0.6$

$$x = \frac{-(0.05) \pm \sqrt{(0.05)^2 - 4(0.009)(-0.6)}}{2(0.009)}$$

$$x = \frac{-0.05 \pm \sqrt{0.0025 + 0.0216}}{0.018}$$

$$x = \frac{-0.05 \pm \sqrt{0.0241}}{0.018} = \frac{-0.05 \pm 0.15524...}{0.018}$$

$$x = \frac{0.10524...}{0.018} = 5.846...$$

$$x = \frac{-0.20524...}{0.018} = -11.402....$$

The negative answer makes no sense in this problem. 1995 + 6 = 2001. There were approximately 1.2 million sales associates in 2001.

49. 6 min is $\frac{6}{60} = \frac{1}{10}$ hr. time $= \frac{\text{distance}}{\text{rate}}$

$$\frac{8}{r} = \frac{8}{r+2} + \frac{1}{10}$$

$$\frac{8}{r} \bullet 10r(r+2) = \frac{8}{r+2} \bullet 10r(r+2) + \frac{1}{10} \bullet 10r(r+2)$$

$80(r+2) = 80r + r(r+2)$

$80r + 160 = 80r + r^2 + 2r$

$r^2 + 2r - 160 = 0$

$a = 1 \qquad b = 2 \qquad c = -160$

$$r = \frac{-(2) \pm \sqrt{(2)^2 - 4(1)(-160)}}{2(1)}$$

$$r = \frac{-2 \pm \sqrt{4 + 640}}{2} = \frac{-2 \pm \sqrt{644}}{2}$$

$$r = \frac{-2 \pm 25.3771...}{2}$$

$$r = \frac{23.3771...}{2} = 11.6885...$$

$$r = \frac{-27.3771...}{2} = -13.6885...$$

The negative answer makes no sense in this problem. $r + 2 = 13.6885$. She bicycled to school at a rate of 13.7 mph.

50. Let t be the time it takes the newer printing press to complete the job alone. In one hour the newer printer completes $\frac{1}{t}$ of the job, the older printer completes $\frac{1}{t+5}$ and together they complete $\frac{1}{2}$ of the job.

$$\frac{1}{t+5} + \frac{1}{t} = \frac{1}{2}$$

$$\frac{1}{t+5} \bullet (t+5)(t)(2) + \frac{1}{t} \bullet (t+5)(t)(2) = \frac{1}{2} \bullet (t+5)(t)(2)$$

$2t + 2(t+5) = t(t+5)$

$2t + 2t + 10 = t^2 + 5t$

$t^2 + t - 10 = 0$

$a = 1 \qquad b = 1 \qquad c = -10$

$$t = \frac{-(1) \pm \sqrt{(1)^2 - 4(1)(-10)}}{2(1)}$$

$$t = \frac{-1 \pm \sqrt{1 + 40}}{2} = \frac{-1 \pm \sqrt{41}}{2}$$

$$t = \frac{-1 \pm 6.4031...}{2}$$

$$t = \frac{-7.4031}{2} = -3.701... \quad t + 5 = 1.298...$$

$$t = \frac{5.4031...}{2} = 2.701.. \quad t + 5 = 7.701...$$

The negative answer makes no sense in this problem. Working alone, it takes the newer printing press about 2.7 hr and the older printing press about 7.7 hr to complete the job.

51. **a.** perimeter $= 2l + 2w = 300$

$2w = 300 - 2l$

$$w = \frac{300 - 2l}{2} = 150 - l$$

b. Area $= lw$

$A(l) = l(150 - l) = 150l - l^2$

51. c.

l	$A(l)=150l-l^2$	(l,A)
25	$150(25)-(25)^2=3125$	$(25,3125)$
50	$150(50)-(50)^2=5000$	$(50,5000)$
75	$150(75)-(75)^2=5625$	$(75,5625)$
100	$150(100)-(100)^2=5000$	$(100,5000)$
125	$150(125)-(125)^2=3125$	$(125,3125)$

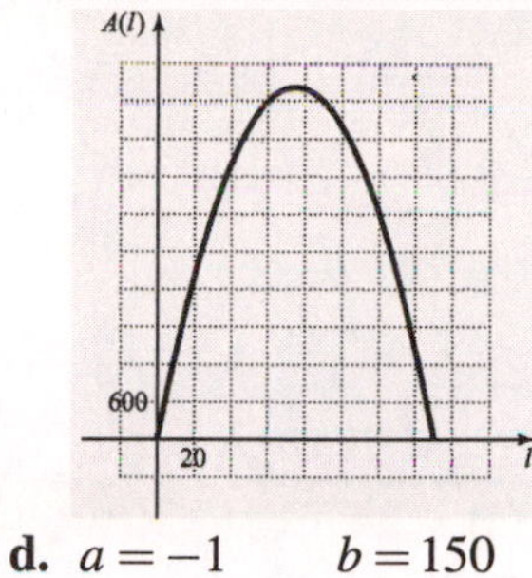

d. $a=-1 \qquad b=150$

Vertex is where $l=-\frac{b}{2a}=-\frac{150}{2(-1)}=75$.

$w=150-l=150-75=75$

$A(75)=150l-l^2=150(75)-(75)^2$

$A(75)=11250-5625=5625$

A park measuring 75 ft by 75 ft will produce a maximum area of 5625 sq ft.

52. a. $C(x)=0.05x^2-2x+100$

x	$C(x)=0.05x^2-2x+100$	(x,y)
0	$0.05(0)^2-2(0)+100=100$	$(0,100)$
10	$0.05(10)^2-2(10)+100=85$	$(10,85)$
20	$0.05(20)^2-2(20)+100=80$	$(20,80)$
30	$0.05(30)^2-2(30)+100=85$	$(30,85)$
40	$0.05(40)^2-2(40)+100=100$	$(40,100)$

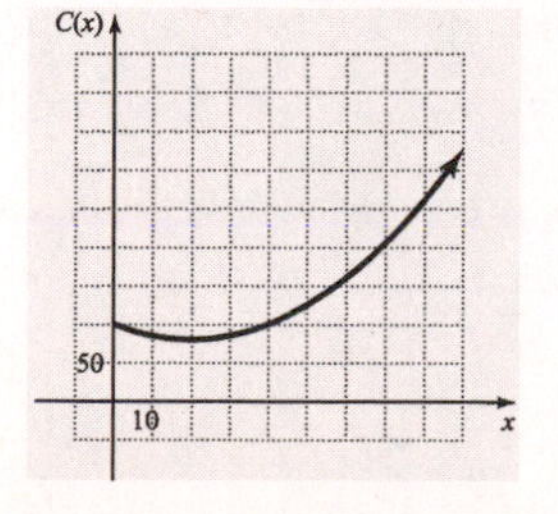

b. Vertex is where $x=-\frac{b}{2a}$

$a=0.05 \quad b=-2$

$x=-\frac{b}{2a}=-\frac{-2}{2(0.05)}=20$

20 units must be fabricated per day in order to minimize the cost.

c. $C(20)=0.05(20)^2-2(20)+100$

$C(20)=0.05(400)-40+100$

$C(20)=20-40+100=80$

The minimum cost is $80.

53. $h(t)=-16t^2+80t+4$

$40=-16t^2+80t+4$

$16t^2-80t+36=0$

$4(4t^2-20t+9)=0$

$4t^2-20t+9=0$

$(2t-1)(2t-9)=0$

$2t-1=0 \quad 2t-9=0$

$2t=1 \qquad 2t=9$

$t=\frac{1}{2} \qquad t=\frac{9}{2}$

The baseball is 40 ft above the ground at $\frac{1}{2}$ sec and $\frac{9}{2}$ sec.

54. Perimeter is $2l+2w=44$

$2l=44-2w$

$l=22-w$

Area $=lw$, $A=(22-w)w$

$120\le 22w-w^2$

$120=22w-w^2$

$w^2-22w+120=0$

$(w-12)(w-10)=0$

$w-12=0 \quad w-10=0$

$w=12 \qquad w=10$

Interval	Test	$120\le 22w-w^2$	Conclusion
$x<10$	9	$22(9)-(9)^2$ $=117$	$120>117$
$10\le x\le 12$	11	$22(11)-(11)^2$ $=121$	$120\le 121$
$12<x$	13	$22(13)-(13)^2$ $=117$	$120>117$

54. (continued)
Any width between and including 10 ft and12 ft will produce an area of at least 120 sq ft.

Chapter 9 Posttest

1.
$$5n^2-11=29$$
$$5n^2=40$$
$$n^2=8$$
$$\sqrt{n^2}=\sqrt{8}$$
$$n=\pm 2\sqrt{2}$$
$$n=2\sqrt{2},\quad n=-2\sqrt{2}$$

2.
$$3p^2-6p=-24$$
$$3\left(p^2-2p\right)=-24$$
$$p^2-2p=-8$$
$$p^2-2p+\left(\frac{-2}{2}\right)^2=-8+\left(\frac{-2}{2}\right)^2$$
$$p^2-2p+1=-8+1$$
$$(p-1)^2=-7$$
$$p-1=\pm\sqrt{-7}$$
$$p=1\pm i\sqrt{7}$$
$$p=1+i\sqrt{7}\quad p=1-i\sqrt{7}$$

3.
$$4x^2+4x-3=0$$
$$a=4\qquad b=4\qquad c=-3$$
$$x=\frac{-(4)\pm\sqrt{(4)^2-4(4)(-3)}}{2(4)}$$
$$x=\frac{-4\pm\sqrt{16+48}}{8}=\frac{-4\pm\sqrt{64}}{8}=\frac{-4\pm 8}{8}$$
$$x=\frac{-4+8}{8}=\frac{4}{8}=\frac{1}{2}$$
$$x=\frac{-4-8}{8}=\frac{-12}{8}=-\frac{3}{2}$$

4.
$$2x^2+7x+9=0$$
$$a=2\quad b=7\quad c=9$$
$$b^2-4ac=(7)^2-4(2)(9)=-23$$
Two complex solutions (containing i).

5.
$$5(3n+2)^2-90=0$$
$$5(3n+2)^2=90$$
$$(3n+2)^2=18$$
$$3n+2=\pm\sqrt{18}=\pm 3\sqrt{2}$$
$$3n=-2\pm 3\sqrt{2}$$
$$n=\frac{-2\pm 3\sqrt{2}}{3}$$
$$n=\frac{-2+3\sqrt{2}}{3},\quad n=\frac{-2-3\sqrt{2}}{3}$$

6.
$$x^2+8x=6$$
$$x^2+8x-6=0$$
$$a=1\qquad b=8\qquad c=-6$$
$$x=\frac{-(8)\pm\sqrt{(8)^2-4(1)(-6)}}{2(1)}$$
$$x=\frac{-8\pm\sqrt{64+24}}{2}=\frac{-8\pm\sqrt{88}}{2}$$
$$x=\frac{-8\pm 2\sqrt{22}}{2}=-4\pm\sqrt{22}$$
$$x=-4+\sqrt{22},\quad x=-4-\sqrt{22}$$

7.
$$x^2-x-2=4x-13$$
$$x^2-5x+11=0$$
$$a=1\qquad b=-5\qquad c=11$$
$$x=\frac{-(-5)\pm\sqrt{(-5)^2-4(1)(11)}}{2(1)}$$
$$x=\frac{5\pm\sqrt{25-44}}{2}=\frac{5\pm\sqrt{-19}}{2}$$
$$x=\frac{5+i\sqrt{19}}{2},\quad x=\frac{5-i\sqrt{19}}{2}$$

8.
$$4x^2+3x=7+6x$$
$$4x^2-3x-7=0$$
$$(x+1)(4x-7)=0$$
$$x+1=0\quad 4x-7=0$$
$$x=-1\qquad 4x=7$$
$$x=\frac{7}{4}$$

9. $2x^2 - 12x + 20 = 1$

$2x^2 - 12x + 19 = 0$

$a = 2 \quad b = -12 \quad c = 19$

$$x = \frac{-(-12) \pm \sqrt{(-12)^2 - 4(2)(19)}}{2(2)}$$

$$x = \frac{12 \pm \sqrt{144 - 152}}{4} = \frac{12 \pm \sqrt{-8}}{4}$$

$$x = \frac{12 \pm 2i\sqrt{2}}{4} = \frac{2(6 \pm i\sqrt{2})}{4}$$

$$x = \frac{6 + i\sqrt{2}}{2}, \quad x = \frac{6 - i\sqrt{2}}{2}$$

10. $\frac{1}{2}x^2 + x - 2 = 0$

$a = \frac{1}{2} \quad b = 1 \quad c = -2$

$$x = \frac{-(1) \pm \sqrt{(1)^2 - 4\left(\frac{1}{2}\right)(-2)}}{2\left(\frac{1}{2}\right)}$$

$$x = \frac{-1 \pm \sqrt{1+4}}{1} x = -1 \pm \sqrt{5}$$

$x = -1 + \sqrt{5}, \quad x = -1 - \sqrt{5}$

11. $x^{1/2} - 2x^{1/4} = -1$

$x^{1/2} - 2x^{1/4} + 1 = 0$

Let $u = x^{1/4}$

$(x^{1/4})^2 - 2(x^{1/4}) + 1 = 0$

$u^2 - 2u + 1 = 0$

$(u-1)^2 = 0$

$u - 1 = 0$

$u = 1$

$x^{1/4} = 1$

$(x^{1/4})^4 = 1^4$

$x = 1$

12. $n = \frac{1}{2} \quad n = \frac{2}{3}$

$2n = 1 \quad 3n = 2$

$2n - 1 = 0 \quad 3n - 2 = 0$

$(2n-1)(3n-2) = 0$

$6n^2 - 7n + 2 = 0$

13. $f(x) = x^2 - 6x + 8$

$$-\frac{b}{2a} = -\frac{-6}{2(1)} = 3$$

$f(3) = (3)^2 - 6(3) + 8 = -1$

Vertex: $(3, -1)$

Axis of symmetry: $x = 3$

$x^2 - 6x + 8 = 0$

$(x-2)(x-4) = 0$

$x - 2 = 0 \quad x - 4 = 0$

$x = 2 \quad x = 4$

x-intercepts: (2, 0) and (4, 0).

$f(0) = (0)^2 - 6(0) + 8 = 8$

y-intercept (0, 8).

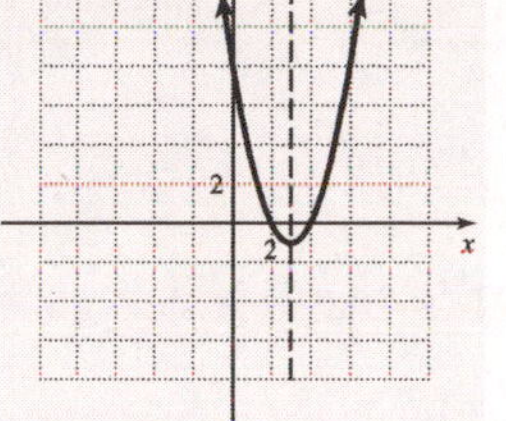

14. $f(x) = -x^2 + 3x + 10$

$$-\frac{b}{2a} = -\frac{3}{2(-1)} = \frac{3}{2}$$

$$f\left(\frac{3}{2}\right) = -\left(\frac{3}{2}\right)^2 + 3\left(\frac{3}{2}\right) + 10 = \frac{49}{4}$$

Vertex: $\left(\frac{3}{2}, \frac{49}{4}\right)$

Axis of symmetry: $x = \frac{3}{2}$

$-x^2 + 3x + 10 = 0$

$(-x-2)(x-5) = 0$

$-x - 2 = 0 \quad x - 5 = 0$

$x = -2 \quad x = 5$

x-intercepts: (–2, 0) and (5, 0).

$f(0) = -(0)^2 + 3(0) + 10 = 10$

y-intercept (0, 10)

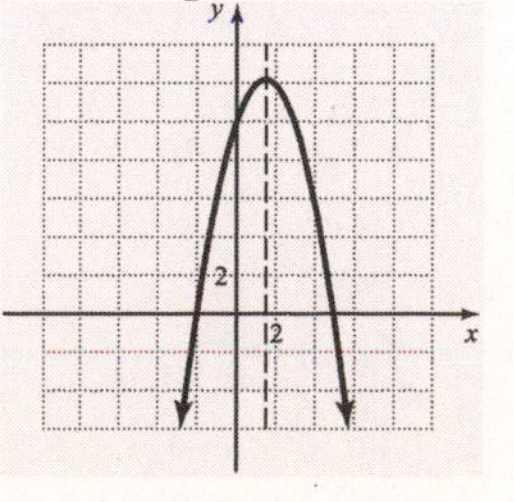

15.

x	$f(x)=2x^2-4x-1$	(x,y)
-1	$2(-1)^2-4(-1)-1=5$	$(-1,5)$
0	$2(0)^2-4(0)-1=-1$	$(0,-1)$
1	$2(1)^2-4(1)-1=-3$	$(1,-3)$
2	$2(2)^2-4(2)-1=-1$	$(2,-1)$
3	$2(3)^2-4(3)-1=5$	$(3,5)$

$$-\frac{b}{2a}=-\frac{-4}{2(2)}=\frac{4}{4}=1$$

$$f(1)=2(1)^2-4(1)-1=-3$$

Vertex: $(1,-3)$

Domain: $(-\infty,\infty)$ Range: $[-3,\infty)$

16. $-x^2-3x+18<0$

$$-x^2-3x+18=0$$
$$x^2+3x-18=0$$
$$(x+6)(x-3)=0$$
$$x+6=0 \quad x-3=0$$
$$x=-6 \quad x=3$$

Interval	Test	$-x^2-3x+18<0$	Conclusion
$x<-6$	-7	$-(-7)^2-3(-7)+18$ $=-10$	$-10<0$
$-6<x<3$	0	$-(0)^2-3(0)+18$ $=18$	$18>0$
$3<x$	4	$-(4)^2-3(4)+18$ $=-10$	$-10<0$

$(-\infty,-6)\cup(3,\infty)$

17. Let r be the average speed of the airplane flying south. Then $r + 20$ is the average speed of the airplane flying west. The planes traveled for one hour.

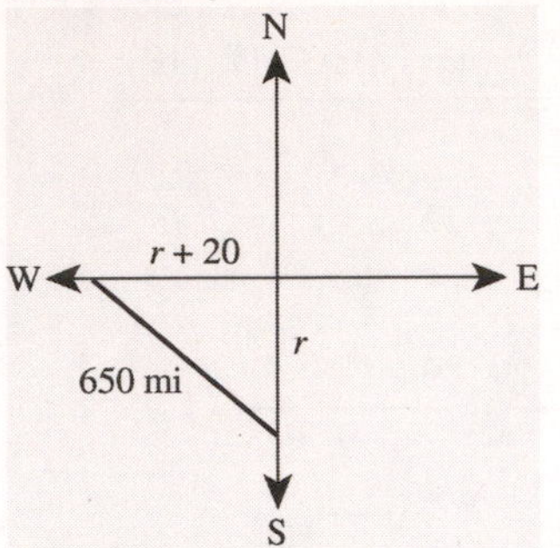

Using the formula $rt = d$ and the Pythagorean theorem, we have

$$(1(r+20))^2+(r)^2=650^2$$
$$r^2+40r+400+r^2=422,500$$
$$2r^2+40r-422,100=0$$
$$2(r^2+20r-211,050)=0$$
$$r^2+20r-211,050=0$$

$$a=1 \quad b=20 \quad c=-211,050$$

$$r=\frac{-(20)\pm\sqrt{(20)^2-4(1)(-211,050)}}{2(1)}$$

$$r=\frac{-20\pm\sqrt{400+844,200}}{2}=\frac{-20\pm\sqrt{844,600}}{2}$$

$$r=\frac{-20\pm 919.0212..}{2}$$

$$r=\frac{-20-919.0212..}{2}=r=\frac{-939.0212..}{2}$$

$$r=-469.51... \Rightarrow r+20=-449.51...$$

$$r=\frac{-20+919.0212...}{2}=\frac{899.0212...}{2}$$

$$r=449.510... \quad r+20=469.510...$$

The negative answer makes no sense in this problem. The average speed of the airplane flying west is approximately 470 mph and the average speed of the airplane flying south is approximately 450 mph.

18. Let t be the time it takes the senior clerk to process the applications alone. In one hour the senior clerk completes $\frac{1}{t}$ of the job, the junior clerk completes $\frac{1}{t+3}$ and together they complete $\frac{1}{2}$ of the job.

$$\frac{1}{t}+\frac{1}{t+3}=\frac{1}{2}$$

$$\frac{1}{t}\bullet 2t(t+3)+\frac{1}{t+3}\bullet 2t(t+3)=\frac{1}{2}\bullet 2t(t+3)$$

$$2(t+3)+2t=t(t+3)$$

$$2t+6+2t=t^2+3t$$

$$t^2-t-6=0$$

$$(t-3)(t+2)=0$$

$$t-3=0 \quad t+2=0$$

$$t=3 \qquad t=-2$$

$$t+3=6 \quad t+3=1$$

The negative answer makes no sense in this problem. Working alone, the senior clerk can process the applications in 3 hr and the junior clerk can process the applications in 6 hr.

19.

$h(t)=-16t^2+96t+3$

$a=-16 \quad b=96$

Vertex is where $t=-\dfrac{b}{2a}=-\dfrac{96}{2(-16)}=3$

$$h(3)=-16(3)^2+96(3)+3$$
$$=-16(9)+288+3=-144+288+3=147$$

The rocket will reach its maximum height of 147 ft in 3 sec.

20. $C(x)=0.025x^2-8x+800<320$

$0.025x^2-8x+800=320$

$0.025x^2-8x+480=0$

$a=0.025 \qquad b=-8 \qquad c=480$

$$x=\frac{-(-8)\pm\sqrt{(-8)^2-4(0.025)(480)}}{2(0.025)}$$

$$x=\frac{8\pm\sqrt{64-48}}{0.05}=\frac{8\pm\sqrt{16}}{0.05}=\frac{8\pm 4}{0.05}$$

$$x=\frac{8+4}{0.05}=\frac{12}{0.05}=240,\quad x=\frac{8-4}{0.05}=\frac{4}{0.05}=80$$

Interval	Test	$0.025x^2-8x+800<320$	Conclusion
$x<80$	60	$0.025(60)^2-8(60)+800$ $=410$	$410>320$
$80<x<240$	100	$0.025(100)^2-8(100)+800$ $=250$	$250<320$
$240<x$	260	$0.025(260)^2-8(260)+800$ $=410$	$410>320$

The factory can produce between, but not including, 80 and 240 units each day.

Chapter 10

EXPONENTIAL AND LOGARITHMIC FUNCTIONS

Pretest

1. $f(x)=2x-1 \quad g(x)=2x^2+5x-3$

a. $(f+g)(x)=2x-1+2x^2+5x-3$
$=2x^2+7x-4$

b. $(f-g)(x)=(2x-1)-(2x^2+5x-3)$
$=2x-1-2x^2-5x+3$
$=-2x^2-3x+2$

c. $(f\bullet g)(x)=(2x-1)(2x^2+5x-3)$
$=4x^3+10x^2-6x-2x^2-5x+3$
$=4x^3+8x^2-11x+3$

d. $\left(\frac{f}{g}\right)(x)=\frac{2x-1}{2x^2+5x-3}=\frac{2x-1}{(x+3)(2x-1)}$
$=\frac{1}{x+3},\ x\neq-3,\frac{1}{2}$

2. $f(x)=\frac{5}{x} \quad g(x)=3x-4$

a. $(f\circ g)(x)=\frac{5}{3x-4};\ x\neq\frac{4}{3}$

b. $(g\circ f)(x)=3\left(\frac{5}{x}\right)-4=\frac{15}{x}-4;\ x\neq0$

c. $(f\circ g)(3)=\frac{5}{3(3)-4}=\frac{5}{5}=1$

d. $(g\circ f)(-5)=\frac{15}{(-5)}-4=-3-4=-7$

3. Yes, it is a one-to-one function as each y-value in the range corresponds to exactly one x-value in the domain.

(x,y) in f	(x,y) in f^{-1}
$(0,4)$	$(4,0)$
$(1,2)$	$(2,1)$
$(3,-2)$	$(-2,3)$

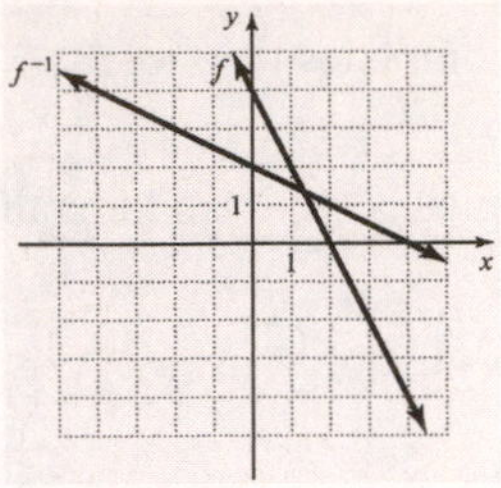

4. $f(x)=3x-7$

$y=3x-7$ Substitute y for $f(x)$
$x=3y-7$ Interchange x and y
$x+7=3y$
$y=\frac{x+7}{3}$
$f^{-1}(x)=\frac{x+7}{3}$

5. a. $f(x)=2^{x-5}$
$f(3)=2^{3-5}=2^{-2}=\frac{1}{2^2}=\frac{1}{4}$

b. $f(x)=-e^{-x}$
$f(-2)=-e^{-(-2)}=-e^2=-7.3890...=-7.389$

6. a. $\log_7(1)=\log_7(7^0)=0$

b. $\log_9\left(\frac{1}{81}\right)=\log_9\left(\frac{1}{9^2}\right)=\log_9(9^{-2})=-2$

7. a. $f(x)=2^x+1$

x	$f(x)=2^x+1$	(x,y)
-3	$2^{-3}+1=\frac{9}{8}$	$\left(-3,\frac{9}{8}\right)$
-2	$2^{-2}+1=\frac{5}{4}$	$\left(-2,\frac{5}{4}\right)$
-1	$2^{-1}+1=\frac{3}{2}$	$\left(-1,\frac{3}{2}\right)$
0	$2^0+1=2$	$(0,2)$
1	$2^1+1=3$	$(1,3)$
2	$2^2+1=5$	$(2,5)$
3	$2^3+1=9$	$(3,9)$

7. (continued)

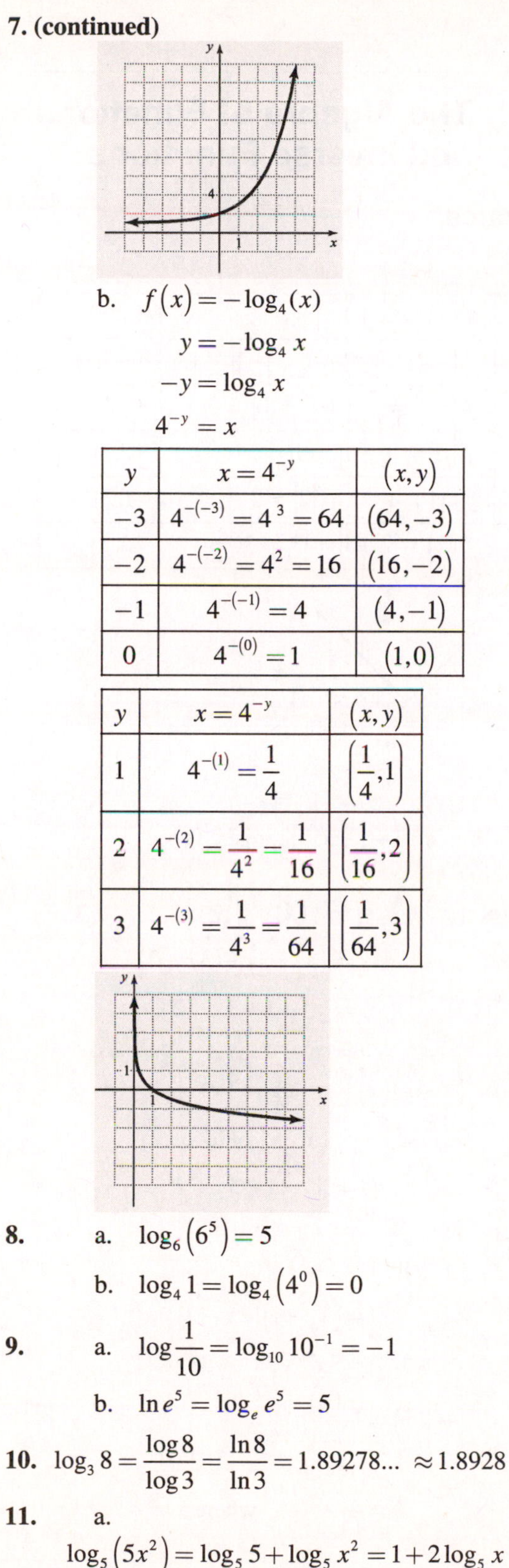

b. $f(x) = -\log_4(x)$

$y = -\log_4 x$

$-y = \log_4 x$

$4^{-y} = x$

y	$x = 4^{-y}$	(x, y)
-3	$4^{-(-3)} = 4^3 = 64$	$(64, -3)$
-2	$4^{-(-2)} = 4^2 = 16$	$(16, -2)$
-1	$4^{-(-1)} = 4$	$(4, -1)$
0	$4^{-(0)} = 1$	$(1, 0)$

y	$x = 4^{-y}$	(x, y)
1	$4^{-(1)} = \frac{1}{4}$	$\left(\frac{1}{4}, 1\right)$
2	$4^{-(2)} = \frac{1}{4^2} = \frac{1}{16}$	$\left(\frac{1}{16}, 2\right)$
3	$4^{-(3)} = \frac{1}{4^3} = \frac{1}{64}$	$\left(\frac{1}{64}, 3\right)$

8. a. $\log_6\left(6^5\right) = 5$

b. $\log_4 1 = \log_4\left(4^0\right) = 0$

9. a. $\log\frac{1}{10} = \log_{10} 10^{-1} = -1$

b. $\ln e^5 = \log_e e^5 = 5$

10. $\log_3 8 = \frac{\log 8}{\log 3} = \frac{\ln 8}{\ln 3} = 1.89278\ldots \approx 1.8928$

11. a.

$\log_5\left(5x^2\right) = \log_5 5 + \log_5 x^2 = 1 + 2\log_5 x$

b. $\log_8 \frac{3x^3}{y} = \log_8 3x^3 - \log_8 y$

$= \log_8 3 + \log_8 x^3 - \log_8 y$

$= \log_8 3 + 3\log_8 x - \log_8 y$

12. a. $3\log_7 2 + \log_7 5 = \log_7 2^3 + \log_7 5$

$= \log_7\left(2^3 \cdot 5\right) = \log_7 40$

b.

$4\log_6 x - 2\log_6(x+2)$

$= \log_6 x^4 - \log_6 (x+2)^2 = \log_6\left(\frac{x^4}{(x+2)^2}\right)$

13. $5^x = 25$

$5^x = 5^2$

$x = 2$

14. $4^{x+1} = 32^x$

$\left(2^2\right)^{x+1} = \left(2^5\right)^x$

$2^{2(x+1)} = 2^{5x}$

$2(x+1) = 5x$

$2x + 2 = 5x$

$-3x = -2$

$x = \frac{2}{3}$

15. $\log_x \frac{1}{2} = -1$

$\frac{1}{2} = x^{-1}$

$\frac{1}{2} = \frac{1}{x}$

$x = 2$

16. $\log_2(x-4) = 3$

$x - 4 = 2^3 = 8$

$x = 12$

17. $\log_2 x + \log_2(x+6) = 4$

$\log_2\left(x(x+6)\right) = 4$

$x(x+6) = 2^4$

$x^2 + 6x = 16$

$x^2 + 6x - 16 = 0$

$(x+8)(x-2) = 0$

$x + 8 = 0 \qquad x - 2 = 0$

$x = -8 \qquad x = 2$

17. (continued)

Check $x = -8$

$$\log_2(-8) + \log_2(-8+6) \stackrel{?}{=} 4$$

$\log_2 x$ is only defined for $x > 0$

-8 is not a solution.

Check for $x = 2$

$$\log_2 2 + \log_2(2+6) \stackrel{?}{=} 4$$

$$1 + \log_2 8 \stackrel{?}{=} 4$$

$$1 + \log_2 2^3 \stackrel{?}{=} 4$$

$$1 + 3 \stackrel{?}{=} 4; \quad 4 = 4 \text{ True}$$

The solution is 2.

18. $C(x) = P(1.04)^x$

$$C(x) = 3.89(1.04)^5 = 3.89(1.21665...)$$

$$C(x) = 4.7327....$$

A gallon of milk will cost about $4.73 in 5 yr.

19. $\ln(r+1) = k$

$$\ln(r+1) = 0.045$$

$$r + 1 = e^{0.045} = 1.04602...$$

$$r = 0.04602...$$

The effective annual interest rate is approximately 4.6%.

20. $C = C_0 e^{-0.21t}$

$$12 = 100e^{-0.21t}$$

$$\frac{12}{100} = e^{-0.21t}$$

$$0.12 = e^{-0.21t}$$

$$\ln(0.12) = -0.21t$$

$$\frac{\ln(0.12)}{-0.21} = t$$

$$t = \frac{-2.12026...}{-0.21} = 10.096...$$

It will take about 10.1 days for 100 mg of gallium-67 to decay to 12 mg.

10.1 The Algebra of Functions and Inverse Functions

Practice

1.

x	$f(x) = \|x+1\|$	(x, y)
-2	$f(-2) = \|-2+1\| = \|-1\| = 1$	$(-2,1)$
-1	$f(-1) = \|-1+1\| = \|0\| = 0$	$(-1,0)$
0	$f(0) = \|0+1\| = \|1\| = 1$	$(0,1)$
1	$f(1) = \|1+1\| = \|2\| = 2$	$(1,2)$
2	$f(2) = \|2+1\| = \|3\| = 3$	$(2,3)$

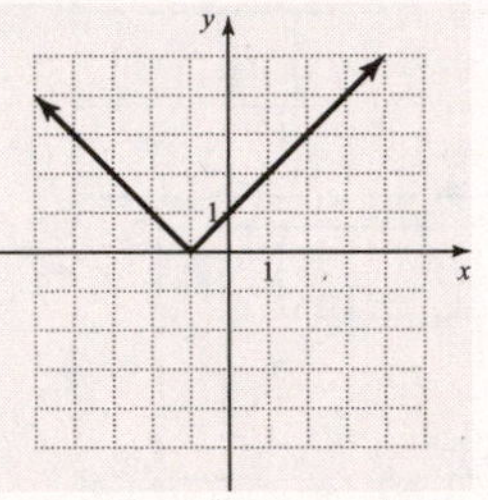

Domain: $(-\infty, \infty)$; range: $[0, \infty)$

2. **a.** $(f+g)(x) = f(x) + g(x)$

$$= (x^2 - 1) + (3x - 2)$$
$$= x^2 - 1 + 3x - 2$$
$$= x^2 + 3x - 3$$

b. $(f-g)(x) = f(x) - g(x)$

$$= (x^2 - 1) - (3x - 2)$$
$$= x^2 - 1 - 3x + 2$$
$$= x^2 - 3x + 1$$

c. $(f \cdot g)(x) = f(x) \cdot g(x)$

$$= (x^2 - 1)(3x - 2)$$
$$= 3x^3 - 2x^2 - 3x + 2$$

d. $\left(\frac{f}{g}\right)(x) = \frac{f(x)}{g(x)}$

$$= \frac{(x^2 - 1)}{(3x - 2)}, \text{ where } x \neq \frac{2}{3}$$

3. **a.**
$$(f+g)(-1) = f(-1)+g(-1)$$
$$= \left(-1-(-1)^2\right)+\left(2(-1)+5\right)$$
$$= -1-1+(-2)+5$$
$$= 1$$

b. $(f-g)(5) = f(5)-g(5)$
$$= \left(5-5^2\right)-\left(2(5)+5\right)$$
$$= 5-25-10-5$$
$$= -35$$

c. $(f\cdot g)(1) = f(1)\cdot g(1)$
$$= \left(1-1^2\right)\left(2(1)+5\right)$$
$$= 0(7)$$
$$0$$

d. $\left(\dfrac{f}{g}\right)(2) = \dfrac{f(2)}{g(2)}$
$$= \frac{2-2^2}{2(2)+5}$$
$$= \frac{-2}{9}$$

4. **a.** $(f\circ g)(x) = f(g(x))$
$$= f(-x+5)$$
$$= 4(-x+5)-3$$
$$= -4x+20-3$$
$$= -4x+17$$

b. $(g\circ f)(x) = g(f(x))$
$$= g(4x-3)$$
$$= -(4x-3)+5$$
$$= -4x+3+5$$
$$= -4x+8$$

c. $(f\circ g)(3) = f(g(3))$
$$= f(-3+5)$$
$$= f(2)$$
$$= 4(2)-3$$
$$= 8-3$$
$$= 5$$

d. $(g\circ f)(3) = g(f(3))$
$$= g(4(3)-3)$$
$$= g(12-3)$$
$$= g(9)$$
$$= -9+5$$
$$= -4$$

5. **a.** No horizontal line intersects the graph at more than one point, so the graph passes the horizontal line test. The function is one-to-one.
b. The graph fails the horizontal line test because the x-axis intersects the graph more than once. The function is not one-to-one.

6. $\{(-1,-3),(0,-1),(2,0),(3,4)\}$

7.
$$g(x) = 3x-1$$
$$y = 3x-1$$
$$x = 3y-1$$
$$x+1 = 3y$$
$$\frac{x+1}{3} = y$$
$$y = \frac{x+1}{3}$$
$$g^{-1}(x) = \frac{x+1}{3}$$

8.
$$p(q(x)) = p(-5x+2)$$
$$= 5(-5x+2)-2$$
$$= -25x+10-2$$
$$= -25x+8$$

Since $p(q(x)) \neq x$, the functions are not inverses of each other.

9. **a.**

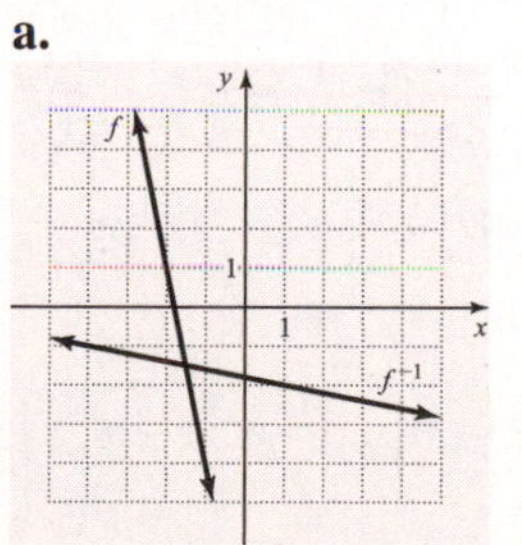

b.

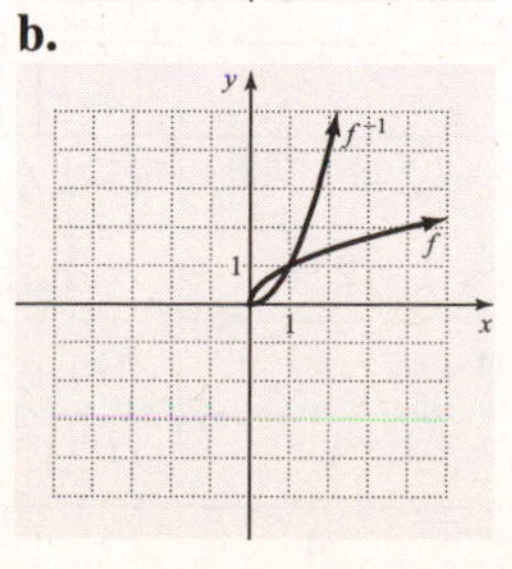

10. **a.** The graph of $C(x)$ is a line and passes the horizontal line test. So C is a one-to-one function.

b.
$$C(x) = 40x$$
$$y = 40x$$
$$x = 40y$$
$$\frac{1}{40}x = \frac{1}{40}\cdot 40y$$
$$y = \frac{1}{40}x$$
$$C^{-1}(x) = \frac{1}{40}x$$

c. The inverse can be used to determine the number of hours that the van was rented based on the amount of money charged.

Exercises

1. The function $f(x) = |x|$ is the <u>absolute value function</u>.

3. If f and g are functions, then the <u>product of f and g</u> is defined as $(f \cdot g)(x) = f(x)\cdot g(x)$.

5. A function f is said to be <u>one-to-one</u> if each y-value in the range corresponds to exactly one x-value in the domain.

7. If f and g are functions and $f(g(x)) = x$ and $g(f(x)) = x$, then $f(x)$ and $g(x)$ are <u>inverse functions of each other</u>.

9. **a.** $f(0) = \left|\frac{1}{2}(0)+3\right| = |0+3| = |3| = 3$

b.
$$f(-8) = \left|\frac{1}{2}(-8)+3\right| = |-4+3| = |-1| = 1$$

c. $f(-4t) = \left|\frac{1}{2}(-4t)+3\right| = |-2t+3|$

d.
$$f(t-6) = \left|\frac{1}{2}(t-6)+3\right| = \left|\frac{1}{2}t-3+3\right| = \left|\frac{1}{2}t\right|$$

11.

x	$g(x) = \|x\|+2$	(x,y)
-2	$g(-2) = \|-2\|+2 = 2+2 = 4$	$(-2,4)$
-1	$g(-1) = \|-1\|+2 = 1+2 = 3$	$(-1,3)$
0	$g(0) = \|0\|+2 = 0+2 = 2$	$(0,2)$
1	$g(1) = \|1\|+2 = 1+2 = 3$	$(1,3)$
2	$g(2) = \|2\|+2 = 2+2 = 4$	$(2,4)$

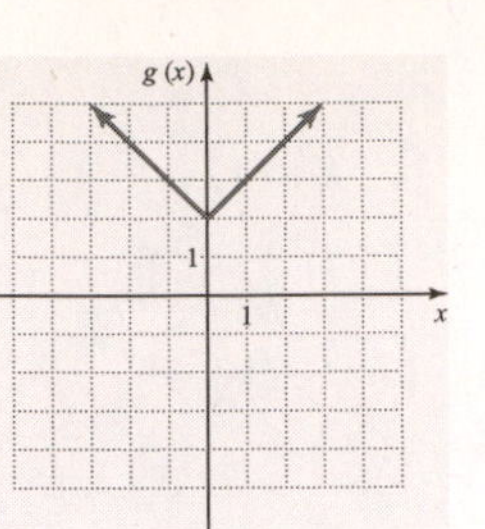

Domain: $(-\infty,\infty)$; range: $[2,\infty)$

13.

x	$h(x) = \|x+2\|$	(x,y)
-2	$h(-2) = \|-2+2\| = \|0\| = 0$	$(-2,0)$
-1	$h(-1) = \|-1+2\| = \|1\| = 1$	$(-1,1)$
0	$h(0) = \|0+2\| = \|2\| = 2$	$(0,2)$
1	$h(1) = \|1+2\| = \|3\| = 3$	$(1,3)$
2	$h(2) = \|2+2\| = \|4\| = 4$	$(2,4)$

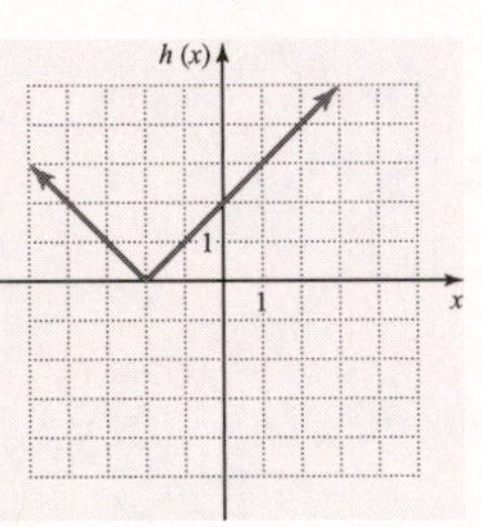

Domain: $(-\infty,\infty)$; range: $[0,\infty)$

15. $f(x) = 2x+3 \quad g(x) = 4x^2-9$

$$(f+g)(x) = 2x+3+4x^2-9 = 4x^2+2x-6$$
$$(f-g)(x) = 2x+3-(4x^2-9)$$
$$= 2x+3-4x^2+9 = -4x^2+2x+12$$
$$(f\bullet g)(x) = (2x+3)(4x^2-9)$$
$$= 8x^3-18x+12x^2-27$$
$$= 8x^3+12x^2-18x-27$$
$$\left(\frac{f}{g}\right)(x) = \frac{2x+3}{4x^2-9} = \frac{2x+3}{(2x+3)(2x-3)}$$
$$= \frac{1}{2x-3};\ x \neq -\frac{3}{2},\frac{3}{2}$$

17. $f(x)=3x^2+11x+6,\quad g(x)=2x^2+5x-3$

$(f+g)(x)=3x^2+11x+6+2x^2+5x-3$

$=5x^2+16x+3$

$(f-g)(x)=3x^2+11x+6-(2x^2+5x-3)$

$=3x^2+11x+6-2x^2-5x+3$

$=x^2+6x+9$

$(f\bullet g)(x)=(3x^2+11x+6)(2x^2+5x-3)$

$$\begin{array}{r} 3x^2+11x+6 \\ \underline{2x^2+5x-3} \\ -9x^2-33x-18 \\ 15x^3+55x^2+30x \\ \underline{6x^4+22x^3+12x^2} \\ 6x^4+37x^3+58x^2-3x-18 \end{array}$$

$\left(\dfrac{f}{g}\right)(x)=\dfrac{3x^2+11x+6}{2x^2+5x-3}=\dfrac{(3x+2)(x+3)}{(x+3)(2x-1)}$

$=\dfrac{3x+2}{2x-1};\ x\neq -3, \dfrac{1}{2}$

19. $f(x)=\dfrac{1}{x+3};\ g(x)=x-1$

$(f+g)(x)=\dfrac{1}{x+3}+x-1$

$=\dfrac{1}{x+3}+(x-1)\bullet\dfrac{x+3}{x+3}$

$=\dfrac{1}{x+3}+\dfrac{(x-1)(x+3)}{x+3}$

$=\dfrac{1+(x^2+2x-3)}{x+3}$

$=\dfrac{x^2+2x-2}{x+3};\ x\neq -3$

$(f-g)(x)=\dfrac{1}{x+3}-(x-1)$

$=\dfrac{1}{x+3}-(x-1)\bullet\dfrac{x+3}{x+3}$

$=\dfrac{1}{x+3}-\dfrac{(x-1)(x+3)}{x+3}$

$=\dfrac{1-(x^2+2x-3)}{x+3}$

$=\dfrac{-x^2-2x+4}{x+3};\quad x\neq -3$

$(f\bullet g)(x)=\left(\dfrac{1}{x+3}\right)(x-1)$

$=\dfrac{x-1}{x+3};\quad x\neq -3$

$\left(\dfrac{f}{g}\right)(x)=\dfrac{\frac{1}{x+3}}{x-1}=\dfrac{1}{x+3}\bullet\dfrac{1}{x-1}$

$=\dfrac{1}{(x+3)(x-1)};\quad x\neq -3, 1$

21. $f(x)=2\sqrt{x};\ g(x)=4\sqrt{x}-1$

$(f+g)(x)=2\sqrt{x}+4\sqrt{x}-1=6\sqrt{x}-1$

$(f-g)(x)=2\sqrt{x}-(4\sqrt{x}-1)$

$=2\sqrt{x}-4\sqrt{x}+1=-2\sqrt{x}+1$

$(f\bullet g)(x)=(2\sqrt{x})(4\sqrt{x}-1)$

$=8(\sqrt{x})^2-2\sqrt{x}=8x-2\sqrt{x}$

$\left(\dfrac{f}{g}\right)(x)=\dfrac{2\sqrt{x}}{4\sqrt{x}-1}=\dfrac{2\sqrt{x}}{4\sqrt{x}-1}\bullet\dfrac{4\sqrt{x}+1}{4\sqrt{x}+1}$

$=\dfrac{8(\sqrt{x})^2+2\sqrt{x}}{16(\sqrt{x})^2-1}$

$=\dfrac{8x+2\sqrt{x}}{16x-1};\ x\neq\dfrac{1}{16}$

23. $(f+g)(3)=(3^2-3(3)-4)+(1-(3)^2)$

$=9-9-4+1-9=-12$

25. $(f-g)(-1)=((-1)^2-3(-1)-4)-(1-(-1)^2)$

$=1+3-4-(1-1)=4-4-0=0$

27. $(f\bullet g)(2)=((2)^2-3(2)-4)(1-(2)^2)$

$=(4-6-4)(1-4)=(-6)(-3)=18$

29. $\left(\dfrac{g}{f}\right)(5)=\dfrac{1-(5)^2}{(5)^2-3(5)-4}=\dfrac{1-25}{25-15-4}$

$=\dfrac{-24}{6}=-4$

31. $f(x)=2x-4;\ g(x)=\frac{1}{2}x-5$

a. $(f\circ g)(x)=2\left(\frac{1}{2}x-5\right)-4$

$=x-10-4=x-14$

b. $(g\circ f)(x)=\frac{1}{2}(2x-4)-5$

$=x-2-5=x-7$

c. $(f\circ g)(3)=(3)-14=-11$

d. $(g\circ f)(-1)=(-1)-7=-8$

33. $f(x)=x-1;\ g(x)=x^2+4x-10$

a. $(f\circ g)(x)=(x^2+4x-10)-1=x^2+4x-11$

b. $(g\circ f)(x)=(x-1)^2+4(x-1)-10$

$=x^2-2x+1+4x-4-10$

$=x^2+2x-13$

c. $(f\circ g)(1)=(1)^2+4(1)-11=1+4-11=-6$

d. $(g\circ f)(-1)=(-1)^2+2(-1)-13$

$=1-2-13=-14$

35. $f(x)=\frac{3}{x};\ g(x)=2x+5$

a. $(f\circ g)(x)=\frac{3}{2x+5};\ x\neq-\frac{5}{2}$

b. $(g\circ f)(x)=2\left(\frac{3}{x}\right)+5=\frac{6}{x}+5;\ x\neq0$

c. $(f\circ g)\left(\frac{1}{2}\right)=\frac{3}{2\left(\frac{1}{2}\right)+5}=\frac{3}{1+5}=\frac{3}{6}=\frac{1}{2}$

d. $(g\circ f)(-2)=\frac{6}{(-2)}+5=-3+5=2$

37. $f(x)=-\sqrt{x};\ g(x)=1-4x$

a. $(f\circ g)(x)=-\sqrt{1-4x};\ x\leq\frac{1}{4}$

b. $(g\circ f)(x)=1-4\left(-\sqrt{x}\right)=1+4\sqrt{x};\ x\geq0$

c. $(f\circ g)(-6)=-\sqrt{1-4(-6)}$

$=-\sqrt{1+24}=-\sqrt{25}=-5$

d. $(g\circ f)(16)=1+4\sqrt{(16)}$

$=1+4(4)=1+16=17$

39. One-to-one function.

41. One-to-one function.

43. Not a one-to-one function.

45. One-to-one function.

47. The function has the point $(3,2)$. The only choice with the point $(2,3)$ is choice c.

49. The function has the point $(3,-5)$. The only choice with the point $(-5,3)$ is choice a.

51. f^{-1}: $\{(5,-4),(3,-2),(1,0),(-1,2),(-3,4)\}$

53. f^{-1}: $\left\{\begin{matrix}(-3,-27),(-2,-8),(-1,-1),(0,0),\\(1,1),(2,8),(3,27)\end{matrix}\right\}$

55. $y=4x$ Substitute y for $f(x)$

$x=4y$ Interchange x and y

$y=\frac{1}{4}x$

$f^{-1}(x)=\frac{1}{4}x$

57. $y=\frac{1}{4}x$ Substitute y for $g(x)$

$x=\frac{1}{4}y$ Interchange x and y

$y=4x$

$g^{-1}(x)=4x$

59. $y=-x-5$ Substitute y for $f(x)$

$x=-y-5$ Interchange x and y

$y=-x-5$

$f^{-1}(x)=-x-5$

61. $y=5x+2$ Substitute y for $f(x)$

$x=5y+2$ Interchange x and y

$x-2=5y$

$y=\frac{x-2}{5}$

$f^{-1}(x)=\frac{x-2}{5}$

63. $y=\frac{1}{2}x-3$ Substitute y for $g(x)$

$x=\frac{1}{2}y-3$ Interchange x and y

$x+3=\frac{1}{2}y$

$2(x+3)=y$

$g^{-1}(x)=2x+6$

65. $y=x^3-4$ Substitute y for $f(x)$

$x=y^3-4$ Interchange x and y

$$x+4=y^3$$
$$y=\sqrt[3]{x+4}$$
$$f^{-1}(x)=\sqrt[3]{x+4}$$

67. $y=\dfrac{3}{x+4}$ Substitute y for $h(x)$

$x=\dfrac{3}{y+4}$ Interchange x and y

$$x(y+4)=3$$
$$y+4=\frac{3}{x}$$
$$y=\frac{3}{x}-4=\frac{3}{x}-\frac{4x}{x}$$
$$h^{-1}(x)=\frac{3-4x}{x}$$

69. $y=\dfrac{x}{2x-1}$ Substitute y for $f(x)$

$x=\dfrac{y}{2y-1}$ Interchange x and y

$2xy-x=y$ or $2xy-x=y$

$2xy-y=x$ $\quad -x=y-2xy$

$y(2x-1)=x$ $\quad -x=y(1-2x)$

$y=\dfrac{x}{2x-1}$ $\quad \dfrac{-x}{1-2x}=y$

$f^{-1}(x)=\dfrac{x}{2x-1}$ or $f^{-1}(x)=-\dfrac{x}{1-2x}$

71. $y=\sqrt[3]{x+4}$ Substitute y for $f(x)$

$x=\sqrt[3]{y+4}$ Interchange x and y

$$x^3=\left(\sqrt[3]{y+4}\right)^3$$
$$x^3=y+4$$
$$x^3-4=y$$
$$f^{-1}(x)=x^3-4$$

73.

$$(f\circ g)(x)=f(g(x))=f(3x+1)$$
$$=\frac{(3x+1)-1}{3}=\frac{3x+1-1}{3}=\frac{3x}{3}=x$$

$$(g\circ f)(x)=g(f(x))=g\left(\frac{x-1}{3}\right)$$
$$=3\left(\frac{(x-1)}{3}\right)+1=x-1+1=x$$

$(f\circ g)(x)=x=(g\circ f)(x)$, the functions are inverses of one other.

75.

$$(p\circ q)(x)=p(q(x))=p\left(\frac{1}{5}x+2\right)$$
$$=5\left(\frac{1}{5}x+2\right)-10=x+10-10=x$$
$$(q\circ p)(x)=q(p(x))=q(5x-10)$$
$$=\frac{1}{5}(5x-10)+2=x-2+2=x$$

$(p\circ q)(x)=x=(q\circ p)(x)$, the functions are inverses of one other.

77.

$$(f\circ g)(x)=f(g(x))=f\left(\frac{x+1}{4}\right)$$
$$=-4\left(\frac{x+1}{4}\right)-1=-(x+1)-1$$
$$=-x-1-1=-x-2;\quad (f\circ g)(x)\neq x$$
$$(g\circ f)(x)=g(f(x))=g(-4x-1)$$
$$=\frac{(-4x-1)+1}{4}=\frac{-4x-1+1}{4}$$
$$=\frac{-4x}{4}=-x$$
$$(g\circ f)(x)\neq x$$

$(f\circ g)(x)\neq x\neq(g\circ f)(x)$, the functions are not inverses of one another.

79. $(g\circ h)(x)=g(h(x))=g\left(\sqrt[3]{x}-5\right)$

$$=\left(\left(\sqrt[3]{x}-5\right)+5\right)^3=\left(\sqrt[3]{x}-5+5\right)^3$$
$$=\left(\sqrt[3]{x}\right)^3=x$$
$$(h\circ g)(x)=h(g(x))=h\left((x+5)^3\right)$$
$$=\sqrt[3]{(x+5)^3}-5=x+5-5=x$$

$(g\circ h)(x)=x=(h\circ g)(x)$, the functions are inverses of one other.

81.

$(p\circ q)(x)=p(q(x))=p(2x^3+7)$

$=\sqrt[3]{2(2x^3+7)-7}=\sqrt[3]{4x^3+14-7}$

$=\sqrt[3]{4x^3+7}\neq x$

$(q\circ p)(x)=q(p(x))=q\left(\sqrt[3]{2x-7}\right)$

$=2\left(\sqrt[3]{2x-7}\right)^3+7=2(2x-7)+7$

$=4x-14+7=4x-7\neq x$

$(p\circ q)(x)\neq x\neq(q\circ p)(x)$, the functions are not inverses of one other.

83.

$(f\circ g)(x)=f(g(x))=f\left(\dfrac{2}{x+3}\right)=\dfrac{2}{\left(\dfrac{2}{x+3}\right)}-3$

$=\dfrac{2}{1}\bullet\dfrac{x+3}{2}-3=\dfrac{2(x+3)}{2}-3$

$=x+3-3=x$

$(g\circ f)(x)=g(f(x))=g\left(\dfrac{2}{x}-3\right)$

$=\dfrac{2}{\left(\dfrac{2}{x}-3\right)+3}=\dfrac{2}{\dfrac{2}{x}-3+3}$

$=\dfrac{2}{\dfrac{2}{x}}=\dfrac{2}{1}\bullet\dfrac{x}{2}=x$

$(f\circ g)(x)=x=(g\circ f)(x)$, the functions are inverses of one other.

85. $(-3,0)$ and $(0,3)$ are on the function, $(0,-3)$ and $(3,0)$ are on the inverse function.

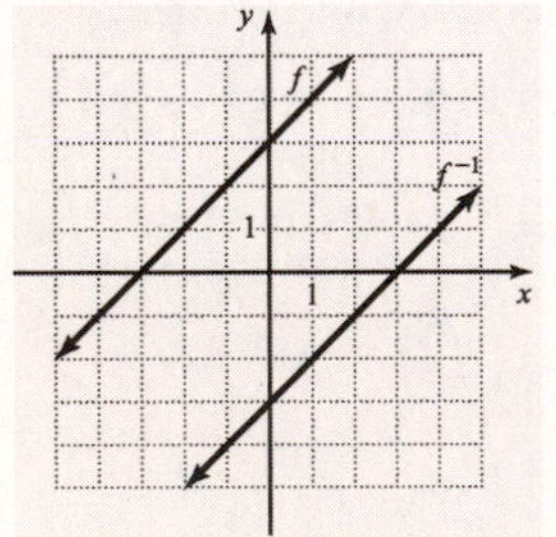

87. $(0,4)$, $(1,3)$, $(2,0)$ and $(3,-5)$ are on the function, $(4,0)$, $(3,1)$, $(0,2)$ and $(-5,3)$ are on the inverse function.

89. $g(x)=-4x+9 \qquad h(x)=\dfrac{9-x}{4}$

$(g\circ h)(x)=-4\left(\dfrac{9-x}{4}\right)+9$

$=-(9-x)+9=-9+x+9=x$

$(h\circ g)(x)=\dfrac{9-(-4x+9)}{4}=\dfrac{9+4x-9}{4}$

$=\dfrac{4x}{4}=x$

$(g\circ h)(x)=(h\circ g)(x)=x$, so the functions are inverses.

91. $f(x)=3x^2-5x+2 \qquad g(x)=x^2-8$

$(g-f)(-3)=\left((-3)^2-8\right)-\left(3(-3)^2-5(-3)+2\right)$

$=(9-8)-(27+15+2)$

$=1-44=-43$

93. $f(x)=-2x+7 \qquad g(x)=\dfrac{4}{x+2}$

a. $(f\circ g)(x)=f(g(x))=f\left(\dfrac{4}{x+2}\right)$

$=-2\left(\dfrac{4}{x+2}\right)+7$

$=-\dfrac{8}{x+2}+7,\ x\neq-2$

b.

$(g\circ f)(x)=g(f(x))=g(-2x+7)$

$=\dfrac{4}{(-2x+7)+2}=\dfrac{4}{-2x+9},\ x\neq\dfrac{9}{2}$

c. $(f\circ g)(6)=-\dfrac{8}{6+2}+7=-1+7=6$

d. $(g\circ f)(0)=\dfrac{4}{-2(0)+9}=\dfrac{4}{9}$

95. The points (–2, 0) and (0, –1) are on the function, so the points (0, –2) and (–1, 0) are on the inverse.

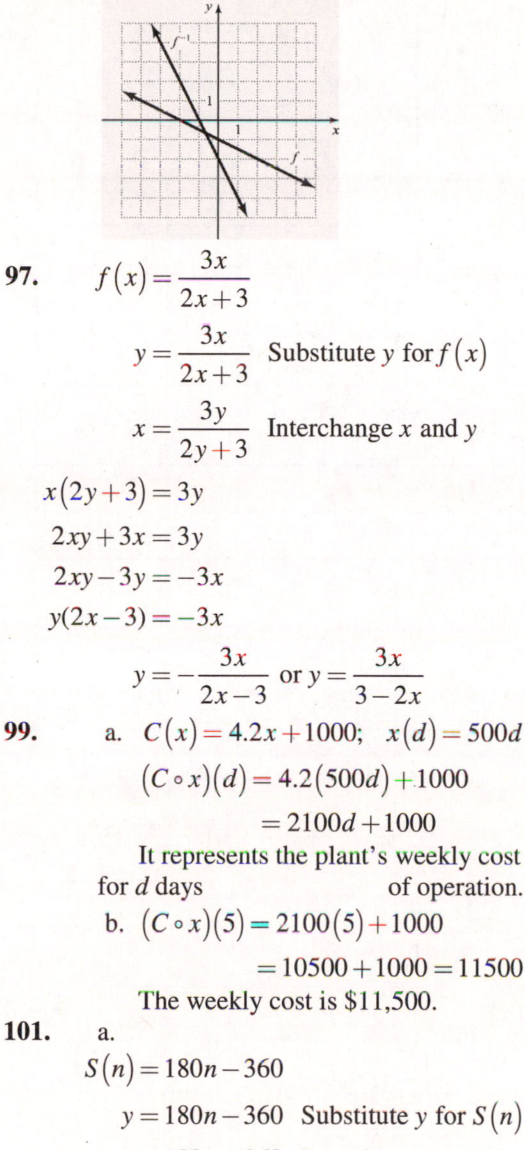

97. $f(x)=\dfrac{3x}{2x+3}$

$y=\dfrac{3x}{2x+3}$ Substitute y for $f(x)$

$x=\dfrac{3y}{2y+3}$ Interchange x and y

$x(2y+3)=3y$

$2xy+3x=3y$

$2xy-3y=-3x$

$y(2x-3)=-3x$

$y=-\dfrac{3x}{2x-3}$ or $y=\dfrac{3x}{3-2x}$

99. a. $C(x)=4.2x+1000;\ \ x(d)=500d$

$(C\circ x)(d)=4.2(500d)+1000$

$=2100d+1000$

It represents the plant's weekly cost for d days of operation.

b. $(C\circ x)(5)=2100(5)+1000$

$=10500+1000=11500$

The weekly cost is \$11,500.

101. a.

$S(n)=180n-360$

$y=180n-360$ Substitute y for $S(n)$

$n=180y-360$ Interchange n and y

$n+360=180y$

$\dfrac{n+360}{180}=y$

$S^{-1}(n)=\dfrac{n+360}{180}$

b. The inverse function can be used to calculate the number of sides of a polygon if the sum of the interior angles is known.

c. $S^{-1}(540)=\dfrac{540+360}{180}=\dfrac{900}{180}=5$

The polygon has five sides.

103. a. $B(t)=0.05t+4.5$

b.

$y=0.05t+4.5$ Substitute y for $B(t)$

$t=0.05y+4.5$ Interchange t and y

$t-4.5=0.05y$

$\dfrac{t-4.5}{0.05}=y$

$y=\dfrac{1}{0.05}t-\dfrac{4.5}{0.05}$

$B^{-1}(t)=20t-90$

c. The inverse can be used to determine the number of minutes of long-distance calling when the monthly bill is known.

10.2 Exponential Functions

Practice

1. **a.** $f(3)=4^3=64$

b. $g(-2)=\left(\dfrac{1}{3}\right)^{-2}=3^2=9$

2. **a.** $f(2)=2^{3(2)-1}=2^5=32$

b. $f(-2)=2^{3(-2)-1}=2^{-7}=\dfrac{1}{2^7}=\dfrac{1}{128}$

3. **a.** $f(-1)=e^{5(-1)}=e^{-5}\approx 0.007$

b. $g(3)=e^{3-1}=e^2\approx 7.389$

4.

x	$f(x)=3^x$	(x,y)
−2	$f(-2)=3^{-2}=\dfrac{1}{9}$	$\left(-2,\dfrac{1}{9}\right)$
−1	$f(-1)=3^{-1}=\dfrac{1}{3}$	$\left(-1,\dfrac{1}{3}\right)$
0	$f(0)=3^0=1$	$(0,1)$
1	$f(1)=3^1=3$	$(1,3)$
2	$f(2)=3^2=9$	$(2,9)$

4. (continued)

x	$g(x)=5^x$	(x,y)
-2	$g(-2)=5^{-2}=\frac{1}{25}$	$\left(-2,\frac{1}{25}\right)$
-1	$g(-1)=5^{-1}=\frac{1}{5}$	$\left(-1,\frac{1}{5}\right)$
0	$g(0)=5^0=1$	$(0,1)$
1	$g(1)=5^1=1$	$(1,5)$
2	$g(2)=5^2=25$	$(2,25)$

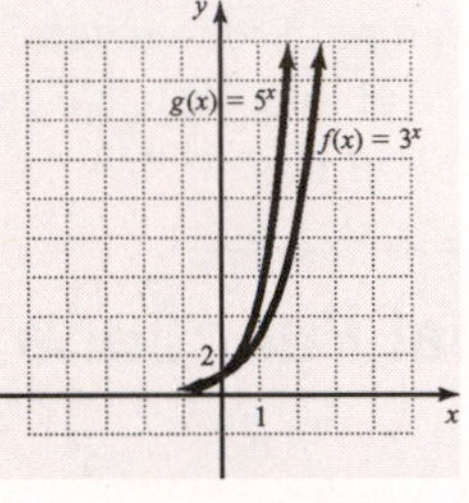

5.

x	$f(x)=\left(\frac{1}{3}\right)^x$	(x,y)
-2	$f(-2)=\left(\frac{1}{3}\right)^{-2}=9$	$(-2,9)$
-1	$f(-1)=\left(\frac{1}{3}\right)^{-1}=3$	$(-1,3)$
0	$f(0)=\left(\frac{1}{3}\right)^0=1$	$(0,1)$
1	$f(1)=\left(\frac{1}{3}\right)^1=\frac{1}{3}$	$\left(1,\frac{1}{3}\right)$
2	$f(2)=\left(\frac{1}{3}\right)^2=\frac{1}{9}$	$\left(2,\frac{1}{9}\right)$

x	$g(x)=\left(\frac{1}{5}\right)^x$	(x,y)
-2	$g(-2)=\left(\frac{1}{5}\right)^{-2}=25$	$(-2,25)$
-1	$g(-1)=\left(\frac{1}{5}\right)^{-1}=5$	$(-1,5)$
0	$g(0)=\left(\frac{1}{5}\right)^0=1$	$(0,1)$
1	$g(1)=\left(\frac{1}{5}\right)^1=\frac{1}{5}$	$\left(1,\frac{1}{5}\right)$
2	$g(2)=\left(\frac{1}{5}\right)^2=\frac{1}{25}$	$\left(2,\frac{1}{25}\right)$

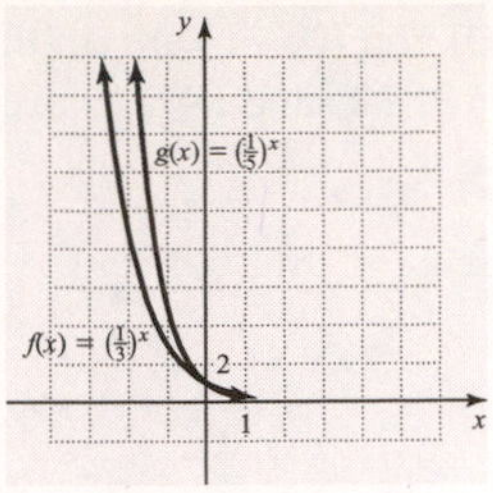

6.

x	$g(x)=3^{x-1}$	(x,y)
-2	$g(-2)=3^{-2-1}=3^{-3}=\frac{1}{27}$	$\left(-2,\frac{1}{27}\right)$
-1	$g(-1)=3^{-1-1}=3^{-2}=\frac{1}{9}$	$\left(-1,\frac{1}{9}\right)$
0	$g(0)=3^{0-1}=3^{-1}=\frac{1}{3}$	$\left(0,\frac{1}{3}\right)$
1	$g(1)=3^{1-1}=3^0=1$	$(1,1)$
2	$g(2)=3^{2-1}=3^1=3$	$(2,3)$

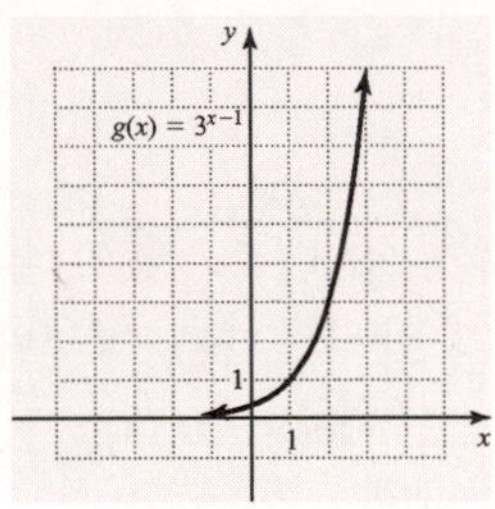

7. **a.**
$$2^x=64$$
$$2^x=2^6$$
$$x=6$$

b.
$$9^x=27$$
$$\left(3^2\right)^x=3^3$$
$$3^{2x}=3^3$$
$$2x=3$$
$$x=\frac{3}{2}$$

8.
$$A(3)=90\left(\frac{1}{2}\right)^{3/18}$$
$$=90\left(\frac{1}{2}\right)^{1/6}$$
$$\approx 80.2$$

Approximately 80 mg will remain after 3 days.

Exercises

1. A(n) exponential function is any function that can be written in the form $f(x)=b^x$, where x is a real number, $b>0$, and $b\neq 1$.\

3. The function defined by $f(x)=e^x$ is called the natural exponential function.

5. An exponential function is decreasing if the base is between 0 and 1.

7. Polynomial function; x has an integer exponent.

9. Radical function; the exponent $\frac{1}{2}$ is $\sqrt{\ }$.

11. Exponential function, it has a variable in the exponent.

13. Polynomial function; x has an integer exponent.

15. $f(x)=2^x$

a. $f(-3)=2^{-3}=\frac{1}{2^3}=\frac{1}{8}$

b. $f(0)=2^0=1$

c. $f(4)=2^4=16$

17. $g(x)=\left(\frac{1}{9}\right)^x$

a. $g\left(-\frac{1}{2}\right)=\left(\frac{1}{9}\right)^{-\frac{1}{2}}=\left(\frac{9}{1}\right)^{\frac{1}{2}}=9^{\frac{1}{2}}=\sqrt{9}=3$

b. $g\left(\frac{1}{2}\right)=\left(\frac{1}{9}\right)^{\frac{1}{2}}=\sqrt{\frac{1}{9}}=\frac{\sqrt{1}}{\sqrt{9}}=\frac{1}{3}$

c. $g(2)=\left(\frac{1}{9}\right)^2=\frac{1^2}{9^2}=\frac{1}{81}$

19. $f(x)=3^{x-4}$

a. $f(1)=3^{1-4}=3^{-3}=\frac{1}{3^3}=\frac{1}{27}$

b. $f(3)=3^{3-4}=3^{-1}=\frac{1}{3}$

c. $f(6)=3^{6-4}=3^2=9$

21. $h(x)=-2^x+3$

a. $h(-3)=-2^{-3}+3=-\frac{1}{2^3}+3$

$=-\frac{1}{8}+\frac{24}{8}=\frac{23}{8}$

b. $h(0)=-2^0+3=-1+3=2$

c. $h(4)=-2^4+3=-16+3=-13$

23. $f(x)=-e^x$

a. $f(-5)=-e^{-5}=-0.00673...\approx -0.007$

b. $f(1)=-e^1=-2.71828...\approx -2.718$

25. $g(x)=e^{-2x}$

a. $g(2)=e^{-2(2)}=e^{-4}=0.01831...\approx 0.018$

b. $g\left(-\frac{1}{2}\right)=e^{-2\left(-\frac{1}{2}\right)}=e^1=2.7182...\approx 2.718$

27. $f(x)=e^{3x-2}$

a. $f\left(-\frac{1}{3}\right)=e^{3\left(-\frac{1}{3}\right)-2}=e^{-1-2}=e^{-3}$

$=0.0497...\approx 0.050$

b. $f(0)=e^{3(0)-2}=e^{-2}=0.13533...\approx 0.135$

29. $f(x)=3^x$

x	$f(x)=3^x$	(x,y)
-3	$(3)^{-3}=\frac{1}{3^3}=\frac{1}{27}$	$\left(-3,\frac{1}{27}\right)$
-2	$(3)^{-2}=\frac{1}{3^2}=\frac{1}{9}$	$\left(-2,\frac{1}{9}\right)$
-1	$(3)^{-1}=\frac{1}{3}$	$\left(-1,\frac{1}{3}\right)$
0	$(3)^0=1$	$(0,1)$
1	$(3)^1=3$	$(1,3)$
2	$(3)^2=9$	$(2,9)$
3	$(3)^3=27$	$(3,27)$

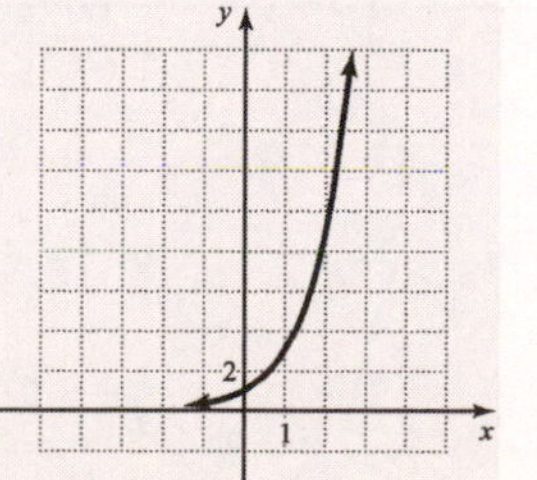

31. $f(x)=\left(\frac{1}{2}\right)^x$

x	$f(x)=\left(\frac{1}{2}\right)^x$	(x,y)
-3	$\left(\frac{1}{2}\right)^{-3}=2^3=8$	$(-3,8)$
-2	$\left(\frac{1}{2}\right)^{-2}=2^2=4$	$(-2,4)$
-1	$\left(\frac{1}{2}\right)^{-1}=2$	$(-1,2)$
0	$\left(\frac{1}{2}\right)^{0}=1$	$(0,1)$
1	$\left(\frac{1}{2}\right)^{1}=\frac{1}{2}$	$\left(1,\frac{1}{2}\right)$

33. $f(x)=-2^x$

x	$f(x)=-2^x$	(x,y)
-3	$-2^{-3}=-\frac{1}{2^3}=-\frac{1}{8}$	$\left(-3,-\frac{1}{8}\right)$
-2	$-2^{-2}=-\frac{1}{2^2}=-\frac{1}{4}$	$\left(-2,-\frac{1}{4}\right)$
-1	$-2^{-1}=-\frac{1}{2}$	$\left(-1,-\frac{1}{2}\right)$
0	$-2^0=-1$	$(0,-1)$
1	$-2^1=-2$	$(1,-2)$
2	$-2^2=-4$	$(2,-4)$
3	$-2^3=-8$	$(3,-8)$

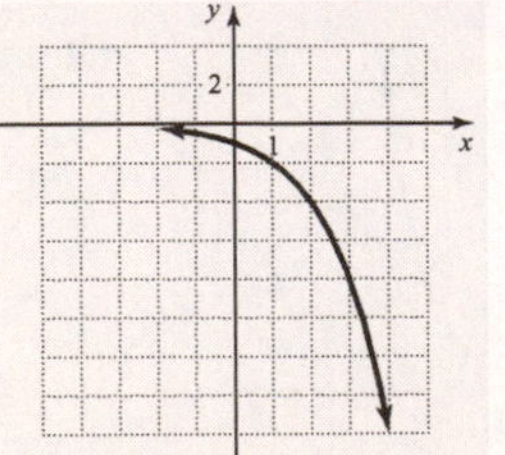

35. $f(x)=2^{x-2}$

x	$f(x)=2^{x-2}$	(x,y)
-1	$2^{-1-2}=2^{-3}=\frac{1}{2^3}=\frac{1}{8}$	$\left(-1,\frac{1}{8}\right)$
0	$2^{0-2}=2^{-2}=\frac{1}{2^2}=\frac{1}{4}$	$\left(0,\frac{1}{4}\right)$
1	$2^{1-2}=2^{-1}=\frac{1}{2}$	$\left(1,\frac{1}{2}\right)$
2	$2^{2-2}=2^0=1$	$(2,1)$
3	$2^{3-2}=2^1=2$	$(3,2)$
4	$2^{4-2}=2^2=4$	$(4,4)$
5	$2^{5-2}=2^3=8$	$(5,8)$

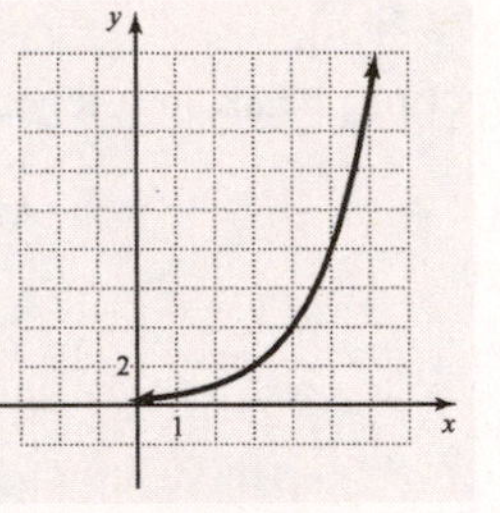

37. $f(x)=\left(\frac{1}{3}\right)^x+2$

x	$f(x)=\left(\frac{1}{3}\right)^x+2$	(x,y)
-3	$\left(\frac{1}{3}\right)^{-3}+2=3^3+2$ $=27+2=29$	$(-3,29)$
-2	$\left(\frac{1}{3}\right)^{-2}+2=3^2+2$ $=9+2=11$	$(-2,11)$
-1	$\left(\frac{1}{3}\right)^{-1}+2=3+2=5$	$(-1,5)$
0	$\left(\frac{1}{3}\right)^{0}+2=1+2=3$	$(0,3)$
1	$\left(\frac{1}{3}\right)^{1}+2=\frac{1}{3}+2=\frac{7}{3}$	$\left(1,\frac{7}{3}\right)$
2	$\left(\frac{1}{3}\right)^{2}+2=\frac{1}{9}+2=\frac{19}{9}$	$\left(2,\frac{19}{9}\right)$
3	$\left(\frac{1}{3}\right)^{3}+2=\frac{1}{27}+2=\frac{55}{27}$	$\left(3,\frac{55}{27}\right)$

37. (continued)

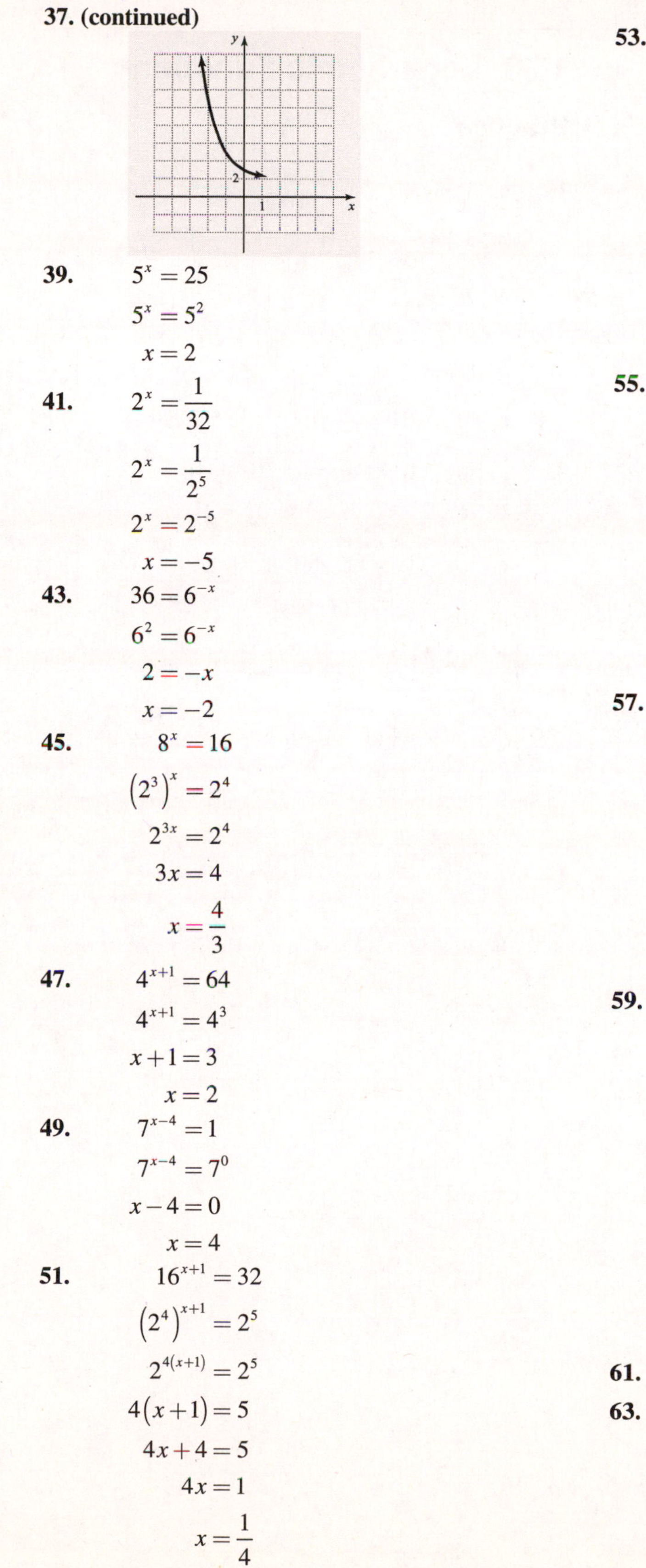

39.
$$5^x = 25$$
$$5^x = 5^2$$
$$x = 2$$

41.
$$2^x = \frac{1}{32}$$
$$2^x = \frac{1}{2^5}$$
$$2^x = 2^{-5}$$
$$x = -5$$

43.
$$36 = 6^{-x}$$
$$6^2 = 6^{-x}$$
$$2 = -x$$
$$x = -2$$

45.
$$8^x = 16$$
$$\left(2^3\right)^x = 2^4$$
$$2^{3x} = 2^4$$
$$3x = 4$$
$$x = \frac{4}{3}$$

47.
$$4^{x+1} = 64$$
$$4^{x+1} = 4^3$$
$$x + 1 = 3$$
$$x = 2$$

49.
$$7^{x-4} = 1$$
$$7^{x-4} = 7^0$$
$$x - 4 = 0$$
$$x = 4$$

51.
$$16^{x+1} = 32$$
$$\left(2^4\right)^{x+1} = 2^5$$
$$2^{4(x+1)} = 2^5$$
$$4(x+1) = 5$$
$$4x + 4 = 5$$
$$4x = 1$$
$$x = \frac{1}{4}$$

53.
$$9^{-x+3} = \frac{1}{27}$$
$$\left(3^2\right)^{-x+3} = \frac{1}{3^3}$$
$$3^{2(-x+3)} = 3^{-3}$$
$$2(-x+3) = -3$$
$$-2x + 6 = -3$$
$$-2x = -9$$
$$x = \frac{9}{2}$$

55.
$$100^{x-5} = 100000^x$$
$$\left(10^2\right)^{x-5} = \left(10^5\right)^x$$
$$10^{2(x-5)} = 10^{5x}$$
$$2(x-5) = 5x$$
$$2x - 10 = 5x$$
$$-3x = 10$$
$$x = -\frac{10}{3}$$

57.
$$64^{x-2} = 128^{x-3}$$
$$\left(2^6\right)^{x-2} = \left(2^7\right)^{x-3}$$
$$2^{6(x-2)} = 2^{7(x-3)}$$
$$6(x-2) = 7(x-3)$$
$$6x - 12 = 7x - 21$$
$$-x = -9$$
$$x = 9$$

59.
$$4^{-x+3} = \frac{1}{32}$$
$$4^{-x+3} = 2^{-5}$$
$$\left(2^2\right)^{-x+3} = 2^{-5}$$
$$2^{2(-x+3)} = 2^{-5}$$
$$2(-x+3) = -5$$
$$-2x + 6 = -5$$
$$-2x = -11$$
$$x = \frac{11}{2}$$

61. $g(x) = x^{3/2}$; radical function

63.
$$g(x) = -5^{x-2}$$
$$g(0) = -5^{0-2} = -5^{-2} = -\frac{1}{5^2} = -\frac{1}{25}$$

65. $f(x)=e^{2x}$

$f(-1)=e^{2(-1)}=e^{-2}=\frac{1}{e^2}\approx 0.135$

67. $S(n)=28000(1.04)^n$

$s(2)=28000(1.04)^2$

$=28000(1.0816)=30284.8$

The employee's salary will be $30,284.80.

69. a. $C(m)=40(0.95)^m$

$C(12)=40(0.95)^{12}$

$=40(0.54036...)=21.6144...$

The concentration will be approximately 21.6 parts per million after 1 yr.

b. $C(0)=40(0.95)^0=40(1)=40$

The initial concentration of pollutants was 40 parts per million.

71.

$A=P\left(1+\frac{r}{n}\right)^{nt}$

$A=8000\left(1+\frac{0.06}{12}\right)^{12(5)}=8000(1.005)^{60}$

$=8000(1.348850...)=10790.801...$

The amount in the account after 5 yr will be $10,790.80.

73. $C(t)=50e^{-0.125t}$

$C(8)=50e^{-0.125(8)}=50e^{-1}$

$=50(0.36787...)=18.393...$

The concentration after 8 hr is approximately 18 mg/L.

75. $N=N_0(2)^{2h}$

$1280=10(2)^{2h}$

$128=(2)^{2h}$

$2^7=2^{2h}$

$7=2h$

$h=\frac{7}{2}=3.5$

1280 bacteria will be present in 3.5 hr.

10.3 Logarithmic Functions

Practice

1. **a.** $8^2=64$

b. $5^{-1}=\frac{1}{5}$

c. $2^{1/3}=\sqrt[3]{2}$

2. **a.** $\log_2 32=5$

b. $\log_7 1=0$

c. $\log_{10}\sqrt{10}=\frac{1}{2}$

3. **a.** $\log_{10} 10=1$ because $10^1=10$.

b. $\log_3 27=3$ because $3^3=27$.

c. $\log_8 2=\frac{1}{3}$ because $8^{1/3}=2$.

4. **a.** $\log_6 6=1$

b. $\log_3 1=0$

5. **a.** $\log_4 x=3$

$4^3=x$

$x=64$

b. $\log_2 x=-1$

$2^{-1}=x$

$x=\frac{1}{2}$

6. **a.** $\log_x 125=3$

$x^3=125$

$x=5$

b. $\log_x 2=\frac{1}{4}$

$x^{1/4}=2$

$x=16$

7. $\log_9 27=x$

$9^x=27$

$(3^2)^x=3^3$

$3^{2x}=3^3$

$2x=3$

$x=\frac{3}{2}$

8.

x	$f(x)=\log_5 x$	(x,y)
1	$f(1)=\log_5 1=0$	$(1,0)$
5	$f(5)=\log_5 5=1$	$(5,1)$
25	$f(25)=\log_5 25=2$	$(25,2)$
$\frac{1}{5}$	$f\left(\frac{1}{5}\right)=\log_5 \frac{1}{5}=-1$	$\left(\frac{1}{5},-1\right)$
$\frac{1}{25}$	$f\left(\frac{1}{25}\right)=\log_5 \frac{1}{25}=-2$	$\left(\frac{1}{25},-2\right)$

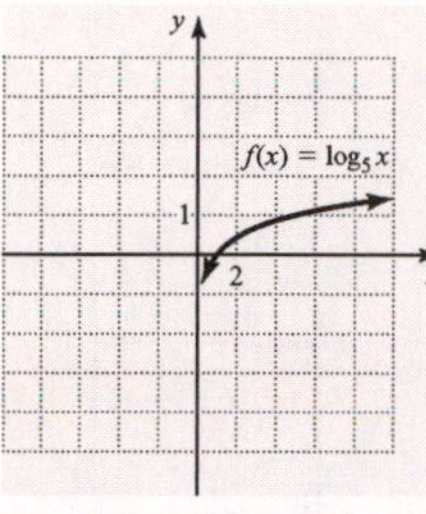

9.

x	$f(x)=\log_{1/5} x$	(x,y)
1	$f(1)=\log_{1/5} 1=0$	$(1,0)$
5	$f(5)=\log_{1/5} 5=-1$	$(5,-1)$
25	$f(25)=\log_{1/5} 25=-2$	$(25,-2)$
$\frac{1}{5}$	$f\left(\frac{1}{5}\right)=\log_{1/5} \frac{1}{5}=1$	$\left(\frac{1}{5},1\right)$
$\frac{1}{25}$	$f\left(\frac{1}{25}\right)=\log_{1/5} \frac{1}{25}=2$	$\left(\frac{1}{25},2\right)$

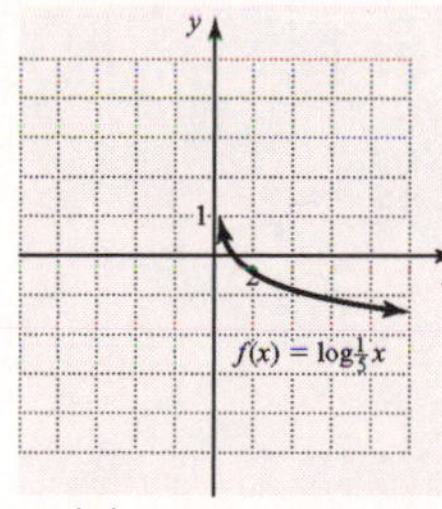

10.
$$\begin{aligned} f(9) &= 27+1.1\log_{10}(9+1) \\ &= 27+1.1\log_{10} 10 \\ &= 27+1.1(1) \\ &= 27+1.1 \\ &= 28.1 \end{aligned}$$

The barometric pressure 9 mi from a hurricane's eye is about 28.1 in. of mercury.

Exercises

1. The inverse of $f(x)=b^x$ is represented by the <u>function</u> $f^{-1}(x)=\log_b x$.

3. A logarithm is a(n) <u>exponent</u>.

5. The logarithmic function is <u>decreasing</u> if the base is between o and 1.

7. $\log_3 81=4 \quad 3^4=81$

9. $\log_{1/2}\frac{1}{32}=5 \quad \left(\frac{1}{2}\right)^5=\frac{1}{32}$

11. $\log_5\frac{1}{25}=-2 \quad 5^{-2}=\frac{1}{25}$

13. $\log_{1/4}4=-1 \quad \left(\frac{1}{4}\right)^{-1}=4$

15. $\log_{16}2=\frac{1}{4} \quad 16^{1/4}=2$

17. $\log_{10}\sqrt{10}=\frac{1}{2} \quad 10^{1/2}=\sqrt{10}$

19. $3^5=243 \quad \log_3 243=5$

21. $\left(\frac{1}{4}\right)^1=\frac{1}{4} \quad \log_{1/4}\left(\frac{1}{4}\right)=1$

23. $2^{-4}=\frac{1}{16} \quad \log_2\frac{1}{16}=-4$

25. $\left(\frac{1}{3}\right)^{-4}=81 \quad \log_{1/3}81=-4$

27. $49^{1/2}=7 \quad \log_{49}7=\frac{1}{2}$

29. $11^{1/5}=\sqrt[5]{11} \quad \log_{11}\sqrt[5]{11}=\frac{1}{5}$

31. $\log_5 125=\log_5 5^3=3$

33. $\log_3 9=\log_3 3^2=2$

35. $\log_6 6=\log_6 6^1=1$

37. $\log_{1/2}\frac{1}{16}=\log_{1/2}\frac{1}{2^4}=\log_{1/2}\left(\frac{1}{2}\right)^4=4$

39. $\log_4\frac{1}{64}=\log_4\frac{1}{4^3}=\log_4 4^{-3}=-3$

41.
$$\begin{aligned} \log_{1/4}16 &= \log_{1/4}4^2=\log_{1/4}\frac{1}{4^{-2}} \\ &= \log_{1/4}\left(\frac{1}{4}\right)^{-2}=-2 \end{aligned}$$

43. $\log_9 1=\log_9 9^0=0$

45. $\log_{27}3=\log_{27}\sqrt[3]{27}=\log_{27}27^{1/3}=\frac{1}{3}$

47. $\log_{1/16}\frac{1}{2}=\log_{1/16}\sqrt[4]{\frac{1}{16}}=\log_{1/16}\left(\frac{1}{16}\right)^{1/4}=\frac{1}{4}$

49. $\log_{36}\frac{1}{6} = \log_{36}\frac{1}{\sqrt{36}} = \log_{36}\frac{1}{36^{1/2}}$

$= \log_{36} 36^{-1/2} = -\frac{1}{2}$

51. $\log_3 x = 3 \Rightarrow 3^3 = x \Rightarrow x = 27$

53. $\log_6 x = -2 \Rightarrow 6^{-2} = x \Rightarrow x = \frac{1}{6^2} = \frac{1}{36}$

55. $\log_{2/3} x = -1 \Rightarrow \left(\frac{2}{3}\right)^{-1} = x \Rightarrow x = \frac{3}{2}$

57. $\log_4 x = \frac{1}{2} \Rightarrow 4^{1/2} = x \Rightarrow x = \sqrt{4} = 2$

59. $\log_x 216 = 3 \Rightarrow x^3 = 216 \Rightarrow x = \sqrt[3]{216} = 6$

61. $\log_x 2 = \frac{1}{3} \Rightarrow x^{1/3} = 2 \Rightarrow x = 2^3 = 8$

63. $\log_x \frac{9}{16} = 2 \Rightarrow x^2 = \frac{9}{16} \Rightarrow x = \sqrt{\frac{9}{16}} = \frac{3}{4}$

65. $\log_x 7 = -1 \Rightarrow x^{-1} = 7 \Rightarrow \frac{1}{x} = 7 \Rightarrow x = \frac{1}{7}$

67. $\log_4 8 = x \Rightarrow 4^x = 8 \Rightarrow \left(2^2\right)^x = 2^3 \Rightarrow$

$2^{2x} = 2^3 \Rightarrow 2x = 3 \Rightarrow x = \frac{3}{2}$

69. $\log_{27} 81 = x \Rightarrow 27^x = 81 \Rightarrow \left(3^3\right)^x = 3^4 \Rightarrow$

$3^{3x} = 3^4 \Rightarrow 3x = 4 \Rightarrow x = \frac{4}{3}$

71. $\log_{1/3} 81 = x \Rightarrow \left(\frac{1}{3}\right)^x = 81 \Rightarrow \left(3^{-1}\right)^x = 3^4 \Rightarrow$

$3^{-x} = 3^4 \Rightarrow -x = 4 \quad x = -4$

73. $y = \log_3(x)$

$x = 3^y$

y	$x = 3^y$	(x, y)
-2	$3^{-2} = \frac{1}{3^2} = \frac{1}{9}$	$\left(\frac{1}{9}, -2\right)$
-1	$3^{-1} = \frac{1}{3}$	$\left(\frac{1}{3}, -1\right)$
0	$3^0 = 1$	$(1,0)$
1	$3^1 = 3$	$(3,1)$
2	$3^2 = 9$	$(9,2)$

75. $y = \log_{1/3} x$

$\left(\frac{1}{3}\right)^y = x$

y	$x = \left(\frac{1}{3}\right)^y$	(x, y)
-3	$\left(\frac{1}{3}\right)^{-3} = 3^3 = 27$	$(27, -3)$
-2	$\left(\frac{1}{3}\right)^{-2} = 3^2 = 9$	$(9, -2)$
-1	$\left(\frac{1}{3}\right)^{-1} = 3$	$(3, -1)$
0	$\left(\frac{1}{3}\right)^0 = 1$	$(1,0)$
1	$\left(\frac{1}{3}\right)^1 = \frac{1}{3}$	$\left(\frac{1}{3}, 1\right)$
2	$\left(\frac{1}{3}\right)^2 = \frac{1}{3^2} = \frac{1}{9}$	$\left(\frac{1}{9}, 2\right)$
3	$\left(\frac{1}{3}\right)^3 = \frac{1}{3^3} = \frac{1}{27}$	$\left(\frac{1}{27}, 3\right)$

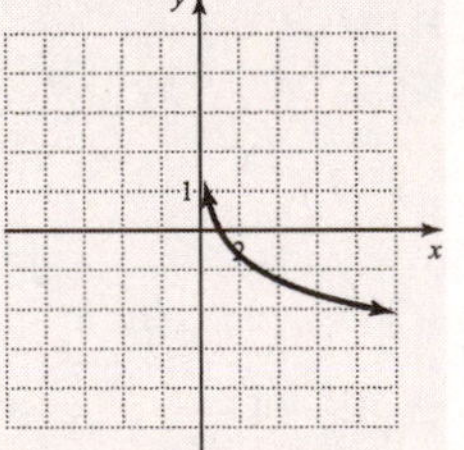

77. $y=-\log_3 x$

$-y=\log_3 x$

$3^{-y}=x$

$x=3^{-y}$

y	$x=3^{-y}$	(x,y)
-3	$3^{-(-3)}=3^3=27$	$(27,-3)$
-2	$3^{-(-2)}=3^2=9$	$(9,-2)$
-1	$3^{-(-1)}=3^1=3$	$(3,-1)$
0	$3^{-0}=1$	$(1,0)$
1	$3^{-1}=\frac{1}{3}$	$\left(\frac{1}{3},1\right)$
2	$3^{-2}=\frac{1}{3^2}=\frac{1}{9}$	$\left(\frac{1}{9},2\right)$
3	$3^{-3}=\frac{1}{3^3}=\frac{1}{27}$	$\left(\frac{1}{27},3\right)$

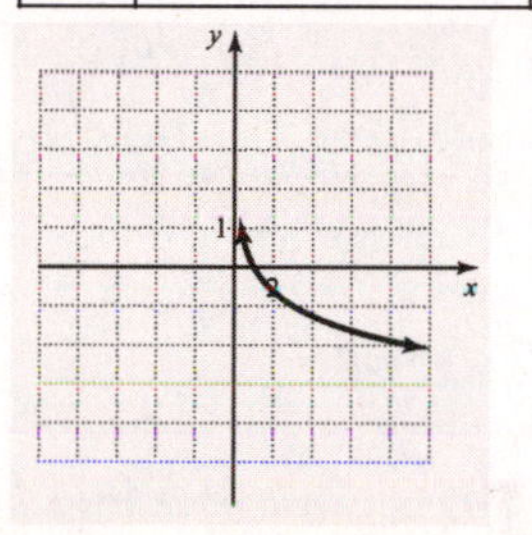

79. $\log_x 5=\frac{1}{3}\Rightarrow x^{1/3}=5\Rightarrow x=5^3=125$

81. $\log_9 27=x\Rightarrow 9^x=27\Rightarrow \left(3^2\right)^x=3^3\Rightarrow$

$2x=3\Rightarrow x=\frac{3}{2}$

83. $\log_2\frac{1}{16}=\log_2 16^{-1}=\log_2\left(2^4\right)^{-1}$

$=\log_2 2^{-4}=-4$

85. $\log_{1/2}16=-4\Rightarrow\left(\frac{1}{2}\right)^{-4}=16$

87. $\left(\frac{1}{7}\right)^{-2}=49\Rightarrow\log_{1/7}49=-2$

89. $P=10\log_{10}\left(\frac{W_2}{W_1}\right)=10\log_{10}\left(\frac{100}{10^{-2}}\right)$

$P=10\log_{10}\left(\frac{10^2}{10^{-2}}\right)=10\log_{10}\left(10^4\right)$

$P=10(4)=40$

The power gain is 40 dB.

91. $h=\log_2\left(\frac{N}{N_0}\right)=\log_2\left(\frac{512}{8}\right)=\log_2(64)$

$h=\log_2\left(2^6\right)=6$

The population reaches 512 bacteria in 6 hr.

93. $\log_2 i=t$

$\log_2 i=3\Rightarrow i=2^3=8$

The amount of the current is 8 amp.

95. $t=N+\log_2 N=64+\log_2 64$

$t=64+\log_2 2^6=64+6=70$

It takes the computer 70 picoseconds to carry out 64 computations.

97.

a. $\log_{36}x=\frac{1}{2}\Rightarrow 36^{1/2}=x\Rightarrow x=\sqrt{36}=6$

The length of the side of the fence is 6 ft.

b. The base represents the area enclosed by the fence.

10.4 Properties of Logarithms

Practice

1. **a.** $\log_6(4\cdot 9)=\log_6 4+\log_6 9$

b. $\log_2 2x=\log_2 2+\log_2 x=1+\log_2 x$

2. **a.** $\log_2 9+\log_2 15=\log_2(9\cdot 15)=\log_2 135$

b.

$\log_7(x+y)+\log_7 x=\log_7[(x+y)x]=\log_7\left(x^2+xy\right)$

3. **a.** $\log_4\frac{3}{4}=\log_4 3-\log_4 4=\log_4 3-1$

b. $\log_b\frac{u}{v}=\log_b u-\log_b v$

4. **a.** $\log_7 49-\log_7 98=\log_7\frac{49}{98}=\log_7\frac{1}{2}$

b. $\log_3(x-y)-\log_3(x+y)=\log_3\frac{x-y}{x+y}$

5. **a.** $\log_2 x^3=3\log_2 x$

b. $\log_5\sqrt[3]{x}=\log_5 x^{1/3}=\frac{1}{3}\log_5 x$

6. **a.** $\log_{10}10^7=7$

b. $3^{\log_3 29}=29$

7. **a.** $2\log_6 4 + 3\log_6 2 = \log_6 4^2 + \log_6 2^3$
$= \log_6 16 + \log_6 8$
$= \log_6 (16 \cdot 8)$
$= \log_6 128$

b. $3(\log_2 u - \log_2 v) = 3\log_2 u - 3\log_2 v$
$= \log_2 u^3 - \log_2 v^3$
$= \log_2 \frac{u^3}{v^3}$

8. **a.** $\log_5 5x^2 = \log_5 5 + \log_5 x^2$
$= 1 + 2\log_5 x$

b. $\log_7 \sqrt{\frac{u}{v}} = \log_7 \left(\frac{u}{v}\right)^{1/2}$
$= \frac{1}{2}\log_7 \frac{u}{v}$
$= \frac{1}{2}\log_7 u - \frac{1}{2}\log_7 v$

c. $\log_u \frac{b^3}{ac^5} = \log_u b^3 - \log_u (ac^5)$
$= \log_u b^3 - \log_u a - \log_u c^5$
$= 3\log_u b - \log_u a - 5\log_u c$

9. **a.** $\log_{10} N = (1)(\log_{10} 6 - \log_{10} 2)$
$\log_{10} N = \log_{10} 6 - \log_{10} 2$
$\log_{10} N = \log_{10} \frac{6}{2}$
$\log_{10} N = \log_{10} 3$
$N = 3$
3 bacteria are present after 1 hr.

b. The number of bacteria in the culture increases by a factor of 3 each hour.

Exercises

1. The logarithm of a product is the sum of the logarithms of the factors.
3. The logarithm of a power of a number is the power times the logarithm of the number.
5. The logarithm with base b of b raised to a power equals that power.
7. $\log_2 (16 \cdot 5) = \log_2 16 + \log_2 5$
$= \log_2 2^4 + \log_2 5 = 4 + \log_2 5$
9. $\log_6 \frac{7}{36} = \log_6 7 - \log_6 36$
$= \log_6 7 - \log_6 6^2 = \log_6 7 - 2$
11. $\log_3 8x = \log_3 8 + \log_3 x$
13. $\log_n uv = \log_n u + \log_n v$
15. $\log_b [a(b-1)] = \log_b a + \log_b (b-1)$
17. $\log_4 \frac{64}{n} = \log_4 64 - \log_4 n$
$= \log_4 4^3 - \log_4 n = 3 - \log_4 n$
19. $\log_7 \frac{v}{u} = \log_7 v - \log_7 u$
21. $\log_a \frac{b-a}{b+a} = \log_a (b-a) - \log_a (b+a)$
23. $\log_2 5 + \log_2 7 = \log_2 (5 \cdot 7) = \log_2 35$
25. $\log_7 10 - \log_7 2 = \log_7 \frac{10}{2} = \log_7 5$
27. $1 + \log_x 6 = \log_x x^1 + \log_x 6 = \log_x 6x$
29. $1 - \log_b 16 = \log_b b^1 - \log_b 16 = \log_b \frac{b}{16}$
31. $\log_5 2 + \log_5 (x-5) = \log_5 2(x-5)$
$= \log_5 (2x-10)$
33. $\log_n x + \log_n z = \log_n xz$
35. $\log_5 k - \log_5 n = \log_5 \frac{k}{n}$
37. $\log_{10}(a+b) - \log_{10}(a-b) = \log_{10} \frac{a+b}{a-b}$
39. $\log_6 8^2 = 2\log_6 8$
41. $\log_3 x^4 = 4\log_3 x$
43. $\log_a \sqrt[5]{b} = \log_a b^{1/5} = \frac{1}{5}\log_a b$
45. $\log_4 a^{-1} = (-1)\log_4 a = -\log_4 a$
47. Since $\log_b b^x = x$, $\log_5 5^4 = 4$
49. Since $\log_b b^x = x$, $\log_2 2^{1/2} = \frac{1}{2}$
51. Since $\log_b b^x = x$, $\log_x x^{-3} = -3$
53. Since $b^{\log_b x} = x$, $8^{\log_8 15} = 15$
55. Since $b^{\log_b x} = x$, $9^{\log_9 x} = x$
57. Since $b^{\log_b x} = x$, $x^{\log_x 10} = 10$
59. $2\log_2 6 + \log_2 3 = \log_2 (6^2 \cdot 3) = \log_2 108$
61. $4\log_7 2 - 3\log_7 4 = \log_7 2^4 - \log_7 4^3$
$= \log_7 \frac{2^4}{4^3} = \log_7 \frac{16}{64} = \log_7 \frac{1}{4}$
63. $-\log_4 x + 6\log_4 y = \log_4 x^{-1} + \log_4 y^6$
$= \log_4 (x^{-1} \cdot y^6) = \log_4 \frac{y^6}{x}$

65. $2\log_b 3+3\log_b 2-\log_b 9$
$=\log_b 3^2+\log_b 2^3-\log_b 9$
$=\log_b\left(\frac{3^2\bullet 2^3}{9}\right)=\log_b\left(\frac{9\bullet 8}{9}\right)=\log_b 8$

67. $\frac{1}{2}(\log_4 x-2\log_4 y)=\frac{1}{2}\log_4 x-\log_4 y$
$=\log_4 x^{1/2}-\log_4 y$
$=\log_4\frac{x^{1/2}}{y}=\log_4\frac{\sqrt{x}}{y}$

69.
$2\log_5 x+\log_5(x-1)=\log_5 x^2+\log_5(x-1)$
$=\log_5 x^2(x-1)$
$=\log_5(x^3-x^2)$

71.
$\frac{1}{3}\left[\log_6(x^2-y^2)-\log_6(x+y)\right]$
$=\frac{1}{3}\log_6\frac{x^2-y^2}{x+y}=\frac{1}{3}\log_6\frac{(x-y)(x+y)}{x+y}$
$=\frac{1}{3}\log_6(x-y)=\log_6(x-y)^{1/3}=\log_6\sqrt[3]{x-y}$

73. $-\log_2 z+4\log_2 x-5\log_2 y$
$=-\log_2 z+\log_2 x^4-\log_2 y^5$
$=\log_2 x^4-\log_2 y^5-\log_2 z=\log_2\frac{x^4}{y^5 z}$

75. $\frac{1}{4}\log_5 a-8\log_5 b+\frac{3}{4}\log_5 c$
$=\log_5 a^{1/4}-\log_5 b^8+\log_5 c^{3/4}$
$=\log_5\frac{a^{\frac{1}{4}}(c^3)^{\frac{1}{4}}}{b^8}=\log_5\frac{\sqrt[4]{ac^3}}{b^8}$

77.
$\log_6 3y^2=\log_6 3+\log_6 y^2=\log_6 3+2\log_6 y$

79. $\log_x x^3y^2=\log_x x^3+2\log_x y=3+2\log_x y$

81. $\log_2 4x^2y^5=\log_2 4+\log_2 x^2+\log_2 y^5$
$=\log_2 2^2+2\log_2 x+5\log_2 y$
$=2+2\log_2 x+5\log_2 y$

83. $\log_8\sqrt{10x}=\log_8(10x)^{1/2}=\frac{1}{2}\log_8 10x$
$=\frac{1}{2}[\log_8 10+\log_8 x]$
$=\frac{1}{2}\log_8 10+\frac{1}{2}\log_8 x$

85. $\log_4\frac{x^2}{y}=\log_4 x^2-\log_4 y=2\log_4 x-\log_4 y$

87. $\log_7\sqrt[4]{\frac{u^3}{v}}=\log_7\left(\frac{u^3}{v}\right)^{\frac{1}{4}}=\frac{1}{4}\log_7\frac{u^3}{v}$
$=\frac{1}{4}\left[\log_7 u^3-\log_7 v\right]$
$=\frac{1}{4}\left[3\log_7 u-\log_7 v\right]$
$=\frac{3}{4}\log_7 u-\frac{1}{4}\log_7 v$

89. $\log_c\frac{a^2c^4}{b^3}=\log_c a^2+\log_c c^4-\log_c b^3$
$=2\log_c a+4-3\log_c b$

91. $\log_6\frac{x^3}{(x-y)^2}=\log_6 x^3-\log_6(x-y)^2$
$=3\log_6 x-2\log_6(x-y)$

93.
$\log_a x^2\sqrt[5]{y^3z^2}=\log_a x^2+\log_a(y^3z^2)^{\frac{1}{5}}$
$=2\log_a x+\frac{1}{5}\log_a(y^3z^2)$
$=2\log_a x+\frac{1}{5}\left[\log_a y^3+\log_a z^2\right]$
$=2\log_a x+\frac{1}{5}\left[3\log_a y+2\log_a z\right]$
$=2\log_a x+\frac{3}{5}\log_a y+\frac{2}{5}\log_a z$

95. $\log_b 8-1=\log_b 8-\log_b b=\log_b\frac{8}{b}$

97. $2\log_a 6-4\log_a 2-3\log_a 3$
$=\log_a 6^2-\log_a 2^4-\log_a 3^3$
$=\log_a 36-\log_a 16-\log_a 27$
$=\log_a 36-(\log_a 16+\log_a 27)$
$=\log_a\frac{36}{16\cdot 27}=\log_a\frac{1}{12}$

99. $\log_p\sqrt[3]{q}=\log_p q^{1/3}=\frac{1}{3}\log_p q$

101. $\log_2 8a^3b^2=\log_2 8+\log_2 a^3+\log_2 b^2$
$=\log_2 2^3+3\log_2 a+2\log_2 b$
$=3\log_2 2+3\log_2 a+2\log_2 b$
$=3+3\log_2 a+2\log_2 b$

103.
$$\begin{aligned}\log_4 64a &= \log_4 64 + \log_4 a\\ &= \log_4 4^3 + \log_4 a\\ &= 3\log_4 4 + \log_4 a\\ &= 3 + \log_4 a\end{aligned}$$

105. $M = \log_{10} I = \log_{10} 10^{8.3} = 8.3$

The magnitude of the earthquake was 8.3.

107.
$$\begin{aligned}S = 10\log_{10}\left(\frac{P}{P_0}\right)^2 &= 2 \cdot 10\log_{10}\frac{P}{P_0}\\ &= 20\left[\log_{10} P - \log_{10} P_0\right]\\ &= 20\log_{10} P - 20\log_{10} P_0\end{aligned}$$

109. **a.**
$$\begin{aligned}\log_{10} C &= \log_{10} C_0 + 0.2t\log_{10}\left(\frac{1}{2}\right)\\ &= \log_{10} C_0 + \log_{10}\left(\frac{1}{2}\right)^{0.2t}\\ &= \log_{10}\left[C_0\left(\frac{1}{2}\right)^{0.2t}\right]\end{aligned}$$
$$C = 10^{\log_{10}\left[C_0(1/2)^{0.2t}\right]} = C_0\left(\frac{1}{2}\right)^{0.2t}$$
$$C = C_0\left(\frac{1}{2}\right)^{0.2t}$$

b.
$$\begin{aligned}C = 50\left(\frac{1}{2}\right)^{0.2(10)} &= 50\left(\frac{1}{2}\right)^2\\ &= 50\left(\frac{1}{4}\right) = 12.5\end{aligned}$$

12.5 g will remain after 10 yr.

10.5 Common Logarithms, Natural Logarithms, and Change of Base

Practice

1. **a.** $\log 6 \approx 0.7782$

b. $\log 27 \approx 1.4314$

2. **a.** $\log 1000 = \log 10^3 = 3$

b. $\log\frac{1}{10} = \log 10^{-1} = -1$

c. $\log\sqrt[3]{10} = \log 10^{1/3} = \frac{1}{3}$

3. **a.** $\ln 4 \approx 1.3863$

b. $\ln 25 \approx 3.2189$

4. **a.** $\ln e^2 = 2$

b. $\ln\frac{1}{e^7} = \ln e^{-7} = -7$

c. $\ln\sqrt[4]{e^3} = \ln e^{3/4} = \frac{3}{4}$

5. $\log_4 2.5 = \dfrac{\log 2.5}{\log 4} \approx 0.6610$

6.
$$\begin{aligned}\ln\frac{\frac{1}{2}}{10} &= -\frac{t}{19}\ln 2\\ \ln\frac{1}{20} &= -\frac{t}{19}\ln 2\\ -19\left(\ln\frac{1}{20}\right) &= t\ln 2\\ t &= \frac{-19\left(\ln\frac{1}{20}\right)}{\ln 2}\\ t &\approx 82.12\end{aligned}$$

It takes approximately 82 yr.

Exercises

1. A(n) <u>common logarithm</u> is a logarithm to the base 10.

3. A(n) <u>natural logarithm</u> is a logarithm to the base *e*.

5. $\log 4 = 0.60205\ldots \approx 0.6021$

7. $\log 18 = 1.25527\ldots \approx 1.2553$

9. $\log\left(\frac{2}{3}\right) = -0.17609\ldots \approx -0.1761$

11. $\log 1.3 = 0.11394\ldots \approx 0.1139$

13. $\ln 3 = 1.09861\ldots \approx 1.0986$

15. $\ln 17 = 2.83321\ldots \approx 2.8332$

17. $\ln\left(\frac{5}{8}\right) = -0.47000\ldots \approx -0.4700$

19. $\ln 1.2 = 0.18232\ldots \approx 0.1823$

21. $\log 1{,}000{,}000 = \log 10^6 = 6$

23. $\log\left(\frac{1}{1000}\right) = \log\left(\frac{1}{10^3}\right) = \log\left(10^{-3}\right) = -3$

25. $\log(0.01) = \log\left(10^{-2}\right) = -2$

27. $\log\sqrt[4]{1000} = \log\sqrt[4]{10^3} = \log 10^{3/4} = \frac{3}{4}$

29. $\log 10^x = x$

31. $\ln e^4 = 4$

33. $\ln\frac{1}{e} = \ln e^{-1} = -1$

35. $\ln\sqrt{e} = \ln e^{1/2} = \frac{1}{2}$

37. $\ln e^b = b$

39. $10^{\log 6} = 6$

41. $e^{\ln 3} = 3$

43.

$$\log_2 7 = \frac{\ln 7}{\ln 2} \text{ or } \frac{\log 7}{\log 2} = 2.80735... \approx 2.8074$$

45.

$$\log_6 21 = \frac{\ln 21}{\ln 6} \text{ or } \frac{\log 21}{\log 6} = 1.69918... \approx 1.6992$$

47.

$$\log_{1/2} 11 = \frac{\ln 11}{\ln\frac{1}{2}} \text{ or } \frac{\log 11}{\log\frac{1}{2}}$$
$$= -3.45943... \approx -3.4594$$

49.

$$\log_3 \frac{1}{2} = \frac{\ln\frac{1}{2}}{\ln 3} \text{ or } \frac{\log\frac{1}{2}}{\log 3} = -0.63092... \approx -0.6309$$

51.

$$\log_5 3.6 = \frac{\ln 3.6}{\ln 5} \text{ or } \frac{\log 3.6}{\log 5} = 0.79588... \approx 0.7959$$

53.

$$\log_7 0.023 = \frac{\ln 0.023}{\ln 7} \text{ or } \frac{\log 0.023}{\log 7}$$
$$= -1.938558... \approx -1.9386$$

55.

$$\log\sqrt[2]{10{,}000} = \log 10{,}000^{1/2} = \frac{1}{2}\log 10{,}000$$
$$= \frac{1}{2}\log 10^4 = \frac{1}{2}\cdot 4\log 10 = \frac{1}{2}\cdot 4 = 2$$

57. $\ln\frac{1}{e^3} = \ln e^{-3} = -3\ln e = -3$

59. $\log_{1/2} 7 = \frac{\log 7}{\log\left(\frac{1}{2}\right)} \text{ or } \frac{\ln 7}{\ln\left(\frac{1}{2}\right)} \approx -2.8074$

61. $\log 2.7 \approx 0.4314$

63.

$$pH = -\log\left[H^+\right] = -\log\left[3.2\times 10^{-13}\right]$$
$$= -\left(\log[3.2] + \log\left[10^{-13}\right]\right)$$
$$= -(0.5051499... - 13)$$
$$= -(-12.49485)... \approx 12.5$$

The pH of household bleach is approximately 12.5.

65.

$$R(x) = 1.45 + 2\log(x+1)$$
$$= 1.45 + 2\log(5+1) = 1.45 + 2\log(6)$$
$$= 1.45 + 2(0.7781...) = 1.45 + 1.5563...$$
$$= 3.006...$$

The company's revenue was approximately $3 million in the year 2005.

67.

$$kd = -(\ln I - \ln I_0)$$
$$kd = -\left(\ln\frac{I}{I_0}\right)$$
$$-kd = \ln\frac{I}{I_0}$$
$$e^{-kd} = \frac{I}{I_0}$$
$$I = I_0 e^{-kd}$$

69.

$$\log_2 60 = \frac{\ln 60}{\ln 2} \text{ or } \frac{\log 60}{\log 2} = 5.90689... \approx 5.9$$

The population will grow to 60 bacteria in approximately 5.9 hr.

10.6 Exponential and Logarithmic Equations

Practice

1.

$$4^{x+3} = 15$$
$$\log 4^{x+3} = \log 15$$
$$(x+3)\log 4 = \log 15$$
$$x+3 = \frac{\log 15}{\log 4}$$
$$x = \frac{\log 15}{\log 4} - 3$$
$$x \approx -1.0466$$

2.

$$\log_4(3x+1) = 2$$
$$4^2 = 3x+1$$
$$16 = 3x+1$$
$$15 = 3x$$
$$x = 5$$

3. $\log_3 x + \log_3(x-6) = 3$

$$\log_3[x(x-6)] = 3$$
$$3^3 = x(x-6)$$
$$27 = x^2 - 6x$$
$$x^2 - 6x - 27 = 0$$
$$(x-9)(x+3) = 0$$
$$x - 9 = 0 \quad \text{or} \quad x + 3 = 0$$
$$x = 9 \qquad\qquad x = -3$$

Since logarithms of negative numbers are undefined, the only solution is $x = 9$.

4. $\log_2(2x+3) - \log_2 x = 3$

$$\log_2 \frac{2x+3}{x} = 3$$
$$2^3 = \frac{2x+3}{x}$$
$$8 = \frac{2x+3}{x}$$
$$8x = 2x + 3$$
$$6x = 3$$
$$x = \frac{1}{2}$$

5. $61{,}102 = 36{,}695e^{10k}$

$$\frac{61{,}102}{36{,}695} = e^{10k}$$
$$\ln\frac{61{,}102}{36{,}695} = \ln e^{10k}$$
$$\ln\frac{61{,}102}{36{,}695} = 10k$$
$$k = \frac{1}{10}\ln\frac{61{,}102}{36{,}695}$$
$$k \approx 0.051$$

The annual rate of population growth was approximately 5.1%.

Exercises

1. $x = \log_3 18 = \dfrac{\ln 18}{\ln 3}$ or $\dfrac{\log 18}{\log 3}$

$$= 2.63092\ldots \approx 2.6309$$

3. $x = \log_{1/2} 9 = \dfrac{\ln 9}{\ln\left(\frac{1}{2}\right)}$ or $\dfrac{\log 9}{\log\left(\frac{1}{2}\right)}$

$$= -3.16992\ldots \approx -3.1699$$

5. $x = \log_4\left(\dfrac{3}{4}\right) = \dfrac{\ln\left(\frac{3}{4}\right)}{\ln 4}$ or $\dfrac{\log\left(\frac{3}{4}\right)}{\log 4}$

$$= -0.20751\ldots \approx -0.2075$$

7. $x = \log_{5.4} 0.0034 = \dfrac{\ln 0.0034}{\ln 5.4}$ or $\dfrac{\log 0.0034}{\log 5.4}$

$$= -3.37048\ldots \approx -3.3705$$

9. $3x = \log_2 4.6$

$$x = \frac{1}{3}\log_2 4.6 = \left(\frac{1}{3}\right)\frac{\ln 4.6}{\ln 2} \text{ or } \left(\frac{1}{3}\right)\frac{\log 4.6}{\log 2}$$
$$= \left(\frac{1}{3}\right)(2.201633\ldots) = 0.73387\ldots \approx 0.7339$$

11. $100^x = 55$

$$(10^2)^x = 55$$
$$10^{2x} = 55$$
$$2x = \log 55$$
$$x = \left(\frac{1}{2}\right)1.74036\ldots = 0.87018\ldots \approx 0.8702$$

13. $x + 4 = \log_3 38 = \dfrac{\ln 38}{\ln 3}$ or $\dfrac{\log 38}{\log 3}$

$$x + 4 = 3.31107\ldots \approx 3.3111$$
$$x = -4 + 3.3111 = -0.6889$$

15. $2x - 3 = \log_5 43 = \dfrac{\ln 43}{\ln 5}$ or $\dfrac{\log 43}{\log 5}$

$$2x - 3 = 2.33696\ldots \approx 2.3370$$
$$2x - 3 = 2.3370$$
$$2x = 5.3370$$
$$x = 2.6685$$

17. $0.25x = \ln(3.1) = 1.13140\ldots \approx 1.1314$

$$x = \frac{1.1314}{0.25} = 4.5256$$

19. $\log_3(x+10) = 4$

$$x + 10 = 3^4$$
$$x = 3^4 - 10 = 81 - 10 = 71$$

21. $\log_5(4x-3) = 1$

$$4x - 3 = 5^1$$
$$4x = 5 + 3 = 8$$
$$x = \frac{8}{4} = 2$$

23. $\log_2(x+1) = -3$

$$x+1 = 2^{-3} = \frac{1}{8}$$

$$x = \frac{1}{8} - 1 = -\frac{7}{8}$$

25. $\log(x^2 - 21) = 2$

$$x^2 - 21 = 10^2 = 100$$

$$x^2 = 100 + 21 = 121$$

$$x = \pm\sqrt{121} = \pm 11$$

$$x = 11, \quad x = -11$$

27. $\log_4 x + \log_4 6 = 2$

$$\log_4 6x = 2$$

$$6x = 4^2$$

$$x = \frac{16}{6} = \frac{8}{3}$$

29. $\log_3 x - \log_3 7 = 2$

$$\log_3\left(\frac{x}{7}\right) = 2$$

$$\frac{x}{7} = 3^2 = 9$$

$$x = 7(9) = 63$$

31. $\log_2 x + \log_2(x-3) = 2$

$$\log_2 x(x-3) = 2$$

$$x(x-3) = 2^2$$

$$x^2 - 3x = 4$$

$$x^2 - 3x - 4 = 0$$

$$(x-4)(x+1) = 0$$

$$x - 4 = 0 \text{ or } x + 1 = 0$$

$$x = 4 \qquad x = -1$$

Check $x = -1$

$$\log_2(-1) + \log_2(-1-3) \stackrel{?}{=} 2$$

Logarithm of values ≤ 0 is undefined.

$x = -1$ is not a solution.

Check $x = 4$

$$\log_2(4) + \log_2(4-3) \stackrel{?}{=} 2$$

$$\log_2 2^2 + \log_2 1 \stackrel{?}{=} 2$$

$2 + 0 = 2$ True

The solution is $x = 4$.

33. $\log_2(x+2) + \log_2(x-5) = 3$

$$\log_2(x+2)(x-5) = 3$$

$$(x+2)(x-5) = 2^3$$

$$x^2 - 3x - 10 = 8$$

$$x^2 - 3x - 18 = 0$$

$$(x-6)(x+3) = 0$$

$$x - 6 = 0 \text{ or } x + 3 = 0$$

$$x = 6 \qquad x = -3$$

Check $x = 6$

$$\log_2(6+2) + \log_2(6-5) \stackrel{?}{=} 3$$

$$\log_2 8 + \log_2 1 \stackrel{?}{=} 3$$

$$\log_2 2^3 + \log_2 2^0 \stackrel{?}{=} 3$$

$3 + 0 = 3$ True

Check $x = -3$

$$\log_2(-3+2) + \log_2(-3-5) \stackrel{?}{=} 3$$

$$\log_2(-1) + \log_2(-8) \stackrel{?}{=} 3$$

Logarithms of values ≤ 0 are undefined.

$x = -3$ is not a solution.

The solution is $x = 6$.

35. $\log_4(2x-1) + \log_4(6x-1) = 2$

$$\log_4[(2x-1)(6x-1)] = 2$$

$$(2x-1)(6x-1) = 4^2$$

$$12x^2 - 8x + 1 = 16$$

$$12x^2 - 8x - 15 = 0$$

$$(2x-3)(6x+5) = 0$$

$$2x - 3 = 0 \text{ or } 6x + 5 = 0$$

$$2x = 3 \qquad 6x = -5$$

$$x = \frac{3}{2} \qquad x = -\frac{5}{6}$$

Check $x = \frac{3}{2}$

$$\log_4\left(2\left(\frac{3}{2}\right) - 1\right) + \log_4\left(6\left(\frac{3}{2}\right) - 1\right) \stackrel{?}{=} 2$$

$$\log_4 2 + \log_4 8 \stackrel{?}{=} 2$$

$$\log_4(2 \bullet 8) \stackrel{?}{=} 2$$

$$\log_4 16 \stackrel{?}{=} 2$$

$$\log_4 4^2 \stackrel{?}{=} 2$$

$2 = 2$ True

35. (continued)

Check $x = -\frac{5}{6}$

$$\log_4\left(2\left(-\frac{5}{6}\right)-1\right)+\log_4\left(6\left(-\frac{5}{6}\right)-1\right)\stackrel{?}{=}2$$

$$\log_4\left(-\frac{8}{3}\right)+\log_4(-6)\stackrel{?}{=}2$$

Logarithms of values ≤ 0 are undefined.

$-\frac{5}{6}$ is not a solution.

The solution is $x=\frac{3}{2}$.

37. $\log_3(7x)-\log_3(x-1)=2$

$$\log_3\frac{7x}{x-1}=2$$
$$\frac{7x}{x-1}=3^2$$
$$7x=9(x-1)$$
$$7x=9x-9$$
$$-2x=-9$$
$$x=\frac{9}{2}$$

39. $\log_2(3x+7)-\log_2(x-1)=3$

$$\log_2\frac{3x+7}{x-1}=3$$
$$\frac{3x+7}{x-1}=2^3$$
$$3x+7=8(x-1)$$
$$3x+7=8x-8$$
$$-5x=-15$$
$$x=3$$

41. $\log_8 x+\log_8(x+1)=\frac{1}{3}$

$$\log_8[x(x+1)]=\frac{1}{3}$$
$$x(x+1)=8^{\frac{1}{3}}=\sqrt[3]{8}=2$$
$$x^2+x=2$$
$$x^2+x-2=0$$
$$(x-1)(x+2)=0$$
$$x-1=0 \text{ or } x+2=0$$
$$x=1 \qquad x=-2$$

Check $x=1$

$$\log_8 1+\log_8(1+1)\stackrel{?}{=}\frac{1}{3}$$
$$\log_8 8^0+\log_8 2\stackrel{?}{=}\frac{1}{3}$$
$$0+\log_8 8^{1/3}\stackrel{?}{=}\frac{1}{3}$$
$$\frac{1}{3}=\frac{1}{3} \text{ True}$$

Check $x=-2$

$$\log_8(-2)+\log_8(-2+1)\stackrel{?}{=}\frac{1}{3}$$
$$\log_8(-2)+\log_8(-1)\stackrel{?}{=}\frac{1}{3}$$

Logarithms of values ≤ 0 are undefined.

$x=-2$ is not a solution.

The solution is $x = 1$.

43. $\log_2(3x-8)-\log_2(x+4)=-1$

$$\log_2\frac{3x-8}{x+4}=-1$$
$$\frac{3x-8}{x+4}=2^{-1}=\frac{1}{2}$$
$$2(3x-8)=1(x+4)$$
$$6x-16=x+4$$
$$5x=20$$
$$x=4$$

45. $\log_3(x^2-16)=2$

$$x^2-16=3^2=9$$
$$x^2=25$$
$$x=\pm 5$$

Check $x = -5$.

$$\log_3\left((-5)^2-16\right)\stackrel{?}{=}2$$
$$\log_3(25-16)\stackrel{?}{=}2$$
$$\log_3 9\stackrel{?}{=}2$$
$$2=2 \text{ True}$$

Check $x = 5$.

$$\log_3(5^2-16)\stackrel{?}{=}2$$
$$\log_3(25-16)\stackrel{?}{=}2$$
$$\log_3 9\stackrel{?}{=}2$$
$$2=2 \text{ True}$$

The solution is $x = -5$ or $x = 5$.

47.
$$\log_2 x - \log_2(x-4) = 2$$
$$\log_2 \frac{x}{x-4} = 2$$
$$\frac{x}{x-4} = 2^2 = 4$$
$$x = 4(x-4)$$
$$x = 4x - 16$$
$$-3x = -16$$
$$x = \frac{16}{3}$$

49.
$$\left(\frac{1}{4}\right)^x = 11$$
$$\log\left(\frac{1}{4}\right)^x = \log 11$$
$$x\log\left(\frac{1}{4}\right) = \log 11$$
$$x = \frac{\log 11}{\log\left(\frac{1}{4}\right)} \approx -1.7297$$

Note that the equation could be solved by using the natural logarithm instead.

51.
$$P = 284(1.007)^t = 350$$
$$(1.007^t) = \frac{350}{284}$$
$$t = \log_{1.007}\frac{350}{284} = \frac{\ln\frac{350}{284}}{\ln 1.007} = 29.955\ldots \text{ or}$$
$$t = \frac{\log\frac{350}{284}}{\log 1.007} = 29.955\ldots$$

The population will be 350 million people in the year 2030.

53.
$$A = P\left(1+\frac{r}{n}\right)^{nt}$$
$$2000 = 1000\left(1+\frac{0.05}{4}\right)^{4t}$$
$$(1.0125)^{4t} = 2$$
$$4t = \frac{\log 2}{\log 1.0125} = 55.797\ldots$$
$$t = \frac{55.797..}{4} = 13.949\ldots$$

It will take approximately 14 years for the amount in the account to double.

55.
$$\text{pH} = -\log\left[\text{H}^+\right] = 7.4$$
$$\log\left[\text{H}^+\right] = -7.4$$
$$\text{H}^+ = 10^{-7.4} = 3.9810\ldots\times 10^{-8}$$

The concentration of hydrogen ions in blood is about 4.0×10^{-8}.

57. **a.**
$$C = 300e^{0.2x}$$
$$\frac{C}{300} = e^{0.2x}$$
$$\ln\left(\frac{C}{300}\right) = 0.2x$$
$$x = \frac{1}{0.2}\ln\left(\frac{C}{300}\right) = 5\ln\left(\frac{C}{300}\right)$$

b. $x = 5\ln\left(\frac{1000}{300}\right) = 5(1.2039\ldots) = 6.019\ldots$

The circulation reached 1,000,000 in about 6 yr after it was launched.

59. **a.** $A = 12000e^{r(0)} = 12000e^0$
$$= 12000(1) = 12000$$

The initial investment was \$12,000; the initial investment was the amount in the account when $t = 0$.

b. $31814 = 12000e^{r(15)}$
$$e^{15r} = \frac{31814}{12000}$$
$$15r = \ln\left(\frac{31814}{12000}\right) = 0.9749997\ldots$$
$$r = \frac{0.9749997\ldots}{15} = 0.06499\ldots$$

The annual interest rate is about 6.5%.

61.
$$1.7 + 2.3\ln(x+1) = R$$
$$1.7 + 2.3\ln(x+1) = 7$$
$$2.3\ln(x+1) = 7 - 1.7 = 5.3$$
$$\ln(x+1) = \frac{5.3}{2.3}$$
$$x+1 = e^{5.3/2.3} = 10.0176\ldots$$
$$x = 9.0176\ldots$$

The company's revenue reached \$7 million 9 months after the action figures were on the market.

Chapter 10 Review Exercises

1.

x	$g(x)=-\|x\|+5$	(x,y)
-2	$g(-2)=-\|-2\|+5=-2+5=3$	$(-2,3)$
-1	$g(-1)=-\|-1\|+5=-1+5=4$	$(-1,4)$
0	$g(0)=-\|0\|+5=0+5=5$	$(0,5)$
1	$g(1)=-\|1\|+5=-1+5=4$	$(1,4)$
2	$g(2)=-\|2\|+5=-2+5=3$	$(2,3)$

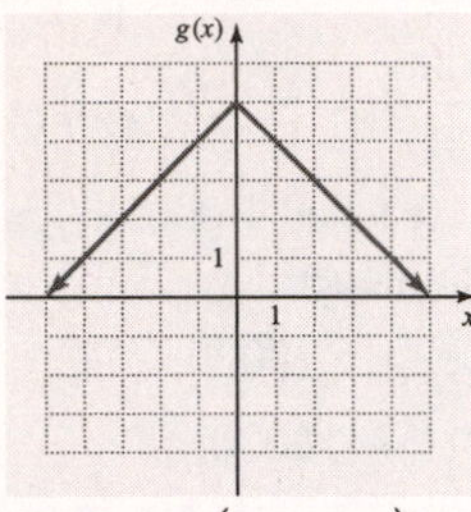

Domain: $(-\infty,\infty)$; range: $(-\infty,5]$

2.

x	$g(x)=\|x+2\|-2$	(x,y)
-2	$g(-2)=\|-2+2\|-2=0-2=-2$	$(-2,-2)$
-1	$g(-1)=\|-1+2\|-2=1-2=-1$	$(-1,-1)$
0	$g(0)=\|0+2\|-2=2-2=0$	$(0,0)$
1	$g(1)=\|1+2\|-2=3-2=1$	$(1,1)$
2	$g(2)=\|2+2\|-2=4-2=2$	$(2,2)$

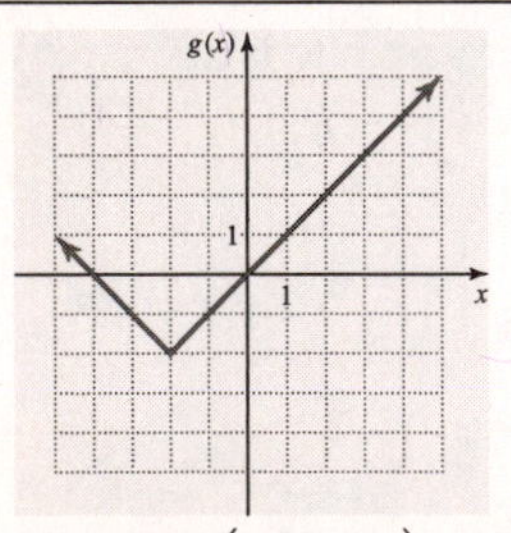

Domain: $(-\infty,\infty)$; range:

3. $(f+g)(x)=5-6x+x-3=-5x+2$

$(f-g)(x)=5-6x-(x-3)=$

$=5-6x-x+3=-7x+8$

$(f\bullet g)(x)=(5-6x)(x-3)$

$=-6x^2+23x-15$

$\left(\frac{f}{g}\right)(x)=\frac{5-6x}{x-3};\ x\neq 3$

4. $(f+g)(x)=3x^2+1+2x^2-4x=5x^2-4x+1$

$(f-g)(x)=3x^2+1-(2x^2-4x)$

$=3x^2+1-2x^2+4x=x^2+4x+1$

$(f\bullet g)(x)=(3x^2+1)(2x^2-4x)$

$=6x^4-12x^3+2x^2-4x$

$\left(\frac{f}{g}\right)(x)=\frac{3x^2+1}{2x^2-4x}=\frac{3x^2+1}{2x(x-2)};\ x\neq 0,2$

5.

$$(f+g)(x)=\frac{2}{x-3}+\frac{3}{x^2-9}$$
$$=\frac{2}{x-3}+\frac{3}{(x+3)(x-3)}$$
$$=\frac{2}{x-3}\bullet\frac{(x+3)}{(x+3)}+\frac{3}{(x+3)(x-3)}$$
$$=\frac{2x+6+3}{(x+3)(x-3)}$$
$$=\frac{2x+9}{(x+3)(x-3)};\ x\neq 3,-3$$

$$(f-g)(x)=\frac{2}{x-3}-\frac{3}{x^2-9}$$
$$=\frac{2}{x-3}-\frac{3}{(x+3)(x-3)}$$
$$=\frac{2}{x-3}\bullet\frac{(x+3)}{(x+3)}-\frac{3}{(x+3)(x-3)}$$
$$=\frac{2x+6-3}{(x+3)(x-3)}$$
$$=\frac{2x+3}{(x+3)(x-3)};\ x\neq 3,-3$$

$$(f\bullet g)(x)=\left(\frac{2}{x-3}\right)\left(\frac{3}{x^2-9}\right)=\frac{6}{(x-3)(x^2-9)}$$

$$\left(\frac{f}{g}\right)(x)=\frac{\frac{2}{x-3}}{\frac{3}{x^2-9}}=\frac{\frac{2}{x-3}}{\frac{3}{(x+3)(x-3)}}$$
$$=\frac{\frac{2}{x-3}\bullet(x+3)(x-3)}{\frac{3}{(x+3)(x-3)}\bullet(x+3)(x-3)}$$
$$=\frac{2(x+3)}{3};\ x\neq 3,-3$$

6.

$$(f+g)(x)=7\sqrt{x}+2+\sqrt{x}-2=8\sqrt{x}$$

$$(f-g)(x)=7\sqrt{x}+2-(\sqrt{x}-2)$$
$$=7\sqrt{x}+2-\sqrt{x}+2=6\sqrt{x}+4$$

$$(f\bullet g)(x)=(7\sqrt{x}+2)(\sqrt{x}-2)$$
$$=7(\sqrt{x})^2-12\sqrt{x}-4$$
$$=7x-12\sqrt{x}-4$$

$$\left(\frac{f}{g}\right)(x)=\frac{7\sqrt{x}+2}{\sqrt{x}-2}\bullet\frac{\sqrt{x}+2}{\sqrt{x}+2}$$
$$=\frac{7(\sqrt{x})^2+16\sqrt{x}+4}{(\sqrt{x})^2-4}$$
$$=\frac{7x+16\sqrt{x}+4}{x-4};\ x\neq 4$$

For exercises 7–10, $f(x)=x^2-6x+5$ and $g(x)=x^2-x$.

7.

$$(f+g)(-2)$$
$$=\left[(-2)^2-6(-2)+5\right]+\left[(-2)^2-(-2)\right]$$
$$=21+6=27$$

8. $(f-g)(3)$
$$=\left[(3)^2-6(3)+5\right]-\left[(3)^2-(3)\right]$$
$$=-4-6=-10$$

9. $(f\bullet g)(1)=\left[(1)^2-6(1)+5\right]\bullet\left[(1)^2-(1)\right]$
$$=0\bullet 0=0$$

10.

$$\left(\frac{f}{g}\right)(-4)=\frac{(-4)^2-6(-4)+5}{(-4)^2-(-4)}=\frac{45}{20}=\frac{9}{4}$$

11. $f(x)=x+2;\ g(x)=x^2-5x+9$

a.

$$(f\circ g)(x)=f[g(x)]$$
$$=[x^2-5x+9]+2=x^2-5x+11$$

b.

$$(g\circ f)(x)=g[f(x)]$$
$$=[x+2]^2-5[x+2]+9$$
$$=x^2+4x+4-5x-10+9$$
$$=x^2-x+3$$

c. $(f\circ g)(0)=(0)^2-5(0)+11=11$

d. $(g\circ f)(-2)=(-2)^2-(-2)+3=9$

12. $f(x)=\sqrt{x};\ g(x)=2x-9$

a. $(f\circ g)(x)=f[g(x)]=\sqrt{2x-9}$

b.

$$(g\circ f)(x)=g[f(x)]=2(\sqrt{x})-9=2\sqrt{x}-9$$

c. $(f\circ g)(9)=\sqrt{2(9)-9}=\sqrt{9}=3$

d. $(g\circ f)(5)=2\sqrt{5}-9$

13. Not one-to-one. It fails the horizontal line test. The line $y=2$ intersects the graph twice.

14. One-to-one. No horizontal line intersects the graph more than once.

15. $f(x)=8x+3$

$y=8x+3$; substitute y for $f(x)$

$x=8y+3$; interchange x and y.

$$x-3=8y$$
$$\frac{x-3}{8}=y$$
$$f^{-1}(x)=\frac{x-3}{8}$$

16. $f(x)=\frac{1}{6}x-1$

$y=\frac{1}{6}x-1$; substitute y for $f(x)$

$x=\frac{1}{6}y-1$; interchange x and y.

$$x+1=\frac{1}{6}y$$
$$6(x+1)=y$$
$$f^{-1}(x)=6x+6$$

17. $f(x)=x^3-5$

$y=x^3-5$; substitute y for $f(x)$

$x=y^3-5$; interchange x and y.

$$x+5=y^3$$
$$(x+5)^{1/3}=(y^3)^{1/3}$$
$$\sqrt[3]{x+5}=y$$
$$f^{-1}(x)=\sqrt[3]{x+5}$$

18. $f(x)=\sqrt[3]{x+10}$

$y=\sqrt[3]{x+10}$; substitute y for $f(x)$

$x=\sqrt[3]{y+10}$; interchange x and y.

$$x=(y+10)^{1/3}$$
$$[x]^3=\left[(y+10)^{1/3}\right]^3$$
$$x^3=y+10$$
$$x^3-10=y$$
$$f^{-1}(x)=x^3-10$$

19. $f(x)=\dfrac{2}{x+3}$

$y=\dfrac{2}{x+3}$; substitute y for $f(x)$

$x=\dfrac{2}{y+3}$; interchange x and y.

$$x(y+3)=2$$
$$xy+3x=2$$
$$xy=2-3x$$
$$y=\frac{2-3x}{x}$$
$$f^{-1}(x)=\frac{2-3x}{x}$$

20.

$f(x)=\dfrac{x}{3x+1}$

$y=\dfrac{x}{3x+1}$; substitute y for $f(x)$

$x=\dfrac{y}{3y+1}$; interchange x and y.

$$x(3y+1)=y$$
$$3xy+x=y$$
$$x=y-3xy$$
$$x=y(1-3x)$$
$$\frac{x}{1-3x}=y$$
$$f^{-1}(x)=\frac{x}{1-3x}$$

21. $f(x)=\dfrac{2}{3}x+2;\quad g(x)=\dfrac{3}{2}x-3$

$$(f\circ g)(x)=f[g(x)]=\frac{2}{3}\left[\frac{3}{2}x-3\right]+2$$
$$=[x-2]+2=x$$
$$(g\circ f)(x)=g[f(x)]=\frac{3}{2}\left[\frac{2}{3}x+2\right]-3$$
$$=[x+3]-3=x$$

$(f\circ g)(x)=x=(g\circ f)(x)$, so the functions are inverses of each other.

22. $f(x)=\dfrac{6}{x}-4;\quad g(x)=\dfrac{6}{x}+4$

$$(f\circ g)(x)=f[g(x)]=\frac{6}{\left[\frac{6}{x}+4\right]}-4$$
$$=\frac{6\bullet x}{\left(\frac{6}{x}+4\right)\bullet x}-4=\frac{6x}{6+4x}-4$$
$$=\frac{6x}{2(3+2x)}-4\bullet\frac{3+2x}{3+2x}$$
$$=\frac{3x}{3+2x}-\frac{12+8x}{3+2x}$$
$$=\frac{-5x-12}{3+2x}\neq x$$
$$(g\circ f)(x)=g[f(x)]=\frac{6}{\left[\frac{6}{x}-4\right]}+4$$
$$=\frac{6\bullet x}{\left(\frac{6}{x}-4\right)\bullet x}+4=\frac{6x}{6-4x}+4$$
$$=\frac{6x}{2(3-2x)}+4\bullet\frac{3-2x}{3-2x}$$
$$=\frac{3x}{3-2x}+\frac{12-8x}{3-2x}$$
$$=\frac{-5x+12}{3-2x}\neq x$$

$(f\circ g)(x)\neq x\neq(g\circ f)(x)$; the functions are not inverses of each other.

23.
$(-2,3)$ and $(0,-3)$ are on the function, $(3,-2)$ and $(-3,0)$ are on the inverse function.

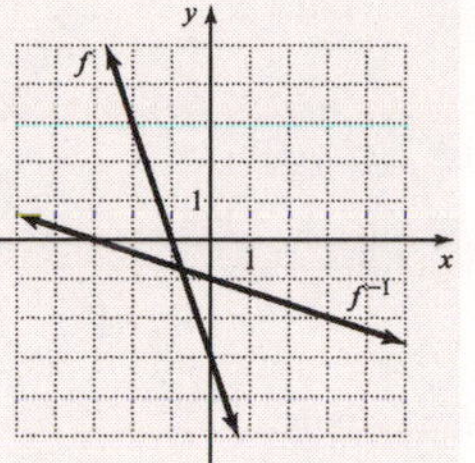

24.
$(0,2)$, $(-1,3)$, and $(-4,4)$ are on the function. $(2,0)$, $(3,-1)$, and $(4,-4)$ are on the inverse function.

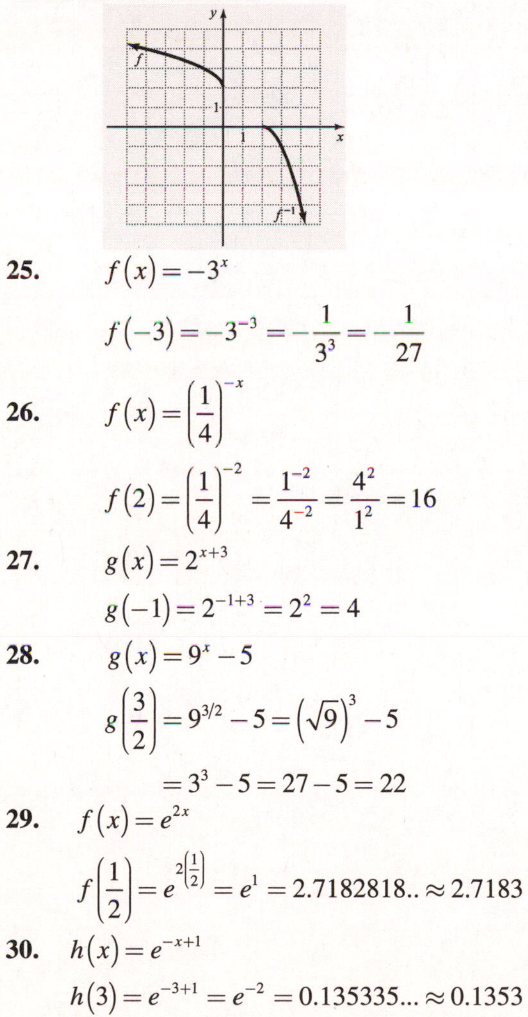

25. $f(x)=-3^x$

$$f(-3)=-3^{-3}=-\frac{1}{3^3}=-\frac{1}{27}$$

26. $f(x)=\left(\frac{1}{4}\right)^{-x}$

$$f(2)=\left(\frac{1}{4}\right)^{-2}=\frac{1^{-2}}{4^{-2}}=\frac{4^2}{1^2}=16$$

27. $g(x)=2^{x+3}$

$$g(-1)=2^{-1+3}=2^2=4$$

28. $g(x)=9^x-5$

$$g\left(\frac{3}{2}\right)=9^{3/2}-5=\left(\sqrt{9}\right)^3-5$$
$$=3^3-5=27-5=22$$

29. $f(x)=e^{2x}$

$$f\left(\frac{1}{2}\right)=e^{2\left(\frac{1}{2}\right)}=e^1=2.7182818..\approx 2.7183$$

30. $h(x)=e^{-x+1}$

$$h(3)=e^{-3+1}=e^{-2}=0.135335...\approx 0.1353$$

31. $f(x)=(3)^{-x}$

x	$f(x)=(3)^{-x}$	(x,y)
-3	$(3)^{-(-3)}=3^3=27$	$(-3,27)$
-2	$(3)^{-(-2)}=3^2=9$	$(-2,9)$
-1	$(3)^{-(-1)}=3$	$(-1,3)$
0	$(3)^{-0}=1$	$(0,1)$
1	$(3)^{-1}$ [illegible]	$\left(1,\frac{1}{3}\right)$
2	[illegible] $\frac{1}{9}$	$\left(2,\frac{1}{9}\right)$
3	$(3)^{-3}=\frac{1}{27}$	$\left(3,\frac{1}{27}\right)$

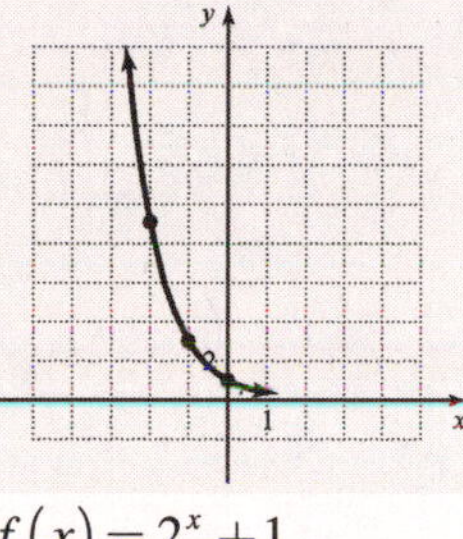

32. $f(x)=2^x+1$

x	$f(x)=2^x+1$	(x,y)
-3	$2^{-3}+1=\frac{1}{2^3}+1$ $=\frac{1}{8}+1=\frac{9}{8}$	$\left(-3,\frac{9}{8}\right)$
-2	$2^{-2}+1=\frac{1}{2^2}+1$ $=\frac{1}{4}+1=\frac{5}{4}$	$\left(-2,\frac{5}{4}\right)$
-1	$2^{-1}+1=\frac{1}{2}+1=\frac{3}{2}$	$\left(-1,\frac{3}{2}\right)$
0	$2^0+1=1+1=2$	$(0,2)$
1	$2^1+1=2+1=3$	$(1,3)$
2	$2^2+1=4+1=5$	$(2,5)$
3	$2^3+1=8+1=9$	$(3,9)$

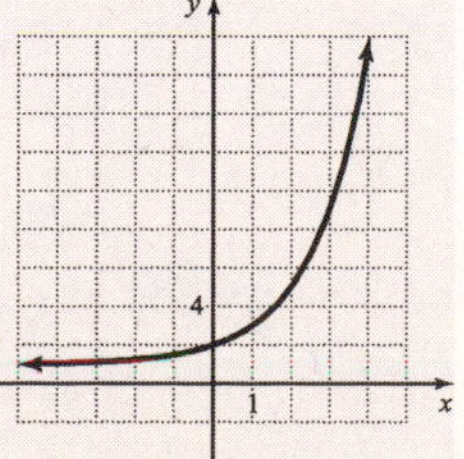

33.
$$\begin{aligned} 4^x &= 32 \\ (2^2)^x &= 2^5 \\ 2^{2x} &= 2^5 \\ 2x &= 5 \\ x &= \frac{5}{2} \end{aligned}$$

34.
$$\begin{aligned} 27^x &= \frac{1}{9} \\ (3^3)^x &= \frac{1}{3^2} \\ 3^{3x} &= 3^{-2} \\ 3x &= -2 \\ x &= -\frac{2}{3} \end{aligned}$$

35.
$$\begin{aligned} 2^{x-3} &= 16 \\ 2^{x-3} &= 2^4 \\ x-3 &= 4 \\ x &= 7 \end{aligned}$$

36.
$$\begin{aligned} 25^{x+2} &= 125^{3-x} \\ (5^2)^{x+2} &= (5^3)^{3-x} \\ 5^{2(x+2)} &= 5^{3(3-x)} \\ 2(x+2) &= 3(3-x) \\ 2x+4 &= 9-3x \\ 5x &= 5 \\ x &= 1 \end{aligned}$$

37.
$$\begin{aligned} \log_6 216 &= 3 \\ 6^3 &= 216 \end{aligned}$$

38.
$$\begin{aligned} \log_{1/3} 9 &= -2 \\ \left(\frac{1}{3}\right)^{-2} &= 9 \end{aligned}$$

39.
$$\begin{aligned} 5^{-2} &= \frac{1}{25} \\ \log_5 \frac{1}{25} &= -2 \end{aligned}$$

40.
$$\begin{aligned} 81^{1/4} &= 3 \\ \log_{81} 3 &= \frac{1}{4} \end{aligned}$$

41. $\log_8 8^1 = 1$

42. $\log_{1/2}\left(\frac{1}{32}\right) = \log_{1/2}\left(\frac{1}{2^5}\right) = \log_{1/2}\left(\frac{1}{2}\right)^5 = 5$

43. $\log_6 1 = 0$

44. $\log_{16} 2 = \log_{16} 16^{1/4} = \frac{1}{4}$

45.
$$\begin{aligned} \log_4 x &= 3 \\ x &= 4^3 = 64 \end{aligned}$$

46.
$$\begin{aligned} \log_2 x &= -6 \\ x &= 2^{-6} \\ x &= \frac{1}{2^6} = \frac{1}{64} \end{aligned}$$

47.
$$\begin{aligned} \log_x 7 &= -1 \\ x^{-1} &= 7 \\ (x^{-1})^{-1} &= (7)^{-1} \\ x &= \frac{1}{7} \end{aligned}$$

48.
$$\begin{aligned} \log_x \frac{4}{9} &= 2 \\ x^2 &= \frac{4}{9} \\ x &= \pm\sqrt{\frac{4}{9}} = \pm\frac{2}{3} \end{aligned}$$

The base of logarithms is defined only for values > 0, so $x = \frac{2}{3}$.

49.
$$\begin{aligned} \log_4 64 &= x \\ \log_4 4^3 &= x \\ x &= 3 \end{aligned}$$

50.
$$\begin{aligned} \log_{1/2} 4 &= x \\ x &= \log_{1/2} 2^2 \\ x &= \log_{1/2}\left(\frac{1}{2}\right)^{-2} = -2 \end{aligned}$$

51. $f(x) = -\log_4 x$

$$y = -\log_4 x$$
$$-y = \log_4 x$$
$$4^{-y} = x$$
$$x = 4^{-y}$$

y	$x = 4^{-y}$	(x, y)
-3	$4^{-(-3)} = 4^3 = 64$	$(64, -3)$
-2	$4^{-(-2)} = 4^2 = 16$	$(16, -2)$
-1	$4^{-(-1)} = 4^1 = 4$	$(4, -1)$
0	$4^{-0} = 1$	$(1, 0)$
1	$4^{-1} = \frac{1}{4}$	$\left(\frac{1}{4}, 1\right)$
2	$4^{-2} = \frac{1}{4^2} = \frac{1}{16}$	$\left(\frac{1}{16}, 2\right)$
3	$4^{-3} = \frac{1}{4^3} = \frac{1}{64}$	$\left(\frac{1}{64}, 3\right)$

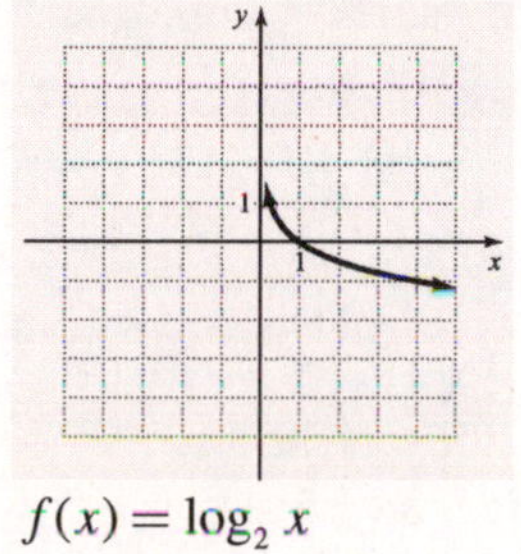

52. $f(x) = \log_2 x$

$$y = \log_2 x$$
$$x = 2^y$$

y	$x = 2^y$	(x, y)
-2	$2^{-2} = \frac{1}{2^2} = \frac{1}{4}$	$\left(\frac{1}{4}, -2\right)$
-1	$2^{-1} = \frac{1}{2}$	$\left(\frac{1}{2}, -1\right)$
0	$2^0 = 1$	$(1, 0)$
1	$2^1 = 2$	$(2, 1)$
2	$2^2 = 4$	$(4, 2)$

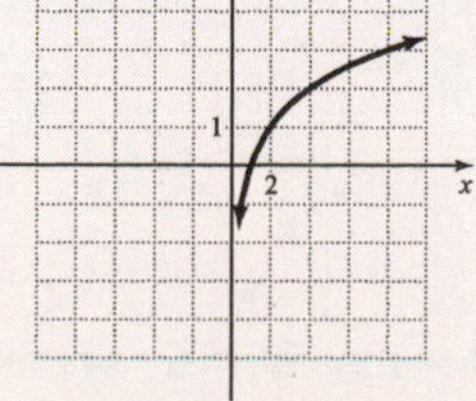

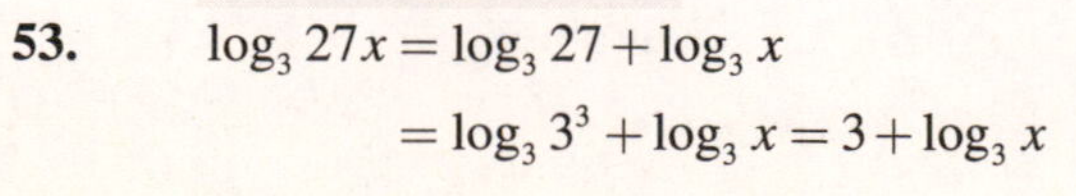

53. $\log_3 27x = \log_3 27 + \log_3 x$

$$= \log_3 3^3 + \log_3 x = 3 + \log_3 x$$

54. $\log_6 \frac{6}{a} = \log_6 6 - \log_6 a = 1 - \log_6 a$

55. $\log_n \frac{x}{x-5} = \log_n x - \log_n (x-5)$

56. $\log_9 [u(u+1)] = \log_9 u + \log_9 (u+1)$

57. $\log_5 x^3 y = \log_5 x^3 + \log_5 y = 3\log_5 x + \log_5 y$

58. $\log_2 \sqrt[5]{x^2 y^3} = \log_2 \left(x^2 y^3\right)^{1/5} = \frac{1}{5}\log_2 x^2 y^3$

$$= \frac{1}{5}\left(\log_2 x^2 + \log_2 y^3\right)$$
$$= \frac{1}{5}\left(2\log_2 x + 3\log_2 y\right)$$
$$= \frac{2}{5}\log_2 x + \frac{3}{5}\log_2 y$$

59. $\log_3 \frac{x^2 z^3}{y^2} = \log_3 x^2 + \log_3 z^3 - \log_3 y^2$

$$= 2\log_3 x + 3\log_3 z - 2\log_3 y$$

60. $\log_b \frac{b^2}{(a-b)^4} = \log_b b^2 - \log_b (a-b)^4$

$$= 2 - 4\log_b (a-b)$$

61. $\log_4 10 + \log_4 n = \log_4 (10n)$

62. $\log_9 x - \log_9 5 = \log_9 \frac{x}{5}$

63. $\log_3 a + 2\log_3 b = \log_3 a + \log_3 b^2 = \log_3 \left(ab^2\right)$

64. $5\log_b a - 4\log_b c = \log_b a^5 - \log_b c^4$

$$= \log_b \left(\frac{a^5}{c^4}\right)$$

65. $\frac{1}{4}(3\log_6 x - 8\log_6 y) = \frac{3}{4}\log_6 x - \frac{8}{4}\log_6 y$

$$= \log_6 x^{3/4} - \log_6 y^2$$
$$= \log_6 \frac{x^{3/4}}{y^2} = \log_6 \frac{\sqrt[4]{x^3}}{y^2}$$

66. $2\log_a x - \log_a y - 4\log_a z$

$$= \log_a x^2 - \log_a y - \log_a z^4 = \log_a \frac{x^2}{yz^4}$$

67. Since $\log_b b^x = x$, $\log_8 8^3 = 3$.

68. Since $\log_b b^x = x$, $\log_x x^{-(1/3)} = -\frac{1}{3}$.

69. Since $b^{\log_b x} = x$, $5^{\log_5 y} = y$.

70. Since $b^{\log_b x} = x$, $a^{\log_a 14} = 14$.

71. $\log 23 = 1.36172... \approx 1.3617$

72. $\log 9.4 = 0.97312... \approx 0.9731$

73. $\ln 48 = 3.87120... \approx 3.8712$

74. $\ln \frac{2}{3} = -0.40546... \approx -0.4055$

75. $\log 0.1 = \log \frac{1}{10} = \log 10^{-1} = -1$

76. $\log \sqrt[4]{1000} = \log \sqrt[4]{10^3} = \log 10^{3/4} = \frac{3}{4}$

77. $\ln e^{100} = 100$

78. $\ln \frac{1}{e^y} = \ln e^{-y} = -y$

79. $\log_5 121 = \frac{\log 121}{\log 5}$ or $\frac{\ln 121}{\ln 5}$

$= 2.97979... \approx 2.9798$

80. $\log_9 6.1 = \frac{\log 6.1}{\log 9}$ or $\frac{\ln 6.1}{\ln 9}$

$= 0.82298... \approx 0.8230$

81. $7^x = 72$

$x = \log_7 72 = \frac{\log 72}{\log 7}$ or $\frac{\ln 72}{\ln 7}$

$= 2.19777... \approx 2.1978$

82.

$6^{2x} = 0.58$

$2x = \log_6 0.58 = \frac{\log 0.58}{\log 6}$ or $\frac{\ln 0.58}{\ln 6}$

$2x = -0.30401...$

$x = \frac{-0.30401...}{2} = -0.15200... \approx -0.1520$

83. $2^{x-5} = 20$

$x - 5 = \log_2 20 = \frac{\log 20}{\log 2}$ or $\frac{\ln 20}{\ln 2}$

$x - 5 = 4.32192...$

$x = 4.32192... + 5 = 9.32192... \approx 9.3219$

84. $3^{2x+3} = 63$

$2x + 3 = \log_3 63 = \frac{\log 63}{\log 3}$ or $\frac{\ln 63}{\ln 3}$

$2x + 3 = 3.77124...$

$2x = 0.77124...$

$x = 0.38562... \approx 0.3856$

85. $\log_7 (x+8) = 0$

$x + 8 = 7^0$

$x + 8 = 1$

$x = -7$

86. $\log_2 12 - \log_2 x = -3$

$\log_2 \frac{12}{x} = -3$

$\frac{12}{x} = 2^{-3} = \frac{1}{8}$

$8(12) = 1(x)$

$x = 96$

87. $\log_6 x + \log_6 (x-5) = 2$

$\log_6 [x(x-5)] = 2$

$x(x-5) = 6^2 = 36$

$x^2 - 5x - 36 = 0$

$(x+4)(x-9) = 0$

$x + 4 = 0$ or $x - 9 = 0$

$x = -4 \qquad x = 9$

Check $x = -4$

$\log_6 (-4) + \log_6 (-4-5) \stackrel{?}{=} 2$

Logarithms of values ≤ 0 are undefined.

$x = -4$ is not a solution.

Check $x = 9$

$\log_6 9 + \log_6 (9-5) \stackrel{?}{=} 2$

$\log_6 [9(4)] \stackrel{?}{=} 2$

$\log_6 36 \stackrel{?}{=} 2$

$\log_6 6^2 \stackrel{?}{=} 2$

$2 = 2$ True

The solution is $x = 9$.

88. $\log_5 (x+2) + \log_5 (x+6) = 1$

$\log_5 [(x+2)(x+6)] = 1$

$(x+2)(x+6) = 5^1$

$x^2 + 8x + 12 = 5$

$x^2 + 8x + 7 = 0$

$(x+7)(x+1) = 0$

$x + 7 = 0$ or $x + 1 = 0$

$x = -7 \qquad x = -1$

Check $x = -7$

$\log_5 (-7+2) + \log_5 (-7+6) \stackrel{?}{=} 1$

$\log_5 (-5) + \log_5 (-1) \stackrel{?}{=} 1$

Logarithms of values ≤ 0 are undefined.

$x = -7$ is not a solution.

88. (continued)

Check $x=-1$

$$\log_5(-1+2)+\log_5(-1+6)\stackrel{?}{=}1$$
$$\log_5(1)+\log_5(5)\stackrel{?}{=}1$$
$$\log_5 5^0+\log_5 5^1\stackrel{?}{=}1$$
$$0+1=1 \text{ True}$$

The solution is $x=-1$.

89. $\log_4(x-3)-\log_4 x=-1$

$$\log_4\left(\frac{x-3}{x}\right)=-1$$
$$\frac{x-3}{x}=4^{-1}=\frac{1}{4}$$
$$4(x-3)=x(1)$$
$$4x-12=x$$
$$3x=12$$
$$x=4$$

90. $\log_3(2x-1)-\log_3(x-4)=2$

$$\log_3\left(\frac{2x-1}{x-4}\right)=2$$
$$\frac{2x-1}{x-4}=3^2=9$$
$$2x-1=9(x-4)$$
$$2x-1=9x-36$$
$$35=7x$$
$$x=5$$

91. **a.**

$$f(c)=2500+c;\quad c(s)=0.008s$$
$$(f\circ c)(s)=f[c(s)]=2500+(0.008s)$$
$$(f\circ c)(s)=0.008s+2500$$

It represents the agent's total monthly salary if his sales totaled s dollars.

b.

$$(f\circ c)(s)=0.008s+2500;\quad s=400000$$
$$(f\circ c)(s)=0.008(400000)+2500$$
$$(f\circ c)(s)=3200+2500=5700$$

The agent's salary is $5,700.

92. **a.**

$$P(x)=10x-150$$
$$y=10x-150;\quad \text{substitute } y \text{ for } P(x)$$
$$x=10y-150;\quad \text{interchange } x \text{ and } y.$$
$$x+150=10y$$
$$y=\frac{x+150}{10}$$
$$P^{-1}(x)=\frac{x}{10}+15$$

b. The inverse can be used to determine the number of jeans the company must sell in order to make a certain profit.

c.

$$P^{-1}(x)=\frac{x}{10}+15$$
$$P^{-1}(1200)=\frac{1200}{10}+15=120+15$$
$$P^{-1}(x)=135$$

135 pairs of jeans must be sold in order to make a weekly profit of $1200.

93. **a.** $A=P\left(1+\frac{r}{n}\right)^{nt}=6500\left(1+\frac{0.054}{2}\right)^{2(5)}$

$$A=6500(1+0.027)^{10}=6500(1.027)^{10}$$
$$A=6500(1.305282...)=8484.3346...$$

The amount in the account after 5 yr will be $8484.33.

b. $2(6500)=(6500)\left(1+\frac{0.054}{2}\right)^{2t}$

$$\left(1+\frac{0.054}{2}\right)^{2t}=2$$
$$(1.027)^{2t}=2$$
$$2t=\log_{1.027}2$$
$$=\frac{\log\ 2}{\log\ 1.027}\text{ or }\frac{\ln 2}{\ln 1.027}$$
$$=26.01715...$$
$$t=13.0085...$$

It will take about 13 yr for this investment to double.

94. **a.** $A=A_0e^{-0.0231t}$

$$A=36e^{-0.0231(60)}=36e^{-1.386}$$
$$=36(0.25007...)$$
$$A=9.0026...$$

9 mg remain after 60 yr.

94. **b.**

$$\frac{1}{2}A_0 = A_0 e^{-0.0231t}$$
$$e^{-0.0231t} = \frac{1}{2} = 0.5$$
$$-0.0231t = \ln(0.5)$$
$$t = \frac{\ln(0.5)}{-0.0231} = \frac{-0.693147...}{-0.0231}$$
$$= 30.006...$$

The half-life of cesium-137 is about 30 yr.

95.
$$t = \frac{2}{3}\log_2 \frac{N}{N_0}$$
$$t = \frac{2}{3}\log_2 \frac{384}{6} = \frac{2}{3}\log_2 64$$
$$t = \frac{2}{3}\log_2 2^6 = \frac{2}{3}(6) = 4$$

384 bacterial will be in the colony in 4 hr.

96.
$$t = 15\log_4 \frac{v}{v_0} = 15\log_4 \frac{24}{3}$$
$$t = 15\log_4 8 = 15\frac{\log 8}{\log 4} \text{ or } 15\frac{\ln 8}{\ln 4}$$
$$t = 15(1.5) = 22.5$$

It will take 22.5 yr for the stamp to have a value of $24.

97.
$$s = \log\frac{10^{90}}{(x+1)^{16}} = \log 10^{90} - \log(x+1)^{16}$$
$$s = 90\log 10 - 16\log(x+1)$$
$$s = 90 - 16\log(x+1)$$

98.
$$\ln(T-A) = \ln(98.6-A) - kt\ln e$$
$$\ln(T-A) - \ln(98.6-A) = -kt\ln e = -kt$$
$$\ln\left(\frac{T-A}{98.6-A}\right) = -kt$$
$$\frac{T-A}{98.6-A} = e^{-kt}$$
$$T-A = (98.6-A)e^{-kt}$$

99.
$$\text{pH} = -\log[\text{H}^+]$$
$$\text{pH} = -\log[2.5\times10^{-7}] = -[\log 2.5 + \log 10^{-7}]$$
$$\text{pH} = -[0.3979400... + (-7)] = -[-6.602059...]$$
$$\text{pH} = 6.602059...$$

The pH of milk is about 6.6

100.
$$t = \frac{\ln x}{r} = \frac{\ln 3}{0.044} = \frac{1.09861...}{0.044} = 24.9684...$$

It will take approximately 25 yr for the investment to triple.

101.
$$V = A(0.85)^t$$
$$15000 = 28500(0.85)^t$$
$$\frac{15000}{28500} = (0.85)^t$$
$$t = \log_{0.85}\frac{15000}{28500}$$
$$= \frac{\log\frac{15000}{28500}}{\log 0.85} \text{ or } \frac{\ln\frac{15000}{28500}}{\ln 0.85}$$
$$t = 3.9494...$$

The value will depreciate to $15,000 about 4 yr after it is purchased.

102.
$$A = Pe^{rt}$$
$$12300 = 8000e^{0.048t}$$
$$e^{0.048t} = \frac{12300}{8000} = 1.5375$$
$$0.048t = \ln(1.5375)$$
$$t = \frac{\ln(1.5375)}{0.048} = \frac{0.430157...}{0.048} = 8.9616...$$

It will take approximately 9 yr for the amount in the account to grow to $12,300.

Chapter 10 Posttest

1. $f(x) = 3x^2 - 4x - 4;\quad g(x) = 3x^2 + 2x$

a. $(f+g)(x) = 3x^2 - 4x - 4 + 3x^2 + 2x$
$$= 6x^2 - 2x - 4$$

b. $(f-g)(x) = 3x^2 - 4x - 4 - (3x^2 + 2x)$
$$= 3x^2 - 4x - 4 - 3x^2 - 2x$$
$$= -6x - 4$$

c. $(f\bullet g)(x) = (3x^2 - 4x - 4)\bullet(3x^2 + 2x)$

$$\begin{array}{r} 3x^2 - 4x - 4 \\ \underline{3x^2 + 2x} \\ 6x^3 - 8x^2 - 8x \\ \underline{9x^4 - 12x^3 - 12x^2 \quad\quad} \\ 9x^4 - 6x^3 - 20x^2 - 8x \end{array}$$

1. **d.**

$$\left(\frac{f}{g}\right)(x)=\frac{3x^2-4x-4}{3x^2+2x}=\frac{(3x+2)(x-2)}{x(3x+2)}$$
$$=\frac{x-2}{x};\ x\neq-\frac{2}{3},0$$

2. $f(x)=\sqrt{x};\ \ g(x)=2x+5$

a.
$$(f\circ g)(x)=f(g(x))=f(2x+5)=\sqrt{2x+5}$$

b.
$$(g\circ f)(x)=g(f(x))=g(\sqrt{x})=2\sqrt{x}+5$$

c. $(f\circ g)(2)=f(g(2))=f(2(2)+5)$
$$=\sqrt{2(2)+5}=\sqrt{9}=3$$

d.
$$(g\circ f)(16)=g(f(16))=g(\sqrt{16})$$
$$=2\sqrt{16}+5=2(4)+5=13$$

3. Yes, it is a one-to-one function as each y-value in the range corresponds to exactly one x-value in the domain.
The function has values:
$(2,0),(1,1),(-2,2)$ and $(-5,2.5)$
The inverse function has values:
$(0,2),(1,1),(2,-2)$ and $(2.5,-5)$

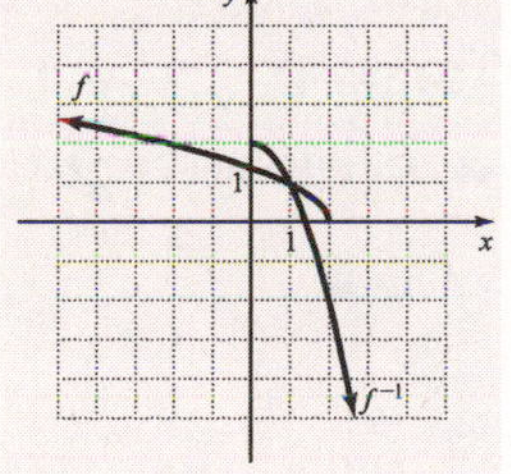

4.
$$f(x)=\frac{1}{x-7}$$
$$y=\frac{1}{x-7};\ \text{substitute } y \text{ for } f(x)$$
$$x=\frac{1}{y-7};\ \text{interchange } x \text{ and } y.$$
$$x(y-7)=1$$
$$xy-7x=1$$
$$xy=7x+1$$
$$y=\frac{7x+1}{x}$$
$$f^{-1}(x)=\frac{7x+1}{x}$$

5. **a.** $f(x)=3^x-2$

$$f(-2)=3^{-2}-2=\frac{1}{3^2}-2$$
$$f(-2)=\frac{1}{9}-\frac{18}{9}=-\frac{17}{9}\text{ or }-1.8889$$

b. $f(x)=e^{x+2}$
$$f(5)=e^{5+2}=e^7=1096.63315...$$
$$f(5)\approx1096.6332$$

6. **a.** $\log_9 1=\log_9 9^0=0$

b. $\log_4\frac{1}{2}=\log_4 1-\log_4 2$
$$=\log_4 4^0-\log_4\sqrt{4}=0-\log_4 4^{1/2}$$
$$=0-\frac{1}{2}=-\frac{1}{2}$$

7. **a.** $f(x)=-\left(\frac{1}{2}\right)^{-x}$

x	$f(x)=-\left(\frac{1}{2}\right)^{-x}$	(x,y)
-2	$-\left(\frac{1}{2}\right)^{-(-2)}=-\frac{1}{4}$	$\left(-2,-\frac{1}{4}\right)$
-1	$-\left(\frac{1}{2}\right)^{-(-1)}=-\frac{1}{2}$	$\left(-1,-\frac{1}{2}\right)$
0	$-\left(\frac{1}{2}\right)^{-0}=-1$	$(0,-1)$
1	$-\left(\frac{1}{2}\right)^{-1}=-2$	$(1,-2)$
2	$-\left(\frac{1}{2}\right)^{-2}=-4$	$(2,-4)$

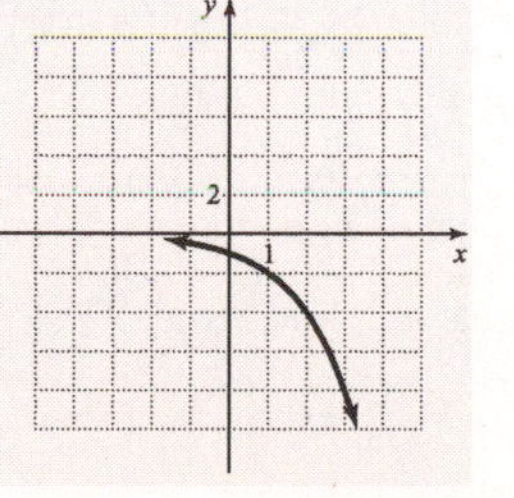

7. **b.** $f(x) = -\log_3 x$

x	$f(x) = -\log_3 x$	(x, y)
$\frac{1}{9}$	$-\log_3 \frac{1}{9} = 2$	$\left(\frac{1}{9}, 2\right)$
$\frac{1}{3}$	$-\log_3 \frac{1}{3} = 1$	$\left(\frac{1}{3}, 1\right)$
1	$-\log_3 1 = 0$	$(1, 0)$
3	$-\log_3 3 = -1$	$(3, -1)$
9	$-\log_3 9 = -2$	$(9, -2)$

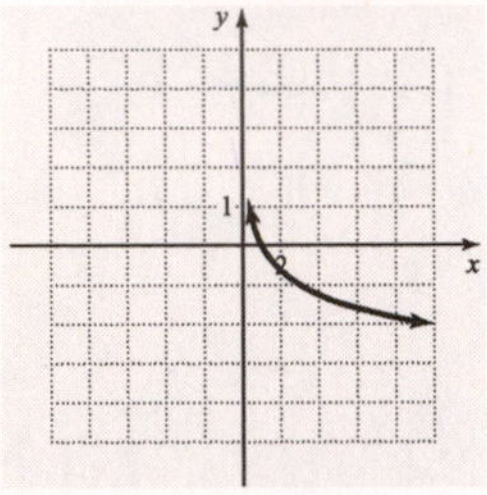

8. **a.**

$$\log_3 (9x^4) = \log_3 9 + \log_3 x^4$$
$$= \log_3 3^2 + 4\log_3 x = 2 + 4\log_3 x$$

b.

$$\log_6 \frac{x^5}{6y^2} = \log_6 x^5 - \log_6 6 - \log_6 y^2$$
$$= 5\log_6 x - \log_6 6^1 - 2\log_6 y$$
$$= 5\log_6 x - 2\log_6 y - 1$$

9. **a.**

$$2\log_5 n + 5\log_5 2 = \log_5 n^2 + \log_5 2^5$$
$$= \log_5 (n^2 \bullet 2^5)$$
$$= \log_5 (32n^2)$$

b. $3\log_b (a-2) - 4\log_b (a+2)$

$$= \log_b (a-2)^3 - \log_b (a+2)^4$$
$$= \log_b \frac{(a-2)^3}{(a+2)^4}$$

10. a. Since $\log_a a^x = x$, $\log_9 9^7 = 7$.

b. Since $a^{\log_a x} = x$, $6^{\log_6 x} = x$.

11. **a.**

$$\log \sqrt{1000} = \log \sqrt{10^3} = \log 10^{3/2} = \frac{3}{2}$$

b. $\ln \frac{1}{e^5} = \ln e^{-5} = -5$

12. $\log_5 42 = \dfrac{\log 42}{\log 5}$ or $\dfrac{\ln 42}{\ln 5} = 2.32234...$

$\log_5 42 \approx 2.3223$

13. $7^x = 49 \Rightarrow 7^x = 7^2 \Rightarrow x = 2$

14.

$$27^{2x-3} = \left(\frac{1}{3}\right)^{x-5}$$
$$\left(3^3\right)^{2x-3} = \left(3^{-1}\right)^{x-5}$$
$$3^{3(2x-3)} = 3^{-1(x-5)}$$
$$3(2x-3) = -(x-5)$$
$$6x - 9 = -x + 5$$
$$7x = 14$$
$$x = 2$$

15.

$$\log_x 4 = -2$$
$$x^{-2} = 4$$
$$\frac{1}{x^2} = 4$$
$$1 = 4x^2$$
$$x^2 = \frac{1}{4}$$
$$x = \pm\sqrt{\frac{1}{4}} = \pm\frac{1}{2}$$

Check $x = -\frac{1}{2}$.

$$\log_{-1/2} 4 \stackrel{?}{=} -2$$

The base of logarithms must be greater than 0.

$-\frac{1}{2}$ is not a solution.

Check $x = \frac{1}{2}$.

$$\log_{1/2} 4 \stackrel{?}{=} -2$$
$$\log_{1/2} 2^2 \stackrel{?}{=} -2$$
$$\log_{1/2} \left(\frac{1}{2}\right)^{-2} \stackrel{?}{=} -2$$
$$-2 = -2 \text{ True}$$

The solution is $x = \frac{1}{2}$.

16.

$$\log_5 (2x+3) = 2$$
$$2x + 3 = 5^2 = 25$$
$$2x = 22$$
$$x = 11$$

17.

$$\log_2(x+3)+\log_2(x-4)=3$$
$$\log_2(x+3)(x-4)=3$$
$$(x+3)(x-4)=2^3$$
$$x^2-x-12=8$$
$$x^2-x-20=0$$
$$(x-5)(x+4)=0$$
$$x-5=0 \text{ or } x+4=0$$
$$x=5 \qquad x=-4$$

Check $x=-4$.

$$\log_2(-4+3)+\log_2(-4-4)\stackrel{?}{=}3$$
$$\log_2(-1)+\log_2(-8)\stackrel{?}{=}3$$

Logarithms of values ≤ 0 are undefined.

$x=-4$ is not a solution.

Check $x=5$

$$\log_2(5+3)+\log_2(5-4)\stackrel{?}{=}3$$
$$\log_2 8+\log_2 1\stackrel{?}{=}3$$
$$\log_2 2^3+\log_2 2^0\stackrel{?}{=}3$$
$$3+0=3 \text{ True}$$

The solution is $x=5$.

18.

$$P(t)=1020(1.014)^t$$
$$t=2009-2001=8$$
$$P(8)=1020(1.014)^8=1020(1.117644...)$$
$$P(8)=1139.997...$$

The population of India in the year 2009 will be approximately 1140 million people.

19.

$$L=10(\log I+12)$$
$$10(\log I+12)=60$$
$$\log I+12=\frac{60}{10}$$
$$\log I+12=6$$
$$\log I=6-12$$
$$\log I=-6$$
$$I=10^{-6}$$

The intensity is $10^{-6}\ \dfrac{\text{W}}{\text{m}^2}$.

20.

$$C=20e^{-0.125t}$$
$$5=20e^{-0.125t}$$
$$e^{-0.125t}=\frac{5}{20}$$
$$e^{-0.125t}=0.25$$
$$-0.125t=\ln 0.25$$
$$t=\frac{\ln 0.25}{-0.125}=11.09035...$$

It takes about 11 hr for the concentration to decrease to 5 mg/L.

Chapter 11

CONIC SECTIONS AND NONLINEAR SYSTEMS

Pretest

1.
$$y = 3x^2 - 12x + 23$$
$$y = 3(x^2 - 4x) + 23$$
$$y = 3\left(x^2 - 4x + \left(\frac{-4}{2}\right)^2\right) + 23 - 3\left(\frac{-4}{2}\right)^2$$
$$y = 3(x^2 - 4x + 4) + 23 - 3(4)$$
$$y = 3(x-2)^2 + 11 \quad \text{Vertex: } (2,11)$$

2. **a.**
$$d = \sqrt{(x_2 - x_1)^2 + (y_2 - y_1)^2}$$
$$d = \sqrt{(-7-(-5))^2 + (13-9)^2}$$
$$d = \sqrt{(-2)^2 + (4)^2} = \sqrt{4+16} = \sqrt{20}$$
$$= 2\sqrt{5} \text{ units}$$

b.
$$d = \sqrt{(x_2 - x_1)^2 + (y_2 - y_1)^2}$$
$$d = \sqrt{(0.8-1.2)^2 + (-0.7-(-1))^2}$$
$$d = \sqrt{(-0.4)^2 + (0.3)^2} = \sqrt{0.16+0.09}$$
$$d = \sqrt{0.25} = 0.5 \text{ units}$$

3.

a. $$\left(\frac{11+(-3)}{2}, \frac{-6+10}{2}\right) = \left(\frac{8}{2}, \frac{4}{2}\right) = (4,2)$$

b. $$\left(\frac{1+0}{2}, \frac{-4+1}{2}\right) = \left(\frac{1}{2}, \frac{-3}{2}\right) = \left(\frac{1}{2}, -\frac{3}{2}\right)$$

4.
$$(x-7)^2 + (y-(-3))^2 = (\sqrt{10})^2$$
$$(x-7)^2 + (y+3)^2 = 10$$

5.
$$x^2 + y^2 - 2x - 5 = 0$$
$$x^2 - 2x + y^2 = 5$$
$$x^2 - 2x + \left(\frac{-2}{2}\right)^2 + y^2 = 5 + \left(\frac{-2}{2}\right)^2$$
$$x^2 - 2x + 1 + y^2 = 5 + 1$$
$$(x-1)^2 + (y-0)^2 = 6$$
Center: $(1,0)$, radius $= \sqrt{6}$

6.
$$x = -2(y+1)^2 + 3$$
$$x = -2(y-(-1))^2 + 3$$
$h = 3,\ k = -1$

Vertex: $(3,-1)$

Axis of symmetry: $y = -1$

y	$x = -2(y+1)^2 + 3$	(x, y)
-3	$-2(-3+1)^2 + 3 = -5$	$(-5,-3)$
-2	$-2(-2+1)^2 + 3 = 1$	$(1,-2)$
-1	$-2(-1+1)^2 + 3 = 3$	$(3,-1)$
0	$-2(0+1)^2 + 3 = 1$	$(1,0)$
1	$-2(1+1)^2 + 3 = -5$	$(-5,1)$

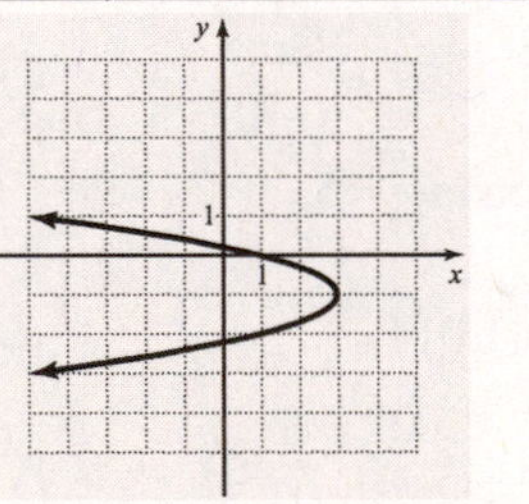

7.
$$x^2 + y^2 = 81$$
$$(x-0)^2 + (y-0)^2 = 9^2$$
Center: $(0,0)$ radius: 9

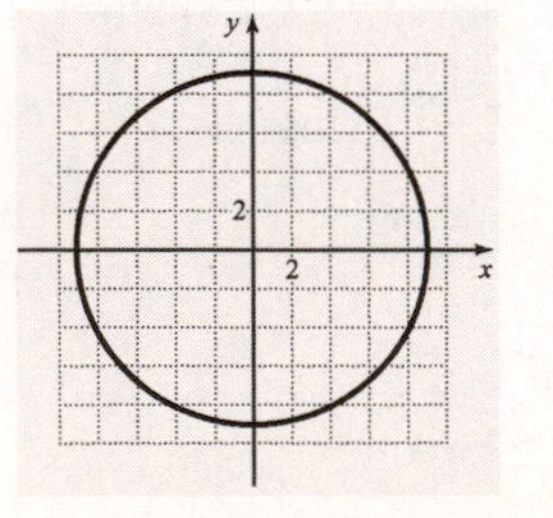

8. $$(x+2)^2+(y-1)^2=16$$
$$(x-(-2))^2+(y-1)^2=4^2$$
Center: $(-2,1)$ radius: 4

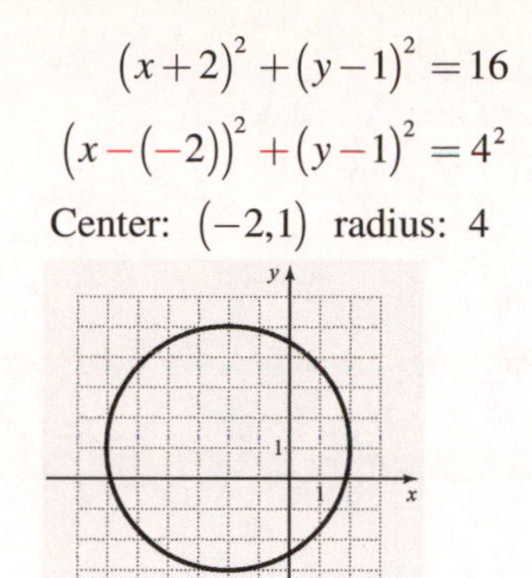

9. $$\frac{x^2}{9}+y^2=1$$
$$\frac{x^2}{3^2}+\frac{y^2}{1^2}=1$$
$a=3,\ b=1$

x-intercepts: $(\pm3,0)$ or $(3,0)$ and $(-3,0)$
y-intercepts: $(0,\pm1)$ or $(0,1)$ and $(0,-1)$

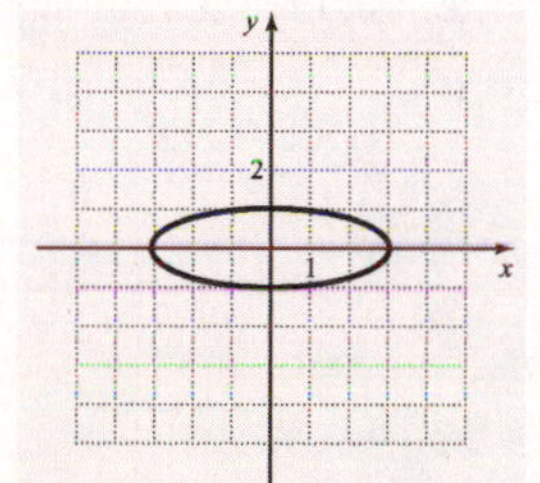

10. $$12x^2+3y^2=48$$
$$\frac{12x^2}{48}+\frac{3y^2}{48}=\frac{48}{48}$$
$$\frac{x^2}{4}+\frac{y^2}{16}=1$$
$$\frac{x^2}{2^2}+\frac{y^2}{4^2}=1$$
$a=2,\ b=4$

x-intercepts are $(\pm2,0)$ or $(2,0)$ and $(-2,0)$
y-intercepts are $(0,\pm4)$ or $(0,4)$ and $(0,-4)$

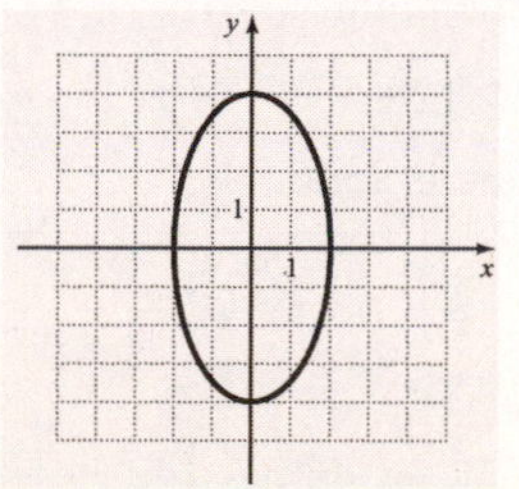

11. $$\frac{y^2}{25}-\frac{x^2}{9}=1$$
$$\frac{y^2}{5^2}-\frac{x^2}{3^2}=1$$
$a=3,\ b=5$
y-intercepts are $(0,\pm5)$ or
$(0,5)$ and $(0,-5)$
Equations of asymptotes are
$y=\pm\frac{5}{3}x$ or $y=\frac{5}{3}x$ and $y=-\frac{5}{3}x$

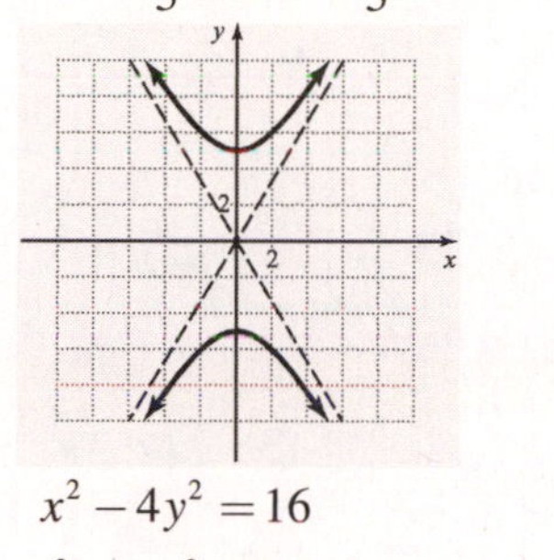

12. $$x^2-4y^2=16$$
$$\frac{x^2}{16}-\frac{4y^2}{16}=\frac{16}{16}$$
$$\frac{x^2}{16}-\frac{y^2}{4}=1$$
$$\frac{x^2}{4^2}-\frac{y^2}{2^2}=1$$
$a=4,\ b=2$
x-intercepts are $(\pm4,0)$ or
$(4,0)$ and $(-4,0)$
Equations of asymptotes are
$y=\pm\frac{2}{4}x$ or $y=\frac{1}{2}x$ and $y=-\frac{1}{2}x$

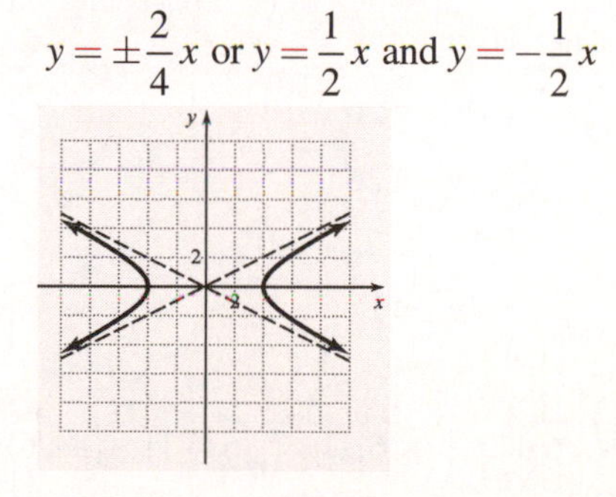

13. $x^2 + y^2 = 9; \quad x = y - 3$

$$(y-3)^2 + y^2 = 9$$
$$y^2 - 6y + 9 + y^2 = 9$$
$$2y^2 - 6y = 0$$
$$2y(y-3) = 0$$
$$2y = 0 \quad \text{or} \quad y - 3 = 0$$
$$y = 0 \qquad y = 3$$
$$x = y - 3 \qquad x = y - 3$$
$$= 0 - 3 \qquad = 3 - 3$$
$$x = -3 \qquad x = 0$$

$(-3,0)$ and $(0,3)$

14. $y = x^2 - 2; \quad y = -x^2 - 6$

$$x^2 - 2 = -x^2 - 6$$
$$2x^2 = -4$$
$$x^2 = -2$$
$$x = \pm\sqrt{-2} = \pm i\sqrt{2}$$
$$y = x^2 - 2$$
$$y = -2 - 2 = -4$$

$(-i\sqrt{2}, -4)$ and $(i\sqrt{2}, -4)$

15.

$$\begin{array}{l} 4x^2 + 2y^2 = 26 \\ \underline{5x^2 - 2y^2 = 28} \\ 9x^2 \qquad\quad = 54 \end{array}$$
$$x^2 = 6$$
$$x = \pm\sqrt{6}$$
$$4x^2 + 2y^2 = 26$$
$$4(6) + 2y^2 = 26$$
$$24 + 2y^2 = 26$$
$$2y^2 = 2$$
$$y^2 = 1$$
$$y = \pm\sqrt{1} = \pm 1$$

$(\sqrt{6}, 1)$, $(-\sqrt{6}, 1)$, $(\sqrt{6}, -1)$, $(-\sqrt{6}, -1)$

16. $x^2 - 9y^2 < 36$

Boundary line is dashed.

$$x^2 - 9y^2 = 36$$
$$\frac{x^2}{36} - \frac{9y^2}{36} = \frac{36}{36}$$
$$\frac{x^2}{36} - \frac{y^2}{4} = 1$$
$$\frac{x^2}{6^2} - \frac{y^2}{2^2} = 1$$

$a = 6, \; b = 2$

x-intercepts are $(\pm 6, 0)$ or $(6,0)$ and $(-6,0)$

Equations of asymptotes are

$y = \pm\frac{2}{6}x$ or $y = \frac{1}{3}x$ and $y = -\frac{1}{3}x$

Test point is $(0,0)$.

$$(0)^2 + 9(0)^2 \overset{?}{<} 36$$

$0 + 0 < 36$ True. Shade the region containing $(0,0)$.

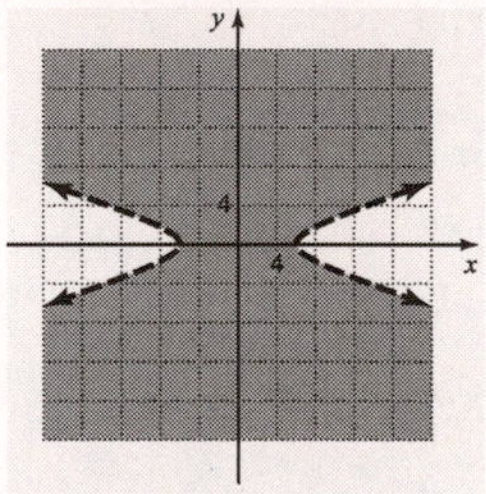

17. $x^2 + (y+3)^2 \le 4$

Boundary line is solid.

$$(x-0)^2 + (y-(-3))^2 = 4$$

Center: $(0,-3)$, radius $= 2$

Test point is $(0,0)$.

$$(0)^2 + (0+3)^2 \overset{?}{\le} 4$$

$9 \le 4$ False, shade the region not containing the point $(0,0)$.

$8x^2 + 32y^2 \ge 128$

Boundary line is solid.

17. (continued)

$$8x^2+32y^2=128$$
$$\frac{8x^2}{128}+\frac{32y^2}{128}=\frac{128}{128}$$
$$\frac{x^2}{16}+\frac{y^2}{4}=1$$
$$\frac{x^2}{4^2}+\frac{y^2}{2^2}=1$$
$$a=4,\ b=2$$

x-intercepts: $(\pm4,0)$ or $(4,0)$ and $(-4,0)$

y-intercepts: $(0,\pm2)$ or $(0,2)$ and $(0,-2)$

Test point is $(0,0)$.

$8(0)^2+32(0)^2\geq128$

$0+0\geq128$ False. Shade the region not containing the point $(0,0)$.

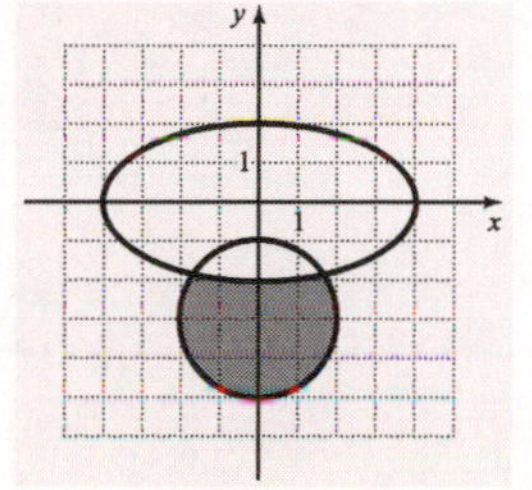

18.

The parabola has vertex at $(0,0)$ and is of the form $x=a(y-0)^2+0.$ $x=ay^2$ It passes through the point $\left(2,\frac{15}{2}\right)$.

$$x=ay^2$$
$$2=a\left(\frac{15}{2}\right)^2$$
$$2=\frac{225}{4}a$$
$$a=2\left(\frac{4}{225}\right)=\frac{8}{225}$$
$$x=\frac{8}{225}y^2$$

19. $\frac{x^2}{354^2}+\frac{y^2}{181^2}=1$

The maximum distance from the sun to the comet (sun to farthest vertex) is 658 million km ($r+r+s$) in the diagram below. That is the distance from the sun to the center of the ellipse plus the distance from the center of the ellipse to the vertex. The distance from the center of the ellipse to the vertex is $a=r+s=354$ million km. The distance from the sun to the center is $r=658-354=304$ million km. The distance from the sun (focus) to the nearest vertex is $s=354-r=354-304=50$. The minimum distance to the sun is about 50 million km.

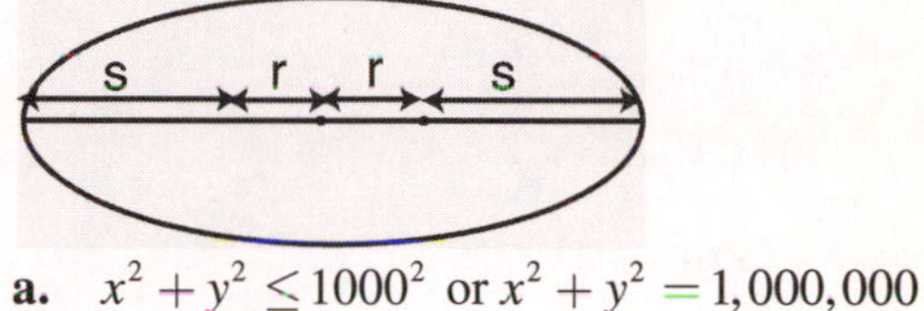

20. **a.** $x^2+y^2\leq1000^2$ or $x^2+y^2=1{,}000{,}000$

Boundary line is solid.

$x^2+y^2=1000^2$

Center: $(0,0)$ radius $=1000$

Test point is $(0,0)$.

$0^2+0^2\overset{?}{\leq}1000^2$

$0\leq1000^2$ True. Shade the region containing the point $(0,0)$.

b.

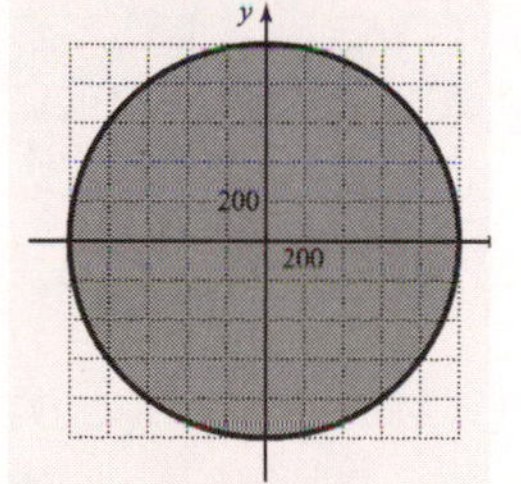

11.1 Introduction to Conics; The Parabola

Practice

1. $y=5(x+1)^2-3$

$h=-1,\ k=-3$

Vertex: $(-1,-3)$

Axis of symmetry: $x=-1$

$y=5(0+1)^2-3$

$=5(1)-3$

-2

y-intercept: $(0,-2)$

$0=5(x+1)^2-3$

$3=5(x+1)^2$

$\frac{3}{5}=(x+1)^2$

$\pm\sqrt{\frac{3}{5}}=x+1$

$x=-1\pm\sqrt{\frac{3}{5}}$

x-intercepts: $\left(-1+\sqrt{\frac{3}{5}},0\right),\left(-1-\sqrt{\frac{3}{5}},0\right)$

2. **a.** $y=4x^2-8x+7$

$y=4(x^2-2x)+7$

$y=4(x^2-2x+1)+7-4$

$y=4(x-1)^2+3$

b. Vertex: $(1,3)$

Axis of symmetry: $x=1$

c. $y=4(0)^2-8(0)+7$

$y=7$

y-intercept: $(0,7)$

$0=4x^2-8x+7$

$b^2-4ac=(-8)^2-4(4)(7)$

$=64-112=-48$

Since the discriminant is negative, the equation has no real solution. So the parabola has no x-intercept.

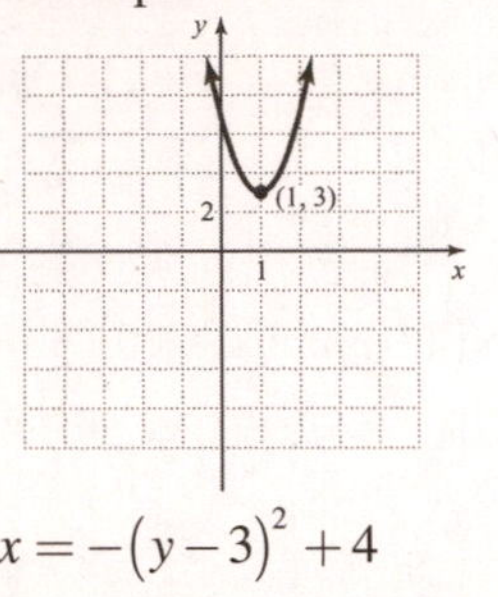

3. $x=-(y-3)^2+4$

$h=4,\ k=3$

Vertex: $(4,3)$

Axis of symmetry: $y=3$

$0=-(y-3)^2+4$

$-4=-(y-3)^2$

$4=(y-3)^2$

$\pm\sqrt{4}=y-3$

$y=3\pm\sqrt{4}$

$y=5$ or $y=1$

y-intercepts: $(0,1)$ and $(0,5)$

$x=-(0-3)^2+4$

$x=-9+4$

$x=-5$

x-intercept: $(-5,0)$

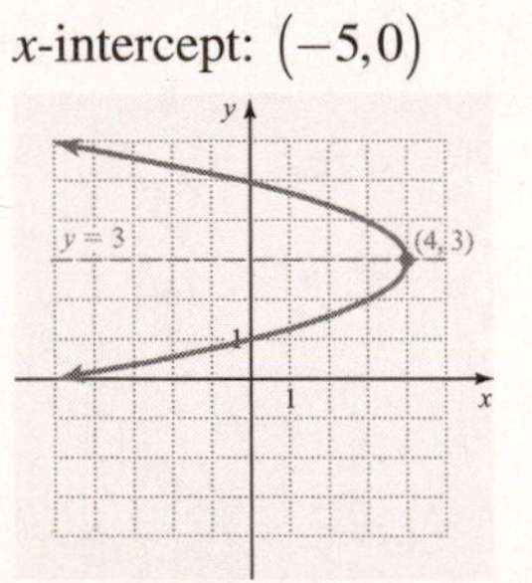

4. $x=-5y^2-20y-9$

$x=-5(y^2+4y)-9$

$x=-5(y^2+4y+4)-9+20$

$x=-5(y+2)^2+11$

Vertex: $(11,-2)$

5. **a.** The vertex is $(2,64)$, so the equation has the form $h=a(t-2)^2+64$. The rocket is launched from the ground, so the point $(0,0)$ is on the graph.

$$0=a(0-2)^2+64$$
$$-64=a(-2)^2$$
$$-64=4a$$
$$a=-16$$

The equation is $h=-16(t-2)^2+64$.

b. $h=-16(1-2)^2+64$

$$=-16(-1)^2+64$$
$$=-16+64$$
$$=48$$

The rocket is 48 ft high 1 sec after it is launched.

Exercises

1. The equation of a parabola that opens upward or downward is in standard form if it is written as $y=a(x-h)^2+k$, where (h,k) is the parabola's vertex and $x=h$ is the equation of the axis of symmetry.

3. The equation of a parabola that opens to the left or to the right is in standard form if it is written as $x=(y-k)^2+h$, where (h,k) is the parabola's vertex and $y=k$ is the equation of the axis of symmetry.

5. $y=(x-3)^2-4$

$h=3,\ k=-4$

Vertex: $(3,-4)$; axis of symmetry: $x=3$

x	$y=(x-3)^2-4$	(x,y)
0	$(0-3)^2-4=5$	$(0,5)$
1	$(1-3)^2-4=0$	$(1,0)$
2	$(2-3)^2-4=-3$	$(2,-3)$
3	$(3-3)^2-4=-4$	$(3,-4)$
4	$(4-3)^2-4=-3$	$(4,-3)$
5	$(5-3)^2-4=0$	$(5,0)$
6	$(6-3)^2-4=5$	$(6,5)$

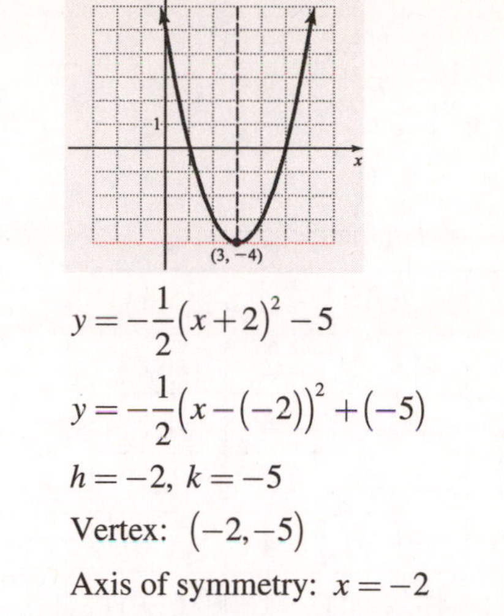

7. $y=-\frac{1}{2}(x+2)^2-5$

$y=-\frac{1}{2}(x-(-2))^2+(-5)$

$h=-2,\ k=-5$

Vertex: $(-2,-5)$

Axis of symmetry: $x=-2$

x	$y=-\frac{1}{2}(x+2)^2-5$	(x,y)
−5	$-\frac{1}{2}(-5+2)^2-5=-\frac{19}{2}$	$\left(-5,-\frac{19}{2}\right)$
−4	$-\frac{1}{2}(-4+2)^2-5=-7$	$(-4,-7)$
−3	$-\frac{1}{2}(-3+2)^2-5=-\frac{11}{2}$	$\left(-3,-\frac{11}{2}\right)$
−2	$-\frac{1}{2}(-2+2)^2-5=-5$	$(-2,-5)$
−1	$-\frac{1}{2}(-1+2)^2-5=-\frac{11}{2}$	$\left(-1,-\frac{11}{2}\right)$
0	$-\frac{1}{2}(0+2)^2-5=-7$	$(0,-7)$
1	$-\frac{1}{2}(1+2)^2-5=-\frac{19}{2}$	$\left(1,-\frac{19}{2}\right)$

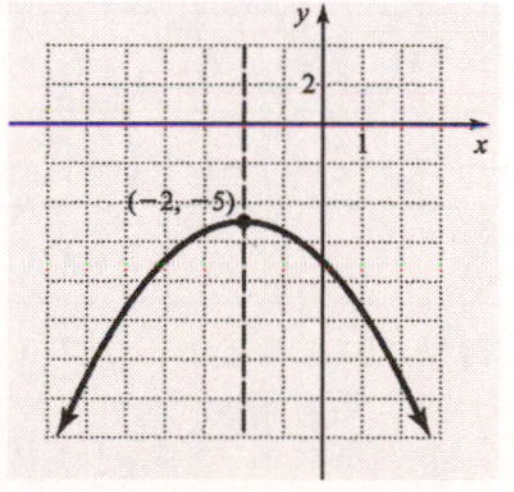

9. $x=4(y-1)^2$

$x=4(y-1)^2+0$

$h=0;\ k=1$

Vertex: $(0,1)$

Axis of symmetry: $y=1$

y	$x=4(y^2-1)$	(x,y)
3	$4(3-1)^2=16$	$(16,3)$
2	$4(2-1)^2=4$	$(4,2)$
1	$4(1-1)^2=0$	$(0,1)$
0	$4(0-1)^2=4$	$(4,0)$
-1	$4(-1-1)^2=16$	$(16,-1)$

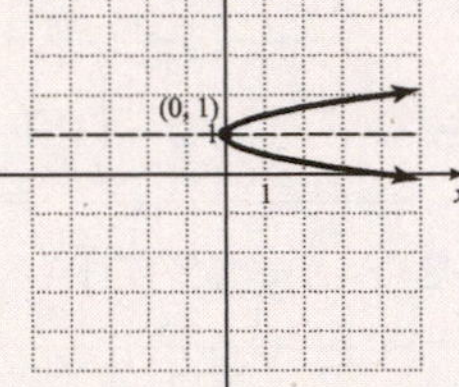

11. $x=-y^2+2y+2$

$$x=-\left(y^2-2y+\left(\frac{-2}{2}\right)^2\right)+2+\left(\frac{-2}{2}\right)^2$$

$x=-(y-1)^2+3$

$h=3,\ k=1$

Vertex: $(3,1)$

Axis of symmetry: $y=1$

y	$x=-y^2+2y+2$	(x,y)
-2	$-(-2)^2+2(-2)+2=-6$	$(-6,-2)$
-1	$-(-1)^2+2(-1)+2=-1$	$(-1,-1)$
0	$-(0)^2+2(0)+2=2$	$(2,0)$
1	$-(1)^2+2(1)+2=3$	$(3,1)$
2	$-(2)^2+2(2)+2=2$	$(2,2)$
3	$-(3)^2+2(3)+2=-1$	$(-1,3)$
4	$-(4)^2+2(4)+2=-6$	$(-6,4)$

13. $y=-2x^2+12x-10$

$y=-2(x^2-6x)-10$

$$y=-2\left(x^2-6x+\left(\frac{-6}{2}\right)^2\right)-10+2\left(\frac{-6}{2}\right)^2$$

$y=-2(x-3)^2+8$

$h=3,\ k=8$

Vertex: $(3,8)$

Axis of symmetry: $x=3$

x	$y=-2x^2+12x-10$	(x,y)
0	$-2(0)^2+12(0)-10=-10$	$(0,-10)$
1	$-2(1)^2+12(1)-10=0$	$(1,0)$
2	$-2(2)^2+12(2)-10=6$	$(2,6)$
3	$-2(3)^2+12(3)-10=8$	$(3,8)$
4	$-2(4)^2+12(4)-10=6$	$(4,6)$
5	$-2(5)^2+12(5)-10=0$	$(5,0)$
6	$-2(6)^2+12(6)-10=-10$	$(6,-10)$

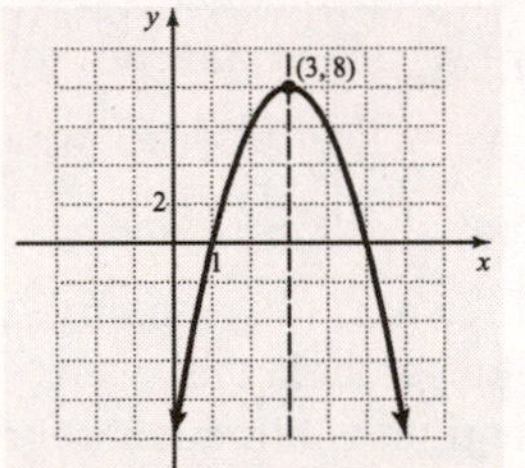

15. $x=3y^2+12y+10$

$x=3(y^2+4y)+10$

$$x=3\left(y^2+4y+\left(\frac{4}{2}\right)^2\right)+10-3\left(\frac{4}{2}\right)^2$$

$x=3(y+2)^2-2$

$x=3(y-(-2))^2-2$

$h=-2,\ k=-2$

Vertex: $(-2,-2)$

Axis of symmetry: $y=-2$

15. (continued)

y	$x=3y^2+12y+10$	(x,y)
1	$3(1)^2+12(1)+10=25$	$(25,1)$
0	$3(0)^2+12(0)+10=10$	$(10,0)$
−1	$3(-1)^2+12(-1)+10=1$	$(1,-1)$
−2	$3(-2)^2+12(-2)+10=-2$	$(-2,-2)$
−3	$3(-3)^2+12(-3)+10=1$	$(1,-3)$
−4	$3(-4)^2+12(-4)+10=10$	$(10,-4)$

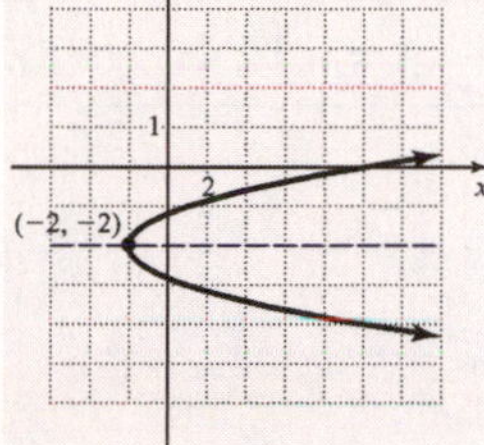

17. $y=x^2-x+1$

$$y=x^2-x+\left(\frac{-1}{2}\right)^2+1-\left(\frac{-1}{2}\right)^2$$

$$y=\left(x-\frac{1}{2}\right)^2+\frac{3}{4}$$

$$h=\frac{1}{2},\ k=\frac{3}{4}$$

Vertex: $\left(\frac{1}{2},\frac{3}{4}\right)$

19. $y=5x^2-50x+57$

$$y=5\left(x^2-10x\right)+57$$

$$y=5\left(x^2-10x+\left(\frac{-10}{2}\right)^2\right)+57-5\left(\frac{-10}{2}\right)^2$$

$$y=5\left(x^2-10x+25\right)+57-5\bullet 25$$

$$y=5(x-5)^2-68$$

$$h=5,\ k=-68$$

Vertex: $(5,-68)$

21. $x=-2y^2-32y-95$

$$x=-2\left(y^2+16y\right)-95$$

$$x=-2\left(y^2+16y+\left(\frac{16}{2}\right)^2\right)-95-(-2)\left(\frac{16}{2}\right)^2$$

$$x=-2(y+8)^2-95+128$$

$$x=-2(y+8)^2+33$$

$$x=-2(y-(-8))^2+33$$

$$h=33, k=-8$$

Vertex: $(33,-8)$

23. $x=3y^2+9y+11$

$$x=3\left(y^2+3y\right)+11$$

$$x=3\left(y^2+3y+\left(\frac{3}{2}\right)^2\right)+11-3\left(\frac{3}{2}\right)^2$$

$$x=3\left(y+\frac{3}{2}\right)^2+11-\frac{27}{4}$$

$$x=3\left(y-\left(-\frac{3}{2}\right)\right)^2+\frac{17}{4}$$

$$h=\frac{17}{4},\ k=-\frac{3}{2}$$

Vertex: $\left(\frac{17}{4},-\frac{3}{2}\right)$

25. $h=-2,\ k=7$

$$y=a(x-(-2))^2+7$$

$$y=a(x+2)^2+7$$

Passes through $(3,12)$

$$12=a(3+2)^2+7$$

$$5=a(5)^2$$

$$5=25a$$

$$a=\frac{5}{25}=\frac{1}{5}$$

$$y=\frac{1}{5}(x+2)^2+7$$

27. $h=4,\ k=0$

$x=a(y-0)^2+4$

Passes through (0, 1)

$0=a(1-0)^2+4$

$-4=a(1)$

$a=-4$

$x=-4(y-0)^2+4$

$x=-4y^2+4$

29. $y=x^2+5x+2$

$y=\left[x^2+5x+\left(\frac{5}{2}\right)^2\right]+\left[2-\left(\frac{5}{2}\right)^2\right]$

$y=\left(x+\frac{5}{2}\right)^2-\frac{17}{4}$

$h=-\frac{5}{2},\ k=-\frac{17}{4}$

Vertex: $\left(-\frac{5}{2},-\frac{17}{4}\right)$

31. $h=2,\ k=-5$

$y=a(x-2)^2-5$

The graph passes through (5, –2), so

$-2=a(5-2)^2-5$

$3=9a$

$\frac{1}{3}=a$

The equation of the graph is

$y=\frac{1}{3}(x-2)^2-5.$

33. $x=-2y^2-4y+1$

$x=-2(y^2+2y)+1$

$x=-2(y^2+2y+1)+(1-(-2)(1))$

$x=-2(y+1)^2+3$

$x=-2(y-(-1))^2+3$

Vertex: $(3,-1)$

Axis of symmetry: $y=-1$

y	$x=-2y^2-4y+1$	(x, y)
–4	$-2(-4)^2-4(-4)+1=-15$	(–15, –4)
–3	$-2(-3)^2-4(-3)+1=-5$	(–5, –3)
–2	$-2(-2)^2-4(-2)+1=1$	(1, –2)
–1	$-2(-1)^2-4(-1)+1=3$	(3, –1)
0	$-2(0)^2-4(0)+1=1$	(1, 0)
1	$-2(1)^2-4(1)+1=-5$	(–5, 1)
2	$-2(2)^2-4(2)+1=-15$	(–15, 2)

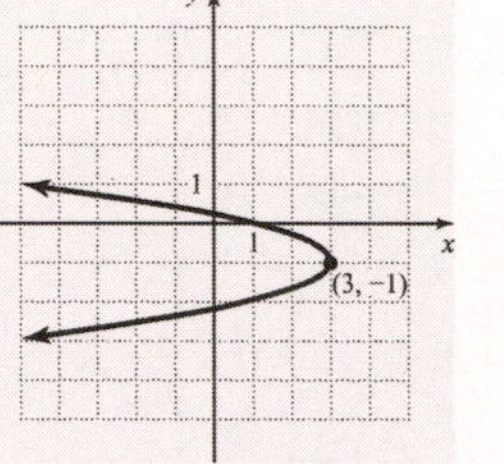

35. $R=-(x-40)^2+1600$

$h=40,\ k=1600$

Vertex: $(40,1600)$

(40, 1600); it shows that the company will have a maximum revenue of $1600 when 40 bottles of perfume are sold.

37. **a.** Perimeter = $2l+2w=500$

$2l=500-2w$

$l=\frac{500-2w}{2}=250-w$

Area = $A=lw=(250-w)w$

$A=-w^2+250w$

$A=-(w^2-250w)$

$A=-\left[w^2-250w+\left(\frac{-250}{2}\right)^2\right]+\left(\frac{-250}{2}\right)^2$

$A=-(w^2-250w+15625)+15625$

$A=-(w-125)^2+15{,}625$

Vertex: $(125,\ 15{,}625)$

$l=250-w=250-125=125$

b. Dimensions of 125 ft by 125 ft will produce a field with a maximum area of 15,625 sq ft.

39. The vertex is at (0, 0), and the parabola passes through (6, 20).
The parabola is of the form

$x = a(y-k)^2 + h; \quad h = 0, \ k = 0$

$x = ay^2$

(6,20) is on the parabola.

$6 = a(20)^2 = 400a$

$a = \frac{6}{400} = \frac{3}{200}$

$x = \frac{3}{200}y^2$

11.2 The Circle

Practice

1. $d = \sqrt{(x_2 - x_1)^2 + (y_2 - y_1)^2}$

$= \sqrt{(0-1)^2 + (-5-(-2))^2}$

$= \sqrt{(-1)^2 + (-3)^2}$

$= \sqrt{1+9}$

$= \sqrt{10} \approx 3.2$ units

2. $\frac{x_1 + x_2}{2} = \frac{3+7}{2} = \frac{10}{2} = 5$

$\frac{y_1 + y_2}{2} = \frac{1+4}{2} = \frac{5}{2}$

The midpoint is $\left(5, \frac{5}{2}\right)$.

3. $x^2 + y^2 = 9$

$x^2 + y^2 = 3^3$

Center: $(0,0)$; radius: 3

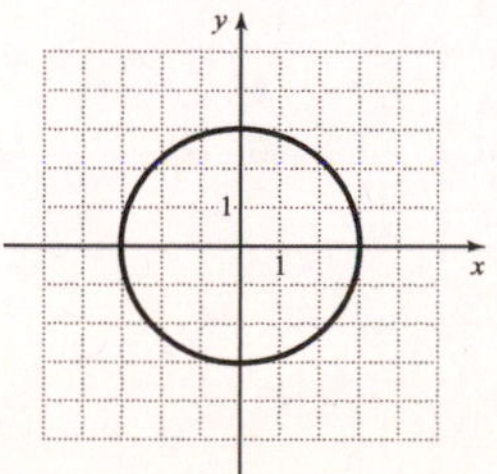

4. **a.** $(x+6)^2 + (y-2)^2 = 100$

$(x+6)^2 + (y-2)^2 = 10^2$

Center: $(-6,2)$; radius: 10

b.

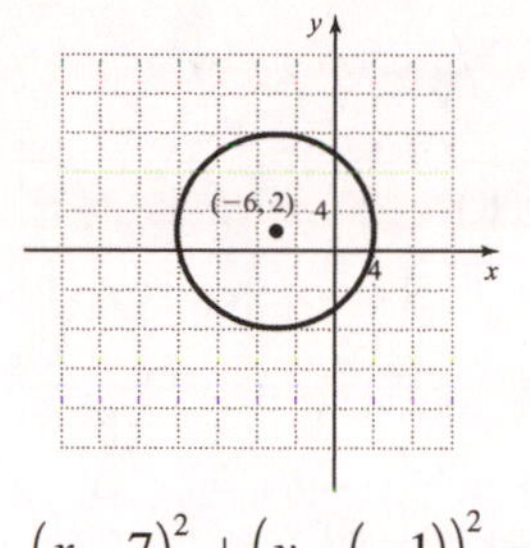

5. $(x-7)^2 + (y-(-1))^2 = 8^2$

$(x-7)^2 + (y+1)^2 = 64$

6. $x^2 + y^2 - 4x - 2y - 19 = 0$

$x^2 + y^2 - 4x - 2y = 19$

$(x^2 - 4x) + (y^2 - 2y) = 19$

$(x^2 - 4x + 4) + (y^2 - 2y + 1) = 19 + 4 + 1$

$(x-2)^2 + (y-1)^2 = 24$

$(x-2)^2 + (y-1)^2 = (2\sqrt{6})^2$

Center: $(2,1)$; radius: $2\sqrt{6}$

7. **a.** Center: $(30,40)$; radius: 10

$(x-30)^2 + (y-40)^2 = 10^2$

$(x-30)^2 + (y-40)^2 = 100$

b. $d = \sqrt{(x_2 - x_1)^2 + (y_2 - y_1)^2}$

$= \sqrt{(35-30)^2 + (18-40)^2}$

$= \sqrt{5^2 + (-22)^2}$

$= \sqrt{25 + 484}$

$= \sqrt{509} \approx 22.6 > 10$

No, the visitor cannot surf the Web.

Exercises

1. The distance between any two points (x_1, y_1) and (x_2, y_2) is equal to the square root of <u>the sum of</u> $(x_2 - x_1)^2$ and $(y_2 - y_1)^2$.

3. A(n) circle is the set of all points on a coordinate plane that are a fixed distance from a given point.

5. The equation of a circle centered at (h,k) with radius r is $(x-h)^2+(y-k)^2=r^2$.

7.
$$d=\sqrt{(x_2-x_1)^2+(y_2-y_1)^2}$$
$$d=\sqrt{(9-3)^2+(10-2)^2}=\sqrt{(6)^2+(8)^2}$$
$$d=\sqrt{36+64}=\sqrt{100}=10 \text{ units}$$

9.
$$d=\sqrt{(x_2-x_1)^2+(y_2-y_1)^2}$$
$$d=\sqrt{(-1-1)^2+(2-6)^2}$$
$$d=\sqrt{(-2)^2+(-4)^2}$$
$$d=\sqrt{4+16}=\sqrt{20}=2\sqrt{5}$$
$$d=4.4721...\approx 4.5 \text{ units}$$

11.
$$d=\sqrt{(x_2-x_1)^2+(y_2-y_1)^2}$$
$$d=\sqrt{(5-8)^2+(-6-(-4))^2}$$
$$d=\sqrt{(-3)^2+(-2)^2}$$
$$d=\sqrt{9+4}=\sqrt{13}=3.6055...\approx 3.6 \text{ units}$$

13.
$$d=\sqrt{(x_2-x_1)^2+(y_2-y_1)^2}$$
$$d=\sqrt{(-1.2-(-3.2))^2+(-5.3-1.7)^2}$$
$$d=\sqrt{(2)^2+(-7)^2}=\sqrt{4+49}=\sqrt{53}$$
$$d=7.2801...\approx 7.3 \text{ units}$$

15.
$$d=\sqrt{(x_2-x_1)^2+(y_2-y_1)^2}$$
$$d=\sqrt{\left(\frac{1}{2}-\frac{1}{4}\right)^2+\left(\frac{2}{3}-\frac{1}{3}\right)^2}$$
$$=\sqrt{\left(\frac{1}{4}\right)^2+\left(\frac{1}{3}\right)^2}$$
$$=\sqrt{\frac{1}{16}+\frac{1}{9}}=\sqrt{\frac{9}{144}+\frac{16}{144}}$$
$$=\sqrt{\frac{25}{144}}=\frac{5}{12} \text{ units}$$

17.
$$d=\sqrt{(x_2-x_1)^2+(y_2-y_1)^2}$$
$$d=\sqrt{\left(\sqrt{3}-(-4\sqrt{3})\right)^2+\left(-3\sqrt{2}-(-2\sqrt{2})\right)^2}$$
$$d=\sqrt{(5\sqrt{3})^2+(-\sqrt{2})^2}=\sqrt{25\bullet 3+2}=\sqrt{77}$$
$$d=8.7749...\approx 8.8 \text{ units}$$

19.
$$\left(\frac{5+7}{2},\frac{9+1}{2}\right)=\left(\frac{12}{2},\frac{10}{2}\right)=(6,5)$$

21.
$$\left(\frac{3+(-12)}{2},\frac{11+(-1)}{2}\right)=\left(\frac{-9}{2},\frac{10}{2}\right)$$
$$=\left(-\frac{9}{2},5\right)$$

23.
$$\left(\frac{-8+(-13)}{2},\frac{-11+(-2)}{2}\right)=\left(-\frac{21}{2},-\frac{13}{2}\right)$$

25.
$$\left(\frac{3.4+0.6}{2},\frac{-1.1+(-3.9)}{2}\right)=\left(\frac{4}{2},\frac{-5}{2}\right)$$
$$=(2,-2.5)$$

27.
$$\left(\frac{\frac{3}{4}+\left(-\frac{1}{3}\right)}{2},\frac{-\frac{2}{5}+\frac{1}{2}}{2}\right)=\left(\frac{\frac{9}{12}-\frac{4}{12}}{2},\frac{-\frac{4}{10}+\frac{5}{10}}{2}\right)$$
$$=\left(\frac{\frac{5}{12}}{2},\frac{\frac{1}{10}}{2}\right)$$
$$=\left(\frac{\frac{5}{12}\bullet\frac{1}{2}}{2\bullet\frac{1}{2}},\frac{\frac{1}{10}\bullet\frac{1}{2}}{2\bullet\frac{1}{2}}\right)$$
$$=\left(\frac{5}{24},\frac{1}{20}\right)$$

29.
$$\left(\frac{-7\sqrt{3}+7\sqrt{3}}{2},\frac{5\sqrt{6}+3\sqrt{6}}{2}\right)=\left(\frac{0}{2},\frac{8\sqrt{6}}{2}\right)$$
$$=(0,4\sqrt{6})$$

31.
$$x^2+y^2=36$$
$$(x-0)^2+(y-0)^2=6^2 \quad h=0,\ k=0$$
Center: $(0,0)$, radius: 6

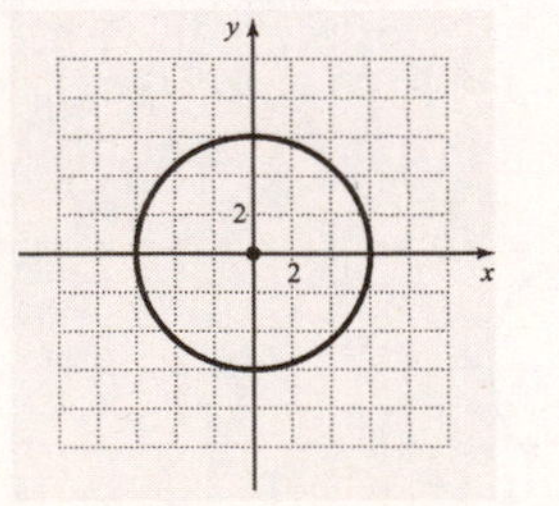

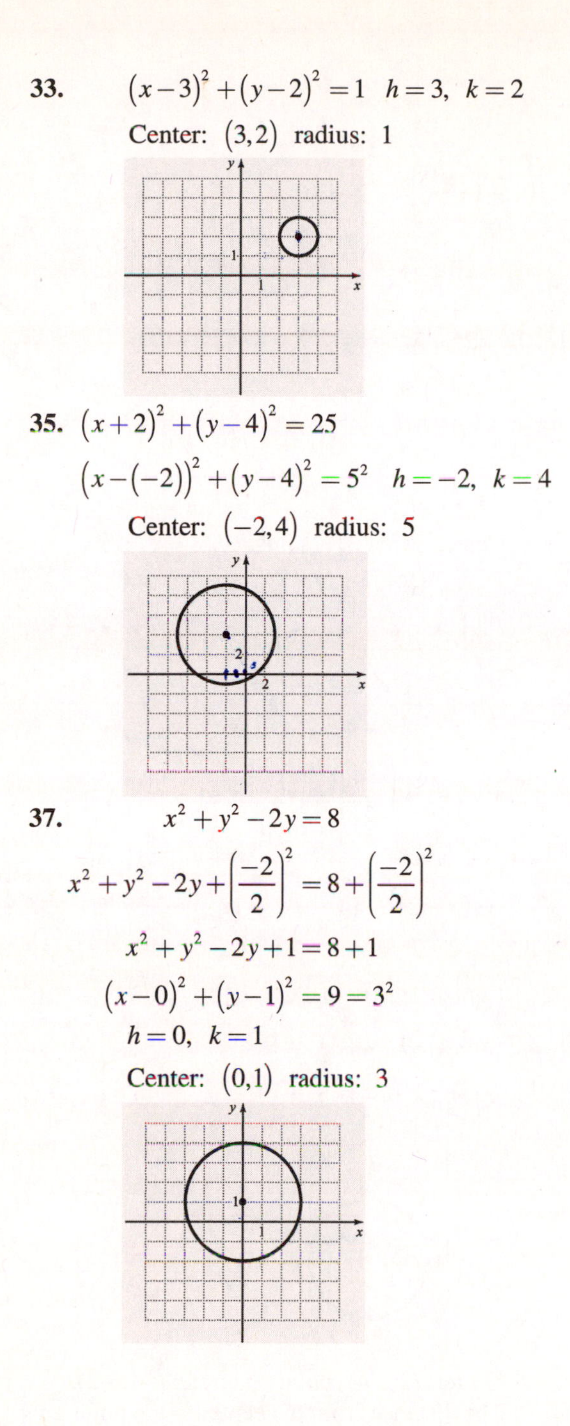

33. $(x-3)^2+(y-2)^2=1 \quad h=3,\ k=2$

Center: $(3,2)$ radius: 1

35. $(x+2)^2+(y-4)^2=25$

$(x-(-2))^2+(y-4)^2=5^2 \quad h=-2,\ k=4$

Center: $(-2,4)$ radius: 5

37. $x^2+y^2-2y=8$

$x^2+y^2-2y+\left(\frac{-2}{2}\right)^2=8+\left(\frac{-2}{2}\right)^2$

$x^2+y^2-2y+1=8+1$

$(x-0)^2+(y-1)^2=9=3^2$

$h=0,\ k=1$

Center: $(0,1)$ radius: 3

39. $x^2+y^2+2x+2y-23=0$

$x^2+2x+y^2+2y=23$

$x^2+2x+\left(\frac{2}{2}\right)^2+y^2+2y+\left(\frac{2}{2}\right)^2$

$=23+\left(\frac{2}{2}\right)^2+\left(\frac{2}{2}\right)^2$

$x^2+2x+1+y^2+2y+1=23+1+1$

$(x+1)^2+(y+1)^2=25$

$(x-(-1))^2+(y-(-1))^2=5^2$

$h=-1,\ k=-1$

Center: $(-1,-1)$ radius: 5

41.

$x^2+y^2-8x-6y+16=0$

$x^2-8x+y^2-6y=-16$

$x^2-8x+\left(\frac{-8}{2}\right)^2+y^2-6y+\left(\frac{-6}{2}\right)^2$

$=-16+\left(\frac{-8}{2}\right)^2+\left(\frac{-6}{2}\right)^2$

$x^2-8x+16+y^2-6y+9=-16+16+9$

$(x-4)^2+(y-3)^2=9=3^2$

$h=4,\ k=3$

Center: $(4,3)$ radius: 3

43. $(x-(-7))^2+(y-2)^2=9^2$

$(x+7)^2+(y-2)^2=81$

45. $(x-0)^2+(y-4)^2=\left(\sqrt{5}\right)^2$

$x^2+(y-4)^2=5$

47. The distance from the center to the point on the circle is the radius.

$$r=\sqrt{(-1-3)^2+(9-5)^2}=\sqrt{(-4)^2+(4)^2}$$
$$r=\sqrt{16+16}=\sqrt{32}$$
$$r^2=\left(\sqrt{32}\right)^2=32$$
$$(x-3)^2+(y-5)^2=32$$

49. The midpoint of the diameter is the center.

$$\left(\frac{-6+2}{2},\frac{1+11}{2}\right)=\left(\frac{-4}{2},\frac{12}{2}\right)=(-2,6)$$

The distance from the center to the endpoint of the diameter is the radius. Using (2, 11),

$$r=\sqrt{(2-(-2))^2+(11-6)^2}=\sqrt{(4)^2+(5)^2}$$
$$r=\sqrt{16+25}=\sqrt{41}$$
$$(x-(-2))^2+(y-6)^2=\left(\sqrt{41}\right)^2$$
$$(x+2)^2+(y-6)^2=41$$

51.
$$x^2+y^2+3x+4y=0$$
$$x^2+3x+y^2+4y=0$$
$$x^2+3x+\left(\frac{3}{2}\right)^2+y^2+4y+\left(\frac{4}{2}\right)^2=\left(\frac{3}{2}\right)^2+\left(\frac{4}{2}\right)^2$$
$$x^2+3x+\frac{9}{4}+y^2+4y+4=\frac{9}{4}+4$$
$$\left(x+\frac{3}{2}\right)^2+(y+2)^2=\frac{9}{4}+\frac{16}{4}=\frac{25}{4}$$
$$\left(x-\left(-\frac{3}{2}\right)\right)^2+(y-(-2))^2=\left(\frac{5}{2}\right)^2$$
$$h=-\frac{3}{2},\ k=-2$$

Center: $\left(-\frac{3}{2},-2\right)$ radius: $\frac{5}{2}$

53.
$$x^2+y^2-10x+6y-4=0$$
$$x^2-10x+y^2+6y=4$$
$$x^2-10x+\left(\frac{-10}{2}\right)^2+y^2+6y+\left(\frac{6}{2}\right)^2=4+\left(\frac{-10}{2}\right)^2+\left(\frac{6}{2}\right)^2$$
$$x^2-10x+25+y^2+6y+9=4+25+9$$
$$(x-5)^2+(y+3)^2=38$$
$$(x-5)^2+(y-(-3))^2=\left(\sqrt{38}\right)^2$$
$$h=5,\ k=-3$$

Center: $(5,-3)$ radius: $\sqrt{38}$

55.
$$3x^2+3y^2-12x-24=0$$
$$3\left(x^2+y^2-4x-8\right)=0$$
$$x^2+y^2-4x-8=0$$
$$x^2-4x+y^2=8$$
$$x^2-4x+\left(\frac{-4}{2}\right)^2+y^2=8+\left(\frac{-4}{2}\right)^2$$
$$x^2-4x+4+y^2=8+4$$
$$(x-2)^2+(y-0)^2=12$$
$$h=2,\ k=0$$

Center: $(2,0)$ radius: $\sqrt{12}=2\sqrt{3}$

57.
$$(x-4)^2+(y+1)^2=9$$
$$h=4,\ k=-1,\ r=\sqrt{9}=3$$

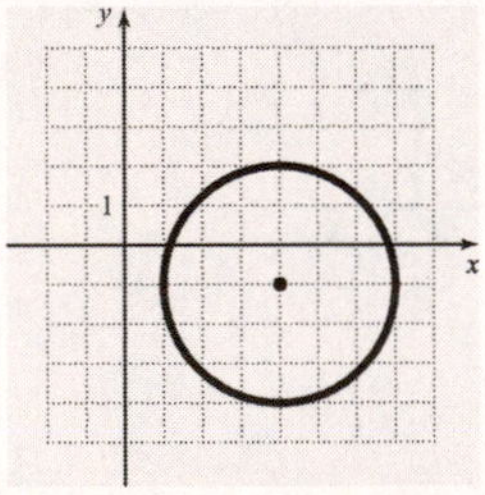

59. Center: (2, −4); point on circle: (−1, −2). The distance from the center to the point on the circle is the radius.

$$r=\sqrt{(-1-2)^2+(-2-(-4))^2}=\sqrt{(-3)^2+(2)^2}$$
$$r=\sqrt{9+4}=\sqrt{13}$$
$$r^2=\left(\sqrt{13}\right)^2=13$$

The equation of the circle is $(x-2)^2+(y+4)^2=13$.

61. $d=\sqrt{(-4-(-6))^2+(7-2)^2}=\sqrt{2^2+5^2}$

$=\sqrt{29}\approx 5.4$ units

63. $\left(\frac{8+(-3)}{2},\frac{5+(-11)}{2}\right)=\left(\frac{5}{2},-\frac{6}{2}\right)=\left(\frac{5}{2},-3\right)$

65. The distance is from (11, 8) to (–1, –1).

$d=\sqrt{(11-(-1))^2+(8-(-1))^2}$

$d=\sqrt{(12)^2+(9)^2}=\sqrt{144+81}=\sqrt{225}=15$

The distance between her home and her office is 15 mi.

67. **a.**

Midpoint of $\overline{AC}$ is D.

$A:\ (2,1)\ C:\ (6,5)$

$\left(\frac{2+6}{2},\frac{1+5}{2}\right)=\left(\frac{8}{2},\frac{6}{2}\right)=(4,3)\quad D:\ (4,3)$

b. Slope of $\overline{BD}$ $B:\ (1,6)\ D:\ (4,3)$

$$m_{\overline{BD}}=\frac{y_2-y_1}{x_2-x_1}=\frac{3-6}{4-1}=\frac{-3}{3}=-1$$

Slope of $\overline{AC}$ $A:\ (2,1)\ C:\ (6,5)$

$$m_{\overline{AC}}=\frac{y_2-y_1}{x_2-x_1}=\frac{5-1}{6-2}=\frac{4}{4}=1$$

The slope of $\overline{BD}$ is –1 and the slope of $\overline{AC}$ is +1. Since the slopes are negative reciprocals $\overline{BD}$ is perpendicular to the base $\overline{AC}$ and is the altitude of the triangle.

c.

Area = $\frac{1}{2}$ base × height.

Area = $\frac{1}{2}\left(\text{length of }\overline{AC}\right)\left(\text{length of }\overline{BD}\right)$.

length of $\overline{AC}=\sqrt{(6-2)^2+(5-1)^2}$

$=\sqrt{(4)^2+(4)^2}$

$=\sqrt{16+16}$

$=\sqrt{32}=4\sqrt{2}$

length of $\overline{BD}=\sqrt{(4-1)^2+(3-6)^2}$

$\sqrt{(3)^2+(-3)^2}=\sqrt{9+9}=\sqrt{18}=3\sqrt{2}$

Area $\frac{1}{2}\left(4\sqrt{2}\right)\left(3\sqrt{2}\right)=\frac{1}{2}\cdot 12\cdot 2=12$

The area of triangle ABC is 12 sq units.

69. Half the diameter of 6 mi. is 3 mi.

$(x-0)^2+(y-0)^2=(3)^2$

$x^2+y^2=9$

71. **a.** Center: (–8, 15) radius: 36

$(x-(-8))^2+(y-15)^2=36^2$

$(x+8)^2+(y-15)^2=1296$

b.

The distance from the center of the circle $(-8,15)$ to the apartment at $(15,-3)$ is $d=\sqrt{(15-(-8))^2+(-3-15)^2}$

$d=\sqrt{(23)^2+(-18)^2}=\sqrt{529+324}=\sqrt{853}$

$d=29.2061...$

Yes, the student can listen to the station's broadcast since $\sqrt{853}\approx 29.2<36$.

11.3 The Ellipse and the Hyperbola

Practice

1. $\frac{x^2}{100}+\frac{y^2}{49}=1$

Center: origin; $a=10$, $b=7$

x-intercepts: $(-10,0)$, $(10,0)$

y-intercepts: $(0,-7)$, $(0,7)$

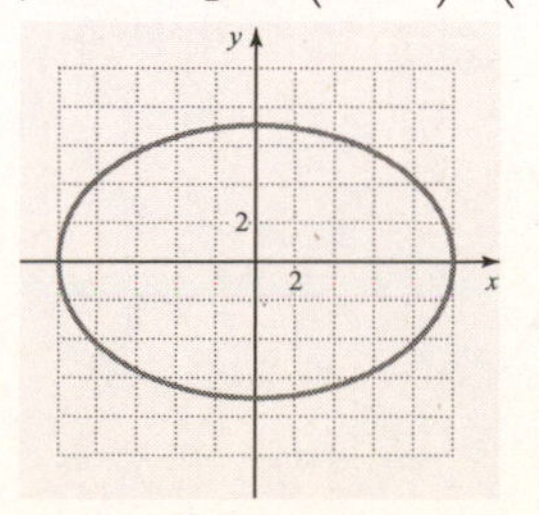

2. $25x^2 + 4y^2 = 100$

$\frac{x^2}{4} + \frac{y^2}{25} = 1$

Center: origin; $a = 2,\ b = 5$

x-intercepts: $(-2,0)$, $(2,0)$

y-intercepts: $(0,-5)$, $(0,5)$

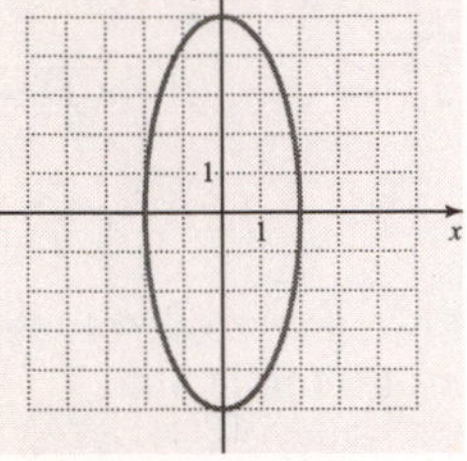

3. $\frac{x^2}{100} + \frac{y^2}{144} = 1$

$a = 10,\ b = 12$

Maximum width: $2 \cdot 10 = 20$ ft

Maximum height: 12 ft

4. $\frac{y^2}{16} - \frac{x^2}{9} = 1$

$a = 3,\ b = 4$

y-intercepts: $(0,-4)$, $(0,4)$

no x-intercepts

asymptotes: $y = \frac{4}{3}x$ and $y = -\frac{4}{3}x$

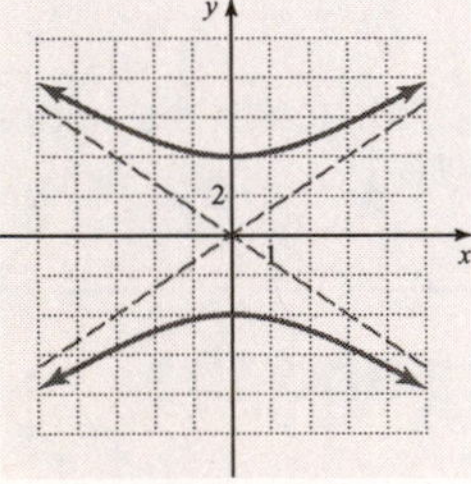

5. $4x^2 - 16y^2 = 144$

$\frac{x^2}{36} - \frac{y^2}{9} = 1$

$a = 6,\ b = 3$

x-intercepts: $(-6,0)$, $(6,0)$

no y-intercepts

asymptotes: $y = \frac{3}{6}x = \frac{1}{2}x$, $y = -\frac{1}{2}x$

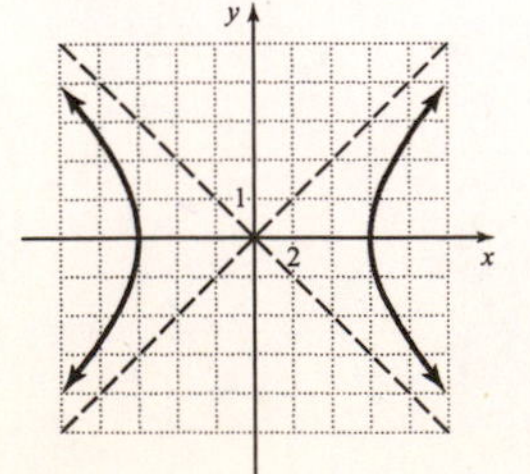

Exercises

1. A(n) ellipse is the set of all points on a coordinate plane the sum of whose distances from two fixed points is constant.

3. A(n) hyperbola is the set of all points on a coordinate plane the difference of whose distances from two fixed points is constant.

5. $\frac{x^2}{16} + \frac{y^2}{4} = 1 \Rightarrow \frac{x^2}{4^2} + \frac{y^2}{2^2} = 1;\quad a = 4,\ b = 2$

x-intercepts: $(\pm 4,0)$ or $(4,0)$ and $(-4,0)$

y-intercepts: $(0,\pm 2)$ or $(0,2)$ and $(0,-2)$

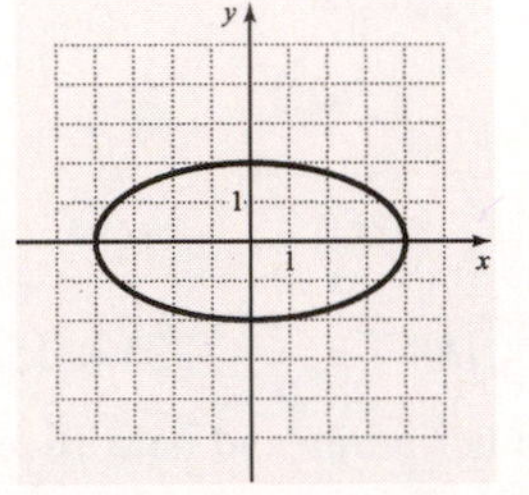

7. $\frac{x^2}{36} + \frac{y^2}{81} = 1 \Rightarrow \frac{x^2}{6^2} + \frac{y^2}{9^2} = 1;\quad a = 6\quad b = 9$

x-intercepts: $(\pm 6,0)$ or $(6,0)$ and $(-6,0)$

y-intercepts: $(0,\pm 9)$ or $(0,9)$ and $(0,-9)$

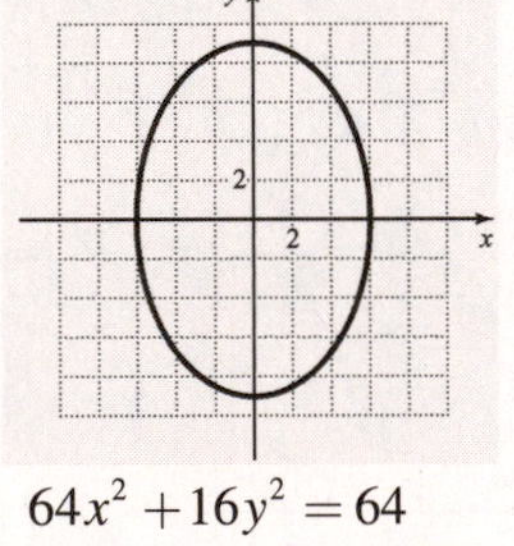

9. $64x^2 + 16y^2 = 64$

$\frac{64x^2}{64} + \frac{16y^2}{64} = \frac{64}{64}$

$\frac{x^2}{1} + \frac{y^2}{4} = 1$

$\frac{x^2}{1^2} + \frac{y^2}{2^2} = 1$

$a = 1, b = 2$

x-intercepts: $(\pm 1,0)$ or $(1,0)$ and $(-1,0)$

y-intercepts: $(0,\pm 2)$ or $(0,2)$ and $(0,-2)$

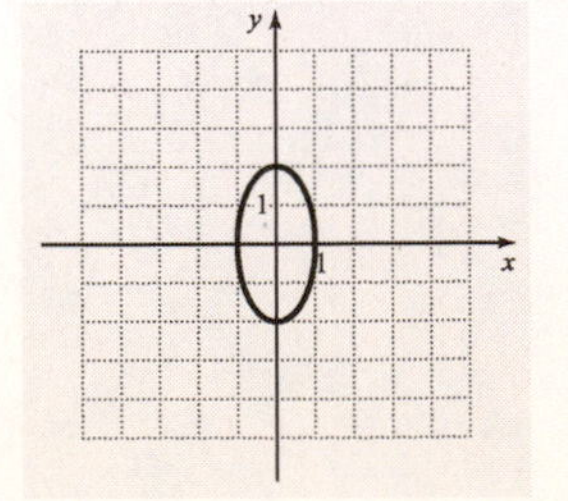

11. $x^2+4y^2=100$

$$\frac{x^2}{100}+\frac{4y^2}{100}=\frac{100}{100}$$

$$\frac{x^2}{100}+\frac{y^2}{25}=1$$

$$\frac{x^2}{10^2}+\frac{y^2}{5^2}=1$$

$a=10;\ b=5$

x-intercepts: $(\pm 10,0)$ or $(10,0)$ and $(-10,0)$

y-intercepts: $(0,\pm 5)$ or $(0,5)$ and $(0,-5)$

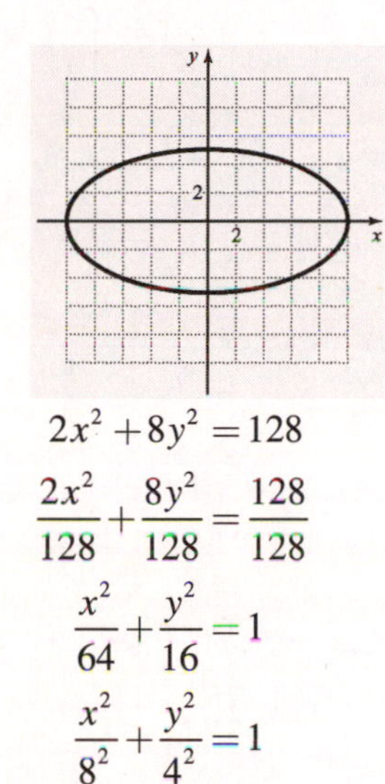

13. $2x^2+8y^2=128$

$$\frac{2x^2}{128}+\frac{8y^2}{128}=\frac{128}{128}$$

$$\frac{x^2}{64}+\frac{y^2}{16}=1$$

$$\frac{x^2}{8^2}+\frac{y^2}{4^2}=1$$

$a=8;\ b=4$

x-intercepts: $(\pm 8,0)$ or $(8,0)$ and $(-8,0)$

y-intercepts: $(0,\pm 4)$ or $(0,4)$ and $(0,-4)$

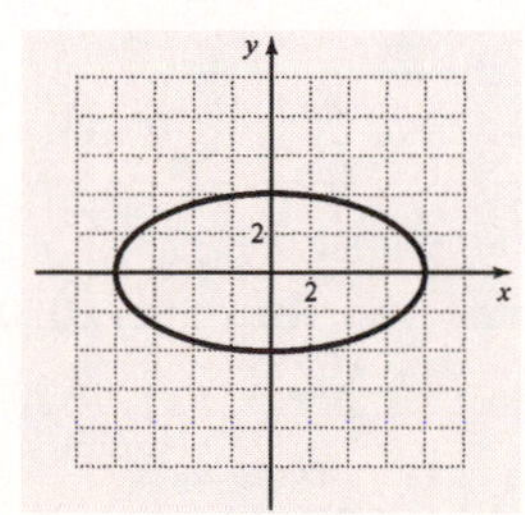

15. $\frac{x^2}{64}-\frac{y^2}{9}=1 \Rightarrow \frac{x^2}{8^2}-\frac{y^2}{3^2}=1;\ a=8,\ b=3$

x-intercepts: $(\pm 8,0)$ or $(8,0)$ and $(-8,0)$

Equation of asymptotes are

$y=\frac{3}{8}x$ and $y=-\frac{3}{8}x$

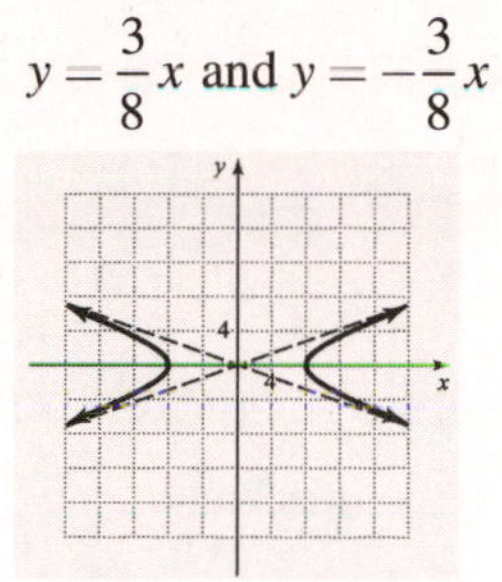

17. $\frac{y^2}{81}-\frac{x^2}{36}=1 \Rightarrow \frac{y^2}{9^2}-\frac{x^2}{6^2}=1;\ a=6,\ b=9$

y-intercepts: $(0,\pm 9)$ or $(0,9)$ and $(0,-9)$

Equation of asymptotes are $y=\frac{9}{6}x=\frac{3}{2}x$

and $y=-\frac{3}{2}x$

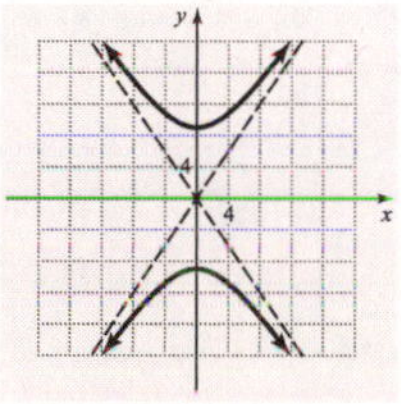

19. $x^2-9y^2=225$

$$\frac{x^2}{225}-\frac{9y^2}{225}=\frac{225}{225}$$

$$\frac{x^2}{225}-\frac{y^2}{25}=1$$

$$\frac{x^2}{15^2}-\frac{y^2}{5^2}=1$$

$a=15,\ b=5$

x-intercepts: $(\pm 15,0)$ or $(15,0)$ and $(-15,0)$

Equation of asymptotes are

$y=\frac{5}{15}x=\frac{1}{3}x$ and $y=-\frac{1}{3}x$

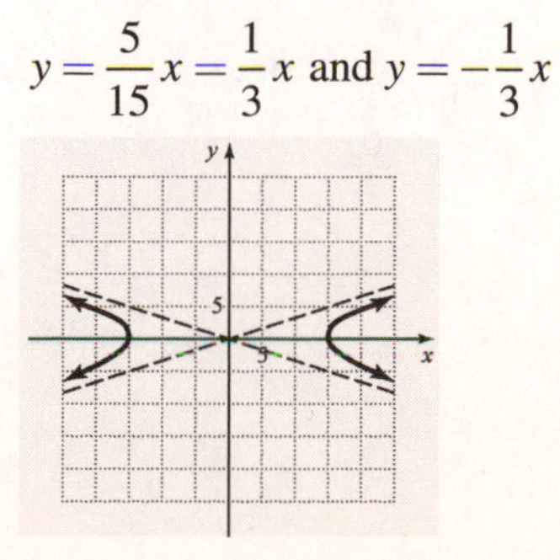

21.
$$3y^2 - 12x^2 = 108$$
$$\frac{3y^2}{108} - \frac{12x^2}{108} = \frac{108}{108}$$
$$\frac{y^2}{36} - \frac{x^2}{9} = 1$$
$$\frac{y^2}{6^2} - \frac{x^2}{3^2} = 1$$
$a = 3,\ b = 6$

y-intercepts: $(0, \pm 6)$ or $(0, 6)$ and $(0, -6)$

Equation of asymptotes are

$y = \frac{6}{3}x = 2x$ and $y = -2x$

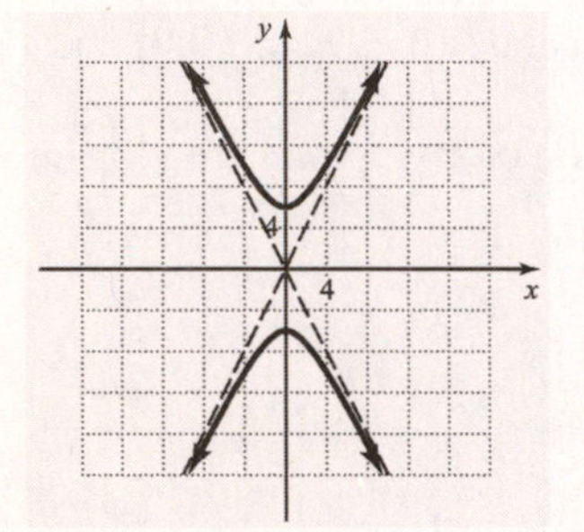

23.
$$8x^2 + 8y^2 = 32$$
$$\frac{8x^2}{8} + \frac{8y^2}{8} = \frac{32}{8}$$
$$x^2 + y^2 = 4 \text{ Circle}$$

25.
$$x^2 - 2y = 10$$
$$2y = x^2 - 10$$
$$y = \frac{1}{2}x^2 - 5 \text{ Parabola}$$

27.
$$10y^2 - 12x^2 = 120$$
$$\frac{10y^2}{120} - \frac{12x^2}{120} = \frac{120}{120}$$
$$\frac{y^2}{12} - \frac{x^2}{10} = 1 \text{ Hyperbola}$$

29.
$a = 11,\ b = 3$
$$\frac{x^2}{a^2} + \frac{y^2}{b^2} = 1 \Rightarrow \frac{x^2}{11^2} + \frac{y^2}{3^2} = 1 \Rightarrow \frac{x^2}{121} + \frac{y^2}{9} = 1$$

31.
$a = \sqrt{7},\ b = 2\sqrt{2}$
$$\frac{x^2}{a^2} + \frac{y^2}{b^2} = 1$$
$$\frac{x^2}{\left(\sqrt{7}\right)^2} + \frac{y^2}{\left(2\sqrt{2}\right)^2} = 1$$
$$\frac{x^2}{7} + \frac{y^2}{4 \bullet 2} = 1$$
$$\frac{x^2}{7} + \frac{y^2}{8} = 1$$

33.
$a = 4;\quad \frac{b}{a} = \frac{1}{4} = \frac{b}{4} \Rightarrow b = 1$
$$\frac{x^2}{a^2} - \frac{y^2}{b^2} = 1 \Rightarrow \frac{x^2}{4^2} - \frac{y^2}{1^2} = 1 \Rightarrow \frac{x^2}{16} - y^2 = 1$$

35.
$$\frac{y^2}{4} - \frac{x^2}{16} = 1 \Rightarrow \frac{y^2}{2^2} - \frac{x^2}{4^2} = 1;\ a = 4,\ \ b = 2$$

y-intercepts: $(0, 2)$ and $(0, -2)$

Equation of asymptotes are $y = \frac{2}{4}x = \frac{1}{2}x$

and $y = -\frac{1}{2}x$

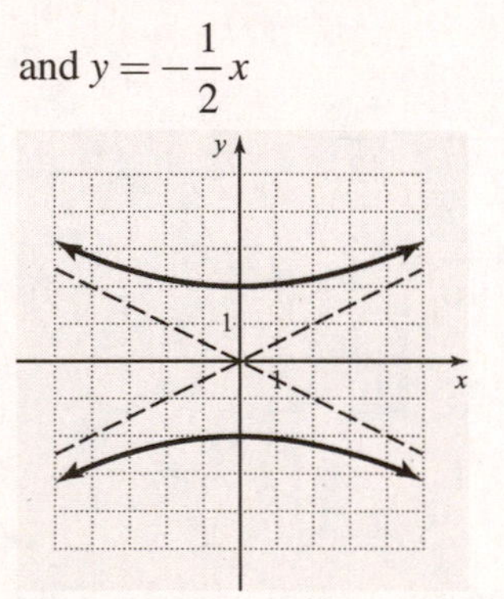

37.
$$3x^2 + 6y^2 = 24$$
$$\frac{3x^2}{24} + \frac{6y^2}{24} = 1$$
$$\frac{x^2}{8} + \frac{y^2}{4} = 1$$
The graph is an ellipse.

39. a is the distance from the center (0, 0) to $\left(3\sqrt{5}, 0\right)$.

$$a^2 = \left(3\sqrt{5} - 0\right)^2 + (0 - 0)^2 = 45$$

b is the distance from the center (0, 0) to $\left(0, \sqrt{3}\right)$. $b^2 = (0 - 0)^2 + \left(\sqrt{3} - 0\right)^2 = 3$

$$\frac{x^2}{45} + \frac{y^2}{3} = 1$$

41. $\frac{x^2}{49}+\frac{y^2}{25}=1 \Rightarrow \frac{x^2}{7^2}+\frac{y^2}{5^2}=1$

$a=7, b=5$

x-intercepts: $(\pm 7,0)$ or $(7,0)$ and $(-7,0)$

y-intercepts: $(0,\pm 5)$ or $(0,5)$ and $(0,-5)$

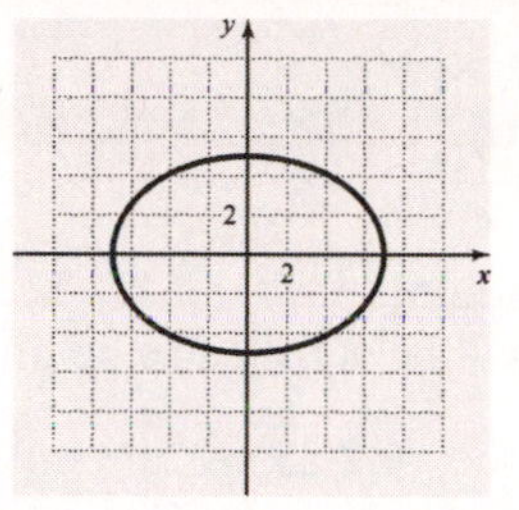

43. **a.**

$\frac{x^2}{8100}+\frac{y^2}{5625}=1 \Rightarrow \frac{x^2}{90^2}+\frac{y^2}{75^2}=1$

$a=90 \Rightarrow 2a=180; \quad b=75 \Rightarrow 2b=150$

The length is 180 m and the width is 150 m.

b. Yes; $135<180<185$ and $110<150<155$

45. **a.** $A=25\pi(R^2-r^2)=100\pi$

$$\frac{25\pi(R^2-r^2)}{100\pi}=\frac{100\pi}{100\pi}$$

$$\frac{R^2-r^2}{4}=\frac{R^2}{4}-\frac{r^2}{4}=1$$

$$a=2 \quad b=2$$

R-intercepts: $(0,\pm 2)$ or $(0,2)$ and $(0,-2)$

Equations of asymptotes are

$R=\frac{2}{2}r=r$ and $R=-r$

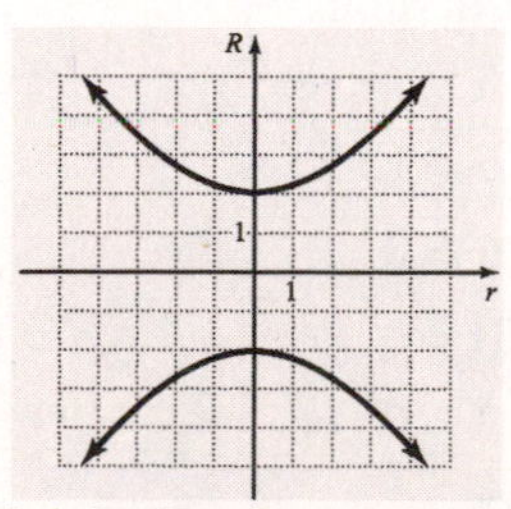

b. In Quadrant I, both the r- and R-coordinates of each point on the graph are nonnegative. Neither radius can be negative.

47. $2a=6 \Rightarrow a=3; \quad 2b=4 \Rightarrow b=2$

$\frac{x^2}{a^2}+\frac{y^2}{b^2}=1 \Rightarrow \frac{x^2}{3^2}+\frac{y^2}{2^2}=1 \Rightarrow \frac{x^2}{9}+\frac{y^2}{4}=1$

11.4 Solving Nonlinear Systems of Equations

Practice

1. **(1)** $x^2+y^2=100$

(2) $y-x=2$

Solve equation **(2)** for y.

$y=x+2$

Substitute this expression for y in equation **(1)**.

$$x^2+(x+2)^2=100$$
$$x^2+x^2+4x+4=100$$
$$2x^2+4x+4=100$$
$$x^2+2x+2=50$$
$$x^2+2x-48=0$$
$$(x+8)(x-6)=0$$
$$x+8=0 \quad \text{or} \quad x-6=0$$
$$x=-8 \qquad x=6$$

Substitute these values for x into either original equation.

$y-x=2$	$y-x=2$
$y-(-8)=2$	$y-6=2$
$y=2-8$	$y=2+6$
$y=-6$	$y=8$

The solutions are $(-8,-6)$ and $(6,8)$.

2. **(1)** $y=x^2-4$

(2) $x^2+y^2=4$

Solve equation **(2)** for x^2.

$x^2=4-y^2$

Substitute this expression for x^2 in equation **(1)**.

$$y=4-y^2-4$$
$$y=-y^2$$
$$y^2+y=0$$
$$y(y+1)=0$$
$$y+1=0 \quad \text{or} \quad y=0$$
$$y=-1$$

Substitute these values for y into either original equation.

$$0 = x^2 - 4 \qquad -1 = x^2 - 4$$
$$x^2 = 4 \qquad x^2 = 3$$
$$x = \pm\sqrt{4} \qquad x = \pm\sqrt{3}$$
$$x = \pm 2$$

The solutions are

$(2,0),(-2,0),(-\sqrt{3},-1),(\sqrt{3},-1)$.

3. **(1)** $x^2 + y^2 = 9$

(2) $2x^2 - y^2 = -6$

Add the equations.

$$x^2 + y^2 = 9$$
$$\underline{2x^2 - y^2 = -6}$$
$$3x^2 \quad = 3$$
$$x^2 = 1$$
$$x = \pm\sqrt{1} = \pm 1$$

Substitute $x^2 = 1$ into either original equation.

$$1 + y^2 = 9$$
$$y^2 = 8$$
$$y = \pm\sqrt{8}$$
$$y = \pm 2\sqrt{2}$$

The solutions are

$(-1,-2\sqrt{2}),(-1,2\sqrt{2})(1,-2\sqrt{2}),(1,2\sqrt{2})$.

4. **(1)** $x^2 - y^2 = -9$

(2) $4x^2 + y^2 = 4$

Add the equations.

$$x^2 - y^2 = -9$$
$$\underline{4x^2 + y^2 = 4}$$
$$5x^2 \quad = -5$$
$$x^2 = -1$$
$$x = \pm\sqrt{-1} = \pm i$$

Substitute $x^2 = -1$ into either original equation.

$$-1 - y^2 = -9$$
$$-y^2 = -8$$
$$y^2 = 8$$
$$y = \pm\sqrt{8}$$
$$y = \pm 2\sqrt{2}$$

The solutions are

$(-i,-2\sqrt{2}),(-i,2\sqrt{2}),(i,-2\sqrt{2}),(i,2\sqrt{2})$.

5. baseball: $y = -16t^2 + 144$

soccer ball: $y = -16t^2 + 64t$

Find t when the heights (y) are the same. Substitute the expression for y in the first equation into the second equation.

$$-16t^2 + 144 = -16t^2 + 64t$$
$$144 = 64t$$
$$t = \frac{9}{4}$$
$$t = 2.25$$

The two balls will be at the same height above the ground $\frac{9}{4}$ sec, or 2.25 sec, after they are released.

Exercises

1. $y = x^2 - 2$

$y = 2x + 1$

Solve by substitution.

$$x^2 - 2 = 2x + 1$$
$$x^2 - 2x - 3 = 0$$
$$(x-3)(x+1) = 0$$
$$x - 3 = 0 \text{ or } x + 1 = 0$$
$$x = 3 \qquad x = -1$$

Find the corresponding y-values:

$$y = (3)^2 - 2 = 9 - 2 = 7$$
$$y = (-1)^2 - 2 = 1 - 2 = -1$$

$(-1,-1),(3,7)$

3. $x^2 + y^2 = 12$

$x = y^2 - 6$

Solve by substitution.

$$y^2 = x + 6$$
$$x^2 + x + 6 = 12$$
$$x^2 + x - 6 = 0$$
$$(x-2)(x+3) = 0$$
$$x - 2 = 0 \text{ or } x + 3 = 0$$
$$x = 2 \qquad x = -3$$

Find the corresponding y-values.

$$y^2 = 2 + 6 = 8 \qquad y^2 = -3 + 6 = 3$$
$$y = \pm\sqrt{8} = \pm 2\sqrt{2} \qquad y = \pm\sqrt{3}$$

$(2,2\sqrt{2}),(2,-2\sqrt{2}),(-3,\sqrt{3}),(-3,-\sqrt{3})$

5. $y = x^2 - 5$

$y = -x^2 + 11$

Solve by substitution.

$x^2 - 5 = -x^2 + 11$

$2x^2 = 16$

$x^2 = 8$

$x = \pm\sqrt{8} = \pm 2\sqrt{2}$

Find the corresponding y-values.

$y = x^2 - 5 = 8 - 5 = 3$

$\left(2\sqrt{2}, 3\right)\left(-2\sqrt{2}, 3\right)$

7. $x^2 + y^2 = 16$

$x - y = 4$

Solve by substitution.

$x = y + 4$

$(y+4)^2 + y^2 = 16$

$y^2 + 8y + 16 + y^2 = 16$

$2y^2 + 8y = 0$

$2y(y+4) = 0$

$2y = 0$ or $y + 4 = 0$

$y = 0 \qquad y = -4$

Find the corresponding y-values.

$x = y + 4$

$x = 0 + 4 = 4 \qquad x = -4 + 4 = 0$

$(4, 0), (0, -4)$

9. $x = 2y^2 - y - 3$

$y = \dfrac{1}{4}x \Rightarrow x = 4y$

Solve by substitution

$x = 2y^2 - y - 3$

$2y^2 - 5y - 3 = 0$

$(2y+1)(y-3) = 0$

$2y + 1 = 0$ or $y - 3 = 0$

$2y = -1 \qquad y = 3$

$y = -\dfrac{1}{2}$

Find the corresponding x-values.

$x = 4y = 4\left(-\dfrac{1}{2}\right) = -2$

$x = 4y = 4(3) = 12$

$\left(-2, -\dfrac{1}{2}\right), (12, 3)$

11. $x^2 + y^2 = 32$

$y = x^2 - 2 \Rightarrow x^2 = y + 2$

Solve by substitution.

$(y+2) + y^2 = 32$

$y^2 + y - 30 = 0$

$(y-5)(y+6) = 0$

$y - 5 = 0$ or $y + 6 = 0$

$y = 5 \qquad y = -6$

Find the corresponding x-values.

$x^2 = y + 2$

$x^2 = 5 + 2 \qquad x^2 = -6 + 2$

$x^2 = 7 \qquad x^2 = -4$

$x = \pm\sqrt{7} \qquad x = \pm\sqrt{-4} = \pm 2i$

$\left(\sqrt{7}, 5\right)\left(-\sqrt{7}, 5\right), (2i, -6), (-2i, -6)$

13.

$$\begin{array}{r} -x^2 + 2y^2 = -8 \\ x^2 + 3y^2 = 18 \\ \hline 5y^2 = 10 \end{array}$$

Solve by elimination

$y^2 = 2$

$y = \pm\sqrt{2}$

Find the corresponding x-values.

$x^2 + 3y^2 = 18$

$x^2 + 3\left(\sqrt{2}\right)^2 = 18$

$x^2 + 6 = 18$

$x^2 = 12$

$x = \pm\sqrt{12} = \pm 2\sqrt{3}$

$\left(2\sqrt{3}, \sqrt{2}\right), \left(-2\sqrt{3}, \sqrt{2}\right)$

$\left(2\sqrt{3}, -\sqrt{2}\right), \left(-2\sqrt{3}, -\sqrt{2}\right)$

15.

$$\begin{array}{rcr} 3x^2 + y^2 = 24 & \longrightarrow & 3x^2 + y^2 = 24 \\ x^2 + y^2 = 16 & \xrightarrow{\times(-1)} & -x^2 - y^2 = -16 \\ \hline & & 2x^2 \quad = 8 \end{array}$$

$x^2 = 4$

$x = \pm\sqrt{4} = \pm 2$

$x^2 + y^2 = 16$

$4 + y^2 = 16$

$y^2 = 12$

$y = \pm\sqrt{12} = \pm 2\sqrt{3}$

$\left(2, 2\sqrt{3}\right), \left(-2, 2\sqrt{3}\right), \left(2, -2\sqrt{3}\right), \left(-2, -2\sqrt{3}\right)$

17.

$$5x^2+6y^2=24 \longrightarrow 5x^2+6y^2=24$$
$$5x^2+5y^2=25 \xrightarrow{\times(-1)} -5x^2-5y^2=-25$$
$$y^2=-1$$
$$y=\pm\sqrt{-1}=\pm i$$
$$5x^2+5y^2=25$$
$$5x^2+5(-1)=25$$
$$5x^2=30$$
$$x^2=6$$
$$x=\pm\sqrt{6}$$

$(\sqrt{6},i),(\sqrt{6},-i),(-\sqrt{6},i),(-\sqrt{6},-i)$

19.

$$5x^2-3y^2=35 \longrightarrow 5x^2-3y^2=35$$
$$3x^2-3y^2=3 \xrightarrow{\times(-1)} -3x^2+3y^2=-3$$
$$2x^2=32$$
$$x^2=16$$
$$x=\pm\sqrt{16}=\pm4$$
$$3x^2-3y^2=3$$
$$3(16)-3y^2=3$$
$$48-3y^2=3$$
$$-3y^2=-45$$
$$y^2=15$$
$$y=\pm\sqrt{15}$$

$(4,\sqrt{15}),(-4,\sqrt{15}),(4,-\sqrt{15}),(-4,-\sqrt{15})$

21.

$$6x^2+6y^2=96 \xrightarrow{\times\left(-\frac{1}{6}\right)} -x^2-y^2=-16$$
$$x^2+9y^2=144 \longrightarrow x^2+9y^2=144$$
$$8y^2=128$$
$$y^2=16$$
$$y=\pm\sqrt{16}=\pm4$$
$$x^2+9y^2=144$$
$$x^2+9(16)=144$$
$$x^2+144=144$$
$$x^2=0$$
$$x=0$$

$(0,4),(0,-4)$

23.

$$\frac{x^2}{4}+\frac{y^2}{16}=1 \xrightarrow{\times16} 4x^2+y^2=16$$
$$\frac{x^2}{2}+\frac{y^2}{24}=1 \xrightarrow{\times(-24)} -12x^2-y^2=-24$$
$$-8x^2=-8$$
$$x^2=1$$
$$x=\pm\sqrt{1}=\pm1$$
$$4x^2+y^2=16$$
$$4(1)+y^2=16$$
$$y^2=12$$
$$y=\pm\sqrt{12}=\pm2\sqrt{3}$$

$(1,2\sqrt{3}),(-1,2\sqrt{3}),(1,-2\sqrt{3}),(-1,-2\sqrt{3})$

25.

$$\frac{y^2}{8}-\frac{x^2}{4}=1 \xrightarrow{\times8} y^2-2x^2=8$$
$$\frac{y^2}{11}+\frac{x^2}{11}=1 \xrightarrow{\times22} 2y^2+2x^2=22$$
$$3y^2=30$$
$$y^2=10$$
$$y=\pm\sqrt{10}$$
$$y^2-2x^2=8$$
$$10-2x^2=8$$
$$-2x^2=-2$$
$$x^2=1$$
$$x=\pm\sqrt{1}=\pm1$$

$(1,\sqrt{10}),(-1,\sqrt{10}),(1,-\sqrt{10}),(-1,-\sqrt{10})$

27.

$$(x-4)^2+y^2=4 \rightarrow x^2-8x+16+y^2=4$$
$$x^2-8x+y^2=-12$$
$$(x+2)^2+y^2=16 \rightarrow x^2+4x+4+y^2=16$$
$$x^2+4x+y^2=12$$
$$x^2-8x+y^2=-12 \xrightarrow{\times(-1)} -x^2+8x-y^2=12$$
$$x^2+4x+y^2=12 \longrightarrow x^2+4x+y^2=12$$
$$12x=24$$
$$x=2$$

27. (continued)

$$(x+2)^2+y^2=16$$
$$(2+2)^2+y^2=16$$
$$16+y^2=16$$
$$y^2=0$$
$$y=0$$

$(2,0)$

29. $x^2+y^2=27$

$$y=x^2+3 \Rightarrow x^2=y-3$$

Solve by substitution.

$$(y-3)+y^2=27$$
$$y^2+y-30=0$$
$$(y+6)(y-5)=0$$
$$y+6=0 \quad \text{or} \quad y-5=0$$
$$y=-6 \qquad y=5$$

Find the corresponding x-values.

$$y=x^2+3 \Rightarrow -6=x^2+3$$
$$-9=x^2$$
$$\pm 3i=x$$
$$5=x^2+3$$
$$2=x^2$$
$$\pm\sqrt{2}=x$$

$(-3i,-6),(3i,-6),(-\sqrt{2},5),(\sqrt{2},5)$

31. $x^2+y^2=25$

$$x=y^2-5 \Rightarrow y^2=x+5$$

Solve by substitution.

$$x^2+(x+5)=25$$
$$x^2+x-20=0$$
$$(x+5)(x-4)=0$$
$$x+5=0 \quad \text{or} \quad x-4=0$$
$$x=-5 \qquad x=4$$

Find the corresponding y-values.

$$x=y^2-5 \Rightarrow -5=y^2-5$$
$$0=y^2$$
$$0=y$$
$$4=y^2-5$$
$$9=y^2$$
$$\pm 3=y$$

$(-5,0),(4,-3),(4,3)$

33.

$$\begin{array}{r} 3x^2+6y^2=24 \\ \underline{3x^2+5y^2=19} \\ y^2=5 \\ y=\pm\sqrt{5} \end{array}$$

Find the corresponding x-values.

$$3x^2+6y^2=24$$
$$3x^2+6\left(-\sqrt{5}\right)^2=24$$
$$3x^2+30=24$$
$$3x^2=-6$$
$$x^2=-2$$
$$x=\pm i\sqrt{2}$$
$$3x^2+6\left(\sqrt{5}\right)^2=24$$
$$3x^2+30=24$$
$$3x^2=-6$$
$$x^2=-2$$
$$x=\pm i\sqrt{2}$$

$\left(-i\sqrt{2},-\sqrt{5}\right),\left(i\sqrt{2},-\sqrt{5}\right),\left(-i\sqrt{2},\sqrt{5}\right),\left(i\sqrt{2},\sqrt{5}\right)$

35. $R=0.25x^2-100x=75x$

$$0.25x^2-175x=0$$
$$0.25x(x-700)=0$$
$$0.25x=0 \quad \text{or} \quad x-700=0$$
$$x=0 \qquad x=700$$

They will have the same revenue if they sell 0 or 700 organizers.

37.

$$x^2+2y^2=7 \longrightarrow x^2+2y^2=7$$
$$x^2-3y^2=2 \xrightarrow{\times(-1)} \underline{-x^2+3y^2=-2}$$
$$5y^2=5$$
$$y^2=1$$
$$y=\pm 1$$

Find the corresponding x-values.

$$x^2+2y^2=7$$
$$x^2+2(1)=7$$
$$x^2=5$$
$$x=\pm\sqrt{5}$$

The comet traveling in the hyperbolic path travels in only one half of the hyperbola $(x \geq 0)$. $x=+\sqrt{5}$ only. The paths will intersect at $\left(\sqrt{5},1\right)$ and $\left(\sqrt{5},-1\right)$.

39.

$$x = y + 3$$
$$x^2 - y^2 = 75$$
$$(y+3)^2 - y^2 = 75$$
$$y^2 + 6y + 9 - y^2 = 75$$
$$6y = 66$$
$$y = 11$$

Find the corresponding x-value.
$x = y + 3 = 11 + 3 = 14$

The smaller square has dimensions 11 in. by 11 in., in contrast to 14 in. by 14 in. for the larger square.

11.5 Solving Nonlinear Inequalities and Nonlinear Systems of Inequalities

Practice

1. $4x^2 + y^2 > 4$

The boundary line is dashed.

$$4x^2 + y^2 = 4$$
$$x^2 + \frac{y^2}{4} = 1$$

Ellipse, centered at the origin with $a = 1,\ b = 2$.

Test point is $(0,0)$.

$$4(0)^2 + (0)^2 \stackrel{?}{>} 4$$
$$0 + 0 = 0 < 4$$

Shade the region not containing $(0,0)$.

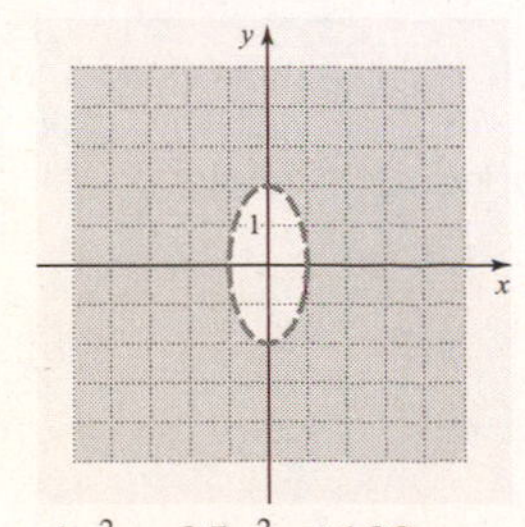

2. $4x^2 - 25y^2 \le 100$

The boundary line is solid.

$$4x^2 - 25y^2 = 100$$
$$\frac{x^2}{25} - \frac{y^2}{4} = 1$$

Hyperbola, centered at the origin with $a = 5,\ b = 4$.

x-intercepts are $(-5,0), (5,0)$.

Asymptotes are $y = \frac{b}{a}x = \frac{4}{5}x,\ y = -\frac{4}{5}x$.

Test point is $(0,0)$.

$$4(0)^2 - 25(0)^2 \stackrel{?}{\le} 100$$
$$0 - 0 = 0 < 100$$

Shade the region not containing $(0,0)$.

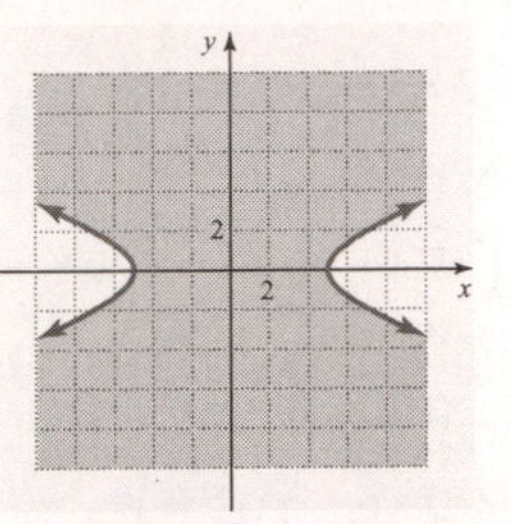

3.

$$x + y \ge 3$$
$$x^2 + y^2 \le 25$$

The boundary lines for both inequalities are solid.

The equation corresponding to the first inequality is a line, $y = -x + 3$. The test point is $(0,0)$.

$$0 + 0 \stackrel{?}{\ge} 3$$
$$0 < 3$$

Shade the region not containing $(0,0)$.

The equation corresponding to the second inequality is a circle centered at the origin with radius 5. The test point is $(0,0)$.

$$0^2 + 0^2 \stackrel{?}{\le} 25$$
$$0 \le 25$$

Shade the region containing $(0,0)$.

The solution is where these shaded regions overlap.

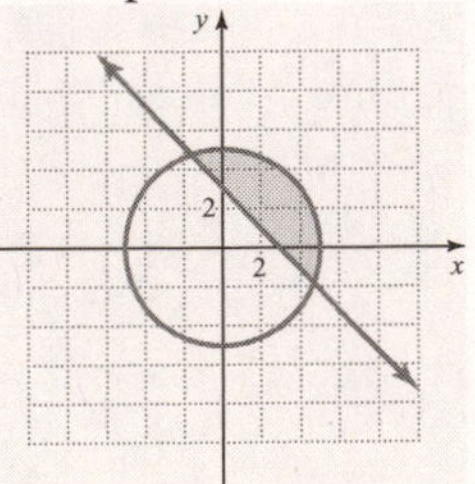

4. $x^2 + y^2 \le 1$

$\frac{x^2}{9} + \frac{y^2}{4} \ge 1$

The boundary lines for both inequalities are solid.

The equation corresponding to the first inequality is a circle centered at the origin with radius 1.

The test point is $(0,0)$.

$0^2 + 0^2 \overset{?}{\le} 1$

$0 \le 1$

Shade the region containing $(0,0)$.

The equation corresponding to the second equation is an ellipse centered at the origin with $a = 3$ and $b = 2$. The x-intercepts are $(-3,0)$, $(3,0)$ and the y-intercepts are $(0,-2)$, $(0,2)$. The test point is $(0,0)$.

$\frac{0^2}{9} + \frac{0^2}{4} \overset{?}{\ge} 1$

$0 + 0 \overset{?}{\ge} 1$

$0 < 1$

Shade the region not containing $(0,0)$.

Since the two shaded regions do not overlap, there are no real solutions.

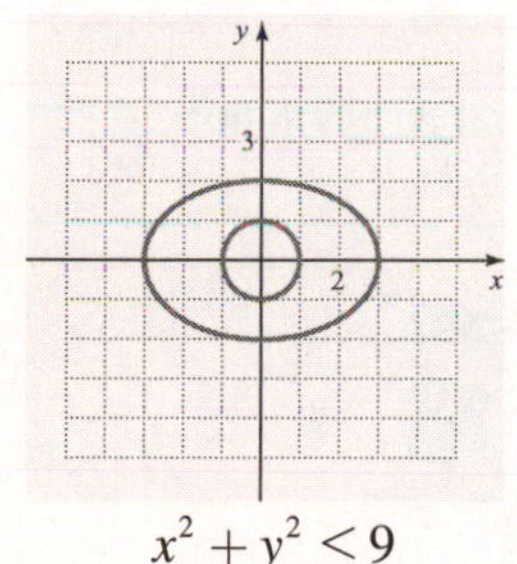

5. $x^2 + y^2 \le 9$

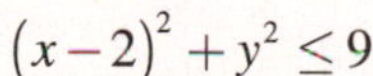

$(x-2)^2 + y^2 \le 9$

The boundary lines for both inequalities are solid.

The equation corresponding to the first inequality is a circle centered at the origin with radius 3.

The test point is $(0,0)$.

$0^2 + 0^2 \overset{?}{\le} 3$

$0 \le 3$

Shade the region containing $(0,0)$.

The equation corresponding to the second equation is a circle of radius 3 centered 2 units to the right of the origin. The test point is $(0,0)$.

$(0-2)^2 + (0)^2 \overset{?}{\le} 9$

$4 + 0 \overset{?}{\le} 9$

$4 \le 9$

Shade the region containing $(0,0)$.

The solution is where these shaded regions overlap.

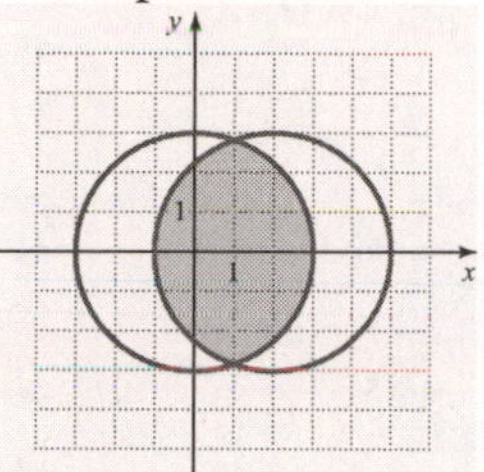

Exercises

1. In graphing a nonlinear inequality that involves either the < or the > symbol, a(n) broken curve is drawn.

3. A solution of a nonlinear system of inequalities must satisfy each inequality in the system.

5. $x^2 + y^2 > 16$

The boundary line is dashed.

$x^2 + y^2 = 16$

$(x-0)^2 + (y-0)^2 = 4^2$

Circle: center: $(0,0)$ radius $= 4$

Test point is $(0,0)$.

$0^2 + 0^2 \overset{?}{>} 16$

$0 > 16$ False, shade the region not containing $(0,0)$.

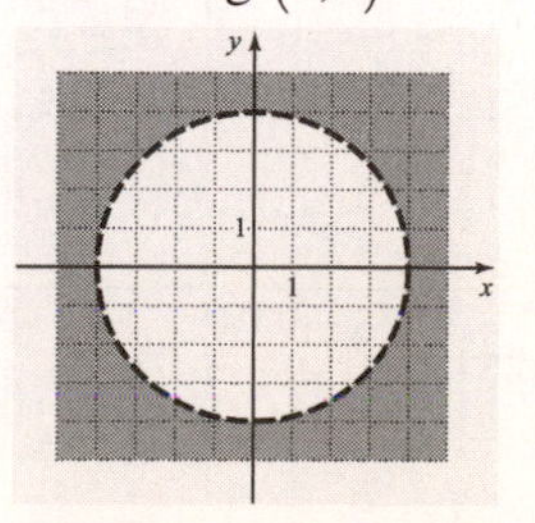

7. $(x-1)^2+(y+1)^2 \le 25$

The boundary line is solid.

$(x-1)^2+(y-(-1))^2=5^2$

Circle: center: $(1,-1)$ radius $=5$

Test point is $(0,0)$.

$(0-1)^2+(0+1)^2 \overset{?}{\le} 25$

$1+1 \overset{?}{\le} 25$; $2 \le 25$ True, shade the region containing the point $(0,0)$.

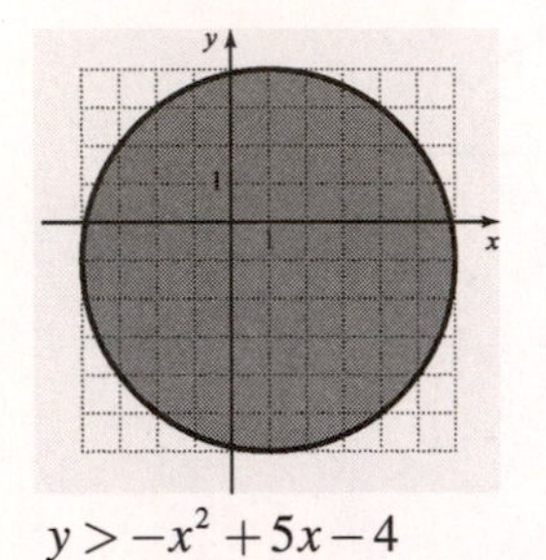

9. $y > -x^2+5x-4$

The boundary line is dashed.

$y=-(x^2-5x)-4$

$y=-\left(x^2-5x+\left(\frac{-5}{2}\right)^2\right)-4+\left(\frac{-5}{2}\right)^2$

$y=-\left(x^2-5x+\frac{25}{4}\right)-\frac{16}{4}+\frac{25}{4}$

$y=-\left(x-\frac{5}{2}\right)^2+\frac{9}{4}$

Parabola opening down with vertex: $\left(\frac{5}{2},\frac{9}{4}\right)$

x	$y=-\left(x-\frac{5}{2}\right)^2+\frac{9}{4}$	(x,y)
0	$-\left(0-\frac{5}{2}\right)^2+\frac{9}{4}=-4$	$(0,-4)$
1	$-\left(1-\frac{5}{2}\right)^2+\frac{9}{4}=0$	$(1,0)$
2	$-\left(2-\frac{5}{2}\right)^2+\frac{9}{4}=2$	$(2,2)$
3	$-\left(3-\frac{5}{2}\right)^2+\frac{9}{4}=2$	$(3,2)$

Test point is $(0,0)$.

$0 \overset{?}{>} -(0)^2+5(0)-4$

$0 \overset{?}{>} 0+0-4$

$0>-4$ True, shade the region containing the point $(0,0)$.

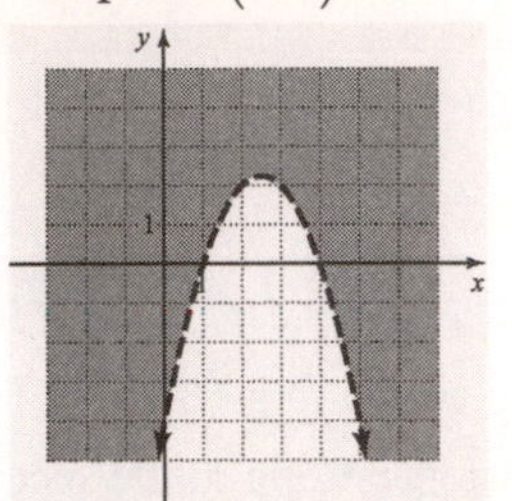

11. $9x^2+y^2 \ge 36$

The boundary line is solid.

$\frac{9x^2}{36}+\frac{y^2}{36}=\frac{36}{36}$

$\frac{x^2}{4}+\frac{y^2}{36}=1 \Rightarrow \frac{x^2}{2^2}+\frac{y^2}{6^2}=1 \Rightarrow a=2,\ b=6$

An ellipse with x-intercepts $(2,0)$ and $(-2,0)$, and y-intercepts $(0,6)$ and $(0,-6)$

Test point is $(0,0)$.

$9(0)^2+(0)^2 \overset{?}{\ge} 36$

$0+0 \overset{?}{\ge} 36$

$0 \ge 36$ False, shade the region not containing the point $(0,0)$.

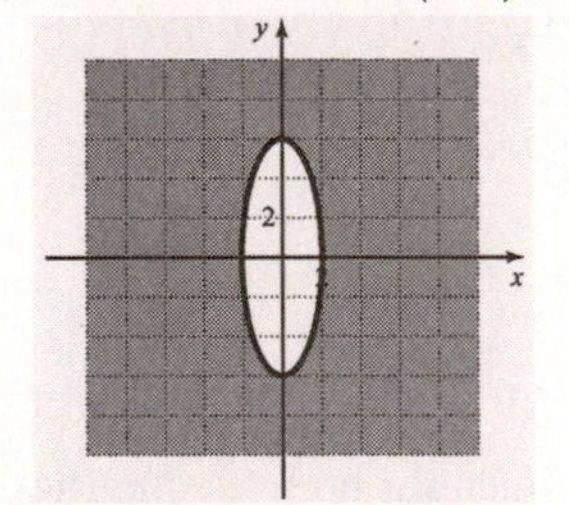

13. $16x^2 + 36y^2 < 144$

The boundary line is dashed.

$$\frac{16x^2}{144} + \frac{36y^2}{144} = \frac{144}{144}$$

$$\frac{x^2}{9} + \frac{y^2}{4} = 1 \Rightarrow \frac{x^2}{3^2} + \frac{y^2}{2^2} = 1 \Rightarrow a = 3, \quad b = 2$$

Ellipse

x-intercepts: $(\pm 3,0)$ or $(3,0)$ and $(-3,0)$

y-intercepts: $(0,\pm 2)$ or $(0,2)$ and $(0,-2)$

Test point is $(0,0)$.

$$16(0)^2 + 36(0)^2 \overset{?}{<} 144$$

$$0 + 0 \overset{?}{<} 144$$

$0 < 144$ True, shade the region containing the point $(0,0)$.

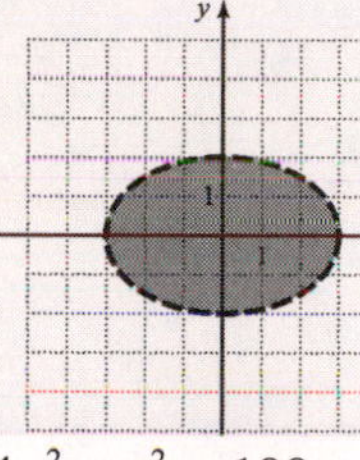

15. $4x^2 - y^2 > 100$

The boundary line is dashed.

$$4x^2 - y^2 = 100$$

$$\frac{4x^2}{100} - \frac{y^2}{100} = \frac{100}{100} \Rightarrow \frac{x^2}{25} - \frac{y^2}{100} = 1 \Rightarrow$$

$$\frac{x^2}{5^2} - \frac{y^2}{10^2} = 1; \quad a = 5, \quad b = 10$$

Hyperbola

x-intercepts: $(\pm 5,0)$ or $(5,0)$ and $(-5,0)$

Equation of asymptotes are

$$y = \frac{10}{5}x = 2x \text{ and } y = -2x$$

Test point is $(0,0)$.

$$4(0)^2 - (0)^2 \overset{?}{>} 100$$

$$0 - 0 \overset{?}{>} 100$$

$0 > 100$ False, shade the region not containing the point $(0,0)$.

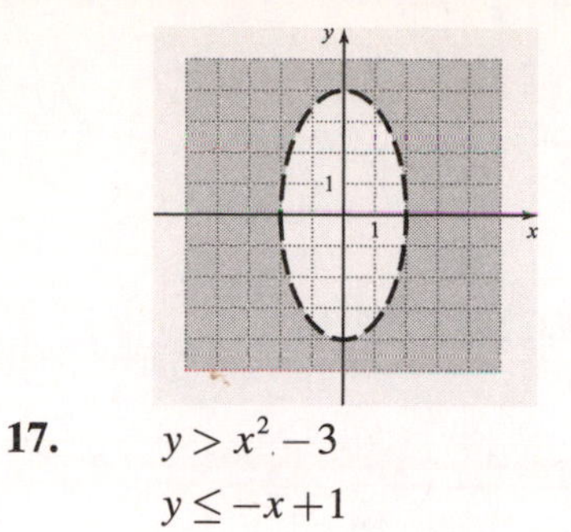

17. $y > x^2 - 3$

$y \le -x + 1$

$y > x^2 - 3$

The boundary line is dashed.

$y = x^2 - 3$

$y = (x - 0)^2 - 3$

Parabola opening up with vertex: $(0,-3)$.

x	$y = x^2 - 3$	(x,y)
-3	$(-3)^2 - 3 = 6$	$(-3,6)$
-1	$(-1)^2 - 3 = -2$	$(-1,-2)$
0	$(0)^2 - 3 = -3$	$(0,-3)$
1	$(1)^2 - 3 = -2$	$(1,-2)$
3	$(3)^2 - 3 = 6$	$(3,6)$

Test point is $(0,0)$.

$$0 \overset{?}{>} 0^2 - 3$$

$0 > -3$ True, shade the region containing the point (0,0).

$y \le -x + 1$

The boundary line is solid.

$y = -x + 1$

x	$y = -x + 1$	(x,y)
0	$-0 + 1 = 1$	$(0,1)$
2	$-2 + 1 = -1$	$(2,-1)$

Test point is $(0,0)$.

$$0 \overset{?}{\le} -(0) + 1.$$

$0 \le 1$ True, shade the region containing the point (0,0).

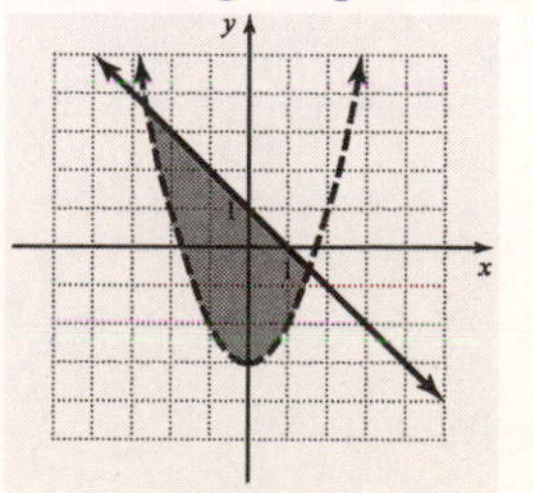

19. $x^2+y^2 \ge 9$
$x^2+y^2 < 25$

$x^2+y^2 \ge 9$
The boundary line is solid.
$x^2+y^2 = 9$
$(x-0)^2+(y-0)^2 = 3^2$
Circle: center: $(0,0)$ Radius $r=3$

Test point is (0,0); $0^2+0^2 \overset{?}{\ge} 9$
$0 \ge 9$ False, shade the region not containing the point (0,0)

$x^2+y^2 = 25$
The boundary line is dashed.
$(x-0)^2+(y-0)^2 = 5^2$
Circle: center: $(0,0)$ radius $r=5$

Test point is $(0,0)$; $0^2+0^2 \overset{?}{<} 25$
$0 < 25$ True, shade the region containing the point $(0,0)$.

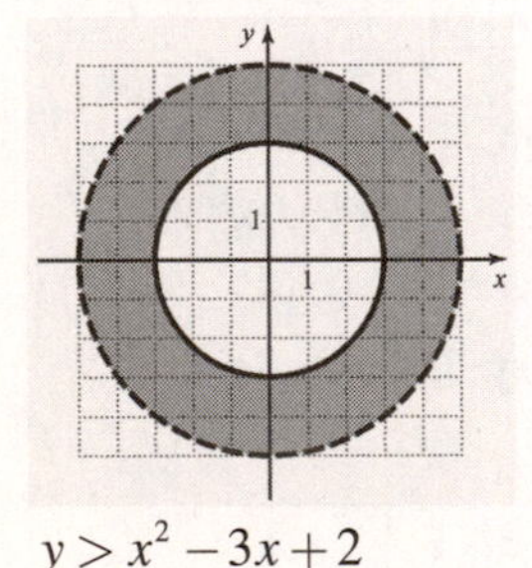

21. $y > x^2-3x+2$
$y \le -x^2-2x-4$
$y > x^2-3x+2$
The boundary line is dashed.
$y = x^2-3x+2$
$y = x^2-3x+2$
Find the vertex by completing the square:

$$y-2+\frac{9}{4} = x^2-3x+\frac{9}{4}$$

$$y+\frac{1}{4} = \left(x-\frac{3}{2}\right)^2 \Rightarrow y = \left(x-\frac{3}{2}\right)^2 - \frac{1}{4}$$

Parabola opening up, vertex: $\left(-\frac{3}{2},\frac{1}{4}\right)$

x	$y=x^2-3x+2$	(x,y)
-1	$(-1)^2-3(-1)+2=6$	$(-1,6)$
0	$(0)^2-3(0)+2=2$	$(0,2)$
1	$(1)^2-3(1)+2=0$	$(1,0)$
2	$(2)^2-3(2)+2=0$	$(2,0)$
3	$(3)^2-3(3)+2=2$	$(3,0)$
4	$(4)^2-3(4)+2=6$	$(4,6)$

Test point is $(0,0)$

$0 \overset{?}{>} (0)^2-3(0)+2$

$0 \overset{?}{>} 0+0+2$

$0 < 1$ False, shade the area not containing the point $(0,0)$, the region above the curve.

$y < -x^2-2x-4$
The boundary line is dashed.
$y = -x^2-2x-4$
$y = -(x^2+2x)-4$

$$y = -\left(x^2+2x+\left(\frac{2}{2}\right)^2\right)-4+\left(\frac{2}{2}\right)^2$$

$y = -(x^2+2x+1)-4+1$
$y = -(x^2+2x+1)-3$
$y = -(x+1)^2-3$
$y = -(x-(-1))^2-3$
Parabola opening down, vertex: $(-1,-3)$

x	$y=-x^2-2x-4$	(x,y)
-3	$-(-3)^2-2(-3)-4=-7$	$(-3,-7)$
-2	$-(-2)^2-2(-2)-4=-4$	$(-2,-4)$
-1	$-(-1)^2-2(-1)-4=-3$	$(-1,-3)$
0	$-(0)^2-2(0)-4=-4$	$(0,-4)$
1	$-(1)^2-2(1)-4=-7$	$(1,-7)$

Test point is $(0,0)$.

$0 \overset{?}{<} -(0)^2-2(0)-4$

$0 \overset{?}{<} 0-0-4$

21. (continued)
$0 < -4$ False, shade the region not containing the point $(0,0)$, the region below the curve.

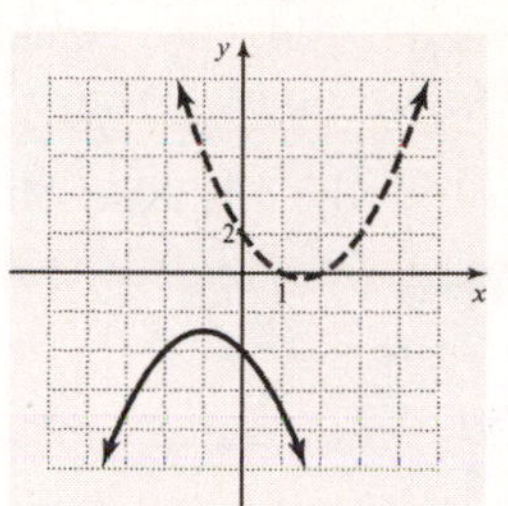

The shaded regions do not overlap, so there are no real solutions for the system.

23. $8x^2 + 2y^2 \le 72$
$x^2 + 4y^2 \le 36$

$8x^2 + 2y^2 \le 72$
The boundary line is solid.

$8x^2 + 2y^2 = 72$

$\frac{8x^2}{72} + \frac{2y^2}{72} = \frac{72}{72}$

$\frac{x^2}{9} + \frac{y^2}{36} = 1 \Rightarrow \frac{x^2}{3^2} + \frac{y^2}{6^2} = 1; \ a = 3, \ b = 6$

Ellipse
x-intercepts are $(\pm 3,0)$ or $(3,0)$ and $(-3,0)$
y-intercepts are $(0,\pm 6)$ or $(0,6)$ and $(0,-6)$
Test point is $(0,0)$.

$8(0)^2 + 2(0)^2 \overset{?}{\le} 72$
$0 + 0 \le 72$ True, shade the region containing the point $(0,0.)$

$x^2 + 4y^2 \le 36$
The boundary line is solid.

$x^2 + 4y^2 = 36$

$\frac{x^2}{36} + \frac{4y^2}{36} = \frac{36}{36}$

$\frac{x^2}{36} + \frac{y^2}{9} = 1 \Rightarrow \frac{x^2}{6^2} + \frac{y^2}{3^2} = 1; \ a = 6, \ b = 3$

Ellipse
x-intercepts are $(\pm 6,0)$ or $(6,0)$ and $(-6,0)$
y-intercepts are $(0,\pm 3)$ or $(0,3)$ and $(0,-3)$
Test point is $(0,0)$.

$(0)^2 + 4(0)^2 \overset{?}{\le} 36$
$0 + 0 \le 36$
True, shade the region containing the point $(0,0.)$

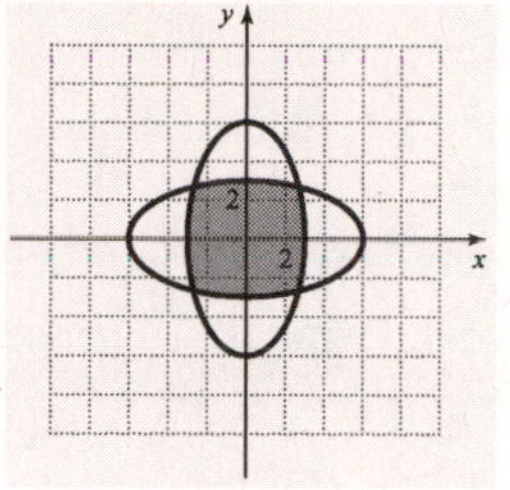

25. $x^2 - 4y^2 < 16$
$x^2 + y^2 \ge 1$

$x^2 - 4y^2 < 16$: The boundary line is dashed.

$x^2 - 4y^2 = 16$

$\frac{x^2}{16} - \frac{4y^2}{16} = \frac{16}{16}$

$\frac{x^2}{16} - \frac{y^2}{4} = 1 \Rightarrow \frac{x^2}{4^2} - \frac{y^2}{2^2} = 1; \ a = 4, \ b = 2$

Hyperbola
x-intercepts are $(\pm 4,0)$ or $(4,0)$ and $(-4,0)$
Equation of asymptotes are

$y = \frac{2}{4}x = \frac{1}{2}x$ and $y = -\frac{1}{2}x$

Test point is $(0,0)$.

$(0)^2 - 4(0)^2 \overset{?}{<} 16$
$0 + 0 < 16$, Shade the region containing the point $(0,0)$.

$x^2 + y^2 \ge 1$: The boundary line is solid.
$x^2 + y^2 = 1$
$(x-0)^2 + (y-0)^2 = 1^2$
Circle, center: $(0,0)$, radius $= 1$
Test point is $(0,0)$.

$0^2 + 0^2 \overset{?}{\ge} 1$
$0 \ge 1$ False, shade the region not containing the point $(0,0)$.

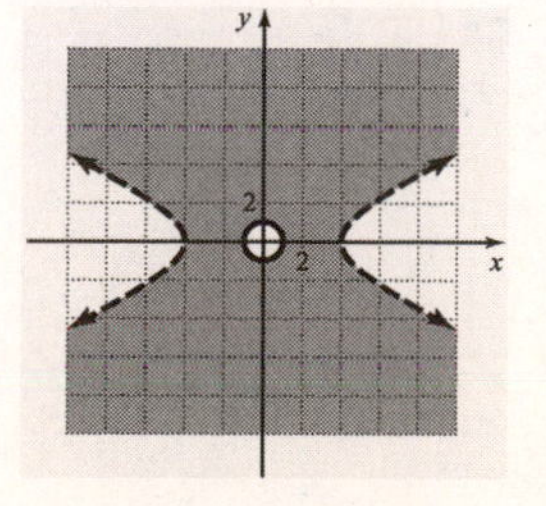

27. $x^2 + y^2 < 4$
$y \geq x^2 + 3$

$x^2 + y^2 < 4$ is a circle with center (0, 0) and radius 2. The boundary is dashed. Shade the area inside the circle.

$y \geq x^2 + 3$ is a parabola with vertex (0, 3). The boundary is solid. Using (0, 0) as the test point, we have $0 \overset{?}{\geq} 0^2 + 3 \Rightarrow 0 \geq 3,$ which is false. Shade the region inside the parabola.

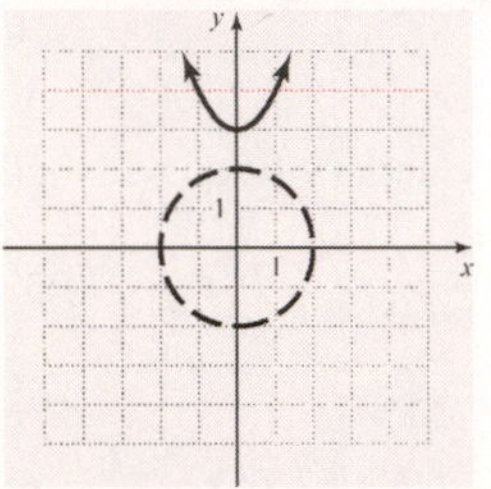

The shaded regions do not overlap, so there are no real solutions for the system.

29. $x^2 + 16y^2 > 16$
$4x^2 + 9y^2 < 36$

$$x^2 + 16y^2 > 16 \Rightarrow \frac{x^2}{16} + \frac{y^2}{1} = 1 \Rightarrow a = 4, b = 1$$

This is an ellipse with x-intercepts (4, 0) and (–4, 0), and y-intercepts (0, 1) and (0, –1). The boundary is dashed. Using (0, 0) as a test point, we have $0^2 + 16(0)^2 \overset{?}{>} 16 \Rightarrow 0 > 16,$ which is false. Shade outside the ellipse.

$$4x^2 + 9y^2 < 36 \Rightarrow \frac{x^2}{9} + \frac{y^2}{4} = 1 \Rightarrow a = 3, b = 2$$

This is an ellipse with x-intercepts (3, 0) and (–3, 0), and y-intercepts (0, 2) and (0, –2). The boundary is dashed. Using (0, 0) as a test point, we have $4(0)^2 + 9(0)^2 \overset{?}{<} 36 \Rightarrow 0 < 36,$ which is true. Shade inside the ellipse.
The overlapping shaded region shows the solution of the system.

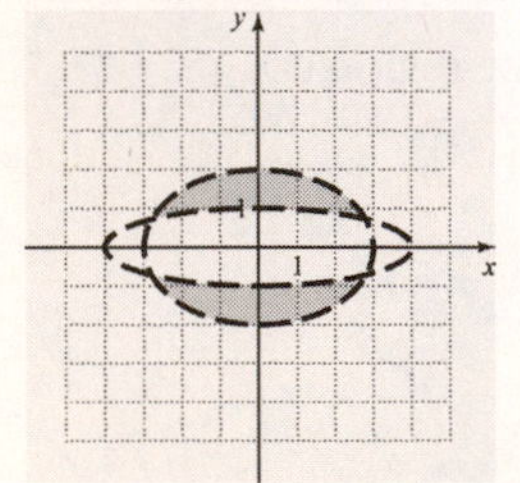

31.

$$4x^2 + y^2 > 16 \Rightarrow \frac{x^2}{4} + \frac{y^2}{16} = 1 \Rightarrow a = 2, b = 4$$

This is an ellipse with x-intercepts (2, 0) and (–2, 0), and y-intercepts (0, 4) and (0, –4). The boundary is dashed. Using (0, 0) as a test point, we have

$4(0)^2 + 0^2 \overset{?}{>} 16 \Rightarrow 0 > 16,$ which is false. Shade the region outside the ellipse.

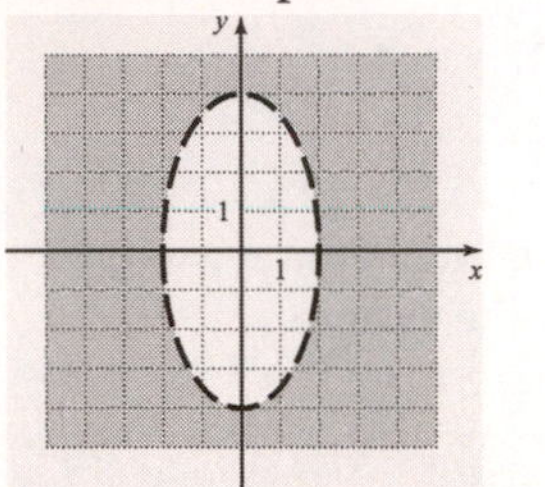

33. $4x^2 + 9y^2 \leq 144$

The boundary line is solid.

$$4x^2 + 9y^2 = 144$$

$$\frac{4x^2}{144} + \frac{9y^2}{144} = \frac{144}{144}$$

$$\frac{x^2}{36} + \frac{y^2}{16} = 1 \Rightarrow \frac{x^2}{6^2} + \frac{y^2}{4^2} = 1; \quad a = 6, \quad b = 4$$

Ellipse.

x-intercepts are $(\pm 6, 0)$ or $(6, 0)$ and $(-6, 0)$

y-intercepts are $(0, \pm 4)$ or $(0, 4)$ and $(0, -4)$

Test point is $(0,0)$.

$$4(0)^2 + 9(0)^2 \overset{?}{\leq} 144$$

$0 + 0 \leq 144$ True, shade the region containing the point $(0,0)$.

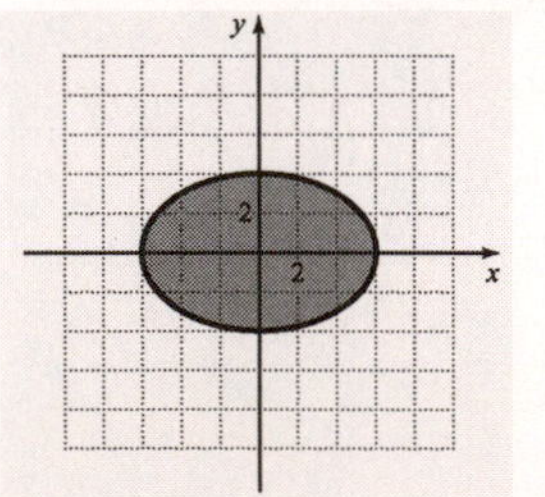

35. **a.** $x^2+y^2\le 2500$

$y\ge x+2$

b. $x^2+y^2\le 2500$

The boundary line is solid.

$x^2+y^2=2500$

$(x-0)^2+(y-0)^2=50^2$

Circle. Center: $(0,0)$, radius $=50$

Test point is $(0,0)$.

$0^2+0^2\overset{?}{\le}2500$

$0\le 2500$ True, shade the region containing the point $(0,0)$.

$y\ge x+2$: The boundary line is solid.

$y=x+2$

x	$y=x+2$	(x,y)
0	$0+2=2$	$(0,2)$
20	$20+2=22$	$(20,22)$

Test point is $(0,0)$.

$0\overset{?}{\ge}0+2$

$0\ge 2$ False, shade the region not containing the point $(0,0)$.

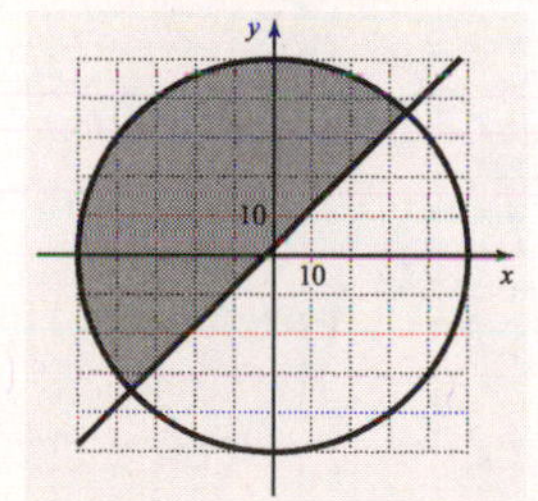

c. No, only points in Quadrant I are possible solutions since the lengths of the gardens cannot be negative.

37. $x^2+y^2\le 900$

$(x-50)^2+y^2\le 1600$

$x^2+y^2\le 900$

The boundary line is solid.

$x^2+y^2=900$

$(x-0)^2+(y-0)^2=30^2$

Circle, center: $(0,0)$ radius $=30$

Test point is $(0,0)$.

$0^2+0^2\overset{?}{\le}900$

$0\le 900$

True, shade the region containing the point $(0,0)$.

$(x-50)^2+y^2\le 1600$

The boundary line is solid.

$(x-50)^2+y^2=1600$

$(x-50)^2+(y-0)^2=40^2$

Circle, center: $(50,0)$ radius $=40$

Test point is $(0,0)$

$(0-50)^2+0^2\overset{?}{\le}1600$

$2500+0\le 1600$ False. Shade the region not containing the point $(0,0)$.

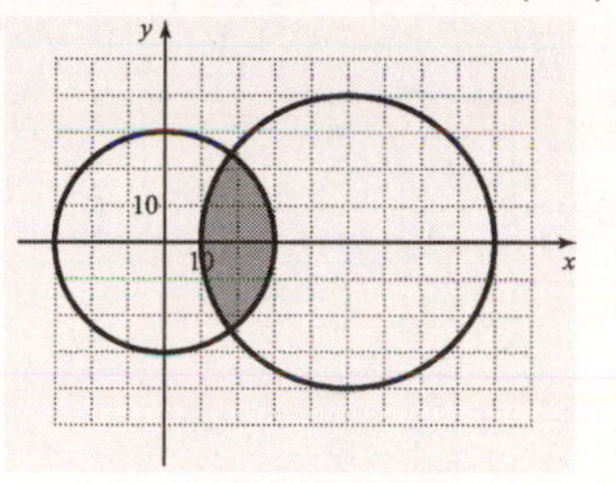

Chapter 11 Review Exercises

1. $y=-2(x+3)^2+5$

$y=-2(x-(-3))^2+5$

$h=-3,\ k=5$

Vertex: $(-3,5)$; Axis of symmetry: $x=-3$

x	$y=-2(x+3)^2+5$	(x,y)
-5	$-2(-5+3)^2+5=-3$	$(-5,-3)$
-4	$-2(-4+3)^2+5=3$	$(-4,3)$
-3	$-2(-3+3)^2+5=5$	$(-3,5)$
-2	$-2(-2+3)^2+5=3$	$(-2,3)$
-1	$-2(-1+3)^2+5=-3$	$(-1,-3)$

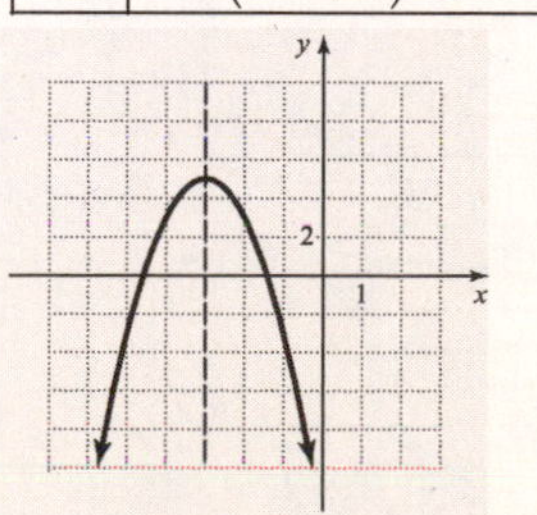

2. $x=3(y-2)^2-10$

$h=-10 \quad k=2$

Vertex: $(-10,2)$

Axis of symmetry: $y=2$

y	$x=3(y-2)^2-10$	(x,y)
5	$3(5-2)^2-10=17$	$(17,5)$
4	$3(4-2)^2-10=2$	$(2,4)$
3	$3(3-2)^2-10=-7$	$(-7,3)$
2	$3(2-2)^2-10=-10$	$(-10,2)$
1	$3(1-2)^2-10=-7$	$(-7,1)$
0	$3(0-2)^2-10=2$	$(2,0)$

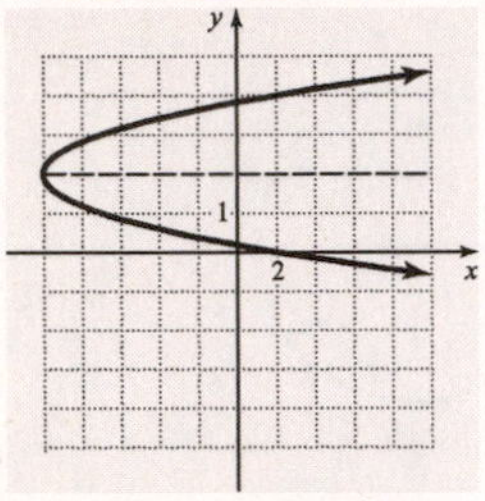

3. $x=4y^2+32y+64$

$x=4(y^2+8y+16)$

$x=4(y+4)^2$

$x=4(y-(-4))^2$

$h=0, \quad k=-4$

Vertex: $(0,-4)$

Axis of symmetry: $y=-4$

y	$x=4y^2+32y+64$	(x,y)
-2	$4(-2)^2+32(-2)+64=16$	$(16,-2)$
-3	$4(-3)^2+32(-3)+64=4$	$(4,-3)$
-4	$4(-4)^2+32(-4)+64=0$	$(0,-4)$
-5	$4(-5)^2+32(-5)+64=4$	$(4,-5)$
-6	$4(-6)^2+32(-6)+64=16$	$(16,-6)$

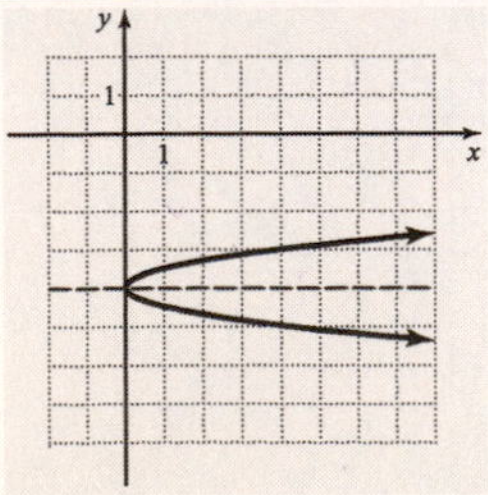

4. $y=-x^2-6x-7$

$y=-(x^2+6x)-7$

$y=-\left(x^2+6x+\left(\frac{6}{2}\right)^2\right)-7+\left(\frac{6}{2}\right)^2$

$y=-(x^2+6x+9)-7+9$

$y=-(x+3)^2+2$

$y=-(x-(-3))^2+2$

$h=-3, \quad k=2$

Vertex: $(-3,2)$

Axis of symmetry: $x=-3$

x	$y=-x^2-6x-7$	(x,y)
-6	$-(-6)^2-6(-6)-7=-7$	$(-6,-7)$
-5	$-(-5)^2-6(-5)-7=-2$	$(-5,-2)$
-4	$-(-4)^2-6(-4)-7=1$	$(-4,1)$
-3	$-(-3)^2-6(-3)-7=2$	$(-3,2)$
-2	$-(-2)^2-6(-2)-7=1$	$(-2,1)$
-1	$-(-1)^2-6(-1)-7=-2$	$(-1,-2)$
0	$-(0)^2-6(0)-7=-7$	$(0,-7)$

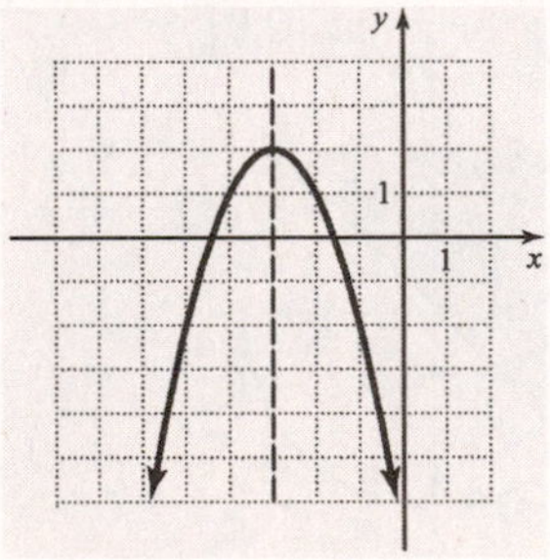

5. $y=8x^2-56x+74$

$y=8(x^2-7x)+74$

$y=8\left(x^2-7x+\left(\frac{-7}{2}\right)^2\right)+74-8\left(\frac{-7}{2}\right)^2$

$y=8\left(x^2-7x+\frac{49}{4}\right)+74-8\left(\frac{49}{4}\right)$

$y=8\left(x-\frac{7}{2}\right)^2+74-98$

$y=8\left(x-\frac{7}{2}\right)^2-24$

$h=\frac{7}{2}, \quad k=-24$

Vertex: $\left(\frac{7}{2},-24\right)$

6.
$$x = 5y^2 + 15y + 7$$
$$x = 5\left(y^2 + 3y\right) + 7$$
$$x = 5\left(y^2 + 3y + \left(\frac{3}{2}\right)^2\right) + 7 - 5\left(\frac{3}{2}\right)^2$$
$$x = 5\left(y^2 + 3y + \frac{9}{4}\right) + 7 - 5\left(\frac{9}{4}\right)$$
$$x = 5\left(y + \frac{3}{2}\right)^2 + \frac{28}{4} - \frac{45}{4}$$
$$x = 5\left(y - \left(-\frac{3}{2}\right)\right)^2 - \frac{17}{4}$$
$$h = -\frac{17}{4} \quad k = -\frac{3}{2}$$
Vertex: $\left(-\frac{17}{4}, -\frac{3}{2}\right)$

7.
$$d = \sqrt{\left(x_2 - x_1\right)^2 + \left(y_2 - y_1\right)^2}$$
$$d = \sqrt{(6-9)^2 + (1-(-5))^2} = \sqrt{(-3)^2 + (6)^2}$$
$$d = \sqrt{9+36} = \sqrt{45}$$
$$d = 3\sqrt{5} = 6.7082... \approx 6.7 \text{ units}$$

8.
$$d = \sqrt{\left(x_2 - x_1\right)^2 + \left(y_2 - y_1\right)^2}$$
$$d = \sqrt{(-2-(-8))^2 + (11-10)^2}$$
$$d = \sqrt{(6)^2 + (1)^2} = \sqrt{36+1}$$
$$d = \sqrt{37} = 6.0827... \approx 6.1 \text{ units}$$

9.
$$d = \sqrt{\left(x_2 - x_1\right)^2 + \left(y_2 - y_1\right)^2}$$
$$d = \sqrt{(-1.8-(-1.3))^2 + (0.5-(-0.7))^2}$$
$$d = \sqrt{(-0.5)^2 + (1.2)^2} = \sqrt{0.25+1.44}$$
$$d = \sqrt{1.69} = 1.3 \text{ units}$$

10.
$$d = \sqrt{\left(x_2 - x_1\right)^2 + \left(y_2 - y_1\right)^2}$$
$$d = \sqrt{\left(-2\sqrt{7} - \sqrt{7}\right)^2 + \left(-5\sqrt{2} - \left(-\sqrt{2}\right)\right)^2}$$
$$d = \sqrt{\left(-3\sqrt{7}\right)^2 + \left(-4\sqrt{2}\right)^2} = \sqrt{9 \bullet 7 + 16 \bullet 2}$$
$$d = \sqrt{63+32} = \sqrt{95} = 9.7467... \approx 9.7 \text{ units}$$

11.
$$\left(\frac{-16+24}{2}, \frac{9+17}{2}\right) = \left(\frac{8}{2}, \frac{26}{2}\right) = (4, 13)$$

12.
$$\left(\frac{-7+10}{2}, \frac{-13+(-5)}{2}\right) = \left(\frac{3}{2}, \frac{-18}{2}\right) = \left(\frac{3}{2}, -9\right)$$

13.
$$\left(\frac{\frac{5}{2} + \left(-\frac{3}{4}\right)}{2}, \frac{\frac{1}{5} + \frac{3}{10}}{2}\right) = \left(\frac{\frac{10}{4} - \frac{3}{4}}{2}, \frac{\frac{2}{10} + \frac{3}{10}}{2}\right)$$
$$= \left(\frac{\frac{7}{4}}{\frac{2}{1}}, \frac{\frac{5}{10}}{\frac{2}{1}}\right)$$
$$= \left(\frac{7}{4} \bullet \frac{1}{2}, \frac{5}{10} \bullet \frac{1}{2}\right) = \left(\frac{7}{8}, \frac{1}{4}\right)$$

14.
$$\left(\frac{-0.35+(-1.05)}{2}, \frac{-2.8+(-4.5)}{2}\right) = \left(\frac{-1.40}{2}, \frac{-7.3}{2}\right)$$
$$= (-0.7, -3.65)$$

15.
$$x^2 + y^2 = 64$$
$$(x-0)^2 + (y-0)^2 = 8^2$$
$$h = 0, \; k = 0$$
Center: $(0,0)$, radius $= 8$

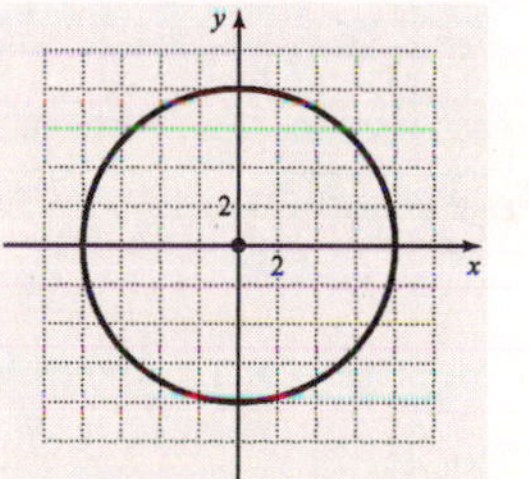

16.
$$(x+2)^2 + (y-4)^2 = 4$$
$$(x-(-2))^2 + (y-4)^2 = 2^2$$
$$h = -2, \; k = 4$$
Center: $(-2,4)$ radius $= 2$

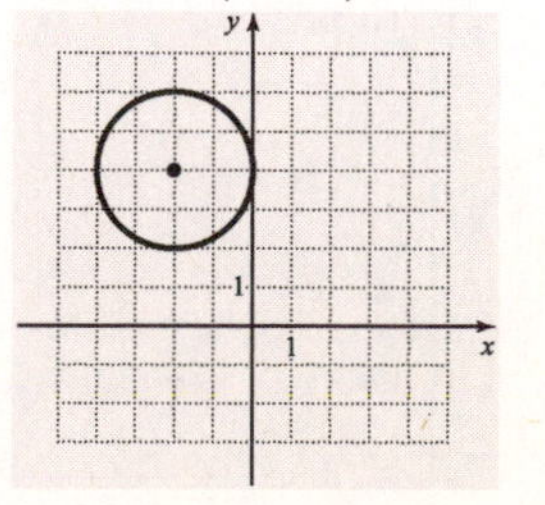

17.

$$x^2+y^2-4x+6y-12=0$$
$$x^2-4x+y^2+6y=12$$
$$x^2-4x+\left(\frac{-4}{2}\right)^2+y^2+6y+\left(\frac{6}{2}\right)^2=12+\left(\frac{-4}{2}\right)^2+\left(\frac{6}{2}\right)^2$$
$$x^2-4x+4+y^2+6y+9=12+4+9$$
$$(x-2)^2+(y+3)^2=25$$
$$(x-2)^2+(y-(-3))^2=5^2$$
$$h=2,\ k=-3$$

Center: $(2,-3)$ radius $=5$

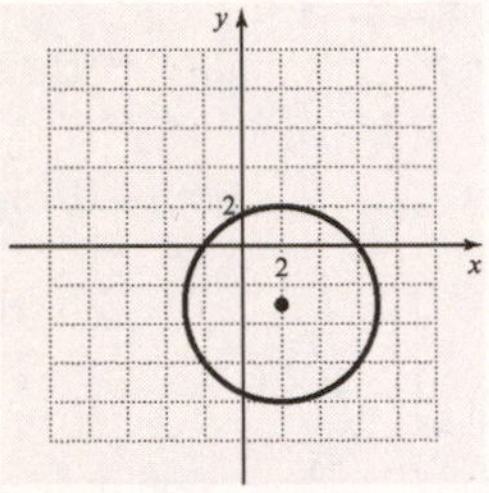

18.

$$x^2+y^2+8x-2y+8=0$$
$$x^2+8x+y^2-2y=-8$$
$$x^2+8x+\left(\frac{8}{2}\right)^2+y^2-2y+\left(\frac{-2}{2}\right)^2=-8+\left(\frac{8}{2}\right)^2+\left(\frac{-2}{2}\right)^2$$
$$x^2+8x+16+y^2-2y+1=-8+16+1$$
$$(x+4)^2+(y-1)^2=9$$
$$(x-(-4))^2+(y-1)^2=3^2$$
$$h=-4,\ k=1$$

Center: $(-4,1)$ radius $=3$

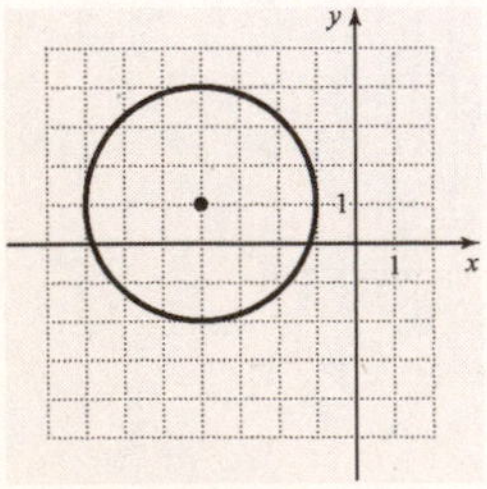

19. $h=9,\ k=0$

$$(x-9)^2+(y-0)^2=13^2$$
$$(x-9)^2+y^2=169$$

20. $h=-6,\ k=10$

$$(x-(-6))^2+(y-10)^2=\left(2\sqrt{5}\right)^2$$
$$(x+6)^2+(y-10)^2=4\bullet 5$$
$$(x+6)^2+(y-10)^2=20$$

21.

$$x^2+y^2-12x-14y-35=0$$
$$x^2-12x+y^2-14y=35$$
$$x^2-12x+\left(\frac{-12}{2}\right)^2+y^2-14y+\left(\frac{-14}{2}\right)^2=35+\left(\frac{-12}{2}\right)^2+\left(\frac{-14}{2}\right)^2$$
$$x^2-12x+36+y^2-14y+49=35+36+49$$
$$(x-6)^2+(y-7)^2=120$$
$$h=6,\ k=7$$

Center: $(6,7)$ radius $=\sqrt{120}=2\sqrt{30}$

22.

$$x^2+y^2+16x+10y+9=0$$
$$x^2+16x+y^2+10y=-9$$
$$x^2+16x+\left(\frac{16}{2}\right)^2+y^2+10y+\left(\frac{10}{2}\right)^2=-9+\left(\frac{16}{2}\right)^2+\left(\frac{10}{2}\right)^2$$
$$x^2+16x+64+y^2+10y+25=-9+64+25$$
$$(x+8)^2+(y+5)^2=80$$
$$(x-(-8))^2+(y-(-5))^2=80$$
$$h=-8,\ k=-5$$

Center: $(-8,-5)$ radius $=\sqrt{80}=4\sqrt{5}$

23. $\frac{x^2}{64}+\frac{y^2}{9}=1\Rightarrow\frac{x^2}{8^2}+\frac{y^2}{3^2}=1;\ a=8,\quad b=3$

x-intercepts: $(\pm 8,0)$ or $(8,0)$ and $(-8,0)$

y-intercepts: $(0,\pm 3)$ or $(0,3)$ and $(0,-3)$

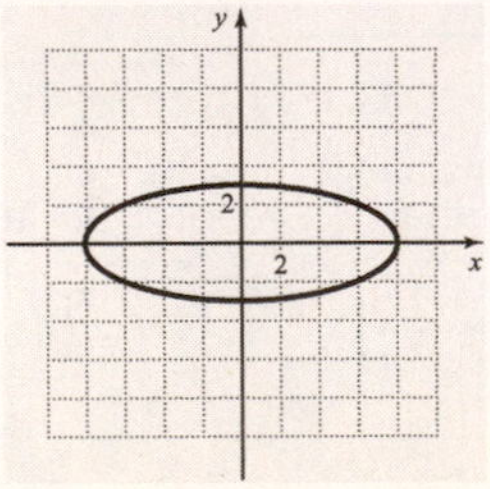

24. $\frac{x^2}{16}-\frac{y^2}{36}=1 \Rightarrow \frac{x^2}{4^2}-\frac{y^2}{6^2}=1;\ a=4,\ \ b=6$

x-intercepts are $(\pm 4,0)$ or $(4,0)$ and $(-4,0)$

Equation of asymptotes are

$y=\frac{6}{4}x=\frac{3}{2}x$ and $y=-\frac{3}{2}x$

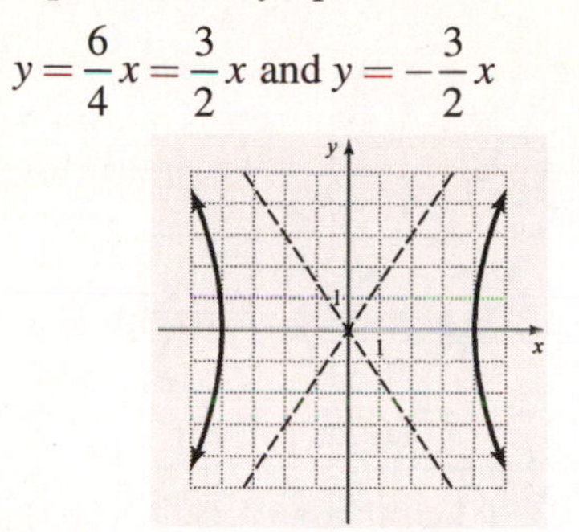

25. $y^2-4x^2=4$

$$\frac{y^2}{4}-\frac{4x^2}{4}=\frac{4}{4}$$

$$\frac{y^2}{4}-\frac{x^2}{1}=1 \Rightarrow \frac{y^2}{2^2}-\frac{x^2}{1^2}=1;\ a=1,\ \ b=2$$

y-intercepts: $(0,\pm 2)$ or $(0,2)$ and $(0,-2)$

Equation of asymptotes are

$y=\frac{2}{1}x=2x$ and $y=-2x$

26. $4x^2+y^2=64$

$$\frac{4x^2}{64}+\frac{y^2}{64}=\frac{64}{64}$$

$$\frac{x^2}{16}+\frac{y^2}{64}=1 \Rightarrow \frac{x^2}{4^2}+\frac{y^2}{8^2}=1;\ a=4\ \ b=8$$

x-intercepts: $(\pm 4,0)$ or $(4,0)$ and $(-4,0)$

y-intercepts: $(0,\pm 8)$ or $(0,8)$ and $(0,-8)$

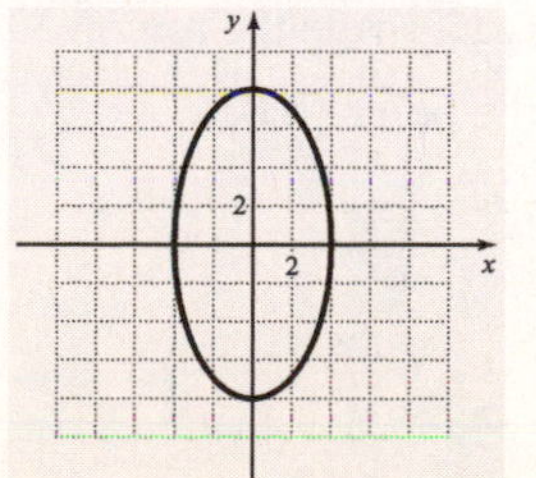

27.
$$y=2x^2+1$$
$$y=2x+5$$

Solve by substitution.

$$2x^2+1=2x+5$$
$$2x^2-2x-4=0$$
$$2(x^2-x-2)=0$$
$$2(x-2)(x+1)=0$$
$$x-2=0 \quad \text{or} \quad x+1=0$$
$$x=2 \qquad\qquad x=-1$$

Find the corresponding y-values.

$$y=2(2)+5=9$$
$$y=2(-1)+5=3$$

$(2,9),(-1,3)$

28.
$$x^2+y^2=10$$
$$y=3x$$

Solve by substitution.

$$x^2+(3x)^2=10$$
$$x^2+9x^2=10$$
$$10x^2=10$$
$$x^2=1$$
$$x=\pm\sqrt{1}=\pm 1$$
$$y=3x$$
$$y=3(1)=3;\ \ y=3(-1)=-3$$

$(1,3),(-1,-3)$

29.
$$x^2+4y^2=16$$
$$x=y^2-4$$

Solve by substitution.

$$y^2=x+4$$
$$x^2+4(x+4)=16$$
$$x^2+4x+16=16$$
$$x^2+4x=0$$
$$x(x+4)=0$$
$$x=0 \quad \text{or} \quad x+4=0 \Rightarrow x=-4$$

Find the corresponding y-values.

$$0=y^2-4 \Rightarrow y^2=4 \Rightarrow y=\pm 2$$
$$-4=y^2-4 \Rightarrow y^2=0 \Rightarrow y=0$$

$(0,2),(0,-2),(-4,0)$

30. $x^2+y^2=8 \xrightarrow{\times(-1)} -x^2-y^2=-8$

$3x^2+y^2=12 \longrightarrow 3x^2+y^2=12$

$2x^2 = 4$

$x^2=2$

$x=\pm\sqrt{2}$

Find the corresponding y-values.

$2+y^2=8$

$y^2=6$

$y=\pm\sqrt{6}$

$(\sqrt{2},\sqrt{6}),(-\sqrt{2},\sqrt{6}),(\sqrt{2},-\sqrt{6}),(-\sqrt{2},-\sqrt{6})$

31. $6x^2+2y^2=16 \xrightarrow{\times\left(\frac{1}{2}\right)} 3x^2+y^2=8$

$2x^2-y^2=17 \longrightarrow 2x^2-y^2=17$

$5x^2 = 25$

$x^2=5$

$x=\pm\sqrt{5}$

Find the corresponding y-values.

$2(5)-y^2=17$

$-y^2=17-10$

$y^2=-7$

$y=\pm\sqrt{-7}=\pm i\sqrt{7}$

$(\sqrt{5},i\sqrt{7}),(-\sqrt{5},i\sqrt{7}),(\sqrt{5},-i\sqrt{7}),(-\sqrt{5},-i\sqrt{7})$

32. $4x^2-9y^2=36 \longrightarrow 4x^2-9y^2=36$

$6x^2+6y^2=54 \xrightarrow{\times\left(\frac{3}{2}\right)} 9x^2+9y^2=81$

$13x^2 = 117$

$x^2=9$

$x=\pm 3$

Find the corresponding y-values.

$4(9)-9y^2=36$

$-9y^2=36-36$

$-9y^2=0$

$y^2=0$

$y=0$

$(-3,0),(3,0)$

33. $y<-x^2+1$

The boundary line is dashed.

$y=-x^2+1$

$y=-(x-0)^2+1$

$h=0,\ k=1$

Vertex: $(0,1)$, axis of symmetry: $x=0$

x	$y=-x^2+1$	(x,y)
-3	$-(-3)^2+1=-8$	$(-3,-8)$
-2	$-(-2)^2+1=-3$	$(-2,-3)$
-1	$-(-1)^2+1=0$	$(-1,0)$
0	$-(0)^2+1=1$	$(0,1)$
2	$-(2)^2+1=-3$	$(2,-3)$

Test point is $(0,0)$.

$0\overset{?}{<}-(0)^2+1$

$0<1$ True, shade the region containing the point $(0,0)$.

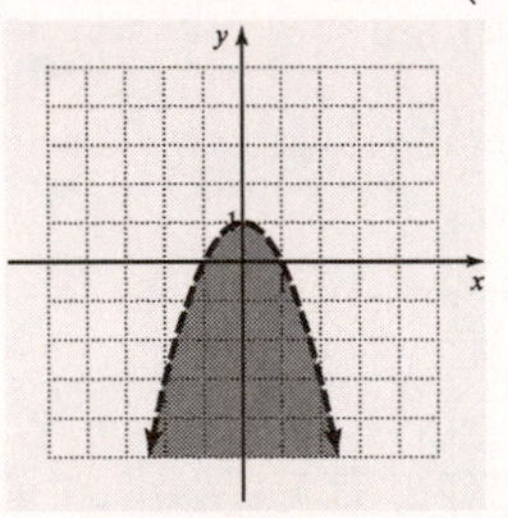

34. $(x+3)^2+y^2\ge 4$

The boundary line is solid.

$(x+3)^2+y^2=4$

$(x-(-3))^2+(y-0)^2=2^2$

$h=-3,\ k=0$

Center: $(-3,0)$ radius $=2$

Testpoint is $(0,0)$

$(0+3)^2+0^2\overset{?}{\ge}4$

$9+0\ge 4$ True, shade the region containing the point $(0,0)$.

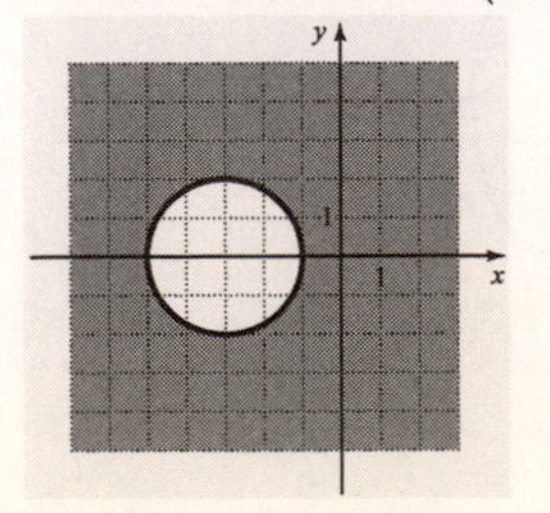

35. $8x^2 + 2y^2 \le 32$

The boundary line is solid.

$$8x^2 + 2y^2 = 32$$

$$\frac{8x^2}{32} + \frac{2y^2}{32} = \frac{32}{32}$$

$$\frac{x^2}{4} + \frac{y^2}{16} = 1 \Rightarrow \frac{x^2}{2^2} + \frac{y^2}{4^2} = 1$$

$a = 2,\ b = 4$

x-intercepts: $(\pm 2,0)$ or $(2,0)$ and $(-2,0)$

y-intercepts: $(0,\pm 4)$ or $(0,4)$ and $(0,-4)$

Testpoint is $(0,0)$.

$$8(0)^2 + 2(0)^2 \overset{?}{\le} 32$$

$0 + 0 \le 32$ True, shade the region containing the point $(0,0)$.

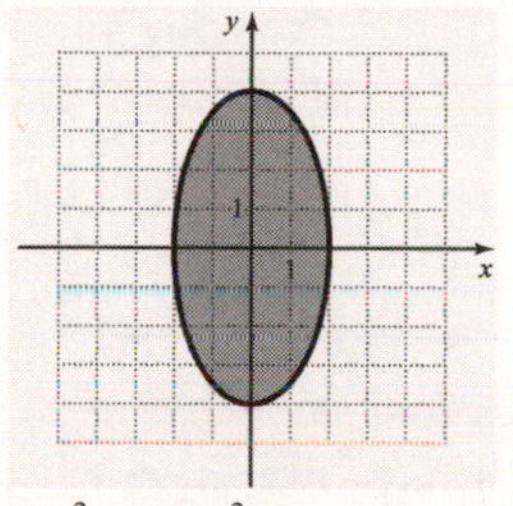

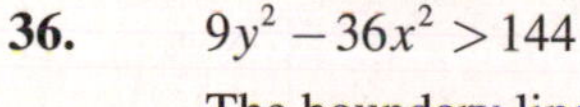

36. $9y^2 - 36x^2 > 144$

The boundary line is dashed.

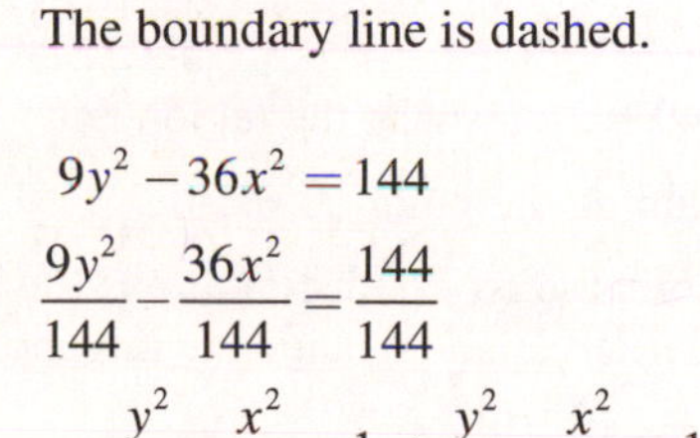

$$9y^2 - 36x^2 = 144$$

$$\frac{9y^2}{144} - \frac{36x^2}{144} = \frac{144}{144}$$

$$\frac{y^2}{16} - \frac{x^2}{4} = 1 \Rightarrow \frac{y^2}{4^2} - \frac{x^2}{2^2} = 1$$

$a = 2,\ b = 4$

y-intercepts: $(0,\pm 4)$ or $(0,4)$ and $(0,-4)$

The equations of the asymptotes are

$$y = \frac{4}{2}x = 2x \text{ and } y = -2x$$

Test point is (0, 0).

$$9(0)^2 - 36(0)^2 \overset{?}{>} 144$$

$0 + 0 > 144$ False, shade the region not containing the point $(0,0)$.

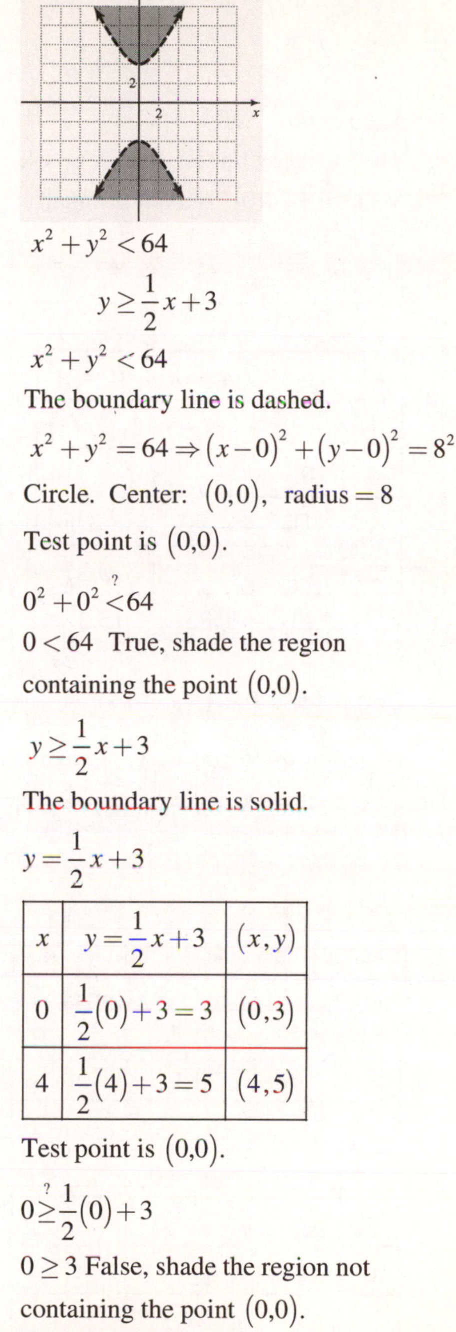

37. $x^2 + y^2 < 64$

$$y \ge \frac{1}{2}x + 3$$

$x^2 + y^2 < 64$

The boundary line is dashed.

$$x^2 + y^2 = 64 \Rightarrow (x-0)^2 + (y-0)^2 = 8^2$$

Circle. Center: $(0,0)$, radius $= 8$

Test point is $(0,0)$.

$$0^2 + 0^2 \overset{?}{<} 64$$

$0 < 64$ True, shade the region containing the point $(0,0)$.

$$y \ge \frac{1}{2}x + 3$$

The boundary line is solid.

$$y = \frac{1}{2}x + 3$$

x	$y = \frac{1}{2}x + 3$	(x, y)
0	$\frac{1}{2}(0) + 3 = 3$	$(0,3)$
4	$\frac{1}{2}(4) + 3 = 5$	$(4,5)$

Test point is $(0,0)$.

$$0 \overset{?}{\ge} \frac{1}{2}(0) + 3$$

$0 \ge 3$ False, shade the region not containing the point $(0,0)$.

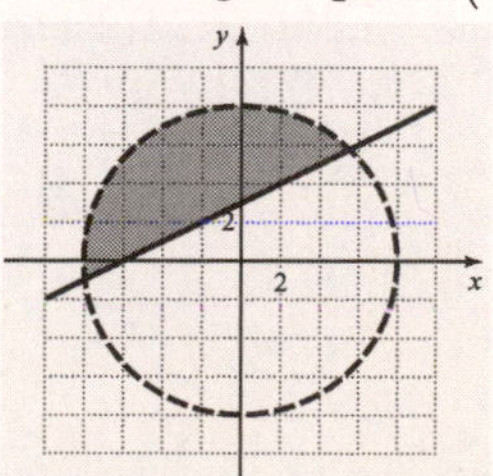

38. $x \geq y^2 - 4$
$x \leq -y^2 + 4$

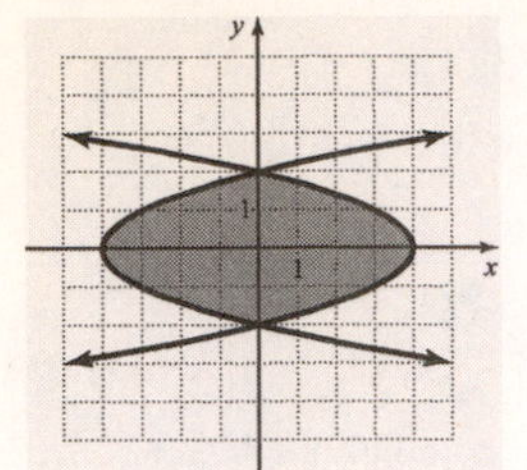

$x \geq y^2 - 4$
The boundary line is solid. $h = -4,\ k = 0$
Vertex: $(-4,0)$ Axis of symmetry: $y = 0$

y	$x = y^2 - 4$	(x,y)
-3	$(-3)^2 - 4 = 5$	$(5,-3)$
-2	$(-2)^2 - 4 = 0$	$(0,-2)$
-1	$(-1)^2 - 4 = -3$	$(-3,-1)$
0	$(0)^2 - 4 = -4$	$(-4,0)$
1	$(1)^2 - 4 = -3$	$(-3,1)$
2	$(2)^2 - 4 = 0$	$(0,2)$
3	$(3)^2 - 4 = 5$	$(5,3)$

Test point is $(0,0)$.

$0 \overset{?}{\geq} (0)^2 - 4;\quad 0 \geq -4$ True,
shade the region containing the point $(0,0)$.

$x \leq -y^2 + 4$
The boundary line is solid. $h = 4,\ k = 0$
Vertex: $(4,0)$ Axis of symmetry: $y = 0$

y	$x = -y^2 + 4$	(x,y)
-3	$-(-3)^2 + 4 = -5$	$(-5,-3)$
-2	$-(-2)^2 + 4 = 0$	$(0,-2)$
-1	$-(-1)^2 + 4 = 3$	$(3,-1)$
0	$-(0)^2 + 4 = 4$	$(4,0)$
1	$-(1)^2 + 4 = 3$	$(3,1)$
2	$-(2)^2 + 4 = 0$	$(0,2)$
3	$-(3)^2 + 4 = -5$	$(-5,3)$

Test point is $(0,0)$.

$0 \overset{?}{\leq} -(0)^2 + 4$
$0 \leq 4$ True, shade the region
containing the point $(0,0)$.

39. $x^2 + y^2 < 16$
$x^2 + y^2 > 49$

$x^2 + y^2 < 16$
The boundary line is dashed.
$x^2 + y^2 = 16 \Rightarrow (x-0)^2 + (y-0)^2 = 4^2$
$h = 0\ \ k = 0$
Center: $(0,0)$ radius $= 4$
Test point is $(0,0)$.

$0^2 + 0^2 \overset{?}{<} 16 \Rightarrow 0 < 16,$ which is true, so shade the region containing the point (0, 0).

$x^2 + y^2 > 49$
The boundary line is solid.
$x^2 + y^2 = 49 \Rightarrow (x-0)^2 + (y-0)^2 = 7^2$
$h = 0\ \ k = 0$
Center: $(0,0)$ radius $= 7$
Test point is $(0,0)$.

$0^2 + 0^2 \overset{?}{>} 49$
$0 > 49$ False, shade the region not containing the point $(0,0)$.

No points satisfy both conditions (inside the smaller circle and outside the larger circle).

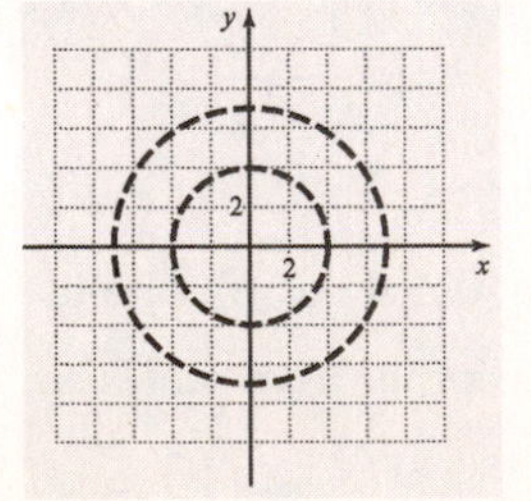

40. $4x^2+25y^2\le 100$
$9x^2-y^2\ge 9$

$4x^2+25y^2\le 100$
The boundary line is solid.
$4x^2+25y^2=100$
$\frac{4x^2}{100}+\frac{25y^2}{100}=\frac{100}{100}$
$\frac{x^2}{25}+\frac{y^2}{4}=1\Rightarrow\frac{x^2}{5^2}+\frac{y^2}{2^2}=1$
$a=5\quad b=2$
x-intercepts: $(\pm 5,0)$ or $(5,0)$ and $(-5,0)$
y-intercepts: $(0,\pm 2)$ or $(0,2)$ and $(0,-2)$
Test point is $(0,0)$.
$4(0)^2+25(0)^2\overset{?}{\le}100$
$0+0\le 100$ True, shade the region containing the point $(0,0)$.

$9x^2-y^2\ge 9$
The boundary line is solid.
$9x^2-y^2=9$
$\frac{9x^2}{9}-\frac{y^2}{9}=\frac{9}{9}$
$\frac{x^2}{1}-\frac{y^2}{9}=1\quad\frac{x^2}{1^2}-\frac{y^2}{3^2}=1\quad a=1\quad b=3$
x-intercepts: $(\pm 1,0)$ or $(1,0)$ and $(-1,0)$

The equations of the asymptotes are
$y=\frac{3}{1}x=3x$ and $y=-3x$
Test point is $(0,0)$.
$9(0)^2-(0)^2\overset{?}{\ge}9$
$0-0\ge 9$ False, shade the region not containing the point $(0,0)$.

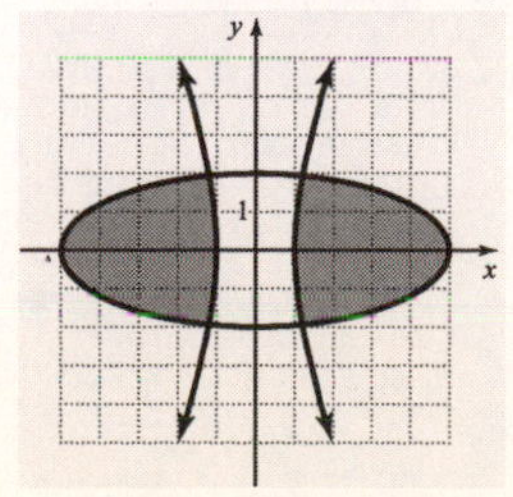

41. **a.**
$R=20x-0.5x^2$
$R=-0.5x^2+20x$
$R=-0.5(x^2-40x)$
$R=-0.5\left(x^2-40x+\left(\frac{-40}{2}\right)^2\right)+0.5\left(\frac{-40}{2}\right)^2$
$R=-0.5(x^2-40x+400)+0.5(400)$
$R=-0.5(x-20)^2+200$

b. $h=20,\ k=200\Rightarrow$ Vertex: $(20,200)$.
This shows that the manufacturing plant will make a maximum revenue of \$200 when it sells 20 units.

42. **a.** The parabola has a vertex at $(0,50)$ and passes through the point $(40,0)$.
$h=0\quad k=50$
$y=a(x-0)^2+50$
$y=ax^2+50$
$0=a(40)^2+50$
$1600a=-50$
$a=-\frac{50}{1600}=-\frac{1}{32}$
$y=-\frac{1}{32}x^2+50$

b. Find y when $x=24$
$y=-\frac{1}{32}(24)^2+50$
$y=-\frac{576}{32}+50=-18+50=32$
The height of the arch 24 ft from the center is 32 ft.

43. The student ends up at
$((8-4),(-15+7))=(4,-8)$

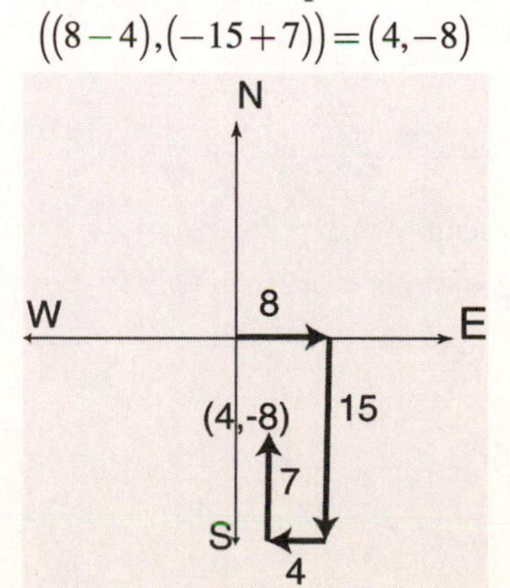

43. (continued)

The distance from $(0,0)$ to $(4,-8)$ is:

$$d=\sqrt{(x_2-x_1)^2+(y_2-y_1)^2}$$
$$d=\sqrt{(4-0)^2+(-8-0)^2}=\sqrt{(4)^2+(-8)^2}$$
$$d=\sqrt{16+64}=\sqrt{80}=4\sqrt{5}=8.9442...$$

The distance between the student's apartment and work is $4\sqrt{5}$ or approximately 8.9 mi.

44. $h=0,\ k=0,\ \text{radius}=60$

$$(x-0)^2+(y-0)^2=60^2$$
$$x^2+y^2=3600$$

45. $h=0,\ k=0,\ \text{radius}=4000+500=4500$

$$(x-0)^2+(y-0)^2=4500^2$$
$$x^2+y^2=20{,}250{,}000$$

46. $\dfrac{x^2}{16}+\dfrac{y^2}{15}=1\Rightarrow\dfrac{x^2}{4^2}+\dfrac{y^2}{\left(\sqrt{15}\right)^2}=1$

$a=4\quad b=\sqrt{15}$

x-intercepts: $(\pm4,0)$ or $(4,0)$ and $(-4,0)$

The distance between x-intercepts is 4 + 4 = 8 billion km. The distance from the sun to one vertex is 3 billion km making the distance from the sun to the other vertex 8 – 3 = 5 billion km. The maximum distance between Pluto and the sun is approximately 5 billion km.

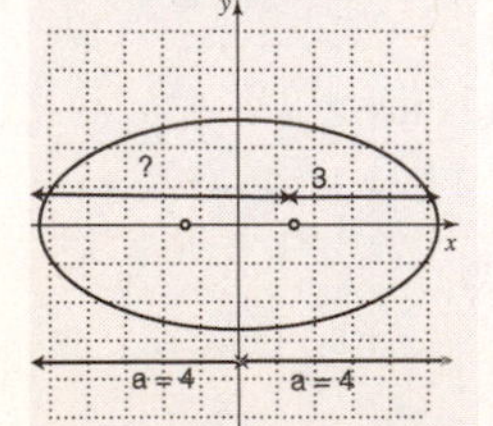

47. The length is $8=2a.\ a=4.$

The width is $4=2b.\ b=2.$

$$\frac{x^2}{4^2}+\frac{y^2}{2^2}=1\Rightarrow\frac{x^2}{16}+\frac{y^2}{4}=1$$

48.
$$-16t^2-32t+120=-16t^2+80$$
$$-32t=-40$$
$$t=\frac{40}{32}=\frac{5}{4}=1.25$$

They will be at the same height $\frac{5}{4}$ sec, or 1.25 sec. after they are released.

49. **a.** Center: $(0,0),\ h=0,\ k=0$

Radius >1

$$(x-0)^2+(y-0)^2>1$$
$$x^2+y^2>1$$

b. The boundary is dashed.

Test point is $(0,0)$.

$$0^2+0^2\overset{?}{>}1$$

$0>1$ False, shade the region not containing the point $(0,0)$.

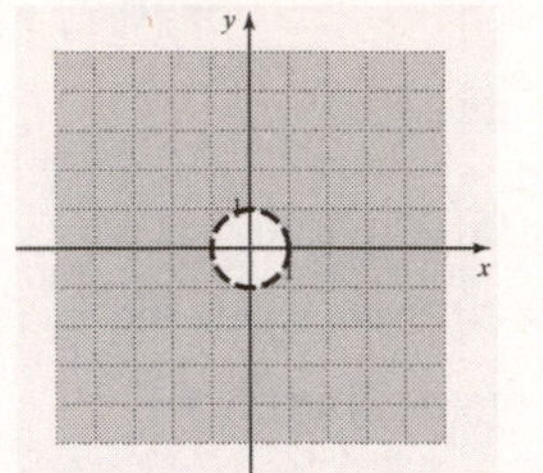

50. **a.** $x^2+y^2\le225$ and $x<y$

$x^2+y^2\le225$

The boundary line is solid.

$x^2+y^2=225$ Circle, center $(0,0)$, radius $=15$.

Test point is $(0,0)$.

$$0^2+0^2\overset{?}{\le}225$$

$0\le225$ True, shade the region containing the point $(0,0)$.

$x<y$

The boundary line is dashed.

$x=y$

x	$y=x$	(x,y)
0	$0=0$	$(0,0)$
2	$2=2$	$(2,2)$

Test point is $(0,2)$.

$0<2$ True, shade the region containing the point $(0,2)$.

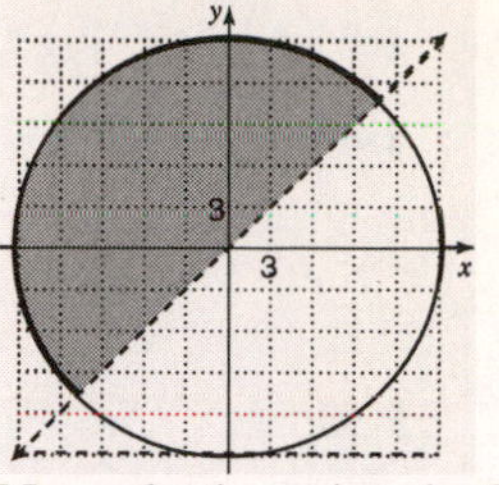

b. No, only the points in Quadrant I are possible solutions since the lengths of the sandboxes must be non-negative quantities.

Chapter 11 Posttest

1.
$$x=-2y^2+16y-91$$
$$x=-2\left(y^2-8y\right)-91$$
$$x=-2\left(y^2-8y+\left(\frac{-8}{2}\right)^2\right)-91+2\left(\frac{-8}{2}\right)^2$$
$$x=-2\left(y^2-8y+16\right)-91+32$$
$$x=-2\left(y-4\right)^2-59$$
$$h=-59,\ k=4$$
Vertex: $(-59,4)$

2. **a.**
$$d=\sqrt{\left(x_2-x_1\right)^2+\left(y_2-y_1\right)^2}$$
$$d=\sqrt{\left(-4-(-2)\right)^2+\left(1-(-1)\right)^2}$$
$$d=\sqrt{(-2)^2+(2)^2}=\sqrt{4+4}=\sqrt{8}$$
$d=2\sqrt{2}$ units

b.
$$d=\sqrt{\left(x_2-x_1\right)^2+\left(y_2-y_1\right)^2}$$
$$d=\sqrt{\left(-4.1-7.9\right)^2+\left(2.6-(-2.4)\right)^2}$$
$$d=\sqrt{(-12)^2+(5)^2}=\sqrt{144+25}=\sqrt{169}$$
$d=13$ units

3. **a.**
$$\left(\frac{-8+(-4)}{2},\frac{-15+9}{2}\right)=\left(\frac{-12}{2},\frac{-6}{2}\right)=(-6,-3)$$

b.
$$\left(\frac{-3+(-6)}{2},\frac{7+3}{2}\right)=\left(\frac{-9}{2},\frac{10}{2}\right)=\left(-\frac{9}{2},5\right)$$

4. $h=-2,\ k=8,$ radius $=3\sqrt{2}$
$$\left(x-(-2)\right)^2+\left(y-8\right)^2=\left(3\sqrt{2}\right)^2$$
$$\left(x+2\right)^2+\left(y-8\right)^2=9\bullet 2$$
$$\left(x+2\right)^2+\left(y-8\right)^2=18$$

5.
$$x^2+y^2+10x-2y+14=0$$
$$x^2+10x+y^2-2y=-14$$
$$x^2+10x+\left(\frac{10}{2}\right)^2+y^2-2y+\left(\frac{-2}{2}\right)^2=-14+\left(\frac{10}{2}\right)^2+\left(\frac{-2}{2}\right)^2$$
$$x^2+10x+25+y^2-2y+1=-14+25+1$$
$$\left(x+5\right)^2+\left(y-1\right)^2=12$$
$$\left(x-(-5)\right)^2+\left(y-1\right)^2=12$$
$$h=-5,\ k=1$$
Center: $(-5,1)$, radius $=\sqrt{12}=2\sqrt{3}$

6. $x=3\left(y-1\right)^2-5$

$h=-5,\ k=1$

Vertex: $(-5,1)$ axis of symmetry: $y=1$

y	$x=3(y-1)^2-5$	(x,y)
3	$3(3-1)^2-5=7$	$(7,3)$
2	$3(2-1)^2-5=-2$	$(-2,2)$
1	$3(1-1)^2-5=-5$	$(-5,1)$
0	$3(0-1)^2-5=-2$	$(-2,0)$
−1	$3(-1-1)^2-5=7$	$(7,-1)$

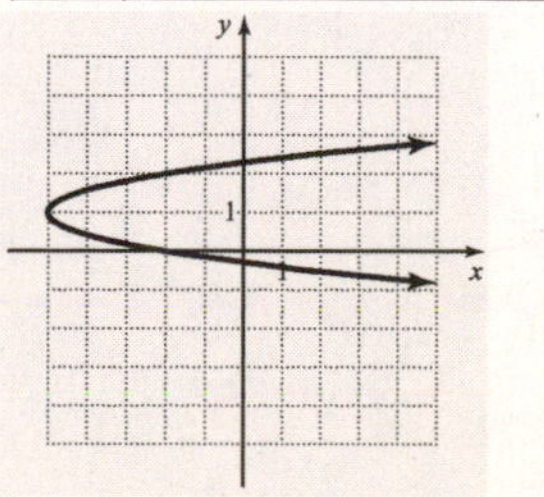

7.
$$x^2+y^2=144$$
$$(x-0)^2+(y-0)^2=12^2$$
$h=0,\ k=0$

Center: $(0,0)$, radius $=12$

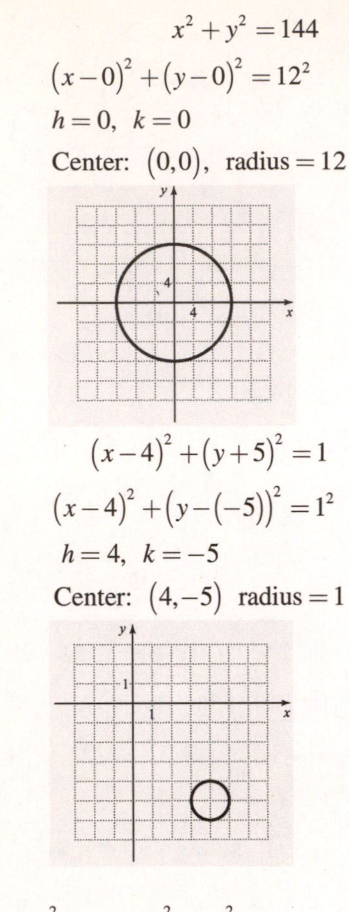

8.
$$(x-4)^2+(y+5)^2=1$$
$$(x-4)^2+(y-(-5))^2=1^2$$
$h=4,\ k=-5$

Center: $(4,-5)$ radius $=1$

9.
$$\frac{x^2}{100}+\frac{y^2}{64}=1\Rightarrow\frac{x^2}{10^2}+\frac{y^2}{8^2}=1$$
$a=10\quad b=8$

x-intercepts: $(\pm10,0)$ or $(10,0)$ and $(-10,0)$

y-intercepts: $(0,\pm8)$ or $(0,8)$ and $(0,-8)$

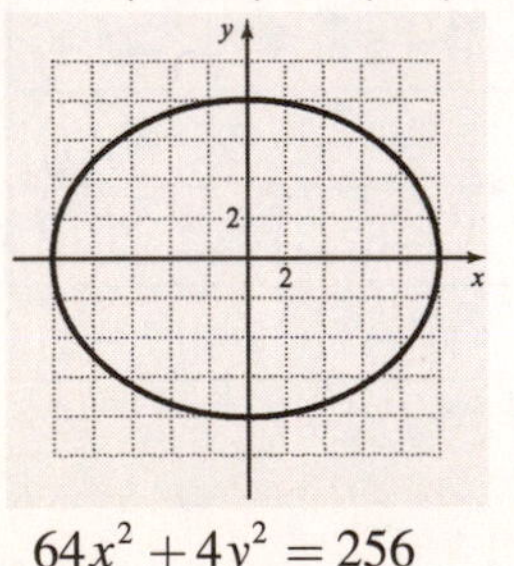

10.
$$64x^2+4y^2=256$$
$$\frac{64x^2}{256}+\frac{4y^2}{256}=\frac{256}{256}$$
$$\frac{x^2}{4}+\frac{y^2}{64}=1\Rightarrow\frac{x^2}{2^2}+\frac{y^2}{8^2}=1$$
$a=2\quad b=8$

x-intercepts: $(\pm2,0)$ or $(2,0)$ and $(-2,0)$

y-intercepts: $(0,\pm8)$ or $(0,8)$ and $(0,-8)$

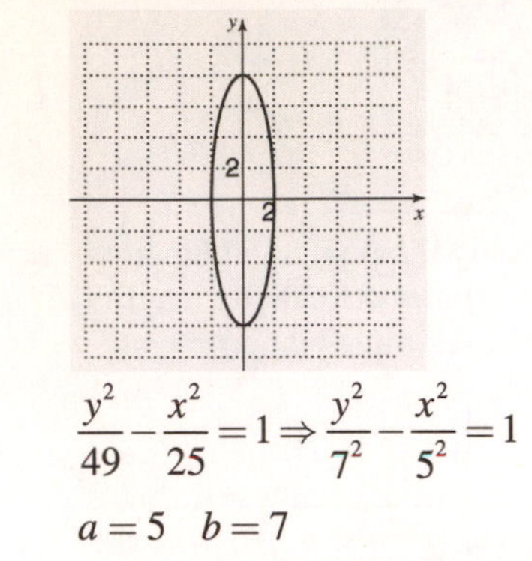

11.
$$\frac{y^2}{49}-\frac{x^2}{25}=1\Rightarrow\frac{y^2}{7^2}-\frac{x^2}{5^2}=1$$
$a=5\quad b=7$

y-intercepts are $(0,\pm7)$ or $(0,7)$ and $(0,-7)$

Equation of asymptotes are

$y=\frac{7}{5}x$ and $y=-\frac{7}{5}x$

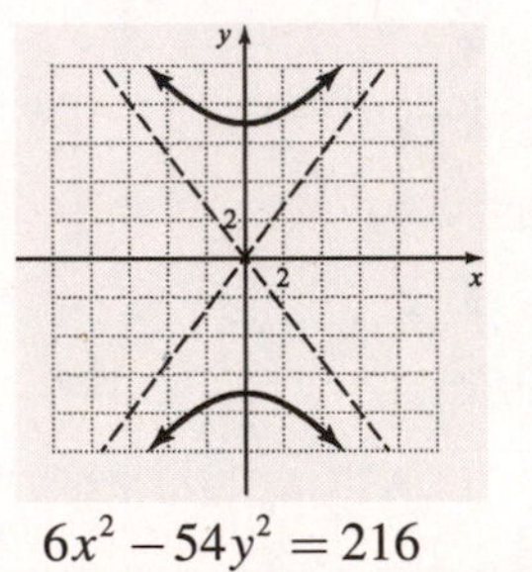

12.
$$6x^2-54y^2=216$$
$$\frac{6x^2}{216}-\frac{54y^2}{216}=\frac{216}{216}$$
$$\frac{x^2}{36}-\frac{y^2}{4}=1\Rightarrow\frac{x^2}{6^2}-\frac{y^2}{2^2}=1$$
$a=6\quad b=2$

x-intercepts: $(\pm6,0)$ or $(6,0)$ and $(-6,0)$

Equation of asymptotes are

$y=\frac{2}{6}x=\frac{1}{3}x$ and $y=-\frac{1}{3}x$

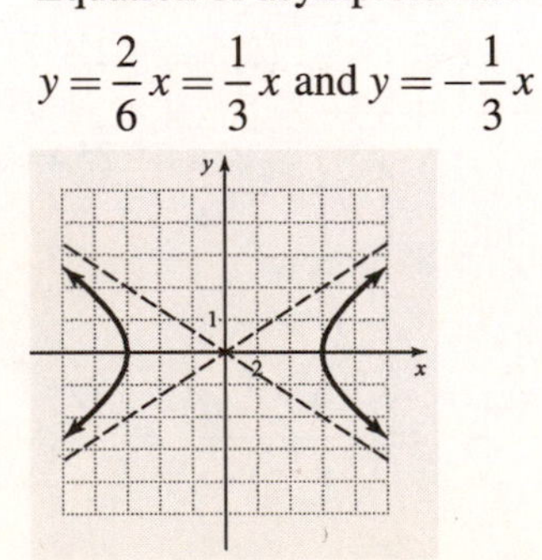

13. $x^2 + y^2 = 25$

$y = x + 5$

Solve by substitution.

$$x^2 + (x+5)^2 = 25$$
$$x^2 + x^2 + 10x + 25 = 25$$
$$2x^2 + 10x = 0$$
$$2x(x+5) = 0$$
$$2x = 0 \quad \text{or} \quad x + 5 = 0$$
$$x = 0 \qquad x = -5$$

Find the corresponding y-values.

$y = x + 5 = 0 + 5 = 5$

$y = x + 5 = -5 + 5 = 0$

$(0,5), (-5,0)$

14. $y = x^2 + 4$

$y = 2x^2 - 6$

Solve by substitution.

$x^2 + 4 = 2x^2 - 6$

$x^2 = 10 \Rightarrow x = \pm\sqrt{10}$

Find the corresponding y-values.

$y = x^2 + 4 = 10 + 4 = 14$

$(\sqrt{10}, 14), (-\sqrt{10}, 14)$

15.

$$3x^2 + y^2 = 9 \xrightarrow{\times 3} 9x^2 + 3y^2 = 27$$
$$9x^2 + 2y^2 = 15 \xrightarrow{\times(-1)} \underline{-9x^2 - 2y^2 = -15}$$
$$y^2 = 12$$
$$y = \pm\sqrt{12}$$

Find the corresponding y-values.

$$3x^2 + y^2 = 9$$
$$3x^2 + 12 = 9$$
$$3x^2 = -3$$
$$x^2 = -1$$
$$x = \pm\sqrt{-1} = \pm i$$

$(i, 2\sqrt{3}), (-i, 2\sqrt{3}), (i, -2\sqrt{3}), (-i, -2\sqrt{3})$

16. $x^2 + 9y^2 > 36$

The boundary line is dashed.

$x^2 + 9y^2 = 36$

$$\frac{x^2}{36} + \frac{9y^2}{36} = \frac{36}{36}$$
$$\frac{x^2}{36} + \frac{y^2}{4} = 1 \Rightarrow \frac{x^2}{6^2} + \frac{y^2}{2^2} = 1$$

$a = 6 \quad b = 2$

x-intercepts: $(\pm 6, 0)$ or $(6,0)$ and $(-6,0)$

y-intercepts: $(0, \pm 2)$ or $(0,2)$ and $(0,-2)$

Test point is $(0,0)$.

$$(0)^2 + 9(0)^2 \overset{?}{>} 36$$

$0 > 36$ False, shade the region not containing the point $(0,0)$.

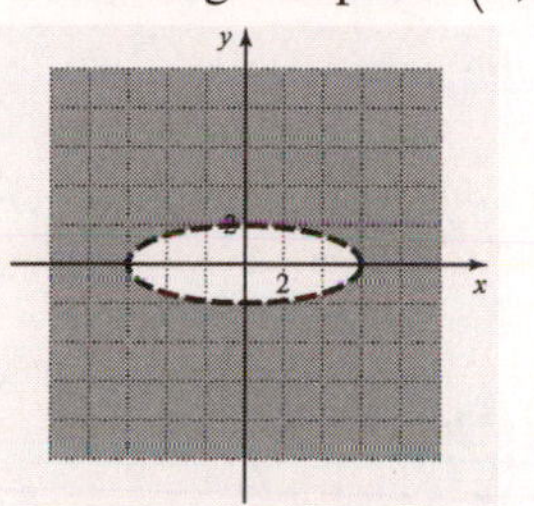

17. $4x^2 + 16y^2 \le 64$

The boundary line is solid.

$4x^2 + 16y^2 = 64$

$$\frac{4x^2}{64} + \frac{16y^2}{64} = \frac{64}{64}$$
$$\frac{x^2}{16} + \frac{y^2}{4} = 1 \Rightarrow \frac{x^2}{4^2} + \frac{y^2}{2^2} = 1$$

$a = 4 \quad b = 2$

x-intercepts: $(\pm 4, 0)$ or $(4,0)$ and $(-4,0)$

y-intercepts: $(0, \pm 2)$ or $(0,2)$ and $(0,-2)$

Test point is $(0,0)$. $4(0)^2 + 16(0)^2 \overset{?}{\le} 64$

$0 + 0 \le 64$ True, shade the region containing the point $(0,0)$.

$4x^2 - y^2 \le 100$

The boundary line is solid.

17. (continued)

$4x^2 - y^2 = 100$

$\frac{4x^2}{100} - \frac{y^2}{100} = \frac{100}{100}$

$\frac{x^2}{25} - \frac{y^2}{100} = 1 \Rightarrow \frac{x^2}{5^2} - \frac{y^2}{10^2} = 1$

$a = 5 \quad b = 10$

x-intercepts: $(\pm 5,0)$ or $(5,0)$ and $(-5,0)$

Equation of asymptotes are

$y = \frac{10}{5}x = 2x$ and $y = -2x$

Test point is $(0,0)$.

$4(0)^2 + (0)^2 \overset{?}{\leq} 100$

$0 + 0 \leq 100$ True, shade the region containing the point $(0,0)$.

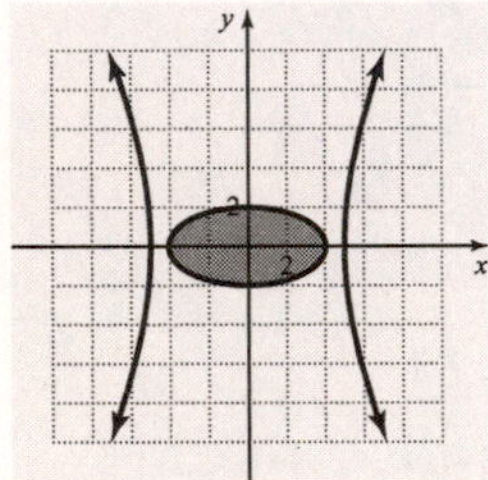

18. The vertex is at $(0,2)$.

$y = a(x-0)^2 + 2 = ax^2 + 2$

It passes through $(450,110)$.

$110 = a(450)^2 + 2$

$108 = a(202,500)$

$a = \frac{108}{202,500} = \frac{1}{1875}$

$y = \frac{1}{1875}x^2 + 2$

19. $2a = 130, \ a = 65$

$2b = 120, \ b = 60$

$\frac{x^2}{65^2} + \frac{y^2}{60^2} = 1$ or $\frac{x^2}{4225} + \frac{y^2}{3600} = 1$

20. **a.** $C \geq 0.5x^2 + 500$

$R \leq 50x$

b. $C \geq 0.5x^2 + 500$

The boundary line is solid.

$C = 0.5x^2 + 500$

x	$C = 0.5x^2 + 500$	(x,C)
0	$0.5(0)^2 + 500 = 500$	$(0,500)$
10	$0.5(10)^2 + 500 = 550$	$(10,550)$
30	$0.5(30)^2 + 500 = 950$	$(30,950)$
50	$0.5(50)^2 + 500 = 1750$	$(50,1750)$
90	$0.5(90)^2 + 500 = 4550$	$(90,4550)$

Test point is $(0,0)$.

$0 \overset{?}{\geq} 0.5(0)^2 + 500$

$0 \geq 0 + 500$ False, shade the region not containing the point $(0,0)$.

$R \leq 50x$

The boundary line is solid.

$R = 50x$

x	$R = 50x$	(x,R)
0	$50(0) = 0$	$(0,0)$
50	$50(50) = 2500$	$(50,2500)$

Test point is $(0,500)$.

$500 \overset{?}{\leq} 50(0)$

$500 \leq 0$ False, shade the region not containing the point $(0,500)$.

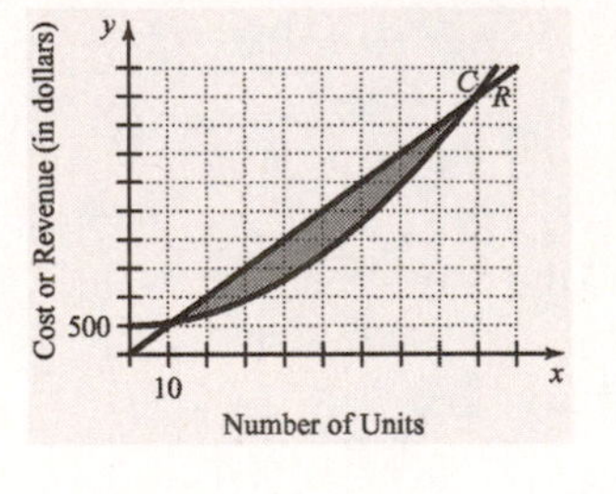